AF330057

Wide-Bandgap Semiconductors for High-Power, High-Frequency and High-Temperature Applications—1999

MATERIALS RESEARCH SOCIETY
SYMPOSIUM PROCEEDINGS VOLUME 572

Wide-Bandgap Semiconductors for High-Power, High-Frequency and High-Temperature Applications—1999

Symposium held April 5–8, 1999, San Francisco, California, U.S.A.

EDITORS:

Steven C. Binari
Naval Research Laboratory
Washington, D.C., U.S.A.

Albert A. Burk
Cree Research Inc.
Durham, North Carolina, U.S.A.

Michael R. Melloch
Purdue University
West Lafayette, Indiana, U.S.A.

Chanh Nguyen
HRL Laboratories
Malibu, California, U.S.A.

Materials Research Society
Warrendale, Pennsylvania

This work was supported in part by the Army Research Office under Grant Number ARO: DAAD19-99-1-0147. The views, opinions, and/or findings contained in this report are those of the author(s) and should not be construed as an official Department of the Army position, policy, or decision, unless so designated by other documentation.

Single article reprints from this publication are available through
University Microfilms Inc., 300 North Zeeb Road, Ann Arbor, Michigan 48106

CODEN: MRSPDH

Published by:

Materials Research Society
506 Keystone Drive
Warrendale, PA 15086
Telephone (724) 779-3003
Fax (724) 779-8313
Web site: http://www.mrs.org/

Library of Congress Cataloging-in-Publication Data

Wide bandgap semiconductors for high-power, high-frequency and high-temperature applications—1999 : symposium held April 5–8, 1999, San Francisco, California, U.S.A. / editors, Steven C. Binari, Albert A. Burk, Michael R. Melloch, Chanh Nguyen
       p.cm.—(Materials Research Society symposium proceedings,
       ISSN 0272-9172 ; v. 572)
       Includes bibliographical references and index.
       ISBN 1-55899-479-3
       1. Wide gap semiconductors—Congresses. 2. Silicon carbide—Congresses.
    3. Gallium nitride—Congresses. I. Binari, Steven C. II. Burk, Albert A.
    III. Melloch, Michael R. IV. Nguyen, Chanh V. Series: Materials Research Society symposium proceedings ; v. 572
TK7871.85.W52 1999   99-40454
621.3815—dc21   CIP

Manufactured in the United States of America

CONTENTS

PART I: <u>SiC DEVICES AND PROCESSING</u>

*Invited Paper

*Invited Paper

*Invited Paper

PART V: <u>GaN DEVICES AND PROCESSING</u>

*Invited Paper

PREFACE

The introduction of the SiC substrate and the demonstration of bright III-N light-emitting diodes were catalysts for a large increase in research and development of wide-bandgap semiconductor materials and devices during the nineties. This symposium, "Wide-Bandgap Semiconductors for High-Power, High-Frequency and High-Temperature Applications—1999," from the 1999 MRS Spring Meeting in San Francisco, California, focused on high-power, high-frequency and high-temperature applications of these wide-bandgap semiconductors. The symposium attracted a wide range of researchers who presented 120 papers in nine different sessions on topics such as bulk crystal growth, epitaxy, materials characterization, processing, and devices.

We would like to thank our invited speakers, J.A. Cooper, S.T. Allen, R. Rupp, V. Balakrishna, S. Seshadri, J.M. Redwing, and G. Sullivan, and the many dedicated anonymous reviewers who made this publication possible.

Generous financial support from the Air Force Research Laboratories, Army Research Office, DARPA, and ODDR&E(R) contributed to the success of this symposium.

Steven C. Binari
Albert A. Burk
Michael R. Melloch
Chanh Nguyen

MATERIALS RESEARCH SOCIETY SYMPOSIUM PROCEEDINGS

MATERIALS RESEARCH SOCIETY SYMPOSIUM PROCEEDINGS

Part I

SiC Devices and Processing

SiC Power Electronic Devices, MOSFETs and Rectifiers

J. A. COOPER, S-H. RYU, Y. LI, M. MATIN, J. SPITZ, D. T. MORISETTE, H. M. McGLOTHLIN, M. K. DAS, M. R. MELLOCH, M. A. CAPANO, AND J. M. WOODALL
School of Electrical and Computer Engineering, Purdue University, West Lafayette, IN
cooperj@ecn.purdue.edu

ABSTRACT

SiC power switching devices have demonstrated performance figures of merit far beyond the silicon theoretical limits, but much work remains before commercial production will be feasible. A significant fraction of the remaining work centers on materials science issues. This paper reviews the current status of unipolar power switching devices in SiC and identifies the major technological and material science barriers that need to be overcome.

1. INTRODUCTION

Since the development of the modified sublimation process for growth of SiC crystals during the 1980's [1, 2] and the commercial availability of single-crystal 6H and 4H SiC wafers during the early 1990's [3], interest and enthusiasm for SiC electronic devices has grown exponentially. SiC is attractive for several reasons: (i) its extreme thermal stability offers the possibility of reliable high-temperature operation; (ii) its high breakdown electric field makes it possible to build power switching devices with resistance-area products 400x lower than silicon [4] and microwave power devices with power densities 100x higher than GaAs; and (iii) its native oxide (SiO_2) enables the whole range of MOS devices and integrated circuits known in silicon [5]. These exciting properties have given impetus to significant device development programs in Europe, Japan, and the United States.

The current device development activities are taking place in an environment where many basic fabrication and material science issues are still unresolved. The situation is reminiscent of the early days of silicon technology, and indeed the SiC industry today is in many ways comparable to the silicon industry of the 1950's. The prospects are exciting, the problems are real, and the future is uncertain. In this review, we focus on one aspect of SiC device and technology development, unipolar (or majority carrier) power switching devices, specifically power MOSFETs and Schottky rectifiers. These devices are expected to be among the first SiC devices to enter commercial production early in the next decade. In the sections that follow, we will outline the basic device designs, review the present status of device development, indicate the relationship between material science issues and device performance, and identify the most critical performance and material science issues still to be resolved.

Mat. Res. Soc. Symp. Proc. Vol. 572 © 1999 Materials Research Society

2. MOTIVATION FOR POWER MOSFETs AND SCHOTTKY RECTIFIERS

Unipolar devices such as the power MOSFET and the Schottky rectifier are attractive for electronic power switching applications for several reasons. Unlike bipolar devices such as the insulated-gate bipolar transistor (IGBT) or the pin diode, the on-state current in unipolar devices does not flow through a forward-biased pn junction. The voltage drop across a forward-biased pn junction in 4H-SiC is about 2.8 V. Assuming a current density of 200 A/cm^2, SiC IGBTs and pin diodes will dissipate 560 W/cm^2 just to establish current flow. This dissipation is absent in power MOSFETs and Schottky diodes. Second, since unipolar devices do not store minority carrier charge in the conducting state, they exhibit minimal reverse recovery transients during switching. In high-frequency switching applications, power dissipated during switching transients can be the dominant power loss in the system.

3. SCHOTTKY RECTIFIERS

3.1 BASIC DESIGN

Figure 1 shows the cross section of a Schottky rectifier in SiC [6]. The starting wafer is 4H-SiC, cut approximately 8° off axis to enable step-controlled epitaxy [7], and polished on the (0001) silicon face. The n-type substrate is doped with nitrogen, with resistivity about 10-20 mΩ-cm and thickness of 300 − 350 μm. A lightly-doped n-type epilayer is grown on the substrate with doping and thickness chosen to provide the desired blocking voltage while minimizing on-resistance. For diodes designed for 1500 V operation, the epilayer is about 10 μm thick, doped 4 - 8x10^{15} cm^{-3}.

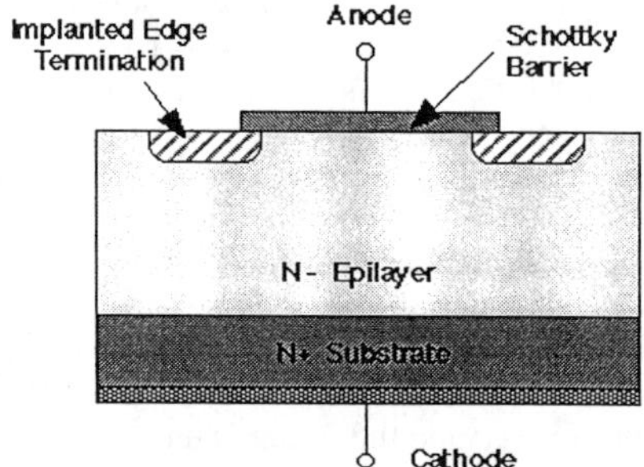

Figure 1. Cross section of a high-voltage SiC Schottky diode.

3.2 SELECTION OF SCHOTTKY METAL

The Schottky metal is chosen to provide the desired barrier height relative to 4H-SiC. Typical Schottky metals are Ni (barrier height 1.3 eV) and Ti (barrier height 0.8 eV). Figure 2 shows I-V characteristics of Ni and Ti Schottky diodes on 4H-SiC. As seen, the lower barrier height of Ti results in a lower turn-on voltage in the conducting state (desirable), but higher leakage current in the blocking state. The reverse leakage current is due to emission of electrons from the Schottky metal into the semiconductor, a process that depends exponentially on barrier height.

The design goal is to obtain the highest possible on current and blocking voltage while minimizing power dissipation in the device. As a general guideline, one would like to support a forward current of at least 100 A/cm^2 in the on-state with a forward drop less than 2 V. This results in a power dissipation of 200 W/cm^2 in the diode. In the blocking state one would like the reverse leakage current to be less than 1 mA/cm^2 at the specified blocking voltage. Although the Ti diode has a lower forward drop than the Ni diode, it has orders of magnitude higher reverse leakage current. Taking 1 mA/cm^2 as the maximum reverse leakage current, the Ti diode could be specified for a maximum reverse voltage of 450 V, while the Ni diode could be rated to 1200 V. For this reason, Ni appears to be the better choice for a Schottky metal on 4H-SiC.

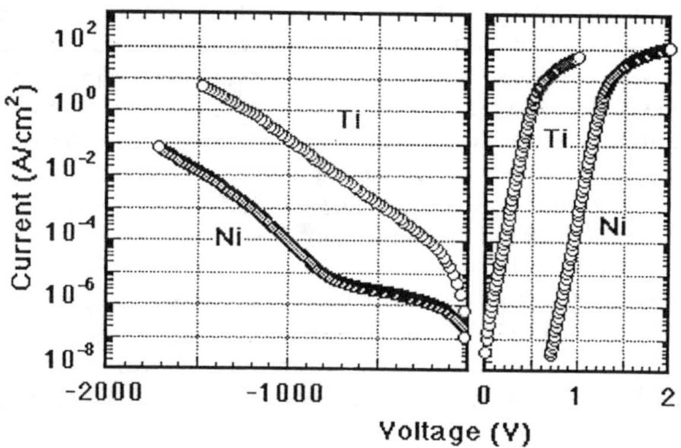

Figure 2. Current-voltage characteristics of Ni and Ti Schottky diodes on 4H-SiC.

3.3 EDGE TERMINATION

The diodes shown in Fig. 2 are fabricated on a 13 μm n-type epilayer that has a theoretical plane-junction breakdown voltage of about 2000 V [6]. The Ni diode exhibits a maximum blocking voltage of 1720 V, about 87% of the theoretical plane-junction breakdown voltage for this epilayer. In order to achieve this result, it is necessary to provide edge termination along the periphery of the Schottky metal to minimize two-dimensional field crowding. The edge termination employed in these diodes makes use of a boron implant under the edge of the Schottky metal, as shown in Fig. 1. This implant is annealed at 1000 °C, a temperature high enough to anneal most of the implant damage without activating the dopants [8]. The effect is to form a resistive layer that carries a small leakage current under reverse bias. This leakage current produces a lateral voltage drop that smoothly tapers the electric field, reducing field crowding at the edges of the Schottky metal. The leakage current is noticeable in the Ni Schottky diode at reverse voltages below about 800 V (see Fig. 2), but is obscured in the Ti diode by the higher Schottky barrier leakage. Other edge terminations have also been used successfully [9, 10].

3.4 YIELD-LIMITING MATERIAL DEFECTS

Although the diodes in Fig. 2 are small, recent work has focused on scaling to larger devices to obtain on-state currents in the range of several A to tens of A [11]. These efforts have made it possible to estimate the density of yield-limiting defects in the material. In one experiment, a 35 mm "low-micropipe" (micropipe density < 30 cm^{-2}) 4H-SiC wafer was fabricated with circular diodes of diameters 1.25, 2, and 3 mm. The yield of good devices for these diodes was 58, 31, and 22 %, respectively, established by testing to 200 V reverse bias. Fitting to a simple yield model indicates a defect density of about 20 cm^{-2}, consistent with the expected density of micropipes in the wafer.

Micropipes are open-core screw dislocations with large Burgers vectors that extend through the substrate and subsequently-grown epilayers. Such defects are obviously fatal to high-voltage power devices. Fortunately, the density of micropipes in commercially-grown wafers has been declining steadily in recent years, and the best reported wafers now have micropipe densities below 1 cm^{-2} [12]. Such wafers would permit fabrication of 50 A Schottky diodes with acceptable yield.

In addition to micropipes, SiC wafers have a much higher density of single-screw dislocations, typically several thousand per cm^2 [13]. Although these defects are not immediately fatal to power devices, they appear to be responsible for "soft breakdown" characteristics, and may possibly give rise to a negative temperature coefficient of breakdown in some devices [14]. A negative temperature coefficient of breakdown is undesirable, since it leads to filamentation and destructive breakdown of the device.

Another problem that becomes apparent when Schottky diodes are scaled to larger diameters is the effect of surface imperfections and irregularities. Since the maximum electric field occurs at the semiconductor surface, the Schottky diode is especially sensitive to surface defects or irregularities. Any macroscopic surface imperfection, such as a pit or asperity, will cause field crowding and premature breakdown. Nomarski microscopy reveals a variety of pits, asperities, and gouges resulting from polishing damage on commercially available SiC wafers. This is an area where the materials science community could make an important contribution, since it is critical to obtaining the large-area high-current Schottky diodes demanded by customers.

3.5 CURRENT STATUS OF EXPERIMENTAL DEVICES

SiC Schottky diodes have achieved impressive performance figures compared to silicon devices. A useful figure of merit for unipolar devices is $\theta = V_B^2 / R_{ON,SP}$, where V_B is the maximum blocking voltage and $R_{ON,SP}$ is the specific on-resistance in Ω-cm^2 ($R_{ON,SP}$ is the product of on-resistance and device area). This figure of merit is appropriate because the minimum achievable $R_{ON,SP}$ in a unipolar device theoretically scales as the square of the blocking voltage. The Ni Schottky diode in Fig. 2, fabricated on a 13 μm 4H-SiC epilayer, has a blocking voltage of 1,720 V and a specific on-resistance of 5.6 mΩ-cm^2, yielding a θ of 528 MW/cm^2 [6]. By comparison, the theoretical limit for silicon unipolar devices is a θ of 4 MW/cm^2. Thus, the diode of Fig. 2 has a figure of merit approximately 130x higher than the theoretical limit for silicon devices. Schottky diodes fabricated in our laboratory on 50 μm epilayers of 4H-SiC have demonstrated blocking voltages as high as 4,200 V [15].

4. POWER MOSFETs

4.1 BASIC DESIGNS

Figure 3 shows cross sections of DMOS and UMOS power transistors in SiC. These MOSFETs are vertical devices with the n+ substrate serving as the drain contact. A lightly-doped n-type epilayer is grown on the substrate to form the drain drift region. As with the Schottky diode, the doping and thickness of the drift region are chosen to obtain the desired blocking voltage while minimizing the on-resistance. In the UMOS structure, a moderately-doped p-type epilayer is also grown on top of the n-type epilayer. This p-type layer is grounded, and acts as the base of the MOSFET (the base is equivalent to the p-type substrate of a planar MOSFET). In the DMOS structure, the p-base is formed by ion implantation of aluminum or boron. N+ source regions are created in selected areas by implantation of nitrogen or phosphorus, and the implants are activated by a high temperature anneal. In the UMOS structure, a trench is defined by reactive ion etching following implant activation. The device is then oxidized to form the gate insulator. Polysilicon is deposited by LPCVD, doped by diffusion of a spin-on dopant, and patterned to form the gate electrode. Ni source contacts and Al base contacts are deposited, defined by liftoff, and annealed.

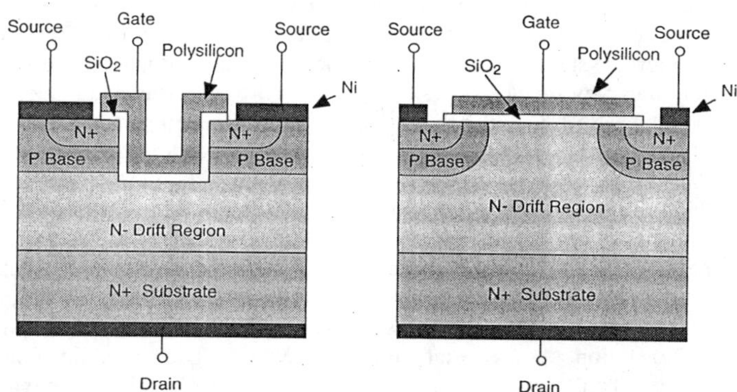

Figure 3. Cross sections of UMOS (left) and DMOS power transistors.

4.2 OPTIMIZATION OF THE MOS INTERFACE

The electrical quality of the oxide/semiconductor interface is of critical importance to the operation of all MOS-based power devices. Oxidation of SiC is both similar to and different from oxidation of silicon [5]. Since the SiC crystal consists of alternating planes of silicon and carbon atoms, the oxidation process involves two reactions: oxidation of silicon to produce silicon dioxide, and oxidation of carbon to produce carbon monoxide and carbon dioxide. The SiO_2 forms the passivating oxide while the carbon byproducts escape in the gas phase.

Oxidation of SiC is much slower than oxidation of silicon, and the oxidation rate depends upon the surface orientation. The fastest oxidation occurs on the (000$\bar{1}$) carbon face, while the slowest oxidation occurs on the (0001) silicon face. The difference in oxidation rates is related to the bonding structure, in which successive planes of silicon and carbon atoms are alternately singly bonded and triply bonded. Oxidation of the carbon face involves oxidation of carbon planes *triply bonded* to silicon planes, followed by oxidation of silicon planes *singly bonded* to carbon planes. In contrast, oxidation of the silicon face consists of oxidation of silicon planes triply bonded to carbon planes, followed by oxidation of carbon planes singly bonded to silicon planes. As might be expected, these two processes exhibit significantly different reaction kinetics.

The SiO_2 - SiC interface is more complex than the SiO_2 - silicon interface, and the role of carbon species in determining the fixed interface charge Q_F and the density of interface states D_{IT} are subjects of current research. At the present time, the optimum oxidation conditions appear to be thermal oxidation in wet O_2 or steam at 1150 °C, followed by a 30 min. in-situ anneal at 1150 °C in argon [16] and a subsequent re-oxidation anneal at 950 °C for 60 min. in wet O_2 [17]. The argon anneal permits carbon species to diffuse out of the oxide under non-oxidizing conditions. The subsequent low temperature re-oxidation forms the final interface, and appears to be beneficial in reducing interface state density. Pre-oxidation cleaning and loading steps are also important. Samples are cleaned using a standard "RCA" clean and loaded into the oxidation tube at 850 °C under flowing O_2. The furnace temperature is then slowly ramped to the 1150 °C oxidation temperature. This procedure minimizes degradation of the surface due to loss of silicon during the furnace insertion [16].

Using the above-described oxidation procedure, MOS interfaces on the (0001) silicon face of 4H and 6H SiC typically have fixed charge densities from $5 - 8 \times 10^{11}$ cm^{-2} and interface state densities that range from around 1.5×10^{11} cm^{-2} eV^{-1} at 0.5 eV above the valence band to around 5×10^{10} cm^{-2} eV^{-1} near midgap [18]. These values are 2 - 3x higher than currently found in silicon devices, but are not high enough to cause problems with MOSFET operation. As a general guideline, a MOSFET in strong inversion is operating with about $6 - 8 \times 10^{12}$ inversion electrons per cm^2, determined by the maximum electric field in the gate oxide (3 – 4 MV/cm). As can be seen, the fixed charge and interface state density numbers quoted above are small compared to the density of inversion electrons in strong inversion.

MOS interfaces formed on the a-axis surfaces of SiC are definitely inferior to those formed on the silicon face. Measurements on the (1$\bar{1}$00) and 11$\bar{2}$0) surfaces indicate interface state densities in the range $5 - 7 \times 10^{11}$ cm^{-2} eV^{-1}, 5 – 10x higher than on the silicon face [19]. MOS interfaces formed on the (000$\bar{1}$) carbon face are even worse, with interface state densities so high as to prevent full modulation of the surface potential. The differences are thought to be due to the presence of carbon bonds at the interface. On the carbon face, essentially all the interface bonds are associated with carbon atoms, while on the a-axis surfaces, approximately half the bonds are associated with carbon. On the silicon face, the vast majority of bonds are associated with silicon atoms, and on an atomic scale this interface resembles the SiO_2 - Si interface.

The breakdown electric field of oxides on 4H and 6H-SiC are comparable to oxides on silicon, typically in the range 8 - 10 MV/cm, although the spread in breakdown voltage across a SiC wafer is larger than for silicon. This is an indication of the relative immaturity of the technology, and probably reflects the difficulty in polishing SiC wafers due to the extreme hardness of the material. As the case with Schottky diodes, any surface imperfection or irregularity will concentrate the surface electric field, leading to premature local breakdown of the oxide.

4.3 MOS OXIDE RELIABILITY

Power switching devices operate at high electric fields, and SiC devices are capable of sustaining much higher internal fields than silicon devices. Indeed, the main advantage of SiC for power devices is the fact that the critical field for avalanche breakdown is 8 − 10x higher than in silicon. In MOS devices, a serious problem occurs at the oxide/semiconductor interface. Because the ratio of dielectric constants between SiO_2 and SiC is about 2.5, the electric field in the oxide is 2.5x higher than the surface electric field in the SiC. Since the critical field for avalanche breakdown in SiC is 2 − 4 MV/cm (depending on doping), the field in the oxide is in the range 5 − 12 MV/cm. These fields are uncomfortably close to the breakdown field of the oxide. Note that if the same oxide were used on a silicon device, the fields in the silicon (and the oxide) would be about 10x lower, safely below the breakdown field of the oxide.

Oxides on SiC are also more susceptible to electron injection from the semiconductor, since the conduction band discontinuity between SiC and SiO_2 is lower than that between silicon and SiO_2, and injection increases exponentially as barrier height is lowered. For these reasons, the long-term reliability of oxides on SiC must be studied carefully to determine the tradeoffs between mean-time-to-failure and oxide field. Although this investigation is still in its early stages, some preliminary conclusions can be drawn. Figure 4 shows mean-time-to-failure of thermal oxides on n-type 6H-SiC as a function of oxide field for three temperatures [20]. Also shown in the figure are the mean-time-to-failure for oxides on silicon at the same temperatures. At high fields, the failure times for oxides on SiC are shorter than for oxides on silicon. However, the field dependence of failure time is stronger for oxides on SiC, with the result that at fields in the range 3 − 4 MV/cm the mean-time-to-failures of SiC and silicon are comparable. These results suggest that acceptable oxide reliability on SiC can be obtained at temperatures up to 150 °C if the field in the oxide is kept below 4 - 5 MV/cm. This can be done without compromising device operation if precautions are taken in the device design to limit the maximum fields in the oxide.

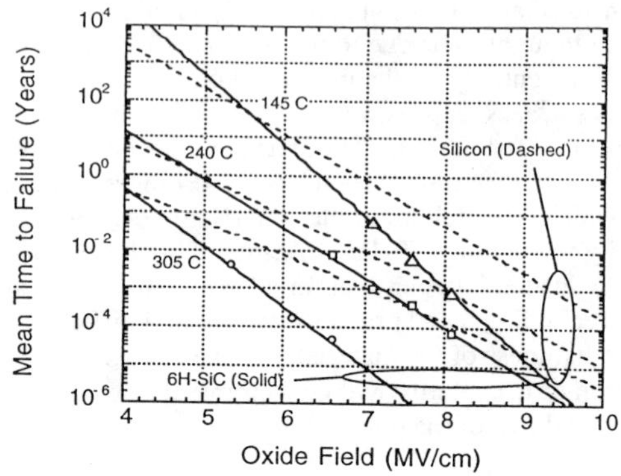

Figure 4. Mean-time-to-failure for oxides on n-type 6H-SiC and on p-type silicon as a function of oxide field and temperature.

4.4 INVERSION LAYER MOBILITY AND IMPLANT ACTIVATION

Since the avalanche breakdown field in SiC is 8 - 10x higher than in silicon, the specific on-resistance (Ω - cm^2) of the drain drift region in power MOSFETs will be about 400x lower than for silicon devices of the same blocking voltage. However, to date no SiC power MOSFET has achieved this ideal. This is because for blocking voltages below about 5,000 V, the resistance of SiC MOSFETs is actually dominated by the channel resistance rather than the drift region resistance. The channel resistance is limited by the inversion layer electron mobility. In silicon, the inversion layer mobility is about half the bulk mobility, but in SiC the inversion layer mobility is only about 10 - 20% of the bulk value. The reason for this is not well understood. Moreover, the inversion layer mobility on 4H-SiC is typically much lower than on 6H, in spite of the fact that the bulk mobility in 4H is higher than in 6H. Typical values for inversion layer mobility reported in the literature are in the range 25 - 50 cm^2/Vs on 4H-SiC and 70 - 90 cm^2/Vs on 6H-SiC. These mobility differences cannot be attributed to differences in fixed oxide charge density or interface state density between 4H and 6H-SiC, since these are fairly comparable between the two interfaces.

It has been recently observed that the inversion layer mobility on 4H-SiC (and probably to a lesser extent on 6H-SiC) can be severely degraded by thermal processing *before* gate oxide formation [21]. This is attributed to step bunching on the surface. Figure 5 shows severe step bunching resulting from a 1700 °C anneal used to activate the boron p-well implant in 4H-SiC [22]. To enable step-controlled epitaxy, 6H and 4H-SiC wafers are routinely cut at 3.5° and 8° off-axis, respectively. During the high-temperature implant anneal, considerable motion of these surface steps occurs and the surface seeks a lower energy state by reducing the density of steps while increasing their height. If this surface is subsequently oxidized to form a MOSFET, the resulting oxide-semiconductor interface can have macroscopic steps whose height is a significant fraction of the oxide thickness. In such a case it may be appropriate to visualize this surface as

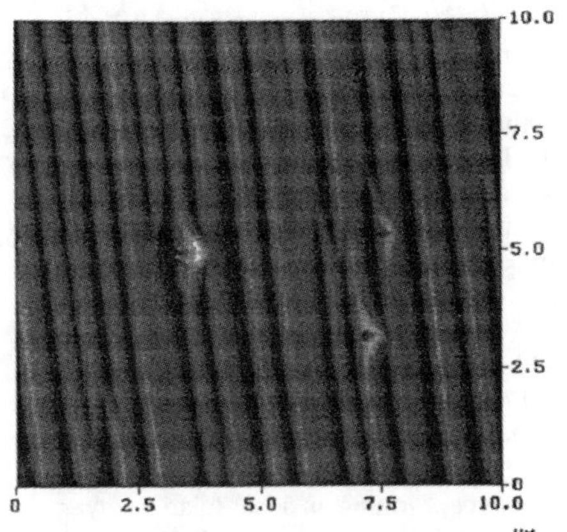

Figure 5. AFM image of the surface of 4H-SiC, implanted with boron to a dose of 4×10^{14} cm^{-2}, typical of a p-well implant, and annealed at 1700 °C for 40 min. in argon. Dimensions of the image are 10 x10 μm.

consisting of alternating horizontal and vertical interface regions, or equivalently, as a series connection of horizontal and vertical MOSFETs. When the surface is brought into inversion by the gate bias, the horizontal regions will typically go into strong inversion while the vertical surfaces are only weakly inverted, or not inverted at all. This occurs because of the relative geometry of the field lines from the gate to the semiconductor, because the oxide on the vertical sidewall is 2-3x thicker than on the horizontal surface, or because the fixed charge and interface state densities on the vertical sidewall are higher than on the horizontal surface. In any event, the steps create potential barriers to electrons flowing from source to drain, significantly degrading the effective surface mobility of the MOSFET. In extreme cases the effective mobility can have fractional values. These extremely low mobilities are correlated with high temperature anneals (1600 - 1700 °C) required to activate p-well implants prior to gate oxidation. Table 1 illustrates the range of inversion layer mobilities observed on 4H-SiC as a function of implant anneal conditions prior to gate oxidation. As seen, in order to obtain high inversion layer mobilities on 4H-SiC it is necessary to keep the maximum implant anneal temperature below about 1200 °C. This temperature is insufficient to activate boron or aluminum used for p-well implants in DMOSFETs, but is high enough to activate nitrogen or phosphorus used for the source implants.

N-type dopants in SiC are nitrogen and phosphorus, and p-type dopants are aluminum and boron. Since diffusion of impurities is impractical in SiC due to the low diffusion coefficient of impurities at any reasonable temperature, ion implantation is used for selective doping. Figure 6 shows sheet resistivity for nitrogen and phosphorus source

Anneal Temperature	1200 °C	1300 °C	1400 °C	1700 °C
No p-well implant	25 cm^2/Vs	7 cm^2/Vs	5 cm^2/Vs	9 cm^2/Vs *
Boron p-well implant				0.32 cm^2/Vs * 0.06 cm^2/Vs

Table 1. Inversion layer electron mobility on 4H-SiC as a function of implant anneal conditions prior to gate oxidation. Asterisks (*) indicate that the implant anneal was conducted under a silane overpressure; all other anneals were conducted in argon.

implants and electron inversion layer mobility in 4H-SiC as a function of implant anneal
temperature [23]. As seen, to obtain inversion layer mobilities greater than 10, it is
necessary to keep the implant anneal temperatures below about 1250 °C. This can be
done successfully if phosphorus is used instead of nitrogen as the source dopant.

The problem becomes more severe if either aluminum or boron implantation is
required in the fabrication process, as for the DMOS structure of Fig. 3. Figure 7 shows
the activation percentage of boron in 4H-SiC as a function of anneal temperature [22].
To obtain good activation, it is necessary to anneal at temperatures in excess of 1600 °C.
Comparison with Fig. 6 indicates that the expected MOS inversion layer mobility
following such an anneal will be in the single digits, at best. The actual results on
DMOSFETs are even worse, since the MOS inversion channel is formed in the region
damaged by the boron implant. The second line in Table 1 shows that inversion layer
mobilities in implanted regions annealed at 1700 °C are fractional. As described in the
previous section, the low effective inversion layer mobility is caused by the dramatic step
bunching that occurs during implant anneal (c.f. Fig. 5). One solution is to convert the
near-surface region of the semiconductor to n-type, forming an accumulation-layer
MOSFET, or ACCUFET [24, 25]. The density of donors in the ACCUFET layer is kept
low so that the layer is completely depleted at zero gate bias, but there the layer has
enough conductivity under positive gate bias that a conductive path is established around
the potential barriers associated with the surface steps, significantly enhancing the
effective mobility of the MOSFET. This solution, while workable, is undesirable
because of the difficulty of controlling the threshold voltage precisely when it is close to
zero, and the likelihood that the threshold voltage will go negative (creating a normally-
on MOSFET) as temperature is raised.

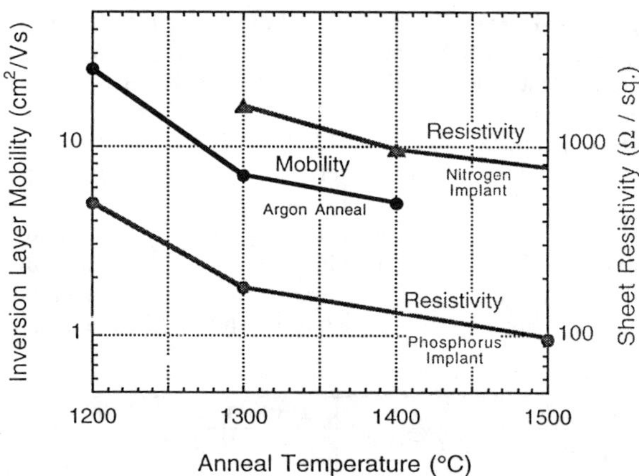

Figure 6. Inversion layer electron mobility and source implant sheet resistivity for
nitrogen and phosphorus in 4H-SiC as a function of implant anneal temperature. The
highest inversion layer mobility (25 cm²/Vs) is obtained for a 1200 °C anneal, where the
phosphorus source implant sheet resistivity is an acceptable 500 Ω per square.

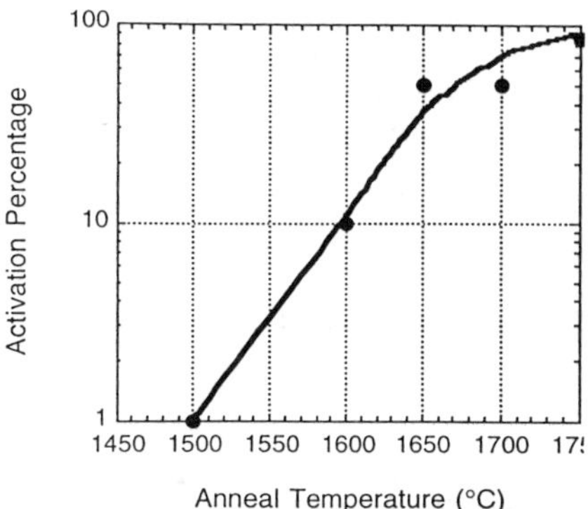

Figure 7. Activation percentage of boron in 4H-SiC as a function of anneal temperature. All anneals are performed for 40 min. in argon.

4.5 CURRENT STATUS OF EXPERIMENTAL DEVICES

In spite of the material and processing difficulties described above, progress in the development of SiC power MOSFETs has been dramatic. Currently, both DMOS and UMOS devices are being developed. UMOSFETs fabricated on 10 μm 4H-SiC epilayers have achieved blocking voltages of 1400 V at a specific on-resistance of 15.7 mΩ-cm^2, for a figure of merit $\theta = V_B^2 / R_{ON}$ of 125 MW/cm^2, approximately 25x higher than the theoretical limit for silicon MOSFETs [25]. Lateral DMOSFETs on insulating 4H-SiC substrates have achieved blocking voltages of 2600 V [26]. With the availability of thicker epilayers (up to 50 μm are currently available), it is anticipated that MOSFETs with blocking voltages in the 3 – 5 kV range will soon be reported.

It is worth noting that the performance of the best reported SiC MOSFETs are still well below the theoretical limit expected for 4H-SiC. For example, the 1400 V UMOSFET reported above with a θ of 125 MW/cm^2 is still a factor of 16 below the 4H-SiC theoretical limit. The main reason for the shortfall is the specific on-resistance R_{ON}, which is dominated by the resistance of the MOSFET channel. In order to reach the SiC theoretical limit, the MOSFET channel resistance needs to be reduced by about an order of magnitude. This is the biggest challenge facing MOSFET device developers today.

5. SUMMARY AND CONCLUSIONS

During the past several years, the development of prototype devices in SiC has been taking place *in parallel* with the optimization of unit fabrication processes and research on fundamental material science issues. This leads to mistakes and false starts in device development, but it also tends to bring into sharp focus the critical material science issues. At the present time, from a device development perspective, the most critical material science issues are: (1) size and cost of SiC wafers, (2) surface morphology of as-received wafers, (3) density of micropipes and single screw dislocations in the material, (4) activation of implants (particularly aluminum and boron p-type implants), (5) low inversion layer electron mobility at the SiO$_2$ – SiC interface and its relationship to previous high-temperature steps, (6) high ohmic contact resistance (particularly to p-type SiC), and (7) the stability and reliability of SiO_2 on SiC under high-field and high-temperature stress. Continued close communication between device developers and material scientists is needed to focus attention on these critical problems and verify their solution.

ACKNOWLEDGEMENTS

This work is supported by ONR under MURI grant. no. N00014-95-1-1302.

REFERENCES

1. G. Ziegler, P. Lanig, D. Theis, and C. Weyrich, *IEEE Trans. Electr. Dev.* **30**, 227 (1983).
2. C. H. Carter, Jr., L. Tang, and R. F. Davis, 4th National Review Meeting on the Growth and Characterization of SiC, Raleigh, NC, USA, 1987; U.S. Patent No. 4,866,005 (Sept. 12, 1989) R. F. Davis, C. H. Carter, Jr., and C. E. Hunter.
3. Cree Research, Inc., Durham, NC.
4. B. J. Baliga, *IEEE Electron Device Lett.* **10,** 455 (1989).
5. J. A. Cooper, Jr., *Physica Stat. Solidi (a)*, **162**, 305 (1997).
6. K. J. Schoen, J. M. Woodall, J. A. Cooper, Jr., and M. R. Melloch, *IEEE Trans. Electr. Dev.* **45**, 1595 (1998).
7. H. Matsunami, T. Ueda, and H. Nishino, *Proc. Mat. Res. Soc. Symp.* **162**, 397 (1990).
8. A. Itoh, T. Kimoto, and H. Matsunami, *IEEE Electron Device Lett.* **17,** 139 (1996).
9. D. Alok, B. J. Baliga, and P. K. McLarty, *IEEE Electron Device Lett.* **15,** 394 (1994).
10. R. Singh and J. W. Palmour, Proc. Int'l. Symp. on Power Semi. Devices and ICs, Weimar, Germany, May 26-29, 1997.
11. G. M. Dolny, D. T. Morisette, P. M. Shenoy, M. Zafrani, J. Gladish, J. M. Woodall, J. A. Cooper, Jr., and M. R. Melloch, *IEEE Device Res. Conf.,* Charlottesville, VA, June 22 - 24, 1998.
12. C. H. Carter, Jr., Cree Research, Inc., Durham, NC, private communication.
13. S. Wang, M. Dudley. C. H. Carter, Jr., and H. S. Hong, *Mat. Res. Soc. Symp. Proc.*, **339**, 735 (1994).
14. P. G. Neudeck and C. Fazi, *IEEE Electron Device Lett.* **18,** 96 (1997).
15. H. M. McGlothlin, D. T. Morisette, J. A. Cooper, and M. R. Melloch, unpublished.
16. J. N. Shenoy, G. L. Chindalore, M. R. Melloch, J. A. Cooper, Jr., J. W. Palmour, and K. G. Irvine, *J. Electronic Mat'ls*, **24**, 303 (1995).
17. L. A. Lipkin and J. W. Palmour, *J. Electr. Mat'ls,* **25**, 909 (1996).
18. M. K. Das, J. A. Cooper, Jr., and M. R. Melloch, *J. Electr. Mat'ls,* **27**, 353 (1998).
19. J. N. Shenoy, M. K. Das, J. A. Cooper, Jr., M. R. Melloch, and J. W. Palmour, *J. Appl. Physics*, **79**, 3042 (1996).
20. M. M. Maranowski and J. A. Cooper, Jr., *IEEE Trans. Electr. Dev.* **46**, (1999).
21. M. K. Das, J. A. Cooper, Jr., M. R. Melloch, and M. A. Capano, *Semi. Interface Specialists Conf.,* San Diego, CA, December 3 - 5, 1998.
22. M. A. Capano, S-H. Ryu, M. R. Melloch, and J. A. Cooper, Jr., *J. Electr. Mat'ls*, **27**, 370 (1998).
23. R. Santharumar, M. K. Das, M. A. Capano, J. A. Cooper, and M. R. Melloch, unpublished.
24. P. M. Shenoy and B. J. Baliga, *IEEE Electron Device Lett.*, **18**, 589 (1997).
25. J. Tan, J. A. Cooper, Jr., and M. R. Melloch, *IEEE Electron Device Lett.*, **19**, 487 (1998).
26. J. Spitz, M. R. Melloch, J. A. Cooper, Jr., and M. A. Capano, *IEEE Electron Device Lett.*, **19**, 100 (1998).

RECENT PROGRESS IN SiC MICROWAVE MESFETs

S.T. ALLEN, S.T. SHEPPARD, W.L. PRIBBLE, R.A. SADLER, T.S. ALCORN, Z. RING, and J.W. PALMOUR
Cree Research, Inc., 4600 Silicon Drive, Durham, NC, 27703; (919) 361-5709

ABSTRACT

SiC MESFET's have shown an RF power density of 4.6 W/mm at 3.5 GHz and a power added efficiency of 60% with 3 W/mm at 800 MHz, demonstrating that SiC devices are capable of very high power densities and high efficiencies. Single devices with 48 mm of gate periphery were mounted in a hybrid circuit and achieved a maximum RF power of 80 watts CW at 3.1 GHz with 38% PAE.

INTRODUCTION

Silicon carbide (SiC) MESFET's are emerging as a promising technology for high- power microwave applications due to a combination of superior properties of SiC, including a high breakdown electric field, high saturated electron velocity and high thermal conductivity. In this paper we report on the substantial progress that has recently been made in microwave SiC MESFET technology. Improvements to device design and fabrication have led to increased breakdown voltages of greater than 150 V for FET's with a channel doping of 3×10^{17} cm^{-3}. Improvements in process control and repeatability have resulted in the ability to yield devices with up to 48 mm of gate periphery.   Cree also recently began fabricating MESFET's on 2-inch diameter semi-insulating SiC substrates, a substantial increase in wafer area over the previous 1-3/8" substrates, preparing the way for producing these devices in large quantity.

ADVANTAGEOUS PROPERTIES OF SILICON CARBIDE

SiC occurs in over 200 different crystal structures, or polytypes, but for semiconductor applications the 6H and 4H polytypes have received the most attention due to the availability of high-quality single crystalline substrates. For microwave MESFET's the 4H-SiC polytype is preferable because it has a larger bandgap and higher electron mobility than 6H-SiC. It is the wide bandgap of 3.2 eV, compared to 1.1 eV for Si and 1.4 eV for GaAs, that gives SiC its primary advantage for high-power devices. This wide bandgap gives rise to a breakdown electric field that is 10 times higher than in GaAs or Si. This is illustrated in Figure 1, which shows the measured breakdown voltage of 4H-SiC p-n junction diodes as well as the theoretical curves for Si and GaAs. This high breakdown field has been exploited to fabricate sub-micron SiC MESFET's with gate-to-drain breakdown voltages exceeding 200 V.

The one drawback to SiC for use in microwave devices is its poor low-field electron mobility, which is in the range of 300 - 500 cm^2/V-sec. for doping levels of interest for MESFET's, i.e., 1×10^{17} cm$^{-3} < N_D < 5\times10^{17}$ cm^{-3}. This results in a larger source resistance and lower transconductance than is typical of GaAs MESFET's, but is partially offset by the ability to operate the SiC devices under extremely high electric fields. The saturated electron velocity in 6H-SiC is 2×10^7 cm/s and has been predicted by Monte Carlo simulations to be 2.7×10^7 cm/s in 4H-SiC [2], almost 3 times higher than in GaAs at high fields. Although the knee voltage of SiC MESFET's is relatively high (typically $\approx$ 9 V), the drain efficiency of the devices is still high because the breakdown voltage is over 100 V. The channel current density is reasonably large, typically around 300 mA/mm for 0.7 μm gate length devices, due to the high saturated velocity. When combined with the high breakdown voltage, this results in the large RF power density of over 4 W/mm that has been measured for SiC MESFET's.

Mat. Res. Soc. Symp. Proc. Vol. 572 © 1999 Materials Research Society

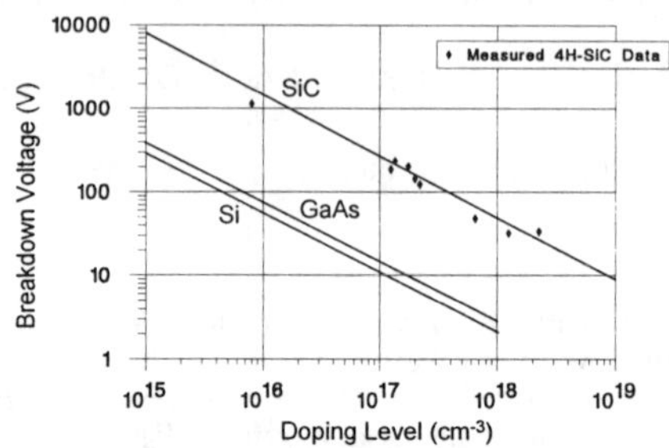

Figure 1: Measured breakdown voltage of 4H-SiC p-n junction diodes as a function of doping, and the theoretical maximums for Si and GaAs.

The other property of SiC that gives it a significant advantage over other semiconductors is its very high thermal conductivity. In order to take advantage of the high electrical power densities available in this material, it is also necessary to dissipate very high thermal power densities, making the thermal conductivity an extremely important parameter. Measured thermal conductivity data for SiC is shown in Table I for low doped n- and p-type, doped n-type and semi-insulating material. The thermal conductivity was measured by creating a temperature difference across a piece of SiC. Thermocouples were inserted into holes drilled 1 cm apart and ΔT and applied power were used to calculate the thermal conductivity at 25°C and 100°C. Copper and Al were used as calibration standards. Thermal conductivity is the product of a material's density, heat capacity and its thermal diffusivity: the latter being dependent on the doping and quality of a material. The very low doped material exhibits a-axis thermal conductivity roughly the same as that reported by Slack [2] for pure Lely platelets. The doped materials and c-axis direction have significantly lower thermal conductivity values. Even the lowest measured value of 3.3 W/cm-K is 7 times higher than the thermal conductivity of GaAs, implying that not only is the power density of SiC high in terms of W/mm of gate periphery but SiC also has extremely high power handling capability in terms of W/mm^2 of die area. For high power, high frequency applications, this is the more important figure of merit since die size becomes constrained by wavelength, and all power devices are operated at their thermal limit.

Table I
Measured Thermal Conductivity Data for SiC

Sample Type	Direction	Carrier (cm^{-3})	Thermal Conductivity 298 K (W/cm-K)	373 K (W/cm-K)
4H Semi Ins.	// c	S.I.	3.3	2.6
4H n	// c	2.0e18 n	3.3	2.5
4H n	$\perp$ c	5e15 n	4.8	2.9
6H n	// c	1.5e18 n	3.0	2.3
6H n	// c	3.5e17 n	3.2	2.3
6H n	$\perp$ c	3.5e17 n	3.8	2.8
6H p	$\perp$ c	1.4e16 p	4.0	3.2
Slack [2]	$\perp$ c	~1e17	~5	~3

RECENT ADVANCES IN SIC SUBSTRATES

The development of SiC electronic devices has been limited in the past by the lack of availability of large, high quality SiC substrates. The primary defects in bulk SiC are micropipes, which are superscrew dislocations in the crystal that have open cores, resulting in pinholes in the wafer. Recently Cree has made advancements in crystal growth technology that resulted in producing 4H-SiC wafers with a micropipe density of < 1 cm^{-2}, which is more than two orders of magnitude less than it was three years ago. Because the active area of microwave MESFET's is very small, limited to the source- drain separation of 4 μm, a micropipe density of < 10 cm^{-2} has a negligible impact on yield. Therefore, with these recent reductions in micropipe defect densities and the increase in wafer diameter size (2-inch in production and 3-inch demonstrated) SiC crystal quality has improved to the point where it would be viable to fabricate SiC MESFET's in production quantities.

The semi-insulating material has been characterized extensively by W.C. Mitchell and R. Perrin at the Air Force Wright Patterson Laboratory. Figure 2 is an Arrhenius plot of the high temperature resistivity of one of these wafers determined from Hall-effect measurements. The extracted activation energy of 1.7 eV is consistent with deep-level compensation in SiC with a bandgap of 3.2 eV and leads to an extrapolated room temperature resistivity of 10^{20} Ω-cm. At a temperature of 500°C, the resistivity exceeds 10^{7} Ω-cm, making the semi-insulating properties of the substrate consistent with a technology capable of operating at extremely high temperatures.

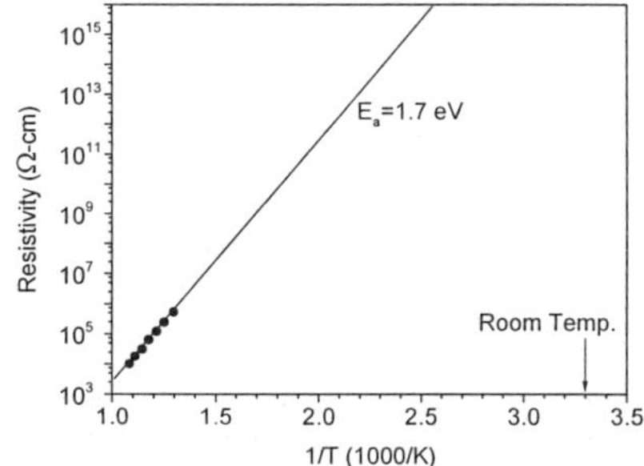

Figure 2: Plot of resistivity vs. 1/T for semi-insulating 4H-SiC as determined with Hall-effect measurements.

MICOWAVE POWER RESULTS

Cree's optimized S-band power FET's have a gate length of 0.7 μm and employ a channel doping of 3×10^{17} cm^{-3}. The details of the fabrication sequence have been discussed previously [3]. The FET's are designed to have a threshold voltage of $V_{gs} = -10$ V, resulting in an I_{dss} of 300 mA/mm at $V_{ds} = 10$ V and a peak transconductance of 45 mS/mm. With this device structure, 1-mm FET's typically have a three-terminal breakdown voltage V_{ds} in the range of 150 - 200 V, defined at the 1 mA/mm point. As determined from small-signal S-parameter measurements, average values for frequency response are $f_T = 9$ GHz and $f_{max} = 20$ GHz.

From these devices a maximum power density of 4.6 W/mm at 3.5 GHz has been measured using an on-wafer load pull system. As shown in Figure 3, a 0.25-mm FET operating at a drain bias of 60 V had a peak power of 1.15 W, a Class A PAE of 39% and an associated power gain of 12.5 dB. A similar device operating at 800 MHz had a power density of 3.0 W/mm and an improved PAE of 60%, as shown in Figure 4, demonstrating that because of the high operating voltages, the intrinsic efficiency of SiC MESFET's is high.

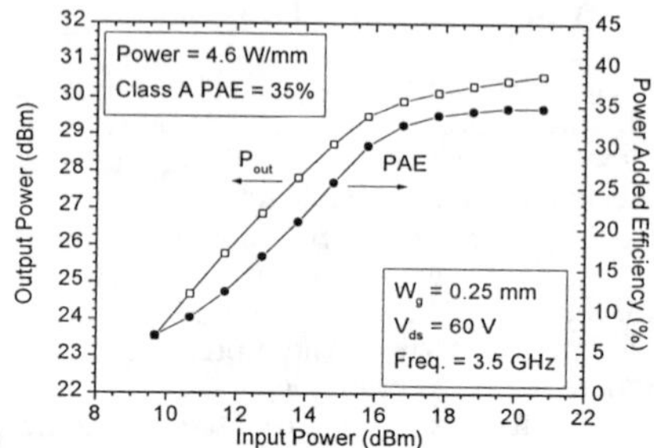

Figure 3: 0.25-mm SiC MESFET with a peak power density of 4.6 W/mm at 3.5 GHz with a drain bias of 60 V.

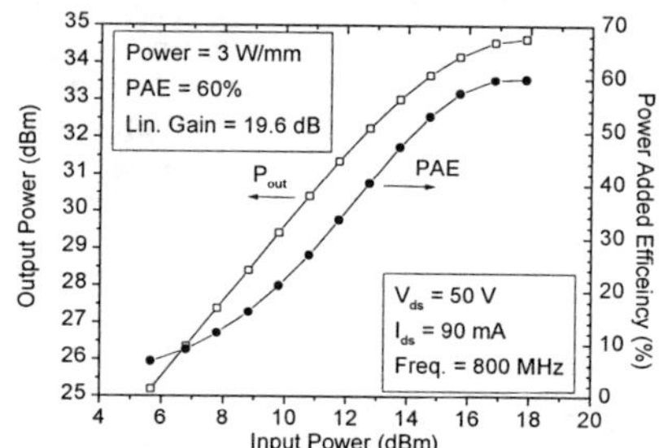

Figure 4: Power sweep of a 0.84-mm SiC MESFET with a PAE of 60% and 3.0 W/mm.

A single device with 48-mm of gate periphery was packaged in a hybrid circuit with input and output matching networks fabricated from alumina, and produced 80 watts CW at 3.1 GHz with 38% PAE, as shown in Figure 5. This is the highest CW power level ever reported for a single device operating at this frequency, and demonstrates the potential of SiC to have a substantial impact on solid state microwave power amplifiers.

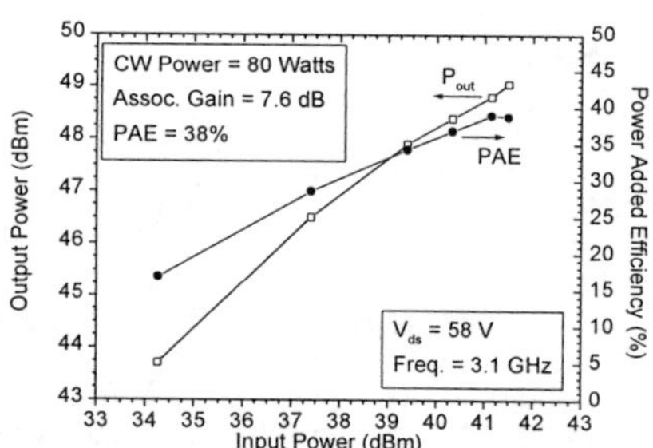

Figure 5: Power sweep of a 48-mm SiC MESFET at 3.1 GHz showing a CW power level of 80 watts.

With an increase in the channel doping and a reduction of the gate length to less than 0.45 µm, SiC MESFET's have shown excellent power performance up to 10 GHz. As illustrated by the power sweep in Figure 6 a power density of 4.3 W/mm was measured from a 0.25-mm MESFET at 10 GHz, with a peak power of 1.1 W, a Class A PAE of 20% and an associated gain of 9 dB. Cree has also recently developed a via hole process for the semi-insulating SiC substrates, making the fabrication of MMIC's possible. Combined with the high power densities shown at X-band, this makes SiC MESFET's an attractive MMIC technology for future systems.

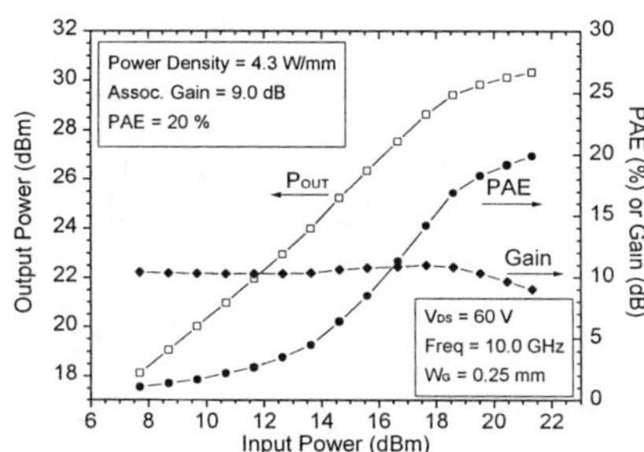

Figure 6: On-wafer load pull measurement of a SiC MESFET showing 4.3 W/mm at 10 GHz.

GAN/ALGAN HEMT's ON SiC SUBSTRATES

Another exciting technology for high power microwave applications that leverages the advantages of semi-insulating SiC is GaN/AlGaN High Electron Mobility Transistors (HEMT's). The close lattice match between SiC and GaN results in a lower defect density epilayer than typically obtained with growth of GaN on sapphire substrates. Additionally, the much high thermal conductivity of the SiC substrate is critical for dissipating the very high power densities possible with the GaN system. GaN devices fabricated on sapphire substrates have achieved relatively high power densities, but the junction temperatures at which these levels are achieved are typically in excess of 300°C due to the very high thermal impedance of the substrate. The structure of the GaN/AlGaN HEMT devices fabricated on SiC substrates is shown in Figure 7. The structure contains an AlN buffer layer, 2 µm of undoped GaN and approximately 27 nm of $Al_{0.14}Ga_{0.86}N$. The AlGaN cap layer comprises a 5 nm undoped spacer layer, a 12 nm, $2x10^{18}/cm^3$ Si-doped donor layer and a 10 nm undoped barrier layer. Device isolation is achieved with shallow mesas dry etched in a chlorine-based plasma.

Typical dc output characteristics of a 1-mm-wide HEMT with L_G = 0.45, L_{GS} = 1.0 and L_{GD} = 1.5 µm show a peak current of 680 mA/mm at V_{GS} = +2 V and a maximum extrinsic transconductance near V_{GS} = -0.5 V of 200 mS/mm. Typical three-terminal gate-drain breakdown voltages range between 60-70 V. Small-signal gain measurements at V_{DS} = 20 V and V_{GS} = -1 V show an extrapolated unity gain frequency f_T of 28 GHz [4]. The maximum available gain (MAG) remained high up to the maximum frequency of the network analyzer, so the f_{MAX} of this device is estimated to be 114 GHz by modeling the power gain above 35 GHz. The effective channel electron velocity, as determined from the slope of the f_T vs. $1/L_G$ data from many devices, lies in the range $6-8x10^6$ cm/s.

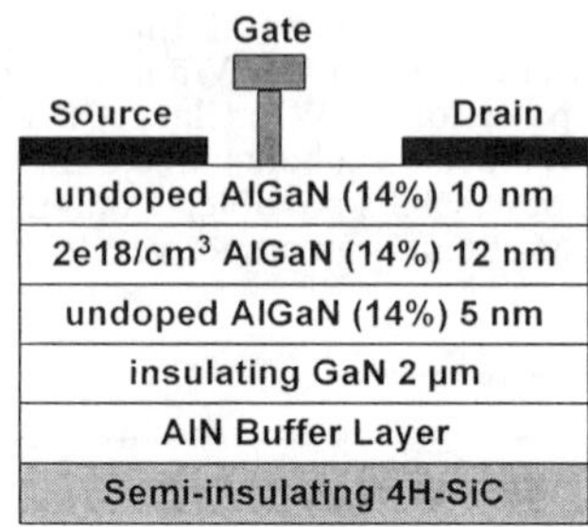

Figure 7: Cross-sectional view of the GaN/AlGaN HEMT structure grown on a semi-insulating 4H-SiC substrate.

On-wafer measurements were performed on a Maury load-pull system at 10 GHz and a drain bias of 30 V. A power sweep for a 0.125 mm HEMT (L_G = 0.45, L_{GS} = 0.5, and L_{GD} = 1.0 μm) is plotted in Figure 8. The most significant result was a record RF power density of 6.9 watts/mm with a PAE of 51 % and an associated gain of 9 dB, that was achieved at -6.0 dB of compression (~4.3 W/mm at -1 dB compression). This power density is about 8 times higher than typical GaAs devices, and is more than twice as high as any other GaN HEMT grown on a sapphire substrate, affirming the potential for AlGaN/GaN HEMT's on SI 4H-SiC substrates for use in high power amplifiers at X-band. When tested at 16 GHz, this device showed a CW power density of 4.4 W/mm with a PAE of 27% and an associated gain of 6.9 dB at the -3 dB compression point, which also demonstrates their advantage for high-power Ku-band operation.

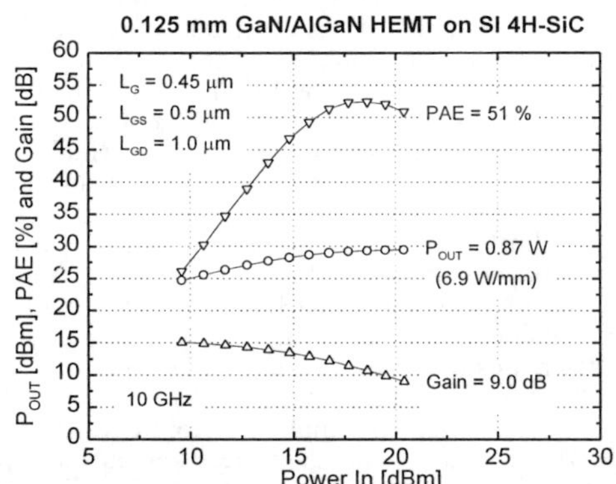

Figure 8: A 10 GHz CW power sweep for a 0.125-mm GaN/AlGaN HEMT on SI 4H-SiC (performed at the Air Force Research Laboratory, SNDM) that shows a record power density of 6.9 W/mm. The device was biased at V_{DS} = 30 V and V_{GS} = -2.25 V.

In order to measure total RF power on larger parts, HEMT's with 3 mm of gate width were packaged with ceramic input and output matching networks. After thinning the SI 4H-SiC substrate to approximately 4 mils, the HEMT was packaged into a hybrid matching network on a carrier that was maintained at 18°C. The power sweep shown in Figure 9 represents a single 3-mm device tested at 7.4 GHz. During the part of the sweep up to 30 dBm input power, the device was biased at V_{DS} = 28.4 V and V_{GS} = -1.8 V. The drain bias was changed to 31 V for the last three data points to achieve a maximum output power of 9.1 watts, a PAE of 29.6 % and an associated gain of 7.1 dB (~2.1 W/mm at -1 dB compression). The high parasitic source inductance introduced into the hybrid circuit by the Au ribbon bonds reduced the gain and frequency response, and the characterization of the 3-mm HEMT at 10 GHz was unwarranted. When source via holes are incorporated in the GaN-on-SiC HEMT process, the total RF power is expected to improve, even at 10 GHz.

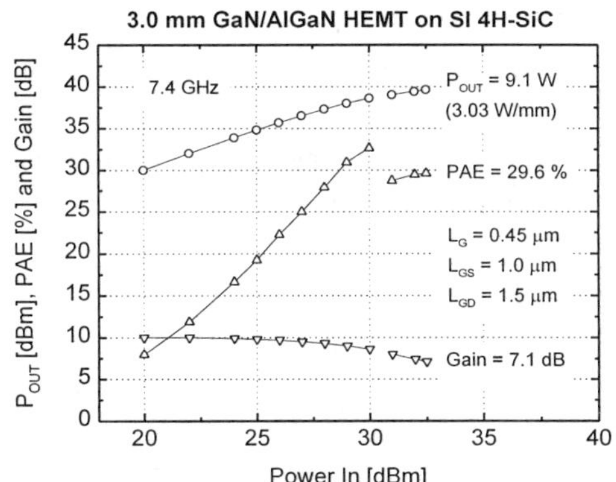

Figure 9: A 7.4 GHz CW power sweep for a 3-mm GaN/AlGaN HEMT on SI 4H-SiC in a hybrid matching circuit. For the initial part of the sweep, the device was biased at V_{DS} = 28.4 V and V_{GS} = -1.8 V. The drain bias was changed to 31 V for the last three data points to achieve a maximum output power of 9.1 watts.

CONCLUSION

The progress in the power performance of SiC MESFET's has been rapid, and these devices are now generating considerable attention for both military and commercial applications. The combination of high electrical power density of the devices and the high thermal conductivity of the SiC substrates enables power levels that are not achievable with other solid state technologies, particularly for CW communication applications. At power levels of 80 W per chip, SiC MESFET's would have a significant power advantage over Si LDMOS or GaAs MESFET's in base station power amplifiers. As SiC MESFET's mature into a MMIC technology, they should have an impact on solid state power applications through X-band. Additionally, recent progress with GaN/AlGaN HEMT's fabricated on semi-insulating 4H-SiC substrates shows this to be an exciting technology with tremendous potential for high power applications at X-band and above.

ACKNOWLEDGMENTS

The SiC material development was sponsored by DARPA, monitored by AFWL on contract no. F33615-95-C-5426. The SiC MESFET work was sponsored by the Naval Research Laboratory, contract no. N00014-96-C-2152, and by a DARPA MAFET Thrust 3 program, contract no. F33615-96-C-1967. The GaN/AlGaN HEMTs were funded on a Phase II SBIR from the Army Research Laboratories, contract no. DAAL01-97-C-0108. The load-pull measurements of Fig. 8 were performed by Lois Kehias and Thomas Jenkins of AFRL.

REFERENCES

[1]   R.P. Joshi, J. Appl. Phys. **78**, 5518 (1995).

[2]   G.A Slack, J. Appl. Phys. **35**, 3460 (1964).

[3]   C.E. Weitzel, J.W. Palmour, C.H. Carter, Jr., K.J. Nordquist, Elect. Dev. Let. **15**, 406 (1994).

[4]   S. T. Sheppard, K. Doverspike, W. L. Pribble, S. T. Allen, J. W. Palmour, L. T. Kehias, T. J. Jenkins, *56th Annual Device Research Conference*, 1998.

CURRENT STATUS OF SIC POWER SWITCHING DEVICES: DIODES & GTOS

S. SESHADRI, A.K. AGARWAL[1], W.B. HALL[2], S.S. MANI[3], M.F. MACMILLAN, R. RODRIGUES[4], T. HANSON[4], S. KHATRI[4] AND P.A. SANGER
ESSS, STC Northrop Grumman Corporation, 1350 Beulah Road, Pittsburgh, PA 15235-5098, sseshadr@aeslcad.essd.pa.northgrum.com
[1]Currently at Cree Research, Inc. 4600 Silicon Drive, Durham, NC 27703
[2]Northrop Grumman Corporation, Baltimore, MD
[3]Currently at Sandia National Laboratory, Albuquerque, NM
[4]Silicon Power Corporation, Malvern, PA

ABSTRACT

The progress that has been made in SiC diodes and GTOs is reviewed. A 100 A/1000 V SiC p-i-n diode package, the highest current rating reported for any SiC device, a 69 A conduction/ 11 A turn-off of a SiC GTO and MTO^{TM}, as well as the first all-SiC, 3 phase Pulse Width Modulated (PWM) inverter are reported, herein, for the first time. The inverter achieves voltage controlled turn off with a high temperature capable, hybrid SiC JFET. Material and process technology issues that will need to be addressed before device commercialization can be realized are discussed.

INTRODUCTION

The potential of SiC devices for power switching applications is well recognized.[1,2] Several markets have been identified for these devices (see Table 1). It is easily seen that there is a greater chance for commercial success of SiC power switching devices over the relatively more mature SiC RF devices. Demonstration of 5.9 kV p-i-n and 1700 V Schottky diodes, 2.6 kV MOSFETs and 1.4 kV Gate Turn-Off (GTO) thyristors are illustrative of the significant advances that have been made in the power area. However, these successes on small, low current devices now need to be followed by demonstration of scaled up devices. While applications of devices with voltage ratings > 9 kV currently appears to be the realm of SiC devices, lower voltage applications face

Table 1

Potential markets for SiC devices (in order of relative size)	
Application area	Typical device
Motor controllers for Electric Vehicles	Schottky diode, MOSFET
Engine and airflow controls for aviation electronics	Schottky diode, MOSFET, sensors
Electric Utility Power applications	P-I-N diode, Thyristors
Industrial High voltage power electronics	P-I-N diode, MOSFET/IGBT
Ballasts for flourescent lamps	MOSFET
Conventional ICE automobile sensors/electronics	Sensors
HDTV transmitters	SIT
Surveillance and tactical radars	MESFET, SIT

significant challenges from the concurrent development of alternative technologies. The expected improvement of silicon technology, illustrated by developments in silicon IGBT and MOSFET devices, as well as interest in other materials for power device applications[3] are cases in point of the need for timely identification and resolution of outstanding technical issues to obtain continued performance and reliability improvements of SiC power devices. This paper focuses only on issues related to the development of SiC diodes and thyristors.

Material quality

Material quality is an important issue for the fabrication of high power devices. Much of the research being performed on high voltage devices are presently aimed at devices having ratings of at least 1-10 A/~1 kV, increasing to 100-1000A/5 kV. The need to minimize on-state voltages limits practical devices to current densities of ~ 500 to 1000 A/cm^2. Such a rating is sufficient to enable conductivity modulation, without significant carrier-carrier and Auger scattering induced resistance and will require device sizes between 5 and 20 mm^2. However, present devices are limited by catastrophic voltage degradation due to hollow-core screw dislocation ("micropipe") bulk defects which exist in densities of as low as 30 cm^{-2}. Thus, device size is limited to ~1 mm^2. Progress in this area is incremental. Because, it is well known that the distribution is NOT uniform one approach to device scaling has been to selectively place devices in sufficiently large micropipe-free areas of the wafer. This approach neglects the contribution of other defects. Another approach would be to parallel many smaller devices. While possible for unipolar devices, the success of this strategy will have to be demonstrated for bipolar devices since current hogging in the on-state and uniform switching to prevent catastrophic destruction are potential problems, especially forGTOs.

Fig. 1 shows the breakdown voltage of devices of various sizes. It is clear from the figure that devices much smaller than 1 mm^2 have severely degraded breakdown characteristics. The observed degradation is consistent not with micropipe defects, but rather with ~10^3 to 10^5 cm^{-2} defect densities associated with closed core screw dislocations in SiC substrates. Such defects have been found to result in higher pre-breakdown reverse leakage currents, softer reverse breakdown I-V curves, and highly localized microplasma breakdown current filaments with space-charge limited conduction in small diameter, low voltage (<250 V) p-i-n diodes. Encouragingly, a positive temperature coefficient of breakdown voltage (PTCBV) -- important to prevent localized self-heating of device to destructive breakdown -- has been observed even for devices with a few screw dislocations. A failure power density of ~2 MW/cm^2, ~5x greater than for silicon rectifiers have been reported for those devices.[5] However, it is not yet clear whether these results apply to higher voltage/large area devices where large numbers of dislocations or the higher power density could prevent true avalanche breakdown. This is important because, if true, it would result in a detrimental, effective negative temperature breakdown voltage coefficient, leading to current filamentation and

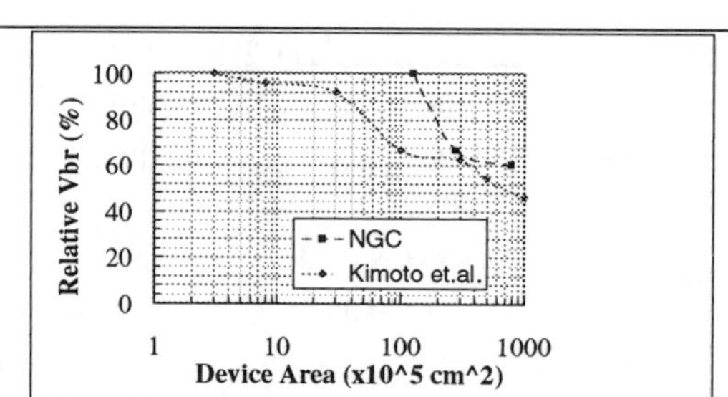

Fig. 1: Dependence of breakdown voltage on device area. Kimoto, et al. Data from ref. 4

catastrophic breakdown. Preliminary measurements on devices made at Northrop Grumman indicate that a positive temperature coefficients can be achieved for multi-kV devices. However, in accord with the data in figure 1, many devices have a negative coefficient, making the paralleling approach a more viable approach than the selective placing of large area devices because smaller devices can be used. Nevertheless, questions about how these dislocations affect diode and bipolar gain devices such as the Bipolar Junction Transistor (BJT), Insulated Gate Bipolar Transistor (IGBT) and GTO thyristors still remain.

Devices with 5 kV ratings will require ~25 to 50 μm thick epitaxial drift regions with doping concentrations < 1×10^{15} cm^{-3}. While such parameters are reasonably achievable with good morphology using current epitaxial growth processes, epitaxial defects such as "tetrahedral pits', "comet tails" and "carrots" do exist and can also degrade device performance. The geometry of these devices locally reduces drift layer thickness, enhancing the peak electric field. Conventional wisdom places their origin at bulk defects or other surface features such as polishing marks. As in the case of bulk defects, de-rating the maximum field (i.e. voltage) rating of a given epitaxially grown drift region generally increases the on-resistance and stored charge resulting in increased forward drop and switching time , respectively, of a device. Moreover, the still uncontrollable minority carrier lifetime of these layers affects the on-state characteristics of high voltage diodes and GTOs. This, combined with the certainty of higher than desirable dislocation densities for the forseeable future makes the reduction of epitaxial defect densities an imperative.

Minority carrier lifetime is an important parameter for bipolar devices. While some measurements have been made on representative material, its value on typical grown layers is not generally well known and more attention is needed in this area. Measurements performed on epitaxially-grown low voltage p-i-n mesa diodes indicates that lifetime is perimeter governed, ostensibly due to poor device passivation. This has important implications for bipolar devices including GTOs where the necessity for high turn-off gains places importance on narrow anode fingers. Most of this problem is undoubtedly due to the poor quality of mesa side walls which forced the abandonment of SiC UMOSFETs in favor of planar DMOS geometries. Planar devices have the advantage of eliminating oxide growth on surfaces containing both Si and C atoms. Since most wafers are grown along the c-axis and prepared with the Si face (as opposed to the C face) as the surface, oxide quality on this surface is expected to more closely resemble the silicon system. However, such an approach results in the need for ion implantation technology.

Activation of ion implants (N and more recently P for n-type, Al or B for p-type) has been shown to occur at temperatures above 1500 C. Low dose (~10^{16} - 10^{17} cm^{-3}) implants are needed for MOSFET, IGBT and GTO gate regions, as well as for the Junction Termination Extension (JTE) edge termination technique favored for high voltage devices. Activation is most critical for JTE, since its implementation requires control of the total charge in the implanted layer. A reduced activation temperature is important to prevent step bunching. In this regard, P implantation shows promise, having been recently shown to result in lower sheet resistance. However, both P and Al have relatively higher masses than their counterparts, resulting in shallower implants with more crystal damage precisely where one does not want it – the device active junction. Indeed, some identified problems related to poor activation include single injection-induced voltage snapback in implanted diodes, poor ohmic contacts and poor turn-on behavior of GTOs. Nevertheless, Al is preferred over B for applications such as guard ring terminations where the greater B diffusion coefficient may result in poor lateral definition. A good ohmic contact is important for optimizing device current and on-state voltage rating. Once again, p-type contacts remain the main problem. High dose implants using Al/C co-implants have been shown to yield sufficiently low contact resistances.

Devices

PIN diodes

SiC p-i-n diodes are expected to be the diode of choice above ~2 kV and in applications requiring temperatures above ~150 C because of the soft breakdown and lower barrier height of Schottky devices. Cree Research has reported an impressive 5.9 kV JTE-terminated implanted diode, the highest blocking voltage yet achieved for this device. While the on-resistance was not reported, this device had a 5.4 V forward drop at 100 A/cm^2. However, the device size was only 100 µm dia. and the authors were

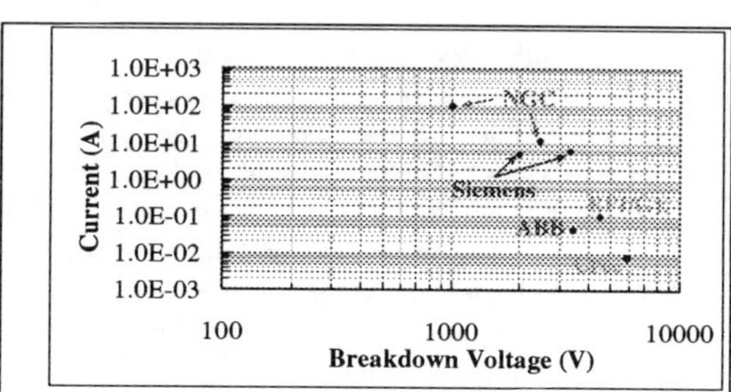

Fig. 2: Current and breakdown voltage rating for some recent SiC p-I-n diodes.

not able to test the breakdown stability. As with the Schottky diodes, large area devices suffer degraded voltage breakdown. Other workers have reported 1 mm^2 planar, Al-implanted JTE-terminated p-i-n diodes of 6 A/3.3 kV (E_c ~ 2.5 MV/cm) and 5 A/2 kV (E_c ~ 2.1 MV/cm) with PTCBV to 150 °C, sustainable avalanche current density of 20 mA/cm^2, 4.8 V @ 70 A/cm^2 and 4.0 V @ 500 A/cm^2, corresponding to on-resistances of 3 mΩ-cm^2 and 2.2 mΩ-cm^2, respectively.[6] The authors estimated a rather low temperature insensitive, lifetime of < 100 ns from turn-off measurements. The results of these and other recent work in the literature are shown in Fig. 2. Also shown in the figure are our most recent results (described below).

Fig. 3 illustrates the first demonstration of a 100 A/1000 V diode package, the highest power rating yet reported for a SiC diode, using a parallel arrangement of 55 400 & 600 µm dia. planar, Al-implanted guard ring terminated diodes with 10 µm drift layers. Measurements performed in a high temperature dielectric fluid showed that PTCBV could be obtained on individual devices up to 100 °C with a sustainable avalanche breakdown current density of 16 mA/cm^2. Additionally, 12 A/ 2.45 kV was achieved in a single (unpackaged) 1 mm dia. diode with 20 µm drift layer. However, in both instances, poor ohmic contacts formed by an oxygen leak during the sintering process were responsible for excessively high on-state voltage at room temperature. The lack of reduction in on-state voltage above 150 °C is thought to be due to non-ideal package and parasitic device resistance. It should be mentioned that no evidence of single injection related snapback has been observed in these devices.

Thyristors

The conductivity modulation possible with bipolar devices such as the Bipolar Junction Transistor (BJT), Insulated Gate Bipolar Transistor (IGBT) and thyristor make these devices more favored than unipolar devices such as the MOSFET for high voltage applications. High power BJTs have not, to date, been developed in SiC. SiC IGBTs have the same temperature

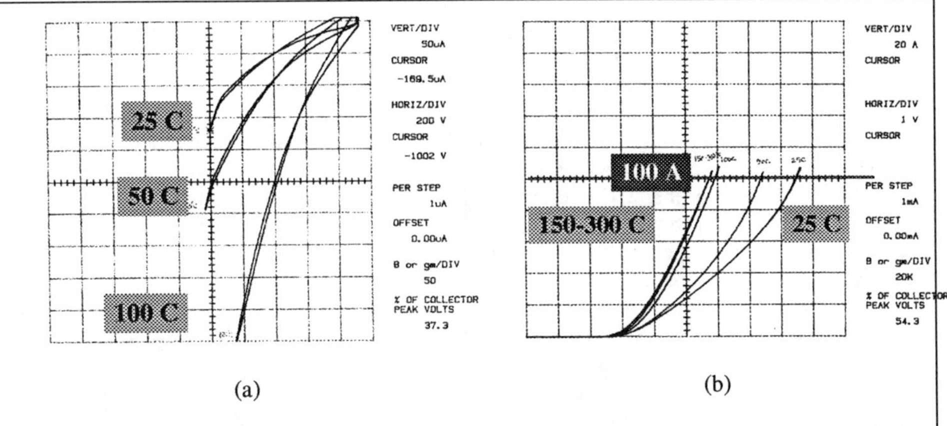

Fig. 3: Temperature dependent (a) reverse and (b) forward I-V characteristic of a 100 A/ 1000 V SiC pin diode package consisting of 15 parallel 600μm dia. diodes.

reliability and high field sensitivity issues faced by conventional SiC MOSFETs.[7,8] These facts make the SiC thyristor presently the best option for high temperature, high voltage applications.

A thyristor is a four layer devices with the emitters of the upper (pnp) and lower (npn) transistors form the anode and cathode contacts, respectively, with the n-base of the pnp forming a gate contact. A common collector/base junction supports the large positive and negative voltages during their blocking state. Thyristors are turned on by a small gate current applied to the upper transistor in the off-state, resulting in low current gain transistor-like characteristics until the sum of the common base gains of the two transistors exceeds 1, at which time regenerative feedback turns the collector base into forward bias, leaving both transistors in voltage saturation. This results in a low on-state voltage of approximately a single diode drop. The devices turn off when the corresponding main current or voltage goes to zero. On-state dissipation considerations due to the larger junction voltage of SiC ensures that thyristor switches become more favorable only at very high voltages or in applications where the high temperature reliability issues of the MOSFET cannot be adequately compensated. A lower voltage range is also possible for applications where the greater current handling capability (with its attendant smaller device size) are at a premium. Thyristors with high voltage (900 V), high power (700 V/6 A), high temperature (500 °C), and increased reliability (> 600 hr lifetime) have already been demonstrated.[9]

Gate Turn-Off thyristors are more favored for DC power switching applications because of their independent gate turn off capability. These devices, therefore, do not need to support large negative voltages in this application. This simplifies their design, resulting in an asymmetric design in which the drift region is made thinner by insertion of a higher doped buffer layer in the lower base layer to prevent the occurrence of punch through breakdown. The problem of terminating the lower junction is also eliminated as only the common base/collector junction need support any voltage. The insertion of the buffer layer increases the turn-off gain by reducing the common base gain of the npn transistor while simultaneously reducing the likliehood of open base breakdown in the npn transistor which limits the off-state voltage. The thinner drift layer of such a structure reduces the on-state voltage drop and the stored charge, decreasing the power dissipation or increasing the switching capability of the device. SiC GTOs are inverted with respect to their silicon counterparts because of the need to avoid using highly resistive p-type SiC substrates.

We have previously demonstrated GTOs with on-state voltages of 2.7 V at 390 C and 3.35 V at room temperature for a current density of 500 A/cm^2. The large decrease in voltage with increasing temperature occurs because of the decreasing SiC band gap, thereby lowering the knee voltage, and because increased ionization of n- and p-type impurities of the anode and substrate lowers the on-resistance of these regions. Room temperature contact resistances for the best devices were a very low 2.4 x 10^{-5} Ω-cm^2. Like the p-i-n diodes, the robustness of the GTO and packaging approach was demonstrated by the conduction of 69 A (4300 A/cm^2) through fifteen GTOs connected in parallel in high power package (see fig. 4). These devices did not show any indication of current hogging for the current densities measured.

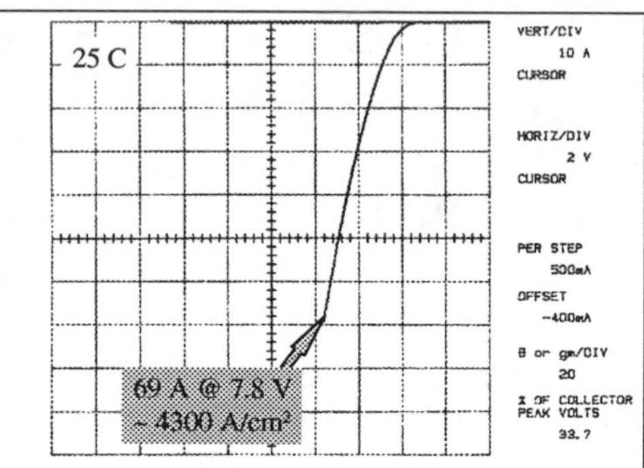

Fig. 4: On-state current through 14 SiC GTOs connected in parallel.

Turn-off gain (I_{ak}/I_g) of GTOs is generally dependent on the exact geometry of the gate and anode. Measurements shown in Fig. 5 on interdigitated devices demonstrate from a relatively high value of 8 at just above the holding current density (~40 A/cm^2), when the device is barely latched on and there is very little minority carrier stored charge in the drift layer, to a minimum of just under 2 at current densities >100 A/cm^2, which occurs once conductivity modulation occurs and the drift region is saturated with minority carriers. Thus, one expects that switching times will be dependent on both cathode current and peak GTO gate current with the slowest turn-off occurring for operation at high turnoff gain and high on-state current densities. The fastest turn-off is then achieved by operating under unity gain conditions with low on-state current density.

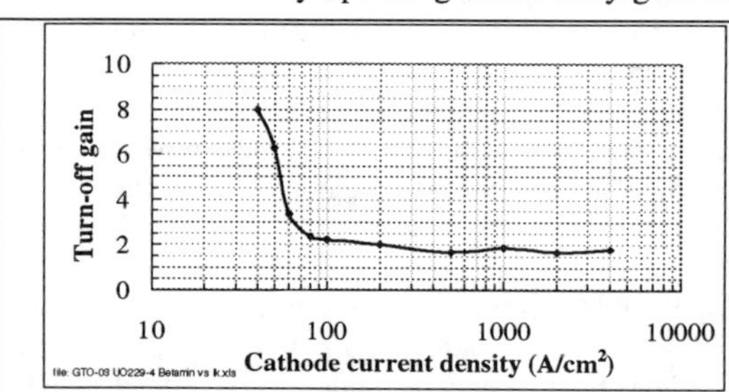

Fig. 5: Turn-off gain versus cathode current for a typical SiC GTO.

However, the desire for fast turn-off is moderated by the desire to minimize device area. The dependence of the turn-off current gain on the cathode current is the reason for the complexity of GTO gate drives. We have chosen a trade-off in which we operate at a current density of 500-1000 A/cm^2. Our demonstration of unity gain switching times of < 200 ns for SiC GTOs at a current density of 1500 A/cm^2, therefore, suggests very strongly that SiC GTOs will live up to their expectations for fast switching.

Storage, rise and fall times of ~113, 43 and 100 ns, respectively, have previously been reported for SiC thyristors at a very low current density of 30 A/cm^2.[10] Maximum turn-off current densities > 5000 A/cm^2 have been achieved both by us and reported by others.[11] However, the inverse relationship between this gain and anode finger width suggests that a trade-off must be made between maximum turn-off gain, maximum current density (on-state voltage) and device foot print within the previously prescribed range of ~500 to 1000 A/cm^2.[12] We have previously reported GTO operation at 325 °C with switching times increasing by a factor of 4 from room temperature due to an increasing minority carrier lifetime. We have also demonstrated repetitive GTO switching up to 350 V (3A) with 3.2 µJ switching losses consistent with the lower stored charge in these devices compared to Si GTOs.

We have now extended the current rating of packaged GTOs to 11 A (@ 800 A/cm^2) (see **Fig. 6** (a)) with switching times < 100 ns using the same package as that demonstrating 69A conduction without degradation in their turn-off current characteristics. This result represents the largest turn-off current reported to date and demonstrates the viability of the paralleling approach with respect to turn-off. Turn-off of 69 A was not expected to be achieved because the structure employed had a peak turn-of current density of ~1000-1500 A/cm^2. Unfortunately, the devices used had poor breakdown voltage so no high voltage measurements could be taken.

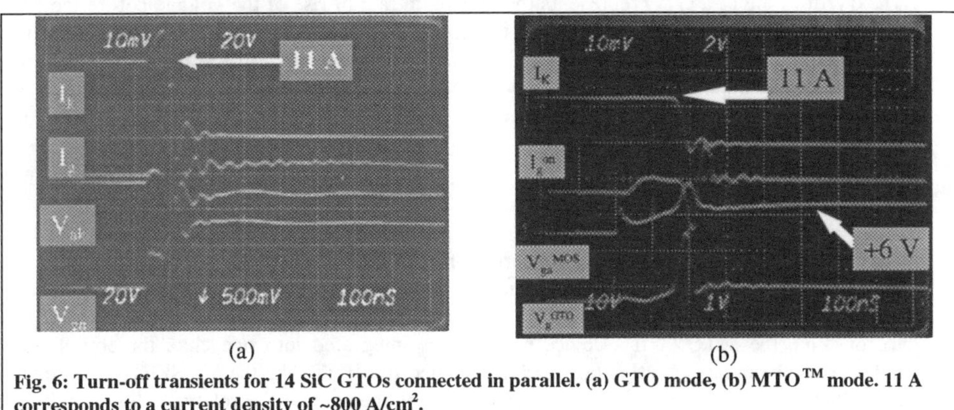

Fig. 6: Turn-off transients for 14 SiC GTOs connected in parallel. (a) GTO mode, (b) MTO^{TM} mode. 11 A corresponds to a current density of ~800 A/cm^2.

We now examine the turn-off characteristics of the 69A GTO package in Fig. 6a shows typical. Before the application of the turn-off gate pulse the device is in its on-state with all three junctions forward biased. Thus, both transistors are in saturation. Application of the turn-off gate pulse induces the gate-to-anode junction into reverse bias, causing V_{AK} to climb towards zero almost instantaneously. After ~80 ns, the storage time (t_s) of the device, the minority carrier concentration at the middle junction decreases below its equilibrium value, reverse biasing it and allowing the voltage to fall to its off-state value. Concurrently, both base and cathode currents decrease rapidly with a characteristic fall time, t_f. The absence of the tail current typically seen in silicon bipolar switches greatly diminishes the switching losses of the SiC GTO and is indicative of the fact that the residual minority carrier density in the base regions is very small, as expected because the 10x larger critical field of SiC enables commensurately thinner drift region, leading to 10x lower stored charge.

Voltage controlled GTO turn-off

Voltage control is especially important during turn-off because of the low turn-off current gain of the GTO. A hybrid MTO™ package containing a silicon power MOSFET with a SiC GTO was assembled to examine the MOS-gated turn-off characteristics of SiC thyristors. The device operates with the MOSFET shunting the gate-to-anode junction in a manner analogous to a removable cathode short in silicon GTOs. The MOSFET in this arrangement is external to the device and thus can be independently optimized without having the oxide subjected to large electric fields. Furthermore, it is only a low voltage device, since it will never see more than the gate/anode breakdown in the worst case. **Fig. 6** (b) illustrates the switching behavior of this hybrid package. Voltage controlled turn-off of an equivalent current to that obtained in the GTO mode switching is clearly demonstrated, indicating that sufficiently large current levels can be easily switched using this device configuration. The turn-off is slower than for the GTO in part because of the fact that in the GTO-mode turn-off, the gate-to-anode junction is strongly reverse biased, whereas in the MTO-mode turn-off this junction is simply held near zero voltage during the storage cycle. The fact that we see no noticeable degradation of turn-off capability while paralleling these many devices under these more restrictive conditions also bodes well for the paralleling approach to device scaling.

The dV/dt rating of a GTO represents the maximum rate of rise of the voltage before the onset of unwanted retriggering of the device due to capacitive displacement currents. Thus, it partially determines the maximum device operating frequency. A high rating also has the added advantage of reducing or eliminating the need for snubber components. Typical silicon GTOs have ratings in the 600-1000 V/μs range. We have measured ratings as high as 700 V/μs in GTO mode, increasing to >1400 V/μs with the use of resistive shunts (as high as 100 Ω) such as the MOSFETs connected across the gate and anode, sufficient to warrant snubberless operation of inverters for motor drive control applications.

The high temperature reliability concerns of MOSFETs has been previously mentioned. Therefore, a thyristor with high temperature, voltage-controlled turn-off capability can be achieved by simply replacing the MOSFET with a SiC JFET, resulting in a Junction Controlled Thyristor. Like the MTO™, this device need not be integrated into one chip, thereby allowing both devices to be independently optimized. Moreover, the JFET, like the MOSFET, need only be a low voltage device, since it is never subjected to voltages above the relatively low gate/anode reverse breakdown voltage.

Turn-off transient was measured on a hybrid device constructed from a 4H-SiC GTO chip and a vertical 4H-SiC JFET (Fig. 7). The JFET had an on-resistance of 2.5 ohm with a gate to source bias, V_{gs}, of +3 V. The JFET can be turned off with V_{gs} = -1 V or less. Turn-on was achieved by forward biasing the anode to gate junction of the GTO while keeping the JFET in off condition through an application of -7.0 V on the gate of the JFET. In order to record the turn-off transient, the JFET was turned on by an application of +3 V on the gate. This presents a 2.5 ohm resistance between the anode and the gate of the GTO thus diverting all of the anode to cathode current through the JFET. The turn-off time was less than 100 ns. These results clearly demonstrate the functionality and potential of the hybrid JCT

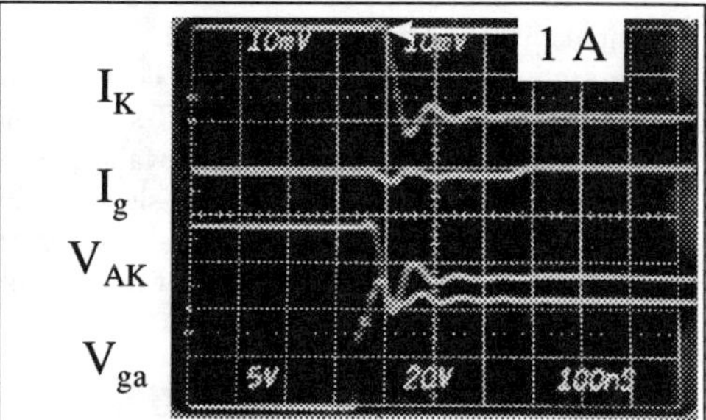

Fig. 7: Turn-off characteristics of a SiC JCT at a current density of 1000 A/cm².

for a higher switching speed up to 50-100 kHz.

All switching tests to date have been performed under resistive loading conditions. However, typical applications require switching under inductive loading conditions, where significant voltage overshoot occurs. Devices have not yet been tested for their ruggedness at the limit of breakdown. Preliminary switching comparisons suggested that SiC GTOs had fall times and turn-off energy losses ~10x lower than for comparable SiMCTs.[13]

SiC inverter

The rapid progress of SiC power device technology has led to the development of the first all-SiC inverter for PWM applications. Fig. 8 shows a schematic of this circuit, which consists of 3 phase switching with each phase consisting of an upper and lower switch containing a SiC GTO with an anti-parallel SiC pin diode and a low voltage SiC JFET connected across the GTO gate and anode. In a significant improvement over silicon

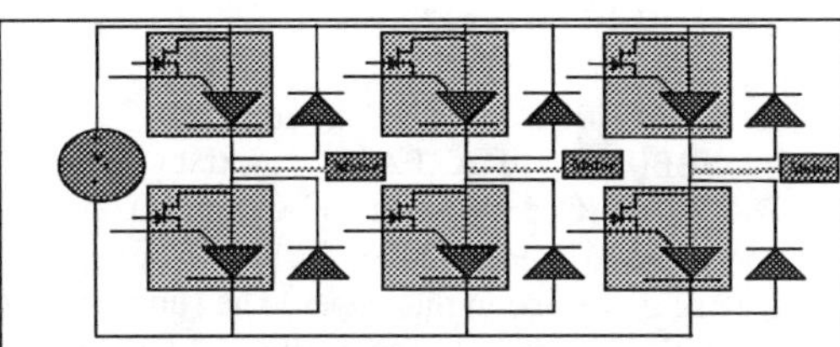

Fig. 8: Schematic diagram of PWM circuit. Three phases each consisting of an upper and lower pole. Each pole contains contains SiC GTO, JFET and diode.

devices, no snubber circuits are used or expected to be necessary because of the high dV/dt of the voltage controlled switch. Fig. 9 illustrates the entire PWM system, including the gate drive circuits and the motor, used in the demonstration.

In the demonstration, a DC bus voltage of 45 V is applied across the load, a 0.25 hp motor of the type used to re-circulate the coolant in an automobile. The GTOs conduct peak and average currents of 1.3 A and 200 mA, respectively, with the upper switches open for a phase angle of 120° and the lower switches being pulsed at a 4 kHz frequency over the same interval. Fig. 10 shows the operation of the inverter, with the phase current (i.e. current through one of the three phases of the motor) integrating to its maximum value with the pulsed bottom switch. The current and voltage levels are presently too low to determine typical switching characteristics of the individual SiC components. Tests are

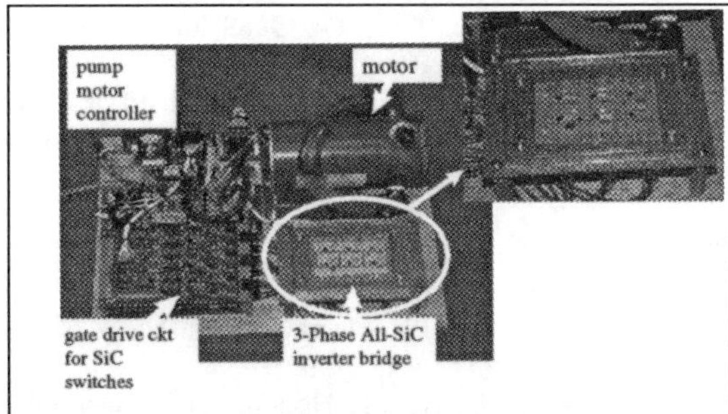

Fig. 9: Photograph of the first all-SiC PWM inverter for motor drive control.

currently underway to examine these issues up to the operational limits in voltage, current, frequency and temperature. We look forward to reporting these results in the near future.

CONCLUSIONS

Progress in SiC p-I-n diodes and GTOs has been reviewed. Among the highlights were demonstration of a 100A/1000V diode package, 69A conduction/11 A turn-off of a GTO package, MOSFET and JFET controlled turn-off of SiC thyristors and all-SiC PWM for motor control applications. The above results are indicative of the fast pace with which these devices are scaling up to power levels needed in typical power switching applications. Materials limitations affecting further improvements were also addressed.

ACKNOWLEDGEMENTS

The authors thank DARPA (contract # 786-DTD-96FEB01), DARPA (contract # F342/14)/EPRI (Order # WO8069-09) and Army (contract # F33615-96-3-2608) for their support of the work described in this paper. The support of V. Hegde and the technicians of the Advanced Electronic Structures Lab, especially J.C. Kotvas, at STC is also gratefully acknowledged.

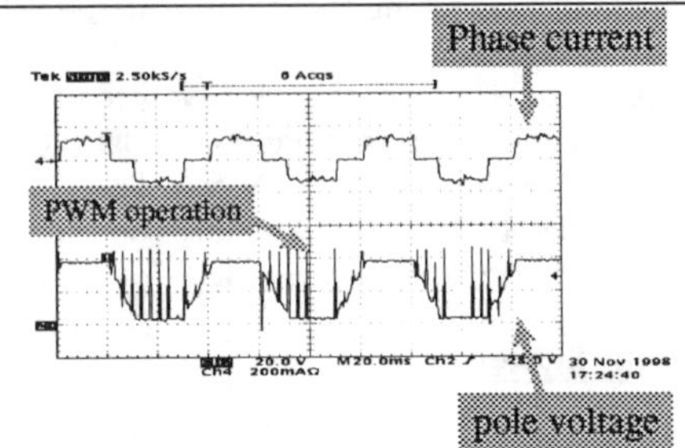

Fig. 10: Phase voltage and current waveforms of the SiC PWM operating at 4 kHz. The narrow spikes in the voltage waveform represents the off-state of on of the lower poles.

REFERENCES

[1] K. Shenai, et al. IEEE Trans. Electr. Dev., V. 36-9, 1989, p. 18111.

[2] M. Bhatnagar and B.J. Baliga, IEEE Trans. Electr. Dev., V.40-3, 1993, p.645.

[3] T. P. Chow, N. Ramungul and M. Ghezzo, Mat. Res. Soc. Symp. Vol. 483, 1998, p. 89.

[4] T. Kimoto, N. Miyamoto and H. Matsunami, IEEE Trans. Electr. Dev. V.46-3, (1999) P. 471.

[5] P. G. Neudeck, W. Huang and M. Dudley, IEEE Trans. Electr. Dev. V.46-3, (1999) p. 478. See also P.G. Neudeck and C. Fazi, IEEE Trans. Electr. Dev. V.46-3, (1999) p. 485.

[6] H. Mitlehener, et al., Proc. Of 19989 Int'l. Symp. On Power Semicond. Devices & ICs (SPSD '98), Kyoto, Japan, June 3-6,1998, p.127.

[7] A.K. Agarwal, S. Seshadri and L.B. Rowland, IEEE Electr. Dev. Lett. V18, (1997) p. 592.

[8] M.M. Maranowski and J.A. Cooper, Jr., IEEE Trans. Electr. Dev. V.46-3, (1999) p. 520.

[9] J.W. Palmour, R. Singh, L.A. Lipkin and D.G. Waltz, Trans. 3rd Int'l. Conf. on High Temp. Elec. (HiTEC) Alb. N.M. Vol 2, XVI-9-14, Jun, 1996.

[10] Z. Xie, J. Fleming and J.Zhao, IEEE Electr. Dev. Lett. 1995.

[11] B. Li, L. Cao and J. Zhao, IEEE Electr. Dev. Lett. (1998).

[12] K. Xie, J.H. Zhao, J.R. Flemish, T. Burke, W.R. Buchwald, G. Lorenzo and H. Singh, IEEE Elec. Dev. Lett. 17, 142 (1996).

[13] J. Mooken, R. Lewis, J.L. Hudgins, A.K. Agarwal, J.B. Casady, S. Siergiej and S. Seshadri, Proc. 32nd IEEE Industry Applications Society (IAS) Conference, Oct. 5-9, New Orleans, LA 1997.

THE EFFECTS OF DAMAGE ON HYDROGEN-IMPLANT-INDUCED THIN-FILM SEPARATION FROM BULK SILICON CARBIDE

R.B. GREGORY *, O.W. HOLLAND **, D.K. THOMAS **, T.A. WETTEROTH *, S.R. WILSON *
*Motorola Inc., Semiconductor Products Sector, Tempe, Arizona
**Oak Ridge National Laboratory, Solid State Division, Oak Ridge, TN. 37831-6048

ABSTRACT

Exfoliation of SiC by hydrogen implantation and subsequent annealing forms the basis for a thin-film separation process which, when combined with hydrophilic wafer bonding, can be exploited to produce silicon-carbide-on-insulator, SiCOI. SiC thin films produced by this process exhibit unacceptably high resistivity because defects generated by the implant neutralize electrical carriers. Separation occurs because of chemical interaction of hydrogen with dangling bonds within microvoids created by the implant, and physical stresses due to gas-pressure effects during post-implant anneal. Experimental results show that exfoliation of SiC is dependent upon the concentration of implanted hydrogen, but the damage generated by the implant approaches a point when exfoliation is, in fact, retarded. This is attributed to excessive damage at the projected range of the implant which inhibits physical processes of implant-induced cleaving. Damage is controlled independently of hydrogen dosage by elevating the temperature of the SiC during implant in order to promote dynamic annealing. The resulting decrease in damage is thought to promote growth of micro-cracks which form a continuous cleave. Channeled H^+ implantation enhances the cleaving process while simultaneously minimizing residual damage within the separated film. It is shown that high-temperature irradiation and channeling each reduces the hydrogen fluence required to affect separation of a thin film and results in a lower concentration of defects. This increases the potential for producing SiCOI which is sufficiently free of defects and, thus, more easily electrically activated.

INTRODUCTION

Hydrogen implantation through an oxide film followed by hydrophilic wafer bonding and a thermal cycle is a process developed to cleave a thin film of silicon-on-insulator (SOI).[1] The process has recently been applied to produce silicon carbide-on-insulator (SiCOI) films for possible use as a wide bandgap semiconductor in power rf and switching devices.[2]

SiC thin films separated from bulk material using this process have measured too resistive, a condition attributed to damage in the SiC thin film caused by the hydrogen implant itself. The experiments described in this work are motivated by the desire to understand the implant damage mechanisms in order to make the separation process more efficient and produce defect-free, low-resistivity SiC.

The problem is illustrated in Figure 1 which shows a schematic of the hydrogen-implant-induced separation process and a channeled RBS spectrum of SiCOI (~500 nm) produced by this process. Backscattered counts from Si in the SiCOI (integrated over channels 540-640) measure 824 greater than similarly measured counts from virgin SiC. Calculating density using the RBS data, one measures 1.27×10^{20} displaced atoms/cm^3. These vacancies have the potential to cause

Mat. Res. Soc. Symp. Proc. Vol. 572 © 1999 Materials Research Society

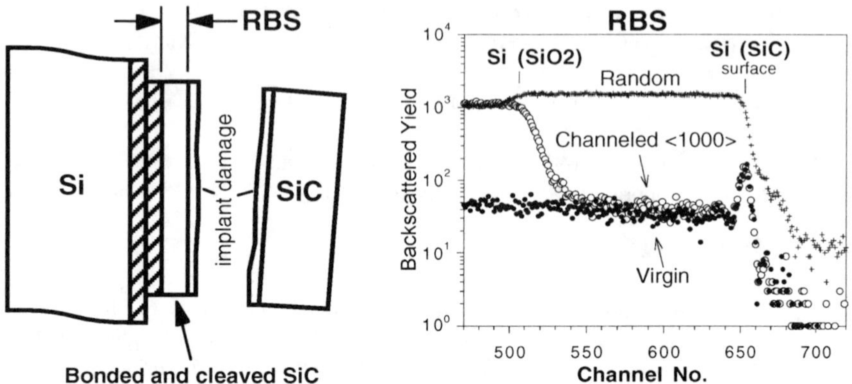

Figure 1. Schematic diagram of transferred and cleaved SiC (left) and RBS-channeling spectra of the transferred SiC film (right) ... H^+ implant damage polished off before analysis.

the deactivation of electrical carriers in material with typical doping concentrations of 10^{17}-10^{18} atoms/cm^3.

Previous work (with Si) shows that the process to cleave a thin film by hydrogen implantation followed by a thermal cycle is a combination of hydrogen chemistry and physical processes. The implant results in the formation of platelet-like microvoids which, during subsequent anneal, expand due to gas pressure of excess hydrogen. This links the microvoids into a continuous fracture, cleaving a thin film from the bulk wafer.[3] The use of H^+-implantation to affect transfer of a thin film from a bulk Si wafer was based upon observations of bubbling and exfoliation of implanted Si wafers after annealing.[4, 5]

The effects of ion-induced damage on the efficiency of the transfer process and its dependence on H^+ dose are demonstrated by observing exfoliation of SiC following H^+-implantation and anneal. Means to control damage independently of H dose are demonstrated with elevated-temperature and channeled implantation. It is proposed that channeled implantation generates less residual damage from the surface to at least half the projected range, $1/2R_p$, of the implant simply because crystalline axes of SiC are aligned with the H^+ beam, decreasing the cross section for ion-solid collisions. The elevated temperature implants affect in-situ, dynamic annealing in order to control H^+ implant damage in SiC.

EXPERIMENTS and RESULTS

Random vs. Channeled H^+-implantation

Experiments to measure damage and exfoliation of SiC as a function of H dose were accomplished using bulk SiC samples, 4H polytype, supplied as research grade material by Cree Research. They were implanted with 60 keV H^+ to doses ranging from 2.5 x10^{16} to 10.5 x10^{16} atoms/cm^2. Samples were tilted 7° from normal to affect random beam alignment. Additional samples were implanted with the H^+ beam aligned to [1000] axes to affect channeled implants over the same dose range. Damage analyses were accomplished by Rutherford backscattering (RBS)-channeling using a 2.3 He^{++} ion beam aligned with [1000] axes normal to the surface of the sample. Backscattered ions were detected at 160° relative to the incident beam using a solid state, surface barrier detector. Samples were then annealed in order to cause exfoliation of thin SiC from the bulk material. The amount of exfoliation was evaluated using optical microscopy.

Figure 2 shows RBS-channeled spectra for three of the samples from the set generated to evaluate the effects of implant damage on exfoliation for random-implanted H^+. The spectra represent as-implanted samples at room - or ambient - temperature (RT). These spectra show that damage to the SiC at the projected range, R_p, increases with the H^+ implant dose. The scattering yields near 1/2 R_p are also progressively greater (as a function of dose) than the yield from the virgin reference sample indicating the presence of ion-induced, residual damage at this location in all the samples. This is possibly due to displaced atoms that either dechannel or directly backscatter the incident He^{++} ions. Analysis by positron annihilation spectroscopy (not shown) indicates the presence of open volume defects, the result of displaced atoms at concentrations below the sensitivity of RBS. Such defectivity may be responsible for deactivating intrinsic carriers in SiC as previously reported for similarly implanted material.[6]

Following RBS characterization, all samples were subjected to 950°C, 15-minute anneals. then optically imaged using a microscope with Nomarski contrast. Figure 3 shows a portion of a series of optical micrographs produced to observe exfoliation of SiC as a function of dose for 60 keV H^+ implants done at RT. During the 950°C, 15-minute anneal, bubble formation occurs as the H^+ dose approaches 4.5 $\times 10^{16}$/cm^2, as seen in the optical micrograph [Figure 3(a)]. (Samples implanted with small increments of dose between 2.5 $\times 10^{16}$ and 4.5 $\times 10^{16}$/cm^2 revealed that the critical dose to produce exfoliation is very near 4.5 $\times 10^{16}$/cm^2.) Evidence for material removal or exfoliation of the 4.5 $\times 10^{16}$/cm^2 sample is clearly seen in Figure 3(a) by the appearance of broken bubbles. The amount of exfoliated surface material maximizes near a dose of 5.5 $\times 10^{16}$/cm^2 [Figure 3(b)], but at higher doses exfoliation decreases as seen in Figures 3(c) and (d), indicating a retrograde effect of the 60 keV H^+ implant to doses greater than 5.5 $\times 10^{16}$/cm^2.

The information conveyed by the images in Figure 3 is represented graphically in Figure 4 which shows the percentage of area that exfoliates following the 950°C anneal. Two sets of data are graphed, one for the randomly implanted samples and one for channel-implanted samples. One sees that channeled implants shift the onset of exfoliation (as well as maximum exfoliation) to approximately 1 $\times 10^{16}$ lower dose than the random implants. Furthermore, the maximum

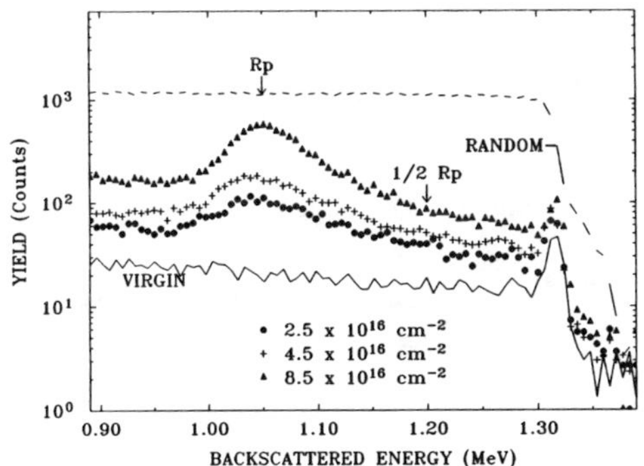

Figure 2. RBS-channeled spectra for Si in 60 keV H^+-implanted SiC. Three dosages shown are 4.5 $\times 10^{16}$, 6.5 $\times 10^{16}$, and 10.5 $\times 10^{16}$/cm^2. Reference spectra include the aligned yield from nonimplanted (virgin) SiC and the randomized yield from an implanted sample.

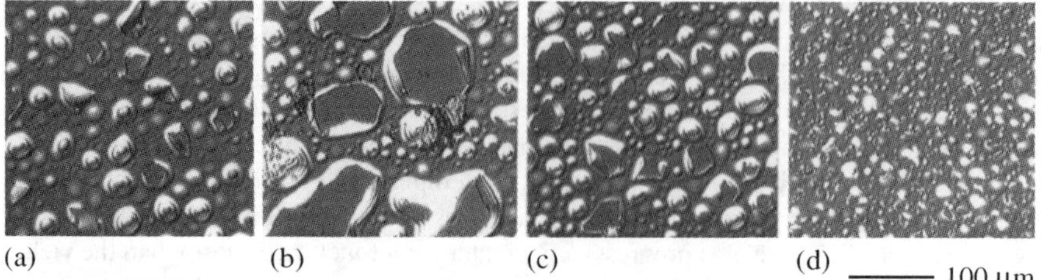

Figure 3. Nomarski optical micrographs of SiC implanted with 60 keV H^+ to dose (a) 4.5×10^{16}, (b) 5.5×10^{16}, (c) 6.5×10^{16}, and (d) $8.5 \times 10^{16}/cm^2$, after furnace annealing at 950°C for 15 minutes.

exfoliated area increases from 37% for the random implant (dosed $5.5 \times 10^{16}/cm^2$) to 69% for the channeled implant (dosed $4.5 \times 10^{16}/cm^2$). The rate of retrograde behavior of exfoliation appears the same for both random and channeled series.

SIMS depth profiles of hydrogen in random and channel-implanted samples are shown in Figure 5. Each of the samples was implanted with 60 keV H^+ to $2.0 \times 10^{16}/cm^2$, then annealed. The dose was held low enough to prevent exfoliation of the SiC during the anneal. The profiles show that the channeled implant has slightly greater range than the random implant. More significant, though, the retained hydrogen concentration measures almost three times greater for the channel-implanted sample.

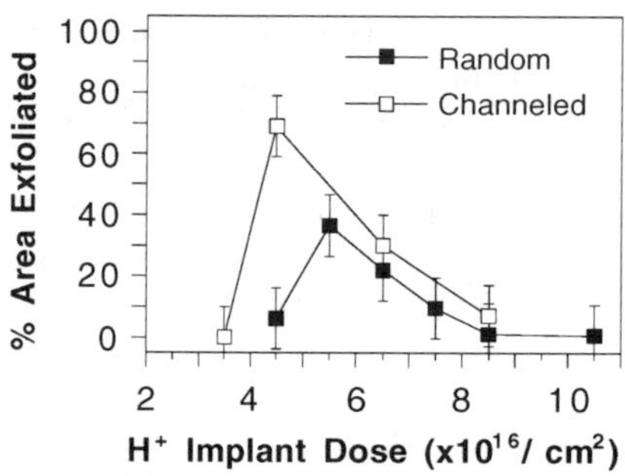

Figure 4. Percentage of area of exfoliated SiC as a function of 60 keV H implant dose. Two sample series graphed, one which was implanted in a random direction and one implanted with samples aligned to [1000] axes.

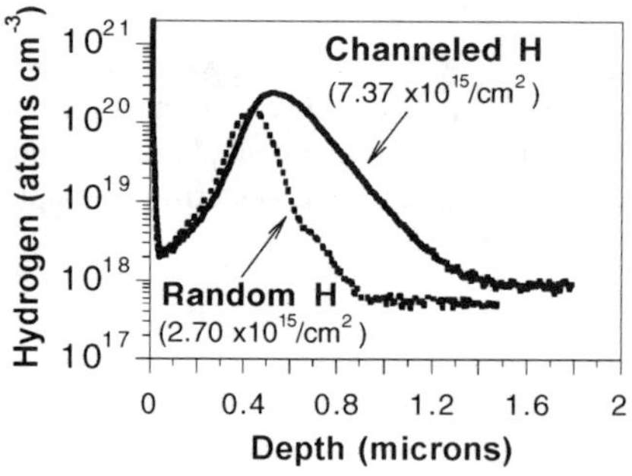

Figure 5. SIMS profiles of 60 keV-implanted hydrogen, dose =2 x10^{16}/cm^2, following anneal at 950°C for 15 minutes.

<u>**Elevated Temperature Implantation**</u>

To learn the effects of elevated temperature implantation for controlling H^+ implant damage, two sets of 4H-SiC samples were implanted with 60 keV H^+. One set was implanted at room temperature to doses ranging from 3.25×10^{16} to 4.5×10^{16}/cm^2. Samples of the second set were heated to 600°C during implant to doses ranging from 2.25×10^{16} to 8.0×10^{16}/cm^2.

Figure 6 shows RBS-channeling spectra for two samples H^+-implanted to a dose of 2.0×10^{16}/cm^2, one implanted at RT and the other with the temperature elevated to 600°C ("hot" implant). It is clear from comparing the scattering yields in the respective samples that the hot implant generated less damage at R_p as well as $1/2\ R_p$. Also evident is a slightly greater R_p for the hot implant. Optical micrographs for the series of hot implants (not shown) indicate the threshold dose for surface exfoliation of SiC during a 950°C anneal is ~2.75×10^{16}/cm^2.

Optical micrographs shown in Figures 7(a) and (b) compare the surface morphology after annealing for hot and RT implants, respectively. The images show about the same degree of bubbling and exfoliation although the H^+ implant dose for the RT implant is much higher, 4.5×10^{16}/cm^2, compared with 2.75×10^{16}/cm^2 for the 600°C implant. Previous work shows that the critical fluence for exfoliation decreases almost linearly with irradiation temperature.[7]

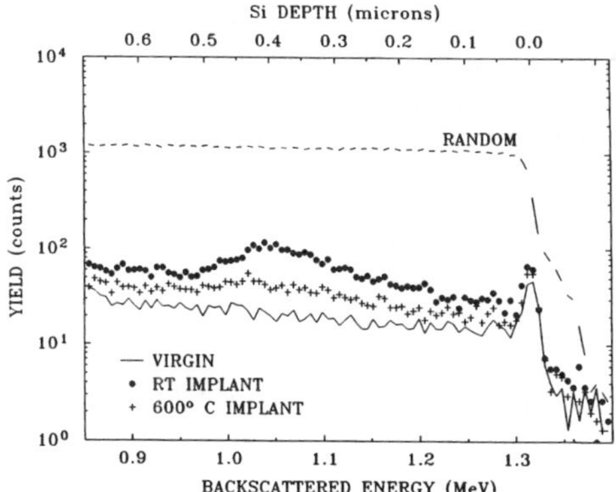

Figure 6. RBS-channeling spectra comparing damage to SiC following 60 keV H^+, 2×10^{16}/cm^2 implants at R.T. and 600°C.

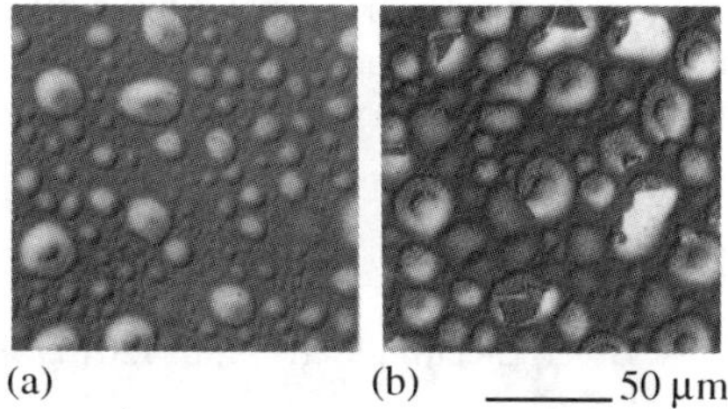

Figure 7. Nomarski optical micrographs of 60 keV H^+-implanted SiC showing similar degrees of bubbling and exfoliation for (a) 2.75×10^{16}/cm^2 @ 600°C and (b) 4.5×10^{16}/cm^2 @ RT.

DISCUSSION and CONCLUSIONS

The dependence of surface exfoliation of SiC on H^+ dose and the retrograde behavior of exfoliation as damage increases beyond a specific dose supports the the following model. Both the hydrogen concentration and the lattice damage affect the degree of exfoliation. Both increase with implantation dose, but damage retards exfoliation. It is clear that more hydrogen available within the lattice will lead to more bubbling and exfoliation, but the role of damage in suppressing the effect is not obvious. As seen from previous work with H^+-implanted Si, the formation of extended defects (i.e., platelets) is critical to the formation of microcracks within the lattice.[3] These microcracks and their ability to expand and interconnect yield the large macroscopic regions within the lattice which become separated from the underlying substrate either during exfoliation or thin-film transfer. One anticipates that substantial lattice damage may inhibit the formation of such macroscopic regions by hindering or stopping the propagation of the microcracks and thus preventing them from forming an interconnecting network.

The present work demonstrates the ability to control ion-induced damage independently from the implant dose by elevating the temperature of a sample to 600°C during the implant in order to dynamically anneal the SiC and potentially reduce the damage relative to an implant performed at room temperature. The RBS data show a significant reduction in residual damage (from the surface to $1/2R_p$) for the hot implant. The optical micrographs indicate that implanting hot also allows a substantial reduction in critical H dose needed for cleaving the thin SiC film, resulting with further decrease in damage.

Channeling the H^+ implant dramatically enhances the process of exfoliation. Measurements of exfoliated area from optical images indicate more robust exfoliation with lower dose relative to random implantation. The SIMS results suggest less out-diffusion of hydrogen during anneal, but increased diffusion into the bulk (below R_p) in the channel-implanted sample. This in turn suggests unique damage morphology at R_p which is not entirely understood.

It appears possible for damage in SiC to reach a concentration great enough to disrupt the formation of a continuous network of cracks. This conclusion is supported when damage is controlled independently of hydrogen concentration, either by elevating the temperature of the SiC during implant, or by channeling the hydrogen, or, quite possibly both. Each of these methods allows a reduction in critical H^+ fluence required to affect separation of a thin film and, therefore, may provide high-quality SiCOI material.

REFERENCES

1. N. Bruel, Electron. Lett. 37, p. 1201 (1995).

2. L. Di Cioccio, Y. Le Tiec, C. Jaussaud, E. Hugonnard-Bruyere, and M. Bruel, Mat. Sci. Forum **264-268,** p. 765 (1998).

3. M. K. Weldon, V. E. Marsico, Y. J. Chabal, A. Agarwal, D. J. Eaglesham, J. Sapjeta, W. L. Brown, D. C. Jacobson, Y. Caudano, S. B. Christman, and E. E. Chaban, J. Vac. Sci. Technol. B **15**, p. 1065 (1997).

4. W. K. Chu, R. H. Kastle, R. F. Lever, S. F. Mader, and B. J. Masters, Phys. Rev. B **16**, p.3851 (1977).

5. C. C. Griffioen, J. H. Evans, P. C. D. Jong, and A. van Veen, Nucl. Instrum. Methods Phys. Res. B**27**, p. 417 (1987).

6. T. Dalibor, C. Peppermuller, G. Pensi, S. Sridhara, R. P. Devaty, W. J. Choyke, A. Itoh, T. Kimoto, and H. Matsunami, Inst. Phys. Conf. Ser. **142**, p. 517 (1996).

7. R. B. Gregory, T. A. Wetteroth, S. R. Wilson, O. W. Holland, and D. K. Thomas, to be published.

CHARACTERIZATION OF SiO₂/SiC SAMPLES USING PHOTOELECTRON SPECTROSCOPY

L. I. JOHANSSON*, P- A. GLANS*, Q. WAHAB*, T.M. GREHK**,
TH. EICKHOFF**, W. DRUBE**
*Department of Physics, Linköping University, S-58183 Linköping
**Hamburger Synchrotronstrahlungslabor HASYLAB am Deutschen Elektronen-
Synchrotron DESY, D-22603 Hamburg

ABSTRACT

The results of photoemission studies of SiO_2/SiC samples for the purpose of revealing presence of any carbon containing by-products at the interface are reported. Two components could be identified in recorded Si 2p and C 1s core level spectra. For Si 2p these were identified to originate from SiO_2 and SiC while for C 1s they were interpreted to originate from graphite like carbon and SiC. The variation in relative intensity of these components with emission angle was first investigated. Thereafter the intensity of the different components were studied after successive Ar^+-sputtering cycles. Both experiments showed contribution from graphite like carbon on top of the oxide but not at the interface.

INTRODUCTION

The defect density at the oxide/semiconductor interface is an important factor for the performance of devices. For SiO_2/SiC the defect densities obtained to date are relatively high and one limiting factor for the formation of a high quality oxide is considered to be a carbon containing by-product at the interface. Studies of oxide layers thermally grown on SiC have earlier been reported using Auger Electron Spectroscopy (AES) [1] , transmission electron microscopy (XTEM) [2] and X-ray Photoelectron Spectroscopy (XPS) [3]. The XTEM results showed a homogeneous SiO_2 layer with a well defined interface and the AES results showed an oxide layer free from carbon related compounds except for a region very close to the interface. In the angle resolved XPS study [3] an interface $Si_4C_{4-x}O_2$ (x<2) oxide was revealed and also presence of $Si_4C_4O_4$ at the surface and in the oxide. These findings motivated our investigation of SiO_2/SiC samples using synchrotron radiation.

By using a high photon energy (3.0 keV) a direct and simultaneous probing of the SiC substrate, the interface and the oxide layer was first made. The probing depth was varied by changing the electron emission angle. Recorded Si 2p and C 1s core level spectra showed two components. The relative intensities of these were extracted and compared to calculated intensity variations. The analysis showed a graphite like carbon layer on top of the oxide but not at the interface. Secondly the composition in the surface region was studied after successive Ar^+-sputtering cycles using lower photon energies, giving a much smaller probing depth.. The result obtained was the same, i.e. that contribution from graphite like

Mat. Res. Soc. Symp. Proc. Vol. 572 © 1999 Materials Research Society

carbon on top of the oxide but not at the interface could be identified. These findings are presented and discussed below.

EXPERIMENT

Two samples with different oxide thicknesses were prepared on n-type Si-terminated 4H-SiC substrates, obtained from CREE Research, with doping concentrations of around 2×10^{18} cm^{-3}. The SiO$_2$ layers were thermally grown via a dry oxidation process at 1100 °C for respectively 30 (Sample A) and 45 minutes (Sample B), followed by annealing in Ar for 10 minutes at the same temperature. The SiO$_2$ layer thickness was measured by ellipsometry and determined to be 58 and 75 Å for samples A and B respectively.

Two different beamlines were utilized for the photoemission experiments. The X-ray wiggler beamline BW2 at HASYLAB [4, 5] and BL 22 at MAX-lab [6]. The end station consisted in both cases of a Scienta hemispherical electron analyzer. At BW2 a photon energy of 3.0 keV was selected with an overall energy resolution of 0.8 eV for the chosen beamline and analyzer settings. The electron emission angle was varied by rotating the sample relative to the fixed analyzer. At BL 22 photon energies of ≤800 eV were utilized and spectra were collected at normal emission after successive Ar$^+$-sputtering cycles. The binding energies were in these latter experiments determined relative to the Fermi edge of a Ta foil mounted on the sample holder.

RESULTS

Si 2p and C 1s core level spectra recorded at 3.0 keV and at different electron emission angles are shown in Fig. 1. Only two prominent peaks are

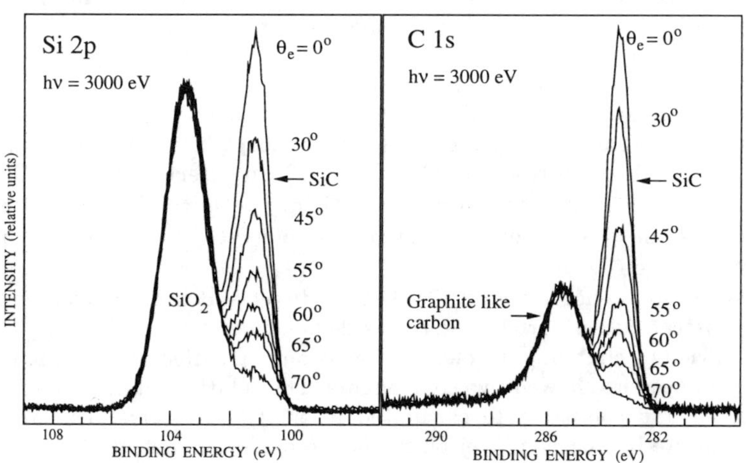

Fig. 1   Si 2p and C 1s spectra recorded at different electron emission angles using a photon energy of 3.0 keV.

clearly visible in both cases. The ones at lower binding energy correspond to the
bulk peaks of SiC. The ones at larger binding energy correspond respectively to
the Si 2p peak from SiO_2 and to a C 1s peak from carbon in a form different from
SiC. The binding energy of this additional C 1s peak correspond fairly well with
graphitic carbon [7,8] but the width of the peak is so large that it cannot originate
from an ordered graphite layer, therefore we refer to it as a graphite like peak.
Only relative intensities are shown in Fig. 1 since the spectra have been
normalized to the high binding energy peak. The variation in relative intensity
with emission angle can be utilized to determine from where in the sample the
graphite like carbon signal originates.

The intensities of the components in the Si 2p and C 1s spectra were
extracted using a curve fit procedure [9]. The intensity ratio between the SiO_2 and
the SiC peak for Si 2p level (labeled Si below) and the ratio between the graphite
like and the SiC peak for C 1s level (labeled C below) were then determined at
each emission angle [10]. The ratios obtained for sample A are shown in Fig. 2,
together with the C/Si ratio multiplied by a factor of 15. The C/Si ratio is seen to
increase monotonically and quite strongly with increasing emission angle. This
indicates that the graphite like carbon signal originates from a carbon containing
layer at the surface and not at the interface, as discussed below.

When applying a simple layer attenuation model to calculate the intensity
ratios assumptions concerning both the elemental distribution and the electron
attenuation length have to be made. Based on earlier reported values [11]
attenuation lengths of 45 Å and 50 Å were assumed for C 1s and Si 2p
photoelectrons at a photon energy of 3.0 keV. Various different models for the
elemental distribution were tried [10] but the observed variations were only

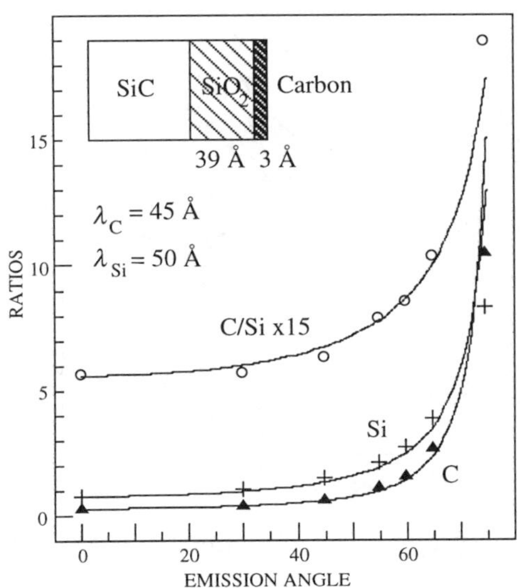

Fig. 2
Experimentally extracted
intensity ratios C (graphite
like carbon/SiC), Si (SiO_2/SiC)
and C/Si (multiplied by 15) for
the sample with a nominal
oxide layer thickness of 58 Å
are shown by filled diamonds,
crosses and open circles
respectively. The solid
lines show calculated
intensity ratios when
assuming the model of the
elemental distribution
shown by the insert.

adequately reproduced when assuming a carbon layer on top of the oxide. The expected intensity ratios are then given by ;

$$C = \frac{c_g^C}{c_{SiC}^C} e^{\frac{d_{ox}}{\lambda_C \cos\theta}} (e^{\frac{d_g}{\lambda_C \cos\theta}} - 1)$$

$$Si = \frac{c_{SiO_2}^{Si}}{c_{SiC}^{Si}} (e^{\frac{d_{ox}}{\lambda_{Si} \cos\theta}} - 1)$$

where c_m^x represents the concentration of element x in matrix m, d_g and d_{ox} the thickness of the graphite like and oxide layers, λ_C and λ_{Si} the electron attenuation lengths and θ the electron emission angle. When assuming elemental concentrations of $c_{SiC}^C = c_{SiC}^{Si} = 1/2$, $c_{SiO_2}^{Si} = 1/3$ and $c_g^C = 1.0$ best agreement between experimental and calculated intensity ratios for sample A was obtained for d_{ox} = 39 Å and d_g = 3.0 Å, which is illustrated by the solid lines in Fig. 2. For sample B layer thicknesses of d_{ox} = 56 Å and d_g = 4.3 Å produced best agreement. These results thus indicate presence of a graphite like carbon layer at the surface but not at the interface for both samples investigated.

We cannot exclude the possibility that the contribution from this layer actually shadows a weaker contribution from graphite like carbon at the interface, however. In order to check this possibility additional experiments were made using lower photon energies in which the Si 2p and C 1s components were monitored after successive Ar⁺-sputtering cycles. Such spectra recorded from sample A are shown in Fig. 3. The emitted photoelectrons have a kinetic energy of around 250 eV in both cases, giving an electron attenuation length of ca. 10 Å. An ion energy of 1.0 keV was utilized which gave a sputter rate of about 1 Å per min

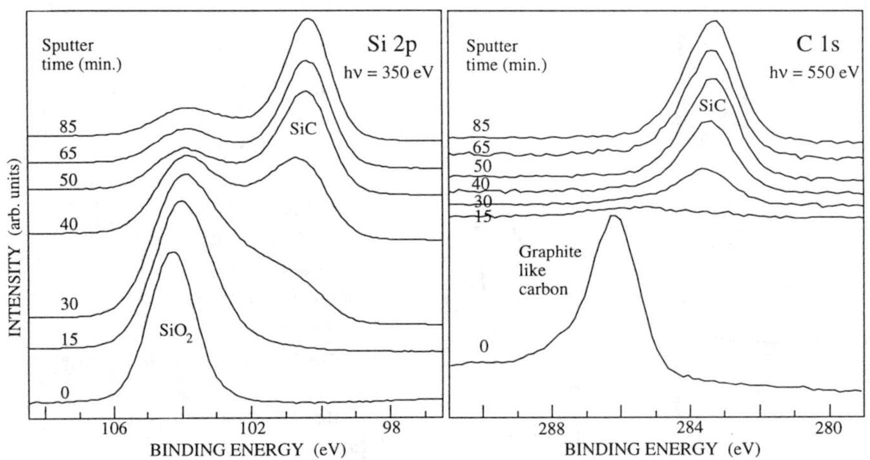

Fig. 3   Normal emission Si 2p and C 1s spectra recorded after different Ar⁺-sputtering times.

as determined from measurements on oxidized Si samples with known oxide thickness values. In Fig. 3 the Si 2p component from SiO_2 is seen to decrease with sputtering time while the contribution from SiC starts to appear after 30 min of sputtering and thereafter increases in relative intensity. For the C 1s components the graphite like carbon component completely dominates the spectrum of the unsputtered surface but is not visible after 30 min of sputtering when the SiC component starts to appear. A weak graphite like C 1s component is still visible after 15 min of sputtering but this should not be interpreted to indicate that there is carbon in the oxide layer. Instead knock on effects during sputtering of the surface carbon layer is believed to give rise to this weak component since measurements made using several different photon energies indicated that this weak component actually originated from the outermost surface region. The important point is that no increased contribution from a graphite like carbon component was possible to identify when the interface between the oxide and SiC was probed. Thus also these experiments showed contribution from graphite like carbon on top of the oxide but not at the interface.

A point worth noticing is that we could not identify more than two components in either the C 1s or Si 2p spectra so no contribution from an interface $Si_4C_{4-x}O_2$ compound or presence of $Si_4C_4O_4$ at the surface and in the oxide, as proposed in an earlier investigation [3], could be identified. Another point to be commented is the discrepancy in the oxide layer thickness values extracted from photoemission and ellipsometry measurements. In this case we believe the main reason to be the carbon layer on top of the oxide since such a layer was not assumed in the analysis of the ellipsometry data.

CONCLUSIONS

The results of photoemission studies of two SiO_2/SiC samples have been reported. Recorded Si 2p and the C 1s core level spectra each showed two components which for the Si 2p level were identified as originating from SiC and SiO_2 respectively while for the C 1s level they were identified as originating from SiC and graphite like carbon. Both angle resolved measurement made using a photon energy of 3.0 keV and normal emission measurements made using lower photon energies and after successive Ar^+-sputtering cycles gave the same results. A graphite like carbon layer on top of the oxide was identified. No contribution of graphite like carbon at the SiO_2/SiC interface or presence of an interface compound, as proposed in earlier studies, could be identified.

ACNOWLEDGEMENTS

Support from the Swedish Natural Science Research Council and SiCEP is gratefully acknowledged. The authors wish to thank Prof. H. Arwin for the help with the ellipsometry measurements.

REFERENCES

1 Q. Wahab, L. Hultman, M. Willander and J.E. Sundgren. J. Electron. Mater. **24**, 1345 (1995).

2 Q. Wahab, R. Turan, L. Hultman, M. Willander and J.-E. Sundgren. Thin Solid Films **287**, 252 (1996).

3 B Hornetz, H.-J. Michel and J. Halbritter, J Mater. Res. 9, 3088 (1994).

4 W. Drube, H. Schulte-Schrepping, H.-G. Schmidt, R. Treusch and G. Materlik, Rev. Sci. Instrum. **66**, 1668 (1995).

5 W. Drube, T. M. Grehk, R. Treusch and G. Materlik, J. Electron Spect. Rel Phenom. **88-89**, 683 (1998) .

6 J.N. Andersen, O. Björnholm, A. Sandell, R. Nyholm, J. Forsell, L. Thånell, A. Nilsson and N. Mårtensson, Synchrotron Radiation News 4, 15 (1991).

7 L. I. Johansson, Fredrik Owman and Per Mårtensson. Phys. Rev. B **53**, 13793 (1996).

8 F.Sette, G.K. Wertheim, Y. Ma, G. Meigs, S. Modesti and C.T. Chen, Phys. Rev. B **41**, 9766 (1990).

9 P.H. Mahowald, D.J. Friedman, G.P. Carey, K.A. Bertness and J.J. Yeah, J. Vac. Sci. Technol. A **5**, 2982 (1987).

10 L. I. Johansson, P.-A. Glans, Q. Wahab T.M, Grehk, Th. Eickhoff and W. Drube, to be published.

11 P.J. Cumpson and M.P. Seah, Surf. and Interf. Analysis **25**, 430 (1997).

ANNEALING OF ION IMPLANTATION DAMAGE IN SiC USING A GRAPHITE MASK

Chris Thomas *, Crawford Taylor *, James Griffin *, William L. Rose *, M. G. Spencer *, Mike Capano **, S. Rendakova ***, and Kevin Kornegay ****

*Materials Science Research Center of Excellence, Howard University, Washington, DC, 20059
** Purdue University, 1285 Electrical Engineering Bldg., West Lafayette, IN, 47907-1285
***TDI, Inc., Gaithersburg, MD, 20877
****Department of Electrical Engineering, Cornell University, Ithaca, New York, 14853

ABSTRACT

For p-type ion implanted SiC, temperatures in excess of 1600 °C are required to activate the dopant atoms and to reduce the crystal damage inherent in the implantation process. At these high temperatures, however, macrosteps (periodic welts) develop on the SiC surface. In this work, we investigate the use of a graphite mask as an anneal cap to eliminate the formation of macrosteps. N-type 4H- and 6H-SiC epilayers, both ion implanted with low energy (keV) Boron (B) schedules at 600 °C, and 6H-SiC substrates, ion implanted with Aluminum (Al), were annealed using a Graphite mask as a cap. The anneals were done at 1660 °C for 20 and 40 minutes. Atomic force microscopy (AFM), capacitance-voltage (C-V) and secondary ion mass spectrometry (SIMS) measurements were then taken to investigate the effects of the anneal on the surface morphology and the substitutional activation of the samples. It is shown that, by using the Graphite cap for the 1660 °C anneals, neither polytype developed macrosteps for any of the dopant elements or anneal times. The substitutional activation of Boron in 6H-SiC was about 15%.

INTRODUCTION

The improvement of the material quality of SiC and the development of its device technology have been, for the past several years, the focus of intense investigation by research groups from around the world. It is projected that its large bandgap, high electron saturation velocity, exceptional thermal conductivity (greater than copper), and large breakdown field strength will improve the standard commercial benchmarks of high-power and high-frequency devices and allow them to operate in caustic environments and at higher temperatures [1]. Of the three methods of doping--ion implantation, thermal diffusion, and in situ doping--ion implantation will play the most important part in the fabrication of these devices. This is because, unlike Silicon technology, thermal diffusion, due to the low diffusion coefficients of the standard dopants below 1800 °C, is not a viable method for doping SiC [2]. In addition, the in situ doping method, while it is the principal method of preparing doped device quality epitaxial material at present, cannot be used for planar device fabrication and other applications where precision is required in small areas. Still, despite its advantages, ion implantation of SiC is not as mature a technology as it is for Silicon, where ion implantation is routinely used in device fabrication. One of the problems associated with the ion implantation of SiC surrounds its post implant anneal. An anneal is necessary to activate the dopant atoms and to reduce the inherent crystal damage created by the ion implantation process. For p-type SiC, this anneal is usually done at temperatures up to 1700 °C [3]. One side effect of this anneal is that, at these high temperatures, the post anneal surface

Mat. Res. Soc. Symp. Proc. Vol. 572 © 1999 Materials Research Society

morphology of the material is dominated by macrosteps. In this paper, we present a technique that eliminates the formation of these macrosteps, and we also demonstrate Boron (p-type) activation. AFM measurements were used to determine the morphology before and after annealing, SIMS data was used to track the Boron ions through the steps of the experiment, and C-V measurements were used to determine the substitutional activation.

EXPERIMENT

The samples used in this experiment were 4 μm thick n-type (3.8 x 10^{15} cm^{-3}) 6H-SiC and (1.5 x 10^{15} cm^{-3}) 4H-SiC epitaxial layers that were ion implanted at 600 °C with low energy Boron. The specific energy/dose schedules for the Boron implants are as follows: 320 keV/3.6 x 10^{14} cm^{-2}, 240 keV/2.8 x 10^{13} cm^{-2}, 160 keV/2.0 x 10 ^{13} cm^{-2}, 100 keV/7.5 x 10^{12} cm^{-2}, 50 keV/5.0 x 10^{12} cm^{-2}, 30 keV/3.75 x 10^{12} cm^{-2}. In addition, n-type (7 x 10^{18} cm^{-3}) 6H-SiC substrates that were ion implanted with low energy Aluminum were also used. The energy/dose schedules for the Aluminum implants are as follows: 260 keV/1.0 x 10^{15}, 150 keV/5.0 x 10^{14}, 80 keV/3.0x 10^{14}, 30 keV/2.0 x 10^{14}.

The samples were first cleaned with a TCE/acetone/methanol degrease, followed by a 1 minute dip in 10:1 HF:H$_2$O and a rinse in DI water. Random 5 x 5 μm^2 AFM scans were taken to determine the surface morphology of the samples before annealing.. The samples were again cleaned in the manner described above, and a 0.8 μm thick carbon mask was fabricated on the surface. For the anneals, an EMCORE resistive heated growth reactor was used. The reactor was calibrated at one point by melting Silicon to verify the temperatures used for the anneals. While Helium was the anneal ambient for most runs, the 6H-SiC substrates were annealed in Argon. The anneal times used were 20 and 40 minutes and the anneal temperature was 1660 °C. There was no data for the 40 minute anneal of the 6H-SiC substrates.

Upon completion of the anneals, the carbon mask was removed by oxidation at 800 °C for one hour in a quartz furnace, and AFM measurements were done again to determine the effect of the anneal on the morphology of the samples. The doping profile for the Boron implanted 6H-SiC was determined from reverse capacitance-voltage measurements of Schottky diodes (Aluminum ohmic and Schottky contacts). The measurements were taken at a frequency of 1MHz.

RESULTS AND DISCUSSIONS

The AFM micrographs shown in figure 1 detail the effect of a 40 minute anneal without a Graphite cap at 1660 °C on the surface morphology of 6H-SiC. Figure 1(a) is the surface just before the anneal is performed. The macrosteps shown in figure 1(b) were developed during the anneal process. The effect is well known and has been reported by groups at Purdue University [3] and Cree Research, Inc. [4]. These macrosteps extend across the entire surface of the sample and are so pronounced that they can be viewed with an optical microscope. Similar results have been reported for ion implanted 4H-SiC [5]. It is known that at about 1400 °C, Silicon containing species (Si, Si$_2$C, and SiC$_2$) sublime from the surface layer of SiC [6]. It is this loss of Si atoms and the ensuing redistribution of the Carbon rich surface layer that leads to the formation of the macrosteps. Figure 2 shows the atomic force micrographs of the final surfaces of the samples that were annealed for 20 and 40 minutes using our Graphite mask as a cap. From the micrographs, it can be seen that no macrosteps were formed for any of the anneal times, even though each anneal was done at 1660 °C. There is, however, a trend for each set of samples: the roughness increases as the time of the anneal increases. This is seen in going from micrographs (b) to (c) and from the

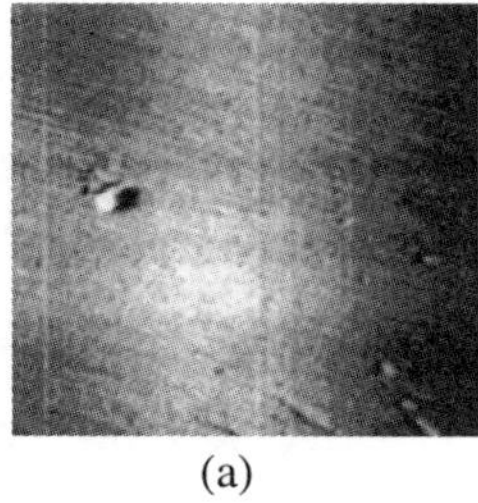
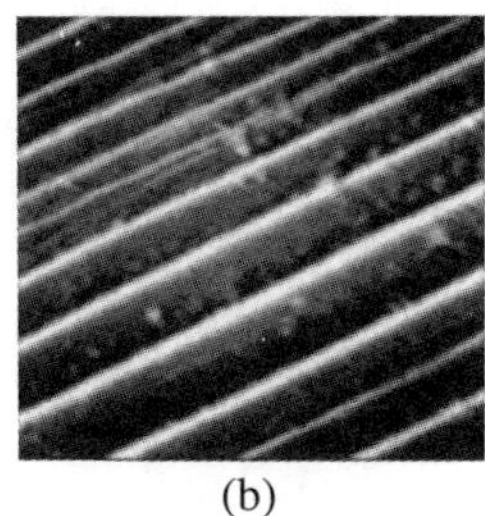

(a)   (b)

Figure 1: AFM picture of Boron implanted (a) 6H-SiC epilayer before anneal (b) 6H-SiC epilayer after 40 minute anneal at 1660 °C. Both micrographs are 5 µm by 5 µm in size.

micrographs (d) to (e), where each pair of micrographs corresponds to time increases from 20 to 40 minutes for 6H-SiC and 4H-SiC respectively. This result is consistent with those reported by Capano et al [3].

Reports of SiC plates [6] and wafers [3] suspended above and almost touching the sample during the annealing process are examples of techniques being used at the moment to reduce the formation of macrosteps. These techniques attempt to create a Si overpressure that allows little to no net movement of Si atoms from the surface layer of the sample, however, while the formation of macrosteps is reduced, the surface roughness is still appreciable. It has also been reported that a Si rich ambient gas such as Silane, if used during the anneal, will maintain the requisite Si

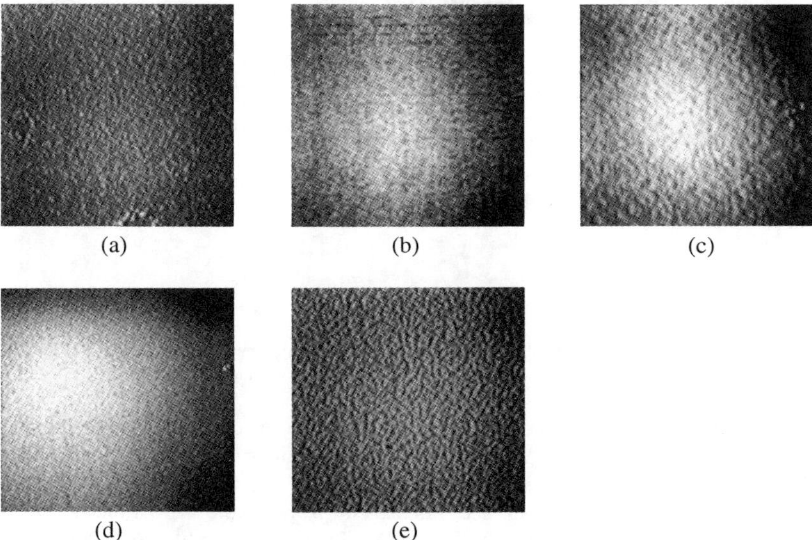

(a)   (b)   (c)

(d)   (e)

Figure 2: Atomic force micrographs of (a) Aluminum implanted 6H-SiC substrate annealed for 20 min, and Boron implanted (b) 6H-SiC epi annealed for 20 min, (c) 6H-SiC epi annealed for 40 min, (d) 4H-SiC epi annealed for 20 min, (e) 4H-SiC epi annealed for 40 min. The edge of all micrographs correspond to 5 µm, and all anneals were done at 1660 °C.

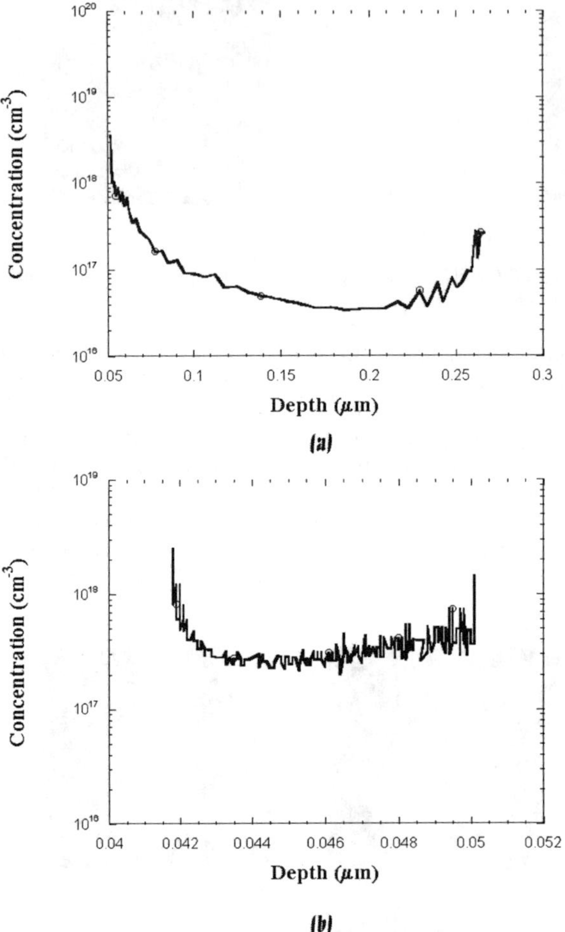

Figure 3: Doping profile for (a) Aluminum implanted 6H-SiC substrate annealed for 20 min, (b) Boron implanted 6H-SiC epi annealed for 20 min., and (c) Boron implanted 6H-SiC epi annealed for 40 min.

overpressure and therefore suppress the formation of macrosteps [4]. Current results with the silane anneal are encouraging, but, from our experiments, the success of the anneal is difficult to reproduce. Our technique, the use of a Graphite mask as a cap, not only suppresses the sublimation of Silicon from the surface, but it also prevents the formation of macrosteps by preventing the redistribution of the surface layer. In addition, unlike AlN, which is another material being used as an anneal cap for SiC, the Graphite mask can operate at temperatures up to 1850 °C.

Along with to the AFM data, C-V measurements were taken to determine the substitutional activation of the the 6H-SiC epilayers that were implanted with Boron. The results are shown in

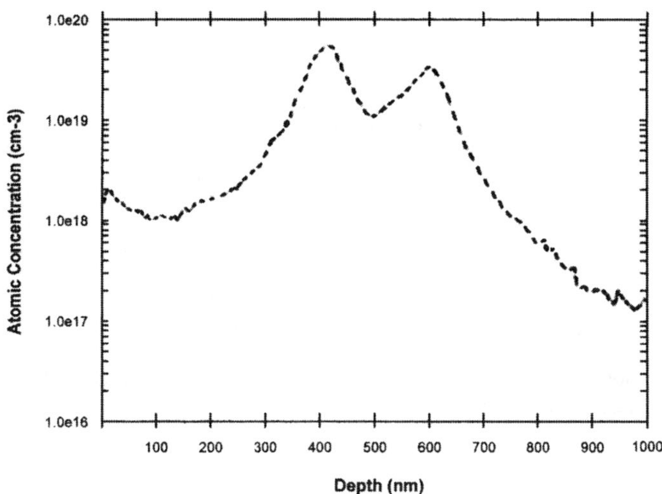

Figure 4: Secondary ion mass spectrometry (SIMS) plot of the Boron implanted 6H-SiC epilayer samples annealed at 1660 °C for 40 minutes.

the doping profiles of figure 3. There is an increase in activation with anneal time as seen by going from profiles (a) to (b) in figure 3, a trend that was also observed by Capano et al [3]. The doping profile data is supported by the SIMS plot shown in figure 4, where up to the depth of 0.5 μm the movement of the Boron atoms produces a concentration of about 2×10^{18}. This means that for the 40 minute anneal corresponding to the doping profile in figure 3(b), the substitutional activation was about 15%.

CONCLUSION

The use of a Graphite mask as an anneal cap was shown to prevent macrostep formation. Temperatures of 1660 °C were used to anneal Aluminum implanted 6H-SiC substrates and Boron implanted 6H- and 4H-SiC epitaxial layers. At these temperatures, macrosteps are usually formed across the surface of the sample. None of the samples annealed with the Graphite cap developed macrosteps during the anneal. There was a general trend of increased roughness for longer anneal times, but the resulting roughness even after a 40 minutes anneal was not appreciable. There was also a trend of increased activation with increased anneal time. Finally, p-type activation was demonstrated for the Boron implanted 6H-SiC epilayer samples.

ACKNOWLEDGMENTS

This work was supported by MURI for manufacturable power switching devices. The contract manager was John Zolper.

REFERENCES

[1]   M. Bhatnagar and B. J. Baliga, IEEE Trans. Electron Devices 40 (1993), p.645.

[2]   M. V. Rao, J. A. Gardner, P. H. Chi, O. W. Holland, G. Kelner, J. Kretchmer, and M. Ghezzo, J. Appl. Phys. Vol. 81, No. 10, p. 6635 (1997).

[3]   M. A. Capano, S. Ryu, M. R. Melloch, J. A. Cooper, Jr., and M. R. Buss, J. Electronic Materials, Vol. 27, No. 4, (1998).

[4]   CREE Research, Inc., Oral presentation at MURI review, (1999).

[5]   M. A. Capano et. al., Oral presentation at MURI review, (1999).

[6]   L. Ottaviani, D. Planson, M. L. Locatelli, J. P. Chante, B. Canut, and S. Ramos, Material Science Forum, Vols. 264-268, p. 709 (1998).

EFFECT OF VARYING OXIDATION PARAMETERS ON THE GENERATION OF C-DANGLING BOND CENTERS IN OXIDIZED SIC

P.J. Macfarlane and M.E. Zvanut, Department of Physics, University of Alabama at Birmingham, 310 Campbell Hall, Birmingham, AL 35294-1170

Abstract

SiC is perhaps the most appropriate material to replace Si in power-metal-oxide-semiconductor-field-effect-transistors (MOSFETs), because, unlike the other wide band-gap semiconductors, SiC can be thermally oxidized similarly to Si to form a SiO_2 insulating layer. In our studies of oxidized SiC, we have used electron paramagnetic resonance (EPR) to identify C-dangling bonds generated by hydrogen release from C-H bonds. While hydrogen's effect on SiC-based MOSFETs is uncertain, studies of Si-based MOSFETs indicate that it is important to minimize hydrogen in MOS structures. To examine the role of hydrogen, we have studied the effects of SiC/SiO$_2$ fabrication on the density of C-related centers, which are made EPR active by a dry heat-treatment. Here we examine the starting and ending procedures of our oxidation routine. The parameter that appears to have the greatest effect on center density is the ending step of our oxidation procedure. For example, samples that were removed from the furnace in flowing O_2 produced the smallest concentration of centers after dry heat-treatment. We report on the details of these experiments and use our results to suggest an oxidation procedure that limits center production.

Introduction

SiC has both a wide band-gap and high thermal conductivity that make it an attractive replacement for Si in high power, high temperature microelectronic devices. For power metal-oxide-field-effect transistors (MOSFETs) in particular, SiC is of interest because unlike the other wide band-gap semiconductors, it can be thermally oxidized similarly to Si in order to create a SiO_2 insulating layer.

In our previous studies of oxidized 3C-SiC, 4H-SiC, and 6H-SiC [1-2], we have observed centers that can be activated by dry heat-treatments at temperatures greater than 800 $^{\circ}$C. The g-values of these centers range from 2.0025 to 2.0029, which is within the range of g-values typical of C-related centers [3-5]. The temperatures at which these centers are generated are significantly greater that those used to generate Si dangling bonds. Thus, we suggest that the centers are unpaired electrons located on C atoms. We have also observed that these centers can be activated by heat-treatment in an ambient that does not contain moisture and passivated in an ambient that contains moisture. Therefore, we suggest that these centers are activated by release of a hydrogenous species from C dangling bonds. Supporting the relation to hydrogen is an experiment in which we heat-treated oxidized samples in dry (<0.6 ppm H_2O) O_2. The concentration of centers activated is similar to that activated by dry heat-treatment in N_2. In addition, we have observed that for the 3C-SiC epilayer samples the density of centers can be reduced by annealing in forming gas. Hydrofluoric acid etching studies of the samples indicated that the centers are not located in the oxide.

In this study of the C-related center in oxidized, heat-treated 3C-SiC, 4H-SiC, and 6H-SiC, we investigate the effects of altering the loading and unloading procedure of our oxidation routine on the concentration of C-related centers produced by dry heat-treatment. In our examinations of the removal steps, we unload the samples in wet O_2 or standard N_2 and compare the results with slowly cooling the samples in air [1-2,6]. We also compare the effects of inserting samples into the furnace in air, wet O_2, and N_2 ambients. The 3C-SiC epilayer samples are included to gain a clear scientific understanding, although the hexagonal polytypes are more applicable to current high power device research.

Experimental Information

The 4H-SiC and 6H-SiC materials are from double-side polished, p-type, 1.5" wafers supplied by Cree Research. The polished surfaces of both wafers are cut approximately 3.5° off the {0001} crystal planes. The 4H-SiC wafer has an epilayer deposited on its Si face. HOYA Corporation supplied the 3C-SiC epilayer layer samples. The approximately 1 μm thick, cubic SiC film was deposited via chemical vapor deposition (CVD) on both the polished and unpolished faces of a (100) oriented Si substrate. Samples are cut into strips 1.5 cm by 0.23 cm, a size suitable for the EPR microwave cavity.

Prior to oxidation, we clean the samples by rinsing in trichloroethane, xylenes, acetone, methanol and deionized water for 5 min time intervals. The samples are then etched for 1 min in 9:1 H_2O:HF (50%). The samples are oxidized for 6 hr at $1150°$ C in O_2 bubbled through deionized water. The loading and unloading steps of the oxidation procedure are described below. After oxidation, the samples are heat-treated in a double-walled quartz tube furnace at 900 °C for 320 min. The heat-treatment is conducted in a dry (<0.3 ppm H_2O) N_2 ambient. At the end of the dry heat-treatment, the samples are quenched to approximately $100°$ C where they are cooled sufficiently before they are removed from the dry ambient.

For the oxidation, five different insert and removal conditions are examined:
1) Insert in air, remove quickly in N_2 ("Pull-out N_2")
2) Insert in air, remove quickly in O_2 ("Pull-out O_2")
3) Insert in air, slowly cool in air ("Air Cooled")
4) Insert in O_2, remove quickly in O_2
5) Insert in N_2, remove quickly in O_2

The slow cool in air is accomplished by gradually pulling the sample to the edge of the furnace tube where the temperature is about 940 °C. The sample remains there about 5 min before being placed in room ambient. For the other removal conditions, the samples are pulled from the furnace with either O_2 or N_2 flowing and immediately placed in room ambient. The insert conditions consisted of a 30 min O_2 or N_2 purge of the furnace tube, followed by insertion of the samples while the O_2 or N_2 gas is flowing. For the latter case, the N_2 gas is terminated and O_2 is used for the oxidation. The oxygen is bubbled through deionized water for all O_2 treatments, and the N_2 as standard grade.

All EPR measurements are conducted at room temperature using an X-band Bruker 200 spectrometer. EPR is a spectroscopic technique able to detect paramagnetic defects in solids. We use the dry heat-treatment to activate the paramagnetic state of a defect already present in the material. The activation process is similar to the one used for the Si P_b center, a Si dangling bond at the interface between SiO_2 and Si [7-8]. The concentration of centers is found by double integration of the EPR spectrum and comparison with the double integral of a spectrum obtained from a known standard. Typically, we integrate the EPR spectrum with the largest amplitude and determine the concentration from the other spectra by scaling their amplitudes. All spectra are measured with 1 G peak-to-peak modulation amplitude and 1 mW microwave power.

Results

Electron paramagnetic resonance spectra were measured for each of the polytypes after oxidation and after dry heat-treatment. From these measurements, the concentration of C-related centers after dry heat-treatment is determined. In Figure 1, we plot the concentration of the centers in each of the three polytypes for three different unloading procedures. The unfilled squares, filled circles, and unfilled triangles represent the concentration of centers in the oxidized 3C-SiC epilayer, 4H-SiC and 6H-SiC samples, respectively. Error in concentration is due to noise in the EPR spectrum. While variation in concentration appears between the different SiC polytypes, a common trend is observed. Unloading the samples in N_2 produced the largest concentration of centers. In contrast, removing the samples from the furnace in wet O_2 produced the smallest concentration of centers.

In Figure 2, we plot the concentration of centers measured in the 6H-SiC samples as a function of their insert ambients. "Air" is used to describe the condition in which the samples were inserted into the furnace in an air ambient. Similarly, "Wet O_2" and "N_2" are used to indicate the conditions in which the samples were inserted into the furnace in flowing wet O_2 and flowing standard N_2 ambients, respectively. The data point for the sample inserted into air is the same as the point in the previous figure of the 6H-SiC sample was removed in an flowing wet O_2 ambient. The error in concentration arises from noise in the EPR spectrum. The concentration of centers is the same for each of the different insert ambients.

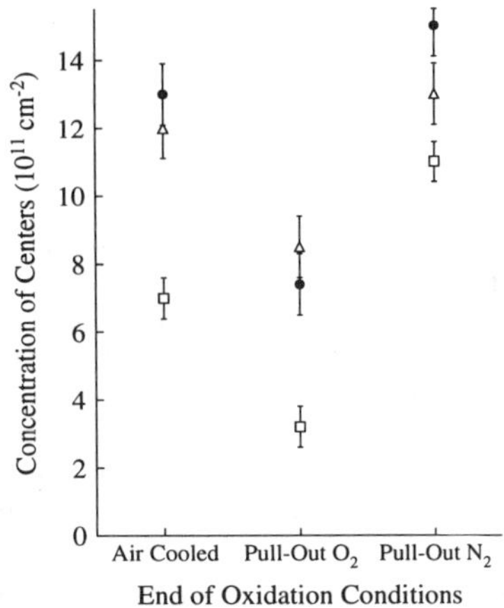

Figure 1: The concentration of centers produced as a function of the ending oxidation conditions. "Air Cooled" describes the condition in which samples were slowly cooled in the furnace in air after the heat was turned off. "Pull-out O_2" and "Pull-Out N_2" describe the conditions in which samples were directly unloaded from the furnace in wet O_2 and standard N_2 ambients, respectively. The □, ●, and Δ indicate the concentration of centers in the oxidized 3C-SiC, 4H-SiC, and 6H-SiC samples.

Discussion

While there is no direct evidence the C-related centers we observe are associated with electrically active defects, centers detected by EPR in oxidized Si have been shown to be related to electrical defects. For example, electrically active interface states were found to be related in part to an interfacial Si dangling bond, called the P_b center [9], and some charge traps in the SiO_2 have been attributed to the E' center, a hole trapped at an O vacancy in SiO_2 [10]. Thus, a center detected by EPR in oxidized SiC could also be related to electrical defects. We observe some similarities between processing procedures that reduce the concentrations of interface states and oxide charge traps and those that reduce the concentration of the centers in our samples. The similar and contrasting results are described below.

In examining the beginning and ending steps of our oxidation procedure, we observe that the final steps of the oxidation have the largest effect on the concentration of centers. Figure 1 indicates that removing the samples in O_2 minimizes the concentration. Compared to removal in N_2, the center density is decreased by 70%, 50%, and 40% in 3C-, 4H- and 6H-SiC samples, respectively. The reduction in center concentration may be related to the 5 min 940 °C wet O_2

"anneal" that the samples receive before they are removed from the furnace. Lipkin and Palmour have demonstrated that both interface states and oxide charge in oxidized 6H-SiC can be reduced by "re-oxidizing" 6H-SiC samples in wet O_2 at 950 °C for 1.5 to 3 hr [11]. Although Lipkin and Palmour's "re-oxidation" anneal was for a significantly longer period of time, it is interesting to note that a brief wet O_2 heat-treatment of our samples also considerably reduced the C-related center concentration. Most likely, unloading the samples in wet O_2 stabilizes the interface between the SiO_2 and the SiC preventing Si evaporation [12].

Assuming that N_2 is primarily responsible for the affects seen in the air annealed samples, comparison of the slowly air cooled and more rapidly N_2 cooled samples suggests that thermal shock may also affect the production of the C-related centers. In Fig. 1, a lower concentration of centers is found for samples that were slowly cooled in air versus samples that were more rapidly removed in N_2. In particular, we observe that the density of centers in 3C-SiC samples that were slowly cooled was 35% lower than the density observed in samples that were rapidly pulled in N_2 to room ambient. Although the density of

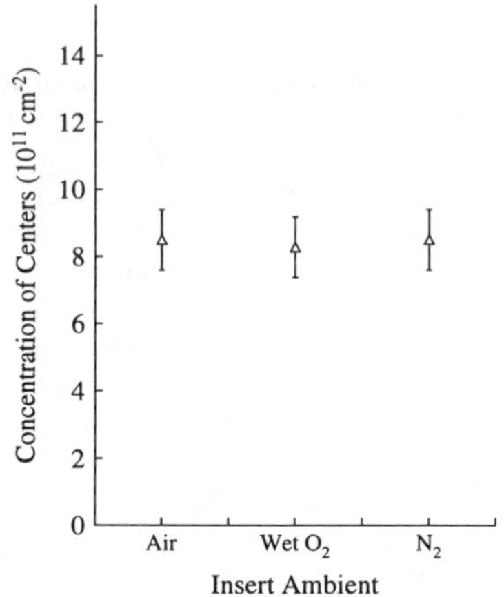

Figure 2: The concentration of centers in the 6H-SiC samples as a function of the insert ambient. "Air," "O_2," and "N_2" indicate that the samples were loaded into the furnace in air, wet O_2, or standard N_2 ambients, respectively.

the centers in the hexagonal polytypes were the same within sample to sample variation, on average the concentration decreased by approximately 10%. The large reduction in center concentration found in the 3C-SiC samples is consistent with the fact that film/substrate samples such as the 3C-SiC epilayers, are more susceptible to thermal shock than bulk substrates. Studying the effect of oxidation conditions on the electrical properties of hexagonal polytypes, Shenoy and coworkers observed a similar behavior [13]. They compared the effects of removing samples using a "slow pull" method, in which the temperature was gradually reduced to 900 °C before the samples were unloaded, to using a "fast pull" technique in which samples were withdrawn from the furnace over approximately 100 sec. Like we observed with the concentration of our C-related center, they found that the density of interface states and oxides charges traps could be reduced by 10% when unloading samples with the "slow pull" procedure versus the "fast pull" method. While the EPR/electrical comparison is intriguing, we must acknowledge that in our experiments, the oxygen in the air may contribute to the reduction of the EPR defect concentration. Future experiments will be conducted in order to determine the role that thermal shock plays in the production of the C-related centers.

Figure 2 indicates that the concentration of the C-related centers does not depend on whether the insert ambient is oxidizing (O_2) or inert (N_2). This is in contrast with electrical studies which show that samples loaded in an inert gas (Ar) have a higher concentration of electrically active defects [13] than samples loaded in O_2. The increase in electrical defects were attributed to Si evaporation from the SiC surface, leaving a C-rich layer on the 6H-SiC substrate. We, however, do not observe this for samples that were inserted in the furnace in flowing N_2 versus flowing O_2. In fact, the concentration of centers indicated by Fig. 2 is approximately the same regardless of whether the samples are inserted in air, N_2 or wet O_2. We speculate that these results indicate that as far as our C-related centers are concerned any initial surface layer is likely consumed during oxidation and converted into the oxidation products. Perhaps, because oxidation occurs at the interface between the SiC substrate and the SiO_2 layer, the ending steps of the oxidation would have more of an effect on the quality of the SiC/SiO_2 interface. This agrees with our observations of sizeable changes in concentration produced in samples oxidized with different ending steps.

Conclusion

In conclusion, the concentration of centers can be affected by altering the ending steps of our oxidation procedure. The ending conditions, ordered from smallest center concentration produced to largest, are unloading the samples in wet O_2, slowly cooling the samples in air, and unloading the samples in standard N_2. The insert ambient did not affect the center concentration. This may indicate that because oxidation occurs at the SiC/SiO_2 interface, any initial effects of the oxidation procedure are effectively removed during the oxidation. Thus, the ending process should have a larger effect on the concentration of centers produced.

Acknowledgements

This work is supported by ONR grant no. N00014-96-2-1238. P.J. Macfarlane is supported by a fellowship from the Alabama Space Grant Consortium.

References

[1]   P.J. Macfarlane and M.E. Zvanut, Appl. Phys. Lett. **71**, 2148 (1997).
[2]   P.J. Macfarlane and M.E. Zvanut, in Hydrogen in Semiconductors and Metals, edited by N.H. Nickel, W.B. Jackson, R.C. Bowman, and R. Leisure (Mater. Res. Soc. **513**, Pittsburgh, PA 1998), pp. 433-438.
[3]   X. Zhou, G. Watkins, K.M. McNamara Rutledge, R.P. Messmer, and S. Chawla, Phys. Rev. B **54**, 7881 (1996).
[4]   G. Gerardi, E. Poindexter, and C. Young, Appl. Spectrosc. **50**, 1427 (1996).
[5]   V.V. Afanas'ev and A. Stesmans, Appl. Phys. Lett. **69**, 2252 (1996).
[6]   P.J. Macfarlane and M.E. Zvanut, J. of Electron. Mater. **28**, 144 (1999).
[7]   K.L. Brower, Physical Review **B42**, 3444 (1990).
[8]   J. Stathis, J. Appl. Phys. **77**, 6205 (1995).
[9]   E. Poindexter, J. Non-Crystalline Solids **187**, 257 (1995).
[10]   P.M. Lenahan and P.V. Dressendorfer, J. Appl. Phys. **55**, 3495 (1984).
[11]   L.A. Lipkin and J.W. Palmour, J. Electron. Mater. **25**, 909 (1996).
[12]   L. Muehloff, W.J. Choyke, M.J. Bozack, and J.T. Yates, J. Appl. Phys. **60**, 2842 (1986).
[13]   J.N. Shenoy, G.L. Chindalore, M.R. Melloch, J.A. Cooper, J.W. Palmour, and K.G. Irvine, J. Electron. Mater. **24**, 303 (1995).

THICK OXIDE LAYERS ON N AND P SiC WAFERS BY A DEPO-CONVERSION TECHNIQUE

Q. Zhang, V. Madangarli, I. Khlebnikov, S. Soloviev and T. S. Sudarshan
Department of Electrical Engineering
University of South Carolina, SC 29208, U.S.A
Tel: 803-777-7302; Fax: 803-777-8045
E-mail: Zhang@engr.sc.edu

ABSTRACT

The electrical properties of thick oxide layers on n and p-type 6H-SiC obtained by a depo-conversion technique are presented. High frequency capacitance-voltage measurements on MOS capacitors with a ~ 3000 Å thick oxide indicates an effective charge density comparable to that of MOS capacitors with thermal oxide. The breakdown field of the depo-converted oxide obtained using a ramp response technique indicates a good quality oxide with average values in excess of 6 MV/cm on p-type SiC and 9 MV/cm on n-type SiC. The oxide breakdown field was observed to decrease with increase in MOS capacitor diameter.

INTRODUCTION

Silicon Carbide (SiC) devices have been proposed for many industrial applications requiring high power, high frequency and hard radiation. Recently, tremendous improvements have been made in the crystal growth and processing techniques including oxidation which will accelerate the commercialization of SiC devices. Oxide layers on SiC not only find application in the fabrication of MOS devices but also as field oxides and passivation layers in high voltage devices. While the oxide layer should have high breakdown strength, low leakage current, and low effective charge density for satisfactory MOS device performance, when oxide layers are used as field oxides in high voltage devices high breakdown strength is most essential. For example, a metal-overlap onto an oxide layer can be used as an edge termination technique for improving the breakdown voltage of high voltage devices by minimizing the electric field enhancement at the contact periphery of devices. In order for this field oxide to be effective it should have good dielectric properties and sufficient thickness to sustain the high breakdown voltage. However, the SiC oxidation rate is too slow to obtain thick oxide layers via the conventional thermal oxidation techniques currently in practice [1,2,3]. Even though alternate techniques such as oxide deposition by LPCVD and poly-silicon conversion have been reported [4,5], the high voltage characteristics of deposited oxide vs a conventional thermal oxide has not been studied in detail. In this paper, we report the possibilities of using simple depo-converison technique for obtaining thick oxide layers with high breakdown strength on n and p-type 6H-SiC wafers.

EXPERIMENT

The 6H-SiC wafers used in our experiments were from Cree Research (substrate doping ~ 1.6×10^{18} cm^{-3}) with a 10 μm thick epilayer of ~ 6×10^{15} cm^{-3} doping concentration for p-type and ~ 3.9×10^{15} cm^{-3} for n-type SiC respectively. The 30 mm diameter wafer was cut into 10 mm x 10 mm square pieces to obtain several samples for experiments. The as-received SiC samples were first cleaned using TCE at 85°C for 15 min. followed by ultrasonic cleaning in acetone and methanol respectively. Then the native surface oxide was chemically etched using a 20% HF solution prior to a modified RCA cleaning process. After the RCA cleaning process the SiC wafers were etched in a 5% HF solution for 10 seconds to remove the surface oxide caused by the RCA process. De-ionized water rinses were used after every step. In order to obtain the depo-converted oxide, a Si film is deposited on the SiC wafer and converted to silicon dioxide by oxidation. The Si films were RF sputtered from a single crystal Si target in a vacuum chamber under Ar ambient and the sputtered Si films were converted to oxide by wet oxidation at 1050°C for 3 hours to ensure total conversion of the deposited Si. All samples with converted oxide had a very uniform surface morphology, indicating that depo-conversion method can be used to obtain thick oxide films. MOS capacitor (MOS-C) structures were fabricated, to characterize the electrical properties of the depo-converted oxide, by evaporating Al on both sides of the sample. Moreover, the usefulness of the depo-converted oxide as a field oxide was demonstrated by fabricating a 1kV 6H-SiC Schottky diode with oxide layer edge termination.

RESULTS AND DISCUSSION

C-V characteristics

Capacitance-voltage measurements were made on MOS capacitors with different oxide thicknesses on *n* and *p* type SiC at room temperature using a Kiethley 590 capacitance-voltage (C-V) analyzer at 100 kHz. Voltage sweeps applied to the MOS capacitor were made from accumulation to depletion/inversion. Typical C-V characteristics of MOS capacitors with thermal and converted oxides fabricated on *n* and *p*-type SiC are shown in Fig. 1.

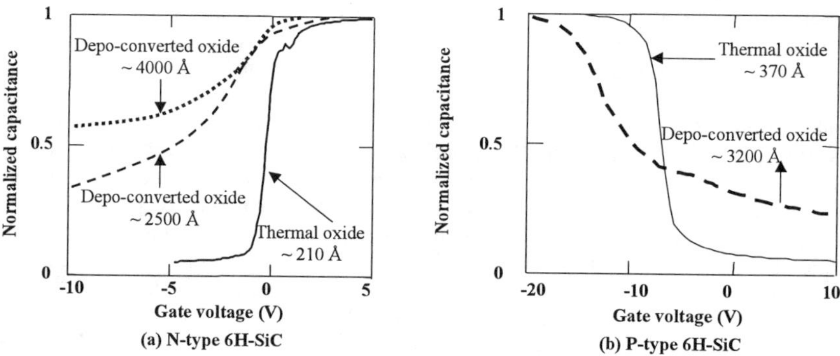

Fig. 1. Comparison of C-V characteristics of MOS capacitors with depo-converted oxide vs thermal oxide on *n* and *p*-type 6H-SiC wafers.

The oxide and SiC parameters obtained from C-V measurements are given in Table 1.

Table 1. Oxide and SiC wafer parameters from C-V measurements

SiC wafer type	Oxide thickness (Å)	Doping concentration (cm^{-3})	Flat-band voltage (V)	Effective charge density (cm^{-2})
N-type	210*	6.5E15	-0.4	7.327E+10
	2500#	4.4E15	-2.4	1.685E+11
	4000#	5.1E15	-1.47	6.114E+10
P-type	370*	4.6E15	-7.25	2.885E+12
	3200#	2.1E15	-11.75	4.156E+12

* Thermal oxide   # Depo-converted oxide

From Table 1, it is observed that the effective charge density of the depo-converted oxide is in the same order of magnitude as that of the thermal oxide. As expected, the flat-band voltage and effective charge density are lower for the oxides grown on N-type SiC compared to those on P-type SiC [6].

Oxide breakdown measurements

In order to determine the breakdown strength of the oxide films, a non-destructive ramp response technique [7] was used to measure the current-voltage characteristics in the accumulation regime of the oxide films on n and p type SiC wafers. From the maximum breakdown voltage and thickness of the oxide films, the critical breakdown field of the oxide films was computed.

Experimental results show that for a given thickness of the oxide, the breakdown field of the oxide reduces with increase in the diameter of the MOS capacitor structure fabricated on the SiC wafer. A typical variation of the oxide breakdown field on a n type SiC with increasing MOS capacitor diameter is illustrated in Fig. 2.

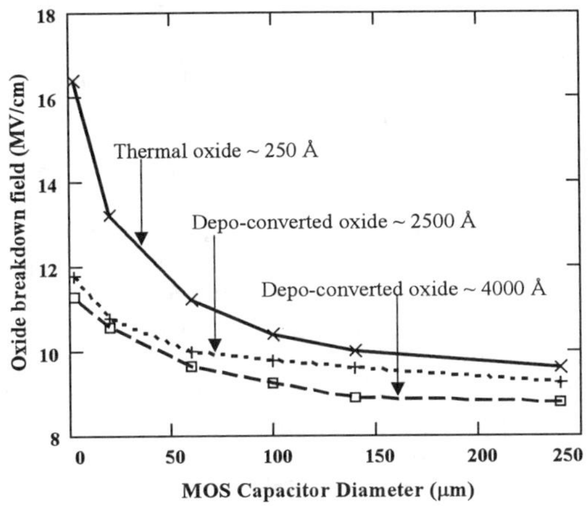

Fig. 2. Dependence of oxide breakdown field on MOS capacitor diameter (n-type SiC).

As observed from Fig. 2, the breakdown field of the thermal oxide increases rapidly with decrease in MOS capacitor diameter below ~ 150 μm. On the other hand both the 2500 Å and 4000 Å thick depo-converted oxides exhibit a gradual reduction in the breakdown field with increase in MOS capacitor diameter upto about 150 μm, after which there is no significant reduction in the breakdown field with further increase in MOS-C diameter. The decrease in the oxide breakdown field with increase in MOS-C diameter could be due to the following reasons. We have observed that oxide breakdown in SiC wafers generally occurs at locations corresponding to the edge of bulk structural defects in the SiC wafer such as polytype inclusions, regions of crystallographic mis-orientation, or different doping concentration [8]. Due to poor SiC material quality the probability of enclosing such defects underneath the gate contact of a MOS-C increases with increase in diameter. This in turn increases the probability of oxide breakdown with increase in MOS-C diameter. Another plausible explanation could be related to the variations in the electric field distribution in the oxide film with increase in MOS-

C diameter. The accumulation layer underneath the gate contact is probably more uniform for small diameter MOS structures compared to large diameter structures due to contact edge effects. Hence, the electric field distribution in the oxide could be more uniform for small diameter structures thus enabling them to withstand higher electric stress compared to large diameter structures where electric field enhancement around the gate contact periphery results in lower oxide breakdown strengths.

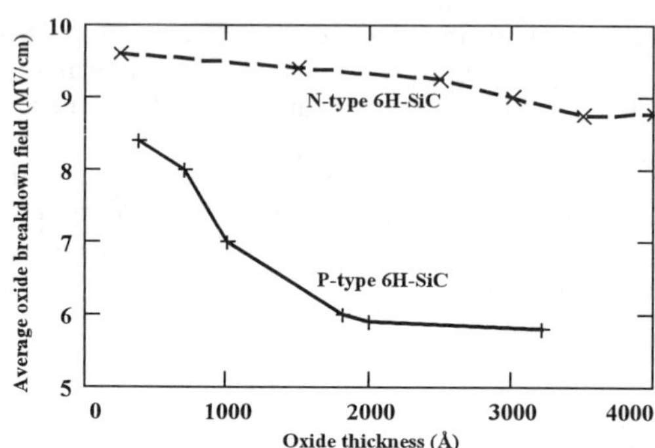

Fig. 3. Dependence of breakdown field on oxide thickness (diameter of MOS-C : 240 μm).

Fig. 3 shows the dependence of the breakdown field on oxide thickness for n and p type SiC wafers. The breakdown field of oxide films in the case of p-type SiC wafers decreases significantly from a value of more than 8 MV/cm for ~ 300 Å thermal oxide to ~ 6 MV/cm for a ~ 3000 Å depo-converted oxide. On the other hand, the breakdown field of oxide films on n-type SiC wafers reduces slightly from ~ 9.6 MV/cm for ~ 300 Å thermal oxide to ~ 8.7 MV/cm for a ~ 4000 Å thick depo-converted oxide. The reduction in oxide breakdown field with increase in thickness is understandable considering the well known fact that the breakdown strength of solid dielectrics decreases with increase in thickness due to a reduction in the heat removal rate which promotes an electro-thermal breakdown at smaller fields [9]. The difference in the extent of reduction of the oxide breakdown field on n-type vs p-type SiC, could be due to poorer oxide quality on p-type SiC with increasing thickness.

The fairly high breakdown strengths of the depo-converted oxide indicates that it is possible to use thick oxide layers obtained by the depo-conversion technique for applications requiring high oxide breakdown strength. For example, the depo-converted oxide can be effectively used for field plate edge termination in Schottky diodes. In fact, we have successfully fabricated high voltage (1kV) p-type SiC Schottky diodes with a ~ 6300 Å depo-converted oxide as edge termination [10].

CONCLUSION

In conclusion, thick oxide films have been successfully formed on n and p-type SiC substrates by converting Si to oxide. C-V measurement indicates that this converted oxide film exhibits an effective charge density comparable to that of a thermal oxide. The depo-converted oxides also indicated fairly high breakdown strengths, in the range of 6-9 MV/cm. The oxide breakdown strength was observed to decrease with increasing oxide thickness especially on p-type SiC. Also, for a given oxide thickness the oxide breakdown strength decreased with increase in MOS capacitor diameter. Thick oxide layers obtained by converting

Si to oxide exhibits promising characteristics for applications in power SiC devices, such as field plate edge termination of Schottky diodes.

ACKNOWLEDGEMENTS

This work was supported by ARO (grant no DAA H04-96-1-0467) via a DEPSCoR program. The authors are pleased to acknowledge many discussions with Dr. G. Gradinaru, and Mr. Y. Gao.

REFERENCE

1. M. Bhatnagar and B.J. Baliga, IEEE Trans. ED **44**, 645 (1993).
2. P.G. Neudeck, J. Electron. Mater. **24**, 283 (1995).
3. J. Schmitt and R. Helbig, J. Electrochem. Soc. **141**, 2262 (1994).
4. J. W. Palmour, U. S Patent No. 5 612 260, (18 March 1997).
5. J. Tan, M. K. Das, J. A. Cooper, Jr., and M. R. Melloch, Appl. Phys. Lett. 70, 2280 (1997).
6. S. Dimitrijev, H. F. Li, H.B. Harrison, and D. Sweatman, IEEE Trans. EDL **18**, 175 (1997).
7. V. P. Madangarli and T.S. Sudarshan, Proc. of the 7th Int. Conf. on SiC, III-Nitrides, and Related Materials (ICSCIII-N'97), 665 (1997).
8. S. Soloviev, I. Khlebnikov, V. Madangarli and T. S. Sudarshan, J. Electron. Mater. **27**, 1124 (1998).
9. B. Tareev, *Physic of Dielectric Materials*, (Mir Publishers, Moscow, 1979).
10. Q. Zhang, V. Madangarli, S. Soloviev and T. S. Sudarshan, *to be presented at 1999 MRS Spring Meeting* (unpublished).

Bias-Temperature-Stress Induced Mobility Improvement in 4H-SiC MOSFETs

K. Chatty†, T. P. Chow†, R. J. Gutmann†, E. Arnold‡, and D. Alok‡

† Rensselaer Polytechnic Institute, Troy, NY 12180-3590, U.S.A.
Tel: 518-276-6044, Fax: 518-276-8761, e-mail: kchatty@unix.cie.rpi.edu

‡ Philips Research, Briarcliff Manor NY 10510, U.S.A.

ABSTRACT

In this work, we report on an instability which affects the field effect mobility in 4H-SiC MOSFETs. The devices (MOSFETs and capacitors) were subjected to a bias-temperature stress (BTS) for 30 minutes at 150°C at stress voltages corresponding to oxide fields upto 1MV/cm. Following a positive BTS(i.e. gate voltage positive), the field effect mobility increased by upto two orders of magnitude from the original value; upon application of a negative BTS to the MOSFET, the device characteristics degraded to the unstressed state. The high mobility state could be recovered by a positive BTS and was reversible with repeated bias stressing. An explanation of this phenomenon is proposed based on the effect of interfacial ions on the dependence of both trapped charge and inversion charge densities on gate bias.

INTRODUCTION

The electrical properties of the current state-of-the-art SiC-SiO_2 interfaces are inferior to those of silicon. The densities of oxide charges and interface states are much higher than those at the Si-SiO_2 interface[1]. The effect of the localized states is seen in the degradation of the transconductance and the increase in the threshold voltage of the MOSFETs[2]. The understanding and control of the characteristics of SiC-SiO_2 interface is crucial to the realization of practical SiC MOS devices. In this work, we report the investigation of an instability which affects the field effect mobility in SiC MOSFETs.

MOSFET FABRICATION

The 4H-SiC wafers used for the fabrication of the MOSFETs had an epitaxial thickness and doping of 10 μm and 4×10^{15} cm^{-3} respectively. The wafers were cleaned before a 800nm thick plasma TEOS oxide (field oxide) was deposited on the wafers. A 100nm thick plasma TEOS oxide was deposited to act as a pad oxide during implantation of the source and drain. The source and drain were then implanted with nitrogen (80keV, 2×10^{15} cm^{-2}; 40keV, 1×10^{15} cm^{-2}) at 650 °C. The samples were then annealed at 1200 °C for 1 hour in an argon to electrically activate the implants.

Mat. Res. Soc. Symp. Proc. Vol. 572 © 1999 Materials Research Society

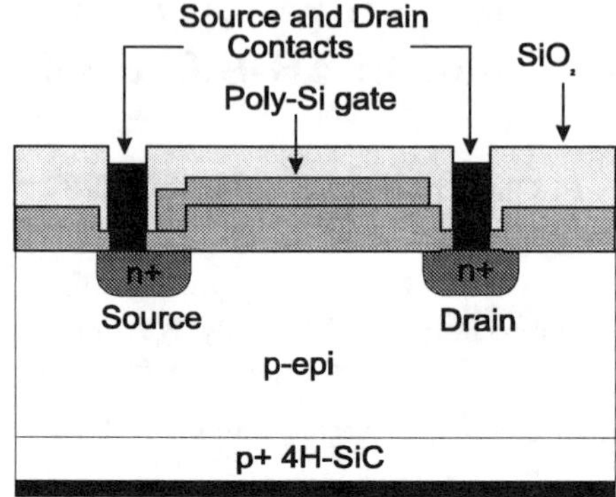

Fig. 1: Schematic of the 4H-SiC lateral MOSFET.

After the implant activation anneal, the field oxide on one of the samples (thin oxide MOSFET) was etched back to 200nm. The samples underwent oxidation in a wet ambient at 1100°C for 6 hours and 40 min, followed by an anneal for 1 hour at the oxidation temperature in argon. The oxide was then subjected to a re-oxidation anneal in a wet ambient at 950°C for 3 hours. The annealing cycles were similar to the work by Sridevan *et al* [3]. Subsequent to the oxidation, polysilicon was deposited and degenerately doped by phosphorus implantation. After definition of the gate, a 600nm thick plasma TEOS oxide was deposited to serve as an interlevel dielectric. The source and drain contacts (Al/Ni/Al) were defined using liftoff technique, and the contacts (source, drain and the substrate) were annealed at 1000°C in argon. The gate contact was patterned and etched, followed by (Ti/Mo) contact metallization(Figure 1).

RESULTS AND DISCUSSION

Experimental results were obtained on two MOSFETs with gate width-to-length (W/L) ratios of 8, and gate oxide thicknesses of 200 nm (thin oxide) and 900 nm (thick oxide). The field effect mobility was calculated from the transconductance dI_D/dV_G at a drain voltage of 25-100mV:

$$\mu_{FE} = \frac{dI_D/dV_G}{C_{ox}V_D(W/L)} \tag{1}$$

where I_{DS} is the drain current, V_G is the gate voltage and C_{ox} is the oxide capacitance per unit area. The mobility values quoted below were taken at the steepest part of the I_D-V_G curve.

The initial field effect mobility of the thin-oxide MOSFET was 0.5 cm^2/V.s. Above room temperature, on repeated gate voltage sweeps, the I_D-V_G transfer characteristic varied. Assuming that these variations were due to mobile ions in the oxide, a bias-temperature-stress(BTS) was applied to stabilize the transfer characteristics

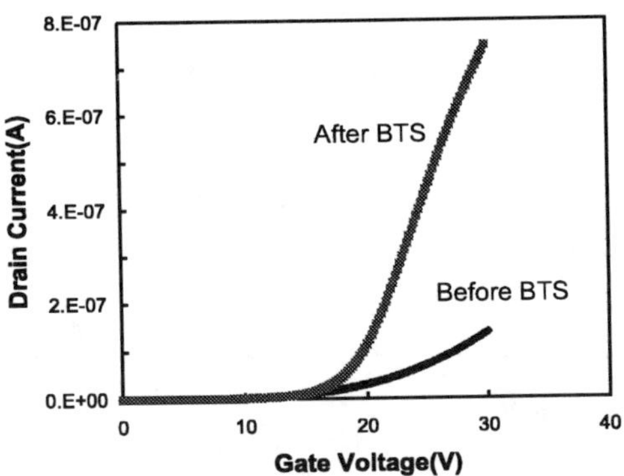

Fig. 2: Transfer characteristics of the thin oxide MOSFET before and after BTS(V_D=25mV).

by drifting the mobile charge towards the semiconductor-insulator interface[4].

The BTS of the thin-oxide MOSFET was done by applying a gate voltage corresponding to an oxide field of 1MV/cm at 150°C for 30 minutes, with the source and drain contacts floating. The device was cooled to room temperature while maintaining the bias voltage on the device. Figure 2 compares the transfer characteristics of the thin-oxide MOSFET before and after BTS. The field effect mobility of the sample increased by a factor of 16, from 0.5 to 8 cm^2/V.s after BTS. Similar instabilities were observed in the transfer characteristics of the thick oxide MOSFET. The BTS consisted of applying a gate voltage corresponding to an oxide field of 0.5MV/cm at 150°C for 30 minutes, with figures 3 and 4 showing the transfer characteristics before and after BTS. In this case, the field effect mobility of the sample increased from 0.1 to 13cm^2/V.s, a two orders-of-magnitude improvement.

Transfer characteristic of another thick-oxide MOSFET before and after BTS is shown in Figure 5, along with the transfer characteristics one month after the bias stressing. While the BTS improves the transconductance, but the characteristics degrade back to the pre-bias-stressed values with time, as the ions diffuse from the SiC-SiO$_2$ interface.

To further investigate this phenomenon, mobile-ion analysis experiments were done on MOS capacitors located on the same wafer. After the initial room temperature capacitance-voltage characteristic was measured using a 1MHz Boonton capacitance meter, a BTS of +20V at 150°C for 30 minutes was applied. The capacitance-voltage characteristic was also measured after a negative BTS.(Following the negative BTS, the MOSFET did not turn on until a subsequent positive BTS). Figure 6 shows the results of the C-V measurements, from which a mobile-ion charge density of 3×10^{12} cm^{-2} was extracted.

Figure 7 compares the gate-voltage dependence of transconductance for a thick

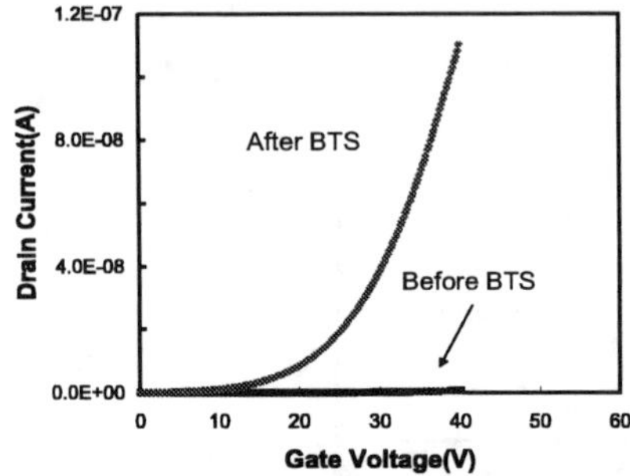

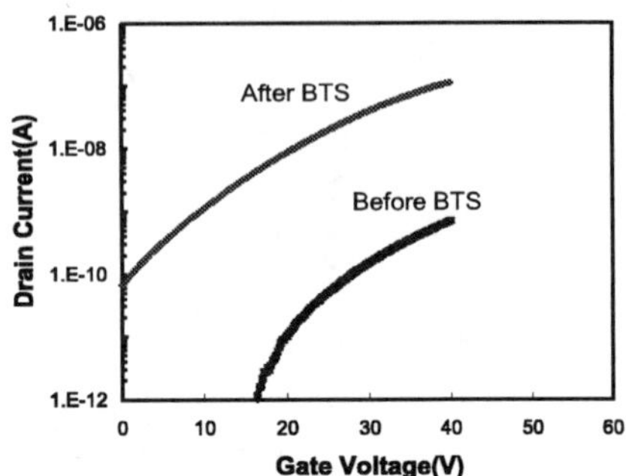

Fig. 3: Transfer characteristics (linear scale) of the thick oxide MOSFET before and after bias stressing (V_D=25mV)

Fig. 4: Transfer characteristics (log scale) of the thick oxide MOSFET before and after bias stressing (V_D=25mV)

oxide MOSFET before and after the BTS. Before BTS, the transconductance increases monotonically with gate bias: after the BTS, the transconductance increases more rapidly, reaches a peak, and decreases with a further increase in gate voltage, as would normally be observed in a conventional MOSFET.

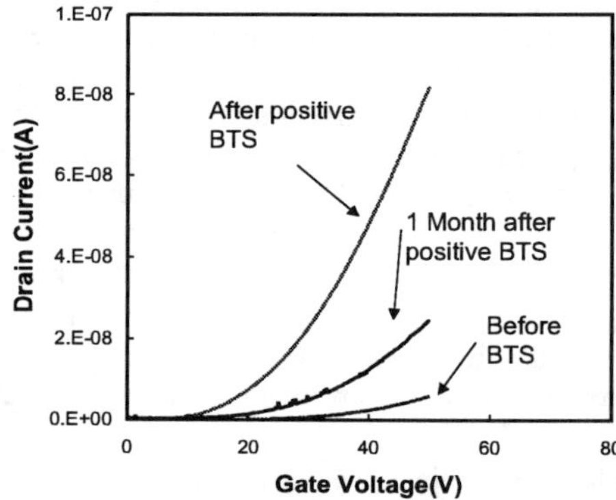

Fig. 5: Transfer characteristics for the thick oxide MOSFET before and after BTS

The unusual SiC MOSFET characteristics are attributed to mobile charges within the gate oxide and their effect on the dependence of interface state occupancy and inversion charge on the gate bias voltage. A possible explanation for this phenomenon is as follows: A change δV_G in gate voltage results in a change δQ_t in the charge trapped in the interface states and a change δQ_{inv} in the inversion charge:

$$\delta V_G = \frac{\delta Q_t + \delta Q_{inv}}{C_{ox}} \tag{2}$$

while the drain current depends only on the inversion charge:

$$\delta I_D = (W/L)V_D \mu_{inv} \delta Q_{inv} \tag{3}$$

where μ_{inv} is the mobility of electrons in the inversion layer. Therefore, $\mu_{FE} < \mu_{inv}$ (Eq. 1), as long as the magnitude of trapped charge increases with gate voltage. The relative change in the inversion charge becomes larger in strong inversion, and the field effect mobility increases with gate voltage. If the interface state density is large, very large gate voltages would be needed for the field effect mobility to approach the inversion layer mobility. With a large positive oxide charge pushed toward the semiconductor interface during BTS, strong inversion can exist at a lower gate bias. Consequently, the field effect mobility increases. The total mobile ion charge is expected to be larger in the thick-oxide MOSFET, which results in a more pronounced mobility improvement.

CONCLUSION

Lateral n-channel MOSFETs with two different gate oxide thicknesses (900nm and 200nm) were fabricated on 4H-SiC. Upon bias stressing the field-effect mobility of the MOSFETs increased by over 2 orders(one order) of magnitude in the case of the thick (thin) oxide MOSFET. The field-effect mobility was degraded upon the application of a negative bias stress but the high mobility state could be recovered by a subsequent positive BTS. The improvement of the transconductance and the field effect mobility of the MOSFETs is attributed to the effect of positive oxide charges on the dependence of interface state occupancy and inversion charge on gate bias.

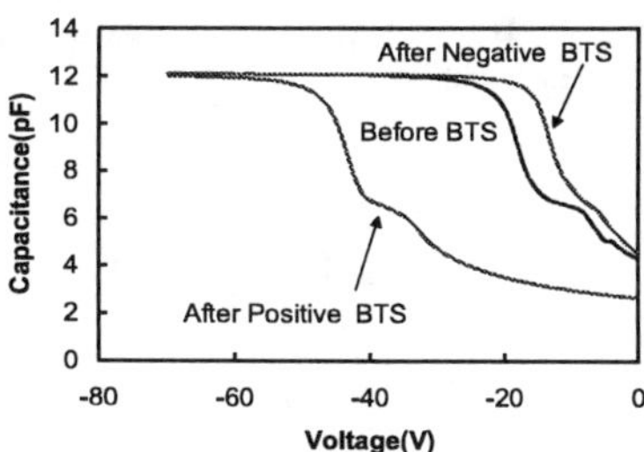

Fig. 6: Capacitance-voltage characteristics before and after BTS.

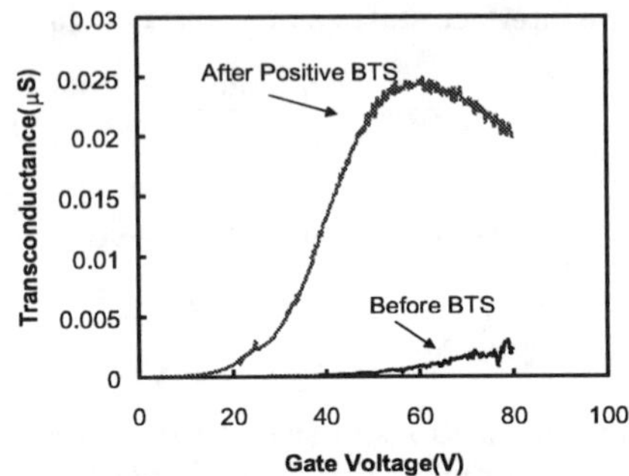

Fig. 7: Gate voltage dependence of transconductance of the thick oxide MOSFET before and after bias stressing (V_D=100mV)

ACKNOWLEDGEMENT

The authors from Rensselaer Polytechnic Institute, Troy, NY, gratefully acknowledge the support of this work by Philips Research, Briarcliff Manor, NY, MURI of the Office of Naval Research under contract no. N0014-95-1-1302 and the Center for Power Electronic Systems (NSF Engineering Research Center under contract no. CR-19229-427756)

REFERENCES

[1] T. Ouisse, Phys. Stat. Sol. (a) **162**, 239 (1997).

[2] E. Arnold, N. Ramungul, T. P. Chow and M. Ghezzo, Proc. 7th Int. Conf. on SiC, III-nitrides and related materials, Stockholm, 1013 (1997).

[3] S. Sridevan and B. J. Baliga, IEEE Electron Device Lett. **19**, 228 (1998).

[4] E. H. Nicollian and J. R. Brews,"MOS(Metal Oxide Semiconductor) Physics and Technology, Ch 15, Wiley 1982.

FULL BAND MONTE CARLO SIMULATION OF SHORT CHANNEL MOSFETs IN 4H AND 6H-SiC

M. HJELM[1,2], H-E. NILSSON[1,2], E. DUBARIC[1,2], C. PERSSON[3], P. KÄCKELL[4], C. S. PETERSSON[2]

[1]Department of Information Technology, Mid-Sweden University, S-851 70 Sundsvall, Sweden, Mats.Hjelm@ite.mh.se

[2]Department of Solid State Electronics, Kungl. Tekniska Högskolan (KTH), Elektrum 229, S-164 40 Kista, Sweden

[3]Department of Physics and Measurement Technology, Linköping University, S-581 83 Linköping, Sweden

[4]Institut für Festkörpertheorie und Theoretische Optik, Max-Wien-Platz 1, 07743 Jena, Germany

ABSTRACT

This is a presentation of a full band Monte Carlo (MC) study, which compares electron transport and device performance for 4H and 6H-SiC 100 nm n-channel MOSFETs. The model used for the electrons is based on data from a full potential band structure calculation using the Local Density Approximation (LDA) to the Density Functional Theory (DFT). For the holes the transport is based on a three band k·p model including spin orbit interaction. The two polytypes are compared regarding surface mobilities obtained with the program, as well as transconductance, unit current gain frequency, carrier velocity, I-V characteristics and energy distribution in the channel for the MOSFETs.

INTRODUCTION

Silicon carbide (SiC) is considered to be a very promising material for high temperature and high power applications due to its high breakdown voltage and high thermal conductivity. One possibility is the fabrication of high speed integrated MOSFETs in 4H-SiC on a semi-insulating substrate, with the expectation of a reliable and stable operation. An advantage for both 4H-SiC and 6H-SiC is their discontinuous energy spectrum in the conduction band along the c-axis direction, which results in limited carrier heating by the electric field. Another factor tending to lower the carrier energy is the strong polar optical scattering. CMOS integrated circuits have been fabricated on 6H-SiC [1,2], showing that commercial SiC MOSFETs will be available in the near future.

This paper presents a comparison, using a full band Monte Carlo model, of 4H-SiC and 6H-SiC regarding the surface mobility and device properties in short channel MOSFETs.

MONTE CARLO MODEL

The full band MC program is based on a large lookup table, stored for the irreducible part of the Brillouin zone and containing the energy and energy gradient versus k-vector from the band structure. Between the points represented in the lookup table, the energy values are calculated using a second order cubic spline, and the energy gradient by linear interpolation. The band data in the simulations presented here are based on the calculations in reference [3] and reference [4].

The following scattering processes are considered: acoustic phonon scattering, polar optical phonon scattering, zero and first order optical intervalley phonon scattering and ionised impurity scattering [5,6]. For the acoustic and optical phonons, the coupling constants are obtained by fitting data from bulk simulations to experimental data from references [7] and [8]. As a result the transport properties in the simulations correspond to those in the material used in the referred experiments.

In the case of surface scattering, we are using the semi-empirical model proposed by E. Sangiorgi and M. R. Pinto [9], which is based on a combination of specular and diffusive reflection in the interface. The diffusive scattering is obtained by the random selection of a k vector in such a way that the energy is conserved and the movement is directed away from the surface. To perform

Mat. Res. Soc. Symp. Proc. Vol. 572 © 1999 Materials Research Society

specular scattering, which is merely an elastic bounce, the normal component of the movement is reversed and the component parallel to the interface is maintained. As both 4H and 6H-SiC are anisotropic the diffusive scattering is also anisotropic with the angular distribution corresponding to the density of states in the k-space. Each time a carrier hits the oxide surface a random number is used to select between the two types of scattering, with a constant (C) defining the probability of diffusive scattering. According to reference [9] a C value of 0.06 results in a good correspondence between MC simulations and experiments for good interfaces between Si and SiO_2. In our simulations we have used this value to simulate a high quality interface. For the simulation of a poor interface we have used a C value of 0.50.

For the overlap integral of electron wave functions in 4H-SiC, we are using equation (1) which is an expression often used for cubic semiconductors [10], where k and k' are the initial and final states, α (= 0.323 for band 1 and 0.8 for band 2, 3 and 4) is the nonparbolicty parameter, ε and ε' are the initial and final energies and θ is the angle between the initial and final states.

$$I^2(k, k') = \frac{[(1 + \alpha\varepsilon)^{1/2}(1 + \alpha\varepsilon')^{1/2} + \alpha(\varepsilon\varepsilon')^{1/2}cos\theta]^2}{(1 + 2\alpha\varepsilon)(1 + 2\alpha\varepsilon')} \tag{1}$$

Due to the high degree of anisotropy, we use another approximate analytic expression, equation (2), for the overlap integral in 6H-SiC.

$$I^2(k, k') = 1 - (0.8732 + 0.0268\ sin\gamma)\{1 - exp[-1.01688 \cdot 10^2(1 - 0.855\,sin\gamma)|k - k'|^2)]\} \tag{2}$$

Here, γ is the angle between the vector $q = k - k'$ and the c-axis. The simulation is self consistent and uses a two dimensional solver for Poisson's equation. All other calculations are performed in three dimensions. There is no need to consider impact ionisation since its threshold energy is very high. The model neither takes into account carrier generation and recombination processes, nor does it consider the carriers at the oxide interface as a two-dimensional gas. All the simulations in the study are made using the same version of the program.

SIMULATION RESULTS

Surface mobilities

The surface mobilities, which are shown in table I are calculated from the diffusion coefficients with the Einstein relation, where the carrier mean energies are sampled values from the simulation. The diffusion coefficients are also obtained from sampled data according to equation (3), where u is the position along the x, y or z axis respectively and D_u is the corresponding diffusion coefficient.

$$D_u = \left(\frac{d}{dt}\right)\langle(u - \langle u \rangle)^2\rangle \tag{3}$$

The simulations are made using a simplified device consisting of two regions with SiC and SiO_2, and with the plane interface between the two materials. We have assumed an electron concentration of 4×10^{18} since the simplified device does not permit calculation of the electron concentration in the channel. The doping level was 1×10^{15} donors/cm^3 and 1×10^{17} acceptors/cm^3. The field perpendicular to the interface was 500 kV/cm.

As can be seen from the table, 6H-SiC has a strong anisotropy with a ratio > 4.8. 4H-SiC is also anisotropic with a ratio of about 0.8 - 0.9. Since the mobility is so low for transport in the c-axis direction for 6H-SiC, we only consider electron transport perpendicular to the c-axis in the remainder of this study. The surface mobilities with C=0 are superior for 4H-SiC. The values for 4H are better than the corresponding bulk mobilities which can be explained by a higher degree of screening of impurities. For good interface qualities, 4H-SiC still has better mobility but with a smaller advantage. In the case of inferior oxide quality, 6H-SiC has an advantage. We interpret

Table I: Surface mobilities in cm^2/Vs. The texts "$\perp$ c-axis" and "c-axis" refer to the transport direction in relation to the c-axis.

Temp. K	C	4H-SiC		6H-SiC	
		$\perp$ c-axis	c-axis	$\perp$ c-axis	c-axis
300	0.00	862	968	358	74
300	0.06	371	454	308	50
300	0.50	74	88	148	14
500	0.00	218	260	115	25
500	0.06	169	193	108	21
500	0.50	61	66	80	11

these characteristics as the result of a higher probability for electrons in 6H-SiC to find final states after scattering, with a large velocity component perpendicular to the c-axis. In figure 1 we show the velocities for 1000 random points in k-space with the energy 0.05 eV for 4H and 6H-SiC. In 4H-SiC the velocities are distributed near the surface of a sphere, while the velocity distribution in 6H-SiC has a flattened form. It can be seen in the figure, that for 6H-SiC the majority of the electrons have a relatively small component parallel to the c-axis. In this context is it also important to consider that the same C constant does not a priori correspond to the same interface quality in the two polytypes.

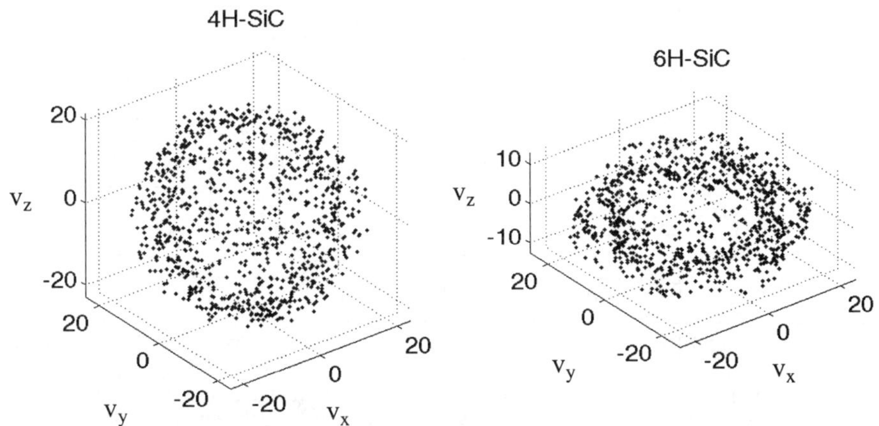

Fig. 1. Velocity distribution for 1000 random points i k-space with energy = 0.05 eV. The velocities are given in cm/s x 10^6. The coordinates are oriented with z parallel to the c-axis.

MOSFET simulations

The simulated MOSFET is shown in figure 2, and in table II the simulation results are shown for transconductance (g_m), total gate capacitance (C_{gtot}) and unit current gain frequency (f_T). These values are for a source potential of 0 V, a drain potential (V_{ds}) of 2.0 V and represent the mean value for gate potentials (V_{gs}) between 1.3 and 2.3 V, with an approximate threshold voltage (V_{th}) of 0.8 V, assuming that the gate is made of aluminium and with no charges in the oxide or on the

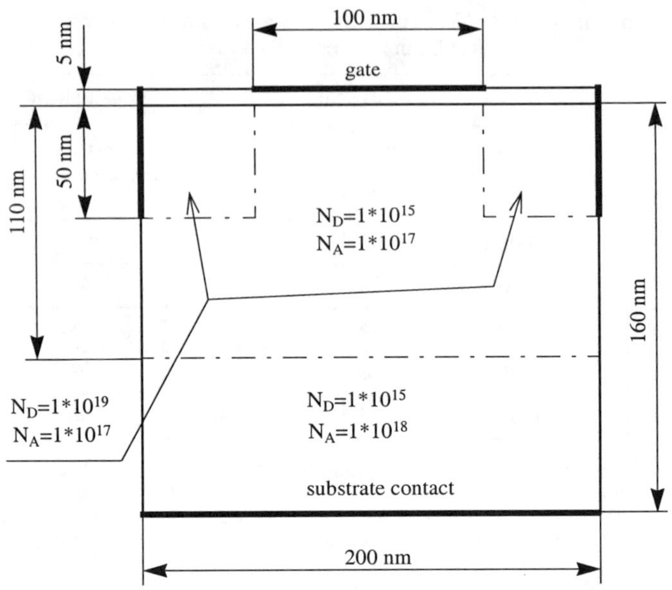

Fig. 2. Simulated MOSFET transistor structure. The width of the device is 1 μm. Doping values are in cm^{-3}.

Table II: Simulation results for the MOSFETs.

Temp. K	C	4H-SiC			6H-SiC		
		g_m S x 10^{-4}	C_{gtot} F x 10^{-16}	f_T GHz	g_m S x 10^{-4}	C_{gtot} F x 10^{-16}	f_T GHz
300	0.06	6.12	5.10	191	4.25	4.86	139
300	0.50	3.80	4.89	123	3.63	4.83	120
500	0.06	5.00	4.98	160	3.10	4.63	105
500	0.50	3.42	4.90	111	2.81	4.62	95

oxide-semiconductor interface. For the calculation of g_m and C_{gtot}, the drain current (I_D) as well as the total charge on the gate (Q_{gtot}) were sampled at equidistant V_{gs} values with a difference of 0.1 V. The g_m was calculated as the quotient of difference in I_D and V_{gs}. Similarly, C_{gtot} was obtained as the quotient of difference in Q_{gtot} and V_{gs}.

In contrast to the bulk mobilities, the MOSFET characteristics are better in all cases for the 4H polytype, although the difference is so small for $C = 0.50$ that it may be considered insignificant. This is not surprising as the MOSFETs work with high electric fields parallel to the channel and the carrier mean velocity reaches saturation or near saturation, i.e. the low field mobilities are not applicable, see figure 3. Furthermore, at the beginning and end of the channel, the electron velocity component perpendicular to the oxide interface is important. Consequently, the very low c-axis mobility for 6H-SiC results in smaller current. An interesting fact, that can also be seen in figure 3, is that 6H-SiC besides lower velocity, has a much lower peak value for the mean energy in

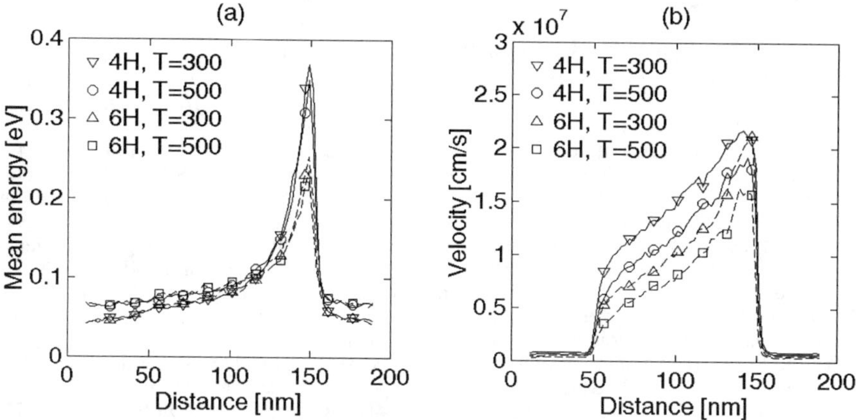

Fig. 3. Mean energy in the channel (a) and mean velocity parallel to the interface (b). V_{gs}=1.8 V, V_{ds}=2.0 V, C = 0.06. The distance is measured from the left border of the component. The mean energy is for a distance up to 2.5 nm from the interface, and the mean velocity for all electrons in the channel.

the channel. The I-V characteristics for C=0.06, are shown in figure 4, together with the corresponding maximum mean energy in the channel. As can be seen, 6H-SiC has an advantage as it has lower energy when $V_{ds} > 0.8$ V, which increases with higher drain potential.

To get a comparison with a corresponding device in silicon, we have made a simulation with the Medici program [11] using the energy balance model. The Si device was identical with the one in SiC, with the exception that the doping in the channel region was increased to 1.3×10^{18} which resulted in a threshold gate potential of 0.8 V. In this case the maximum carrier energy with V_{gs}=1.8 V and V_{ds}=2.0 V, was 0.54 eV, which may be compared with the carrier energies 0.39 eV and 0.25 eV for 4H-SiC and 6H-SiC respectively, obtained with the MC program. The approximate maximum fields were: 5.4×10^5 V/cm parallel to the interface in the Si device and 6.5×10^5 in the SiC device, 6.8×10^5 V/cm and 4.2×10^5 perpendicular to the interface in Si and SiC respectively.

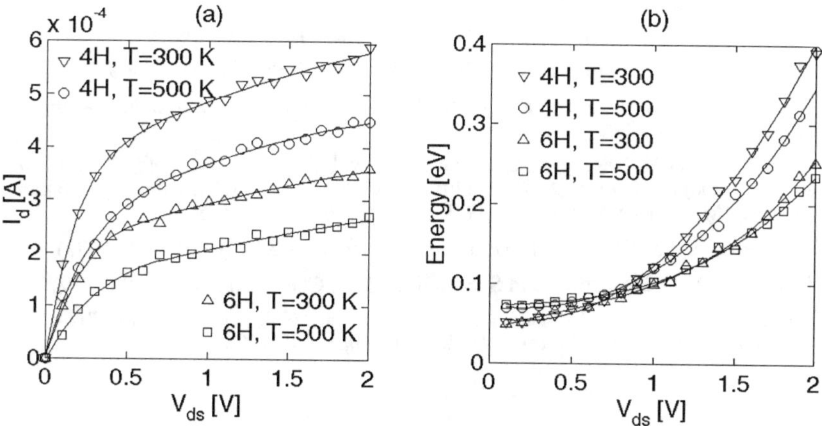

Fig. 4. I-V characteristics (a) and maximum mean energy (b). V_{gs} = 1.8 V, C = 0.06.

CONCLUSIONS

We have presented a full band Monte Carlo simulation, comparing both surface mobilities and device performance for 100nm n-MOSFETs in 4H and 6H-SiC. We have used the model of Sangiorgi and Pinto [9] for the oxide semiconductor interface with a diffuse scattering factor (C) modelling the oxide quality.

The surface mobility simulations show a difference in surface mobility in favour of 4H-SiC for good interface quality (C=0.06), but for bad interfaces (C=0.50) 6H-SiC has higher mobility. We consider this to be an effect of the strong anisotropy in 6H-SiC, with higher probabilities for carriers to find final states after scattering in the channel direction.

The simulated MOSFETs show better transconductance and unit current gain frequency for 4H-SiC in all cases. For instance, g_m was 6.12 x 10^{-4} S and f_T 191 GHz for the 4H-SiC device with a gate width of 1 μm at 300 K and C=0.06. The corresponding values for 6H were 4.25 x 10^{-4} S and 139 GHz. However, for the poor interface quality, the difference is small and the f_T values are 123 and 120 GHz for the same conditions as above. This means that in order to benefit from the higher mobility in 4H-SiC a good oxide quality is necessary. The electrons in the channel have lower energies in the SiC MOSFETs than in a corresponding Si device. At high V_{ds} values, 6H-SiC has an advantage having lower carrier energies than 4H-SiC. For instance with V_{gs}=1.8 V, V_{ds}=2.0 V, T= 300 K, and C=0.06, the maximum mean energy in the channel is 0.39 eV for 4H-SiC and 0.25 eV for 6H-SiC.

The difference in carrier energy in favour of 6H-SiC is a potential that could be utilized in device designs, leading us to the conclusion that 6H-SiC may be as good as, if not better than, 4H-SiC in MOSFET devices.

ACKNOWLEDGMENTS

Financial support from ISS Foundation and Mid-Sweden University is gratefully acknowledged.

REFERENCES

1. J. N. Shenoy, J. A. Cooper, Jr. and M. R. Melloch, IEEE Dev. Lett. **18**, 93 (1997).

2. J. Spitz, M. R. Melloch, J. A. Cooper, Jr. and A. Capano, IEEE Dev. Lett **19**, 100 (1998).

3. P. Käckell, B. Wenzien and F. Bechstedt, Phys. Rev. B **50** (15) 10761 (1994).

4. C. Persson and U. Lindefelt, J. Appl Phys. **82** (11), 5,496 (1997).

5. K. Tsukioka, D. Vasileska and D. K. Ferry, Physica B **185**, 466 (1993).

6. H-E. Nilsson, U. Sannemo and C. S. Petersson, J. Appl. Phys. **80** (6), 3365-3369 (1996).

7. M. Schadt, G. Pensl, R. P. Devaty, W. J. Choyke, R. Stein and D. Stephani, Appl. Phys. Lett. **65**, 3120 (1994).

8. W. J. Schaffer, G. H. Negley, K. G. Irvine and J. W. Palmour in *Diamond, Silicon Carbide and Nitride Wide Bandgap Semiconductors*, edited by C. H. Carter, G. Gildenbalt, S. Nakamura, and R. Nemanich, (Mater. Res. Soc. Proc. **339**, Pittsburgh, PA 1994) pp. 595-600.

9. E. Sangiorgi and M. R. Pinto, IEEE Trans. on Elec. Dev. **39** (2), 356-361 (1992).

10. W. Fawcett, A. D. Boardman and S. Swain, J. Phys. Chem. Solids, **31**, 1963 (1970).

11. Avant! Corporation, TCAD Business Unit, Fremont, California. *Medici, Two-Dimensional Device Simulation Program, Version 4.1, Users Manual*, July 1998.

High Voltage Schottky Barrier Diodes on P-Type SiC using Metal-Overlap on a Thick Oxide Layer as Edge Termination

Q. Zhang, V. Madangarli, S. Soloviev and T. S. Sudarshan
Department of Electrical Engineering
University of South Carolina, SC 29208, U.S.A
Tel: 803-777-7302; Fax: 803-777-8045
E-mail: Zhang@engr.sc.edu

ABSTRACT

P-type 6H SiC Schottky barrier diodes with good rectifying characteristics upto breakdown voltage as high as 1000V have been successfully fabricated using metal-overlap over a thick oxide layer (~ 6000 Å) as edge termination and Al as the barrier metal. The influence of the oxide layer edge termination in improving the reverse breakdown voltage as well as the forward current – voltage characteristics is presented. The terminated Schottky diodes indicate a factor of two higher breakdown voltage and 2-3 times larger forward current densities than those without edge termination. The specific series resistance of the unterminated diodes was ~228 mΩ-cm^2, while that of the terminated diodes was ~84 mΩ-cm^2.

INTRODUCTION

Silicon carbide (SiC) has recently been given renewed attention because of its fine characteristics such as stability at high temperatures, wide bandgap, high breakdown field and high thermal conductivity. In the case of high-voltage devices, edge termination plays a very critical role in determining the breakdown voltage. High breakdown voltage (~ 730 V) SiC Schottky barrier diodes have been reported using edge termination techniques [1,2]. However, in these processes ion implantation is required to obtain a high resistivity layer on the surface at the edges of the device [1]. A metal-overlap onto an oxide layer at the edge of the device can be used to minimize the electric field enhancement so that the breakdown voltage can approach the ideal plane parallel value of the SiC wafer [3]. But due to the slow oxidation rates on SiC it is difficult to get a thick oxide layer on SiC by conventional thermal oxidation of SiC [4]. Moreover, although there have been a lot of publications on high voltage Schottky diodes on n-type SiC, there is very little work reported on high voltage Schottky contact on p-type SiC. To the best of our knowledge the only significant study upto now on p-type SiC Schottky diodes has been by R.Raghunathan and B.J.Baliga, who reported p-type 4H- and 6H-SiC Schottky diodes with breakdown voltage upto 600V in 1998 [5]. In this paper, we compare the forward and reverse I-V characteristics of a conventional p-type 6H-SiC Schottky diode with no edge termination to that of a 1 kV Schottky diode with a thick oxide layer edge termination.

EXPERIMENT

Schottky diodes were fabricated on a p-type 6H-SiC wafer from Cree Research (substrate doping ~ 1.6$\times10^{18}$ cm^{-3}) with a 10 micron thick epilayer of ~ 6$\times10^{15}$ cm^{-3} doping concentration, and ideal parallel plane breakdown voltage of ~ 1200 V. The 30 mm diameter wafer was cut into 10 mm x 10 mm square pieces to obtain several samples for experiments. All the samples were cleaned by RCA procedure to obtain a SiC surface with minimal surface contamination prior to diode fabrication. Unterminated Schottky diodes of approximately 140 μm diameter

Mat. Res. Soc. Symp. Proc. Vol. 572 ©1999 Materials Research Society

were fabricated on few samples following a conventional photolithographic process (Fig. 1(a)), while Schottky diodes with oxide layer edge termination were fabricated as follows.

In order to obtain a thick oxide layer for edge termination, instead of sputter depositing SiO_2 on top of the thermal oxide as reported by other groups [3], we have adopted a depo-conversion technique which is discussed in a companion paper [6]. This technique essentially involves sputter deposition of a thick Si layer on the SiC wafer and conversion of Si to SiO_2 by oxidation. After conversion, the sample was annealed in Ar for half an hour in order to improve the oxide quality.

The thick oxide layer was then selectively etched using a first mask to form the Schottky contact window with a diameter of 140 microns. After a cleaning process, Al was evaporated onto both sides of the sample in high vacuum ($<10^{-5}$ Torr). A Schottky contact was formed on the polished epilayer while an ohmic contact is obtained on the roughened backside of the sample. The Al on the epilayer side was selectively etched using a second mask with a diameter of 250 microns to fabricate individual Schottky diode structures with a metal overlap length of 55 micron over the thick oxide layer as shown in Fig.1 (b). It has to be noted that, even though generally it is sufficient to have an overlap approximately equal to the epilayer thickness [3], we have used a larger overlap because of the availability of a photomask with 250 µm diameter circular features.

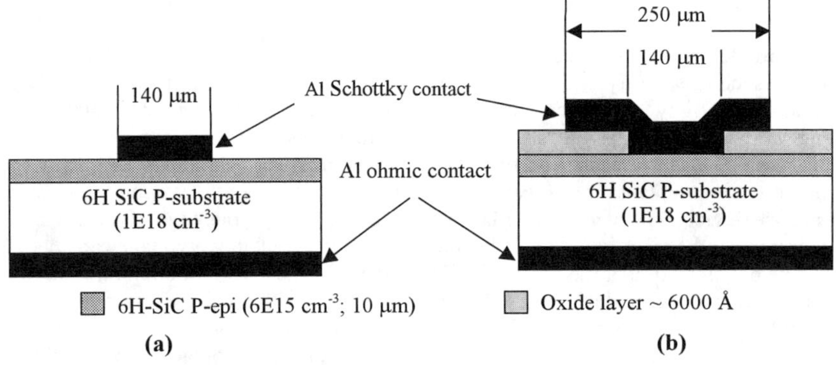

Fig. 1. Cross section of Al/6H-SiC Schottky diode (a) without edge termination, and (b) with thick oxide edge termination.

After the fabrication of the diodes, their forward and reverse I-V characteristics were measured using a DC voltage source (*Kiethley 237 High Voltage SMU*) as well a pulse measurement system. The pulse measurement system [7] comprised of a Tektronix *FG540* function generator inputting a ramp pulse of ~ 500 µs duration into a Trek *50/750* High Voltage Pulse Amplifier, and a Tektronix TDS 540 Digitizing oscilloscope used for recording the applied voltage and measuring the current through the diode structure.

RESULTS AND DISCUSSION

(a) Forward characteristics

Typical forward current density vs voltage (J-V) characteristics of Al/6H-SiC Schottky diodes with and without edge termination under DC bias conditions are shown in Fig. 2. The current density through the diodes with edge termination was observed to be approximately 2-3 times higher than that through the diodes without edge termination. The ideality factor (n) for both the diodes were calculated to be between 1.1 and 1.8. The forward voltage drop at a current density of 50 A/cm^2 was ~ 11.7 V for the diodes without edge termination, while it was ~ 6.5 V for the diodes with edge termination. The series resistance ($R_{on,,sp}$) calculated from a plot of *IdV/dI vs I* was found to be ~228 mΩ-cm^2 for diodes without edge termination, while it was ~ 84 mΩ-cm^2 for diodes with edge termination. These values are significantly higher than the ideal series resistance value of ~ 13 mΩ-cm^2 calculated using the doping concentration and thickness of the epitaxial layer and substrate provided to us by the manufacturer and the mobility of holes in the epi-layer and substrate [5]. The discrepancy between the ideal and experimental $R_{on,,sp}$ values could be due to a significantly lower substrate doping concentration and / or a large series resistance of the backside ohmic contact. Also, inherently the epitaxial and substrate region series resistance is high for p-type SiC due to the large ionization energy of the dopant atom (Al) in SiC [5,8]. The rather high forward voltage drop of the p-type SiC Schottky diodes could be attributed to the large series resistance values. The barrier height calculated from the forward I-V measurement was between 1.2-1.7 eV, which is comparable to that reported by Raghunathan et al [5] for P-type 6H-SiC Schottky diodes.

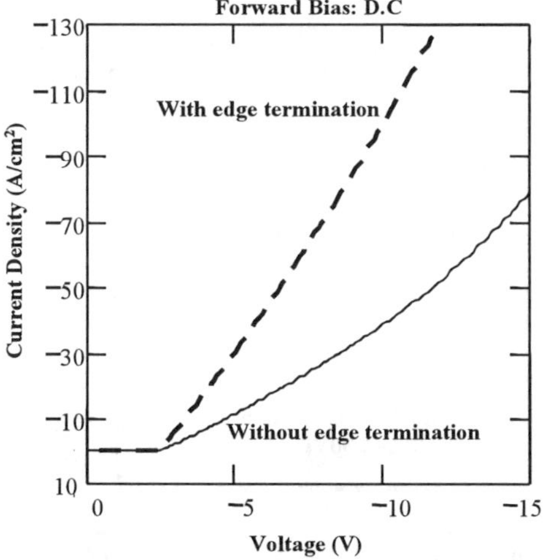

Fig. 2. Typical forward current density vs voltage (J-V) characteristics of Al/6H-SiC Schottky diodes with and without edge termination under DC excitation.

The higher forward current density through the diode with termination could be attributed to the formation of a low resistivity accumulation layer in the semiconductor below the MOS structure surrounding the Schottky contact. This can result in an effective increase in the current conduction area (Fig. 3), which will lead to a lower series resistance, and hence higher currents and a lower forward voltage drop.

It has to be noted that the current density in Fig. 2 was calculated assuming uniform current conduction across the 140 μm dia. (D1) Schottky contact area, while the effective current conduction area in case of the terminated diodes could be larger (D2 >140 μm) due to the presence of a low resistivity accumulation layer surrounding the Schottky contact area. We are in the process verifying this hypothesis by 2D numerical simulation using ATLAS, and a detailed analysis of this phenomena will be presented in a future paper.

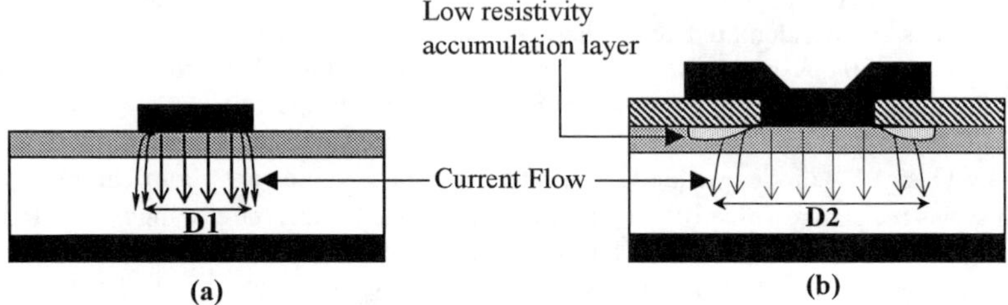

Fig. 3. Schematic of expected current flow pattern through the Al/6H-SiC Schottky diode (a) without edge termination, and (b) with thick oxide edge termination.

(b) Reverse characteristics

A comparison of reverse bias current density vs voltage characteristics of the Schottky diodes with and without edge termination under DC bias conditions is shown in Fig. 4.

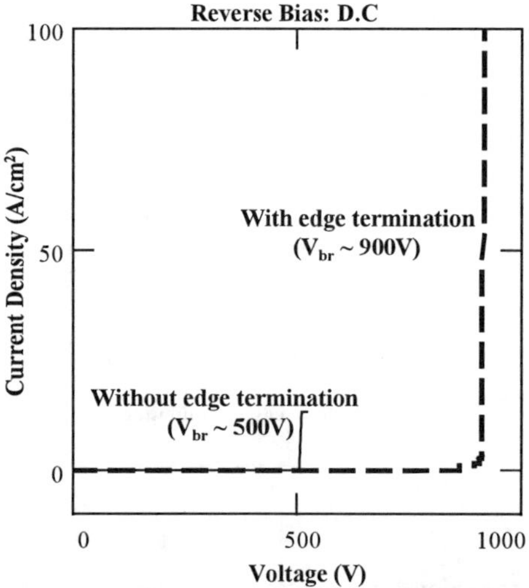

Fig. 4. Typical reverse current density vs voltage (J-V) characteristics of Al/6H-SiC Schottky diodes with and without edge termination under DC excitation.

The reverse breakdown voltage of the Schottky diodes with edge termination is observed to be approximately two times larger than that of the diodes without edge termination, while the reverse leakage current densities of both diodes were < 100 $\mu A/cm^2$ upto 400 V. Both the diodes also indicated a reasonably high ON/OFF ratio ($J_{F,-5V}$ / $J_{R,200V}$) of about 4 x 10^5. Extensive 2D numerical simulation using ATLAS (to be presented in a future paper) confirms that the increase in the breakdown voltage of the diodes with edge termination is due to an ~ 30% reduction in the peak electric field at the Schottky contact periphery as a result of the field plate surrounding the Schottky contact. In fact, the numerical simulation clearly indicates the expansion of the depletion region below the MOS region surrounding the Schottky contact, and hence the high field region is shifted away from the Schottky contact edge to inside the oxide layer corresponding to the edge of the metal overlap on top of the oxide layer, resulting in a higher breakdown voltage.

(c) Effect of oxide thickness on reverse breakdown voltage

Under reverse bias conditions of the terminated Schottky diodes, the oxide layer will sustain a portion of total applied voltage. As mentioned earlier, the high field region now shifts away from the Schottky contact edge into the oxide layer, near the edge of the metal overlap. If the electric field in the oxide layer exceeds the breakdown field of SiO_2, the Schottky diode will fail due to breakdown of the oxide layer. In order to investigate the influence of the thickness of the oxide layer on the Schottky diode breakdown voltage, Schottky diodes with different oxide thicknesses were fabricated. Fig. 5 shows the normalized pulse J-V characteristics of Schottky diodes, with different oxide thicknesses, under reverse bias.

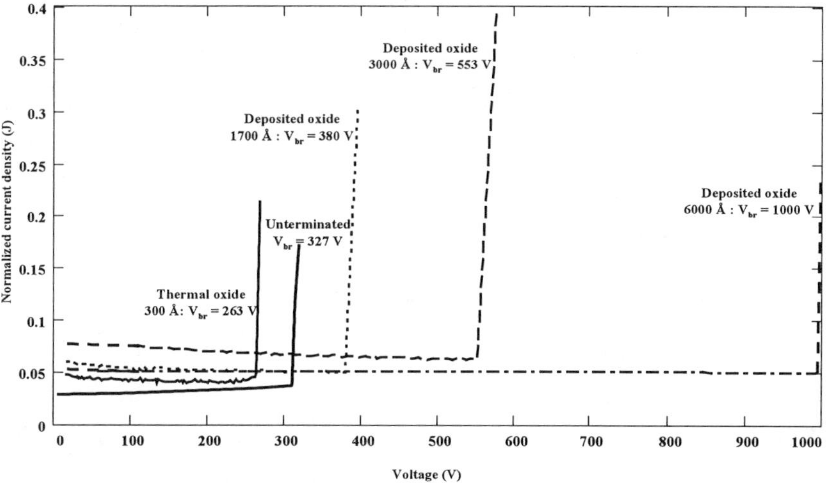

Fig. 5. Normalized pulse J-V characteristics of Schottky diodes with different oxide thicknesses, under reverse bias.

It is interesting to note that Schottky diodes with a thin thermal oxide (300 Å) failed at a lower voltage than the unterminated Schottky diode, indicating possible oxide breakdown when the oxide layer is too thin. But it has to be noted, however, that due to poor SiC material quality, a large scatter was observed in the breakdown voltage of both the unterminated and terminated diodes (> 20% difference between $V_{br,min}$ and $V_{br,max}$). Hence the anomalous reduction in the diode breakdown voltage with a thin oxide could also be due to poor material quality. A detailed measurement of the breakdown voltage of several unterminated and thin oxide terminated Schottky diodes is necessary to pin point the exact reason for the anomalous breakdown of a thin oxide terminated diode. Still, from Fig. 5 it is clearly evident that the breakdown voltage increases with increase in the oxide thickness. ATLAS simulation results (to be presented in a future paper) indicates that if the oxide thickness less than a minimum value, the field plate edge termination is not effective, and a thin oxide terminated diode structure can possibly fail at a lower voltage than an unterminated diode. Also, while the breakdown voltage is observed to increase with increase in the oxide thickness, theoretically beyond a certain value of oxide thickness there is no significant increase the breakdown voltage. For the Schottky diodes investigated in our laboratory, increasing the oxide thickness beyond 6000 Å did not result in further improvement in the breakdown voltage (up to 10000 Å thick oxide layers were investigated in the present experiment).

CONCLUSION

High voltage p-type 6H SiC Schottky diodes up to 1000 V have been successfully fabricated using a thick oxide layer for edge termination. The forward current density of the Schottky diodes with edge termination was 2-3 times larger than those for diodes without edge termination, while the reverse leakage current density of both diodes were $< 100 \ \mu A/cm^2$ upto 400V. The difference in forward current densities could be due to a larger effective conduction area for diodes with edge termination, as a result of an accumulation layer underneath the MOS region surrounding the Schottky contact. Also, the Schottky diodes with edge termination indicated a factor of two higher breakdown voltages than those without edge termination as a result of electric field relief at the Schottky contact edge. A minimum oxide thickness of ~ 6000 Å was necessary to attain 1 kV reverse breakdown voltage for the Schottky diodes fabricated on p-type 6H-SiC in this experiment. Further increase in oxide thickness did not result in any improvement in the breakdown voltage.

ACKNOWLEDGEMENTS

This work was supported by ARO (grant no DAA H04-96-1-0467) via a DEPSCoR program. The authors also wish to acknowledge Dr. I. Khlebnikov for his valuable suggestions and Mr. Marc Tarplee for performing the ATLAS simulation.

REFERENCE

1. D. Alok and B.J. Baliga, IEEE Trans. ED **44**, 1013 (1997).
2. A. Itoh, T. Kimoto and H. Matsunami, IEEE Trans. EDL **17**, 130 (1996).
3. V. Saxena, J. N. Su, and A.J. Steckl, IEEE Trans. ED **46**, 456 (1998).
4. M. Bhatnagar and B.J. Baliga, IEEE Trans. ED **40**, 645 (1993).
5. R. Raghunathan and B.J. Baliga, IEEE Trans. EDL **19**, 71 (1998).
6. Q. Zhang, V. Madangarli, I. Khlebnikov, S. Soloviev and T. S. Sudarshan, *to be presented at 1999 MRS Spring Meeting* (unpublished).
7. T. S. Sudarshan, V.P. Madangarli, G. Gradinaru, C.C. Tin, R. Hu, and T. Isaacs-Smith, MRS Proc. **423**, 99 (1996).
8. S.M. Sze, *Physics of Semiconductor Devices*, 2nd. Ed. (John Wiley & Sons Publishers, New York, 1981).

HIGH VOLTAGE P-N JUNCTION DIODES IN SILICON CARBIDE USING FIELD PLATE EDGE TERMINATION

R.K. CHILUKURI, P. ANANTHANARAYANAN, V. NAGAPUDI, B.J. BALIGA
Power Semiconductor Research Center, North Carolina State University
Raleigh, NC 27606
ravi@apollo.psrc.ncsu.edu

ABSTRACT

In this paper, we report the successful use of field plates as planar edge terminations for P^+-N as well as N^+-P planar ion implanted junction diodes on 6H- and 4H-SiC. Process splits were done to vary the dielectric material (SiO_2 vs. Si_3N_4), the N-type implant (nitrogen vs. phosphorous), the P-type implant (aluminum vs. boron), and the post-implantation anneal temperature. The nitrogen implanted diodes on 4H-SiC with field plates using SiO_2 as the dielectric, exhibited a breakdown voltage of 1100 V, which is the highest ever reported measured breakdown voltage for any planar ion implanted junction diode and is nearly 70% of the ideal breakdown voltage. The reverse leakage current of this diode was less than $1x10^{-5}$ A/cm^2 even at breakdown. The unterminated nitrogen implanted diodes blocked lower voltages (~840V). In contrast, the unterminated aluminum implanted diodes exhibited higher breakdown voltages (~800V) than the terminated diodes (~275V). This is attributed to formation of a high resistivity layer at the surface near the edges of the diode by the P-type ion implant, acting as a junction termination extension. Diodes on 4H-SiC showed higher breakdown than those on 6H-SiC. Breakdown voltages were independent of temperature in the range of 25 °C to 150 °C, while the leakage currents increased slowly with temperature, indicating surface dominated components.

INTRODUCTION

SiC has become an attractive semiconductor in recent years for high speed, high power and high temperature applications because of its wide band-gap, high critical electric field and high thermal conductivity. However, its widespread application has been limited due to the fact that the device fabrication technology for SiC is still in its stage of infancy. Due to low impurity diffusion coefficients, doping in SiC is usually obtained either by epitaxial growth or by ion implantation. The former is not attractive because it requires mesa etching for edge termination, and hence the device topology becomes non-planar making passivation difficult. Hence, in order to improve the commercial viability of SiC devices, the successful development of planar ion implantation technology for SiC for microelectronic technology is of major importance. Dopant activation, surface morphology control and ohmic contacts have been the major obstacles in the development of a reproducible ion implantation technology. Until recently, P-N junctions in SiC have been formed using multiple epitaxial layers or unmasked ion implantation. However, these approaches compromise on surface planarity and are not preferred [1-3].

Although there have been quite a few reports on ion implantation, 6H-SiC has received most of the attention [2-6]. In most cases, the dopants used have been nitrogen, aluminum and boron. There have been only a couple of reports on other dopants like phosphorous and beryllium [7,8] where breakdown voltages of 675 V were reported for phosphorous implanted non-planar junctions. The diodes fabricated in the above work have been mesa-etch terminated which are not compatible with IC technology. Various planar edge terminations such as floating metal rings and resistive Schottky barrier field plates have been explored for 6H-SiC devices [9] but only 50% of ideal breakdown voltage was achieved. A planar, near ideal, edge termination

Mat. Res. Soc. Symp. Proc. Vol. 572 © 1999 Materials Research Society

using Argon implantation [10,11] has been reported for 6H-SiC Schottky barrier diodes but the leakage current is greatly increased by the implanted region. This method was also demonstrated for 4H-SiC with breakdown voltages exceeding those reported for 4H-SiC mesa-etch terminated diodes [12]. Most recently, a 3.4 kV ion implanted PIN-rectifier has been implemented on 4H-SiC with low leakage currents obtained by using junction termination extension created by boron ion implantation [13]. In this paper, reverse blocking characteristics of diodes made using phosphorous, nitrogen, aluminum and boron implanted layers and having different edge terminations are discussed.

EXPERIMENT

The starting wafers were research grade 6H- and 4H-SiC wafers (both n-type and p-type) consisting of a 10 μm thick epitaxial layer (0.85-1.0×10^{16} cm^{-3}) on a 300 μm thick substrate (1×10^{18} cm^{-3}), obtained from CREE. Devices were fabricated using two process sequences differing mainly in the dielectric used to mask the implants, the temperature of post implantation anneal and the final dielectric used in device termination.

The first fabrication process was a 2-mask process in which the post-implantation anneal was at <1400 °C and the dielectric used for passivation was silicon dioxide. In this process, after cleaning the wafers by standard techniques, a 7000 Å thick oxide was deposited using Low Pressure Chemical Vapor Deposition (LPCVD). Next, windows for ion implantation were opened in the oxide, and a 500 Å pad oxide was deposited by LPCVD. This was to ensure that the peak implant doping occurs at the surface, which would result in good ohmic contacts. High temperature (1000 °C) nitrogen and aluminum implantations were performed using multiple implant sequences with energies 25, 50, 75 and 120 keV to give box-profiles with varying total doses. Monte Carlo simulations performed using SUPREM showed that these implant conditions would yield depths of uniform concentration or junction depths in the range of 0.25-0.3 μm in the case of the nitrogen implants and 0.17-0.2 μm in the case of the aluminum implants. Post-implantation proximity anneals for dopant activation were done at 1300°C and 1400°C for the nitrogen and aluminum implanted samples, respectively, in Argon for 30 minutes. Aluminum contacts were formed by lift-off on the device side, after etching the pad oxide. Blanket evaporation of aluminum was done on the back side to form ohmic contacts to the substrate.

For post-implantation anneals at temperatures >1600 °C, which is desirable for high dopant activation, the above process was not feasible due to reflow of the oxide at such temperatures. Hence, a 4-mask process was designed for device fabrication. In this process, the silicon nitride was used as the dielectric for passivation. Diodes were also made from phosphorous and deep high energy boron implants, and planar edge terminations such as field rings were also incorporated. After cleaning the wafers, the first mask level was used to etch alignment marks on bare SiC surface using RIE. Following an RCA clean, 1 μm each of oxide and polysilicon were deposited by LPCVD. Such a thick stack of polysilicon and oxide was found, based on simulations, to be required to mask the high energy boron implants. Windows were opened in the stack of polysilicon and oxide for implantation and a 500 Å of pad oxide was deposited by LPCVD on both wafers for obtaining peak doping of the implantation profile at the surface. Based on simulations, the implantation profiles yielded uniform concentration upto a depth of 0.3, 0.2, 0.2 and 0.6 μm in case of nitrogen, phosphorous, aluminum and deep boron/aluminum implants, respectively, with net doses of 4.6×10^{15}, 4.6×10^{15}, 3.4×10^{16} and 6.8×10^{14} cm^{-2}. After ion implantation, blanket etch was done to remove both the oxide and the polysilicon. Next, post-implantation proximity anneals were performed for all wafers at 1600 °C in Argon for 30 min in a SiC crucible. The SiC crucible helped maintain a vapor pressure of SiC

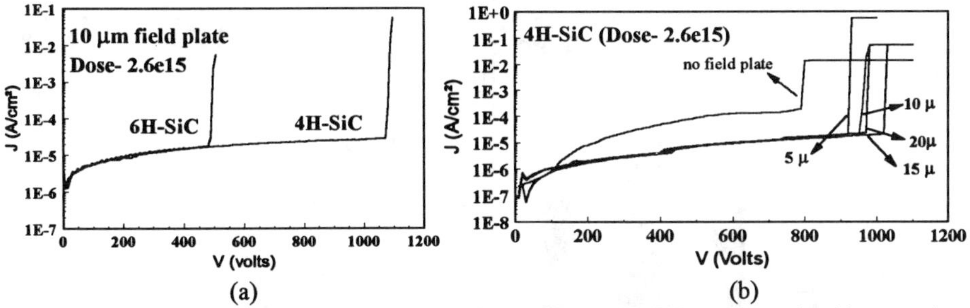

Fig. 1. Reverse J-V characteristics of nitrogen implanted diodes with SiO_2 as dielectric : (a) comparison of 6H- and 4H-SiC diodes with field plates; (b) effect of field plate length.

and hence completely prevented surface degradation and pitting [4]. Following the post implantation anneal, a 0.55 μm thick nitride layer was deposited by LPCVD. A 0.4 μm LPCVD oxide was used as the mask to open contact windows in the nitride film. Aluminum contacts were formed by lift-off for the device side, and by blanket evaporation on the back side.

RESULTS & DISCUSSION

A. Diodes with SiO_2 as dielectric (post-implantation anneal at < 1400 °C)

The diode reverse I-V measurements were performed using Keithley 251 I-V setup and the system leakage was found to be a few pico-amperes. It is known that edge termination plays an important role in determining the breakdown voltage of a p-n junction diode. In the case of nitrogen implanted diodes on 6H-SiC, field plates served to increase the breakdown voltage from 350 ± 50 V to 500 ± 20 V. The breakdown voltage of nitrogen implanted diodes on 4H-SiC increased from 800 ± 40 V to 1100 ± 60 V with field plates. This indicates that a simple planar field plate edge termination can be used to obtain higher breakdown voltages. Extremely low leakage current densities in the range of 1 x 10^{-5} A/cm^2 were obtained in both 6H- and 4H-SiC nitrogen implanted diodes terminated with field plates even just before breakdown. A comparison of the reverse I-V characteristics of these nitrogen implanted diodes on 6H- and 4H-SiC with field plates are shown in Fig. 1(a). It can be observed that much higher breakdown voltages were obtained in 4H-SiC than in 6H-SiC. The variation of the breakdown characteristics with the dose of nitrogen implant was studied in 6H-SiC and no significant trend or variation was found. The nitrogen implanted junction diodes without field plates on 4H-SiC also showed good reverse blocking characteristics with a maximum breakdown voltage of 840 V and average leakage currents in the range of 1x10^{-4} A/cm² prior to breakdown as shown in Fig. 1(b). A breakdown voltage of 1100 V obtained in this study on 4H-SiC is the highest ever reported measured breakdown voltage for any planar ion implanted junction diode on SiC. The breakdown voltage is nearly 70 % of the theoretical parallel plane ideal breakdown voltage of about 1600 V for a diode with the given epitaxial layer specifications. A total of 30 devices were measured, 15 with field plates and 15 without field plates, and it was found that the presence of field plate improved the reverse I-V characteristics of the diodes consistently and as expected from theory [14]. The breakdown voltage was found to be almost independent of the of field plate length as shown in Fig. 1(b). The distribution of the leakage currents just before breakdown for the nitrogen implanted diodes on 4H SiC across the sample is shown in Fig. 2. The average leakage current for diodes with field plates, reduced by an order of magnitude when compared to that of diodes without field plates, to a value as low as 1x10^{-5} A/cm² just before breakdown.

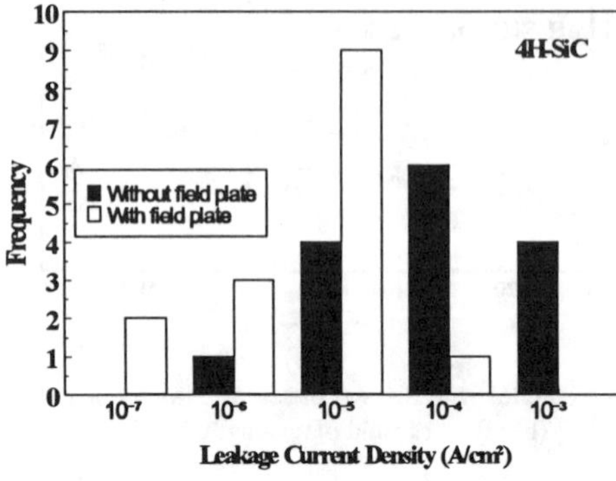

Fig. 2. Reverse leakage current density histogram of nitrogen implanted diodes near breakdown.

Fig. 3. Comparison of reverse J-V characteristics of nitrogen and aluminum implanted diodes.

The aluminum implanted junction diodes on 4H-SiC exhibited higher breakdown voltages without field plates than with field plates (Fig. 3). For a dose of 6.8×10^{15} cm^{-2}, while the aluminum implanted diodes without field plates showed breakdown voltages of 800 ± 10 V, the ones with field plates supported only 275 ± 40 V. This is attributed to the poor activation of the aluminum implanted region due to which, the actual doping in the ion implanted region is much lower than the implanted aluminum concentration. The sheet resistance of the aluminum implanted region, measured using Kelvin test elements on the same wafers, and was found to be 25 kΩ/square which supports the above argument. This results in the presence of a high resistivity layer at the surface near the edges of the diode (between points A and B in Fig. 4(a)) in case of the diodes without field plates. It is believed that this region then acts like a junction termination extension and promotes the spreading of the potential along the surface laterally which results in reduced electric field [13] at a given voltage for the diodes without field plates. Hence, a higher breakdown voltage is observed for diodes without field plates when compared to diodes with field plates. The breakdown voltage of aluminum implanted diodes was found to be higher for a higher dose. However, like in the nitrogen implanted diodes, no significant variation of breakdown voltage with the length of field plate was observed.

B. Diodes with Si₃N₄ as dielectric (post-implantation anneal at 1600 °C)

The reverse I-V characteristics were measured for all diodes with different edge terminations and the results are summarized in Fig. 5. The unterminated diodes gave the lowest

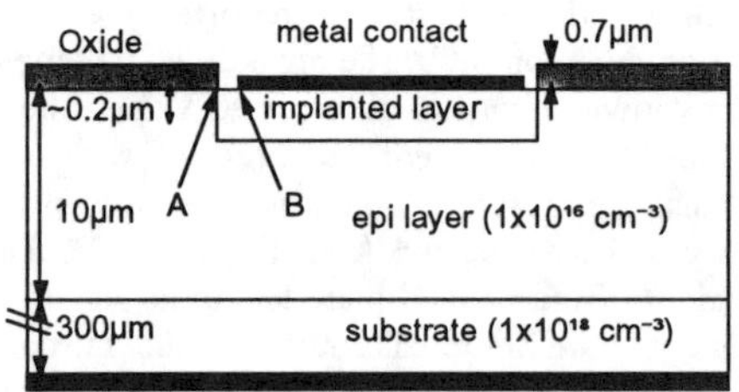

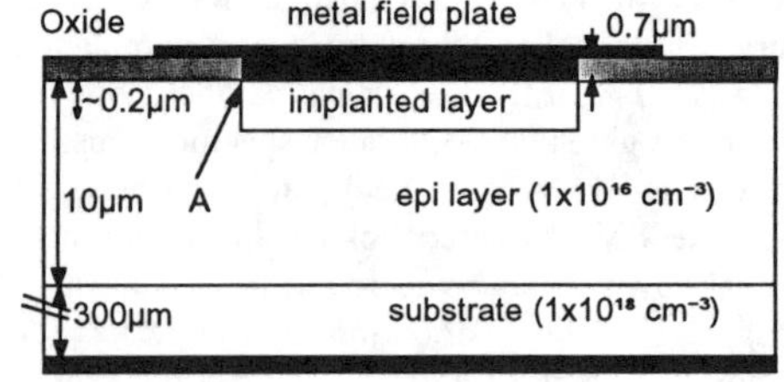

Fig. 4. Cross-section of fabricated planar diodes with SiO₂ as dielectric : (a) without field plate (b) with field plate.

Implanted Species	No edge termination	Single field plate	Single field ring	1 field ring + 1 field plate	2 field rings + 1 field plate
N	300 +/-25 V	550 +/-20 V	350 +/-100 V	350 +/- 30 V	500 +/-60 V
P	400 +/-50 V	700 +/-40 V	400 +/-100 V	400 +/- 50 V	475 +/- 100 V
Al	150 +/-20 V	600 +/-50 V	260 +/-40 V	300+/- 30 V	300+/- 100 V
B and Al	75+/-50 V	350+/-50 V	200 +/-20 V	200 +/- 50	175+/-25 V

Fig. 5. Variation of breakdown voltages with edge terminations of 4H-SiC diodes using Si_3N_4 as dielectric.

breakdown voltages in all cases. The highest breakdown voltages were obtained for the diodes with a single field plate for all the different dopants. The least variation of breakdown voltages across the wafer was observed for these diodes. This confirmed that a simple planar field plate edge termination can be used to obtain higher breakdown voltages. However, no other trend could be established for the variation of breakdown voltage with the edge terminations. The higher effectiveness of the field plate termination than the guard ring termination has also been reported with 4H-SiC Schottky diodes [15]. Further, in our study, it was observed that phosphorous is a better choice than nitrogen for making donor implanted diodes in SiC. Not only did phosphorous implanted diodes show higher breakdown voltages, but they also exhibited lower leakage currents for the same voltage when compared to nitrogen implanted diodes. Leakage currents were found to be $1x10^{-4}$, $1x10^{-3}$, $1x10^{-6}$ and $1x10^{-7}$ Amperes for phosphorous, nitrogen, aluminum and boron/aluminum implanted diodes, respectively, just before breakdown. The aluminum implanted diodes consistently supported higher voltages than boron/aluminum implanted diodes. The acceptor implanted diodes showed much lower leakage currents than the donor implanted diodes. The leakage currents obtained for nitrogen implanted diodes on 4H-SiC were found to be higher in this process when compared to the previous process. The high leakage currents in the diodes was attributed to excessive leakage at the periphery of the junction due to residual ion implantation damage [4,7] caused by the higher dose. Hot implantation is also known to bring about inferior junction characteristics due to formation of dislocation loops [5].

High temperature (25 °C to 150 °C) characterization was also performed for all the diodes. The variation of breakdown voltage with temperature is shown in Fig. 6. The breakdown voltage was found to be almost invariant with temperature which is an encouraging result when compared to previously reported negative temperature coefficient for the breakdown voltage [1]. The leakage currents also remained in the same order of magnitude when the temperature was increased from room temperature to 150 °C (~423 K), which is again highly desirable. We believe that the leakage currents increased very slowly with temperature because the leakage current was predominantly due to surface leakage components and not due to diffusion or space charge generation components which are extremely low in SiC.

CONCLUSIONS

Electrical properties of P^+-N as well as N^+-P planar ion implanted junction diodes on 6H- and 4H-SiC with different edge terminations were studied. The single field plate edge termination resulted in higher breakdown voltages than those with other planar edge terminations such as the floating field rings. The nitrogen implanted diodes on 4H-SiC with field plates using SiO_2 as the dielectric, exhibited a breakdown voltage of 1100 V, which is the highest ever reported measured breakdown voltage for any planar ion implanted junction diode and is nearly 70% of the ideal breakdown voltage. The reverse leakage current of this diode was less than $1x10^{-5}$ A/cm^2 even at breakdown. Diodes on 4H-SiC showed higher breakdown than those on 6H-SiC. Breakdown

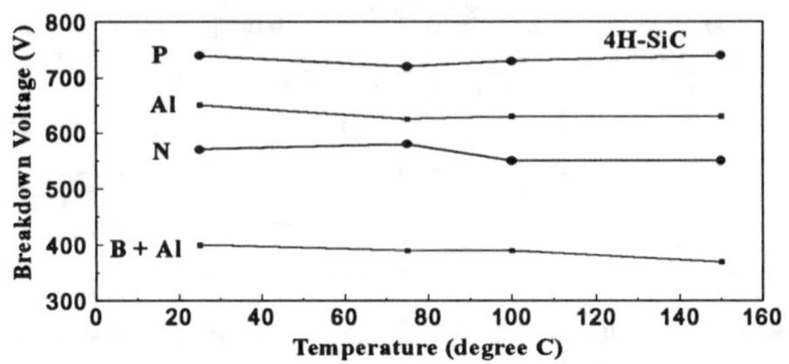

Fig. 6. Variation of breakdown voltages with temperature of field plate terminated 4H-SiC diodes using Si_3N_4 as dielectric.

voltages were independent of temperature in the range of 25 °C to 150 °C; while, the leakage currents increased slowly with temperature, but stayed within the same order of magnitude. These results are of relevance to design and fabrication of high voltage devices such as MOSFETS containing reverse blocking junctions.

ACKNOWLEDGEMENTS

The authors would like to acknowledge the industrial sponsors of Power Semiconductor Research Center for supporting this work.

REFERENCES

1. J.W. Palmour and L.A. Lipkin, Trans. of 2nd International High Temperature Electronic Conf., 1, XI-3 (1994).
2. L.G. Matus, J. A. Powell, and C.S. Salupo, Appl. Phys. Lett., 59 (14) 1770 (1991).
3. P.G. Neudeck, D.J. Larkin, J. A. Powell, C.S. Salupo, and L.G. Matus, IEEE Trans. Electron Devices, 41 (5), 826 (1994).
4. M.V. Rao, J. Gardner, P. Griffiths, O.W. Holland, G. Kelner, P.H. Chi, and D.S. Simons, Nucl. Instr. and Meth. in Phys. Res., B (106), 333 (1995).
5. T.Kimoto, A. Itoh, H. Matsunami, T. Nakata, and M. Watanabe, J. of Electronic Materials, 24 (4), 235 (1995).
6. M. Ghezzo, D.M. Brown, E. Downey, and J. Kretchmer, Appl. Phys. Lett., 63 (9), 1206 (1993).
7. Y. Zheng, N. Ramungul, R. Patel, V. Khemka, and T.P. Chow, Proc. of International Conference for SiC and Related Materials, MoP-06 (1997).
8. R. Patel, V. Khemka, N. Ramungul, T.P. Chow, M. Ghezzo, and J. Kretchmer, Proc. of International Symposium on Power Semiconductor Devices & ICs, 122 (1998).
9. M. Bhatnagar, H. Nakanishi, S. Bothra, P.K. McLarty, and B.J. Baliga, Proc. of International Symposium on Power Semiconductor Devices and ICs, 89 (1993).
10. D. Alok, B.J. Baliga, and P.K. McLarty, IEEE Electron Device Lett., 15, 394 (1994).
11. D. Alok and B.J. Baliga, IEEE Trans. on Electron Devices, 44 (6), 1013 (1997).
12. D. Alok, R. Raghunathan, and B.J. Baliga, IEEE Trans. on Electron Devices, 43 (8), 1315 (1996).
13. K. Rottner, Proc. of International Conference for SiC and Related Materials, 136 (1997).
14. B.J. Baliga in *Power Semiconductor Devices*, (PWS publishing company, 1995).
15. R. Singh and J.W. Palmour, Proc. of International Symposium on Power Semiconductor Devices & ICs, 157 (1997).

CARBON AND SILICON RELATED SURFACE COMPOUNDS OF PALLADIUM ULTRATHIN FILMS ON SIC AFTER DIFFERENT ANNEALING TEMPERATURES

W.J. LU*, D.T. SHI, T. CRENSHAW, A. BURGER, W.E. COLLINS

Department of Physics and the Center for Photonic Materials and Devices

Fisk University

Nashville, TN 37208

ABSTRACT

Pd/SiC Schottky diode has triggered interest as a chemical sensor to be operated at high temperatures. Various surface compounds formed at high temperatures are known to alter the device performance. In this work, the carbon and silicon related compounds and morphology of Pd ultra-thin film on 6H-SiC and 4H-SiC are investigated after thermal annealing using X-ray photoelectron spectroscopy (XPS) and atomic force microscopy (AFM). The Pd ultra-thin films of about 3 nm in thickness are deposited by RF sputtering. The XPS analysis reveals the presence of silicon oxycarbides (SiC_xO_y) as deposited. After being annealed above 300°C, the atomic ratio of C to O in SiC_xO_y decreases with increasing the annealing temperatures, and the Pd film becomes a Pd silicide nanofeatured layer on SiC. When the annealing temperature is at 500°C, the majority of the SiC_xO_y is converted into SiO_2. An amorphous Si phase exists after annealing at 200 to 400°C, which indicates that the Si-C bonds in SiC are broken at lower temperatures due to the presence of Pd. Graphite and C=O are found on the as deposited samples and also after annealing at temperatures up to 600°C. The formations of the carbon and silicon related compounds on Pd/4H-SiC are very similar to those on Pd/6H-SiC.

INTRODUCTION

SiC-based device developments require that the metal contacts and interconnects are physically, chemically, and electrically stable under severe conditions, such as at high temperatures. Metal contact studies on SiC have resulted in commercially available SiC devices which can be operated at high temperatures [1,2]. The diffusions and reactions between metal thin films and the SiC substrate result in the formation of various interfacial compounds, and alter the electrical properties. Therefore, it is one of the most critical issues to investigate the interfacial compositions of metal/SiC at elevated temperatures in SiC based device research.

Pd/SiC Schottky diodes have been successfully demonstrated as a chemical sensor for hydrogen and hydrocarbon [3,4], which can be operated at high temperatures. The heat treatment significantly promotes interfacial diffusion and chemical reactions, and broadens the interface region. The two-dimensional surface diffusion and surface segregation of Si from dissociated SiC result in a thin silicon oxide layer on the top of the Pd film [3]. The Pd chemical states are various co-existing palladium silicides (Pd_xSi, x = 1,2,3,4) for the Pd thickness of about 400Å [4] after annealing at 425°C. Lu *et al.* [5] found the Pd exists as PdSi and Pd_2Si for the ultra-thin Pd film (~30Å thickness) at the elevated temperatures. To the best of our knowledge, the formations of the silicon and carbon related compounds in Pd/SiC after annealing at high temperatures have not been systematically

Mat. Res. Soc. Symp. Proc. Vol. 572 © 1999 Materials Research Society

investigated .

In this study, the interfacial composition and the morphological features of Pd ultra-thin films on 6H-SiC and 4H-SiC at different annealing temperatures were investigated using XPS and AFM. The Pd ultra-thin films were prepared by the RF sputtering method, and the thickness was about 30Å. The silicon and carbon related compounds in Pd/SiC were determined by XPS after annealing.

EXPERIMENTAL

1. Samples and Pd Ultra-thin Film Fabrication

n-type, Si-face 6H-SiC and 4H-SiC wafers with 3.5^0 off-axis on Si (0001) substrates were purchased from Cree Research Inc. The doping concentration was 2.6 x 10^{18} cm^3. An RF sputtering system (Kurt J. Lesker Company) was used for the Pd ultra-thin film preparation. The SiC wafer cleaning procedure and the Pd ultra-thin preparation procedure are in Ref.[5]. The thickness of Pd thin film on SiC was about 30 Å, so that XPS can effectively detect the Pd/SiC interfacial compositions.

The Pd/SiC samples were analyzed and annealed consecutively from 100°C to 600°C in 100°C increments in air for 30 minutes each time. The Pd/6H-SiC and Pd/4H-SiC samples were prepared and annealed at the same time.

2. Characterization

The XPS experiments were performed on a Kratos X-SAM 800 spectrometer with an energy resolution of 0.1 eV. The base pressure was at 10^{-9} Torr. The Mg Kα (hv=1253.6 eV) radiation was used. The energy scale of the analyzer was calibrated by the Cu ($2p_{3/2}$) XPS peak at 932.67 eV, and the Au ($4f_{7/2}$) XPS peak at 84.00 eV.

The exit angle of the photoelectron was 90° to the sample surface. The data were smoothed and satellite peaks were subtracted before background subtraction and deconvolution were performed. The background of the XPS was subtracted using the Sherley algorithm. The Gaussian-Lorentzian method was applied to deconvolute the XPS.

A NanoScope E (Digital Instruments, Inc.) was used for AFM measurements. The standard silicon nitride cantilever supplied by Digital Instruments, Inc. was employed. The force constant of the silicon nitride cantilever is 0.12 N/m. At least three different regions on the surface were measured by AFM for each sample.

RESULTS

1. Carbon related compounds in Pd/6H-SiC

XPS is used to examine the surface composition changes on Pd/6H-SiC and Pd/4H-SiC samples from the as deposited and after each annealing step. The surface species formed on the Pd/SiC surface are quite complex, and two kinds of surface compounds are co-existing on Pd/SiC samples after annealing; (a) the Pd silicides and Pd element, (b) the silicon and carbon related compounds. In this paper, we present the results for the silicon and carbon related compounds on 6H-SiC and 4H-SiC substrates.

The studies on the Pd-related compounds on Pd/SiC samples after different annealing temperatures has been presented at another report [5]. There are two reactions between Pd and SiC in the range from room temperature to 600°C; (a) at 300°C, the Pd reacts to SiC to form Pd_2Si on both 6H-SiC and 4H-SiC substrates, and (b) Pd_2Si reacts with SiC to form PdSi at 500°C for Pd/4H-

SiC, and at 600°C for Pd/6H-SiC.

Table I lists the XPS peaks of the C(1s) and Si($2p_{3/2}$) after deconvolution for Pd/6H-SiC and Pd/4H-SiC as deposited and after each annealing step. Figure 1 shows the deconvolution of the C(1s) XPS for Pd/6H-SiC as deposited and after consecutive annealing at 300°C and 600°C. As shown, the strong peak at the binding energy of 282.9 eV represents the C(1s) XPS from the Si-C bonds in SiC substrate [5-8]. The peak at 284.1 eV exists in the Pd/SiC samples as deposited, and it is assigned to the carbon oxycarbide (SiC_xO_y)compounds [6,7]. The SiC_xO_y is formed during the Pd film deposition. The 285.1 eV peak is from the C-C bond (graphite), which is commonly found in almost any sample due to the contaminants. The broad peaks at 287.2-287.8 eV and ~286.5 eV are assigned to the C=O and C-O bonds, respectively [9]. Table II shows the chemical shifts in XPS of various silicon and carbon compounds compared with the XPS for the Si-C bond in SiC.

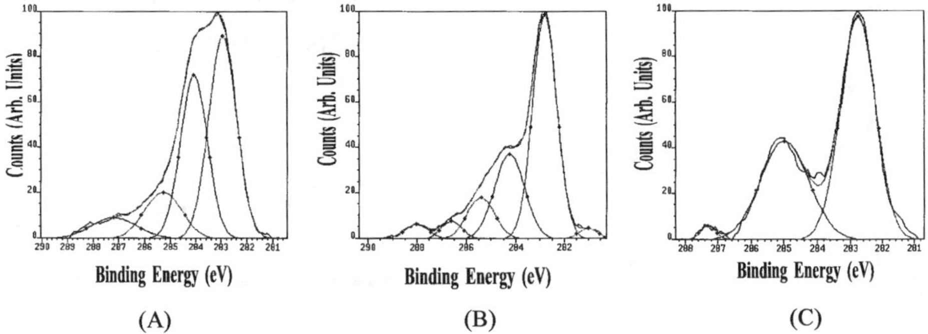

(A)   (B)   (C)

Figure 1. Deconvolution of the C (1s) XPS for Pd/6H-SiC as deposited (A) and after consecutive annealing at 300°C (B) and at 600°C (C).

TABLE I. Deconvolution of the XPS peaks of the C (1s) and Si ($2p_{3/2}$) on Pd/SiC as deposited and after consecutive annealing in air for 30 minutes.

Samples	C (1s) peaks	Si ($2p_{3/2}$) peaks
as deposited	282.9, 284.1, 285.1, 287.2	100.8, 102.3, 103.6
annealed at 100°C	282.9, 284.1, 285.1, 287.3	100.8, 102.3,
annealed at 200°C	282.9, 284.1, 285.2, 287.4	99.0, 100.8, 102.3
annealed at 300°C	282.9, 284.1, 285.2, 286.5, 287.8	99.0, 100.8, 102.3
annealed at 400°C	282.9, 284.1, 285.2, 287.8	99.0, 100.8, 102.6
annealed at 500°C	282.9, 284.2, 285.1	100.8, 103.2, 103.6
annealed at 600°C	282.9, 285.1, 287.6	100.8, 103.6

The concentrations of the carbon related compounds on Pd/SiC change with increasing annealing temperatures. The C(1s) XPS peak of the C-C bond also increases with increasing annealing temperatures. The surface graphite comes not only from the contaminants, but results from the products of the reactions of the Pd and the SiC substrate [5]. The SiC_xO_y and C=O groups on the surface decrease with increasing the annealing temperatures. A small amount of the C-O group is also found after annealing at 300°C. At the annealing temperature of 600°C, the main carbon compound is graphite although small amounts of SiC_xO_y and C=O still exist on the surface. The C=O group is also found on the substrate at room temperature, and the concentration of the C=O is low through the annealing temperature up to 600°C. The C=O group is difficult to remove, and the thermal cleaning of the SiC substrates usually requires at least 800°C to obtain the oxygen-free SiC surface under ultrahigh vacuum (UHV) environment [10]. It seems that the oxygen in air directly attacks the surface C of the SiC substrate to form C=O.

TABLE II. Chemical shifts in XPS of various silicon and carbon compounds compared with the Si-C bond in SiC.

Species	C (1s) in SiC (282.9 eV)	Si (2p$_{3/2}$) in SiC (100.8 eV)
C-C	2.0 [6,7], 1.6 [9]	
Si-Si		-1.8 [11]
Si-Pd		0 [4]
SiC_xO_y	1.2 [6,7]	0.5 - 2.4 [6,7]
SiO_2		>2.4 [6,7], >1.8 [9], 2.5 [8]
C-O	3.6 [9]	
C=O	4.9 [9]	

2. Silicon related compounds in Pd/6H-SiC

Figure 2 shows the deconvolution of the Si (2p$_{3/2}$) XPS for Pd/6H-SiC after consecutive annealing at 300°C, and 600°C. As shown, the strong peak at 100.8 eV is assigned to the Si (2p$_{3/2}$) XPS from the SiC substrate and the Pd silicides [4]. The chemical shifts of the Si (2p$_{3/2}$) peak from the silicon oxycarbides (SiC_xO_y) vary with the stoichiometry of C and O [6,7]. The binding energy of SiC_xO_y is 102.3 eV for the sample as deposited. It increases to 102.7 eV after consecutive annealing at 400°C, and 103.2 eV after annealing at 500°C. After being annealed at 600°C, the majority of the silicon compounds on Pd/6H-SiC is SiO_2 instead of silicon oxycarbides.

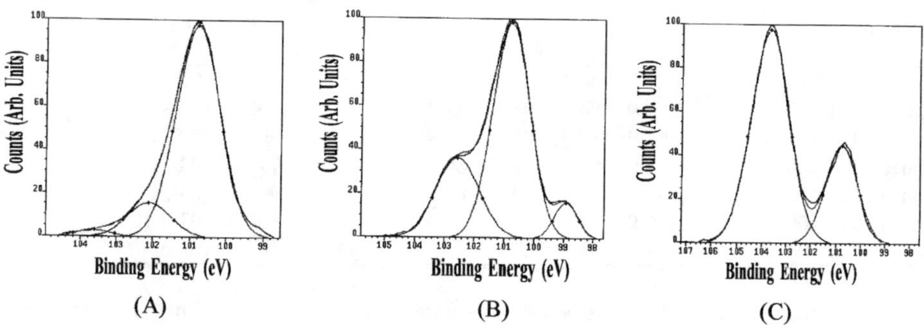

Figure 2. Deconvolution of the Si (2p$_{3/2}$) XPS for Pd/6H-SiC as deposited (A) and after consecutive annealing at 300°C (B) and at 600°C (C).

After annealing at 200°C, a small XPS peak at 99.0 eV appears. This peak is assigned to Si-Si bond [11], and diminishes after annealing at 400°C. As in our previous work [5], Pd starts to react with SiC to form Pd_2Si and C at 300°C. The presence of Si indicates that the Si-C bonds in SiC break in to C and Si phases on Pd/SiC at 200°C. Obviously, the existing of Pd seems to act as a catalyst to break the Si-C bonds. At higher temperatures, Si is consumed by the reactions with Pd silicides [5]. Table III summarizes the various carbon and silicon compounds on Pd/SiC after consecutive annealing from 100 to 600°C.

TABLE III. Carbon and silicon related compounds on Pd/SiC as deposited and
after consecutive annealing in air for 30 minutes.

Annealing Temperatures (°C)	Carbon and silicon related compounds
as deposited	C, SiC_xO_y, C=O
100°C	C, SiC_xO_y, C=O
200°C	Si, C, SiC_xO_y, C=O
300°C	Si, C, SiC_xO_y*, Pd_2Si, C=O, C-O
400°C	Si, C, SiC_xO_y*, Pd_2Si, C=O
500°C	C, SiC_xO_y*, Pd_2Si, C=O, SiO_2
600°C	C, SiO_2, PdSi, C=O

* the atomic ratio of C to O decreases with increasing annealing temperatures.

Figure 3 shows the morphological changes for Pd/6H-SiC as deposited and after consecutive annealing at 400°C and 600°C using AFM. As shown, the surface of the Pd/SiC sample as deposited has a good uniformity (Figure 3A). After annealing at 300°C, the Pd film becomes a nanofeatured layer. The average size of the nanofeatures is about 20-30 nm (Figure 3B). With increasing the annealing temperature to 600°C, the nanofeatured layer has been broken into nanosize rounded clusters on 6H-SiC surface (Figure 3C). The changes in the morphological features on the Pd/SiC surface are related to the surface compositions. At the annealing temperature of 300°C, Pd_2Si is formed and the nanofeatured layer appears [5]. Meanwhile, the atomic ratio of C to O in the SiC_xO_y decreases. When the annealing temperature is 500 to 600°C, the majority of the SiC_xO_y is converted into SiO_2, and Pd_2Si reacts with SiC to form PdSi [5].

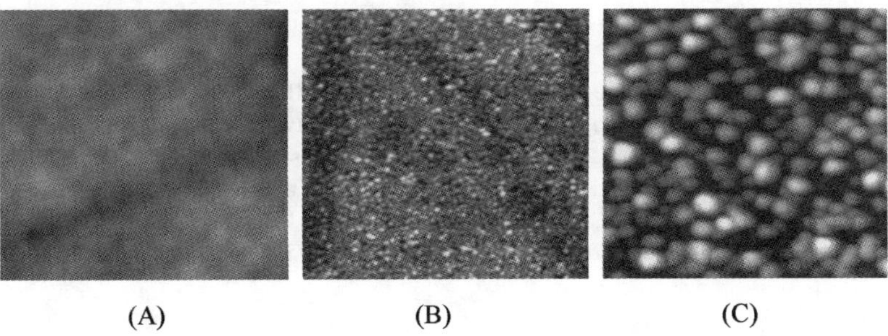

(A)   (B)   (C)

Figure 3. AFM images for Pd/6H-SiC as deposited (A, (1 x 1 μm^2 x 5 nm)), after consecutive annealing at 300°C (B, (1 x 1 μm^2 x 10 nm)), and 600°C (C, (1 x 1 μm^2 x 40 nm)).

3. Comparison of Pd/6H-SiC and Pd/4H-SiC

6H-SiC and 4H-SiC have very similar structures and reactivity with Pd at high temperatures. The formations of Pd silicides on Pd/4H-SiC is slightly enhanced compared to those on Pd/6H-SiC [5]. However, the formations of the carbon and silicon related compounds on 6H-SiC and 4H-SiC substrates are almost identical.

CONCLUSIONS

This study reports that various carbon and silicon compounds are formed with the presence of Pd ultra-thin film on SiC surface after different annealing temperatures. Silicon oxycarbides (SiC_xO_y) are found and the atomic ratio of C to O decreases with increasing annealing temperatures. SiO_2 is formed after annealing at 500°C and above. The amorphous Si is formed after annealing in the temperature region of 200-400°C, and Pd seems to act as a catalyst to break the Si-C bonds in SiC into Si and C phases. A low concentration of the C=O group exists on Pd/SiC through the annealing temperature up to 600°C. We found no significant differences in the formation of carbon and silicon compounds between the Pd/6H-SiC and Pd/4H-SiC samples.

ACKNOWLEDGMENTS

The work was supported by NASA grant Nos NAG3-2126 and NCC8-133. The authors are thankful for the discussion with Dr. Gary W. Hunter at NASA Lewis Research Center.

REFERENCES

[1]   V. Saxena, and A.J. Steckl, *"SiC Materials and Devices"*, Semiconductors and Semimetals, **Vol. 52,** edited by Y.S. Park, Academic Press, San Diego, CA, USA, 1998, Chapter 3, p77.

[2]   L.M. Porter, and R.F. Davis, *Mater. Sci. and Engin.*, **B34**, 83 (1995).

[3]   L.-Y. Chen, G.W. Hunter, P.G. Neudeck, G. Bansal, J. B. Petit, and D. Knight, *J. Vac. Sci. Technol.* **A 15(3)**, 1228 (1997).

[4]   L.-Y. Chen, G.W. Hunter, P.G. Neudeck, and D. Knight, *J. Vac. Sci. Technol.* **A 16(5)**, 2890 (1998).

[5]   W.J. Lu, D.T. Shi, A. Burger, and W.E. Collins, accepted by *J. Vac. Sci. Technol., A* for publication in 1999.

[6]   C. Önneby, and C.G. Pantano, *J. Vac. Sci. Technol., A*, **15(3),** 1597 (1997).

[7]   C. Önneby, and C.G. Pantano, *J. Vac. Sci. Technol., A*, **16(4),** 2742 (1998).

[8]   S. Contarini, S.P. Howlett, C. Rizzo, and B.A. De Angelis, *Applied Surf. Sci.*, **51**, 177(1991).

[9]   B. Hornetz, H.-J. Michel, and J. Halbritter, *J. Vac. Sci. Technol., A*, **13(3),** 767 (1995).

[10]   S. Tanaka, R. S. Kern, R.F. Davis, J. F. Wendelken, and J. Xu, *Surf. Sci.*, **350**, 247 (1996).

[11]   J. R. Shallenberger, *J. Vac. Sci. Technol., A*, **14(3),** 693 (1996).

A MATERIALS INVESTIGATION OF NICKEL BASED CONTACTS TO n-SiC SUBJECTED TO OPERATIONAL THERMAL STRESSES CHARACTERISTIC OF HIGH POWER SWITCHING

M.W. COLE*, C.W. HUBBARD*, C.G. FOUNTZOULAS*, D.J. DEMAREE*, F. REN**
* US Army Research Laboratory, WMRD, Aberdeen Proving Ground, MD 21005
** Dept. of Chem. Eng., University of FL, Gainesville, FL 21213

ABSTRACT

This study developed and performed Laboratory experiments which mimic the acute cyclic thermal loading characteristic of pulsed power device switching operation. Ni contacts to n-SiC were the device components selected for cyclic thermal testing. Modifications of the contact-SiC materials properties in response to cyclic thermal fatigue were quantitatively assessed via Rutherford backscattering spectrometry (RBS), scanning electron microscopy (SEM), atomic force microscopy (AFM), surface profilometry, transmission electron microscopy (TEM), nanoindentation testing and current-voltage measurements. Decreases in nanohardness and elastic modulus were observed in response to thermal fatigue. No compositional modifications were observed at the metal-semiconductor interface. Our results demonstrated that the majority of the material changes were initiated after the first thermal pulse and that the effects of subsequent thermal cycling (up to 10 pulses) were negligible. The stability of the metal-semiconductor interface after exposure to repeated pulsed thermal cycling lends support for the utilization of Ni as a contact metallization for pulsed power switching applications.

INTRODUCTION

Much attention has focused on SiC as a material for high power, high temperature and high radiation tolerance device applications[1-3] It is the exceptional properties of SiC, such as high breakdown field, large bandgap, high thermal conductivity and high electron saturation velocity, which are responsible for these device application interests [1]. To date, SiC electronic materials research efforts have focused predominantly on growth, processing science and packaging issues. However, in order to promote, design and realize reliable SiC power devices it is important to assess the performance of device components under the influence of their potential operational stress regimes. This is particularly critical for pulsed power device applications, namely palpitated high power switching, where the operational environment is dominated by acute cyclic pulsed power actions which ultimately translate into severe thermal, electrical and mechanical cyclic stresses in the device materials. In order to fully explore SiC's utilization for pulsed power switching applications it is necessary to determine the effects of such cyclic stress regimes, both individually and as combined effects, on the fundamental pulsed power device components. In this paper we report results of a unidimensional, that is, non-combined effects, investigation which evaluated the reliability of Ni-SiC ohmic contact device components in response to acute cyclic thermal loading. Our results demonstrate that the electrical, compositional and structural integrity of the metal-SiC interface strongly influences the reliability of the Ni-SiC ohmic contacts under acute cyclic thermal stress.

It is well documented that device performance is often limited by the electrical and materials integrity of the ohmic contacts [4,5]. Since ohmic contacts are a fundamental component of all pulsed power devices the ohmic contact-SiC device structure was selected for cyclic thermal testing. A number of different metals have been proposed as suitable ohmic contacts to n-SiC. Specifically, metals such as Ni, Al/Ni/Al, Cr, Al, Au-Ta, $TaSi_2$, W, Ta, Ti, Ti/Au, $TiSi_2$, Co and WSi have been studied with the Ni based metallization systems suggested as superior candidates due to their low specific contact resistance, (ρ_c), less than 5.0×10^{-6} ohms-cm^2 [6-9]. Additionally, published data suggests that annealed Ni contacts, which react with SiC to form Ni_2Si, exhibit excellent static thermal stability at temperatures as high as $500°C$ [7]. Based on this information Ni was selected as the contact metallization for cyclic thermal fatigue testing.

Mat. Res. Soc. Symp. Proc. Vol. 572 © 1999 Materials Research Society

EXPERIMENTAL

200 nm of Ni was deposited on 4H n-type SiC substrates purchased from CREE Research. The SiC substrates were non-research grade with a micropipe density of greater than 100 cm^{-2}. The substrates were Si faced and the donor density was 2.0 x 10^{19} cm^{-3}. Prior to the metal deposition the wafers were cleaned in warm electronic grade trichoroethane (TCA), boiling acetone and methanol followed by a rinse in deionized water. The Ni deposition was accomplished via electron beam evaporation with a base pressure of 5x10^{-7} Torr. The Ni on SiC samples, were annealed at 950°C for 5 min. in a N_2 ambient in order to produce ohmic behavior. Cyclic thermal fatigue experiments were conducted using a 10.6 μm IR pulsed CO_2 laser. The pulsed thermal fatigue design was configured for a 3 second heating interval followed by a 60 second cooling interval. The heating and cooling intervals were chosen to mimic military switching requirements.   Laser power levels were tailored to maintain a temperature of 650° C for 1 cycle and 10 consecutive cycles. Temperature verification was obtained with the aid of both a thermocouple and pyrometer.

Current-voltage measurements, were performed on the as-deposited, annealed and thermally fatigued Ni-SiC samples. The electrical measurements were internally consistent and were used solely to assess the electrical changes resulting from thermal fatigue relative to the non-fatigued sample. All samples were analyzed for surface morphology modification via atomic force microscopy (AFM) and surface profilometry. A Topometrix Discoverer Scanning Probe system was used to obtain the AFM data and the profilometry measurements were obtained via a Tencore P-2 low scan profiler. Elemental diffusion and phase formation were tracked using Rutherford backscattering spectroscopy (RBS). RBS measurements were obtained on a National Electronics Corp. 55DH-2 Accelerator using 2 MeV He$^+$ ions with a scattering angle of 170° and a solid angle of 5.5 msr. Experiments were performed at a tilt angle of 10° away from the detector in IBM geometry (beam, surface normal and detected beam are coplanar). Simulations were produced using the computer code RUMP [10]. The structural durability of the thin film metallization on SiC was evaluated via nanoindentation testing. The nanoindentation measurements were made using a Nano Instruments XP nanoindenter, with a Berkovich 3-sided pyramid diamond indenter with controlled penetration depths of 300 nm. The instrument allowed a indenter penetration vs. force curve to be determined allowing the determination of both nanohardness and the effective elastic modulus from the slope of the hysteresis curve. Microstructural analyses were obtained via cross-sectional TEM using a JEOL 3010 STEM operated at 300 keV.

RESULTS AND DISCUSSION

The I-V characteristics of the as-deposited Ni films displayed rectifying behavior suggestive of a large barrier height, typically 1 eV or greater. The RBS analysis (figure 1) revealed no reaction between the Ni and SiC substrate and showed no evidence of an interfacial oxide upon Ni deposition. The I-V curve for the annealed e-beam deposited Ni on SiC possessed near-linear characteristics and was symmetric with reversal of the voltage polarity. Due to lack of dry etching facilities, no values for specific contact resistance were measured, thus a quantitative value for specific contact resistance is not available. However, the Ni-SiC annealing protocol used in this study is well documented in the literature to produce good ohmic behavior with reproducible results,[7-9] thus, it is not unreasonable to assume that our samples also possess good ohmic behavior, that is, low specific contact resistance . The fact that the I-V curve for the annealed Ni-SiC structure possessed linear characteristics and was symmetric with reversal of the voltage polarity, is indicative of ohmic contact formation [11]. The RBS analysis of the e-beam deposited Ni on SiC (figure 2) showed a thin layer of NiO present at the contacts surface and Ni silicide formation adjacent to the SiC. The NiO is most likely due to the presence of oxygen during the annealing cycle. However, the near surface nature of the NiO layer combined with the linear I-V characteristics suggests that the oxygen contamination was not severe enough to impede ohmic contact formation. The RBS depth profile (figure 2) and TEM analyses (not shown) revealed a non-abrupt contact-SiC interface in which the thickness of the contact increased from 200 nm to 400 nm as a result of interfacial reactions during the anneal. This large increase in contact thickness has been documented in the literature for Ni contacts on SiC [6, 8, 12]. For Ni contacts to SiC, formation of Ni silicides upon annealing appears to be a requirement for ohmic behavior. The critical step in silicide formation requires the continual supply of Si atoms through breaking bonds in the substrate. The Si-C bonds can be broken in several ways; sufficient thermal energy to break the SiC bonds (high temperatures) and/or rapid interstitial migration of the metal through the SiC lattice which assists bond breaking and aids silicide formation. Since our samples were annealed at a fairly high temperature (950°C) and the fact that Ni is very small (.69 Å ionic radius) compared to that of Si (2.71 Å) and C (2.60) suggests that both these mechanisms may have contributed to the formation of the silicide phase detected by RBS.

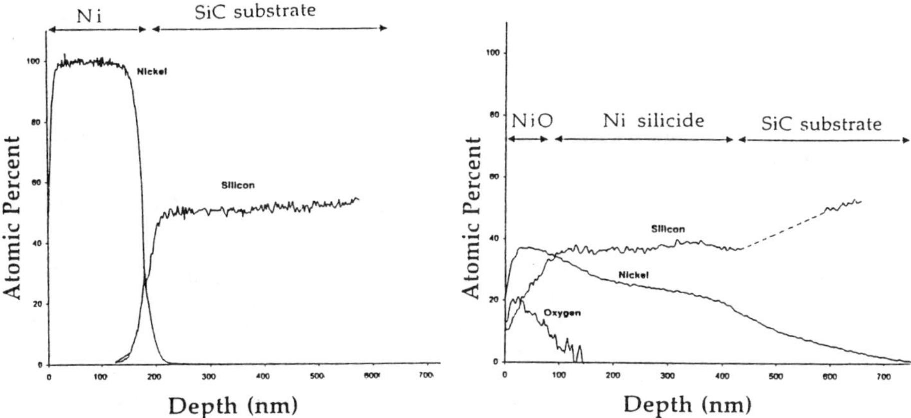

Fig. 1. RBS depth profile for the e-beam as-deposited Ni on SiC.

Fig. 2. RBS depth profile for annealed Ni contact to SiC.

The I-V curve for the 10 cycle thermal fatigued contact on SiC was notably similar to that of the unfatigued sample, that is, linear, symmetric with reversal of voltage polarity and possessed no deviation in the slope of the I-V curve relative to that of the unfatigued sample. The fact that ten cycles of thermal fatigue did not significantly degrade the I-V curve speaks well for the electrical integrity of this contact metallization in response to pulsed thermal stress. Figure 3 displays the RBS depth profile for this sample. The results revealed no compositional changes at the metal semiconductor interface. However, oxygen appears to have penetrated a bit deeper into the sample. It is speculated that attenuated nanofractures, resultant from thermal shock, confined to the upper portion of the contact may be responsible for this indiffusion of oxygen [13]. It must be kept in mind that the interfacial properties of the metal-semiconductor contact strongly influences electrical performance [4,5]. Thus, the fact that the metal-semiconductor interfacial region remained compositionally and electrically unchanged lends support to the excellent reliability of this contact in response to acute pulsed thermal stress.

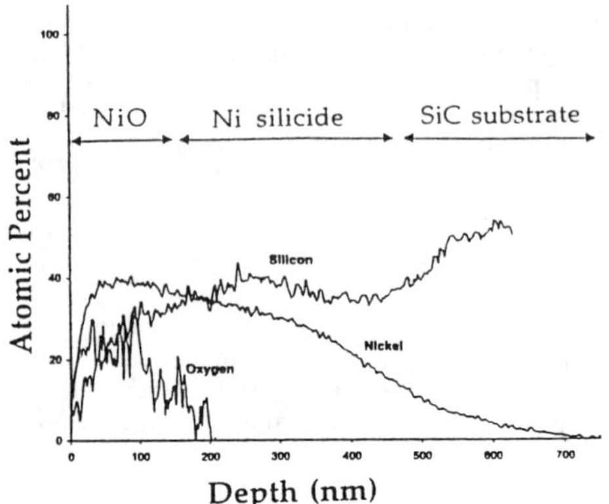

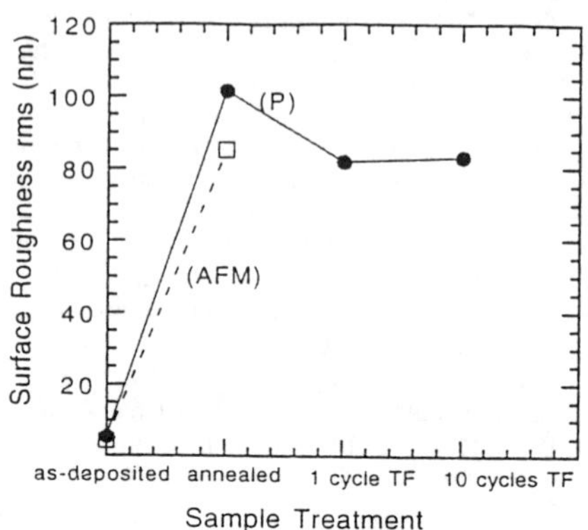

Fig.3. RBS depth profile for the annealed Ni contact to SiC after 10 cycles of thermal fatigue.

Fig. 4. Surface roughness as a function of sample treatment. Atomic force microscopy (AFM) & profilometry (P) data parallel one another.

Optical and SEM analysis showed the surfaces of the as-deposited e-beam evaporated Ni thin film on SiC to be smooth and mirror like in appearance. However, after annealing at 950°C the surface changed drastically. The mirror-like metallic luster changed to a lusterless dull gray color and the film was no longer smooth. The surface morphological changes were quantified via profilometry and AFM analyses, and are displayed in figure 4. The magnitude of the profilometry data is larger than that of the AFM data but the general trends parallel one another. It is evident from figure 4 that annealing augmented the surface roughness of the metal film. Specifically, the surface roughness increased by an order of magnitude. The films surface remained lusterless after the thermal fatigue however the surface became smoother. This surface smoothening occurred during the first cycle of the pulsed thermal fatigue and maintained at steady state through the 10th cycle, that is, the rms roughness value remained constant throughout the duration of the thermal cycling. We suggest that the initial thermal shock of the first laser pulse caused the removal of a thin layer of loosely bound particles from the films surface which resulted in a lower rms roughness value. The fact that the film did not delaminate and the rms roughness value remained constant throughout the thermal cycling bodes well for the films strong adhesion and cohesion properties.

Mechanical properties such as nanohardness and Young's modulus were used to assess the changes in the physical durability of the metal-SiC component in response to thermal fatigue. Simply defined the hardness of a thin film is the resistance of the film to penetration of its surface, that is, resistance to local plastic deformation [14]. Hardness is a complex macroscopic property related to the strength of interatomic forces and depends on several variables at the nanoscopic level. Grain size, area and grain boundary structure, film composition (new phase formation), impurities, defects and film texture all influence and/or control the nanohardness or durability of thin films [15]. Thus, there is a strong relationship between a films microstructure and its mechanical properties.

The magnitude of Young's (elastic) modulus is determined by the strength of the atomic bonds in the film. The stronger the atomic bonding, the greater the stress required to increase the interatomic spacing and thus the larger the value of the modulus of elasticity and the more durable the film [14]. Like hardness, the modulus is a macroscopic property which depends on many different variables at the nanoscopic-atomistic level. Specifically, modification of the films chemical composition strongly influences the elastic modulus value since the various phases are composed of different atomic species with different bond strengths. Thus, changes in the contacts mircrostructure and composition will be reflected in the values of the contacts nanohardness and elastic modulus.

Figure 5 displays the nanohardness and Young's modulus values for the as-deposited, annealed and thermal fatigued contact metallization on SiC. The nanohardness and modulus values changed significantly with sample treatment, showing a decrease in magnitude in response to annealing and cyclic thermal fatigue. The largest change in both the hardness and modulus occurs after the first thermal pulse. Additional thermal cycling caused negligible changes in these properties. The compositional change from Ni to Ni-silicide (and the thin surfacial NiO layer) accounts for the change in the durability values in response to annealing. TEM results on Ni-SiC ohmic contacts has shown the grain size of the Ni-Silicide to be smaller that that of the as-deposited Ni on SiC [13]. Thus, the diminishment of the nanohardness after annealing is easily explained by the increase of grain boundary area of the Ni-silicide with respect to that of Ni. The decrease in durability values in response to the first thermal pulse is most likely due to the promotion of thermally induced nanofractures (figure 6) within the upper portion of the Ni-Silicide contact layer [13]. The fact that the hardness and modulus do not change significantly after 10 cycles suggests that multiple or repeated thermal cycling has a negligible affect on the films durability. This speaks well for the reliability of the Ni contacts endurance to repeated pulsed thermal fatigue. In addition, the fact that the metal film did not delaminate in response to the cyclic thermal stress bodes well for the films strong adhesion and cohesion properties.

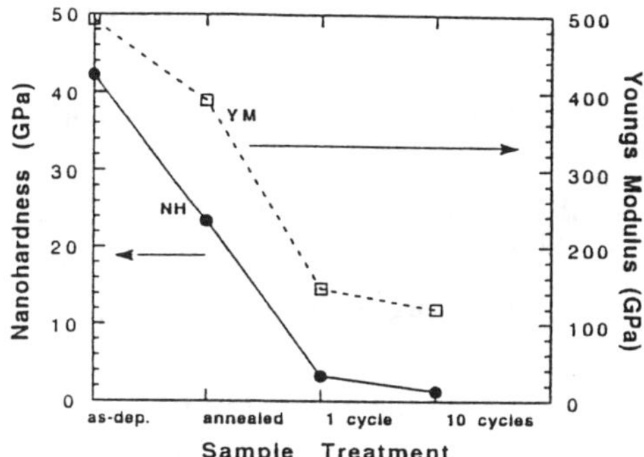

Fig. 5. Nanohardness and Young's modulus values as a function of sample treatment.

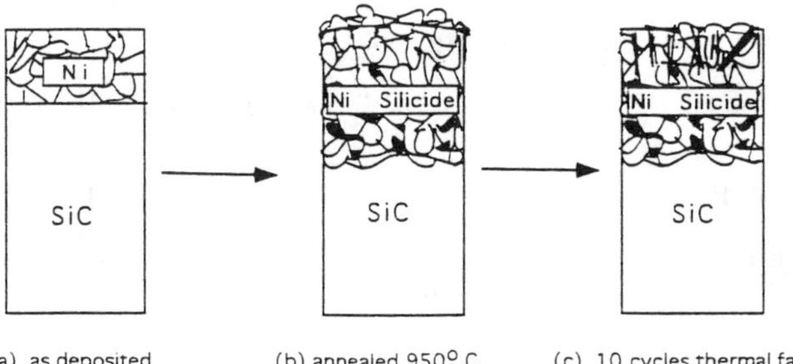

Fig. 6. Schematic representation of the microstructural changes as function of sample treatment. The dark black areas in the annealed sample represent voids (voids were determined via X-TEM) and the dark lines at the surface of the thermally fatigued sample represent nanofractures.

CONCLUSIONS

We have developed and performed laboratory experiments which simulate the acute thermal cyclic fatigue incured as a result of pulsed power switching operation. Evaluation of the reliability of Ni contacts to SiC in response to cyclic thermal fatigue was investigated. Our results demonstrated that most of the material changes occurred in response to the first thermal pulse and that further pulsing (up to 10 pulses) inflected negligible changes in the contact-SiC durability, compositional and electrical properties. The stability of the metal-semiconductor interface after acute thermal cyclic fatigue lends support for for the utilization of Ni as a contact metallization for pulsed power applications.

REFERENCES

1. S.J. Pearton, F. Ren, R.J. Shul and J.C. Zolper, Proceedings of The Electrochemical Society, **97-1**, 138 (1997).

2. Philip G. Neudeck, J. of Electronic Materials, **24**, 283 (1995).

3. J.W. Palmour and C.H. Carter, Proceedings of 1993 International Semiconductor Device Research Symposium, 695 (1993).

4. W.Y. Han, Y. Lu, H.S. Lee, M.W. Cole, L.M. Casas, A. DeAnni and K.A. Jones, J. Appl. Phys. **74**, 754 (1993).

5. M.W. Cole, W.Y. Han, L.M. Casas, D.W. Eckart and K.A. Jones, J. Vac. Soc. Technol. A **12**, 1904 (1994).

6. C. Hallin, R. Yakimova, B. Pecz, A. Georgieva,T.S. Marinova, L. Kasamakova, R. Kakanakov, E.Janzen, J. of Electronic Materials, **26**, 119 (1997).

7. J. Crofton, P.G. McMullin, J.R. Williams, M.J. Bozack, J. Appl. Phys., **77**, 1317 (1995).

8. J. Crofton, L.Beyer, T. Hogue, R.R. Siergiej, S. Mani, J.B. Casady, T.N. Oder, J.R. Williams, E.D. Luckowski, T. Isaacs-Smith, V.R. Iyer and S.E. Mohney, Proceedings of The Fourth International High Temperature Electronics Conference, 84 (1998).

9. M.R. Melloch and J.A. Cooper, MRS Bulletin, **23**, 42 (1997).

10. L.R. Doolittle, Nucl. Instrum. Methods **B9**, 344 (1985).

11. S.M. Sez, *Semiconductor Devices Physics and Technology*, (John Wiley & Sons, New York, (1985).

12. M.I. Chaudhry, W.B. Berry and M.V. Zeller, Mat. Res. Soc. Proc., **162**, 507 (1990).

13. M.W. Cole, C. Hubbard, D. Demaree, C.G. Fountzoulas, D. Harris, A. Natarajan, P. Searson, R.A. Miller and D. Zhu, Proceedings of the Army Science Conference, **22**, 35 (1998).

14. W.D. Nix, Metallurgical Transactions A, **20A**, 2217 (1989).

15. J.E. Sundgren and H.T. G. Hentzell, J. Vac Sci. Tech. A, **4**, 2259 (1986).

PREPARATION OF CONDUCTIVE TUNGSTEN CARBIDE LAYERS FOR SIC HIGH TEMPERATURE APPLICATIONS

H. ROMANUS[*], V. CIMALLA[*], S.I. AHMED[*], J.A. SCHAEFER[*], G. ECKE[**], R. AVCI[+], L. SPIESS[***]

[*] Institute of Physics, Technical University of Ilmenau, Germany , D-98693 Ilmenau
[**] Institute of Solid State Electronics, Technical University of Ilmenau, Germany
[***] Institute of Materials Engineering, Technical University of Ilmenau, Germany
[+] Department of Physics, Montana State University, Bozeman, Montana 59717

ABSTRACT

Thin tungsten carbide films of different compositions were prepared by DC magnetron sputtering of tungsten and carbon and subsequent annealing in different environments. The onset of carbide formation was around 800°C. Annealing in a pure hydrogen ambient generally results in carbon depletion in the layers with the formation of a dominant W_2C phase. Adding propane enhances the carbon content in the layers and stimulates the formation of the WC phase. On silicon nitride substrates, variation of the propane concentration in an annealing environment allows a continuous alteration of the layer structure between polycrystalline single phase WC and a mixed layer with dominant W_2C and with it, the adjustment of different values of the electrical resistance. In contrast, on thin (100)SiC layers a textured W_2C phase was grown after annealing in propane/hydrogen at 900°C whereas at higher temperatures the formation of silicides was observed. In addition, the chemical composition and the temperature dependence of the electrical specific resistance were investigated and are also discussed.

INTRODUCTION

In recent years, increasing interest has been directed to wide band-gap semiconductors like silicon carbide (SiC) due to its potential for applications in power and high temperature electronics [1]. SiC devices show very good high temperature, chemical and mechanical stability and high breakdown voltages. Recent improvements in crystal growth techniques now provide industry and research with high quality bulk and epilayer material. However, there are still a variety of factors limiting the commercialization of devices in this area [2]. One of the most important factors with respect to high temperature devices is the requirement of a metallization, which should maintain a low contact resistivity, have good adhesion to the substrate material and have high stability at elevated temperatures, particularly at the interface. Several attempts have been made to achieve good Ohmic contacts to SiC. However, at elevated temperatures most metals are not stable on SiC and form rather good rectifying contacts with a barrier height > 1 eV [3].

The formation of good stable Ohmic contacts to p-type SiC remains an important technological problem that is hindering the industrial production of high temperature and power devices. A theoretical study based on an ideal band bending model of a semiconductor-metal interface has shown the impossibility of forming enhancement contacts to p-type SiC [4], because no contact material exists with the required work function. In all cases, to form Ohmic contacts the width of the depletion zone has to be minimized to increase thermionic emission. This can be attained

Mat. Res. Soc. Symp. Proc. Vol. 572 ©1999 Materials Research Society

by a high doping of the surface region. Nevertheless, a high value of the contact material work function is a precondition for the formation of Ohmic contacts.

Metal carbides are promising contact materials that could fulfill the above mentioned requirements. Transition-metal carbides like tungsten carbide possess an unusual combination of physical properties: They are hard and very refractory as many ceramics and have good electrical conductivity that is comparable to the parent transition metals [5]. The extreme stability and hardness are already being exploited in many industrial applications like wear-resistant coatings, cutting and drilling tools. However, the electrical properties make tungsten carbide also an attractive material for contacts in high temperature electronics or as conductive protective layers in sensor applications. Tungsten carbide exists in different phases, most important are the WC and W_2C phases. The close match of the lattice constants of the hexagonal phase to the (0001)SiC plane and the high work function of 3.6 eV and 4.58 eV, respectively, [5] favors them in applications as highly stable epitaxial contacts to SiC.

Several attempts have been made to prepare thin tungsten carbide layers on different substrates including sputtering, chemical vapor deposition (CVD), solid-phase reaction, laser ablation and ion beam synthesis. In most cases a mixed phase or amorphous tungsten carbide was grown, mainly for hard and wear resistant coatings. The only techniques that have successfully prepared single phase WC layers are the solid state reaction of tungsten on diamond [6] and CVD on tantalum substrates [7] at high temperatures. However, both methods are strongly influenced by the substrate material and CVD was shown to be inappropriate for SiC substrates. Up to now neither have the contact properties of thin tungsten carbide layers to SiC been investigated nor has epitaxial growth been possible.

EXPERIMENTAL

The tungsten-carbon layers were deposited by DC magnetron sputtering of a sintered stoichiometric WC target in argon at a pressure of 3×10^{-2} Torr at room temperature or *in situ* annealing at 500°C, respectively. The substrates were silicon wafers covered by 250 nm silicon nitride or by thin cubic (100)SiC formed by carbonization [8]. Prior to sputter deposition they were ultrasonically cleaned in methanol. The tungsten-carbon layer thickness was varied between 50 and 300 nm. Annealing experiments were performed *in situ* or *ex situ* in a conventional horizontal quartz tube in argon for 20 min or in a rapid thermal processing (RTP) equipment [9] between 10 and 60 s in argon/ hydrogen/ propane mixtures with a propane content that was varied between 0 and 5%. The layer structure was analyzed by X-ray diffraction (XRD) with CuK_α radiation, the surface morphology by atomic force microscopy (AFM), the electrical conductivity by a linear four-point method at room temperature and at elevated temperatures up to 470°C and the chemical composition by Auger electron spectroscopy (AES) and X-ray photoelectron spectroscopy (XPS).

RESULTS

The as-deposited layers were amorphous or micro crystalline with a homogeneous, smooth surface without any distinct features. A compositional analysis revealed a carbon depletion around 10 %. Annealing *in situ* resulted in micro crystalline thin films. *Ex situ* annealing in pure hydrogen by RTP lead to strong carbon depletion via reaction to hydrocarbons, and the formation of the W_2C and other high temperature phases (Fig. 1). This reaction can be suppressed by adding propane to the ambient. With an enhancement of the propane concentration up to 0.025% on the silicon nitride covered substrates, the content of the W_2C phase decreased while content of the WC phase increased (Fig. 2). Only WC was detected by XRD above 0.025% at 1200°C and in the AFM images small grains appeared at the surface. In contrary, annealing of the tungsten carbon layers on the (100)SiC covered substrates at 900°C under the same conditions enhances the formation of the W_2C phase. Only one tungsten carbide diffraction peak was detected indicating the formation of a texture. At higher temperatures (i.e. 1200°C in Fig. 3) tungsten silicide was additionally observed. This phase is probably formed by the reaction of tungsten with silicon from the underlying substrate which diffuses through the thin SiC layer.

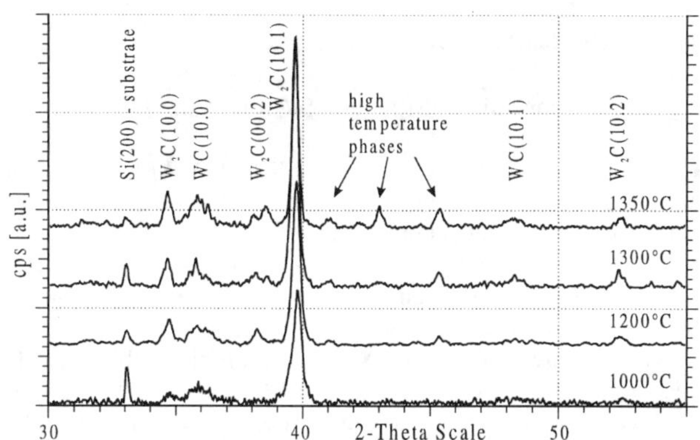

Fig. 1 XRD patterns, RTP annealing in pure hydrogen between 1000°C and 1350°C

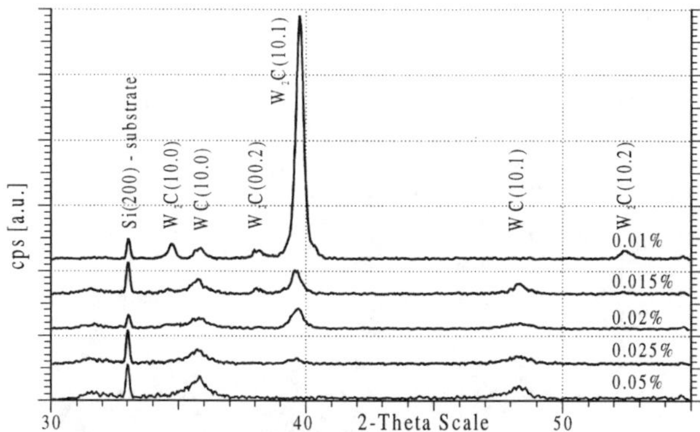

Fig. 2 XRD patterns, RTP annealing at 1200°C, of 0.01% to 0.05% propane in hydrogen

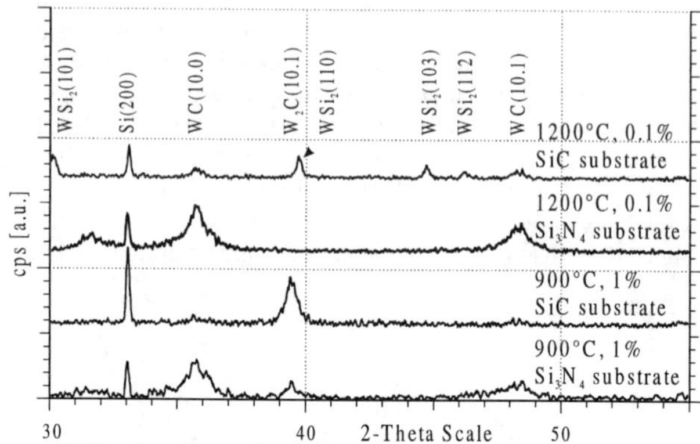

Fig. 3 XRD patterns, RTP annealing - Si$_3$N$_4$ and SiC substrates

The chemical analysis of the thin films was accomplished by AES depth profiling using a sputter rate of around 1 to 2 nm/min. For the as-deposited films as well as for the RTP (Fig. 4) and *in situ* annealed samples the tungsten and carbon amount is constant over the entire thickness, independent of the annealing time (Table I). A low oxygen content of approximately 1% to 3 % was detected. A higher concentration was found only after annealing in the quartz tube where already low traces of oxygen or water are sufficient to oxydize the layers. Similar results were found by XPS depth profiling. XPS spectra indicate the formation of carbides. The binding energy was estimated to be 32.1 eV and 31.36 eV for the WC and W$_2$C, respectively, where the latter is located very close to the value for elemental tungsten (31.32 eV [10]). The calculated content of the two carbidic phases is in good agreement with XRD observations.

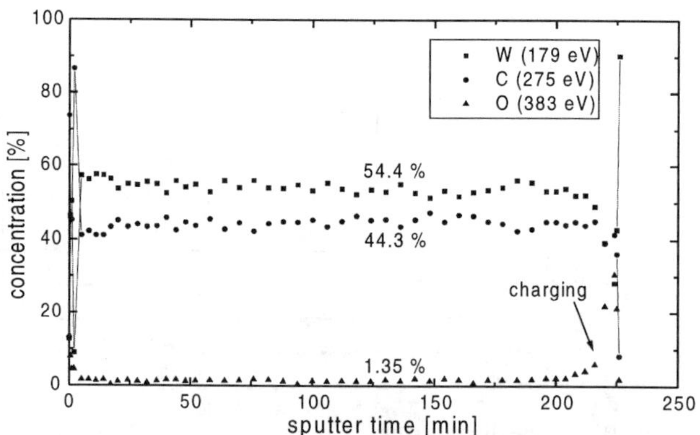

Fig. 4 AES depth profiling, RTP at 1200°C and 0.02% propane (The changing concentrations near the interface are due to charging effects at the isolating substrate.)

Table I Comparison between XRD, AES and XPS

Sample	XRD	AES			XPS
		W	C	O	
in situ annealed	micro crystalline	53.3%	43.3%	<3.2%	
Annealed in quartz tube	mixture of W/W_2C and a low content WC	40..77%	2..24%	59..5%	approximately 10% WC, 10% W_2C, 80% W, oxide only near surface
RTA, 1200°C, 0.02%, 60s	mixture of WC and W_2C	54.4%	44.3%	1.35%	approximately 40%WC, 60% W_2C
RTA, 1200°C, 0.02%, 10s	mixture of WC and W_2C	55.2%	43.2%	1.58%	
RTA, 1200°C, 0.05%, 60s	W_2C and a low content WC	65.5%	32.8%	1.66%	approximately 10% WC and 90% W_2C

At room temperature the as-deposited tungsten-carbon layers had a specific resistance of around 200 $\mu\Omega$cm. With increasing content of the W_2C phase at lower propane concentrations the specific resistance decreased continuously to 100 $\mu\Omega$cm (Fig. 5). As a result of the temperature dependent measurements the as-deposited and in situ annealed layers revealed a negative temperature coefficient which is well known for amorphous or microcrystalline metal films. The specific resistivity of the crystalline tungsten carbide layers showed a temperature behavior depending on their composition (Fig. 6): a linear increase with the temperature for pure WC layers similar to pure metals and a non linear behavior in the case of temperature dependence of the WC / W_2C mixtures. Above 450°C surface oxidation affected the measurements.

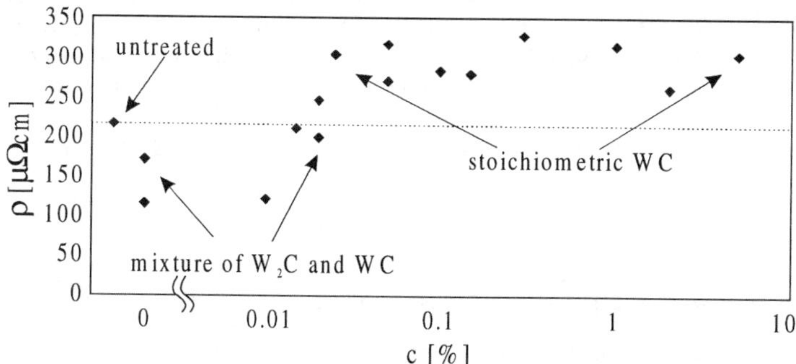

Fig. 5 Electrical resistivity on propane concentration measured at room temperature, untreated and rapid thermal processing at 1200°C

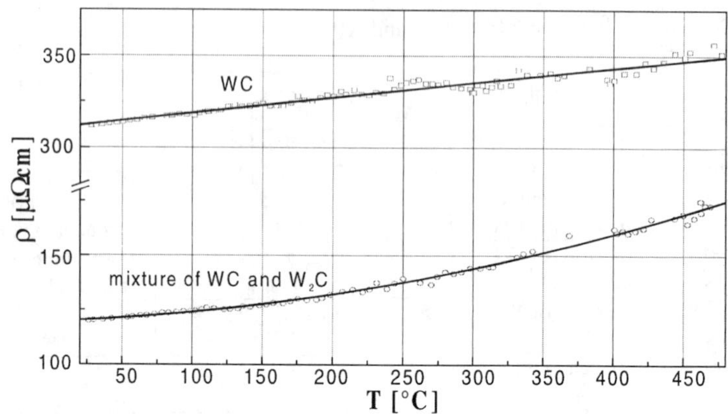

Fig. 6 Electrical specific resistance on the measurement temperature

CONCLUSIONS

Single phase polycrystalline WC layers were prepared by sputtering of WC targets and subsequent short time annealing in a propane-hydrogen ambient. Thermal treatment in pure hydrogen resulted in carbon depletion in the layers and in the formation of W_2C. Propane diluted in the annealing ambient stimulated a transformation of the tungsten-carbon layers to a stoichiometric WC phase. On amorphous silicon nitride polycrystalline tungsten carbide layers with adjustable ratio between the WC and the W_2C phase were formed and therefore different values of the electrical resistance in dependence on the propane concentration emerged. In contrary on thin cubic (100)SiC layers a preferred formation of a textured W_2C phase was observed after annealing in propane/hydrogen at 900°C. At higher temperatures the presence of additional silicon resulted in the formation of a tungsten silicide by diffusion of silicon through the thin SiC layer or by an interface reaction. The origin of this silicon as well as the influence of orientation of the SiC layer ((100) vs. (111)) will be the subject of further investigations.

REFERENCES

[1]   J. W. Palmour, L. A. Lipkin, R. Singh, D. B. Slater, A.V. Suvorov, C. H. Carter, jr., Diam. Rel. Mater. **6,** 1400-4 (1997).
[2]   V. E. Chelnokov, A. L. Syrkin, V. A. Dmitriev, Diam. Rel. Mater. **6,** 1480-4 (1997).
[3]   L. M. Porter, R. F. Davis, Mater. Sci. Eng. B (Solid-State Materials for Advanced Technology) **34** (2-3), 83-105 (1995).
[4]   Spieß, L.; Nennewitz, O.; Weishart, H.; Lindner, J.; Skorupa, W.; Romanus, H.; Erler, F.; Pezoldt, J.: Aluminum implantation of p-SiC for Ohmic contacts; Diam. Rel. Mater. **6,** 1414-8 (1997).
[5]   *Carbide, Nitride and Boride Materials - Synthesis and Processing*, edited by A. W. Weimer, Chapman & Hall, London, (1997).
[6]   A. Bächli, J. S. Chen, R.P. Ruiz, M.A. Nicolet, MRS Symp. Proc. **339,** 247-52 (1994).
[7]   P. Tägtström, H. Högberg, U. Jansson, J. O. Carlsson, J. de Phys. IV, **5** (C5, pt.2) 967-74 (1995).
[8]   V. Cimalla, J.K. Karagodina, J. Pezoldt, G. Eichhorn, Mater. Sci. Eng. **B29,** 170-175 (1994).
[9]   G. Leitz, J. Pezoldt, I. Patzschke, J.-P. Zöllner, G. Eichhorn, MRS Symp. Proc. **303,** 171-176 (1993).
[10] *Practical Surface Analysis*, edited by D. Briggs and M.P. Seah, John Wiley & Sons, New York, (1990).

A FORMATION OF SiO$_2$/4H-SiC INTERFACE BY OXIDIZING DEPOSITED POLY-SI AND HIGH TEMPERATURE HYDROGEN ANNEALING

K. Fukuda (NEDO industrial researcher)*, K. Sakamoto**, K. Nagai**, T. Sekigawa*,**
S. Yoshida*,**, and K. Arai*,**
*Ultra-Low Loss Power Device Technologies Research Body, Electrotechnical Laboratory, 1-1-4,
Umezono, Tsukuba, Ibaraki, 305-8568 Japan, kfukuda@etl.go.jp
**Electrotechnical Laboratory, 1-1-4, Umezono, Tsukuba, Ibaraki, 305-8568 Japan

ABSTRACT

A formation of SiO$_2$/4H-SiC interfaces by oxidizing deposited poly-Si on a 4H-SiC substrate and high temperature hydrogen annealing at low pressure (8.5×10^2 Pa) has been investigated.   The oxidation rate of deposited poly-Si was approximately 100 times faster than that of a SiC. Hydrogen annealing more effectively reduced the flat band voltage shift (ΔV_{fb}) of the 4H-SiC MOS structure than argon and vacuum annealing. Moreover, the good SiO$_2$/4H-SiC interface was formed because ΔV_{fb} decreased as the oxidation temperature increased.

INTRODUCTION

Recently, many studies on high-temperature, high-power and high-frequency electronic devices fabricated from 6H- and 4H-SiC have been reported because SiC has excellent physical properties such as high electric field breakdown strength, high saturated electron velocity and high thermal conductivity [1-3]. Metal-oxide-semiconductor field-effect-transistors (MOSFETs) are as important in power SiC devices as in power Si devices. However, SiC MOSFETs have not been realized yet for practical use because of two main problems. One is the difficulty of thermally oxidizing SiC due to the anisotropy and very small value of the oxidation rate. Another is a large amount of SiO$_2$/SiC interface state. Especially, the latter is thought to be one of the origins of low channel mobility of SiC MOSFETs.

In Si MOS technology, the SiO$_2$ film formed by oxidizing poly-Si is used, for example, as an insulator film between a floating gate and a control gate in erasable programmable read only memory (EPROM).   The oxidation rate of poly-Si is faster than that of single crystal Si and exhibits no anisotropy.   It is also well known that hydrogen annealing terminates dangling bonds of Si at the SiO$_2$/Si interface and decreases the interface state density (D_{it}).

However, Afanasev et al. reported that hydrogen annealing at the pressure of 1.1×10^5 Pa occurred positive charge at the SiO$_2$/SiC interface, resulting in the shift of C-V characteristics of 6H-SiC MOS structures toward negative voltage from the ideal C-V characteristics, and that hydrogen annealing seemed to be insignificant for the decrease of dangling bonds [4,5]. In

Mat. Res. Soc. Symp. Proc. Vol. 572 © 1999 Materials Research Society

contrast, we found that hydrogen annealing at low pressure (8.5×10^2 Pa) did not shift C-V characteristics of 4H-SiC MOS structures toward negative voltage from the ideal C-V characteristics and decreased D_{it} [6,7]. The positive charge which occurs at the SiO_2/4H-SiC interface in hydrogen annealing at the pressure of 1.1×10^5 Pa is considered to be due to the excess hydrogen. Therefore, if the combination of oxidizing deposited poly-Si and high temperature hydrogen annealing at low pressure (8.5×10^2 Pa) is applied to the fabrication of SiC MOSFETs, two problems as mentioned above will be solved.

In this study, we have investigated a formation of the SiO_2/4H-SiC interface by oxidizing deposited poly-Si and high temperature hydrogen annealing at low pressure (8.5×10^2 Pa) and effect of the oxidation temperature on high-frequency C-V characteristics.

EXPERIMENTAL

The 8° off-angled n-type 4H-SiC (0001) substrates from Cree Research with a 4.9-μm -thick n-type epitaxial layer were used for this study. The n-type dopant was nitrogen. The effective carrier density (N_d-N_a) was 1.7×10^{16} cm^{-3}. After the standard RCA cleaning, 10 nm thick sacrifice oxide films were grown at 1100°C in dry O_2, and then they were removed by 5% HF solution before loading in ultra-high-vacuum (UHV). The 22 nm thick poly-Si layers were deposited on 4H-SiC substrates in a Si molecular-beam-epitaxy (MBE) apparatus. The poly-Si layers were thermally oxidized at 1050°C, 1150°C and 1250°C for 30 min in dry O_2. The samples were moved from the hot zone to the cool zone and cooled down to 20°C rapidly. The post-oxidation annealing (POA) of samples oxidized at 1050°C and 1150°C were performed in hydrogen (8.5×10^2 Pa) at 1000°C for 30 min and POA of samples oxidized at 1250°C were performed in hydrogen (8.5×10^2 Pa) , argon (8.5×10^2 Pa) and vacuum (1×10^{-4} Pa) at 1000°C for 30 min. Aluminum on top of oxide films and on the backs of the samples was evaporated to make gate electrodes and ohmic contacts of MOS structures, respectively. High-frequency C-V measurements were performed using an HP 4274 LCR meter in a shielded dark box at room temperature. The thickness of SiO_2 films grown on 4H-SiC substrates by oxidizing deposited poly-Si as shown in Fig. 1 were measured using the surface profilometer (Dektak IIA). The thickness of SiO_2 films of MOS structures, estimated from capacitance-voltage (C-V) characteristics, was 60 ± 5nm.

RESULTS AND DISCUSSION

<u>Oxidation rate</u>

Figure 1 shows the oxidation time dependence of the SiO_2 thickness formed by oxidizing deposited poly-Si on SiC substrates, single crystal Si [8] and single crystal SiC [9] at 1050°C. The oxidation rate of a poly-Si is approximately 100 times faster than that of single crystal SiC

[9] and approximately 1.2 times faster than that of single crystal Si as expected.   This value is available for practical use.

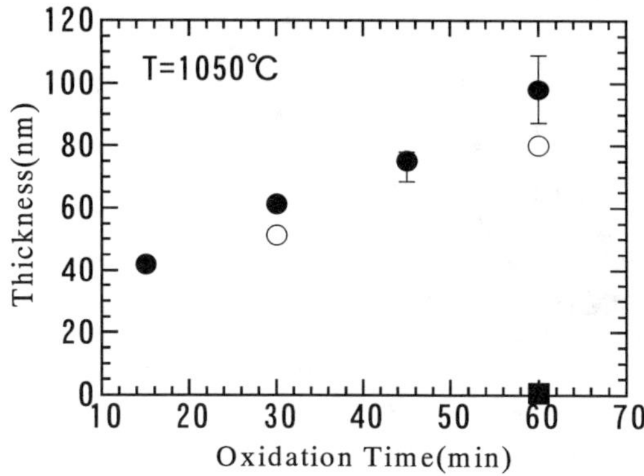

Figure 1 Oxidation time dependence of the thickness of SiO_2 films formed by oxidizing deposited poly-Si (closed circles) on 4H-SiC substrates, single crystal Si [8](open circles) ,and single crystal SiC [9] (closed square) at 1050°C.

POA effect on high-frequency CV characteristics of 4H-SiC MOS structures

Figure 2 shows effect of POA on high-frequency C-V characteristics of 4H-SiC MOS structures. The dotted line was calculated using the oxide capacitance (C_{ox}) and N_d-N_a of sample with hydrogen annealing. The measured capacitance at Vg<-3V are lower than the calculated values due to a small amount of minority carriers generated at room temperature because of the wide-gap of 4H-SiC.   POA condition dependence of ΔV_{fb} and the surface state density (N_{ss}) + the fixed charge density (N_f) is summarized in Table 1.   The values of ΔV_{fb} and N_{ss}+N_f of sample without any annealing are 14V and -5.0×10^{12} cm^{-2}. Even the vacuum and argon annealing decrease ΔV_{fb} and N_{ss}+N_f. Especially, ΔV_{fb} and N_{ss}+N_f were reduced to 2.2V and −7.4×10^{11} cm^{-2} by argon annealing, respectively. Hydrogen annealing decreases more effectively ΔV_{fb} and N_{ss}+N_f. C-V characteristics of the sample with hydrogen annealing is very close to the ideal C-V characteristics. We have already revealed that the density of hydrogen which accumulates at the SiO_2/SiC interface formed by thermal oxidation of an SiC substrate increases as hydrogen annealing temperature increases using secondary ion mass spectroscopy (SIMS) [6,7]. As it is thought that the same phenomena occurs in this experiment, hydrogen terminates dangling bonds of Si or C atoms at the SiO_2/SiC interface more strongly than oxygen because the binding energy of Si-H is lower than that of Si-O.   As a result, the values of ΔV_{fb} and N_{ss}+N_f of the sample with hydrogen annealing are smaller than those of samples with argon or vacuum annealing, which are 1.1 V and -4.0×10^{11} cm^{-2}, respectively. Hydrogen annealing is the most available for the good SiO_2/4H-SiC interface.

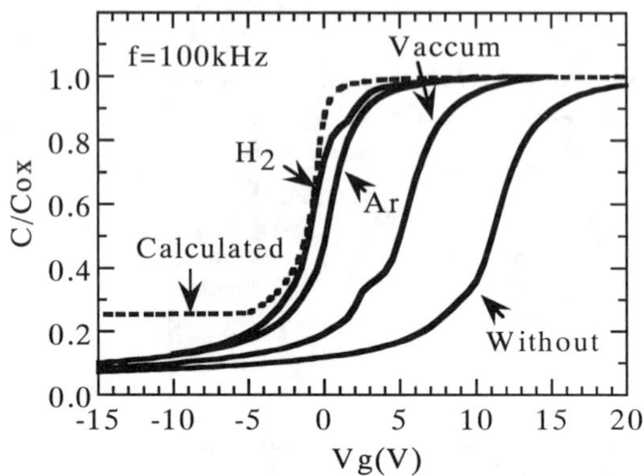

Figure 2　POA effect on high-frequency C-V characteristics of 4H-SiC MOS structures. The gate oxide films were grown at 1250°C. POA in hydrogen (8.5×10^2 Pa), argon(8.5×10^2 Pa) and vacuum(1×10^{-4} Pa) was performed at 1000°C for 30min after oxidation.

Table 1　POA condition dependence of ΔV_{fb} and $N_{ss}+N_f$

POA condition	ΔV_{fb}(V)	$N_{ss}+N_f$ (cm^{-2})
Without	14	-5.0×10^{12}
Vacuum (1×10^{-4} Pa)	7.4	-2.8×10^{12}
Argon　(8.5×10^2 Pa)	2.2	-7.4×10^{11}
Hydrogen (8.5×10^2 Pa)	1.1	-4.0×10^{11}

<u>Oxidation temperature effect on high-frequency CV characteristics of 4H-SiC MOS structures</u>

Figure 3 shows oxidation temperature effect on high-frequency C-V characteristics of 4H-SiC MOS structures. All samples were annealed in hydrogen (8.5×10^2 Pa) at 1000°C for 30 min after oxidation. The dotted line and C-V characteristics at 1250°C are the same as those in Fig.2. ΔV_{fb} and N_f decrease with increasing temperature as shown in Fig.4. This suggests that dangling bonds of Si or C atoms are terminated by oxygen atoms strongly as temperature increases. The SiO$_2$/4H-SiC interface becomes better as oxidation temperature increases.

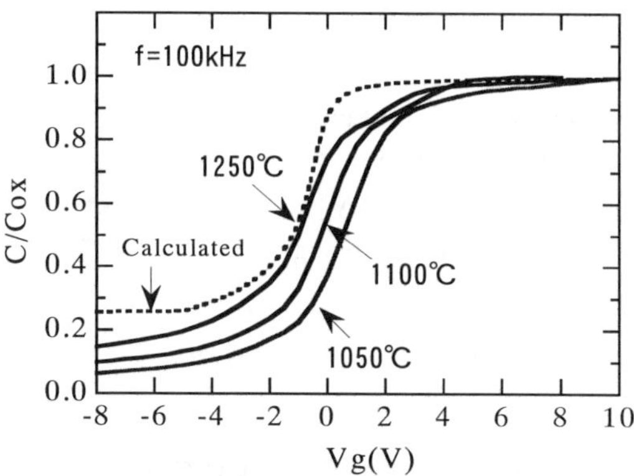

Fig. 3 Effect of oxidation temperature on high-frequency C-V characteristics of 4H-SiC MOS structures. They were annealed in hydrogen (8.5×10^2 Pa) at 1000°C for 30 min after oxidation.

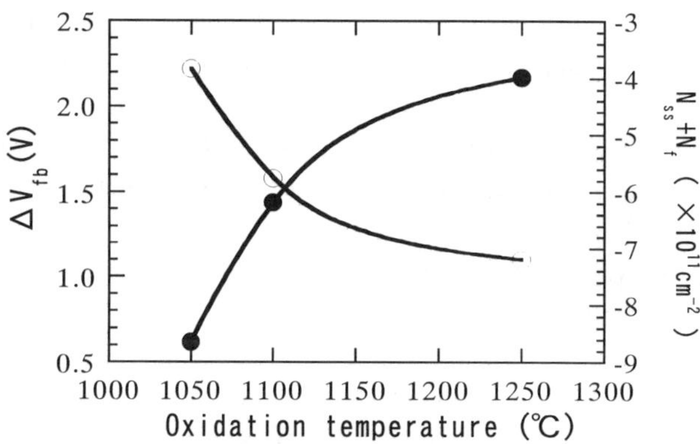

Fig.4 Oxidation temperature dependence of ΔV_{fb} and $N_{ss}+N_f$ estimated from C-V characteristics of 4H-SiC MOS structures as shown in Fig. 3.

CONCLUSION

The oxidation rate of deposited poly-Si is approximately 100 times faster than that of single crystal SiC, which is available for practical use. Hydrogen annealing at low pressure (8.5×10^2 Pa) more effectively reduced ΔV_{fb} and $N_{ss}+N_f$ than argon and vacuum annealing, resulting in ΔV_{fb} and $N_{ss}+N_f$ of 1.1V and -4.0×10^{11} cm^{-2}, respectively. The value of ΔV_{fb} is also reduced as oxidation temperature increases. Finally, the problems of slow oxidation rate of single crystal SiC and the bad SiO_2/4H-SiC interface can be solved using these technologies.

ACKNOWLEDGEMENTS

This work was carried out by the Ultra-Low Loss Power Device Technologies Project under the management of the R&D Association for Future Electron Devices (FED) as a part of the Ministry of International Trade and Industry (MITI) R&D of Industrial Science and Technology Frontier Program supported by New Energy and Industrial Technology Development Organization (NEDO).

REFERENCES

1. J. A. Cooper, Jr., M. R. Melloch, J. M. Woodall, J. Spitz, K. J. Schoen and J. P. Henning : III-Nitrides and Related Materials, eds. G. Pensil, H. Morkoç, B. Monemar and E. Janzén (Trans Tech Publications Ltd, Switzerland·Germany·UK·USA, 1998), p. 895.

2. K. Hara : III-Nitrides and Related Materials, eds. G. Pensil, H. Morkoç, B. Monemar and E. Janzén (Trans Tech Publications Ltd, Switzerland·Germany·UK·USA, 1998), p. 901.

3. C. E. Weitzel : III-Nitrides and Related Materials, eds. G. Pensil, H. Morkoç, B. Monemar and E. Janzén (Trans Tech Publications Ltd, Switzerland·Germany·UK·USA, 1998), p. 907.

4. V. V. Afanasev, M. Bassler, G. Pensl and M. Schulz : Silicon Carbide, eds. W. J. Choyke, H. Matsunami and G. Pensl (Akademie Verlag, Berlin, 1997) Vol. 2, p. 321.

5. V. V. Afanasev, A. Stesmans and C. I. Harris : III-Nitrides and Related Materials, eds. G. Pensil, H. Morkoç, B. Monemar and E. Janzén (Trans Tech Publications Ltd, Switzerland· Germany·UK·USA, 1998), p. 857.

6. K. Fukuda, K. Nagai, T. Sekigawa, S. Yoshida, K. Arai and M. Yoshikawa : Extended Abstract of the 1998 International Conference on Solid State Device and Materials (the Japan Society of Applied Physics, Tokyo, 1998), p100.

7. K. Fukuda, K. Nagai, T. Sekigawa, S. Yoshida, K. Arai and M. Yoshikawa : to be published in Jpn. J. Appl. Phys.

8. Y. Kamigaki and Y. Itoh : J. Appl. Phys.,**48**, 2891 (1977).

9. A. Gölz, G. Horstmann, E. Stein von Kamienski and H. Kurz : Silicon Carbide and Related Materials, eds. S. Nakashima, H. Matsunami, S. Yoshida and H. Harima (IOP Publishing Ltd, London, 1995),p.634.

HIGH TEMPERATURE STABLE WSi$_2$-CONTACTS ON p-6H-SILICON CARBIDE

Frank ERLER [*], Henry ROMANUS [*], Jörg K.N. LINDNER [**], Lothar SPIESS [*]
[*] Technical University of Ilmenau, Institute of Applied Materials Science, Germany
[**] University of Augsburg, Institute of Physics, Germany

ABSTRACT

Amorphous tungsten-silicon layers were deposited by DC co-sputtering and subsequently annealed in an argon atmosphere up to 1325 K to form tetragonal crystalline WSi$_2$. Al-implanted p-6H-SiC exhibits a small depletion area forming an ohmic contact with low specific contact resistance. A modified Circular Transmission Line Model (CTLM), introduced by Marlow & Das [1] and Reeves [2], was used to characterize the electrical properties of the prepared contacts in the range between 300 K and 650 K. Deviations between calculated field-emission contact resistances and measured contact resistances (ρ_C=2·10^{-2} Ωcm^2, T=650 K) could be explained by TEM-cross section investigations. These deviations are caused by inhomogeneous contact interfaces originating from technological difficulties during contact preparation.

INTRODUCTION

Silicon Carbide as a wide band gap semiconductor is suitable for high temperature devices with different applications. For this purpose it is necessary to obtain low-resistivity ohmic contacts on active semiconductor areas which do not degrade at high temperatures. Therefore, the contact material must have a high melting point and should not allow chemical reactions with SiC which may influence the electrical properties of the contact area.

Due to theoretical calculations of a p-depletion contact on 6H SiC [3], WSi$_2$ is favourable as metallization material because of the high work function of 4.62 to 4.70 eV [4, 5]. The high acceptor activation energy (> 0.24 eV) causes an ionization of only approximately 1% of available acceptors at room temperature [6]. Dopant incorporation via diffusion is not feasible in SiC. However, ion implantation is a possibility to increase the acceptor concentration at the surface. This causes a narrower depletion zone which allows for thermionic-field or field emission. Fig. 1 demonstrates the influence of the hole concentration on the specific contact resistance [7, 8], using Eq. (1).

$$\rho_C = \frac{k}{qA^*T}\exp\left\{\frac{4\pi\sqrt{m^*\varepsilon_S}}{qh}\left(\frac{\Phi_B}{\sqrt{N_A^-}}\right)\tanh\left[\frac{qh}{4\pi\sqrt{m^*\varepsilon_S}}\left(\frac{\sqrt{N_A^-}}{kT}\right)\right]\right\} \tag{1}$$

Thus an ohmic contact with a low specific contact resistance should be feasible.

Further research work [9] focused on the preparation and patterning of tungsten silicide metallization and on measurement of contact resistivity, applying the Circular Transmission Line Model [1, 2].

Mat. Res. Soc. Symp. Proc. Vol. 572 © 1999 Materials Research Society

EXPERIMENT

In order to increase the acceptor concentration near the surface, an Al-implantation was performed using different doses between $5 \cdot 10^{14}$ and $6 \cdot 10^{15}$ cm^{-2} at room temperature and an implantation energy of 50 keV. A theoretical calculation of the implantation profiles (TRIM) is shown in Fig. 2. The recrystallization of SiC was achieved by annealing at 1925 K in argon atmosphere for 20 min.

To obtain high temperature stable ohmic contacts consisting of WSi$_2$, tungsten and silicon were DC co-sputtered in a stoichiometric ratio Si:W of 2.1:1, resulting in an amorphous layer of 400 nm thickness. During the formation of crystalline WSi$_2$ at 1325 K in an argon atmosphere for 20 min a theoretical shrinkage to 300 nm is expected to appear [10].

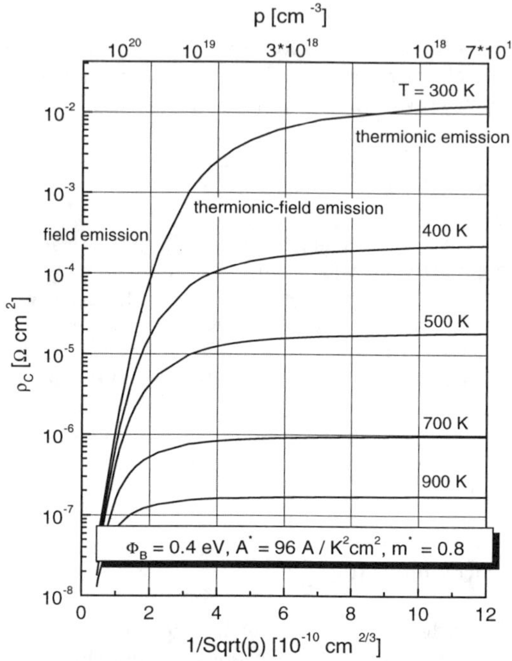

Fig. 1   Calculated specific contact resistance versus hole concentration

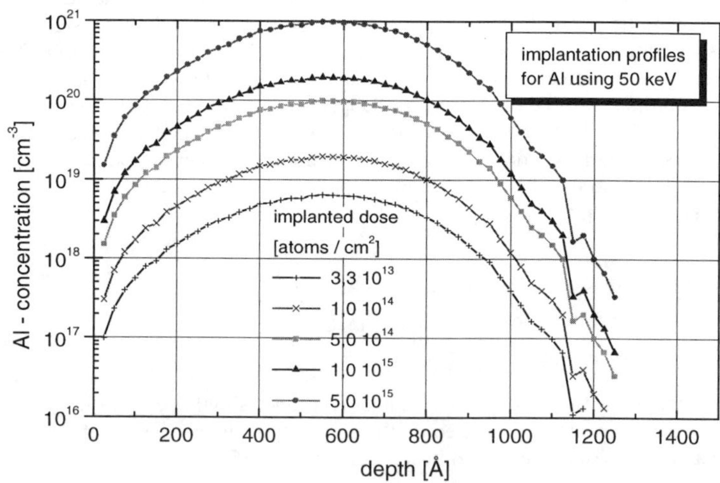

Fig. 2   Implantation profiles calculated using TRIM

The formation of a crystalline WSi_2-layer is demonstrated by x-ray diffraction using different angles of incidence and Bragg-Brentano-geometry (Fig. 3).

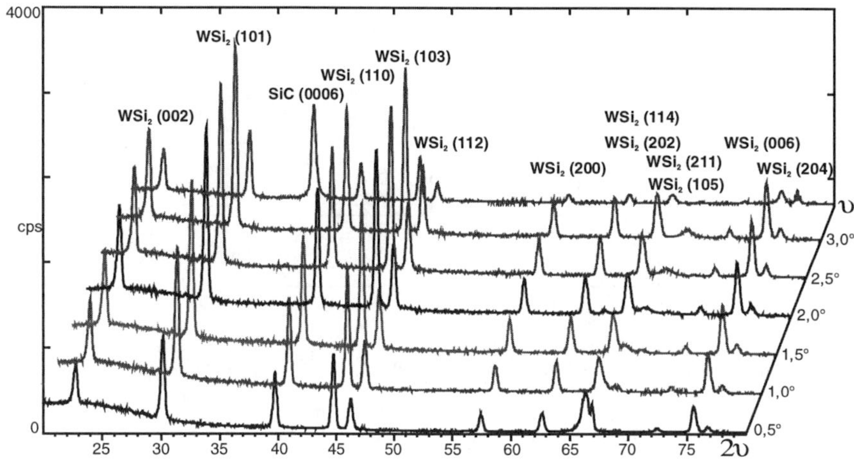

Fig. 3   X-ray diffraction patterns of annealed WSi_2

Measurement structures were patterned by a plasma enhanced chemical etching process using a photo resist mask. The used gas fluxes were 20 sccm Cl_2 and 20 sccm SF_6, the chamber pressure amounted to $3 \cdot 10^{-2}$ mbar and the RF-power was 50 W. Etching efficiency was controlled by optical microscopy.

An example of contact resistance measurement structures is shown in Fig. 4. A modified Circular Transmission Line Model according to Marlow & Das [1] and Reeves [2] was applied for obtaining specific contact resistances at different temperatures up to 650 K.

RESULTS

Fig. 4   Contact resistance measurement structures according to Marlow & Das (a) and Reeves (b)

The summarized results for the measured specific electrical resistivity and the temperature coefficient of the WSi_2-layer are shown in Fig. 5. The obtained resistivity of approximately 47 µΩcm is slightly larger than the value of 40 µΩcm for WSi_2-layers given in literature [4, 5]. This deviation indicates complete transformation into crystalline WSi_2 with a small amount of pores.

Atomic Force Microscopy was used to determine the thickness of the annealed WSi_2-metallization layer. It amounts to approximately 350 nm (Fig. 6), thus it is thicker than the expected value of 300 nm. One possible explanation for this is the formation of pores in the metallization layer during the annealing process.

Specific contact resistances of different samples were calculated from measured current-voltage-plots. One example of such current-voltage-plots for different temperatures is shown in Fig. 7. The equation systems, developed using CTLM and regression analysis, were solved by a mathmatical software package. Table I shows some obtained values of specific contact resistances for different temperatures, upper epi-layer concentrations and implantation doses.

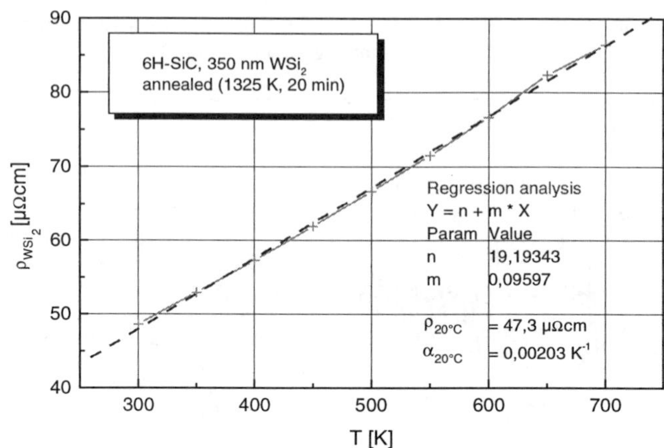

Fig. 5   Specific electric resistance and temperature coefficient of WSi$_2$-metallization

Table I   Measured specific contact resistivity of implanted and WSi$_2$-coated p-6H SiC
(N_{Al} upper epi-layer concentration, $N_{D\,imp}$ implanted dose)

p-6H-SiC	$\rho_{C\,300K}$ [Ωcm^2] no WSi$_2$-annealing	$\rho_{C\,300K}$ [Ωcm^2] annealed	$\rho_{C\,650K}$ [Ωcm^2] annealed
$N_{Al} = 3\cdot10^{18}$ cm^{-3}, $N_{D\,imp} = 5\cdot10^{14}$ cm^{-2}	Schottky-contact	Schottky-contact	Schottky-contact
$N_{Al} = 7\cdot10^{16}$ cm^{-3}, $N_{D\,imp} = 1\cdot10^{15}$ cm^{-2}	Schottky-contact	Schottky-contact	0.2 - 0.8
$N_{Al} = 3\cdot10^{18}$ cm^{-3}, $N_{D\,imp} = 6\cdot10^{15}$ cm^{-2}	Schottky-contact	0.25 - 1.7	$(2 \dots 5)\cdot10^{-2}$

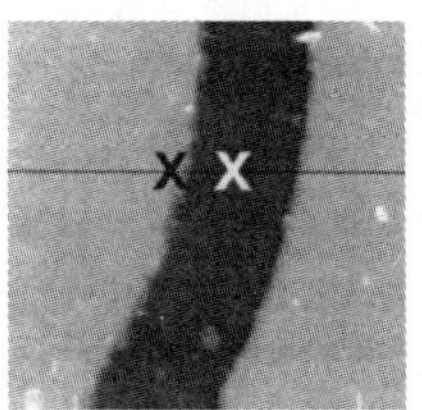
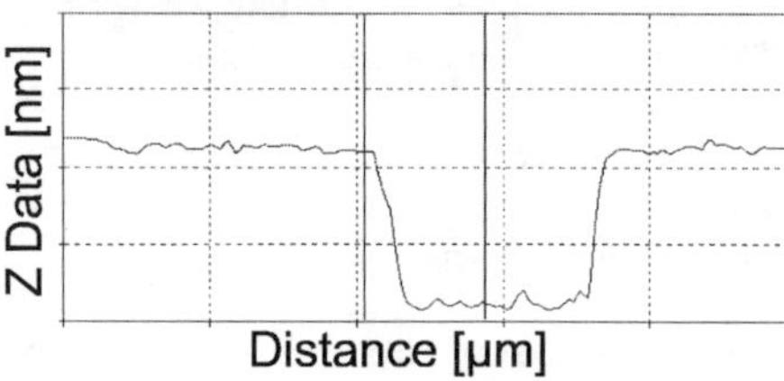

Fig. 6   Atomic Force Microscope image of an annealed and patterned ring structure; depth-measurement

At higher temperature more of the implanted acceptors are ionized. Therefore, a lower ohmic contact resistivity was obtained.

A high-resolution cross-sectional TEM (XTEM) micrograph of metallized 6H-SiC (implantation: $N_{D\,imp} = 6\cdot10^{15}$ cm^{-2}, 50 keV) is shown in Fig. 8. The grainy structure of the metallization layer (denoted as M) is clearly visible, the grain size being in the range of a few hundred nanometers. The metallic layer contains various pores, which are characterized by bright areas in Fig. 8 and which are most likely the result of the thermal treatment within the metallization process rather than of the sputtering technique used in specimen preparation. The interface (small arrow) between the metal layer and the SiC substrate shows a wavy morphology with trenches reaching up to 100 nm down into the substrate. These trenches are partially filled with metal grains.

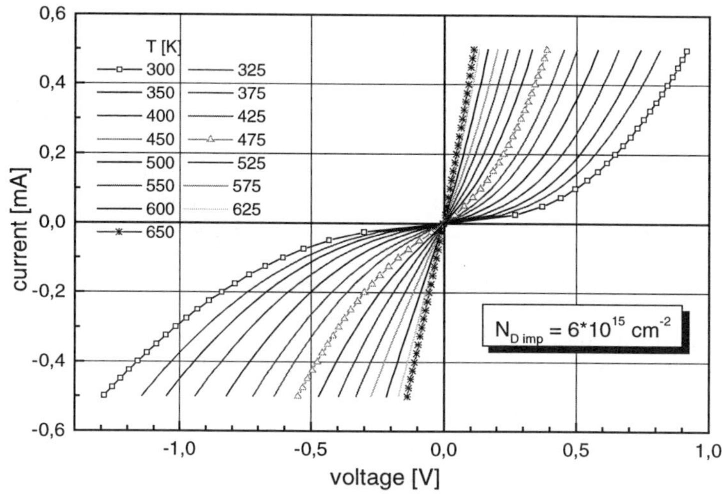

Fig. 7   Measured current-voltage-plot of a 3-ring contact structure, Reeves pattern, temperature as parameter

The SiC substrate is subdivided into four regions, the undamaged 6H substrate followed by a 6H-layer with small defects causing grainy contrasts in Fig. 8. The 6H layers are covered with epitaxially grown 3C-SiC $((111)_{3C} \parallel (0001)_{6H})$, the interface being located approximately 130 nm beneath the metal/ semiconductor interface. The 3C-SiC part is subdivided into a defect rich, approximately 80 nm thick lower zone showing stripy contrasts (planar defects) in Fig. 8 and a 50 nm thick nearly defect free upper zone.

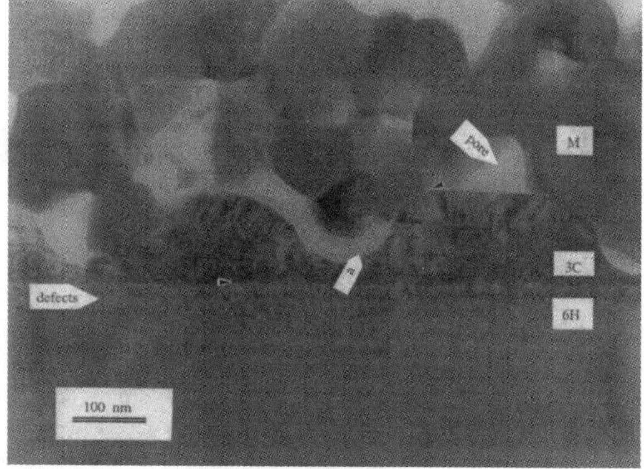

Fig. 8   TEM image of the interface region WSi$_2$-metallization layer / p-6H Silicon Carbide (shows different defects, e.g. pores, 3C SiC)

The results indicate that the metal semiconductor contact mainly consists of a metal/3C-SiC/6H-SiC contact, even though due to the trenches formed there may be locally also some small metal/6H-SiC contact regions.

CONCLUSIONS

Doping of 6H-SiC by implantation of 50 keV aluminium ions using a dose of $6 \cdot 10^{15}$ cm^{-2} probably leads to the formation of an amorphous surface layer. This amorphous layer may recrystallize as a twinned 3C-SiC layer during the subsequent 20 min anneal at 1925 K. In addition it has to be clarified whether the trenches observed are the result of the annealing procedure, the subsequent HF and plasma cleaning or the reaction of deposited W or Si with the SiC surface during metallization. Since the defect layers in SiC are flat, any implantation related reasons can be ruled out.

Due to different difficulties during the implementation of implantation and metallization process it was not possible to obtain contact resistances less then 10^{-3} Ωcm^2. However, the high implanted samples lead to an ohmic contact particularly at higher temperatures. This indicates a field emission conducting process which is demonstrated in the left part of Fig. 1. Therefore, WSi$_2$ should be an available contact material for high temperature applications.

The aim for further research work has to be, firstly, the prevention of solid-state transformation to 3C-SiC during the implantation annealing process. Secondly, to avoid the growth of pores during the annealing for the formation of the tungsten disilicide layer.

ACKNOWLEDGMENTS

The authors gratefully acknowledge the advice and support of Olaf Nennewitz, Jörg Pezoldt, Jakob Kriz and Wolfgang Skorupa.
Parts of this work were supported by the BMBF under contract no. 01 BM 303/8 (1993-1996).

REFERENCES

1.   G.S. Marlow, M. Das, Solid-State Electronics **25**, 91-94 (1982).
2.   G.K. Reeves, Solid State Electronics **23**, 487-490 (1980).
3.   L. Spiess, O. Nennewitz, H. Weishart, J. Lindner, W. Skorupa, H. Romanus, F. Erler, J. Pezoldt, Diamond and Related Materials **6**, 1414-1419 (1997).
4.   S.P. Muraka, Metallization-*Theory and practice for VLSI and ULSI*, (Butterworth-Heinemann, Reed Publishing Inc., USA, 1993), pp. 41-64.
5.   H.F. Hadamovsky, *Werkstoffe der Halbleitertechnik*, 1st ed. (Deutscher Verlag für Grundstoffindustrie, Leipzig, DDR, 1985), pp. 262-291. [German]
6.   L. Spiess, O. Nennwitz, J. Pezoldt, Inst. Phys. Conf. Ser. **142**, 585-588 (1996).
7.   Bergmann, Schäfer, Freyhardt, *Festkörper - Lehrbuch der Experimentalphysik*, Band 6 (Walter de Gruyter & Co., Berlin, Germany, 1992), pp. 531-539. [German]
8.   S.S. Cotten, G.S. Gildenblat, Metal Semiconductor contacts and devices, in *VLSI Electronics*, edited by N.G. Einspruch (Academic Press Inc., London, 1986)
9.   J. Kriz, K. Gottfried, C. Kaufmann, T. Gessner, Diamond and Related Materials **7**, pp. 77-80 (1998).
10. A. Fabricius, O.Nennewitz, L. Spiess, V. Cimalla, J. Pezoldt, Mater. Res. Soc. Symp. Proc. **402**, Boston, 1995, pp. 625-630

For further information, please feel free to contact:
Frank Erler, TU Ilmenau, PF 100565, 98684 Ilmenau, Germany
phone: +49 36 77 / 69 31 11, fax: +49 36 77 / 69 31 04
email: erler@e-technik.tu-ilmenau.de, www: http://phase.e-technik.tu-ilmenau.de

STRUCTURAL AND ELECTRICAL PROPERTIES OF BERYLLIUM IMPLANTED SILICON CARBIDE

T. HENKEL*, Y. TANAKA*, N. KOBAYASHI*, H. TANOUE*,
M. GONG**, X.D. CHEN**, S. FUNG** and C.D. BELING**
*   Electrotechnical Laboratory, Tsukuba, Ibaraki 305-8568, Japan, henkel@etl.go.jp
**   Physics Department, The University of Hong Kong, Hong Kong, China

ABSTRACT

Structural and electrical properties of beryllium implanted silicon carbide have been investigated by secondary ion mass spectrometry, Rutherford backscattering as well as deep level transient spectroscopy, resistivity and Hall measurements. Strong redistributions of the beryllium profiles have been found after a short post-implantation anneal cycle at temperatures between 1500 °C and 1700 °C. In particular, diffusion towards the surface has been observed which caused severe depletion of beryllium in the surface region. The crystalline state of the implanted material is well recovered already after annealing at 1450 °C. However, four deep levels induced by the implantation process have been detected by deep level transient spectroscopy.

INTRODUCTION

Silicon carbide (SiC) is a promising material for high temperature, high frequency, and high power device applications. Group III elements are commonly used as acceptor dopants for this semiconductor. Investigations on p-type SiC doped with these elements, however, revealed poor electrical characteristics due to high acceptor ionization energies and low hole mobilities [1-5]. Therefore, alternative dopants with a higher electrical activation are highly desirable.

Due to the low mass and high solubility in the SiC lattice [6], beryllium (Be) can be used for the production of thick p-type layers applying ion implantation. However, there have been very few reports on Be doped SiC [7-12]. Be is known to be an electrically active impurity, i.e., a doubly charged acceptor in SiC [10]. Two acceptor levels at 0.42 and 0.6 eV, respectively, were determined by Hall measurements [8]. Further, one deep level at 0.38 eV was obtained by I-V measurements on p-n junctions produced by Be implantation [12]. However, the precision of these results is questionable because that data analysis is based on simplified model assumtions. Despite the fact that Be has been successfully applied in the fabrication of diodes [11,12], much is still unknown about the structure of this dopant in the SiC lattice.

In this paper, we report on structural and electrical properties of Be implanted SiC. Samples were characterized by secondary ion mass spectrometry (SIMS), Rutherford backscattering spectrometry / channeling (RBS/C), deep level transient spectroscopy (DLTS), resistivity and Hall measurements.

EXPERIMENT

Epitaxial layers ([0001] orientation, n-type, off-axis, thickness: 10 μm, carrier concentration: 1×10^{16} cm^{-3}) grown on 6H-SiC substrates as purchased from Cree Research [13] were used as starting material. ^{9}Be^{+} was implanted applying energies between 50 and 590 keV (see Table I) in order to obtain a box-shaped profile. Samples were maintained at room temperature (RT), and

Mat. Res. Soc. Symp. Proc. Vol. 572 © 1999 Materials Research Society

were tilted 7° with respect to the ion beam to minimize channeling effects during implantation. Ion range and nuclear energy distributions were obtained by Monte Carlo (MC) simulations using the TRIM code (SRIM-98, full cascade) [14]. A mean displacement energy of 25 eV for both Si and C atoms was applied. The concentration profile and the nuclear energy density distribution are shown in Fig. 1. The critical energy density for the amorphization of SiC crystal at RT (2×10^{21} keV/cm^3 [15]) is indicated by the dashed line.

Ion energy (keV)	Ion fluence (10^{14} cm^{-2})
50	0.61
75	0.62
100	0.63
130	0.84
170	0.74
210	0.85
260	0.80
320	1.00
400	1.03
490	0.94
590	1.52

Table I. Schedule used for Be implantation.

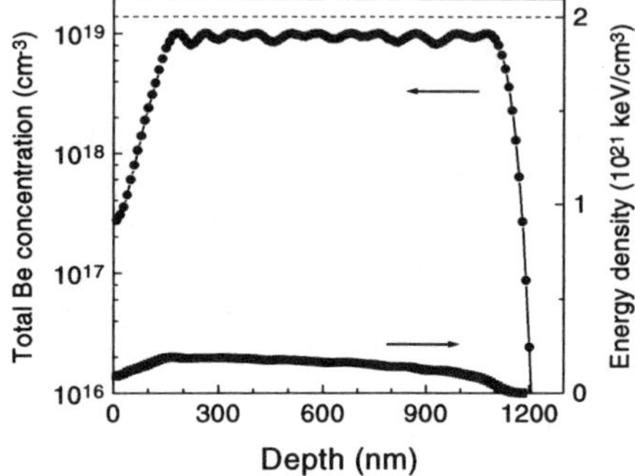

Fig. 1. Total nuclear energy distribution and total dopant concentration for Be implanted SiC as calculated using TRIM

To repair the crystal damage and activate the implanted dopant, samples were annealed in flowing argon (Ar) gas (1-2 atm) at temperatures between 1450 °C and 1700 °C for 1 min using a rapid thermal annealing (RTA) system. The temperature rise and fall rates were about 50 °C/s and 20 °C/s. Each sample was covered with another SiC crystal to protect the sample surface from Si dissociation during annealing. Before the electrical measurements, a planar layer of about 0.6 μm was removed from the top of the epilayer by applying a combination of ion implantation and wet chemical etching [16,17]. Ohmic contacts were prepared by metal deposition (titanium on the top, nickel on the backside) using an e-beam evaporation system and a post-deposition RTA cycle at 1100 °C for 5 min in flowing Ar gas.

To evaluate the thermal stability of the Be profile, SIMS measurements were conducted using a CAMECA ims 4f instrument with a 14 keV NO$_2^-$ primary ion beam. The depth conversion was performed by measuring the total crater depth with a surface profiler and assuming a constant sputtering rate during depth profiling. The concentration calibration was performed using a standard sample fabricated by a single energy ^{9}Be$^+$ implantation. A sensitivity factor was then derived by comparing the area under the SIMS profile obtained from this sample with that one of a profile obtained by a MC simulation.

Resistivity and Hall measurements employing the van der Pauw geometry were performed at RT. To study deep level defects, DLTS measurements were conducted using an equipment described elsewhere [18]. Ionization energies and capture cross-sections were evaluated from the temperature dependence of the emission rates. Finally, damage distributions were determined by 3 MeV ^{4}He$^+$ RBS/C along the [0001] axis using a scattering angle of 150°. The depth scale given in the RBS spectra below was calculated using the mean energy approximation [19] and the density of the crystalline material (9.66×10^{22} at/cm^3 [15]).

RESULTS

SIMS measurements were performed on both the as-implanted and the post-implantation annealed samples to investigate the Be distribution before and after the high temperature treatment. As can be seen in Fig. 2, the near surface tail and the plateau concentration of the as-implanted atom distribution are well reproduced by the MC simulations. However, the slope of the tail towards the substrate is lower compared with the theoretical profile. Channeling effects can be responsible for the observed discrepancy which are not considered in SRIM-98. Since vacancy-type defects far beyond the nuclear energy deposition profile were detected by Positron annihilation spectroscopy [20], defect-enhanced diffusion could be another reason.

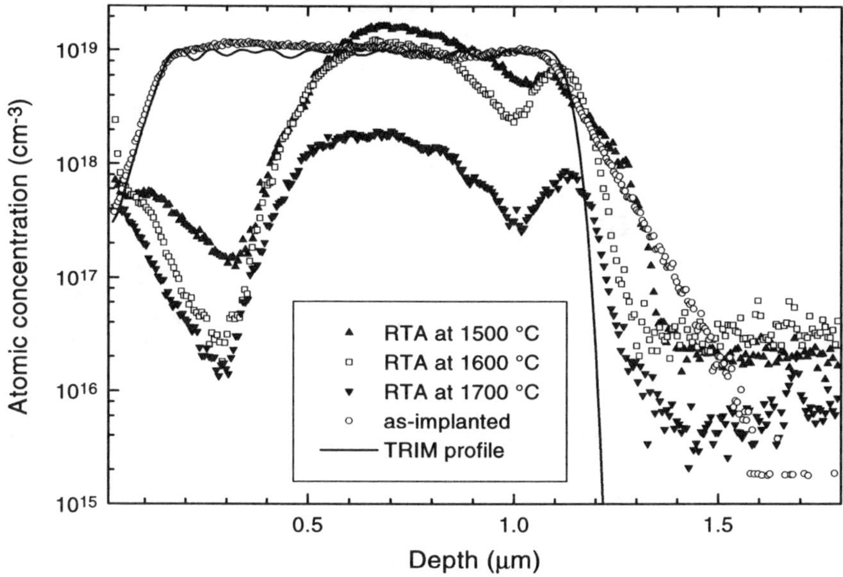

Fig. 2.   SIMS ^{9}Be depth profiles in SiC before and after annealing at the temperatures indicated. A simulated depth profile as obtained using TRIM is shown for comparison.

Strong redistributions were found in the annealed samples. The dopant profile is already thermally unstable after a short RTA cycle at 1500 °C. In particular, the heat treatment caused severe depletion of Be in the surface region up to two orders of magnitude below the as-implanted atom concentration. The integral of the depth profile, which is a measure for the implanted ion fluence, results in a fluence around 30 % lower compared with the as-implanted profile. This indicates out-diffusion of the dopant which increases with increasing temperature, i.e., 43 % and 92 % Be loss were found at 1600 °C and 1700 °C, respectively. Implantation induced defects may govern the diffusion process. However, because of the small atomic size of Be, migration via an interstitial mechanism is also anticipated.

Additionally, in-diffusion into the bulk of the epilayer, although less pronounced, was also observed. The higher the anneal temperature the stronger is the redistribution in the tail region.

Moreover, after RTA at 1500 °C a peak in the Be profile can be seen at a depth of about 1.1 μm which became more pronounced at higher anneal temperatures. It is assumed that Be atoms were trapped by thermally stable defects formed during the anneal process in this (end-of-range) region.

To remove the Be depleted surface layer, all samples (except the as-implanted specimen) to be examined in the following by RBS and electrical measurements were etched as described above and processed at temperatures ≤1600 °C.

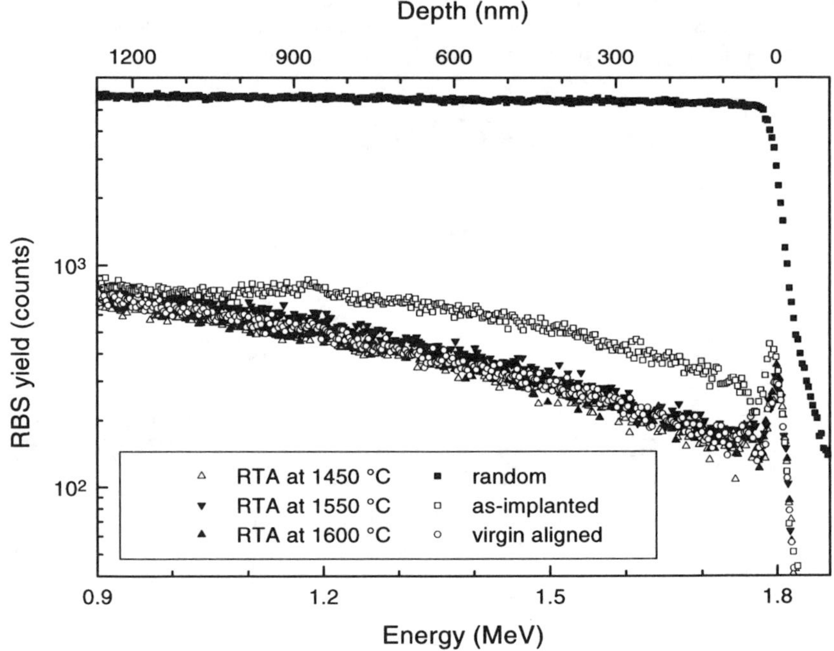

Fig. 3.   RBS/C spectra (Si sublattice) of 3 MeV ^{4}He$^+$ backscattered from Be implanted SiC before and after annealing at the temperatures indicated

To characterize the as-implanted state and to evaluate the recovery of the crystal lattice, RBS/C measurements were performed (see Fig. 3). A random and an aligned spectrum from virgin unimplanted material are also shown for comparison. The scattering yield in the aligned spectrum obtained from the as-implanted sample is far below the yield in the random spectrum. This means that the crystal was not amorphized during implantation, in agreement with the nuclear energy deposition calculated (see Fig. 1). This is a necessary condition for complete annealing of damage induced by ion implantation since defect-free recrystallization of amorphous SiC is not possible [15]. Further, the yield in the as-implanted sample approaches the yield in the virgin material at about 1.2 μm, i.e., the damaged region extends up to this depth which is consistent with the calculated energy deposition profile. The peak observed at an energy of 1.18 MeV is due to an oxide layer on the SiC surface which is confirmed by SIMS.

As can be further seen in Fig. 3, the scattering yield in the aligned spectra from the annealed samples almost coincides with the yield in the aligned spectrum from the virgin sample at all depths indicating a good lattice quality within the sensitivity limit of RBS. Thus, it can be assumed that the crystalline state is well recovered after a RTA cycle at temperatures $\geq$1450 °C.

Resistivity and Hall measurements were performed at RT to obtain electrical properties of the doped SiC layers. For comparison, a virgin unimplanted n-type sample was etched and provided with contacts as described above. Although a weak p-type conduction was detected in the implanted epilayers (free hole concentrations in the 10^{16} cm^{-3} range), well reproducible Hall voltages could not be obtained. Obviously, the Hall measurements were strongly affected by the substrate due to insufficient isolation of the p-type layer. However, resistivities were found to be about 0.5 Ωcm, i.e., one magnitude lower compared to the virgin sample. Within the limits of these measurements, a dependence on the post-implantation anneal temperature in the range from 1450 to 1600 °C was not detected.

Finally, a typical DLTS spectrum as obtained after RTA at 1600 °C is shown in Fig. 4. Four peaks labeled BE$_1$, BE$_2$, BE$_3$, and BE$_4$, respectively, were observed in the temperature range from -150 to +100 °C. Since no DLTS signals corresponding to deep levels were observed in virgin unimplanted samples, all the deep levels were therefore introduced by Be implantation. Ionization energies and capture cross-sections of these levels are given in Table II. For the calculation of the capture cross-sections, these levels were assumed to be electron traps.

Since the free hole concentration of the p-type layer produced by Be implantation was found to be of the same magnitude compared to the free electron concentration of the epilayer as stated above, the width of the p-type depletion layer is assumed to be comparable to that of the n-type one. Thus, the DLTS signals observed may arise from either electron traps at the n-side or hole traps at the p-side of the p-n junction. The question, whether these levels are donor- or acceptor-like, cannot be answered from the present results but possibly by DLTS measurements on Be implanted p-type material provided with Schottky contacts on top of the samples. Further investigations are necessary in order to understand the origin of the deep levels and the structure of the corresponding defects.

Deep level	Ionization energy (eV)	Capture cross-section (cm^2)
BE$_1$	0.34	5×10^{-13}
BE$_2$	0.46	5×10^{-14}
BE$_3$	0.52	5×10^{-14}
BE$_4$	0.66	4×10^{-16}

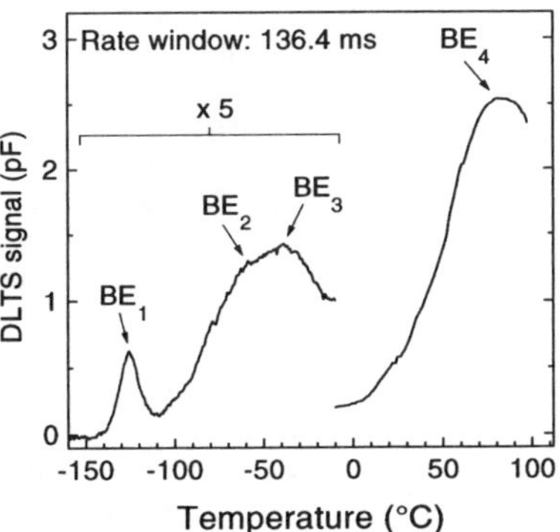

Table II. Ionization energies and capture cross-sections of the deep levels as determined from DLTS data

Fig. 4. DLTS spectrum recorded on Be implanted n-type SiC after annealing at 1600 °C

CONCLUSIONS

The doping behaviour of Be implanted into n-type 6H-SiC epitaxial layers was investigated by SIMS, RBS/C, DLTS, resistivity and Hall measurements. Strong redistributions of the as-implanted Be profiles were found after RTA at temperatures ≥ 1500 °C. In particular, Be diffused towards the surface on a higher level than into the epilayer. As a consequence, severe Be depletion in the surface region occured. Moreover, it was shown by RBS/C that the crystalline state is well recovered after a short RTA cycle at 1450 °C. However, four deep levels labeled BE_1, BE_2, BE_3, and BE_4 were observed by DLTS which were generated by Be implantation. The energy position of these levels as well as the nature of the corresponding defects is still an open question.

ACKNOWLEDGMENTS

We thank Y. Ishida for the help during the Hall measurements. One of the authors (T.H.) gratefully acknowledges the support of this work by Science and Technology Agency of Japan.

REFERENCES

1. T. Troffer, M. Schadt, T. Frank, H. Itoh, G. Pensl, J. Heindl, H.P. Strunk, M. Maier, phys. stat. sol. (a) **162**, 277 (1997).
2. M.V. Rao, J.A. Gardner, P.H. Chi, O.W. Holland, G. Kelner, J. Kretchmer and M. Ghezzo, J. Appl. Phys. **81**(10), 6635 (1997).
3. T. Henkel, Y. Tanaka, N. Kobayashi, I. Koutzarov, H. Okumura, S. Yoshida and T. Ohshima, Mat. Res. Soc. Symp. Proc. **512**, 163 (1998).
4. M. Gong, S. Fung, C.D. Beling, G. Brauer, H. Wirth and W. Skorupa, J. Appl. Phys. **85**(1) 105 (1998).
5. T. Troffer, G. Pensl, A. Schöner, A. Henry, C. Hallin, O. Kordina, E. Janzen, Mater. Sci. Forum **264-268**, 557 (1998).
6. G.L. Harris, *Properties of Silicon Carbide*, (INSPEC, London, 1995), pp.153.
7. A.A. Kalnin, Y.M. Tairov and D.A. Yaskov, Sov. Phys. - Solid State **8**(3), 755 (1966).
8. Y.P. Maslakovets, E.N. Mokhov, Y.A. Vodakov and G.A. Lomakina, Sov. Phys. - Solid State **10**(3), 634 (1968).
9. O.J. Marsh and H.L. Dunlap, Rad. Eff. **6**, 301 (1970).
10. P.G. Baranov, Mater. Sci. Forum **264-268**, 581 (1998).
11. N. Ramungul, Y. Zheng, R. Patel, V. Khemka and T.P. Chow, Mater. Sci. Forum **264-268**, 1049 (1998).
12. N. Ramungul, V. Khemka, Y. Zheng, R. Patel and T.P. Chow, IEEE Trans. Electron Devices **46**(3), 465 (1999).
13. Cree Research, Inc., 4600 Silicon Drive, Durham, NC 27703
14. J.F. Ziegler, J.P. Biersack, and U. Littmark, *The Stopping and Range of Ions in Solids*, (Pergamon, New York, 1985), pp. 1.
15. V. Heera, W. Skorupa, Mat. Res. Soc. Symp. Proc. **438**, 241 (1997).
16. J.A. Edmond, J.W. Palmour and R.F. Davis, J. Electrochem. Soc. **133**(3), 650 (1986).
17. D. Alok and B.J. Baliga, J. Electron. Mater. **24**(4), 311 (1995).
18. C.V. Reddy, S. Fung, and C. D. Beling, Rev. Sci. Instrum. **67**(1), 257 (1996).
19. W. K. Chu, J. W. Mayer, and M. A. Nicolet, *Backscattering Spectrometry*, (Academic, New York, 1978), pp. 64.
20. H. Wirth, W. Anwand, G. Brauer, M. Voelskow, D. Panknin, W. Skorupa and P.G. Coleman, Mater. Sci. Forum **264-268**, 729 (1998).

ELEVATED TEMPERATURE SILICON CARBIDE CHEMICAL SENSORS

M.A.GEORGE*, M. A. AYOUB*,
D. ILA** and D. J. LARKIN***
*Department of Chemistry, University of Alabama in Huntsville
**Center for Irradiation of Materials, Alabama A&M University
***NASA Lewis Research Center

ABSTRACT

In this study, the (I-V) properties of the sensors were measured as a function of hydrogen, propylene and methane exposure at temperatures up to 400° C and sensor responses were observed for each gas. The response to hydrogen and propylene had a rapid increase and leveling off of the current followed by the subsequent decrease to the baseline when the gas was switched off. However, exposure to methane resulted in a rapid spike in the current followed by a gradual increase with continued exposure. X-ray photoelectron (XPS) studies of methane exposed SiC sensors revealed that this behavior is attributed to the oxidation of methane at the Pd surface.

INTRODUCTION

The Study of SiC has focused on methods to grow high quality SiC for a wide range of applications including high temperature, high power devices [1] as well as optoelectronic devices [2]. Recently, it has also been shown that SiC can be employed as both an oxygen and a hydrogen sensor that operates in a temperature regime considerably higher than conventional sensors such as tin oxide (SnO_2) or silicon. Because of its outstanding thermal stability, silicon carbide can be employed as a hydrogen and hydrocarbon sensor that can potentially operate at temperatures up to 1000 °C [3-7]. Potential uses of elevated temperature SiC sensors include automotive applications, process gas monitoring, aeronautics, and aerospace applications.

The deposition of a catalytic metal such as Palladium (Pd) onto silicon carbide (SiC) results in Schottky diode behavior. The adsorbing gas changes the space charge region under the metal clusters which in turn affect the conductivity of the crystal. This change in conductivity is measured and can be correlated to surface concentrations and to the levels of the sampled gas in the ambient. The high sensitivity for hydrogen containing combustible gases is enhanced by the presence of catalytic metals. Hydrogen containing species dissociatively adsorb to the metal and hydrogen atoms migrate to the Pd/SiC interface where they affect the current-voltage (I-V) properties of the SiC [7].

EXPERIMENTAL

For this study of the response of silicon carbide sensors to various gases, 5 x 7 mm samples of epitaxial silicon carbide films deposited on bulk silicon carbide substrates were examined. Both silicon face and carbon face samples were produced, however the preliminary studies were performed on silicon faced silicon carbide films only. The sensors were prepared by depositing 100 nm thick aluminum films to the backside of the sample, while palladium films were deposited to

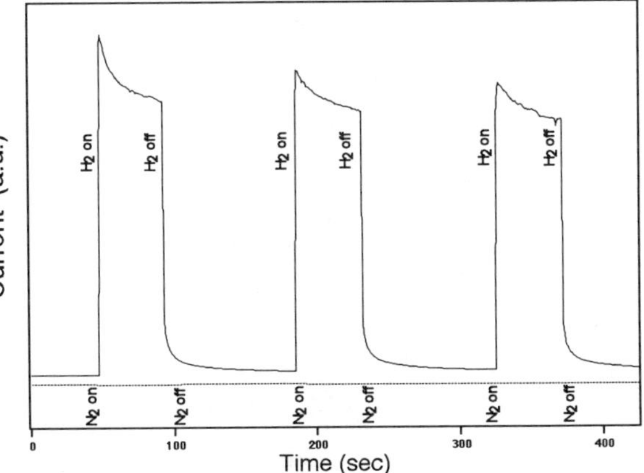

Figure 1 Current response at 0.7 V to H_2 and N_2 exposure.

Mat. Res. Soc. Symp. Proc. Vol. 572 © 1999 Materials Research Society

the epitaxial silicon carbide film side. The Pd deposits ranged from 0.1 mm in diameter to 1.5 mm. Up to four palladium deposits were applied to one silicon carbide sample. The silicon carbide sensors were then mounted in a Flatpax sample holder and contacts were attached using conductive adhesive, colloidal silver.

In order to establish a baseline for the studies of silicon carbide chemical sensors, the current-voltage response of the sensors was examined upon exposure to three specific gases, hydrogen, propylene and methane. Experiments involved exposing the sensors to the respective gases in both air and nitrogen carrier gases. The sensors were kept in ambient room air during all of the tests. This enabled the establishment of preliminary baseline responses to the gases under ambient conditions, the conditions that may be closer to the actual conditions of operational sensors. The responses to the gases were carried out at room temperature, 100, 200, 300 and 400° C. Responses at room temperature were not significant and therefore will not be discussed in this paper. The response to methane was measured at 200, 300 and 400° C. Both current-voltage and current-time (at constant voltage) measurements were obtained using a Keithley 2400 Source-meter interfaced to a Pentium computer. All of the experiments are presented as current-time plots where the sensor has been exposed to the various gases of interest. All of these measurements were performed with a forward bias at 0.7 V.

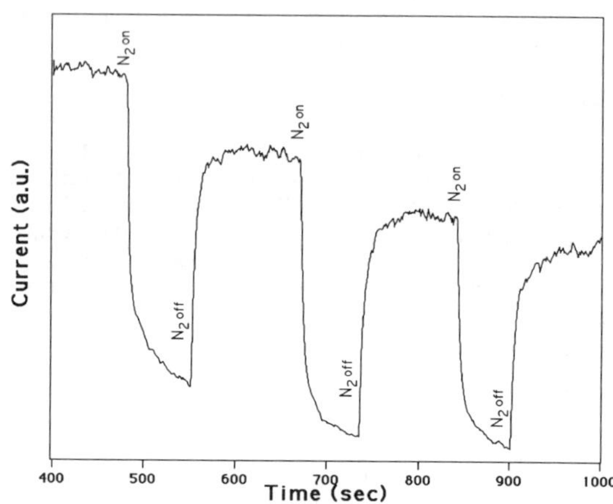

Figure 2 Current response to nitrogen at 200 °C.

RESULTS

The sensor response to hydrogen and nitrogen at 200 ° C is shown in Figure 1. The measurements were obtained using 1000 ppm H$_2$ in N$_2$. The figure contains two plots, one for pure N$_2$ (99.999%) and one for the H$_2$-N$_2$ mix. As can be seen in the plots, there is a good response to the H$_2$ upon turning the H$_2$-N$_2$ mix on, with a rapid decrease in current upon turning the gas off. Three on-off cycles are shown in the plot demonstrating reproducibility for the response. The same process was performed for the pure N$_2$. As can be seen in the plot, the behavior of the pure N$_2$ did not follow that of the H$_2$-N$_2$ mix. There was a change in current observed for the pure N$_2$. However, there was a slight decrease in current rather then the increase that occurs with the H$_2$-N$_2$ mix. We attribute this change to thermal fluctuations of the sensor when the N$_2$ gas is switched on and off. Both the N$_2$ and the H$_2$-N$_2$ mix were done at same flow rates, therefore the magnitude of the decrease in current

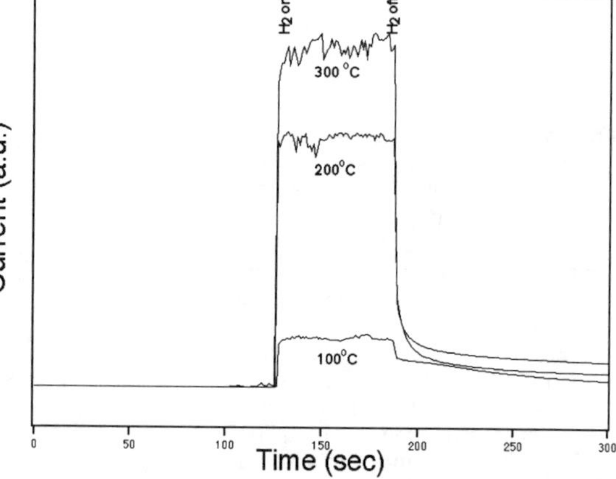

Figure 3 Current response to H$_2$ at 100 ° C, 200 ° C and 300 ° C

124

from the thermal fluctuations due to gas flow are insignificant when compared to the increase in the current from the Schottky sensor response to H_2. In Figure 2, the sensor response to pure N_2 is plotted and scaled by itself. The plot shows the decrease in current as the N_2 is turned on and the subsequent increase in current when the gas is turned off. The magnitude of the decrease is around 10 microamps, while the magnitude of the H_2 response is an increase of 30 miliamps. In order to establish the behavior of the sensor H_2 response to temperature, I-t plots at 100, 200 and 300 ° C are shown in Figure 3. As can be seen in the figure, the sensor shows an increase in sensitivity with increasing temperature consistent with published results by Hunter *et-al* [3, 4].

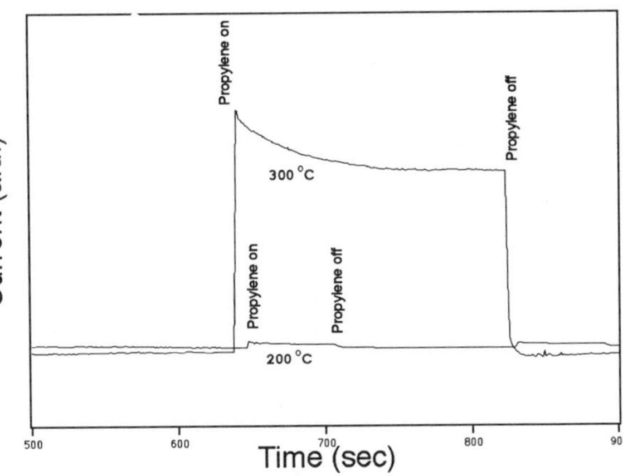

Figure 4 Current response to propylene at 200 °C and 300 °C.

The current-time response to propylene was measured at 100, 200 and 300° C. Consistent with the results of Chen et-al [5], the measurement at 100 ° C was too low for a sensor response. The plots at 200 °C and 300 °C are shown in Figure 4, and clearly exhibit a temperature dependent increase in current with exposure to propylene. This may be attributed to several factors including sensor preparation, contacts for the leads and the method of gas dosing. Our experiments were carried out in ambient air, while the Chen work involved systematic purging of the sensor-sampling chamber in air and nitrogen prior to exposure.

In order to understand the nature of the sensor response to methane, both current time measurements and x-ray photoelectron spectroscopy experiments were performed on silicon carbide sensors. As reported in the work by Chen, the response to methane did not follow the behavior of H_2 or the other two hydrocarbons examined in their work. They attributed the decreased response to poisoning of the Pd surface by the methane. Pronounced responses to propylene and ethylene were observed in their study for temperatures ranging from 200-400 ° C, however the methane had an initial increase in current followed by a gradual decrease down to the original baseline.

The responses to methane at 300 °C and 400 °C observed in our study are shown in Figure 5. The response at 300 ° C reveals that, upon turning the methane on, there is an initial fast response followed by a decrease as with the Chen study. However in our

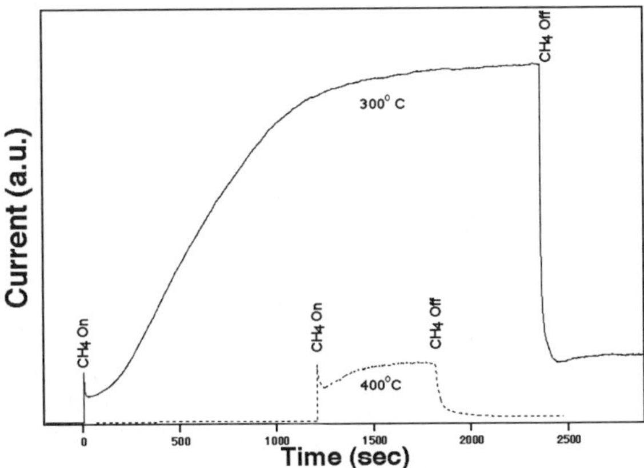

Figure 5 Current response to methane at 300° C and 400° C.

work, the decrease is immediately followed by a gradual increase in current over a period of a few thousand seconds. In Figure 5, it is seen that the current continues to increase until at around 2400 seconds where the methane is turned off. Upon switching the methane flow off, there is a rapid decrease as the current returns to near the original baseline.

Unlike the response to H_2 where there is a rapid increase up to saturation, followed by generally a constant current, with the methane, there is the initial response, decrease, and then a gradual increase in current. Saturation does not occur immediately, but depending on the sensor temperature may take several minutes to occur. This process appears to involve surface reactions on the Pd metal, rather than the silicon carbide, and is dependent on the temperature and the concentration of methane vapor that is near the surface that subsequently adsorbs and/or reacts on the Pd

3. XPS Study of CH_4 adsorption on Pd/silicon carbide

In our experiments, exposure to methane was carried out in ambient air, and as such, CO, CO_2 and O_2 were present to participate in surface reactions at the Pd surface. Several recent studies have addressed the catalytic behavior of Pd upon methane adsorption and discuss the reaction processes that occur in the presence of these gases [8-11].

In an effort to examine the surface chemistry for CH_4 interactions with Pd, XPS was performed on Pd/silicon carbide sensors prior to and after exposure to CH_4. Both the unexposed and exposed samples were heated at 300°C for 30 min. In the case of the exposed sample, methane was introduced during the heating process. Figure 6 shows the XPS spectra of these samples in the C1s region. The C1s spectra are from the silicon carbide surface of a CH_4 exposed sample, shown in Figure 6(a), from the Pd surface on a sample that was not exposed to CH_4, Figure 6(b) and from the Pd surface for a CH_4 exposed sample, Figure 6(c). The peak shapes on silicon carbide were the same for both CH_4 exposed and unexposed samples.

The C1s XPS spectra in Figure 6 obtained from the silicon carbide surface contains two peaks: silicon carbide at around 282 eV and CH_x at around 284 eV. The CH_x peaks observed on the silicon carbide surface is not attributed to the CH_4 sample gas; rather it is due to pre-adsorbed carbon from exposure to ambient conditions. The Pd surface for both exposed and unexposed samples contains at least three components: CH_x at around 284 eV, CO at around 288 eV and another unassigned peak at around 292 eV.

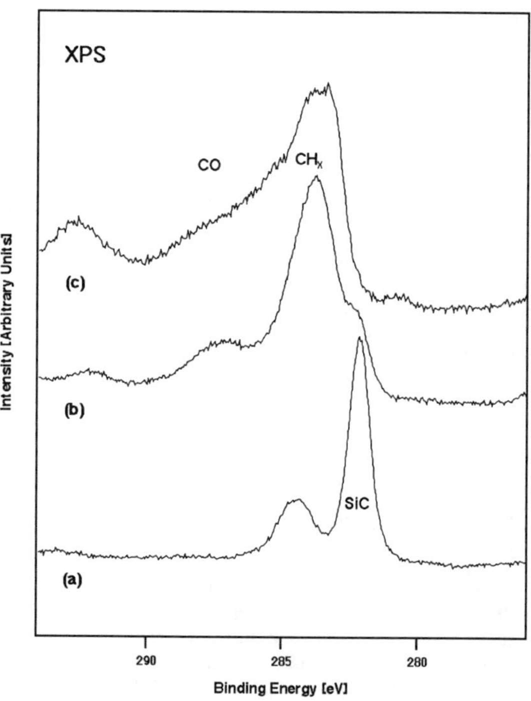

Figure 6 XPS spectra in the C1s region for the (a) unexposed surface, (b) unexposed Pd/SiC Surface and (c) methane exposed Pd/SiC surface.

The peak at 292 eV may correspond to CO_2, however, it has been reported that CO_2 dissociatively adsorbs on Pd in a temperature range of 200-400 °C [8]. Therefore, unless CO_2 adsorption on silicon carbide supported Pd results in a stable adsorbed CO_2 species, we will not assign the 292 eV peak to CO_2. The spectra show the exposed sample with an obvious increase in CO as well as the unassigned component at 292 eV. The heating process at 300° C in ambient air therefore may have resulted in the dissociative adsorption of CO_2 and the resulting CO peaks observed on the unexposed sample.

This process requires the presence of hydrogen and follows the reaction:

$$CO_2 + H_2 ====> CO + H_2O$$

This implies that some adsorbed hydrogen was present on the unexposed sample for this reaction to occur. The source of this hydrogen may be from the hydrocarbon layer that exists on the sample from exposure to the ambient.

For the CH_4 exposed sample, the occurrence of adsorbed CH_x as a result of CH_4 exposure is not evident in the XPS spectra since there was no obvious increase in the CH_x XPS peaks after exposure to CH_4. There is however, a slight broadening of the C1s peak in the area that CH_x occurs and therefore some CH_x is present. It has been reported that the adsorption of CH_4 on Pd when pre-adsorbed CO is present can result in a stabilized adsorbed CH_x species [9]. The CH_x peaks observed in Figure 6 might also be due to the pre-adsorbed hydrocarbon species observed on the unexposed sample that originates from the adsorption of carbon from the ambient. The determination of the adsorption of a stable CH_x species on Pd from methane exposure requires more study in a controlled system where sample cleaning and gas dosing can be performed *in-situ.*

It is evident in the XPS spectra that there is an increase in the occurrence of the oxidized form of carbon, CO. The interaction of CH_4 at the Pd surface is most likely then resulting in the oxidation of the CH_4 leaving the adsorbed carbon monoxide. This reaction has been reported to occur as [8]:

$$CO_2 + CH_4 =====> 2CO + 2H_2$$

This reaction results in the format ion of hydrogen which should of course, in the absence of competing processes, in turn dissociate and affect the surface potential at the Pd-SiC interface leading to the increase in current observed in the I-V behavior. However, the experiment as performed occurs in ambient air with an abundance of CO_2 present to induce the oxidation of the CH_4. The response to hydrogen therefore may be inhibited due to the increase in surface CO. This inhibited response to the dissociated H_2 from the CH_4 molecule is seen as the spike followed by the immediate decrease in current in the I-t plot in Figure 5.

It is apparent that both the carbon and oxygen XPS peaks should give some indication of the overall process. The XPS spectra for the same samples involved in the carbon XPS study were examined in the oxygen region of the spectrum. These are shown in Figure 7. Again, (a) corresponds to the silicon carbide region of the exposed sample while (b) is on the Pd of the unexposed sample and (c) on the Pd of the CH_4 exposed sample.

The O1s peak occurs at around 533 eV while Pd has a peak representing Pd3d very near the

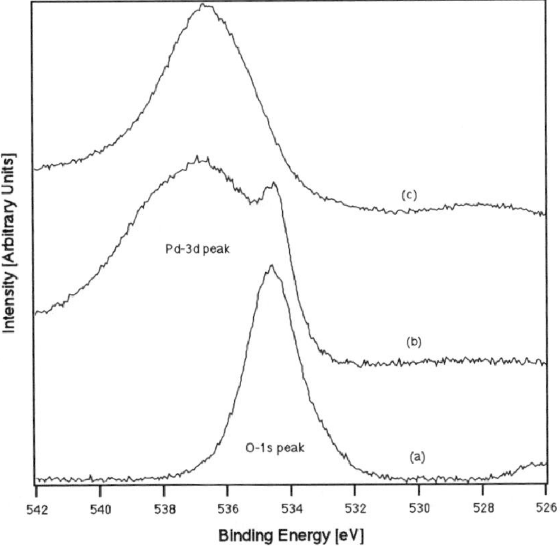

Figure 7 XPS spectra in the O1s region for (a) the methane exposed SiC, (b) the unexposed Pd/SiC surface and (c) the exposed Pd/SiC surface.

oxygen peak at 537 eV. As would be expected, the spectrum on the silicon carbide shows no Pd3d peak, while both of the samples on the Pd film has the Pd3d peak. There is a pronounced difference in the peak shapes between the two Pd sample spectra. The Unexposed sample shows both the O1s peak and the Pd3d peak, while the CH_4 exposed sample has no O1s peak or it has "vanished" below the background produced by the Pd3d peak. This indicates that exposure to methane at 300 °C somehow involves the pre-adsorbed oxygen. In the absence of CH_4, heating the sample to 300 °C had no apparent affect on the O1s peak as seen with the unexposed sample.

CONCLUSIONS

The surface reaction involving the adsorbed oxygen species is perhaps the mechanism that leads to the gradual increase in current observed in methane-exposed SiC sensors over time. This reaction is certainly temperature and time dependent and may follow kinetics that yields the gradual increase in current. The oxidation of methane involving surface oxygen would result in the increase in binding energy for O1s on the exposed films if the surface oxygen were due to CO. This would bury the O1s peak in the Pd3d background. It is also reasonable to suggest that the net effect of decreasing the surface oxygen from Pd-oxide may lead to the gradual increase in current as seen in the I-t curve of Figure 5. Clearly, this is a bit speculative at this point and a more detailed systematic study would provide the information needed to better understand the mechanism for the observed behavior of the I-t curves for methane on the silicon carbide sensors.

ACKNOWLEDGEMENTS

This work was supported by a grant from the NASA Lewis Research Center NAG3-2020.

REFERENCES

1. K. Shenai, R.S. Scott, B.J. Baliga, IEEE trans. Electron. Dev. 36(1989)1811.
2. J.R. Waldrop and R.W. Grant, Appl. Phys. Lett., 62(1993)2685.
3. G.W. Hunter, P.G. Neudeck, G.D. Jefferson, G.C. Madzsar, C.C. Liu and Q.H. Wu, *Report E-7773 NASA, (1993)*.
4. G.W. Hunter, P.G. Neudeck, C.C. Liu and Q.H. Wu, *Conference on advanced Earth-To-Orbit Propulsion Technology, (1994)*.
5. L. –Y Chen, G.W. Hunter , P.G. Neudeck, D. Knight, C.C. Liu and Q.H. Wu, *Proceedings 190ᵗʰ Meeting of Electrochemical society (1996)*.
6. L. A. Spetz, A. Baranzahi, P. Tobias, I. Lundstrom, High temperature sensors based on metal-insulator-silicon carbide devices, Physica Status Solidi (A) Applied Research, v 162, 1 (1997) 493-511.
7. G. Muller, G. Krotz, E. Niemann, SiC for sensors and high-temperature electronics, Sensors and Actuators, A, 43, 1-3 (1994) 259-268
8. A. Erdöhelyi, J. Cserényi, E. Papp and F. Solymosi, *Applied Catalysis A, 108 (1994) 205-219*.
9. J. –J. Chen and N. Winograd, *Surface Science 314 (1994) 188-200*.
10. A.K. Bhattacharya, J.A. Breach, S. Chand, D. K. Ghorai, A. Hartridge, J. Keary and K.K. Mallick, *Applied Catalysis A: General, 80 (1992)L1-L5*.
11. W. Lisowski, *Surface Science, 312 (1994) 157-166*.

THE EFFECT OF ANNEALING ON ARGON IMPLANTED EDGE TERMINATIONS FOR 4H-SIC SCHOTTKY DIODES

A P KNIGHTS*, D J MORRISON**, N G WRIGHT**, C M JOHNSON**, A G O'NEILL**, S ORTOLLAND** , K P HOMEWOOD*, M A LOURENÇO*, R M GWILLIAM*, AND P G COLEMAN***,
*School of Electronic Engineering, Information Technology and Mathematics, University of Surrey, Guildford GU2 5XH, United Kingdom. a.knights@ee.surrey.ac.uk
**Department of Electrical and Electronic Engineering, University of Newcastle, Newcastle-upon-Tyne, United Kingdom, NE1 7RU.
***School of Physics, University of East Anglia, Norwich, NR4 7TJ, United Kingdom.

ABSTRACT

The edge termination of SiC by the implantation of an inert ion species is used widely to increase the breakdown voltage of high power devices. We report results of the edge termination of Schottky barrier diodes using 30keV Ar^+ ions with particular emphasis on the role of post-implant, relatively low temperature, annealing. The device leakage current measured at 100V is increased from 2.5nA to 7µA by the implantation of 30keV Ar^+ ions at a dose of $1\times10^{15}cm^{-2}$. This is reduced by two orders of magnitude following annealing at 600°C for 60 seconds, while a breakdown voltage in excess of 750V is maintained. The thermal evolution of the defects introduced by the implantation was monitored using positron annihilation spectroscopy (PAS) and deep-level-transient spectroscopy (DLTS). While a concentration of open-volume defects in excess of $1\times10^{19}cm^{-3}$ is measured using PAS in all samples, electrically active trapping sites are observed at concentrations $\sim1\times10^{15}cm^{-3}$ using DLTS. The trap level is well-defined at $E_c\text{-}E_t = 0.9eV$.

INTRODUCTION

SiC Schottky barrier diodes are ideal for power switching applications as they can be operated at higher voltages and higher temperatures than equivalent Si or GaAs devices. The requirements for a high power Schottky barrier diode are a low forward voltage drop, low reverse leakage current and high breakdown voltage. Premature voltage breakdown often occurs because of electric field crowding at the periphery of the device. In order to achieve high breakdown voltages near to the theoretical limits expected for SiC, it is necessary to employ an edge termination.

One technique used to terminate Schottky diodes is via the implantation of inert ions. The main area of the device is protected using a mask and inert ions are implanted into the sample. This process forms an area of high resistivity which allows the potential to spread across the surface of the biased sample. Unfortunately, a significant and undesirable increase in leakage current accompanies the increase in breakdown voltage. For example, devices exhibiting near-ideal breakdown voltages on 6H-SiC have been reported using Ar^+ ion implantation [1], however the reverse leakage of these devices was $5\times10^{-2}A/cm^2$ at <100V.

In this study we report an improvement in reverse leakage current of Ar^+ implanted 4H-SiC Schottky barrier diodes. Following implantation the devices are annealed at temperatures up to 600°C with a resulting decrease in the reverse leakage current of almost two orders of magnitude while a breakdown voltage in excess of 750V is maintained.

Mat. Res. Soc. Symp. Proc. Vol. 572 © 1999 Materials Research Society

In addition, the implantation induced defects have been monitored using positron annihilation spectroscopy (PAS) and deep-level-transient spectroscopy (DLTS). These two complementary techniques provide information on (1) the size and distribution of open-volume defects (such as vacancies and voids), and (2) the position of the defect related mid-gap trap position respectively.

EXPERIMENTAL DETAILS

A ready-diced n-type 4H-SiC epitaxial wafer ($N_{Dsub}= 1x10^{18}cm^{-3}$, $N_{Depi}=1x10^{16}cm^{-3}$) supplied by Cree Inc. was used in this study. Each 5mm by 5mm section was given a thorough solvent clean using acetone and IPA. After a 5 minute soak in 10:90 $HF:H_2O$, 1000Å of Ni was deposited on the highly doped backside and alloyed at 1000°C to form a large area backside ohmic contact.

Four samples were front-patterned with 312µm diameter dots using standard photolithography and were placed in 10:90 $HF:H_2O$ for 5 minutes immediately before the deposition of 1000Å Ni Schottky contacts. The samples were implanted at room temperature with $1x10^{15}cm^{-2}$ Ar^+ ions at 30keV , with the contact pads acting as implant masks for the 30keV ions. Three samples were subsequently annealed at either 400°C, 500°C, or 600°C for a duration of 60 seconds. The remaining sample was left unannealed. The resulting diodes were characterised using forward and reverse I-V measurements.

Four further samples were implanted and annealed under the same conditions, before the fabrication of Schottky contact pads. These samples were first analysed using the entirely non-destructive technique of positron annihilation spectroscopy and subsequently, Ni Schottky contact pads were deposited on top of the irradiated area and the samples were subjected to deep-level-transient spectroscopy analysis.

RESULTS AND DISCUSSION

Current-Voltage Measurements

Figure 1 shows the I-V characteristics for the Schottky diodes. The diodes exhibit classic Schottky behavior with excellent forward conduction. For the reverse-biased, unimplanted sample, a leakage current at 100V of 2.5nA is measured with a subsequent increase to 7µA following edge-termination via Ar^+ implantation. These values are consistent with previous measurements [1]. Upon annealing, the leakage current is seen to decrease significantly with increasing temperature. Following an anneal at 600°C for 1 minute it is reduced to a value of 0.09µA. At the same time, the breakdown voltage for the 600°C device was found to be consistently greater than 750V, or ~50% that of the theoretical, ideal breakdown.

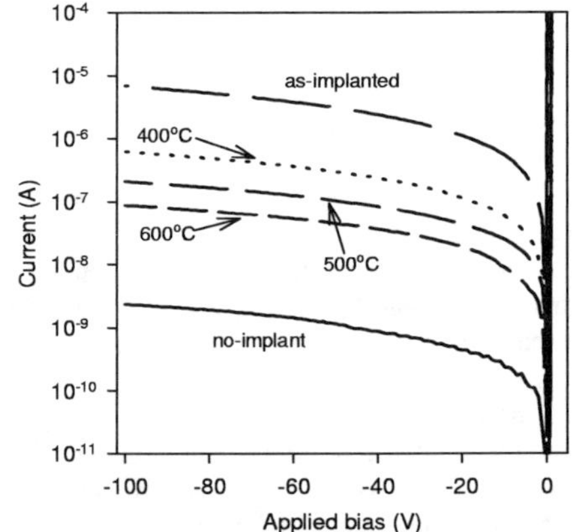

Figure 1 - I-V curves for reverse biased, edge-terminated Schottky diodes. The anneals were performed for 1 minute at temperatures indicated on the plot.

This improvement in leakage current is analogous to the situation found during the formation of high resistivity regions in GaAs via implant isolation. Although high levels of resistivity can be induced in doped GaAs following inert ion implantation, a low temperature anneal is required to obtain maximum resistivity, close to the intrinsic value [2]. The anneal selectively removes shallow levels responsible for hopping conduction leaving only the deep-levels required for efficient isolation.

Positron Annihilation Spectroscopy (PAS)

The application of positron annihilation to the study of ion implantation induced defects has been comprehensively described elsewhere [3]. Monoenergetic positrons were implanted into the samples in the energy range 0-30keV and at each energy the Doppler-broadened annihilation gamma photopeak linewidth was measured and described by the parameter S. Depth resolved information is obtained by varying the energy of the incident positron beam with the mean depth in Angstroms, $z = 130\ E^{1.6}$, where E is the incident positron energy in keV. The implantation profile becomes broader with increasing E; its FWHM is approximately equal to the mean implantation depth.

The size of the samples (5mm x 5mm) required aperturing of the positron beam to 4mm diameter immediately after leaving the source region. The samples were attached to thin tungsten wires to minimise the surface area of the backing material and thereby the probability of detecting radiation from positron annihilation at sites other than in the sample under study. The positron beam was centred on the target at each incident energy by direct observation of the beam and sample shadow on a microchannelplate-phosphor screen-CCD camera assembly at the end of the beam line. The total photopeak count rate was ~ 800 s^{-1} and run times at each incident energy were 2-3000s.

Figure 2 shows the S parameter (representative of open-volume defect size/concentration) versus incident positron energy (depth) for the as-implanted and annealed samples, together with the spectrum for an unimplanted control sample. Information on the size and distribution of the defects is obtained from fitted

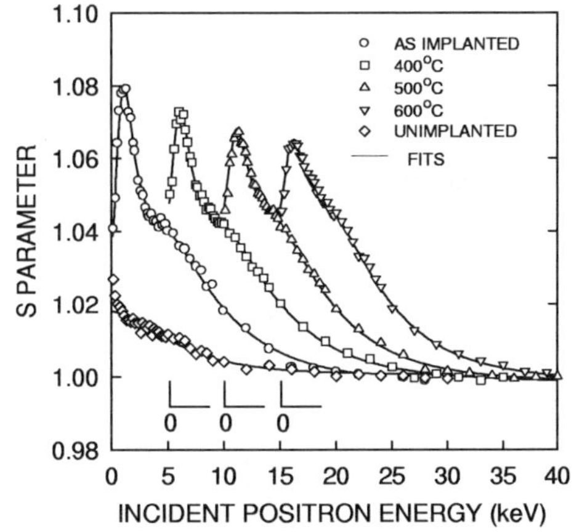

Figure 2 - Positron S-parameter (open-volume defect size/concentration) versus incident positron energy (depth) for implanted and annealed samples.

models to the data shown as solid lines in the figure. All of the implanted samples show two defected regions containing a large concentration of defects ($>1\times10^{19}$cm^{-3}). The first defected layer extends to a depth consistent with the implanted Ar$^+$ ions of ~20nm, with a dominant positron trap, saturated S parameter of >1.085. The second layer extends well beyond the range of the ions to a depth of ~250nm. The saturated S parameter for these defects is ~1.045. This type of deep layer has also been observed in 200keV Ge$^+$ implanted SiC [4] and probably results from a combination of implanted ion channelling and defect diffusion.

In a recent publication Brauer et al [5] presented a plot illustrating the dependence of the S parameter (normalised to its value, S_b, in undefected bulk SiC) on the size of the open-volume defect in SiC in which a positron is trapped at the moment of annihilation. They found that for the Si-C divacancy $S/S_b \sim 1.05$. This value rises to ~ 1.08 for a cluster of four divacancies, and approaches an asymptotic value of ~ 1.15 for large clusters or voids. This would suggest that in the present study the shallow layer comprises, in the main, vacancy clusters or voids, while the deeper layer contains point-type defects such as divacancies.

Upon annealing, the larger sized (void) defects are observed to decrease in size/concentration. However, the smaller, point-type defects are unchanged in either distribution, size or concentration. If the open-volume defects are playing a significant role in the determination of the electrical characteristics of the diodes it is likely that it is related to the break-up of the void defects introduced up to the range of the implanted ions.

<u>Deep-level-transient spectroscopy (DLTS)</u>

Deep-level-transient spectroscopy (DLTS) experiments were performed under dark conditions using a Bio-Rad DL4600 system. Samples were mounted on a stage in a liquid nitrogen cryostat. The temperature was monitored using a platinum resistance thermometer attached directly to the stage, giving an uncertainty of ± 0.5 K on the measured value. All DLTS spectra were taken twice (ramping the temperature up and then down) to account for temperature lag. Arrhenius plots were obtained from the average values of the ramp-up and ramp-down peak positions. Prior to the DLTS measurements capacitance-voltage (CV) and current-voltage (IV) measurements were performed at various temperatures, and standard Schottky junction characteristics were obtained. The IV and CV characteristics of all samples did not show any significant differences.

DLTS spectra were obtained under the same bias and pulse conditions (reverse voltage, V_R = -2.0 V; forward voltage, V_F = 0.0 V; fill pulse, t = 2 ms) to enable direct comparison between the different treated samples. Under these excitation conditions, around 0.5 to 1 μm of bulk SiC layer is sampled. For all samples, a typical DLTS spectrum indicated the presence of a dominant majority carrier trapping centre, associated with a single exponential positive peak present in the temperature range of 300 - 450 K, with activation energy around 900 meV and apparent capture cross section of order of 10^7 s^{-1}K^{-2}.

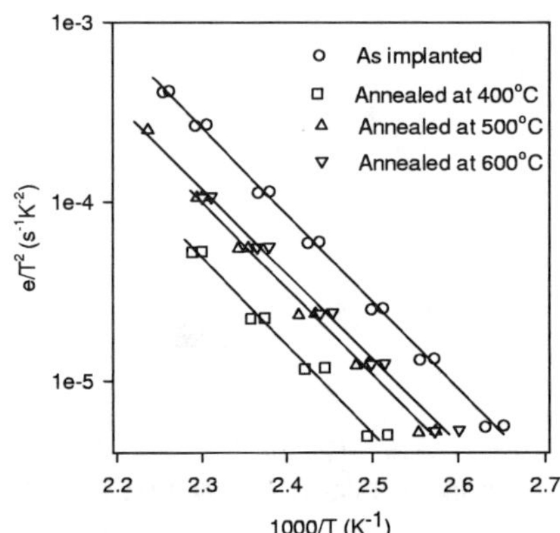

Figure 3 - Arrhenius plots of the irradiation induced deep level obtained from the as-implanted and annealed samples.

The samples in this study compare favourably with those from Alok et al. [6] where DLTS measurements of similarly implanted SiC yielded traps between 0.2 and 0.81eV below the conduction band. A value of 900meV is consistent with efficient edge termination [6].

Figure 3 shows the Arrhenius plots of this deep level obtained from the as-implanted and annealed samples. An estimation of the density of this trap gives a concentration of $\sim 1 \times 10^{15} cm^{-3}$ for the as-implanted sample, reduced to 40% of this value following annealing. The same trap concentration was obtained from all annealed samples and was independent of the annealing temperature, within the experimental errors. Comparison with the high concentration of open-volume defects observed by PAS would suggest that if the carrier traps are vacancy-type, they form a small subset of the open-volume defects present in the sample. The absence of a strong dependence on annealing temperature of any shallow trapping sites makes it difficult to make a direct correlation between the observed reduction in diode leakage current and the optimisation of implant isolation of GaAs. Hence, it is implied that the removal of shallow defects by the relatively low temperature anneal is not the explanation for the reduction in leakage current.

CONCLUSIONS

We have described results of the edge termination of Schottky barrier diodes using 30keV Ar^+ ions with particular emphasis on the role of post-implant, relatively low temperature, annealing. The device leakage current measured at 100V was increased from 2.5nA to 7µA by the implantation of 30keV Ar^+ ions at a dose of $1 \times 10^{15} cm^{-2}$, and reduced by two orders of magnitude following annealing at 600°C for 60 seconds, while a breakdown voltage in excess of 750V was maintained. Positron annihilation spectroscopy showed the concentration of open-volume defects to be in excess of $1 \times 10^{19} cm^{-3}$ in all samples. They are contained in two distinct defect bands. The first is consistent with the implanted ion range and contains voids. The second extends to >10 times the ion range and is dominated by point-type defects. A reduction in void size/concentration with annealing is correlated with the reduction in leakage current. Electrically active trapping sites are observed at concentrations $\sim 1 \times 10^{15} cm^{-3}$ using DLTS. The trap level is well-defined at E_c-E_t = 0.9eV.

ACKNOWLEDGEMENTS

This work is supported as part of the SCEPTRE project under EPSRC grant no. GR/L62320.

REFERENCES

1. D Alok, B J Baliga, P K McLarty, IEEE Electron Device Letters, **15**, 394 (1994).

2. S J Pearton, International Journal of Modern Physics B., **7**, 4687 (1993).

3. P Asoka-Kumar, K G Lynn, and D O Welch, J. Appl. Phys., **76**, 4935 (1994).

4. G Brauer, W Anwand, P G Coleman, A P Knights, F Plazaola, Y Pacaud, W Skorupa, J Stormer and P Willutski, Phys Rev B **54** 3084 (1996)

5. G Brauer, W Anwand, P G Coleman, J Stoermer, F Plazaola, JM Campillo, Y Pacaud and W Skorupa, J. Phys. Condens. Matter **10**, 1147 (1998)

6. D. Alok, B J Baliga, M Kothandaramam, and P K McLarty, *Proceedings of Silicon Carbide and related materials*, Kyoto, Japan, (1995) pp.565-568.

OXIDATION MODELLING FOR SIC

N.G. WRIGHT, C.M. JOHNSON AND A.G. O'NEILL
Dept. Of Electrical and Electronic Engineering, The University of Newcastle upon Tyne, Newcastle UK, NE1 7RU, n.g.wright@ncl.ac.uk

ABSTRACT

A simple mechanistic model of the oxidation of SiC is presented and analysed using Monte-Carlo simulation techniques. The model explains the observed anisotropic oxidation rate of SiC in terms of the effect of weakening/strengthening of Si-C bonds arising from the on-going incorporation of highly electronegative oxygen atoms into the crystal lattice. The extraction of key process metrics (such as oxide thickness, interface roughness and oxide defect density) from the Monte-Carlo simulations is discussed.

INTRODUCTION

High voltage MOS devices are one of the most attractive classes of switch in SiC technology. Such devices offer the possibility of switching up to 10kV loads with low gate drive requirements and low on-state losses. For good control and safety, normally-off devices are generally required in power switching applications and so there has been substantial international research into developing good quality SiO_2 layers on SiC for use in inversion/accumulation mode SiC power MOSFETs. To date attempts at producing such oxide layers have not been particularly successful. Oxides grown on SiC (both the 4H and 6H polytypes) exhibit high fixed charge densities and poor oxide-semiconductor interfaces with significant roughness [1]. Systematic study of oxidation of the 6H and 4H-SiC polytypes by various groups has however produced a wealth of information about oxidation processes in SiC [2]. For example, it is now well established that different crystal faces of SiC oxidise at different rates resulting (for example) in uneven oxide thickness around an etched trench [3, 4]. The lowest/highest oxidation rates are observed on the so-called silicon/carbon faces (the $0001/000\bar{1}$ planes respectively) with a corresponding increase in oxidation rate for planes between the two extremes. Dependency of oxidation rate on crystallographic plane is also observed in silicon where it is often explained by arguments based on the number of silicon-silicon bonds exposed to various crystal faces. As $0001/000\bar{1}$ planes have the same number of surface bonds but widely differing oxidation rates, such an explanation is not sufficient for SiC. This paper presents recent work in the development of a simple oxidation model for SiC proposed by the authors in an earlier paper [5]. The consequences of the model are explored using Monte Carlo based simulation techniques and new results on the oxidation of trench structures presented.

THEORY

Mechanistic Oxidation Model

As full details of the proposed oxidation model have been presented in an earlier paper [5], we present only a brief discussion of the main issues here. The crystal structure and proposed bonding of 4H-SiC is shown schematically in Figure 1 (with the z-axis scale greatly exaggerated for clarity). From the point of view of the proposed oxidation model, the important structural characteristic of all the hexagonal SiC structures is the absence of

Mat. Res. Soc. Symp. Proc. Vol. 572 © 1999 Materials Research Society

inversion symmetry along the z-axis of the chemical bonding between Si and C atoms. To explore the consequences of this on oxidation, consider oxygen molecules interacting with the crystal structure at the Si-face.

The first stage of oxidation is oxygen atoms bonding to the Si atoms outside the crystal. The high electronegativity of the oxygen atoms will make the bond between the surface Si atoms and the back-surface C atoms (labelled α in Figure 1) less ionic and hence lower the bond energy (see Figure 2) [6]. At the second stage of oxidation, the α-bond will be broken by the incoming oxygen resulting in a Si-O-C bond. The high electronegativity of the oxygen atom

now in the α-bond position will increase the ionic nature of C-Si β-bonds (see Figure 2) which will be consequently strengthened. Further oxidation proceeds via attack of these strengthened β-bonds - a process which will be slower than the oxidation of the weakened α-bond. Once the bonds around the carbon atom are fully oxidised, the release of a CO molecule can then occur - with subsequent re-ordering of the Si-O bonding structure. At the Si-face then, there are one weakened and three strengthened bonds resulting in an overall increase in the amount of energy that must be supplied by the oxidation process (i.e. the average activation energy of the process is increased). Subsequent oxidation of other layers can proceed by the same mechanism with CO being released from the structure as the oxidation of the bonds around each C atom is complete.

At the C-face however, such a mechanism results in three weakened and one strengthened bond resulting in a lowering of the activation energy of the oxidation process. At other crystal faces, the proportion of weakened to strengthened bonds will be between these two extremes producing consequent intermediate oxidation rates. Adapting the model for other oxidising ambients (e.g. wet oxidation) is achieved by using chemical bond data to modify the reactions assumed to occur during oxidation

Monte Carlo Approach

The proposed oxidation model has been implemented in the simulator, OXYSIM, which is based on using a Monte Carlo type approach to model the oxidation process at an

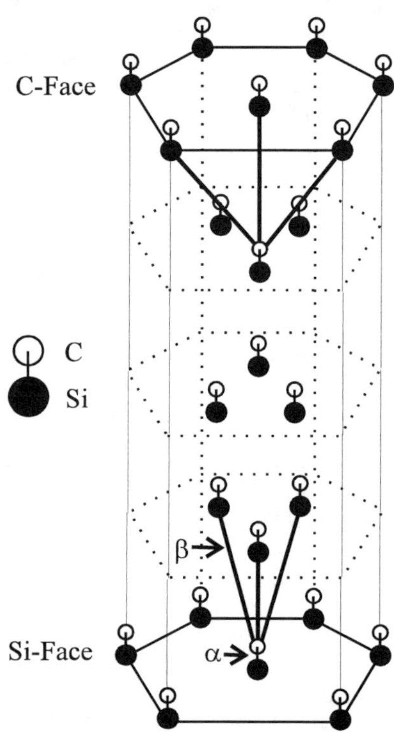

Figure 1: a schematic diagram of the crystal structure of 4H-SiC in which the z-axis (shown vertical in the diagram) is greatly exaggerated. The layer nature of the crystal is obvious with each layer corresponding approximately to one of the three stacking positions (traditionally labelled ABC) of hexagonal layer structures. The 4H-SiC structure shown is given by the stacking sequence ABCB with other polytypes built up from different stacking sequences (e.g. 6H-SiC $\equiv$ ABCACB).

atomic level. OXYSIM works by considering the interaction between the semiconductor crystal lattice and any oxidising species (note: the model can be adapted to all common oxidising reactants. For simplicity of explanation however, the description is given in terms of dry oxidation, i.e. oxygen atoms reacting with the semiconductor). After the structure of the lattice has been initialised by placing atoms in their correct position and setting relevant bond energies, oxygen atoms (in this case) enter the lattice on calculated trajectories. The oxidation temperature of the simulated process can be used to determine the number and trajectories of the incoming oxygen atoms according to the well-known laws of statistical mechanics governing gases. Those oxygen atoms that come within pre-defined distances of lattice atoms are considered as possible candidates for interaction with the lattice atoms to form bonds (and thus disrupt the original lattice). The program then uses a simple Monte Carlo type decision process to decide whether the candidate interaction is allowed [7, 8]. The probability of interaction, P_{int}, is calculated using the formula:

$$P_{int}=\exp[-E_{bond}/kT] \tag{1}$$

where E_{bond} is the energy of the lattice bond, k is the Boltzmann constant and T is the temperature (in Kelvin). The decision to allow or not allow the interaction is then made by comparing this probability to a random number and acting according to the criterion:

$$P_{int} \geq \text{random number} \Rightarrow \text{interaction allowed} \tag{2}$$

$$P_{int} < \text{random number} \Rightarrow \text{interaction not allowed} \tag{3}$$

If the interaction is allowed then the oxygen atom enters the lattice and the bonding and structure of the lattice modified according to pre-set structural rules (as specified in the model described below). The process then continues for a specified time/number of incoming oxygen atoms (similarly if the original interaction is not allowed).

Following the completion of the oxidation process, the resulting structure is analysed to determine a number of simple process metrics. The thickness of the oxide layer at a given point in the lattice is determined by simply calculating the perpendicular distance from the lattice surface to the oxygen atom furthest from the surface. The mean roughness of the oxide/lattice interface is defined to be the standard deviation of the thickness of the oxide at a given point from the mean oxide thickness across the whole lattice. Information about the defect density of the simulated oxide layer can be extracted by examining the bonding of both the original lattice atoms and the incorporated oxygen atoms. The density and nature of mis-bonding can then be used to predict the density of electrically active defects from published models of trap formation [9].

RESULTS

A quantitative test of the proposed model can be made by comparing the effect of the proposed bond weakening with the activation energies for oxidation extracted from experimental data using a linear rate model. Figure 2 shows such a comparison for wet oxidation of SiC in which the surprisingly good agreement between the proposed model and observed data can be clearly seen (despite the obvious simplifications of the model). Such information could be of great importance in the process optimisation of SiC trench devices such as UMOSFETs (in which a gate oxide is grown on the exposed surfaces of a deep trench). Under the high field conditions of a SiC UMOSFET, the high electric field at the trench corner can cause device breakdown in the oxide layer (c.f. breakdown at the device periphery in well designed silicon devices). This can of course be alleviated by growing a thicker oxide but only at the expense of increasing the device threshold voltage (for a given doping level in the SiC). As illustrated by Figure 2, oxide thickness will be far from uniform around an etched trench and so device optimisation can be hindered by the conflicting requirements of requiring a thick oxide at the trench corner (to increase device breakdown voltage) with the desire for a thin oxide on the trench side-wall (for an acceptable threshold voltage). In fact, this requirement is of key importance in deciding whether to fabricate UMOSFETs on the silicon face (0001) or the carbon face ($000\bar{1}$). The carbon face produces a trench oxide will the required thin/thick

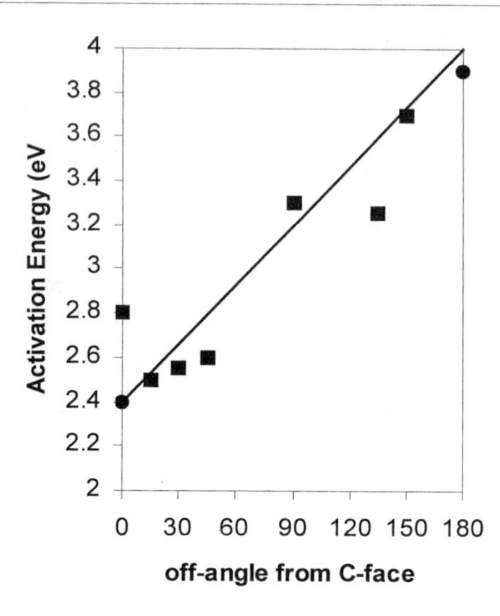

Figure 2: a graph illustrating a comparison between the observed dependency of activation energy (extracted from a linear oxidation rate model) with angle of crystallographic plane (from the C-Face) and that predicted by the proposed model (shown as solid line). Experimental data from [3] ■ symbol and [4] ● symbol.

characteristics on the side/bottom walls of the trench but generally results in a lower quality oxide/SiC interface compared to oxides grown on the silicon face [10]. In order to examine whether a trench oxidation process on the silicon face could be optimised to reduce the differences in thickness between the side and bottom walls of the trench (and thus comer closer to the desirable characteristics of the carbon face trench), we have examined the relative oxidation rates of different crystallographic faces as a function of temperature.

Figure 3 shows data from simulating the relative dry oxidation rates of the silicon and carbon faces between 800K and 1500K (the approximate limits on practical oxidation processes). The relative oxidation rates of the C- and Si-faces are clearly temperature dependent (in agreement with experiment [3, 4]) – offering the potential for process optimisation according to the desired criteria. Such data suggests the oxidation temperature plays an important role in determining the relative thickness of the side-wall oxide compared to the oxide on the trench bottom. The effect of process temperature on the predicted oxide shape in a 1.5μm wide by 1.5μm deep trench is explored in figure 4. For an oxide grown on a trench etched into a silicon face wafer, the anisotropy in the oxide thickness of the side-wall and trench bottom is predicted to fall with temperature (although the mean oxidation rate will of course rise). This suggests that a short oxidation time at high temperature is the optimal process condition and will produce the most even oxidation thickness around the trench in such a device.

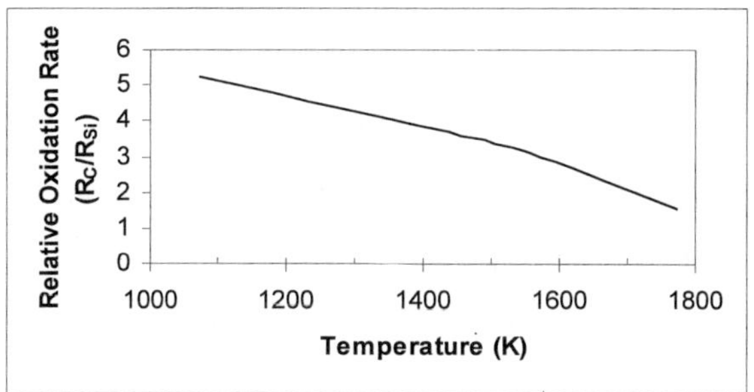

Figure 3: a graph showing the simulated relative oxidation rates of the C- and Si-faces (R_C and R_{Si} respectively) as a function of temperature (K).

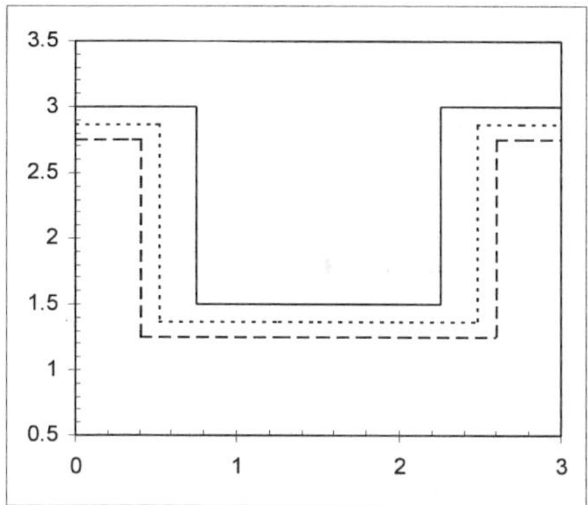

Figure 4: Oxidation of a 1.5µm wide x 1.5µm deep trench etched into the silicon face of 4H-SiC. The solid line represents the original trench shape, the dotted/dashed lines the resulting oxide/SiC interfaces after oxidation at 1200K and 1500K respectively.

CONCLUSION

A mechanistic model for the oxidation of 4H-SiC has been presented and explored using both a simple analytic approach and the OXYSIM simulator. It has been shown that the model reproduces well the anisotropic oxidation rates observed in 4H-SiC and, when used in conjunction with OXYSIM, can predict important oxide quality metrics such as oxide thickness, interface roughness and defect density. Such an approach can thus offer insights into process optimisation and be useful in improving 4H-SiC oxide quality.

ACKNOWLEDGEMENTS

This work was supported by the UK Engineering and Physical Sciences Research Council. The authors would also like to thank the sponsors of the SCEPTRE project for their generous contributions to this work.

REFERENCES

1. J.A.Cooper Jr., Phys. Stat. Sol. **162**, p. 305 (1997)

2. D. Alok, P. K. McLarty and B. J. Baliga, Appl. Phys Lett, **64**, p. 2845 (194)

3. K. Ueno, Phys. Stat. Sol. **162**, p. 290 (1997)

4. A. Rhys, N. Singh and M. Cameron, J. Electrochem. Soc **142**, p. 1318 (1995)

5. N.G. Wright, C. M. Johnson and A.G. O'Neill, Mat. Sci. Eng. B**56** (1999).

6. L. Pauling, "The Nature of the Chemical Bond" 3rd ed., Cornell University Press, New York 1960

7. H.J. Herrman in "The Monte-Carlo Method in Condensed Matter Physics" (e.d. K.Binder), Springer 1995

8 G. Bhanot, Rep Prog Phys **51**, p. 429 (1988)

9. V. F. Afanasev, M. Bassler, G. Pensl and M. Schulz, Phys. Stat. Sol. **162**, p. 321 (1997)

10 S.Onda, R. Kumar and K. Hara, Phys. Stat. Sol. **162**, p. 369 (1997)

ANNEALING EFFECTS OF SCHOTTKY CONTACTS ON THE CHARACTERISTICS OF 4H-SIC SCHOTTKY BARRIER DIODES

S.C. KANG, B.H. KUM, S.J. Do[*], J.H. Je[*], and M.W. SHIN
Department of Ceramic Materials Engineering, Myong Ji University 38-2 Nam-Dong, Yongin-Si, Kyunggi-Do, Korea 449-728 Phone: 82-335-330-6465 Fax: 82-335-33-6457
*Department of Materials Science and Engineering, Pohang Institute of Science and Tehnology P O. BOX 125, POHANG 790-600, Korea. Tel: 82-562-79-2139 Fax: 82-562-79-2399

ABSTRACT

This paper reports on the relationship between the microstructure and the device performance of Pt/4H-SiC schottky barrier diodes (SBDs). The evolution of microstructure in the metal/SiC interfaces annealed at different temperatures was characterized using X-ray scattering techniques. The reverse characteristics of the devices were degraded with annealing temperatures. The maximum breakdown voltages of as-deposited devices and 850 ℃ annealed devices are 1300 V and 626 V, respectively. However, the forward characteristics of the devices were found out to improve with annealing temperatures. X-ray scattering analysis showed that Pt-silicides were formed by annealing performed at or higher than 650 ℃. The formation of silicides was shown to increase the roughness of the Pt/SiC interface. It is believed that the forward characteristics of the SBDs be strongly dependent on the crystallity of silicides formed in the Pt/SiC interface during the annealing process.

INTRODUCTION

SiC has been given significant attention as a potential material for high-frequency, high-power, and high-temperature applications due to its unique electrical and thermal properties. These properties include a high electric field at breakdown (2×10^6 V/cm), a high electron velocity (2×10^7 cm/sec), a large band gap (2. 86 eV for 6H and 3.2 eV for 4H), and a high thermal conductivity (4 W/K cm) [1]. In particular, the extremely high critical electric field of SiC makes it a prime candidate for high-voltage applications, such as high-power rectifiers. Rectifiers utilize SBDs to suppress high-voltage transients induced on the power line during current switching [2]. For a negligible dissipation of power during the switching, the reverse current transient of the SBD must be suppressed, maintaining a high reverse voltage without breakdown. There have been a lot of reports on the design and the fabrication of a SiC SBD to achieve its theoretical breakdown voltage [3]. However, there are only a few reports found on the relationship between

Mat. Res. Soc. Symp. Proc. Vol. 572 © 1999 Materials Research Society

the microstructure and the device performance of 4H-SiC SBDs. In this study, we attempted to establish the relationship between the current-voltage characteristics of Pt/4H-SiC SBDs and the microstructure of the Pt/4H-SiC interface annealed at various temperatures. It was found out that the annealing conditions for the schottky contact have a significant impact on the device performance. The evolution of microstructure in the Pt/4H-SiC interface was characterized using X-ray scattering techniques.

EXPERIMENT

Single crystal 4H-SiC wafer with a nitrogen doped ($N_D \sim 1.2$ x 10^{16}/cm^3) epi-layer with 10 μm thickness was used to fabricate the SBDs. The substrates were cleaned according to the standard chemical cleaning procedure. The processing details for the fabrication of 4H-SiC SBDs are found elsewhere [Baliga] except that Al shadow mask was employed as the shadow mask for the Boron implantation [4]. Pt schottky contact (t=3000 Å) was deposited by the sputtering method through a metal shadow mask in a vacuum ($\sim 1 \times 10^{-6}$ torr). The implant dose and energy were 1.0×10^{15} cm^{-2} and 30 keV, respectively. Ohmic contacts were formed by evaporation of Ni (t=3000 Å) in a vacuum($7 \sim 9 \times 10^{-6}$ torr) for the backside blanket. The ohmic contact was annealed at around 1050 ℃ for 30 min in Ar ambient to remove the implant damage. To investigate the annelaing effects on the device performance, the devices were annealed at various temperature (RT, 500 ℃, 650 ℃, 750 ℃, and 800 ℃) right after the formation of schottky contact. The microstructure of the Pt/SiC interfaces was characterized using X-ray scattering techniques. The I-V characteristics of device were measured using a Sony tektronix 370 programmable curve tracer.

RESULTS AND DISCUSSIONS

The structure of the fabricated device and its typical current-voltage characteristics are shown in Figure 1. The maximum blocking voltage obtained through this study was 1300 V (for samples with as-deposited schottky contact).

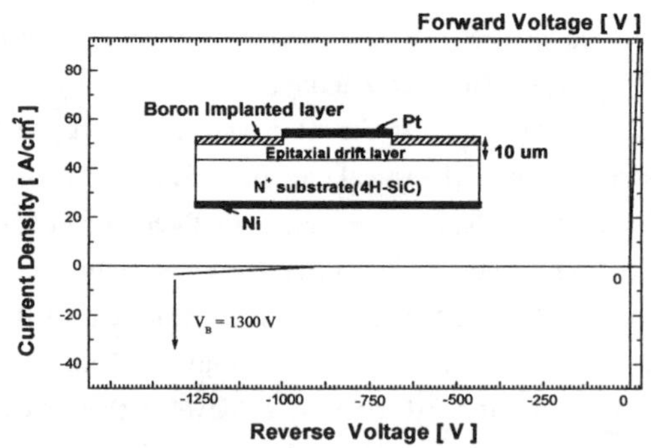

Figure 1. Schematic device structure of 4H-SiC SBD and its typical I-V characteristics showing the maximum breakdown voltage of 1300 V (device with as-deposited schottky contact)

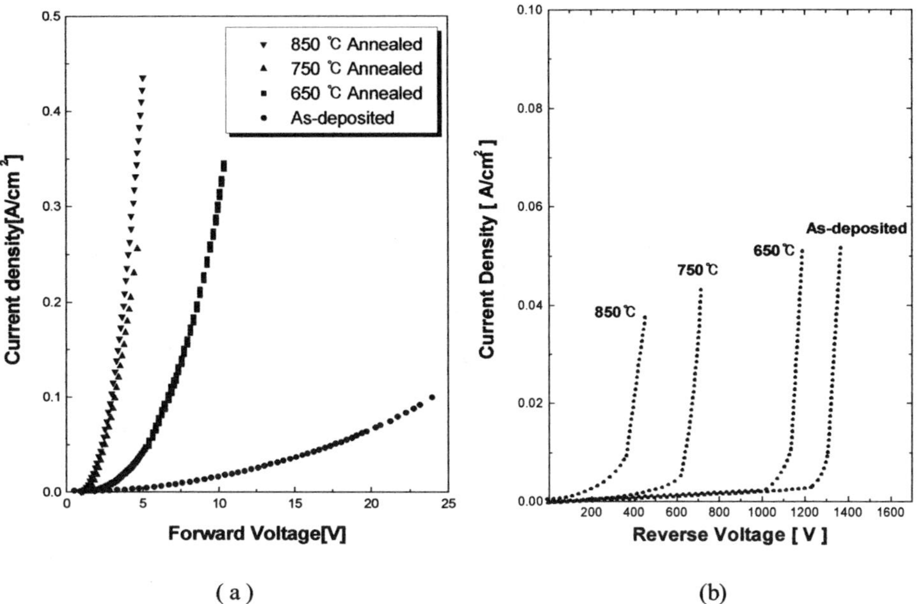

Figure 2. The variation of forward and reverse bias I-V characteristics of Pt/SiC schottky barrier diodes annealed at different temperatures; (a) forward and (b) reverse characteristics

Figures 2 (a) and (b) show the variation of forward and reverse bias I-V characteristics of Pt/SiC schottky barrier diodes annealed at different temperatures. It is shown that the forward current density of diodes is enhanced with the annealing temperature. The current density of devices annealed at 850 ℃ is as high as 420 mA/cm^2 at 5 V. The current density of devices with the as-deposited schottky contact is about 140 times lower than the value for the samples that were annealed at 850 ℃. Apart from the forward characteristics, the breakdown voltage of devices is found out to decrease with the annealing temperature. The ideality factor of the samples annealed were shown to vary in a range 1.2 (850 ℃) to 4.5 (as-deposited). The barrier heights are expected to be higher for samples annealed at lower temperature. The degradation of breakdown voltage of devices annealed at high temperature can be explained by the calculated specific-on resistance. Figure 3 compares the specific-on resistance (R_{on}) of as-deposited samples and of samples annealed at 850 ℃. The distribution of specific-on resistance of as-deposited samples is shown to be about 1 order higher than that of annealed samples. It is known that the breakdown voltage is reversely proportional to R_{on} [5].

The electrical characteristics of devices are examined in view of the microstructure of the interface between schottky contact and 4H-SiC. Figure 4 shows the intensity of reflected X-ray from the surface and interface of Pt/4H-SiC layers annealed at different temperatures.

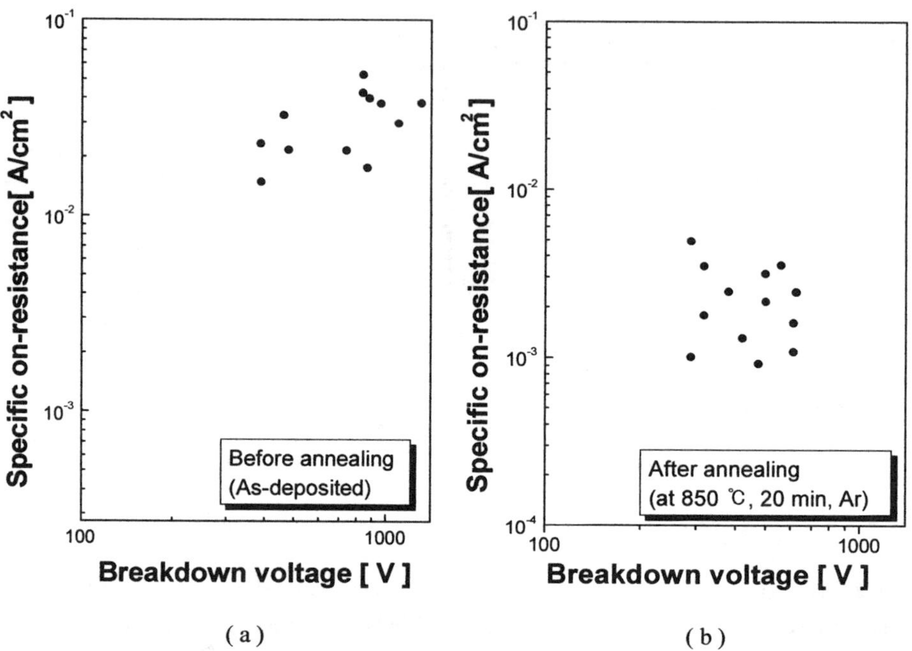

Figure 3. Breakdown dependence of specific-on resistance; (a) as-deposited and (b) annealed at 850 ℃.

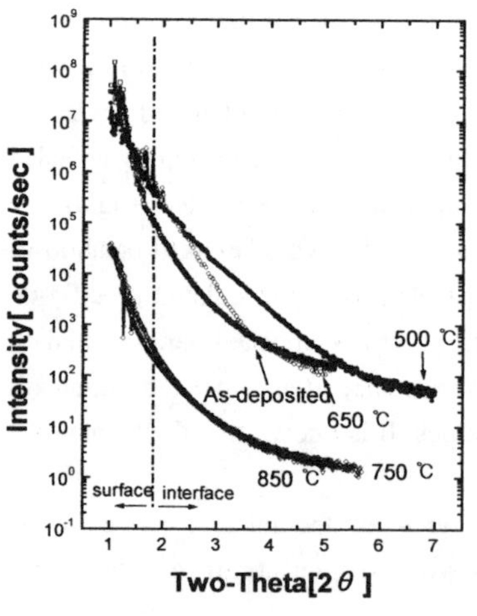

It is evident that the roughness of the interface is suddenly increased for the samples annealed at temperatures higher than 650 ℃. The higher reflectivity for the samples annealed at 500 ℃ compared to the as-deposited samples can be attributed to the improvement of crystallity of Pt. From X-ray scattering analysis it was shown that Pt silicides form at temperatures higher than 650 ℃, while the crystallity of Pt itself is improved with temperature below 650 ℃.

Figure 4. Intensity of reflected X-ray from the surface and interface of Pt/4H-SiC layers annealed at different temperatures.

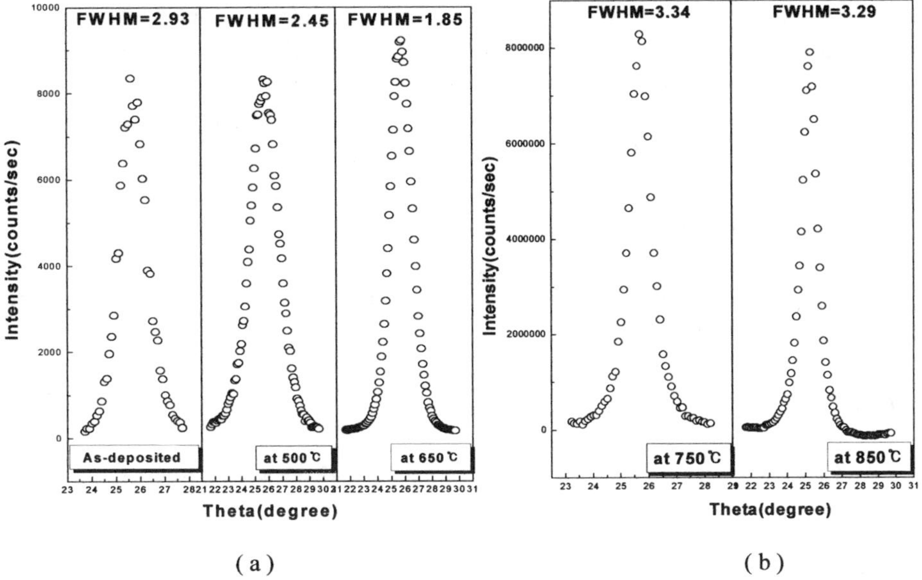

Figure 5. Rocking curves of (a) Pt (111) diffraction when the Pt/4H-SiC layer are annealed at different temperatures (as-deposited, 500 ℃ annealed, and 650 ℃ annealed) and (b) Pt$_2$Si annealed at 750 ℃ and 850 ℃.

Figure 5 (a) compares the rocking curves of Pt (111) diffraction when the Pt/4H-SiC layer are annealed at different temperatures (as-deposited, 500 ℃ annealed, and 650 ℃ annealed). Fig. 5 (b) shows the rocking curves for Pt$_2$Si. Long-scan X-ray diffraction exhibited that there is no Pt phase left after annealing the sample at 850 ℃ [6]. It is concluded that the higher current density of samples annealed at higher temperature stem from the better crystallity of Pt (below 650 ℃) and Pt2Si (above 650 ℃). It is worthwhile noting that the formation of silicides results in the increase of roughness of Pt/4H-SiC interface [7]. This can be easily understood by comparing Figure 4 and Figure 5. The forward and reverse characteristics in Pt/4H-SiC SBDs are found to be dominated by the interface state which are controlled by the thermal annealing.

CONCLUSIONS

Electrical characteristics of Pt/4H-SiC SBDs were interpreted in light of the evolution of microstructure of the Pt/4H-SiC interface by using X-ray scattering techniques.   Devices with the as-deposited Pt schottky contact exhibited the maximum breakdown voltage of 1300 V. The reverse characteristics of the devices were degraded with annealing temperatures. It was shown that the forward characteristics of the devices improve with annealing temperatures. X-ray scattering analysis showed that Pt-silicides were formed by annealing at or higher than 650 ℃. The formation of silicides was shown to increase the roughness of the Pt/SiC interface. The forward and reverse characteristics in Pt/4H-SiC SBDs are found to be dominated by the interface state which are controlled by the thermal annealing.

ACKNOWLEDGMENTS

The authors wish to acknowledge the financial support of the Korea Research Foundation made in the program year of 1998.

REFERENCES

1. Charles E. Weitzel, John W. Palmour, Calvin H. Carter, Jr., Karen Moore, Kevin J. Nordquist, Scott Allen, Christine Thero, and Mohtt Bhatnagar, *IEEE transactions on Electron Devices* **41**(10), 1732-1741, (1996)

2. Iver Lauermann, Rudiger Memming, and Dieter Meissner, J. Electronchem. Soc., **144**(1), p.73-80. (1997)

3. A. Itoh, T. Kimoto, and H. Matsunami., *Member, IEEE, IEEE Electron Device letters,* **16**(6), p. 280-282. (1995)

4. Dev. Alok and B.J. Baliga, *Proceeding of 1995 International Symposium on Power Semiconductor Devices & Ics, p. 96-100,* (1995)

5. J. Crofton, E. D. Luckowski, et al, *Inst. Phys. Conf. Ser.* No. 142 : Chapter 3, *Paper presented at Silicon Carbide and Related Materials 1995 Conf. Kyoto, Japan,* (1995)

6. L.M. Porter, R.F. Davis, J.S. Bow, M.J. Kim, and R.W.Carpenter, *Inst. Phys. Conf. Ser.* No 137, p. 581-584, (1993)

7. N. A. Papanicolaou, A. Christou, and M. L. Gipe, *J. Appl. phys.* **65**(9), p. 3526-3530, (1989)

Part II

SiC Epitaxy and Characterization

EPITAXIAL GROWTH OF SIC IN A VERTICAL MULTI-WAFER CVD SYSTEM: ALREADY SUITED AS PRODUCTION PROCESS?

Roland Rupp,[1] Christian Hecht,[2] Arno Wiedenhofer[2], and Dietrich Stephani[2]

[1]Siemens AG, Semiconductor Components Group, HL PS E SiC, D-80312 Munich, Germany

[2]Siemens AG, Corporate Technology, Department ZT EN, Box 3220, D-91050 Erlangen, Germany

ABSTRACT

Results about a new CVD system suited for epitaxial growth on six 2 inch SiC-wafers at a time are presented. Excellent gas flow stability is achieved for this new reactor type as shown by in- situ observations of the gas flow dynamics in the reactor chamber. These experimental results agree favorably with numerical process simulation results.

The epitaxial layers grown in the multi-wafer system so far show a by an order of magnitude higher background impurity level ($\leq 10^{15}$ cm^{-3}) as reported previously for layers grown in single-wafer systems by the authors and other groups ($\leq 10^{14}$ cm^{-3}). On the other hand, the doping homogeneity achieved until today is very encouraging. The variation on a 2 inch wafer is less than $\pm$ 20% at about $1*10^{16}$ cm^{-3}. The wafer to wafer variation of the average doping value both within a run and from run to run is within 15 %. The reproducibility and uniformity of the layer thickness is even better (total thickness variation $\leq$ 5% on a 2 inch wafer). The surface of the epitaxial layers is very smooth with a typical growth step height of 0.5 nm (4H, 8° off orientation). First measurements on Schottky diodes build on these layers show low leakage current values indicating low point defect density in the epitaxial layers.

INTRODUCTION

In the last 10 years significant progress in SiC epitaxial growth took place. Advancements cover control, reproducibility and homogeneity of doping and thickness but also background doping. This was enabled by better understanding of the deposition process (step control [1], influence of graphite parts [2], site competition [3,4]) and by the development of commercially available epitaxial equipment [5,6,7], which allows accurate control of the relevant process parameters. Nevertheless the costs of the epitaxial process - apart from the still extremely high wafer prices - are a major drawback for a wide range commercialization of SiC devices. This would hold even in the case of zero micropipe density. A pragmatic way to achieve a substantial potential in cost reduction is the use of multiple-wafer instead of single-wafer processing. On the other hand, this leads to new challenges in direction of process control and homogeneity adjustment.

Today there exist three major commercial suppliers for SiC epi systems, all using different basic setups. These companies are EPIGRESS in Sweden (hot wall tube reactor [6]), AIXTRON in Germany (planetary reactor with independent rotation of each wafer [5]) and EMCORE in New Jersey, USA (vertical cold wall reactor [7]). To the knowledge of the authors multi-wafer epitaxial processing of SiC is only reported either on AIXTRON ([8], seven 2" wafers) or on EMCORE ([9], six 2" wafers) systems.

Mat. Res. Soc. Symp. Proc. Vol. 572 © 1999 Materials Research Society

In the following paper we shall describe how and how far we have been able to solve the difficulties of this technology and to report about growth results recently achieved with this tool. A comparison will be made with results formerly achieved with a single-wafer system.

EXPERIMENTAL

The principal construction of the SiC multi-wafer CVD-system is based on the EMCORE D180 type reactors besides a special chamber and heater design necessary for the high temperature operation (see Fig. 1a,b), i. e. an RF-heating of the wafers is applied, which allows wafer temperatures up to 1600°C.

A palladium cell is used to provide high purity hydrogen to the process and a loadlock equipped with a turbo pump is attached to the process chamber to reduce the unproductive time between processes. Silane and propane (2% / 5% diluted in hydrogen) are used as reactive gases added to the main gas stream of hydrogen. Controlled n-type doping is achieved by feeding nitrogen to the process gas. Intentional p-doping is not applied in the system to avoid memory effects and to enable the growth of low nitrogen doped layers with compensation as small as possible. The process temperature is measured by 3 optical pyrometers pointing at different radial positions and allowing control of the temperature homogeneity across the wafer (ΔT typically < 10 K across 2').

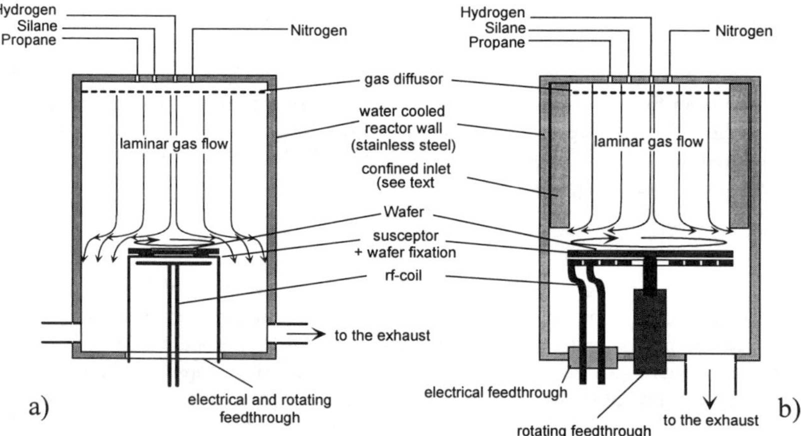

Fig. 1: Sketch of the EMCORE vertical epi reactor conception
a) single-wafer system [10, 17]   b) multi-wafer system

The inner diameter of the reactor chamber is reduced in the upper part (see Fig. 1b) by a water cooled insert attached to the top flange („confined inlet"). In this way, the amount of hydrogen flow necessary to adjust stable flow without recirculation can be decreased significantly. This is mainly due to an increased flow velocity in this area with reduced cross section.

The RF-coil is exposed to the process atmosphere and therefore a condensation of undefined Si/C compounds from the process gas mixture occurs at the coil, primarily determining the maintenance cycles by the need for coil cleaning. Presently the length of these cycles is 80-100 hours of growth time (corresponding to 300-400 µm total growth).

The RF-susceptor / wafer holder assembly consists of a Mo-Plate with six SiC-coated graphite pies as inserts, which have pockets for placing the wafers in it. The whole assembly is displayed in Fig. 2.

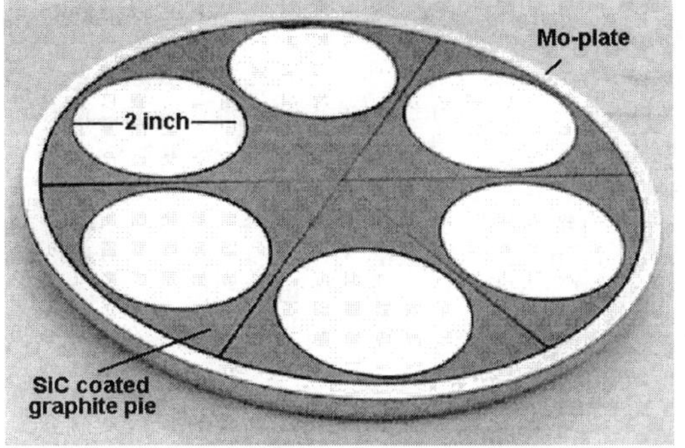

Fig. 2:
RF-susceptor / wafer holder assembly used in this study. The total diameter of the Mo-plate is 185 mm.

4H-SiC-wafers with a diameter of 35 and 51 mm purchased from Cree Res. Inc. (Durham NC) were used as substrates for most of our growth experiments. These wafers are oriented in the (0001)-direction (Si-face) with an off-angle of 8 degrees towards (11-20).
Thickness and growth rate of the epitaxial layers were determined by weight difference [10] or - spatially resolved - either optically (room temperature infrared reflectance measurement [11,12]) or electrically (**c**apacitance-**v**oltage-measurement; **CV**). This CV-method (contact size $1*1$ mm^2) was also employed for determination of lateral and vertical doping profiles. Further measurements were made at the University of Pittsburgh (**l**ow **t**emperature **p**hoto **l**uminescence; **LTPL**) and at the Fraunhofer Institute for Integrated Circuits in Erlangen (**a**tomic **f**orce **m**icroscopy; **AFM**).

For the evaluation of inhomogeneities of properties (thickness and doping of layers) we used the following scheme:
The measurements are performed on typically ≥ 300 points equally spread on the wafer surface without edge exclusion. The derived data are accumulated and the delta between the 10% and the 90 % value of the accumulated distribution is used as measure for the inhomogeneity. Percentage values are then calculated by dividing this delta by the 50 % value of the accumulated distribution (see Fig. 3). This technique is suited to eliminate the influence of measuring points affected by crystal defects on the homogeneity assessment .

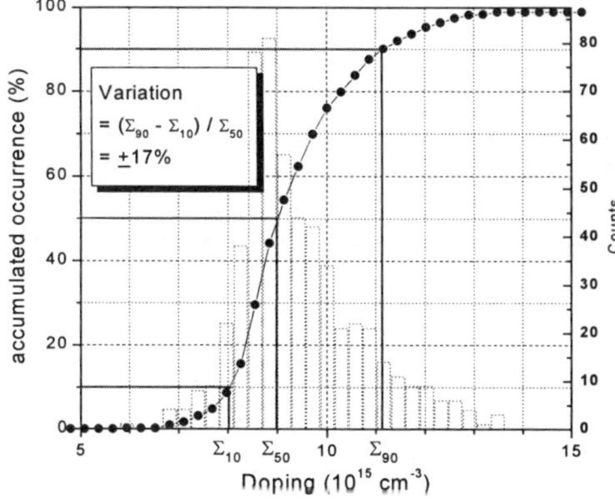

Fig. 3:
Example of deriving a measure for the total variation of a property (in this case doping) from an accumulated occurrence plot. The right axis belongs to the histogram (dashed).

151

RESULTS AND DISCUSSION

Flow dynamics

In the past we employed numerical process simulations to predict both flow stability and deposition behavior in a single-wafer SiC-VPE system as reported elsewhere [13,14]. Encouraged by the good correlation between numerical and experimental results, these tools were also applied during the design phase of the new multi-wafer reactor chamber to ensure that stable gas flow conditions are achievable. The results showed that a total gas flow in excess of 100 slm (at 300 mbar chamber pressure) should be necessary to avoid gas recirculation in case of a cylindrical reactor wall. On the other hand, with a *confined inlet* (see Fig. 1b) the flow was expected to be stable with less than half of this amount of hydrogen. A typical flow pattern and temperature distribution as derived by the numerical simulation is shown in Fig. 4. This picture shows completely laminar flow without any recirculation in the reaction chamber above the wafer plate.

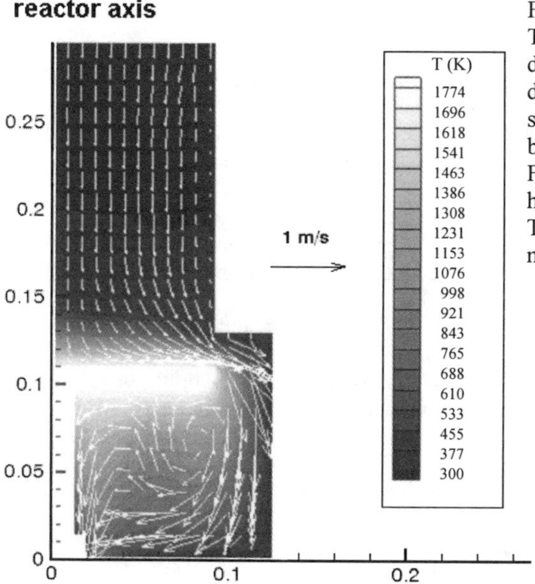

Fig. 4:
Typical flow pattern and temperature distribution in the reactor chamber as derived by a 2-dimensional numerical simulation. The susceptor is assumed to be isothermal as a boundary condition. For symmetry reasons only the right half of the chamber is plotted.
The two axes give the coordinates in meters.

As reported previously for a single-wafer reactor[13,14,15], it is possible to observe the stability of gas flow in the reactor due to a silicon cluster formation. This holds also for the multi-wafer system and Fig. 5 gives an example of the appearance of the resulting irradiant layer. Therefore, we were able to test the numerical prediction after installation of the system by observing the behavior of this layer. These observations confirmed that the flow remains stable even at a hydrogen flow of only 40 slm (250 mbar, 350 rpm, 1500°C wafer temperature), i. e. no swirls are visible moving upwards in the reactor chamber.

The formation of the Si clusters is caused by a local supersaturation of Si generated by the complete dissociation of SiH_4 and the relatively steep T-gradient directly above the wafer plate as explained in [15].

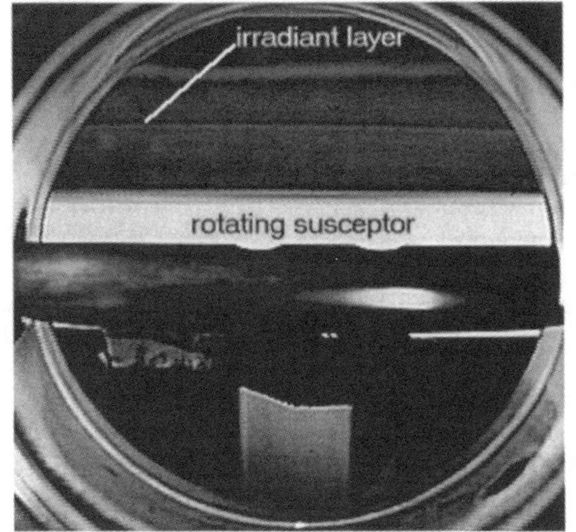

Fig. 5:
Picture of an irradiant layer due to Si supersaturation and cluster formation in the gas phase. The picture is taken through a viewport, which is normally closed by a manual shutter.
A stable, not time dependent and strictly localized layer is an indication for stable flow conditions and no recirculation in the gas phase.

Background doping:

The electrically active impurity concentration was mainly quantified by voltage-capacitance measurements on undoped layers with Ti Schottky contacts. Usually the undoped layers are of p-type conductivity similar to what we reported earlier for single-wafer epitaxy [16,17].
Fig. 6 gives a comparison between the lateral impurity distribution in an epitaxial layer grown on a 35 and a 51 mm diameter wafer. Both layers show a background level below 10^{15} cm^{-3} besides an edge area with several mm width. The impurity distribution is not complete rotational symmetry. It shows an eccentric minimum (shifted to the left) and a higher concentration on the right.

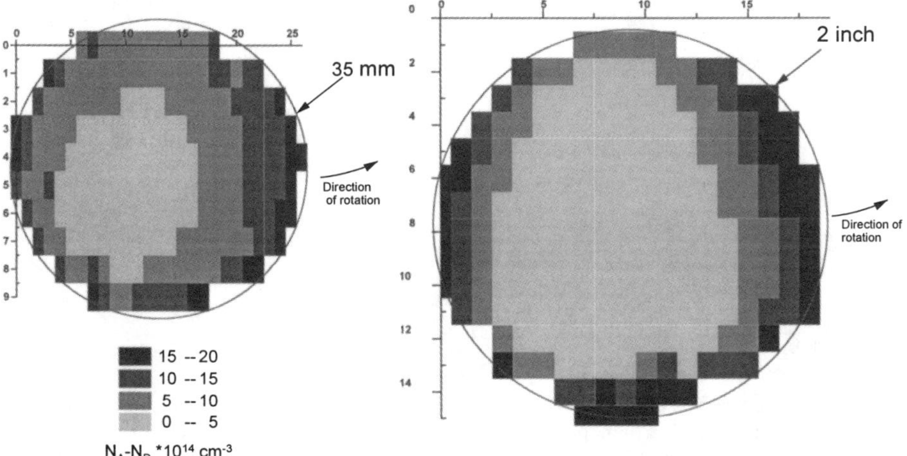

N_A-N_D $*10^{14}$ cm^{-3}

Fig. 6: Background impurity mapping (CV) of epitaxial layers upon wafers with different diameter. A clear edge effect is visible. The axes just give relative coordinates of the metal contact points.

This can be explained by the impurity generation and transport mechanisms. The main source for p-type impurities is outgasing of susceptor and wafer-holder materials during the epitaxial growth process. These contaminants can be transported to the growing surface by diffusion and forced convection due to the high-speed rotation. On the leading edge of the wafer with respect to the direction of rotation, the convection acts in the same direction as the diffusion (and against diffusion on the other edge). This leading edge corresponds to the right hand edge of the wafer maps displayed in Fig. 6 which shows a wider zone of increased impurity concentration. The average impurity concentration drops for increasing wafer diameter, due to a reduced importance of those edge effects. In the center of the 2" wafer it has a value below 10^{14} cm^{-3}. The electrically determined impurity concentration steadily decreases during the growth process, i. e. the impurity level is highest at the interface to the substrate wafer and lowest at the surface. This decrease starts to saturate for layer thickness > 8-10 µm. A possible explanation for this effect is that impurity-releasing surfaces in the neighborhood of the wafers become more and more coated during the growth process and therefore the amount of impurities in the gas phase is continuously decreasing. Comparing the electrically active impurity concentration between different wafers processed within the same run, the background level can vary up to 30 %. This also indicates that the direct surrounding of the wafer and probably mechanical tolerances of the wafer in the respective pocket play the dominant role for the achievable purity.

The type of impurities acting as acceptors was determined by LTPL [18]. A typical spectrum revealed by this technique from a nominally undoped epitaxial layer is displayed in Fig. 7. It shows boron being the dominating impurity and some traces of Al. In addition, a well developed nitrogen line is visible. The Ti-line (not displayed in Fig. 7) is very weak, intrinsic point defect related lines (e. g. L-lines) are hardly detectable. On the other hand, a well developed intrinsic line (I_{75}) shows the good crystalline quality of the epitaxial layer.

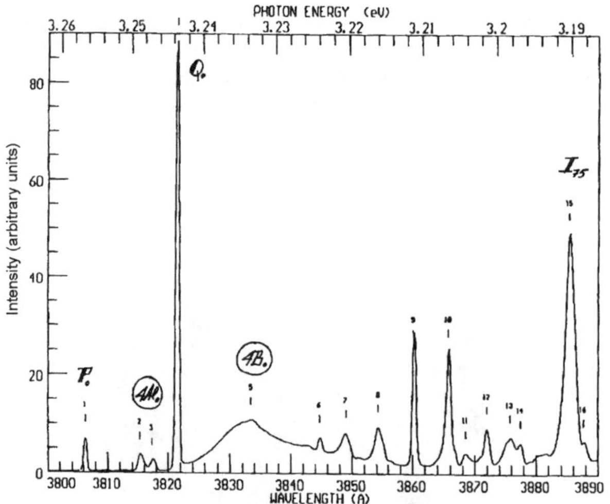

Fig. 7:
A typical LTPL spectrum (excitation wavelength: 244 nm) revealed from a nominally undoped epitaxial layer with 10 µm thickness.
(intrinsic free exciton line I_{75}; impurity related bound exciton lines: P_0, Q_0: nitrogen related; $4Al_0$: aluminum related; $4B_0$: boron related [18])

The purity described above can only be achieved as long as the SiC-coating of the graphite pies is not damaged. Such a damage may take place by thermal stress induced crack formation or via a reaction between the Mo plate and these parts and marks the end of their life time. In case of such a damage an increase in acceptor impurity concentration of more than one order of magnitude occurs. The length of the life time of the pies we worked with so far varied between 20 and 70 h of total process time. The reason of this big variation seems to be both quality and thickness of the SiC coating. Further investigation on this topic is necessary because this life time is a very important factor for process stability and total epi costs.

<u>**Homogeneity:**</u>

Contrary to typical vertical single-wafer systems [10] the wafers in the new equipment have to be placed outside the center of rotation (see Fig. 1) obviously. Without additional precautions this usually leads to a significant radial gradient in flow velocities and probably also in chemical gas composition at the wafer location. Therefore our system allows control of the radial distribution of silane, propane and dopants. Thereby we were able to achieve a reasonable spatial uniformity for both thickness and doping as displayed in Figs. 8 and 9. The thickness mainly varies in the radial direction, whereas the intentional nitrogen doping distribution shows a typical maximum at the position of the trailing edge of the wafer. This distribution holds only for optimized gas flow and growth conditions. For homogeneous gas introduction along the whole diameter of our top flange we typically get a significantly higher variation of the nitrogen incorporation in the range of about a factor of 2 even on a 35 mm wafer. In this case the maximum of the doping is at the outmost part of the wafer with respect to the axis of the wafer plate.

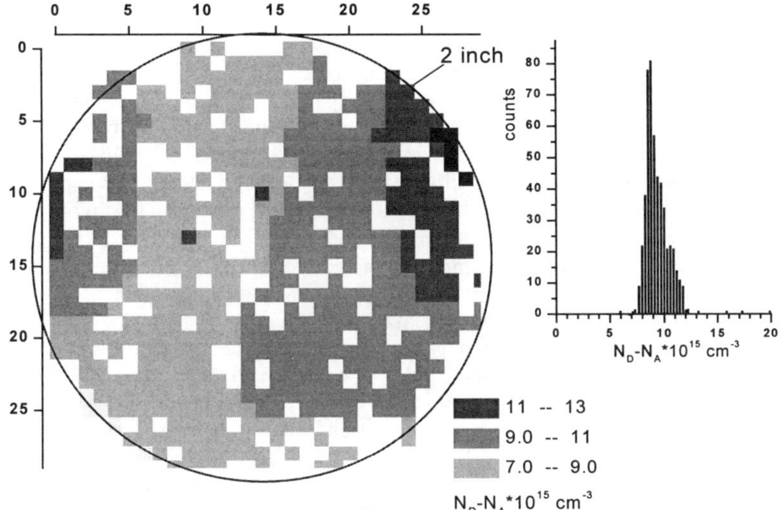

Fig. 8: CV-map (contact size 1*1 mm^2) and distribution function of intentional N-doping of an epitaxial layer upon a 2 inch wafer. At the voltage necessary to measure in a depth of 3.5 μm (>150 V) all contact areas containing a defect have already failed (white areas). The axes in the map give relative coordinates of the metal contact points.

The improved homogeneity shown in Fig. 8 was partly achieved by careful adjustment of the above mentioned radial distribution of process gases but also by reducing the silicon supersaturation, i. e. reduction of Si cluster formation (Fig. 5) by means of decreasing the silane flow. This effect of Si cluster formation on the doping homogeneity probably can be explained by a radially unequal reevaporation of these Si-clusters, leading to a radial variation of the local Si/C ratio and – therefore – to a dopant incorporation efficiency which also depends on the radial coordinate.

Unfortunately, this reduction of Si supersaturation additionally causes a reduction in growth rate. Thus, the homogeneity shown in Fig. 8 can only be maintained at growth rates up to 3.5 μm/h so far.

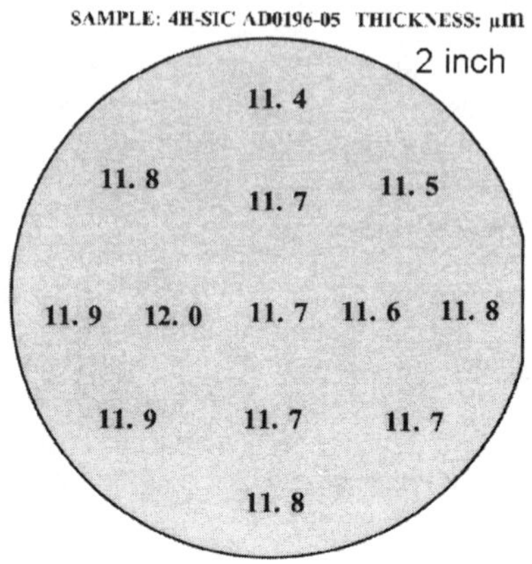

Fig. 9:
Thickness uniformity map of an epitaxial layer upon a 2" wafer measured by room temperature infrared reflectance [11,12]. The total delta of the thickness variation is < 5%.

The remaining mainly azimuthal orientated doping inhomogeneity can be dedicated to the influence of the surrounding of the wafer: Usually the temperature of the wafer surface is about 30 degrees lower than the temperature of the surrounding coated graphite surface. This is due to the additional heat transfer boundary between this graphite part and the wafer. Correlated to this temperature difference there is a difference in the local chemical equilibria on the two surfaces. In practice, one has to expect a higher Si vapor pressure above the hotter surrounding surfaces than above the SiC wafer surface. Thus, the trailing edge of the wafer is always moving in this area of increased Si vapor pressure, it faces a higher Si/C ratio in the gas phase leading to a higher nitrogen incorporation. The resulting variation of doping in azimuthal direction of about ± 15 % can only be reduced by means of reducing the temperature difference between wafer and surrounding material, i. e. a completely different growth set up would be necessary.

Besides the homogeneity of a singular epitaxial layer we also have to deal with the variation of properties from wafer to wafer during one run (inter wafer homogeneity) and from run to run. In Fig. 10 the typical scatter of the doping concentration within one run is shown in dependence of the wafer position (full load of the system with 6 two inch wafers). The data points give the average doping concentration on each wafer, the size of the error bar corresponds to the inhomogeneity on each wafer. It is clearly visible that there is not a random scatter of the average doping, but a sinusoidal fluctuation along the azimuthal coordinate of the wafer plate (Fig. 2) with an amplitude of 10 to 15 %. This behavior is very reproducible and can be explained by a not completely leveled plane of rotation of the wafer plate with respect to the coil plane. In this case temperature may vary slightly in azimuthal direction leading to the observed variation of doping along this coordinate.

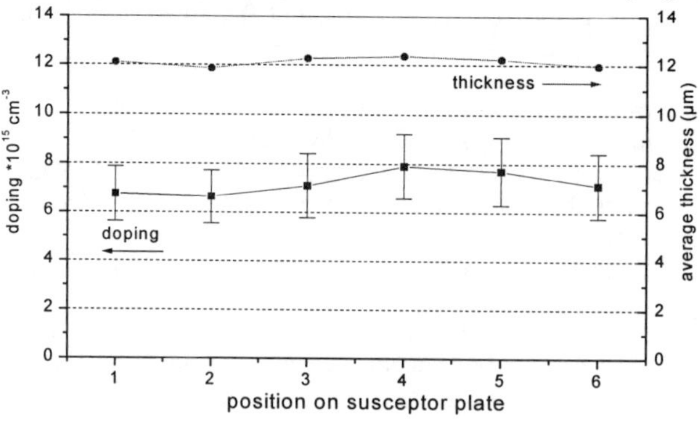

Fig. 10:
Variation of thickness and doping within one full loaded run. For this plot the average thickness is evaluated by weight difference, doping is measured by CV.

As also shown in Fig. 10 the layer thickness is varying in the same manner from wafer to wafer but with a much smaller amplitude of less than 5 %. The thickness inhomogeneity on one wafer usually is very small, typically about 4 to 8 % on two inch wafers like depicted in Fig. 9 (measured by room temperature infrared reflectance [11,12]).

Both thickness and doping seem to be reproducible with an alteration of less than 5 % from one run to the next. This holds at least for the short time range, we do not have enough data to make statements about the long term stability so far. In this case one has to take into account aging effects of helping materials like the wafer plate assembly, which may cause systematic shifts of epi-layer properties even in the case of macroscopic equal control parameters.

Surface morphology

Similar to what we have reported for a single-wafer epitaxial system formerly [16], epilayers grown in the new multi-wafer system show very good surface morphology apart from wafer related defects. An AFM image of a typical layer grown with 3.5 µm/h is depicted in Fig. 11. The surface shows steps corresponding to its off orientation, which have a height of about 0.5 nm (twice the distance between two Si-C layers). This compares favorably with values reported by other groups on 4H wafers with 8° off orientation [19]. A surface as smooth as shown in Fig. 11 can be maintained over a wide range of process parameters and external Si/C ratios (0.6 – 5 at 250 mbar, calculated from the macroscopic silane and propane flow). Exceeding the limit of stable growth by further increase of propane flow leads to rough surface formation first at the inner (pointing to the center of rotation) edge of the wafer. This is also an indication for a increasing local Si/C ratio along the radial coordinate as discussed in the chapter on *doping homogeneity*.

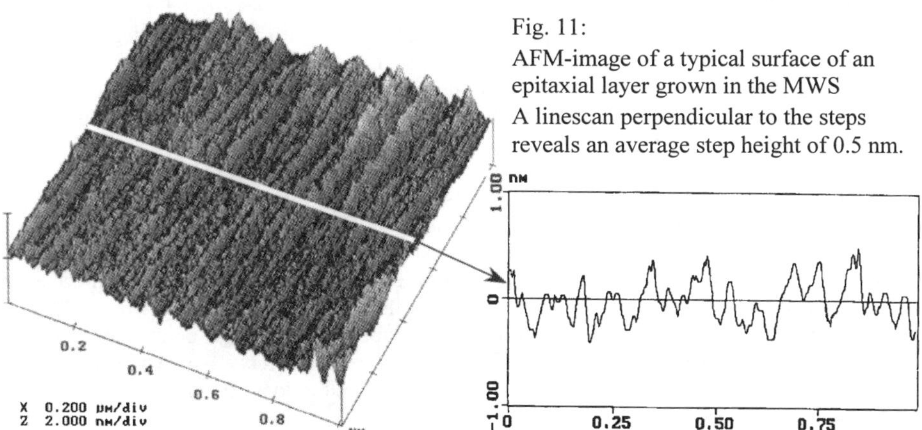

Fig. 11:

AFM-image of a typical surface of an epitaxial layer grown in the MWS

A linescan perpendicular to the steps reveals an average step height of 0.5 nm.

Electrical characteristics and device yield

Recently we have started to use epitaxial layers grown in the multi-wafer system in our device program. Typical electrical characteristics of edge-terminated [20,21] Schottky diodes manufactured on such layers are shown in Fig. 10. At a reverse voltage level of 600 V these diodes show a reverse current of $\ll 10^{-3}$ A/cm^2 with a yield up to 75 % (active area 1.2 mm^2). This yield is in good correlation with the specified micropipe density of the wafers used as substrate (≤ 30 cm^{-2}). Whereas this result was achieved by using 35 mm diameter wafers, similar observations are made on 2 inch substrates but with a slightly lower yield.

Packaged diodes have a specific differential on-resistance of 1.2-1.5*10^{-3} Ωcm^2, indicating a carrier mobility in the epitaxial layer of about 700 – 900 Vs/cm^2 (substrate resistivity: 0.02 Ωcm). This value together with the reasonably low leakage current density indicates a good quality of the epitaxial layers and – specifically – very little defects generated in addition to the wafer defects during the epi process and the subsequent process steps.

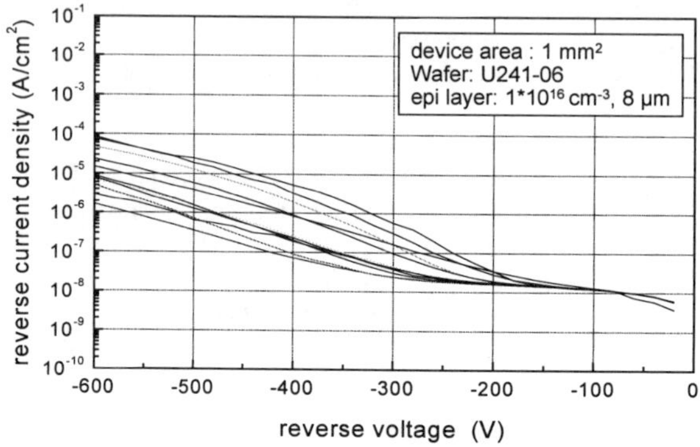

Fig. 12:
Typical reverse characteristics of Ti-Schottky diodes with implanted *junction terminated extension* [18,19] manufactured on an epitaxial layer grown in the multi wafer system.
(specific differential on-resistance: 1.2-1.5*10^{-3} Ωcm^2)

CONCLUSIONS

First results with a vertical multi-wafer VPE system showed that reasonable purity and homogeneity of SiC epitaxial layers can be achieved. The attained background impurity level < 10^{15} cm^{-3} is already sufficient for the manufacturing of devices with a reverse voltage up to 1000 V and there is potential for further improvement. The intra-wafer homogeneity of both doping and thickness is acceptable, the variation of doping from wafer to wafer within one run and from run to run is in a range of about 20 % and therefore suited for some starting applications of SiC like Schottky diodes in the above mentioned voltage range.

A key issue for such devices are the process costs for the epi layer growth. The system described in this paper allows growth on 6 wafers at a time and the necessary cooling, purging and reloading time between two runs is less than one hour (i. e. time between end of growth and start of growth on the next set of wafers). This is enabled by the avoidance of thermal isolations in our reactor and by the use of a loadlock. A first estimation of possible manufacturing costs of 10 µm thick epitaxial layers on two inch wafers show that a value of significantly below 200 US$ will be possible in a production environment (including capital usage, consumables, lab cost, personal etc). It is a necessity to meet this value before SiC epitaxy can be named "production suited", because otherwise it will be very difficult to identify volume applications, which can accept the still very high price per unit area of SiC devices.

In our case we made a trade-off between quality and costs of the epitaxial layers: Compared to results already reported for single-wafer epi systems [16,17,22,23], the properties of layers as described above are not very impressive, but they can be achieved at quite low costs!

In summary, if we can stabilize and reproduce the results described above over a longer time, we then have a production-suited epitaxial process at least for SiC devices like Schottky diodes dedicated to the < 1000 V range.For higher blocking voltages and/or for applications less tolerant towards doping variations further work on process and hardware optimization is necessary.

ACKNOWLEDGEMENT

The authors would like to thank W. J. Choyke (Univ. Pittsburgh) for the photoluminescence measurement on our epitaxial layers and C. Q. Chen (Univ. Erlangen) for the determination of epi layer thickness by infrared reflectance. Further thanks go to Yu.N. Makarov (Univ. Erlangen) for his work on numerical simulation of our epitaxial process. We also want to acknowledge the engagement of Paul Fabiano, Alex Guarry, David Voorhees and Dennis Stucky (all of EMCORE) in setting up the epi system and solving initial problems.

REFERENCES

1. H. Matsunami, T. Ueda, H. Nishino, Mater. Res. Soc. Symp. Proc. **162** (1990) p. 397

2. K. Rottner, R. Helbig, J. Cryst. Growth. **144** (1994) p. 258

3. D.J. Larkin, Mat. Res. Soc. Symp. Proc., **410** (1996) p. 337

4. D.J. Larkin, P.G. Neudeck, J.A. Powell, L.G. Matus, Appl. Phys. Lett. **65** (1994) p. 1659

5. Aixtron AG, Kackertstrasse 15-17, D52072 Aachen, Germany

6. Epigress AB,Ideon Science & Technology Park, SE-223 70 Lund, Sweden

7. EMCORE Corporation, 394 Elizabeth Avenue, Somerset, NJ 08873, USA

8. A.A. Burk, M.J. O'Loughlin, S.S. Mani, Mat. Sci. Forum **264-268** (1998) p. 83

9. R. Rupp, A. Wiedenhofer, D. Stephani, Proc. of the 2nd Europ. Conf. on SiC and Rel. Mat. ECSCRM`98, Sept. 2-4 1998 Montpellier, France, in press

10. R. Rupp, P. Lanig, J. Voelkl, D. Stephani, J. Cryst. Growth, **146** (1995) p. 37

11. M.F. McMillan, U. Forsberg, P.O.Å. Perssons, L. Hultman, E. Janzén, Material Science Forum, **264-268** (1997) p. 245

12. C.Q. Chen, F. Engelbrecht, C. Pepperemüller, N. Schulze, R. Helbig, R. Rupp, Institute of Appl. Physics, Univ. Erlangen-Nürnberg, to be published

13. R. Rupp, Yu.N. Makarov, H. Behner, A. Wiedenhofer, phys stat sol (b), **202** (1997) p. 281

14. Yu.N. Makarov, Yu.E. Egorov, A.O. Galyukov, A.N. Vorob`ev, A.I. Zhmakin, R. Rupp, Proc. of the 2nd Europ. Conf. on SiC and Rel. Mat. ECSCRM`98, Sept. 2-4 1998 Montpellier, France, in press

15. A.N. Vorob'ev, S.Yu. Karpov, A.E. Komissarov, Yu.N. Makarov, A.S. Segal, A.I. Zhamakin, R.Rupp, Proc. of the 2nd Europ. Conf. on SiC and Rel. Mat. ECSCRM`98, Sept. 2-4 1998 Montpellier, France, in press

16. R. Rupp, A. Wiedenhofer, P. Friedrichs, D. Peters, R. Schörner, D. Stephani, Material Science Forum, **264-268** (1998) p. 89

17. R. Rupp, P. Lanig, J. Völkl, D. Stephani, Mat. Res. Soc. Proc. **423** (1996) p. 253

18. R.P. Devaty, W.J. Choyke, phys stat sol (a), **162** (1997) p. 5

19. T. Kimoto, A. Itoh, H. Matsunami, Appl. Phys. Lett. **66** (1995) p. 3645

20. H. Mitlehner, W. Bartsch, M. Bruckmann, K.O. Dohnke, U. Weinert, Proc. of the ISPSD´97 Weimar 1997 IEEE p. 165

21. H. Mitlehner, P. Friedrichs, D. Peters, R. Schörner, U. Weinert, B. Weis, D. Stephani, Proc. of the ISPSD´98 Kyoto 1998 IEEE p. 127

22. A.A. Burk, L.B. Rowland phys stat sol (b), **202** (1997) p. 263

23. O. Kordina, A. Henry, E. Janzén, C.H. Carter, Material Science Forum, **264-268** (1998) p. 97

MULTI-WAFER VPE GROWTH OF HIGHLY UNIFORM SiC EPITAXIAL LAYERS

M.J. O'Loughlin, H.D. Nordby, Jr., and A.A. Burk, Jr. *
Advanced Technology Laboratories, Northrop Grumman ES[3], PO Box 1521 Baltimore, MD
21203, michael_j_oloughlin@md.northgrum.com
*current address, Cree Research, Durham, NC

ABSTRACT

A multi-wafer silicon carbide vapor phase epitaxy reactor is employed that features full
planetary motion and is capable of high quality epitaxy on seven, two-inch diameter substrates.
We are currently performing preproduction growths of static induction transistor (SIT) and metal
semiconductor field effect transistor (MESFET) active layers. On a 2-inch diameter substrate,
layer uniformity is typically ±5% (standard deviation/mean) for both dopant concentration and
layer thickness (for 1 3/8-inch substrates, layer uniformity is around ±3%). For the seven wafers
within a run, interwafer uniformity has been dramatically improved to approximately ±8% for
dopant concentration and ±3% for layer thickness. Process control charts will be presented
which exhibit that interrun (run-to-run) variation in both thickness and doping can be kept within
±10% of the desired values.

INTRODUCTION

Prototype silicon carbide (SiC) static induction transistors (SIT) and metal semiconductor
field effect transistors (MESFET) have been demonstrated with performance exceeding that
typically available from both silicon and gallium arsenide transistors [1,2,3]. To realize
production of SiC based devices, a multi-wafer SiC vapor phase epitaxy (VPE) reactor has been
developed in collaboration with Aixtron, AG [4]. The reactor, based on the design of Frijlink
and coworkers [5], features full planetary motion and, as currently configured, is capable of
simultaneous, high quality epitaxy on seven, two-inch diameter substrates.

To achieve satisfactory uniformity and reduce run-to-run variability, it was necessary to
modify many of the original reactor components [6]. The improvements in uniformity and
variability resulting from those modifications were sufficient to allow modest production goals
for device quality SiC epitaxial layers to be exceeded in the previous year. Significantly higher
volumes of epitaxial layers with more stringent specifications for dopant concentration and layer
thickness values and uniformities will be required to meet current and future production goals.
Indeed, production specifications are likely to require that all areas of all epitaxial layers be
within ±10% of target values for both dopant concentration and thickness. Several more reactor
modifications will be summarized which have allowed those production specifications to be
realized with high yield.

EXPERIMENT

The VPE reactor used to grow the SiC epitaxial layers discussed within has been described
in detail elsewhere [3,6,7]. A cross sectional view of the reactor is shown in figure 1. The
reactor consists of an inductively heated, mechanically rotated susceptor with seven wafer
holders (satellites). The satellites are themselves rotated on the susceptor by means of gas foil
levitation [5]. The satellite rotation is desirable to reduce upstream/downstream variation in
growth. The susceptor rotation averages inhomogeniety in the RF induction heating and
geometry of the non-rotating reactor components. However, susceptor rotation does not directly

affect inhomogenieties that result from anisotropic conductivity in the graphite susceptor, dimensional nonuniformity of the machined and coated susceptor, and misalignment, *i.e.* tilt or eccentricity, of the susceptor on its support. All of those sources of inhomogeniety can be minimized but are difficult to eliminate.

A graphite foil-on-foam ceiling [8] separates the growth region from the water-cooled reactor chamber. The ceiling is passively heated by radiation from the ~1600°C susceptor creating a pseudo hot-wall effect. The hot ceiling reduces supersaturation and condensation of silicon vapor in the reactor. In a previous communication [6], it was reported that the graphite foil-on-foam ceilings could only be used for tens of runs. However, recently, ceiling lifetime has been extended to nearly 100 runs, which has increased epitaxial layer production efficiency.

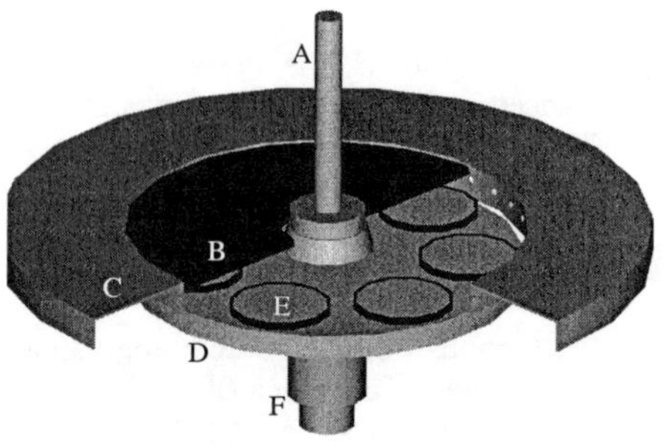

Figure 1. Stylized cross section of VPE reactor. A) Gas Injector, B) Ceiling, C) Gas Collector, D) Susceptor, E) Wafer Holder, F) Rotation Axis

Finally, a perforated gas collector terminates the outer diameter of the susceptor, the growth region, and the ceiling. The gas collector, made from graphite foam or molybdenum, ensures laminar flow of the gases in the growth chamber.

Typical conditions for SiC epitaxial layer growth in this reactor have been described in detail elsewhere [6]. All of the growths reported in this study were performed at reduced pressure, susceptor temperatures of ~1600°C, growth rates of ~3 µm/hr, and silicon-to-carbon ratios around 0.88. The mass of each wafer is measured with an analytical balance both before and after each growth to determine epitaxial layer thickness, growth rate, and wafer-to-wafer uniformity. After growth, all epitaxial layers, except highly doped layers used for ohmic contacts, are characterized by capacitance versus voltage (CV) measurements on 200 µm diameter photolithographically defined Cr-Au Schottky diodes. CV measurements are performed at regular intervals along one radius of each wafer and used to extract relative thickness and dopant concentration values and uniformities. The rotating wafers have been found to be highly axially symmetric making the measurement of one radius sufficient. To minimize the affect of long-term drift in growth rates and background impurity concentration, the measurements from each run are used as a calibration for the next run.

RESULTS

All epitaxial layers were characterized by mass and CV measurement. Three levels of uniformity were examined; intrawafer uniformity, interwafer (or intrarun) uniformity, and interrun (run-to-run) uniformity.

A five-point overlay of dopant concentration as a function of depth, extracted from CV measurements, is presented in figure 2. The 4.6 µm thick epitaxial layer, grown on a two-inch diameter, low resistivity, n-type substrate, was intentionally doped with nitrogen to a concentration of 4×10^{15} cm^{-3}. For the epitaxial layer shown in figure 2, the uniformities (standard deviation divided by the mean) of the dopant concentration and thickness were 4.9% and 5.1% respectively. In both cases, the range (maximum-minimum) of measurements as a percentage of the mean is approximately 12%. The uniformity exhibited by this layer does not represent a "best-effort", but is representative of typical two-inch epitaxial layers. For example, the average uniformity for a recent series of runs with two-inch wafers was 4.7% for dopant concentration and 5.2% for thickness. Excellent intrawafer uniformity for 1 3/8-inch wafers has also been observed. For those smaller diameter wafers, the average uniformity, based on over 200 wafers, was 3.6% for dopant concentration and 2.4% for thickness.

After extensive reduction of susceptor induced inhomogeniety [6,7], interwafer (wafer-to-wafer within a run) uniformity had only been reduced to 10% at best. With that large of a standard deviation, at least one-third of the wafers would fall outside the ±10% absolute specifications for production epitaxial layers. However, by replacing the graphite foam gas collector with a molybdenum one with smaller perforations, the interwafer thickness uniformity improved to approximately 3%. The time evolution of interwafer uniformity is presented in figure 3. The molybdenum gas collector was installed at point A on that chart. It is evident that the improvement in thickness uniformity was not matched by a corresponding improvement in dopant concentration uniformity. The dopant concentration of the individual wafers still exhibited a systematic distribution, *i.e.* the highest doped position was opposite the lowest doped position. The highest doped position correlated with the hottest side of the susceptor. By rearranging the susceptor hardware to improve the temperature balance (point B in figure 3), the dopant concentration uniformity has been reduced to 7%. Most of the remaining non-uniformity is randomly distributed and arises from small differences in local temperature due to variations in; emissivity of components, thermal contact between wafer and satellite, satellite levitation, *etc.* An overlay of the dopant profiles at five positions along the radius of each of seven wafers in a single run is presented in figure 4. For the run represented by that figure, both the thickness and dopant concentration uniformity at 2×10^{17} cm^{-3} are approximately 5%. At 1×10^{16} cm^{-3} the dopant concentration uniformity is slightly degraded to 7%.

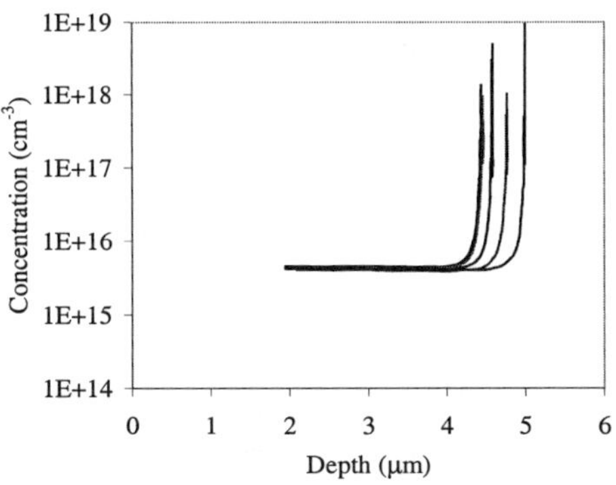

Figure 2. CV dopant profile measured at five positions along a radius of a 2" diameter wafer. The epitaxial layer is thinnest in the center. The thickest profile is measured at 4.6 mm from the edge.

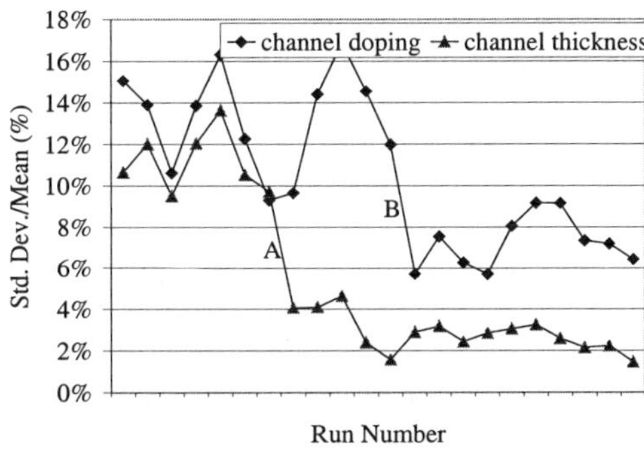

Figure 3. Inter wafer uniformity for 22 runs. A new gas collector was installed at A. Susceptor hardware was rearranged at B to improve the temperature balance.

Having demonstrated good uniformity on a 2-inch diameter wafer and for each wafer in a run, the remaining obstacle to high yield production epitaxy is interrun variation. A control chart representing our ability to meet target values for thickness and dopant concentration as a function of run number is presented in figure 5. The data points represent the percentage deviation of the average value for a run from the target value. The error bars represent the interwafer uniformity within each run. Where the error bars (1σ) fall outside the $\pm 10\%$ of target lines, there will be a yield loss. It is evident, from figure 5, that the improvements that were effected to decrease the interwafer variation also permitted better control of run-to-run variation. In fact, after point B (corresponding to improved gas collector geometry and susceptor temperature uniformity), the average absolute deviation from target values were 2.8% for dopant concentration and 3.6% for thickness.

CONCLUSIONS

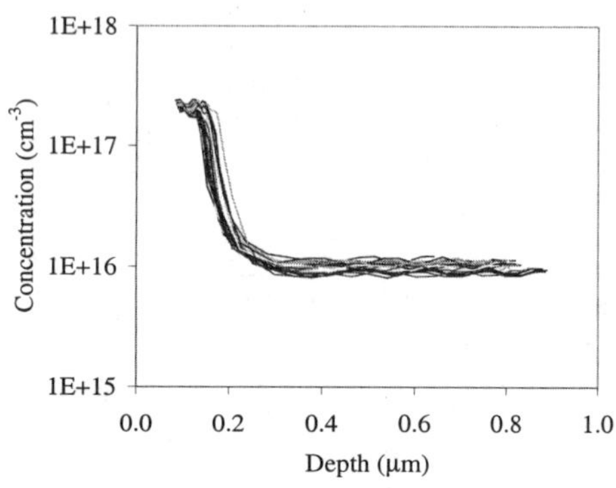

Figure 4. An overlay of the CV dopant profile measured along the radius of each of seven wafers in a run.

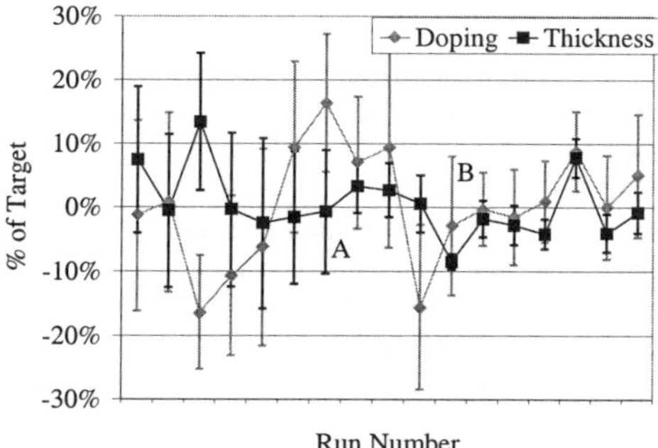

Figure 5. Control chart for doping and thickness. Symbols represent percent deviation of the average value for a run from the target value.

Sufficient uniformity has been demonstrated for SiC epitaxy on seven at-a-time two-inch diameter wafers for high yield in production. Intrawafer uniformity of approximately 5%, similar to what was previously reported for 1 3/8-inch wafers, can be routinely achieved on two-inch diameter wafers. The most significant improvement has been the reduction of interwafer uniformity to 3% for thickness and 7% for dopant concentration. The combination of good intrawafer and interwafer uniformity has been coupled with better control of run-to-run variation such that a very high percentage of epitaxial layers meet stringent production specifications.

ACKNOWLEDGMENTS

The authors gratefully acknowledge David Stanley, Gil Rykiel, Sam Ponczak, Ollie Gildow, and Rich Siergiej for their assistance in reactor modifications, epitaxial growth, and characterization. Development of this reactor was supported in part by the Department of the Air Force under contracts F33615-92-C-5912 and F33615-95-5427 (Tom Kensky, contact monitor and Laura Rea, program direction leader).

REFERENCES

1. S. Sriram, G. Augustine, A. A. Burk, Jr., R.C. Glass, H. M. Hobgood, P. A. Orphanos, L.V. Rowland, T. J. Smith, C. D. Brandt, M. C. Driver, and R. H. Hopkins, IEEE Electron Device Lett., **EDL-17**, p. 369 (1996).

2. R. R. Siergiej, S. Sriram, R. C. Clarke, A. K. Agarwal, C. D. Brandt, A. A. Burk, Jr., T. J. Smith, A Morse, and P. A. Orphanos, *Tech. Digest Int. Conf. SiC and Rel. Mat'95*, (Kyoto, Japan 1995), p. 321.

3. A. A. Burk, Jr., M. J. O'Loughlin, R. R. Siergiej, A. K. Agarwal, S. Sriram, R. C. Clarke, M. F. MacMillan, V. Balakrishna, and C. D. Brandt, J. Solid State Electronics, accepted for publication.

4. Aixtron Inc. Kackerstr. 15-17, D-52072 Aachen, Germany.

5. P. M. Frijlink, J. Crystal Growth, **93**, p. 207 (1988).

6. A. A. Burk, Jr., M. J. O'Loughlin, and H. D. Nordby, Jr., J. Crystal Growth, accepted for publication.

7. A. A. Burk, Jr., M. J. O'Loughlin, and S. S. Mani, in *Silicon Carbide, III-Nitrides, and Related Materials,* , edited by G. Pensl, H. Morkoç, B. Monemar, and E. Janzén (Materials Science Forum, **264-268**, Trans Tech Publications, Switzerland 1998), p. 83-88.

8. Calcarb, Inc., Rancocas, New Jersey, USA.

CHARACTERIZATION OF THICK 4H-SiC HOT-WALL CVD LAYERS

M.J. Paisley*, K.G. Irvine, O. Kordina, R. Singh, J.W. Palmour, and C.H. Carter, Jr.
Cree Research, Inc., 4600 Silicon Drive, Durham, NC 27703-8475, USA
*Mike_Paisley@Cree.com

ABSTRACT

Epitaxial 4H-SiC layers suitable for high power devices have been grown in a hot-wall chemical-vapor deposition (CVD) system. These layers were subsequently characterized for many parameters important in device development and production. The uniformity of both thickness and doping will be presented.

Doping trends vs. temperature and growth rate will be shown for the p-type dopant used. The n-type dopant drops in concentration with increasing temperature or increasing growth rate. In contrast, the p-type dopant increases in concentration with decreasing temperature or increasing growth rate. A simple descriptive model for this behavior will be presented.

The outcome from capacitance-voltage and SIMS measurements demonstrate that transitions from n to n^-, or p to p^-, and even n to p levels can be made quickly without adjustment to growth conditions. The ability to produce sharp transitions without process changes avoids degrading the resulting surface morphology or repeatability of the process. Avoiding process changes is particularly important in growth of thick layers since surface roughness tends to increase with layer thickness.

Device results from diodes producing two different blocking voltages in excess of 5 kV will also be shown. The higher voltage diodes exhibited a breakdown behavior which was near the theoretical limit for the epitaxial layer thickness and doping level grown.

INTRODUCTION

The properties of silicon carbide, such as wide bandgap, high breakdown electric field strength, and high thermal conductivity are characteristics that make the material highly desirable for high power devices. Recent device demonstrations and measurements of the superior electrical and physical properties of silicon carbide (SiC) have shown that it is the premier semiconductor material for fabrication of high power and power microwave electronic devices. In addition, SiC devices can operate at elevated temperatures allowing them to be used at either high ambient temperature or with reduced cooling requirements in nominal ambients. In recent years substrate and CVD developments have made great progress, thereby allowing for the development of SiC-based high power high frequency devices such as Metal-Semiconductor Field Effect Transistors (MESFETs)[1]. Other high power devices such as diodes, MOSFETs and GTOs have also been demonstrated [2,3,4]. While current levels of uniformity of layer thickness and doping are sufficient for production use, the current wafer diameters are not. Thus work needs to continue on layer uniformity for increasing wafer sizes as they become available to be ready for production when it becomes commercially viable.

EXPERIMENTAL SETUP

The wafers were grown in a horizontal hot-wall CVD reactor. The key component of the reactor is the graphite susceptor which is similar in construction to those described in Ref. 5 and 6. The susceptor is heated inductively and is designed to obtain good heating uniformity over a large area. The temperature is measured by a pyrometer focused on a position at the leading edge of the susceptor. The susceptor is also tapered to compensate for gas reactant depletion which may be quite severe in hot-wall systems. The precursors, silane and propane, are diluted in a high flow of purified hydrogen. Growth temperatures in the range of 1500-1700°C were used. Nitrogen was used as the n-type dopant and trimethyl aluminum (TMA) was used as the p-type dopant.

The thicknesses and thickness uniformities were measured by observing the cleaved edge of the sample in a scanning electron microscope (SEM). The difference in doping between the highly doped substrate and the epitaxial layer gives sufficient contrast for determining the

Mat. Res. Soc. Symp. Proc. Vol. 572 © 1999 Materials Research Society

thickness. The doping and doping uniformities were measured by a mercury probe capacitance-voltage (C-V) profiler.

RESULTS AND DISCUSSION

Epitaxial Layer Uniformities

Epitaxial layers were deposited on prototype 75 mm diameter wafers, cleaved and then measured along the flow direction for layer thickness by cross-section scanning electron microscopy (SEM). The results are shown in Figure 1. The uniformity is calculated as the standard deviation divided by the mean. The "epi-crown" (the raised ridge along the wafer circumference) was not included in the calculations.

Additional wafers were grown and characterized by C-V measurements to determine the doping uniformity. For a doping level of 7.5×10^{15} cm^{-3} n-type the resulting uniformity was 7%, also calculated as standard deviation divided by the mean. Similar growth runs on 75 mm wafers for p-type layers resulted in a uniformity of 9%. The results for various diameter wafers are shown in Table 1.

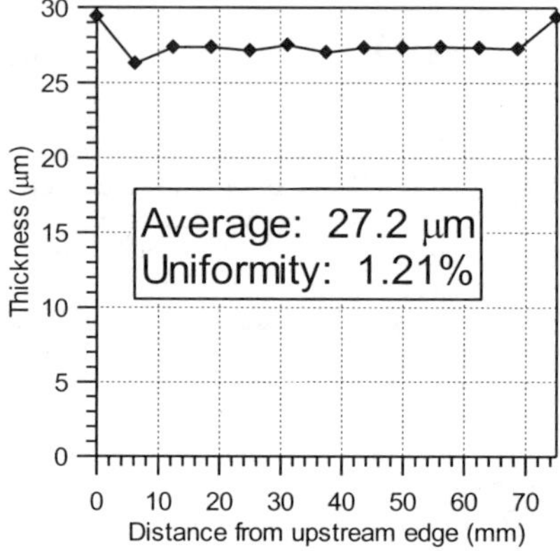

Figure 1. Epi thickness of 75 mm wafer.

Achieving the desired doping level is also very important to the final device performance. However, reaching the desired levels of around 1×10^{15} cm^{-3} and below for thick layers of SiC is often difficult. Nearly all reactor parameters can affect the doping at these levels to easily shift the results outside allowable limits. Additionally, C-V measurements in this range are less reliable when performed with a mercury probe profiler. We have collected data from over 80 reactor runs with different dopant types and doping. All runs were for low doped thick layers (>20 µm). The results show that if we accept levels of within ±20% of the target doping, the reactor yield is 93%. It should be noted that this high yield applies only to standard product offerings which are well characterized. Non-standard combinations of thickness and doping levels will significantly impact the resulting reactor yields.

Table 1. Uniformities of epitaxial layers on various wafer sizes.

Wafer	Thickness	p-type	n-type
75 mm	1.2%	9%	7%
50 mm	<1%	n/a	1.6%
35 mm	<1%	2%	2%

Dopant Incorporation Model

The incorporation behavior of various dopants in SiC has been studied for some time. The behavior of nitrogen incorporation in SiC has been established as decreasing with both increasing growth temperature and increasing growth rate [7]. This has been our experience as well, though it does appear that susceptor geometry can affect the magnitude of these changes, at least in our laboratory. It turns out that the incorporation behavior of Al is remarkably different where it still decreases with increasing temperature, but also increases with increasing growth rate. Figure 2 shows the Al concentration measured by C-V as a function of temperature offset from an arbitrary value. These temperatures were maintained for the duration of a single deposition run. Temperature offset is used because the actual wafer surface temperature is not known as the control temperature is measured at some distance from the wafer. Figure 3 shows the growth rate of the layers shown in Figure 2, which was also held constant during each individual run. As can be seen in the figure, the growth rate drops at higher temperatures due to increased etching by the hydrogen carrier gas.

Figure 2 shows that the N_a–N_d concentration drops rapidy with increasing deposition temperature as was also observed by Kimoto [7]. Figure 3 shows the growth rate dependence for the temperature range. It is reasonable to assume that the sticking coefficient of the Al will drop as the wafer temperature is increased. At a constant growth temperature offset of 20°C, if the growth rate is increased by 50% (from ■ point to ▼ point in both figures), the doping level increases by 38%. But if the growth rate is increased again by 60%, then the doping level increases by an additional 230%. At the highest temperature offset of 120°C, an increase in growth rate of 83% (▲ point), results in a doping increase of 100%. So as the deposition temperature is changed by 120°C while the deposition rate is held constant within 8% and the TMA flow is held constant, the resulting Al concentration drops by a factor of 250. The stoichiometry of the input precursors was kept the same though it is likely to assume that the effective stoichiometry changes when the temperature is varied over this range. While we think that stoichiometry may play some minor role in the observed behavior [8], it can not explain a change of this magnitude.

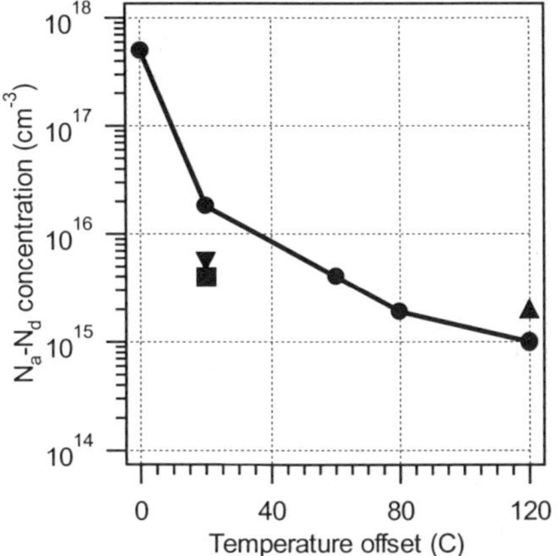

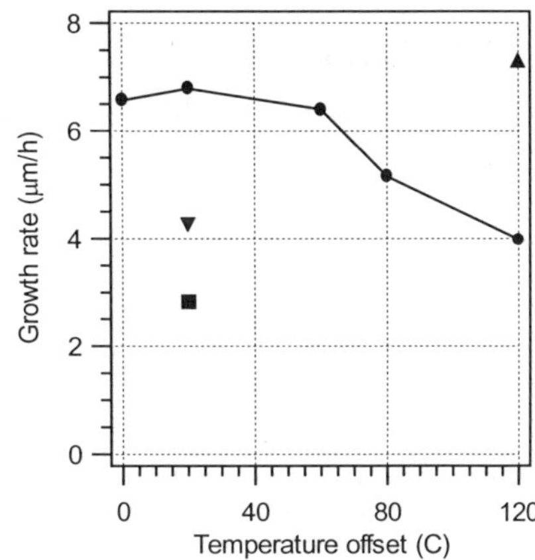

Figure 2. N_a–N_d concentration (from C-V) as a function of adjusted deposition temperature (at constant TMA flow).

Figure 3. Epitaxial growth rate as a function of adjusted deposition temperature.

The data points where the growth rate was increased indicates an initial proportional change in doping levels with growth rate that becomes non-linear. This appears to us to indicate that the p-type doping is influenced by surface roughness or defects. Thus as the growth rate increases, the surface roughens, and/or defect concentrations increase which provide additional sites for Al incorporation. It may also explain the high incorporation rate of Al at lower deposition temperatures where many more defects can act as bonding sites for Al. Nitrogen, in contrast, decreases with increasing growth rate.

Doping Transitions

One component of device performance is the ability to make transitions from n to n^-, or p to p^-, and even n to p levels. More abrupt transitions permit devices to be fabricated with layer thicknesses closer to the ideal design thickness. Also, transitions from n to p or p to n levels will result in potential recombination regions that can be minimized by the abruptness of the transition. We chose a typical p^- layer structure that consisted of an n^+ substrate followed by an n^+ buffer layer, then a p^+ buffer layer and finally the p^- region. A sample of this structure was profiled using both C-V measurements and secondary ion mass spectrometry (SIMS). The SIMS profile is shown below in Figure 4.

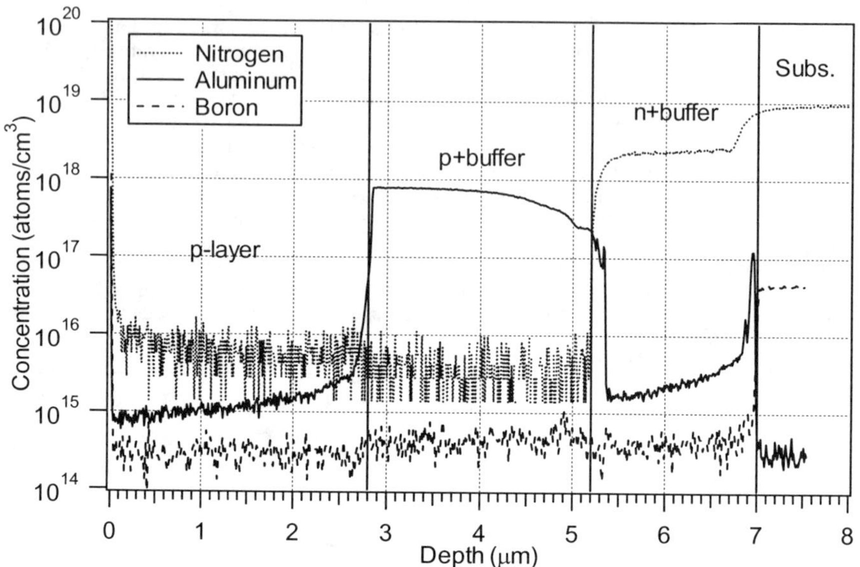

Figure 4. Secondary ion mass spectrometry (SIMS) depth profile of a multi-layer structure indicating sharp transitions between adjacent regions.

It should be noted that the profiles observed with SIMS were also observed with C-V profiling. These data are not shown since C-V profiles are subject to depletion effects and a complete profile of the entire structure was not possible on a single structure. The SIMS data itself is also subject to "layer-mixing" effects which can smear out transitions that occur at greater depths. The Al profile shows a "spike" at the substrate interface which has been previously observed by Burk and Rowland [9]. While removal of this "spike" is a matter of some further research, doping levels are controlled such that no "buried interfaces" occur. The transition from the n^+ buffer to the p^+ buffer shows that this change is possible within much less than 0.5 μm. The Al profile also shows a "rounding up" behavior which is not present in layers grown more recently, due to injection of larger quantities of trimethyl aluminum at the start of such regions. Finally, the transition to the final p^- level which falls three orders of magnitude takes less than 3 μm, with the majority of the transition requiring only 0.2 μm to fall to 3×10^{15} cm^{-3}. The entire structure was deposited without changes in Si/C ratios or other deposition conditions that might affect the quality of the layers. The surface morphology was equivalent to single layers grown under these conditions.

High Voltage Device Behavior

Low doped, thick 4H-SiC epitaxial layers are essential for the realization of high voltage devices. A simple high-power device structure that can use these epitaxial layers is a high voltage P-i-N diode as shown in Figure 5. To ease the gradient of the electric field at the edge of the device, an edge termination suitable for very high voltage must be used. A batch of wafers using a 4H-SiC n^- layer with a thickness of 85 μm (also called the drift layer) and a later batch with a layer thickness of approximately 50 μm were fabricated using junction termination extension (JTE) as the edge termination technique. The termination region has a low dose implanted p-type ring surrounding the high voltage anode junction. While the edge termination was not designed to fully exploit the capability of the epitaxial layer used, a high blocking voltage in excess of 5.5 kV was achieved on one of the diodes of the first batch. With improvements in the processing, a blocking voltage of 5.9 kV was achieved on one diode in the second batch. The reverse current-voltage characteristics are shown in Figure 6. The first diode batch gave a varying leakage current that went from mid-10^{-7} A to mid-10^{-5} A. The marked improvement in leakage behavior of the second batch resulted in leakage current of 10^{-8} A over most of the voltage range and demonstrates the much improved processing. Modeling of the ideal breakdown behavior for SiC at a given layer thickness and doping level showed that the second diode with a 50 μm epitaxial layer (*vs.* 85 μm for the first batch) closely approached the maximum theoretical voltage.

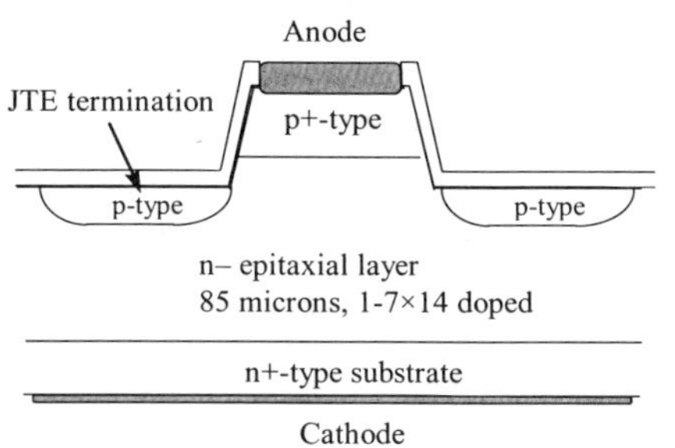

Figure 5. The structure of the 5.5 kV P-i-N diodes in Figure 6.

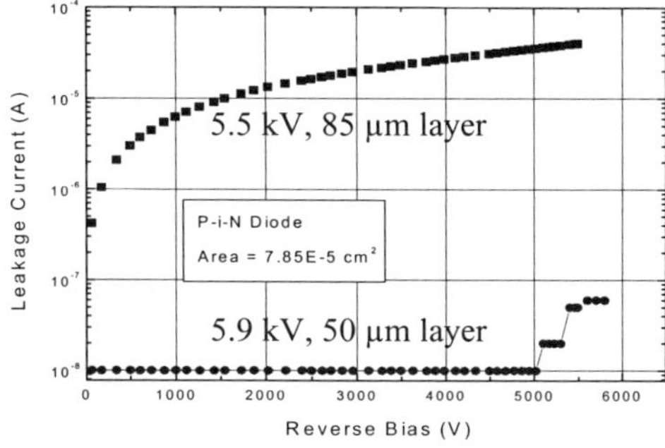

Figure 6. Reverse bias I-V characteristics of diodes.

In addition to a high reverse blocking voltage, a diode must also have a low on-state voltage drop. In the diodes of the first batch achieving 5.5 kV, the forward voltage drop at 100 A/cm^2 was 5.4 V. In the second batch of diodes, the contacts were poorly annealed and had a much higher resistivity which resulted in a larger voltage drop. However, we have recently produced diodes with a 50 μm drift layer which have a forward voltage drop of only 6.5 V at a current density of 1000 A/cm^2. This layer had a blocking voltage of 5 kV.

The formulation presented in Ref. 10 was used along with the on-state voltage drop observed in the low doped drift region in the device to estimate a carrier lifetime. The results indicated that the carrier lifetime of the low doped epitaxial layers used in these diodes was to be greater than 1 μs [11].

SUMMARY

Highly uniform epitaxial layers of 4H-SiC have been achieved by optimizing growth conditions in a hot–wall CVD reactor. The thickness and doping uniformity for 3" diameter wafers have been shown for the first time at <2% and <10%, respectively. The thickness and doping uniformities for other wafer sizes were also shown with results at or below 2% in all cases.

A simple model was shown for the incorporation of Al in 4H-SiC showing that increased growth rate resulted in higher doping levels. It was proposed that this behavior was due to increased surface defects and/or roughness. SIMS data was presented that showed sharp transitions across different doping levels and even dopant types. These changes were possible without changes in deposition conditions which maintains the resulting surface morphology.

High voltage P-i-N diodes were fabricated on low doped epitaxial layers of two different thicknesses. The best forward voltage drop of 6.5 V at 1000 A/cm^2 was achieved using an epitaxial layer thickness of 50 μm. Improved processing resulted in both a higher blocking voltage of 5.9 kV and dramatically improved leakage current behavior. Modeling showed that the higher voltage diode approached theoretical limits for breakdown behavior in SiC for its thickness and doping level. From the on-state resistance the carrier lifetime in the thick, low doped blocking layer was estimated to be greater than 1 μs.

ACKNOWLEDGEMENTS

This work was partially supported by DARPA through Air Force Contract No. F33615-C-5426 and by ONR through its MURI program Contract No. N00014-95-1-1302.

REFERENCES

1. S.Sriram, G.Augustine, A.A. Burk, Jr., R.C. Glass, H.M. Hobgood, P.A. Orphanos, L.B. Rowland, R.R. Siergiej, T.J. Smith, C.D. Brandt, M.C. Driver, and R.H. Hopkins, IEEE Electron Device Lett. **17**, 369, (1996).

2. O. Kordina, J.P. Bergman, A. Henry, E. Janzén, S. Savage, J. Andre, L.P. Ramberg, U. Lindefelt, W. Hermansson, and K. Bergman, Appl. Phys. Lett. **67**, 1561, (1995).

3. J.N. Shenoy, J.A. Cooper, Jr., and M.R. Melloch, IEEE Electron Device Lett. **18**, 93 (1997).

4. A.K. Agarwal, J.B. Casady, L.B. Rowland, S. Seshadri, W.F. Valek, and C.D. Brandt, Submitted to *IEEE Electron Device Lett.*

5. O. Kordina, C. Hallin, A. Henry, J. P. Bergman, I. Ivanov, A. Ellison, N. T. Son, and E. Janzen, Phys. Stat. Sol. B **202**, p. 321 (1997).

6. O. Kordina, A. Henry, E. Janzen, and C.H. Carter, Jr., Silicon Carbide, III-Nitride and Related Materials **2**, p. 107 (1997).

7. T. Kimoto, A. Itoh, N. Inoue, O. Takemura, *et al*, Mater. Sci. Forum **264-8**, 675 (1998).

8. D.J. Larkin, Phys. Stat. Sol. (b) **202**, 305 (1997).

9. A.A. Burk, Jr., and L.B. Rowland, Appl. Phys. Lett. **68**, 382 (1996).

10. B. J. Baliga, *Power Semiconductor Devices* (PWS Publishing, 1996).

11. Paul Chow, Rensselaer Polytechnic Institute, private communication.

HOMO-EPITAXIAL AND SELECTIVE AREA GROWTH OF 4H AND 6H SILICON CARBIDE USING A RESISTIVELY HEATED VERTICAL REACTOR

Ebenezer Eshun*, Crawford Taylor*, M. G. Spencer*, Kevin Kornegay**, Ian Ferguson***, Alex Gurray*** and Rick Stall***
*Howard University, Materials Science Research Center of Excellence, Washington, DC 20059
**Department of Electrical Engineering, Cornell University, Ithaca, NY 14853
***EMCORE Corporation, SOMMERSET, NJ 08873.

ABSTRACT

Silicon carbide technology is rapidly developing into a production process. This is due to rapid progress in the development of high quality epitaxy and substrates. We report on the development of a resistively heated vertical reactor and it's application to homo-epitaxy and selective area growth. Epitaxial growth of 4H and 6H-SiC requires high temperatures (in excess of 1500°C). In this work we investigate resistive heating which offers advantages in cost, temperature uniformity and power efficiency of heating. However, resistive heating presents major technological challenges. Due to the power efficiencies possible with resistive heating we are able to obtain temperatures in excess of 1750°C. Using this system we have grown "state of the art" 4H and 6H-SiC. At 1580°C our background doping is p-type at a level of $3\text{-}5 \times 10^{15} \text{cm}^{-3}$ as measured by capacitance techniques in agreement with earlier results presented by investigators from Siemens Corp using a similar system. The background concentration increases by about an order of magnitude at 1680°C. This system has also been used to perform experiments with selective area growth of SiC using a graphite mask. This masking technology allows for the growth of SiC in specific regions at elevated temperature in excess of 1600°C.

INTRODUCTION

Silicon Carbide (SiC) has several superior properties as compared to other semiconductor materials. These properties include high thermal conductivity, inertness to chemical reactions, hardness, high breakdown field, high saturated electron drift velocity. These superior properties, makes SiC an excellent candidate for high power and high temperature electronic devices. This and the availability of high quality epitaxy and SiC substrates have resulted in the rapid development of the SiC technology. In the epitaxial growth of 6H and 4H-SiC, high temperatures in excess of 1500°C are required. High voltage, bipolar devices require thick, low doped epitaxial layers with long carrier life times, and are most conveniently grown by Chemical Vapor Deposition (CVD)[1].

The reactor used in this study is a EMCORE Corporation system, equipped with a vertical chamber and a rotating disk technology. Based on the experience of EMCORE in the field of III-V epitaxy, our reactor is a combination of the rotating disk technology and the high temperature needs of homoepitaxial SiC growth [2]. In conjunction with EMCORE Corporation, we have been able to develop a usable resistive heating system as an alternative to RF heating. The construction of the reactor is shown in figure 1.

Mat. Res. Soc. Symp. Proc. Vol. 572 © 1999 Materials Research Society

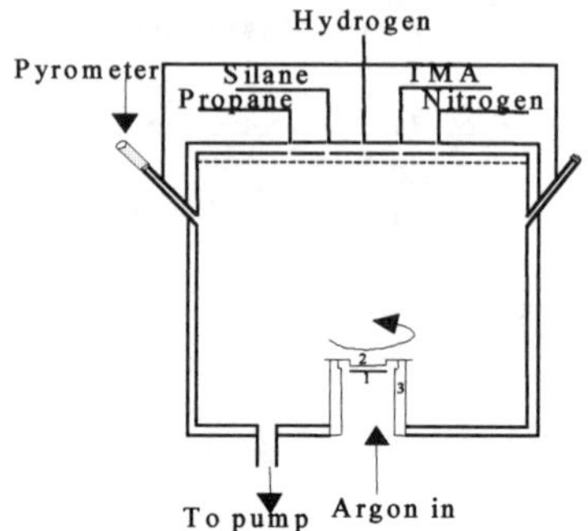

Figure1: Diagram showing the construction of the above-mentioned resistively heated CVD reactor. (1) Resistively heated filament (2) wafer carrier or susceptor and (3) graphite cup.

It consists of a double walled, water-cooled vertical chamber, made out of stainless steel. It has three view ports one of which is used for pyrometric temperature measurement. It is also equipped with a RHEED gun for in-situ RHEED measurements to characterize the grown layers immediately after growth. The susceptor is resistively heated with a graphite filament with power consumption at 1700°C of 8kW.

The challenge in the development of the resistive heating system has been the decrease of the filament lifetime due to hydrogen etching. Hydrogen is present in the reactor as a carrier gas and decomposition product of the reactant gases (SiH_4 and C_3H_8). Hydrogen etching is a function of the filament temperature and the partial pressure of hydrogen in the region of the filament. Studies at EMCORE indicate that filament failure is likely after a 5% reduction in the filament cross sectional area. In order to increase the filament lifetime, hydrogen should be eliminated from the filament region and/or the filament temperature minimized. Our current design involves the use of a thin (20mil thick) graphite susceptor which drops into an open cup resulting in a 20mil to 40mil air gap between the filament and the susceptor. This configuration is used because in minimizes the temperature offset between the filament and susceptor. The base of the cup is purged with argon as shown in figure 1 to reduce the partial pressure of hydrogen in the vicinity of the filament. If the hydrogen can be eliminated the filament lifetime will be determined by carbon evaporation. For our geometry the predicted filament lifetime is over 1000 hours at a filament temperature of 2000°C. Because we have not been able to completely exclude hydrogen from the filament the current filament life is still determined by hydrogen etching. We have been able to achieve filament life times of 15 to 20 hours by this method.

In our bid to achieve higher filament lifetimes (>100 hours), we are developing a growth process, which uses argon as the shroud, thus reducing the volume of hydrogen significantly. We are also designing a new cup and susceptor, which will lock in place to reduce the flow of hydrogen into the cup and allowing a higher flow of argon for purging.

EXPERIMENTS

Homoepitaxial Growth / Reactor Calibration

For the growth of α- SiC, we used (0001) SiC substrates with the polished growth surface tilted at angles between 3° and 8° from the basal plane . This tilt angle ensures that "step-flow" homoepitaxial growth occurs and that good morphology is obtained [3],[4],[5],[6]. In order to grow high quality α- SiC epitaxial layers and also to calibrate our CVD reactor, we have been working on optimization of the growth parameters including growth temperature, chamber pressure, reactant gas flows, growth rates, carrier gas flows and rotation speed. In our system a shroud flow of 21-30 slm is required to stabilize the reactor. Typical growth parameters are shown in table I.

Table I: Typical growth parameters for the homoepitaxial growth of α- SiC using a resistively heated CVD reactor.

Growth Temperature	Chamber Pressure	C_3H_8 Flows	SiH_4 Flows	Shroud (H_2) Flows	Rotation Speed
1500°C	50 Torr	8.4sccm	275 sccm	21 slm	750 rpm
to	to	to	to	to	
1700°C	250 Torr	100 sccm	550 sccm	30 slm	

Propane (C_3H_8) and silane (SiH_4) were used as the reactant gasses with hydrogen as the carrier gas. The substrates were heated resistively as discussed previously. Initial calibration growths were performed without in-situ doping of the epitaxial layers thus we were able to find out about the levels of background impurities during the growth process. We have performed growth experiments of α- SiC at 1580°C and elevated temperatures (1680°C). We calibrate our reactor at 1420°C by observing the melting of silicon.

Selective Area Growth of α- SiC

In order to selectively grow α- SiC, we mask parts of the substrate using our high temperature graphite mask[7]. The graphite mask after growth was removed by oxidizing it and then etching it in an HF/HNO_3 (1:2) etch solution. The growth of the selective epi is verified with a scanning electron microscope (SEM).

RESULTS AND DISCUSSIONS

In the characterization of our epitaxial layers, we employ several methods to access the quality of the grown layers. The surface morphology is assessed using an atomic force microscope (AFM) and an optical microscope .Other characterization methods include two point probe to check the breakdown voltage and capacitance-voltage (CV) measurements (by depositing Al schottky diodes) using a doping profiler which gives a quick assessment of the background dopant type, background doping levels and hence a doping profile. Figure 2 shows the doping profiles for two 4H samples grown under the same conditions except temperature.

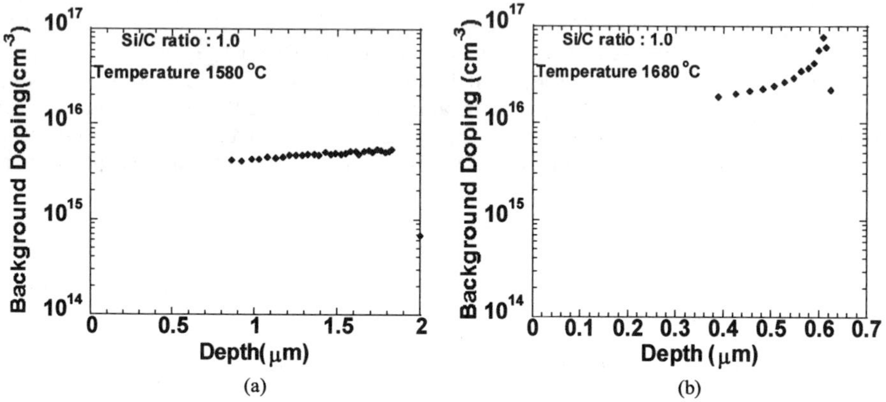

Figure 2: Doping profiles for unintentionally doped p-type 4H SiC epitaxial layers at (a) 1580°C and (b) 1680°C

These background doping levels are due to the fact that we do not use SiC coated parts, hence although they are reasonably low, the use of SiC coated parts will enable us to achieve even lower backgrounds.

High temperature growth is advantageous because higher growth rates could be achieved to satisfy the thick layer requirements of SiC power devices. Results at the elevated temperatures show significant step bunching as shown in figure 3 and also higher background doping levels which we believe comes from the graphite out gassing at the higher temperature. Experiments with the selective area growth show promising results. We see nucleation on the mask after growth including some etching as shown in figure 4. The SEM picture shows the step from substrate to epi. The rough edges are due to SiC growing into the undercut in the graphite mask made during fabrication.

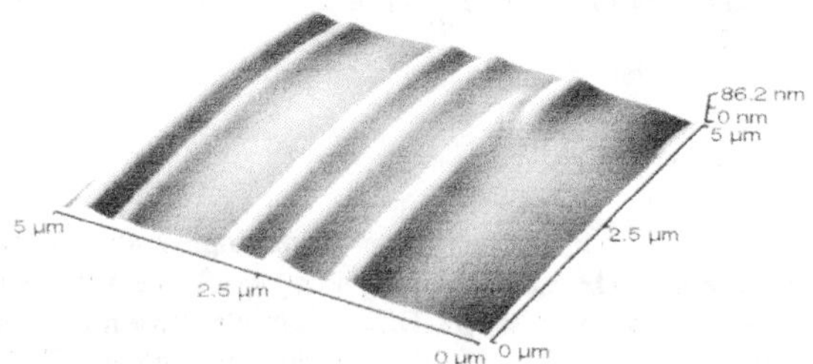

Figure 3: AFM picture of 4H SiC epitaxial layer, grown at 1680°C. Significant step bunching is observed.

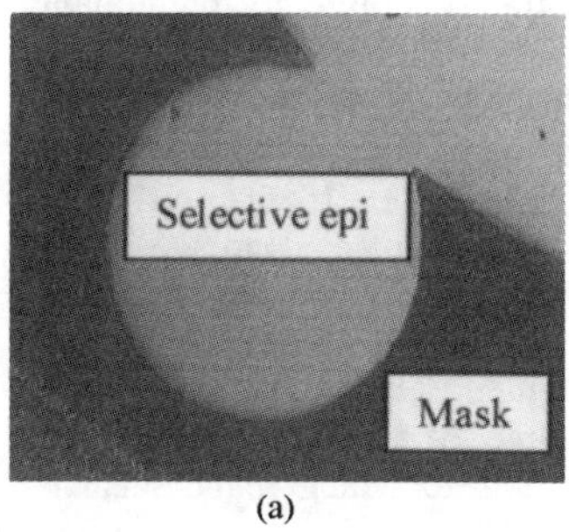

(a)

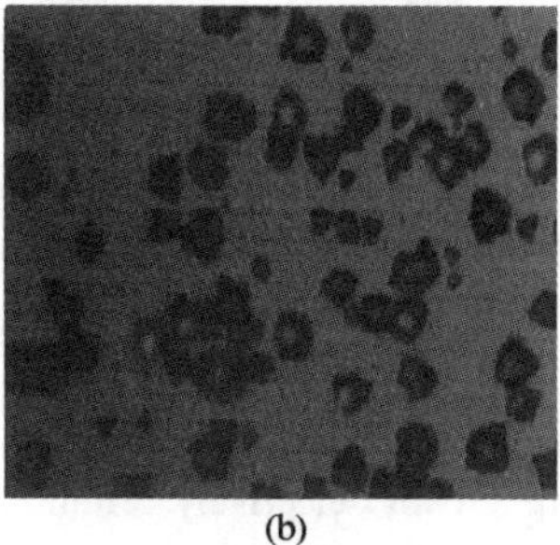
(b)

Figure 4: Optical microscope picture of (a) selectively grown 4H SiC epitaxial layer(magnification 162x) (b)graphite mask surface showing nucleation during growth (magnification 1618x).

In order to determine the growth rates, we determine the layer thickness. We have developed the first direct measure of SiC epi thickness measurement by measuring the height of our selectively grown epi as shown in figure 5, this is a unique advantage of our selective epi technique.

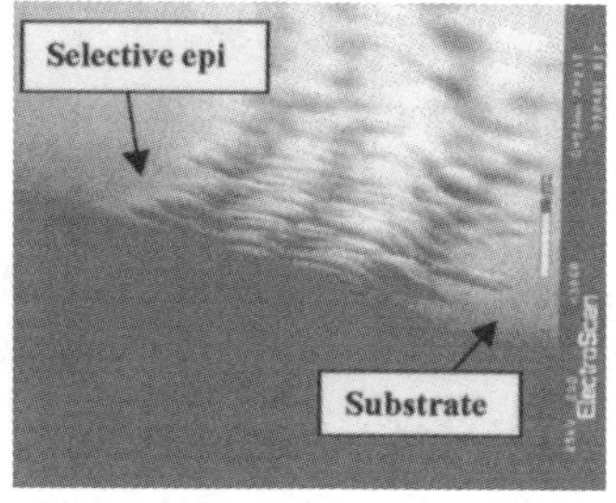

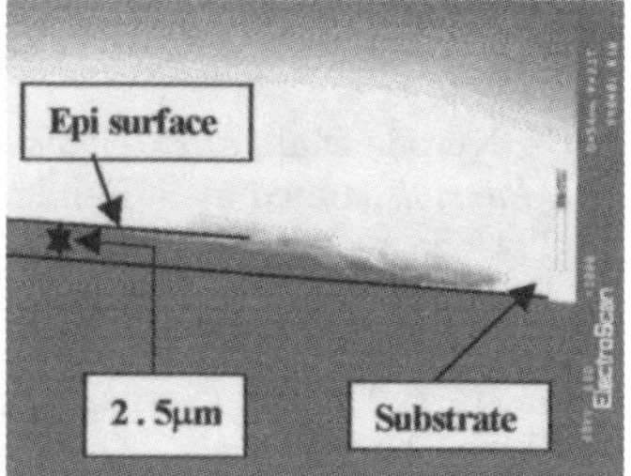

Figure 5: SEM picture showing selectively grown 4H SiC and a direct measure of epi thickness. Rough edge is due to growth of SiC into the undercut in the graphite mask.

The graphite mask is easily removed with a silicon etch, leaving the masked area clean without degradation.

CONCLUSION

We have successfully developed a usable resistively heated vertical CVD reactor for the growth of SiC. Filament lifetimes of 15 to 20 hours have been achieved. These lifetimes are due to the fact that we have not been able to totally eliminate hydrogen from the vicinity of the filament. Because we know the degradation mechanism, we have various plans to increase the lifetime further. We have grown "state of the art" α- SiC epi layers with this reactor, with a growth rate of 5μm/hr for Si/C ratio of 1.0. Higher growth rates can be achieved with higher

Si/C ratios, this is possible because we have a wide window for the Si/C ratios for high quality epi growths. We have found the background doping to be p-type ~ 5×10^{15} cm^{-3} as reported in [2]. The high temperature capability of resistive heating has enabled us to perform growth experiments at elevated temperatures (1680°C). Significant step bunching is observed with an order of magnitude higher background doping relative to growths at 1580°C. Process optimization is underway to obtain quality epi layers at faster growth rates, an advantage of high temperature growth. We have successfully grown high quality selective epi (using a graphite mask) comparable to that grown without masking. Using our selective epi technique, we have developed the first direct measure of SiC epi thickness measurement, resulting in an accurate determination of growth rates. Critical device areas like the channel of a power MOSFET can therefore grown selectively, eliminating the ion implantation damage which reduces carrier mobilities.

ACKNOWLEDGEMENTS

The authors acknowledge the support of the MURI on Manufacturable Power Switching Devices, contract manager John Zolper.

REFERENCES

[1]   O. Kardina et al., in *Growth and Characterization of SiC Power Device Material*, (Linköping Studies in Science and Technology. Dissertations No. 352 1994) pp. 47

[2]   R. Rupp et al., in *Journal of Crystal Growth*, 146(1994) 37-41

[3]   N. Kuroda et al., *in Extended Abstracts of the 19th Conference of Solid State Devices and Materials*, edited by S. Furukawa, (Tokyo, Japan 1987) pp. 227

[4]   T. Ueda et al, in *J. Crystal Growth*, 104, 695 (1990)

[5]   H. Matsunami et al., Mater. Res. Soc. Symp. Proc. 162, 397 (1990)

[6]   H. Matsunami et al., in *Amorphous and Crystalline Silicon Carbide*, edited by G.L Harris and C.Y.W Yang, (Springer Proceedings in Physics 1989, Vol.34) pp. 34-39

[7]   Chris Thomas et al., *Annealing of Ion Implantation Damage in SiC using a Graphite Mask*, presented at the 1999 MRS Spring Meeting, San Francisco, CA, 1999 (unpublished).

Properties of 4H-SiC by Sublimation Close Space Technique

S.Nishino, K.Matsumoto, Y.Chen and Y.Nishio
Department of Electronics and Information Science
Faculty of Engineering and Design, Kyoto Institute of Technology
Matsugasaki, Sakyo-ku, Kyoto 606-8585, Japan

ABSTRACT

SiC is suitable for power devices but high quality SiC epitaxial layers having a high breakdown voltage are needed and thick epilayer is indispensable. In this study, CST method (Close Space Technique) was used to rapidly grow thick epitaxial layers. Source material used was 3C-SiC polycrystalline plate of high purity while 4H-SiC(0001) crystals inclined 8° off toward <1120> was used for the substrate. Quality of the epilayer was influenced significantly by pressure during growth and polarity of the substrate. A p-type conduction was obtained by changing the size of p-type source material. The carrier concentration of epilayer decreased when a lower pressure was employed. Schottky diode was also fabricated.

INTRODUCTION

SiC is suitable for high power devices because of its wide-bandgap and high thermal conductivity. To make a SiC device having high breakdown voltage, thick epitaxial layers are needed. In conventional CVD method (in which silane and propane are normally used), epitaxial growth rate is about 3 μm/h . From industrial point of view, rapidly grown thick epilayers are required. Sublimation epitaxy (CST : close space technique) has been demonstrated to grow thick epilayers at higher growth rate [1-5].

A close space technique has two advantages compared to other growth methods. One is that the growth apparatus has a simple configuration. The other is the ability to keep the growth system pure since only two materials are needed, one for the source and the other for the substrate. This growth method had been applied to grow various semiconductors such as Ge [6]. This technique has also been applied to SiC [7,8]. It is basically the same as conventional sublimation method [9]. A merit of this configuration is that it minimizes the area of graphite wall in the crucible. Normally in the conventional sublimation method unwanted carbon species comes from the wide area of the graphite wall , and then the quality of epilayers degrades. In CST, unwanted C species are minimized and sublimed vapor is transferred to the substrate in quasi-thermal equilibrium condition.

We have previously studied homoepitaxy of 4H-SiC by CST[5]. The surface morphology was observed by optical microscope in the Nomarsky mode and AFM (Atomic Force Microscope). Crystallinity and purity of the epilayer were characterized by Raman spectroscopy. Smooth layers without step-bunching were obtained at a pressure lower than 40 Torr. The surface morphology of the epilayers was mainly influenced by the off-angle of the substrate and the ambient pressure.

EXPERIMENT

3C-SiC polycrystalline plates of semiconductor grade (MUHSIC:Admap Ltd.) were used as SiC source material while 4H-SiC (Si-face, C-face) substrates of commercially

Mat. Res. Soc. Symp. Proc. Vol. 572 ©1999 Materials Research Society

obtained   wafers inclined   8º off (0001) towards <1120> were used as substrates. Size of the substrate was 5 mm x 5 mm. In order to have an accurate spacer height between source and substrate, a graphite spacer with a square opening size of 4 mm x 4 mm was used. The source and substrate were set in a graphite crucible. A crucible with outer diameter of 34 mm, inner diameter of 24 mm and 33 mm long including cap was used. The crucible was thermally shielded by graphite foam. A vertical quartz tube cooled by an air-fan was used as the reaction tube. The temperatures of top and bottom of the crucible were measured simultaneously by two-color pyrometers(Chino IR-AQ). The experiments were carried out in Ar atmosphere. The crucible in the reaction tube was heated by an rf generator at a frequency of about 45kHz. Our typical growth conditions were as follows: growth temperature of source material (Tg) 1900-2100℃, temperature gradient (grad T) 3.5℃/mm, growth pressure (P) 0.3-100Torr, distance ($\triangle$X) between source and substrate 1.5 mm.

The 3C-SiC plate was rinsed in HF to remove native oxide from the surface. The substrate was also cleaned with HF before setting. The crucible was baked at around 2000℃ in Ar atmospheric pressure before setting the substrate and source plate.

RESULTS AND DISCUSSION

The growth rate of epitaxial layers (Vg) was investigated by varying the substrate temperature with $\triangle$X~1.5 mm, P~10 and 100 Torr, and grad T~ 3 ℃/mm . The growth rate increased exponentially with growth temperature. The activation energies calculated by using an Arrhenius plot was 162 kcal/mol at 100 Torr and 149 kcal/mol at 10 Torr as shown Fig 1. Those two values are in the range of published data [10, 11].   According to the literature, growth rate is limited by two mechanisms , one is surface reaction limited and the other is

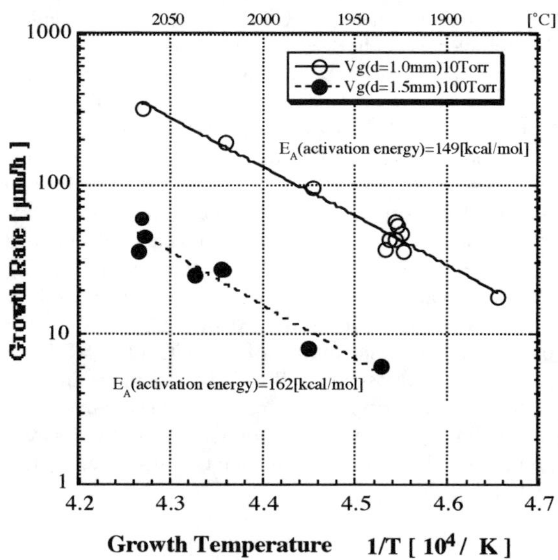

Fig.1 Temperature dependence of the growth rate

diffusion limited [7]. Growth rate at 100 Torr was limited by diffusion , whereas, it was limited by surface reaction at pressure of 10 Torr. To verify those rate limited regime, growth rate dependence on Ar pressure was measured. Growth rate dependence on pressure is shown in Fig.2. When the pressure was low , the growth rate increased. At lower pressure, the source was activated and the concentration of chemical species increase. Generally speaking, growth rate (Vg) is determined by three factors: etching (Ve), diffusion (Vd) and crystallization (Vc) velocities. Ve and Vc are related to surface reaction rate(Vs). Consequently, Vg is expressed by the competing processes of Vs and Vd. In Fig.2, growth rate decreased at higher pressure indicating that the growth rate is limited by diffusion process [7]. However, when the pressure was low, growth rate increased indicating that growth rate is limited by surface reaction. This rate limiting process might be related to the different value of activation energy at different growth pressure as shown in Fig.1.

Surface morphology of the epilayer on (0001)Si-face grown at a pressure of 10 Torr was smooth as observed by optical microscope. Even though surface scratches remained on the substrate , smooth surface was obtained after the epi-growth. In case of (0001)C-face, the

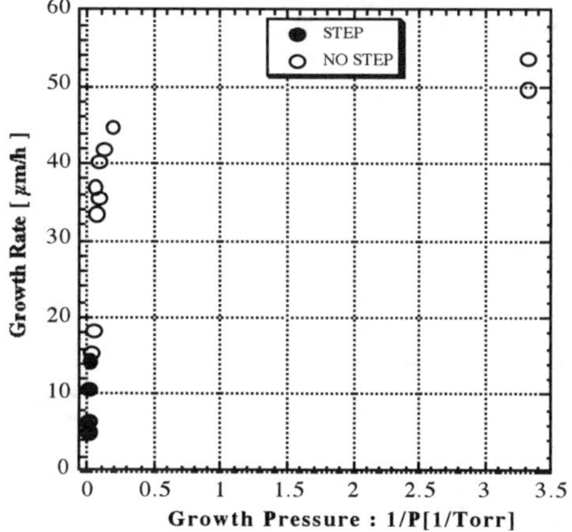

Fig. 2 Pressure dependence of the growth rate

situation was quite different. The surface of as-received C-face was often rough compared to the (0001)Si-face. A lot of polishing scratches were observed on it. To create a smooth surface of the substrate, two kinds of treatments were carried out. One method was to use molten-KOH etching and the other was RIE etching using CF_4+O_2. When the as-received substrate was used for the growth, the scratched pattern remained in the epilayer even though 40 mm thick epilayer was grown as shown in Fig.3 (a). Morphology of epilayer on

molten-KOH treated substrate was smooth but micropipes appeared which originally existed in the substrate as shown in Fig.3(b). Morphology of the epilayer on RIE treated substrate was also smooth and step-bunching was not observed as shown in Fig.3(c). In these experiments, surface pre-treatment was a key to obtaining smooth surface. In comparing the surface morphology of Si-face and C-face, C-face was very sensitive to pretreatment. However, it is confirmed that the C-face gives a smoother surface, if surface treatment is carefully done.This difference might be related with surface energies of those faces.

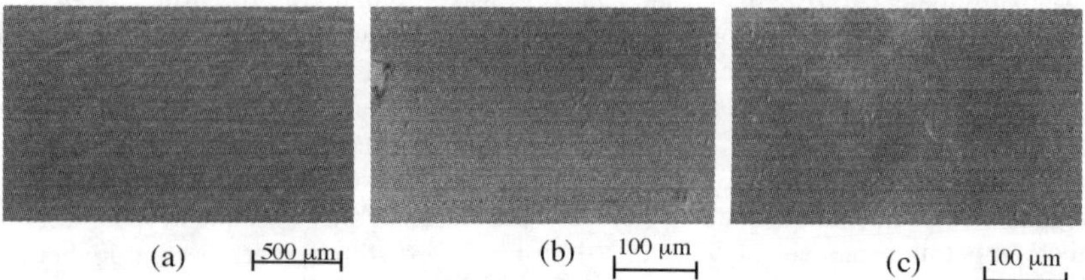

Fig.3   Surface morphoplogy of the epilayers depending various pretreatments.
(a) epilayer on as-received substrate,   (b) epilayer on molten KOH etching substrate and (c) epilayer on RIE treated substrate.

Normally, surface energy of Si-face is high compared with C-face[12].

The purity of the epitaxial layers was investigated by Raman spectroscopy. Free carrier concentration of the epilayer was characterized by measuring the LO-phonon-plasmon-coupled mode in Raman spectroscopy [13]. LO-phonon peaks at around 970 cm^{-1} shifted to lower wave numbers, when the substrate temperature was increased. This shift indicates that carrier concentration also decreased by increasing the substrate temperature. This is related to the site-competition mechanism [14]. As higher concentration of C species was produced at high temperatures, nitrogen was not incorporated in high C ambient, as reported in homoepitaxial CVD growth.

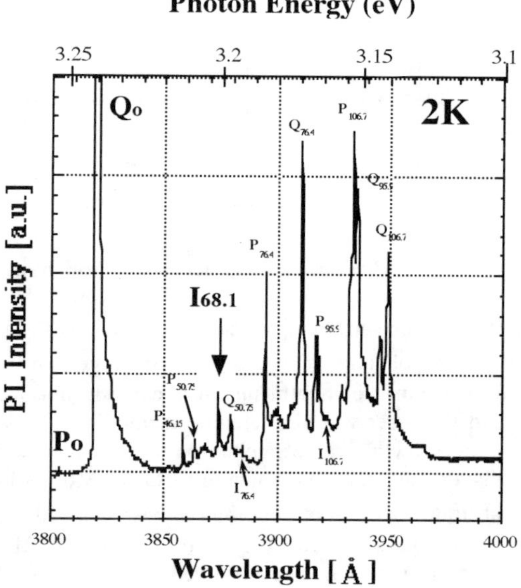

Fig.4 Photoluminescence spectrum of epilayer prepared at 0.3 Torr.

We also investigated the purity of the epilayers by performing photoluminescence measurement at 11K. Several sharp peaks, P-series and Q-series due to recombination of excitons bound to neutral nitrogen donors were observed. Peaks of Po and Qo are zero-phonon lines. The Qo peak, which is due to excitons bound to neutral nitrogen donors substituting C atom on the cubic sites, was observed in the epilayer grown at 0.3 Torr. However, this peak was not observed in the epilayer prepared at a pressure greater than 40 Torr. Those observations indicate that purer samples were grown at lower pressure. Phonon replica of free exciton emission is also observed at 2K as shown in Fig.4. Phonon replica of free-exciton band is also observed. The relative intensity of I-series (free exciton) and Q-series (bound exciton) suggests concentration of nitrogen donors in epilayers [15]. Peak intensity of the ratio between I series and Q-series was 3×10^{-2}, from which the carrier concentration of epilayer is estimated to be the order of 10^{17} cm^{-3} [15].

P-type epilayer was made using p-type source material. A small amount of p-type chip wafer (2mm x 2mm) was placed on the source plate of SiC (undoped n-type). The growth condition was the same as for the undoped case. Photoluminescence of the epilayer is shown in Fig.5. D-A pair emission peaks (Bo,B$_{TO, LO}$,,C$_{TO, LO}$)were observed and indicate that Al was incorporated in the epilayer in this growth method. When a small amount of p-type platelet was used, D-A pair emission band shifted to much lower energy side compared to when entire p-type source was used. This is the result of large separation between donor and acceptor achieved by adding a small amount of p-type source. This suggests the a possibility of controlling doping concentration by changing the size of the p-type source materials.

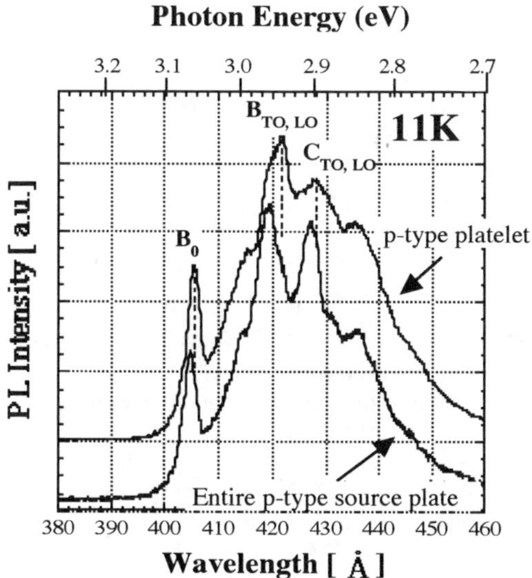

Fig.5   Photoluminescence spectra of p-type epilayers
prepared by using different size of source plates.

To verify the usefulness of CST for large size wafer, we used 4H-SiC substrate of 1 inch size. One inch diameter spacer was made and the periphery (1 mm) of the substrate was held by the spacer. Growth conditions were the same as described previously, and 40 micron-thick epilayer was obtained. Surface morphology was smooth and the entire area was the same polytype as the substrate. However, central part was thicker than the periphery due to the temperature gradient. The temperature uniformity will be optimized by changing the crucible shape.

Schottky diode was also made on the n-type epilayer. Ohmic contact was made by Al evaporation and annealing at 400 ℃. Au was evaporated as Schottky metal. I-V characteristic showed breakdown voltage of 40 V, built-in voltage of 2.3 V and n-value of 1.6. Carrier concentration estimated by C-V measurement was about 7×10^{16} cm^{-3} .

CONCLUSION

We have obtained smooth 4H-SiC homoepitaxial layers at a high growth rate of 40 μm/h. Surface morphology depended on the polarity of the substrate. When careful pretreatment was done for the C-face, smooth and step-bunch-free surface was obtained on C-face. Excitons bound to neutral nitrogen donors were observed by photoluminescence measurement at 2 K, indicating high quality epitaxial layers. A low carrier concentration of 5×10^{16} cm^{-3} was obtained at low pressure. P-type epilayer was obtained by adding small amount of p-type source . The CST is applicable for large size epitaxial layer growth also.

ACKNOWLEDGEMENT
The authors would like to thank to Dr. H. Harima and Dr. S. Nakashima at Osaka University for comment on Raman data. The authors also thank to Dr. A. Henry and Dr. E. Janzen in Linkoping University for LTPL measurement. This work was partially supported by a Grant-in-Aid for Science Research No.09450011 from the Ministry of Education, Science and Culture, Japan and Kansai Electric Power Company, Ion-Engineering Ltd, and NEDO /FED.

REFERENCES
1) T.Yoshida et al. Proceedings of ICSC Ⅲ-N97, Trans Tech Publications (1997), 155.

2) A.K.Georgierva et al.Proceedings of ICSC Ⅲ-N97, Trans Tech Publications (1997)147.

3) M.Syvajarvi et al. Proceedings of ICSC Ⅲ-N97, Trans Tech Publications (1997)143.

4) S.Nishino et al. Mat.Res.Soc.Symp.Proc.vol.483,(1988)307.

5) S.Nishino et al. ECSCRM98, Abstract (1998, Montpellie, France), .

6) F.H.Nicoll, J.Electrochem.Soc.110,(1963)1165.

7) S.K.Lilov et al. Phys.Stat.Sol.(a)37(1976)pp.143.

8) M.M.Anikin et al. Material Science and Engineering,B11(1992)113.

9) Yu.M.Tairov et al. J. Cryst. Growth 36 (1976),147.

10) L.K.Kroko, J.Electrochem.Soc.113(1966)801.

11) J.Drowart, G.de.Maria, J.Chem.Phys.41(1958)1015.

12) E.Pearson et al J.Crystal Growth, 70(1984)33.

13) H.Harima, S.Nakashima, and T.Uemura, J.Appl.Phys.78, (1995)1996.

14) D.J.Larkin et al. Appl.Phys.Lett. 65,(1994)1659.

15) A.Itoh "Control of electrical Properties of 4H-SiC Brown by Vapor Phase Epitaxy for Power Electronic Aplication" Ph.D. Thesis, Kyoto University, Kyoto (1995).

EFFECT OF Ge ON SiC FILM MORPHOLOGY IN SiC/Si FILMS GROWN BY MOCVD

W.L. Sarney[1], L. Salamanca-Riba[1], P. Zhou[2], M.G. Spencer[2], C. Taylor[2], R.P. Sharma[3], and K.A. Jones[4]

[1]Dept. of Materials & Nuclear Engineering, University of Maryland, College Park, MD
[2]Materials Science Research Center of Excellence, Howard University, Washington D.C.
[3]Center for Superconductivity, University of Maryland, College Park, MD
[4]U.S. Army Research Laboratory, Adelphi, MD

ABSTRACT

SiC/Si films generally contain stacking faults and amorphous regions near the interface. High quality SiC/Si films are especially difficult to obtain since the temperatures usually required to grow high quality SiC are above the Si melting point. We added Ge in the form of GeH_2 to the reactant gases to promote two-dimensional CVD growth of SiC films on (111) Si substrates at 1000°C. The films grown with no Ge are essentially amorphous with very small crystalline regions, whereas those films grown with GeH_2 flow rates of 10 and 15 sccm are polycrystalline with the 3C structure. Increasing the flow rate to 20 sccm improves the crystallinity and induces growth of 6H SiC over an initial 3C layer. This study presents the first observation of spontaneous polytype transformation in SiC grown on Si by MOCVD.

INTRODUCTION

Growing high quality SiC films on Si is challenging due to large differences in their thermal expansion coefficients and lattice constants. SiC grown on Si usually forms in the cubic polytype, typically with high densities of stacking faults and dislocations. Films that have a large lattice mismatch with their substrate grow by either the Volmer-Weber or the Stranski-Krastanov mode, both of which lead to island formation.[1] Growth of SiC on Si is limited by the Si species. It is believed that the Si is first absorbed on the surface, and then it migrates to the proper site and is subsequently carbonized on either terraces or at step edges.[2] In an attempt to improve the quality of SiC films grown on Si at 1000° C, we added Ge in the form of GeH_2 during growth. We find that adding Ge promotes two-dimensional growth, improves film quality, and at high concentrations may induce a cubic-to-hexagonal polytype transformation.

EXPERIMENT

The SiC/Si films were grown in a commercial vertical rotating disk CVD reactor. All of the samples were grown with a constant Si/C ratio under identical conditions with the exception of the Ge concentration. SiH_3, C_3H_8, GeH_2, and H_2 were reacted at 1000°C on Si (111) substrates. After growth the samples were characterized ex-situ by AFM, X-ray, RBS, and high resolution transmission electron microscopy (HRTEM). The TEM samples were prepared using tripod polishing and ion milling, and were observed in a JEOL 4000 FX operated at 300 KV.

RESULTS

Transmission electron microscopy was used to obtain high resolution lattice images and diffraction patterns from each sample. The first sample was grown with 0 sccm Ge, but due to

185

the presence of GeH$_2$ in the CVD reactor, it was unintentionally doped with a trace amount of Ge. Figure 1 is the HRTEM image and diffraction pattern from this sample. The distinct spots in the (1$\bar{1}$0) diffraction pattern correspond to the Si substrate. The very light, diffuse ring seen in the diffraction pattern corresponds to the SiC film and indicates that the film is primarily amorphous. The elongated spots over the diffuse ring, however, indicate that some crystallites are formed and they have a preferred orientation with respect to the substrate. The SiC/Si interface in the lattice image is visible and denoted by the arrowheads. The amorphous-like contrast of the film in Fig. 1 could not be the result of ion milling since SiC is much harder than Si and the Si lattice in the image is intact.

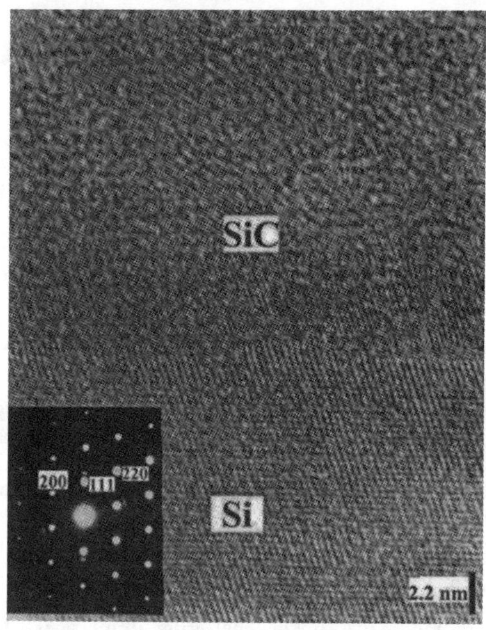

Fig. 1   HRTEM image and diffraction pattern from the sample grown with a trace amount of Ge.

Figure 2 shows that the crystallite size increases and the preferred orientation in the film improves with increasing GeH$_2$ flow. The lattice image of the film with 10 sccm Ge in Fig 2. shows regions with lattice fringes for SiC on a background of amorphous-like SiC. The elongated spots and rings in the (112) diffraction pattern shown as inset to Fig. 2 are less diffuse and sharper than the spots in Fig. 1. This result indicates that the film is polycrystalline 3C SiC with some preferred orientation of the crystallites with respect to the substrate. The SiC/Si interface in the lattice image is very abrupt and obvious. The TEM lattice image of the sample grown with 15 sccm Ge showed slightly improved crystallinity, but otherwise was not appreciably different from the sample with 10 sccm.

Fig. 2   HRTEM image from the sample grown with 10 sccm Ge. The diffraction pattern indexes denoted with the subscript s refer to the Si substrate.

Adding 20 sccm GeH_2 to the reactor led to the multi-polytype structure shown in Figs. 3 and 4. The diffraction pattern (Fig. 5) shows the presence of the 3C and 6H polytypes, and corresponds to the $(1\bar{1}0)$ zone axis for the cubic structure and the $(0\bar{1}10)$ zone axis for the hexagonal structure. The initial structure is 3C SiC, which ends abruptly after approximately 14 nm from the interface and is followed by a 19 nm transistion region consisting of mixed cubic and hexagonal polytypes. The remaining 33 nm of film is predominately 6H SiC with some cubic regions, as shown in Fig. 3. The hexagonal structure is evident by the zig-zag stacking sequence seen in the high resolution image (upper part of Fig. 3) and the presence of the (0001) spots in the diffraction pattern (Fig. 5).

Although the SiC/Si interface is abrupt in all of the samples, it became more visibly obvious for increasing amounts of Ge. Figure 4 shows that the sample containing 20 sccm Ge has a white band at the interface while Figs. 1 and 2 do not. Although surfactants are intended to float at the growth front without incorporating into the film, our RBS results indicate that a small amount of Ge is present at the interface. The white band may represent an interfacial layer of SiGe. Although all of the Ge does not float to the surface, it certainly improves the crystalline quality of the film. Similarly, Hatayama, et al, observed Ge at the SiC/Si interface in samples grown by dimethylgermane source molecular beam epitaxy.[3] In the case of MBE films, however, no transition to 6H SiC was observed.

Fig. 3   HRTEM image from the sample grown with 20 sccm Ge. The image shows from bottom to top the Si substrate, the 3C SiC region, the transition region of predominately 6H SiC, and the remaining region of 6H SiC.

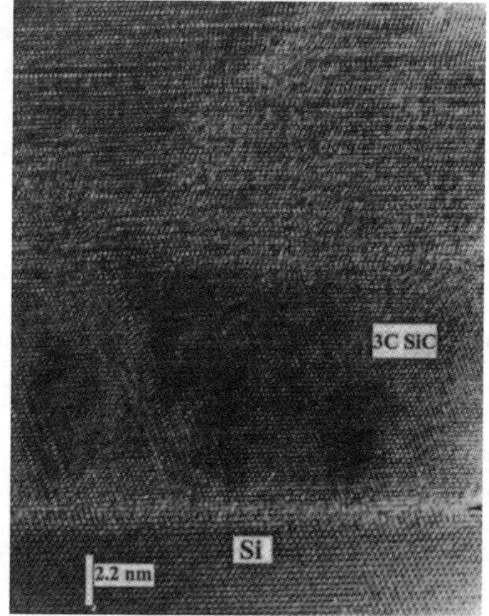

Fig. 4   Higher magnification HRTEM image from the region close to the film/substrate interface of the sample grown with 20 sccm Ge. The upper part of the figure is the transistion region.

188

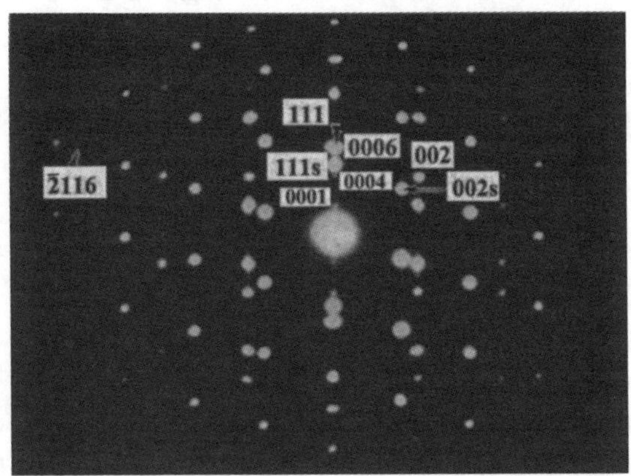

Fig. 5 Diffraction pattern from the sample grown with 20 sccm Ge. The indexes denoted with the subscript s refer to the Si substrate. The spots labeled with three and four indexes represent the cubic and hexagonal polytype of SiC, respectively.

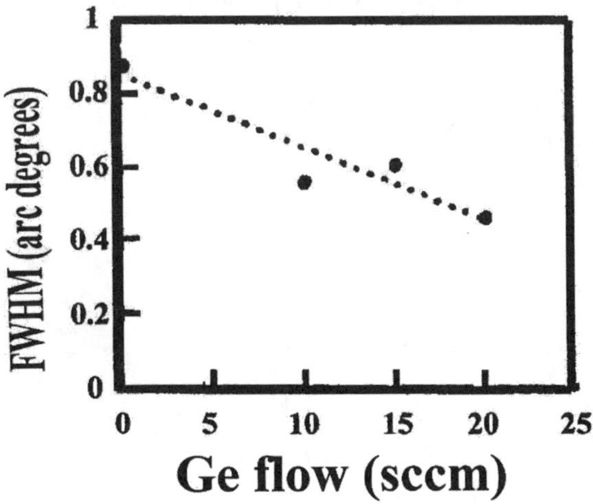

Fig. 6 X-ray FWHM from SiC films vs. GeH$_2$ flow in the reactor.

Ex-situ AFM results indicate that the surface roughness of the film grown with a trace amount of Ge is approximately 16 nm. The roughness decreases with increasing Ge content up to 15 sccm and then increases again for the film grown with 20 sccm Ge. The increase in roughness probably results from the presence of some cubic SiC in the 6H SiC layer. The samples with 10 and 15 sccm Ge contain larger amorphous regions, which would tend to be smoother than the crystalline regions.

X-ray experiments show that the rocking curve FWHM narrows as a function of Ge concentration, as shown in Fig. 6. The x-ray linewidth is reduced by almost a factor of 2, indicating that the crystallite size increases with increasing Ge, in agreement with the TEM results.

CONCLUSION

To our knowledge, there are no reports of SiC films grown on Si with a sudden abrupt transition from the 3C to the 6H polytype. The TEM images show that the crystallinity of 3C SiC films grown on Si(111) by CVD is improved by adding small amounts of Ge during film growth. We have obtained fairly high-quality 6H-SiC films and improved the crystalline quality of 3C SiC films grown on (111) Si at 1000° C by adding 20 sccm Ge. The quality of the SiC films may be improved further by optimizing the growth conditions, particularly the Ge concentration. This growth method could have very important implications in the fabrication of substrates of SiC for the growth of GaN.

ACKNOWLEDGEMENTS

This work was supported by MRCP Army Grant No. DAAL 019523530.

REFERENCES

1. E. Tournie, K. Ploog, Thin Solid Films **231**, 43 (1993).
2. T.Kimoto, H. Nishino, W-S.Yoo, H. Matsunami, J. Appl. Phys., **73** , 726 (1993).
3. T. Hatayama, N. Tanaka, T. Fuyuki, H. Matsunami, J. Elec. Mater., **26**, 160 (1997).

Properties of Heteroepitaxial 3C-SiC Layer on Si Using $Si_2(CH_3)_6$ by CVD

Y. Chen, Y. Masuda, Y. Nishio, K. Matsumoto, and S. Nishino
Department of Electronics and Information Science ,
Faculty of Engineering and Design, Kyoto Institute of Technology,
Matsugasaki, Sakyo-ku, Kyoto 606, JAPAN

ABSTRACT

Single crystal cubic silicon carbide (3C-SiC) has been deposited on Si(100) by atmospheric CVD at $1350^{o}C$ using $Si_2(CH_3)_6$. The 3C-SiC epilayers were characterized by XRD, Raman scattering and photoluminescence (PL). The 3C-SiC distinct TO near 796 cm^{-1} and LO near 973 cm^{-1} were recorded by Raman measurement. The PL spectra of SiC films at 11K included the nitrogen-bound exciton (N-BE) lines, the "defect-related" W band near 2.15eV, and 2.13eV peak corresponding to D-A pair recombination as well as the "divacancy-related" D1 peak at 1.97eV. The thickness dependences of Raman and PL measurement were made and it was observed that tensile stress and strain in films decrease with increasing film thickness. Electrical properties of the films were measured by making schottky diodes and using Van der Pauw method. Above 300K, the electron mobility changed as $\mu_H \sim T^{-1.45 \sim -1.56}$ and the highest mobility was about 400 $cm^2V^{-1}s^{-1}$ at room temperature. In 3C-SiC the scattering processes are affected prominently by acoustic scattering in this temperature range.

INTRODUCTION

The heteroepitaxial growth of 3C-SiC on Si substrates by CVD method has been investigated intensively. This is because commercially available 6H and 4H-SiC wafers still have some problems such as small substrate size and defects like micro-pipes. Atmospheric pressure CVD (APCVD) method has been widely used for the growth of 3C-SiC on Si substrate. It has been reported that SiC films can be grown on Si substrate at a relatively low temperature in APCVD system by pyrolyzing HMDS (hexamethyldisilane: $Si_2(CH_3)_6$), a single-source organosilane precursor which contains direct Si-C bonds. Most of the work on the growth of SiC films using HMDS reported SiC film thickness of about 5μm on Si(111) substrate[1, 2]. However, because the thermal stress in the films on Si(111) is larger than that of Si(100), a number of microcracks exit easily in 3C-SiC films on Si(111)[3]. It is clear that using a (100) Si substrate instead of a (111) Si substrate will reduce thermal stress. A higher quality crystallinity 3C-SiC films can be obtained on Si(100).

In this work, we report the growth of high quality single crystalline 3C-SiC films on Si(100) up to 17.5μm using $Si_2(CH_3)_6$ + H_2 gases by APCVD. The resulting film surfaces are all mirror-like. The growth rate of 3C-SiC films was about 4.3μm/h. This growth rate is rather high compared with SiH_4 + C_3H_8 system. The 3C-SiC epilayers were characterized by XRD, Raman scattering and PL. We successfully obtained high quality single crystal 3C-SiC films and observed the complete line structure of PL at 11K. Thickness dependences of the Raman and PL measurement were determined and stress-related peaks were obtained. Electrical properties of

Mat. Res. Soc. Symp. Proc. Vol. 572 ©1999 Materials Research Society

the films were also measured by making Schottky diode and using the Van der Pauw method.

EXPERIMENTAL

The CVD system used in this study was made from a 50mm diameter air-cooled horizontal quartz tube connected to a rotary pump. A SiC-coated graphite susceptor, supported by a quartz boat, was heated by RF-induction, operating at 450KHz and 5KW. For Raman and PL measurements, n-type Si (100) substrate with a resistivity of 0.01 ~ 0.02 $\Omega \cdot$ cm were used, while films grown on p-type Si(100) substrate was used for Hall measurement. The silicon substrate, cut into 25×15 mm pieces, was placed on the graphite susceptor. The temperature of the substrate was measured by means of an optical pyrometer focused at a hole in the susceptor through a window at the end of the tube. HMDS flow rate was controlled by bubbling H_2 gas through the liquid HMDS. Manual flow control valves were used to control the flow rate. The overall growth process consisted of three steps. (a): Si substrate was etched at 1175°C in HCl(63.1sccm) + H_2(1slm). (b): Si substrate was carbonized for 3 minutes at 1350°C in C_3H_8(1.0sccm) + H_2(1slm). This procedure provides a so-called "buffer layer " for subsequent CVD. This is important to obtain a good crystallinity of SiC films. (c): After carbonization, HMDS gas was quickly introduced for growth at 1350°C. HMDS flow rate was 0.5 sccm and H_2 flow rate was 2.5 slm. Usually, for a low flow rate of HMDS, the final surface of the 3C-SiC films was mirror-like with some haze, and hillocks in the films appeared and became denser with increase of the HMDS flow rate. The 3C-SiC epilayers on Si(100) were characterized by X-ray diffraction (XRD), Raman scattering (excitation by argon-ion laser: 5145 $\text{\AA}$) and photoluminescence (PL: excitation by He-Cd laser with a line at 3250 $\text{\AA}$) at 11K. Electrical properties of the films were also measured by making Au - Schottky diode and Van der Pauw method.

RESULTS AND DISCUSSION

The epilayer surface orientations of each sample were characterized by XRD, and were the same as the substrate Si(100). Only one peak appeared at 2θ=41.5 degree corresponding to the SiC(200) peak. The full width at half maximum(FWHM) of X-ray rocking curves changed from 1.04 to 0.19 degree with increasing film thickness from 0.5 to 17.5μm. This indicates that the crystallinity of the 3C-SiC epilayers improved with increasing film thickness.

Raman scattering measurements were performed at room temperature. For every sample, both the 3C-SiC TO(Γ) phonon near 796cm^{-1} and the 3C-SiC LO(Γ) near 973cm^{-1} were recorded. The intensity of TO phonon peak is less than that of the LO phonon peak. With increase of film thickness, the TO(Γ) and LO(Γ) peaks become sharper and stronger, which is characteristic of improved crystal quality[4].

Fig.1 shows the Raman spectra of 9.1μm and 17.5μm 3C-SiC films on Si(100) and 9.1μm free standing films removed from the Si substrate. For 9.1μm films, the line position of the 3C-SiC LO(Γ) phonon shifts from 973.2cm^{-1} to 974.6cm^{-1} after the Si substrate was removed. This indicates that the existence of the tensile biaxial stress in CVD films on Si substrate, probably due to the thermal coifficent mismatch. A tensile stress in the 3C-SiC films on Si(100) shifts the LO(Γ) phonon to lower energy. The Raman shift between the 3C-SiC on Si and free 3C-SiC films is $\leq 2cm^{-1}$. This corresponds to a biaxial stress of 0.4 -1.0 GPa and an inplane strain of about 0.1% - 0.2% [4]. After removal of Si by HF/HNO3 1:1 etch, the intensity of the TO and LO phonon peaks increase dramatically, and the TO phonon peak becomes much more intense relative to the LO phonon peak. Both effects are the result of the strain relief in the film[5]. Additionally, the LO phonon peak of 17.5μm films on Si shifts to a higher frequency relative to that of 9.1μm films on Si. This means that the tensile stress in films decreases gradually with increase of the film thickness.

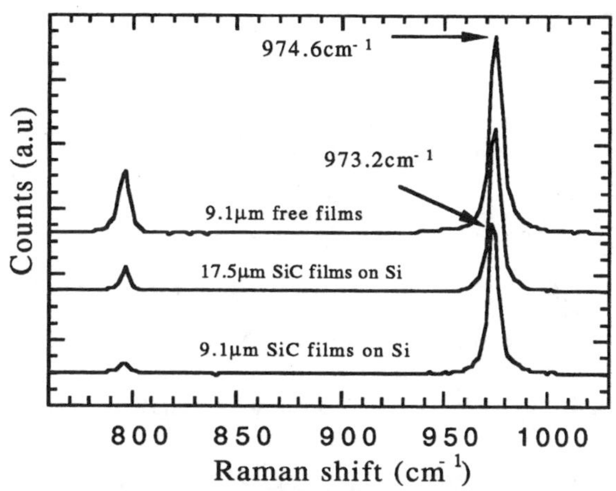

Fig.1 The Raman spectra of 3C-SiC film

Fig.2 shows the CVD 3C-SiC/Si PL spectra for SiC films(3.1 to 17.5μm) at 11K. The spectra have a nitrogen-bound exciton (N-BE) zero phonon line N_0, one-, and two-phonon lines and D-A (donor-acceptor: N-Al: B_0) pair recombination peak as well as D1 line[6, 7]. The D1 center no-phonon line at 1.97 eV, which is caused by divacancy related center, appears and becomes stronger with increasing thickness. This is similar to that reported by Powell et al. [6]. It indicates that a large number of dislocations and other extended defects near interface region are reduced, and recombination due to point defect complexes becomes more prominent gradually in the thicker films.

A peak at 2.13eV is observed in all samples. The peak energy corresponds to D-A pair recombination in 3C-SiC[7]. The peak was confirmed by dependences of spectra on the excitation intensity and on the measurement temperature. The peak energy of D-A recombination shifts to higher energy side with increasing excitation intensity or temperature.With increases of the excitation intensity and temperature, the number of neutral donors decreases more quickly than that of neutral acceptors, because the donor level (53.6meV) is shallower than the acceptor level (269meV). The recombination ratio of close D-A pairs to distant D-A pairs increases, so that the contribution of Coulomb energy between D-A pairs becomes stronger and the peak energy shifts to the high energy side.

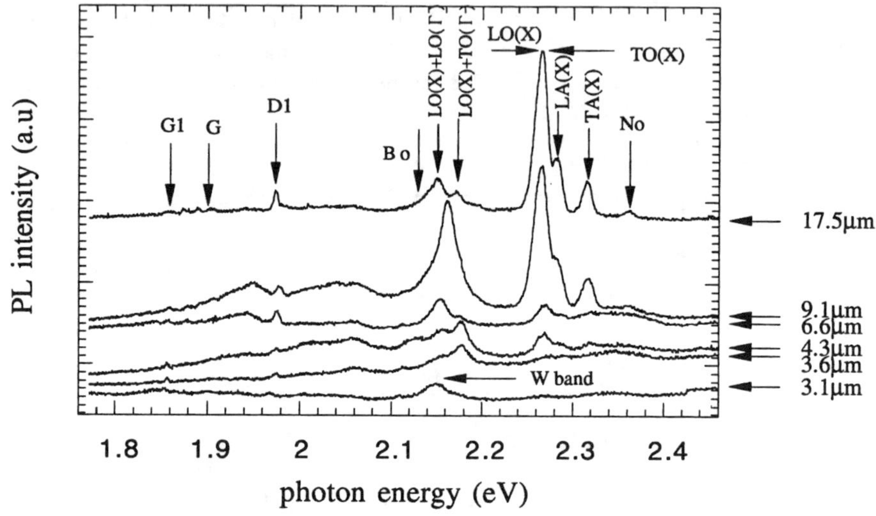

Fig.2 PL spectra of the 3C-SiC films on Si(100) at 11K

A broad band at $\sim$ 2.15eV(W band) is also observed when the SiC film thickness is thinner than about 3.1μm, which is caused by the dislocation and extended defects near the interface region and heavy deformation in these films[6]. The energy region of the W band around 2.15eV overlaps the two phonon energy region of the N-BE. But it is seen that the two phonon replica of N-BE is enhanced and the W band decreases at greater film thickness. This fact shows that a stronger strain exists in the thinner films. The strain is relaxed with increase of the film thickness, corresponding to the Raman result. The N-BE line structure is developed and N_0 and one phonon replicas becomes stronger in the thicker films. The photoluminescence spectra of the thicker 3C-SiC films consist of much stronger N-BE lines and stronger D1 line. The result shows that the density of dislocations and extended defects decreases and the point-defect complexes become more prominent, the crystalline perfection of CVD 3C-SiC films has been improved with increasing film thickness [6].

Hall measurements were done by using the Van der Pauw method. SiC films(5x5 mm) removed from the Si substrate were used. Four Al electrodes of 0.5mm were deposited on the epilayers and ohmic contact was obtained. Hall measurement were performed from 80K to 500K. All the epilayers obtained showed n-type conduction. The carrier concentrations were 1.5 $\times 10^{17}$ at 17.5μm and 3×10^{17} cm^{-1} at 8.3μm.

In Fig. 3, the Hall mobilities of 8.3μm and 17.5μm free films are shown as a function of temperature. The Hall mobility obtained was about 400 cm^{-2}V^{-1}s^{-1} at room temperature. Above

300K, the mobility decreases and changes as $\mu_H \sim T^{-1.45} \sim {}^{-1.56}$. It has been suggested that in 3C-SiC the scattering processes are affected prominently by acoustic phonon scattering in this temperature range[8].

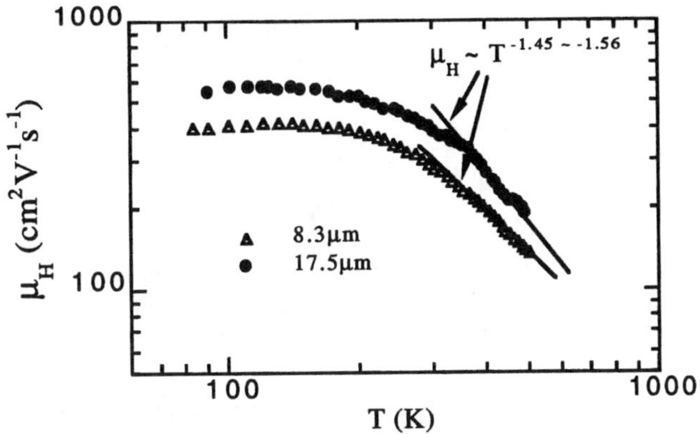

Fig.3 Temperature dependencies of Hall mobilities of non-doped crystals.

The capacitance-voltage characteristics of schottky barrier contacts were measured at a frequency of 1MHz. The voltage dependence of the Au-3C-SiC diode capacitance is shown in Fig.4. As shown in Fig.4, the plots are almost linear. The carrier concentration calculated decreases from 7.3×10^{17} to 3.9×10^{17} for 2.6μm and 6.6μm thickness, respectively, which is in good agreement with Hall measurement. The barrier height calculated from the C-V measurement is about 0.9 ~ 1.35 V.

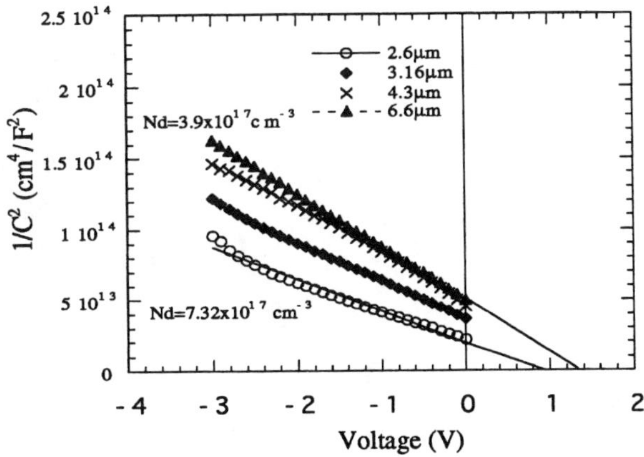

Fig.4 C -V characteristics of Au - 3C-SiC schottky barrier contact

CONCLUSION

To summarize, a high quality single crystal films of 3C-SiC have been obtained using HMDS by APCVD. The results indicated that the defects and tensile stress in the films on Si decrease and the crystalline perfection has been improved with increase of film thickness. The highest mobility was 400 $cm^2V^{-1}s^{-1}$ at room temperature. The electron mobility changed as $\mu_H \sim T^{-1.45} \sim -1.56$ above 300K. Compared with the SiC polytypes ($\mu_H \sim T^{-2.0} \sim -2.6$), the mobility of 3C-SiC is still fairly large at high temperature. It suggests that 3C-SiC is a promising material for the devices operated at high temperature.

REFERENCES

[1] K. Takahashi, S. Nishino, J. Saraie, J. Electrochem. Soc. 139 (1992) 3565-3571.

[2] C. H. Wu, C. Jacob, X. J. Ning, et al., J. Cryst. Growth 158 (1996) 480-490.

[3] H. P. Liaw and R.F. Davis, J. Electrochem. Soc.:SOLID-STATE SCIEVCE AND TECHNOLOGY, Vol. 131 (1984) 3014-3018

[4] Z. C. Feng, A. L. Mascarenhas, W. J. Choyke, et al., J. Appl. Phys. 64(6) (1988) 3176-3186.

[5] J. A. Freitas, S. G. Bishop, A. Addamiano, et al., Mat. Res. Soc. Symp. Proc.46. 581- 586.

[6] W. J. Choyke, Z. C. Feng, J. A. Powell, J. Appl. Phys. 64 (6) (1988) 3163-3175.

[7] W. J. Choyke, L. Pattick, Phys. Rev. B2 (1970) 4959.

[8] K. Sasaki, E. Sakuma, S. Misawa, S. Yoshida, S. Gonda, App. Phys. Lett. 45(1), (1984) 72-73.

CHARACTERIZATION OF P-TYPE BUFFER LAYERS FOR SiC MICROWAVE DEVICE APPLICATIONS

A.O. KONSTANTINOV, S. KARLSSON, P.-Å. NILSSON, A.-M. SAROUKHAN,
J.-O. SVEDBERG, N.NORDELL, C.I. HARRIS, J. ERIKSSON[*] and N. RORSMAN[*].
Industrial Microelectronics Center, IMC, S-164 40 Kista, SWEDEN
[*]Chalmers University, Dept. of Microwave Technology, Göteborg, SWEDEN

ABSTRACT

Low-loped p-type silicon carbide buffer layers are grown by chemical vapor deposition on conducting and semi-insulating substrates. Capacitance-voltage and electrical admittance techniques are developed for accurate non-destructive characterization. The electrical admittance techniques suggested are capable of measuring the resistivity in a very wide range, up to 7 orders of magnitude. MESFET devices using thick buffer layers on conducting substrates are reported with F_i=8.4 GHz and F_{max}=32 GHz.

INTRODUCTION

Low-doped epitaxial p-type SiC is commonly used as buffer-layer material in power microwave transistors. Silicon carbide microwave MESFETs formed on conducting substrates employ low-doped buffer layer to minimize conductive and capacitive losses [1,2]. MESFET devices formed on semi-insulating substrates often use p-type buffer layers to minimize the influence of substrate impurities and of short-channel effects [2,3]. In the present work we report on the growth and electrical characterization of low-doped p-type layers. We also report on the application of low-doped p-buffer layers in SiC microwave device technology.

EXPERIMENT

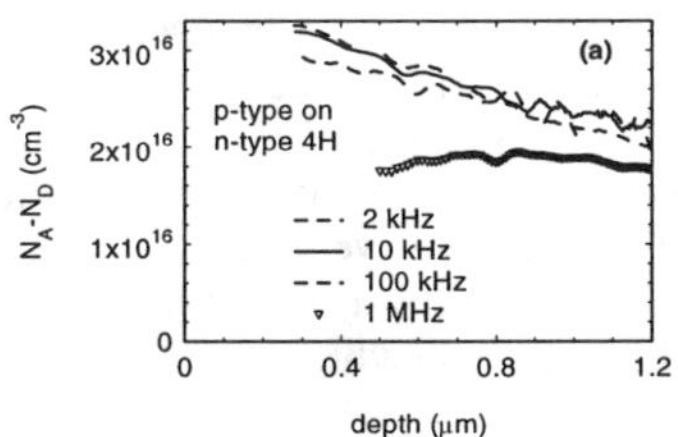

Figure 1. Acceptor concentration profiles for a p-type 4H SiC layers grown on a conducting n-type substrate.

Epitaxial growth has been performed by chemical vapor deposition (CVD) using two types of CVD reactors: a single-wafer horizontal reactor built by Epigress AB [4] and a multiwafer vertical TurboDisk reactor supplied by EMCORE, Inc. Both reactors are operated at a low pressure of about 200-400 mBar. The silicon and carbon precursors used are silane and propane. Trimethylaluminum (TMA) is used as Al dopant source. Nickel Schottky barriers 0.4 mm in diameter were deposited onto the front layer side. A large-area metal contact on the front side or metallized wafer backside was used as an Ohmic contact. Multifrequency capacitance-voltage measurements and measurements of the electrical admittance were performed using the HP 4284A Precision LCR Meter either in a shielded probe station or in a cryostat.

The results of room-temperature C-V measurements were processed using the standard technique to extract the doping profiles. The resulting profiles for a p-type layer grown on a conducting n-type substrate are plotted in Fig. 1. It is seen from the plots that the data tend to converge for low measurement frequencies. By contrast, 1 MHz measurements yield a very large

Mat. Res. Soc. Symp. Proc. Vol. 572 © 1999 Materials Research Society

error for the layers grown on n-type wafer. Lower frequencies are therefore preferable for accurate measurement of doping profiles.

The frequency dispersion of electrical admittance can provide information on the layer conductivity. Plotted in Fig. 2 are the measurement results for the parallel capacitance C_p and the normalized conductance $G_p/2\pi F$ for the layer with the doping profile plotted in Fig. 1. The measurements were performed at zero bias. Only about a half of actual experimental plots are shown in the figure. The peaks of two clearly different types are seen on the plots. The first type appears in the low-temperature range, close to the freezeout point of aluminum acceptors, Fig. 2a. The other type can be observed even at room temperature, as those seen in Fig. 2b.

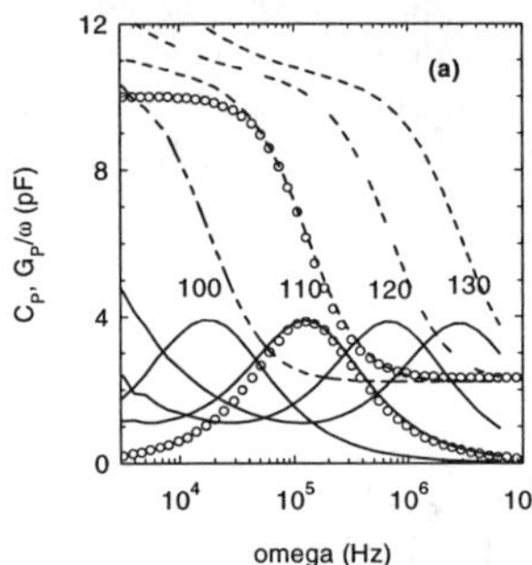

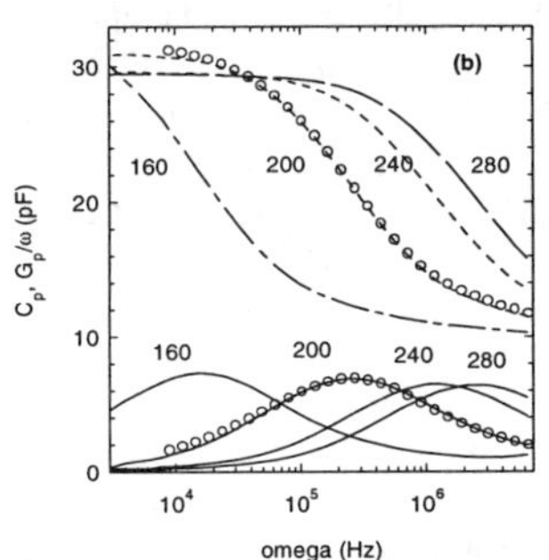

Figure 2. Parallel capacitance (dashed lines) and conductance (solid lines) spectra of a low-doped p-type layer on n-type 4H SiC substrate measured at 100-280K. Circles are the simulation results.

SIMULTATION RESULTS AND DISCUSSION

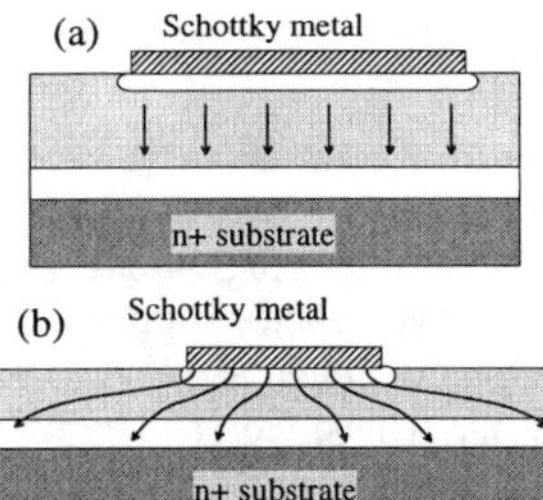

Figure 3 Current pathways for the regimes corresponding to low- and high-temperature admittance peaks.

A simulation has been performed in order to understand the origin of admittance spectra and to extract material parameters from the measurement data. According to the simulation results, the conductance peak originates from the conduction losses in the high-resistivity layer. Two clearly different regimes of AC current transport are observed depending on the layer resistivity. At low temperatures the layer resistivity is very high. The conduction losses are determined by vertical current transport through the layer, whereas the lateral spreading of AC current is negligible. The conductance peak appears close to the dielectric relaxation frequency, as a result of the transition of the p-type material from conductor to insulator. The conductivity σ can be determined from the peak position F_{peak}^{lo} using the relationship:

$$\sigma = 2\pi F_{peak}^{lo} \varepsilon\varepsilon_0 (w_0 + w_{MS} + w_{PN})\left(\frac{1}{w_{MS}} + \frac{1}{w_{PN}}\right), \tag{1}$$

where ε is the permittivity, w_{MS}, w_{PN} and w_0 are the thickness of the Schottky-barrier and p-n junction depletion regions and of the non-depleted layer portion respectively. The parallel capacitance will increase from the low-temperature dielectric limit, $\varepsilon S /(w_0 + w_{MS} + w_{PN})$ to approximately ½ of the Schottky-barrier capacitance, $\varepsilon S /(w_{MS} + w_{PN})$, where S is the contact area. The factor of ½ originates from the series connection of the p-n junction and the Schottky barrier, as it is shown in Fig. 3a.

The high-temperature conductance peak originates from the conduction losses due to lateral current transport. At a high temperature the lateral transport effectively eliminates the contribution of the p-n junction to the device impedance, as it is shown in Fig. 3b. The capacitance increases approximately twofold as result, and a peak of conduction losses appears in the transition region. The layer conductivity can be determined from the conductance peak position F_{peak}^{hi} using the results of the numerical calculation performed:

$$\sigma = \frac{2\pi F_{peak}^{hi} \varepsilon \varepsilon_0}{K} \frac{r_C^2}{w_0 w_{MS}} \tag{2}$$

where r_C is Schottky contact radius w_0 the non-depleted layer thickness, w_{MS} is the depletion region width for the Schottky barrier and the p-n junction respectively. Numerical factor K depends on the ratio of the p-n junction to the Schottky-barrier depletion layer thickness. The simulation yields $K=1.29$ for $w_{PN}/w_{MS}=1$, $K=1.61$ for $w_{PN}/w_{MS}=1.25$ and $K=1.93$ for $w_{PN}/w_{MS}=1.5$.

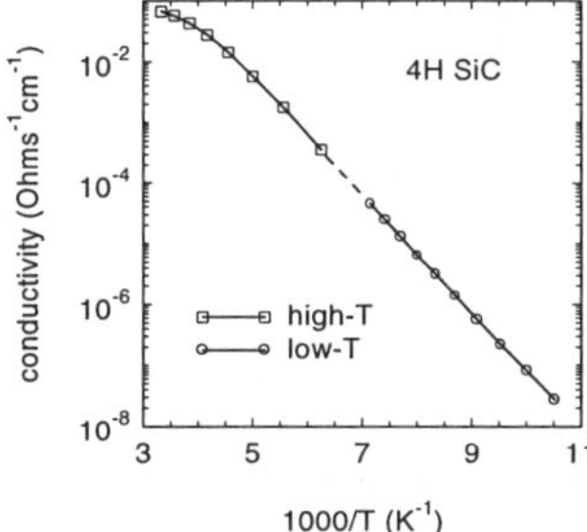

Figure 4. The temperature dependence of conductivity extracted from the admittance data using Eqs. (1) and (2)

A good quantitative agreement is observed between theoretical and experimental spectra for both capacitance and conductance, as one can see it from Fig. 2. Using the data plotted in Fig. 2 we have determined the temperature dependence of the layer conductivity for the temperature range from 100 to 300K. The results are plotted in Fig. 4. The slope of the dependence corresponds to the apparent activation energy of 185 meV, which is close to the activation energy of aluminum acceptors in 4H SiC. The admittance results are in agreement with the data obtained for this sample using the SIMS technique.

Electrical admittance measurements appear to be a convenient characterization tool for low-doped p-type epitaxial layers of silicon carbide. No processing apart from Schottky barrier deposition is required to determine the major electrical properties such as the dopant concentration and conductivity. The technique can therefore be used as a tool for non-destructive characterization and process monitoring.

MATERIAL APPLICATION IN MESFET TECHNOLOGY

Low-doped p-type buffer layers were used for fabrication of microwave MESFET devices. The layers were unintentionally doped in the mid-10^{15} cm^{-3} range, grown in the multiwafer EMCORE reactor to a thickness of 15 µm. The dominant acceptor is boron as it was determined by electrical admittance and SIMS measurements. The active layer was formed by nitrogen implantation, and the capping n$^+$ contact layer was grown by CVD. The wafers were processed to form recessed-gate microwave MESFETs. The devices had a split gate configuration, which is shown in Fig. 4. Sintered nickel is used as an Ohmic contact to source and drain regions, and Ti/Au metal stack was selected as Schottky-barrier contact.

The gate length is varied in the range 0.25-2 µm, the drift region length is 1.5-2.5 µm. The threshold voltage appeared to be lower than expected, only about 2 Volts. The reason for this is presumably related to incomplete nitrogen activation. The on-state currents are correspondingly low, about 40 mA/mm for the devices with a short gate length. The MESFETs, nevertheless, show a good microwave performance. The cut-off frequencies F_t and F_{max} are as high as 8.4 and 32 GHz for the gate length of 0.25 µm. Only a marginal difference is observed between the 0.25 and 0.5 µm-gate devices; the latter have F_t=7.8 GHz and F_{max}=31 GHz. The cut-off frequencies, however, decrease more rapidly with further increase of the gate length. Summary of the frequency performance measured at a drain bias of 40 V is given in Table I. The frequency performance achieved is comparable with the best results reported for conducting substrates [5], in spite of the fact that the channel doping is yet below the optimal value. Further performance improvement can be achieved through optimization of the channel implant dose.

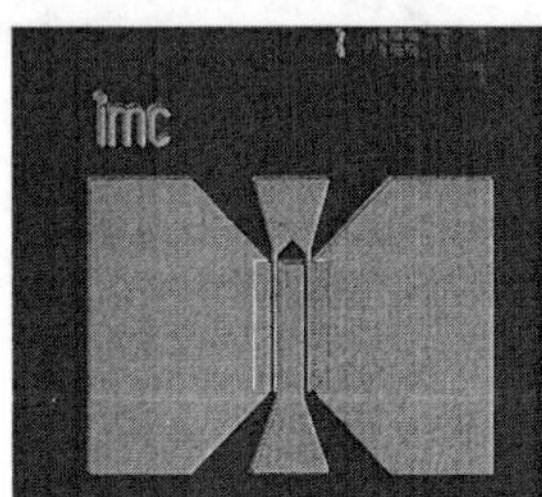

Figure 5. Layout of Microwave
MESFETs

Table I.

Dependence of the frequency
performance on gate length.
Channel width is 200 µm.

gate length (µm)	F_t (GHz)	F_{max} (GHz)
0.25	8.4	32.0
0.5	7.8	30.7
0.75	7.3	28.3
1	5.9	21.2

SUMMARY

In conclusion we have developed the techniques for electrical characterization of low-doped p-type buffer layers for SiC microwave device applications. A very wide range of layer resistivity is available for characterization with the admittance techniques suggested, up to 7 orders of magnitude. Implanted-channel MESFET devices with high operation frequencies have been fabricated using thick p⁻ buffer layers on conducting substrates.

ACKNOWLEDGEMENTS

The authors would like to thank M. Linnarsson for performing the SIMS measurements and to ABB Corporate Research for access to the EMCORE TurboDisk deposition system. The work was supported by the EU TELSiC program and by the Swedish Board for Industrial and Technical Development (NUTEK). The Swedish Nanometric Facilities at Chalmers University were used for electron-beam lithography.

REFERENCES

1. J.W. Palmour, C.E. Weitzel, K.J. Nordquist and C.H. Carter, IOP Conf. Ser. **137**, 495 (1994).
2. S.T. Allen, R.A. Sadler, T.S. Alcorn, J.Sumakeris, R.C. Glass, C.H. Carter, and J.W. Palmour, Materials Science Forum **264-268**, 953 (1998).
3. O. Noblanc, C. Arnodo, E. Chartier and C. Brylinski, Materials Science Forum **264-268**, 949 (1998)
4. Epigress AB, Ideon Science Park, S-223 70 Lund, Sweden.
5. C.E. Weitzel, Materials Science Forum **264-268**, 909 (1998).

OPTICAL CHARACTERIZATION OF SiC WAFERS

J.C. BURTON*, M. POPHRISTIC*, F.H. LONG*, I. FERGUSON**
*Chemistry Dept., Rutgers University, Piscataway, NJ 08854, fhlong@rutchem.rutgers.edu
**EMCORE Corp., Somerset, NJ 08873

ABSTRACT

Raman spectroscopy has been used to investigate wafers of both 4H-SiC and 6H-SiC. The two-phonon Raman spectra from both 4H- and 6H-SiC have been measured and found to be polytype dependent, consistent with changes in the vibrational density of states. We have observed electronic Raman scattering from nitrogen defect levels in both 4H- and 6H-SiC at room temperature. We have found that electronic Raman scattering from the nitrogen defect levels is significantly enhanced with excitation by red or near IR laser light. These results demonstrate that the laser wavelength is a key parameter in the characterization of SiC by Raman scattering. These results suggest that Raman spectroscopy can be used as a noninvasive, *in situ* diagnostic for SiC wafer production and substrate evaluation. We also present results on time-resolved photoluminescence spectra of n-type SiC wafers.

INTRODUCTION

Silicon carbide (SiC) has recently been the subject of renewed interest as an important material for a wide variety of high-power and high-temperature electronic applications. SiC exhibits a large number (250) of polytypes with different structural and physical properties. [1] The polytypes have the same chemical composition but exhibit different crystallographic structures and stacking sequences along the principal crystal axis. Several important polytypes of SiC such as 4H- and 6H- have C_{6v} crystallographic symmetry. In the a-direction 4H- and 6H-SiC are almost identical ($< 1 \%$ change); however, the 4H- polytype consists of 4 units in the c-direction and the 6H- consists of 6 units. Different polytypes have different band gaps, electron mobilities, and other physical properties; for example, 4H-SiC has attracted significant attention due to its high electron mobility and excellent thermal properties. Recently high quality wafers of both 4H- and 6H-SiC have been grown.[2] Wafers of SiC are also a promising substrate for nitride semiconductor growth due to their compatible lattice structure and similar thermal expansion coefficients. In this paper we discuss second-order Raman scattering and resonance enhancements of the electronic Raman scattering from SiC in the near IR.

EXPERIMENT

Raman spectra were recorded using both confocal Raman microscopy and a bulk Raman spectrometer. The bulk Raman system consisted of a Coherent Model INNOVA 90 Ar/Kr laser, a SPEX Model 1877E triple monochromator, and a CCD (charge-coupled-detector) cooled with liquid nitrogen. A liquid nitrogen cryostat was used to take low temperature data. This system has been described elsewhere.[3] All confocal data was collected at room temperature using both a Dilor LabRam system and a Renishaw Series 1000 Raman microscope. Micro-Raman spectra were obtained with laser excitation at 785 nm (1.58 eV) and 633 nm (1.96 eV), which was compared with data taken at 514 nm (2.41 eV), 568 nm (2.18 eV) and 647 nm (1.92 eV) with the bulk Raman spectrometer. The spectral resolution is approximately 1 cm^{-1} for all spectra. The samples were aligned such that the collection of scattered light was in the near back-scattered

Mat. Res. Soc. Symp. Proc. Vol. 572 © 1999 Materials Research Society

geometry perpendicular to the (0001) face of the sample. The polarization of the laser light, both incident and collected, was unspecified. Confocal Raman spectra taken 5-10 microns below the surface of the SiC wafer yielded results very similar to the surface of the SiC; therefore we conclude that the effects we have observed are not surface specific.

4H- and 6H-SiC samples were examined which came from a variety of sources. Most of the samples studied were *n*-type nitrogen doped, except for one semi-insulating and one *p*-type (aluminum) doped sample of each polytype.

RESULTS

Overtone Spectra

Many groups have studied the vibrational spectroscopy and dynamics of 4H- and 6H-SiC.[4] We have examined the two-phonon Raman spectra of 4H- and 6H-SiC. The optical branch of the two-phonon Raman spectra for 4H- and 6H-SiC are shown in figures 1 and 2. Two-phonon Raman spectra are known to be measures of the

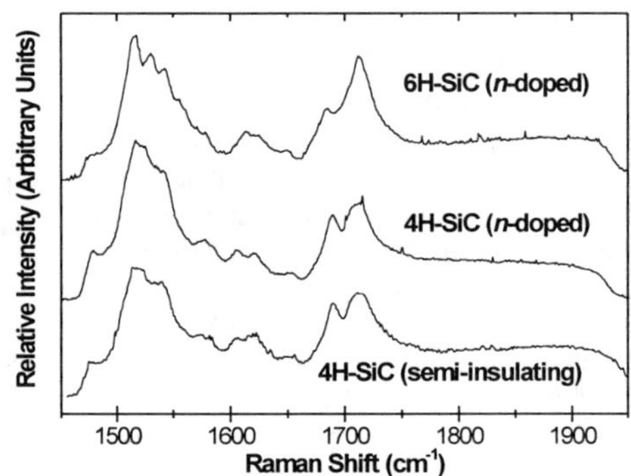

Figure 1: Optical branch of the second-order Raman spectra of 6H- and 4H-SiC at room temperature.

vibrational density of states and therefore these are essential to improved lattice dynamics models of SiC. The two-phonon Raman spectra of the wurtzitic forms of the SiC were found to be far more complex than their 3C (cubic) counterpart. The two-phonon Raman spectra are not simply twice the one-phonon spectra, but reflect the entire three dimensional band structure. The two-phonon Raman spectra for 4H- and 6H-SiC are quite different from each other. A detailed discussed of our peak assignments for the overtone spectra can be found elsewhere.[5]

Electronic Raman Scattering

Figure 2 shows room temperature Raman spectra of a single nitrogen doped 4H-SiC wafer (n= 5.5 x 10^{18} cm^{-3}) which were taken at different laser excitation wavelengths: 514 nm (2.41 eV), 633 nm (1.98 eV) , and 785 nm (1.58 eV). There is a very significant change in the Raman spectra upon changing the laser wavelength.   When the laser excitation is in the red or near IR, clear resonant enhancements are observed for peaks at approximately 400, 530, and 570 cm^{-1}, labeled N_a, N_b, and N_c. At 620 cm^{-1}, a Fano resonance is clearly observed with 785 nm (1.58 eV) excitation. The strongest peak at 530 cm^{-1} (N_b) is broad and asymmetric; this asymmetry is possibly due to another peak near 500 cm^{-1}. The peaks at 530 and 570 cm^{-1} are consistent with previous measurements of n-type 4H-SiC at low temperature.[6] At low temperatures, a sharp peak at 57 cm^{-1} has also been measured.[6] The three high frequency peaks and the low frequency mode at 57 cm^{-1} make for a total of four peaks that can be attributed to nitrogen donors in 4H-SiC. This is more than the number of inequivalent sites, two, for 4H-SiC.

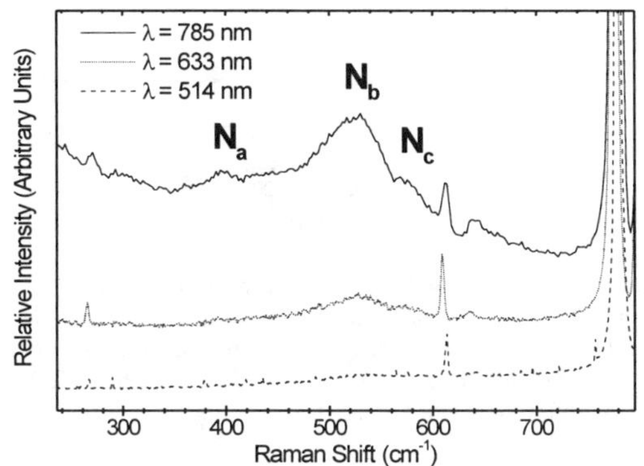

Figure 2: Raman spectra from a single n-type 4H-SiC sample taken at room temperature with different laser excitation wavelengths: 514 nm (2.41 eV), 633 nm (1.98 eV) , 785 nm (1.58 eV). Note the clear appearance of several additional peaks, labeled N_a, N_b, and N_c, as the laser wavelength is tuned to the near IR. We attribute these peaks to electronic Raman scattering from the nitrogen defect levels. The spectra are normalized to the peak at 777 cm^{-1}.

Figure 3 shows Raman spectra for *n*-type 4H-SiC at different doping concentrations, taken at room temperature with 785 nm (1.58 eV) excitation. As the *n*-type nitrogen doping concentration increases from semi-insulating to 7.1×10^{18} cm^{-3}, we clearly see an increase in the intensity of the peaks at approximately 400, 530 and 570 cm^{-1}. The absence of peaks N$_a$, N$_b$, and N$_c$ in the semi-insulating sample demonstrates that these peaks are associated with nitrogen doping. Careful inspection of the peak near 530 cm^{-1} determines that the absolute peak position shifts to smaller values of Raman shift as the nitrogen concentration is increased.

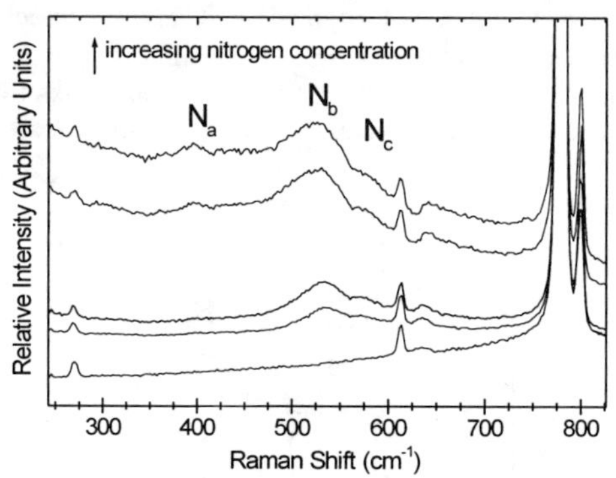

Figure 3: Raman spectra for 4H-SiC taken at room temperature at 785 nm (1.58 eV) for different *n*-type doping concentrations. The peaks under discussion, N$_a$, N$_b$, and N$_c$, are clearly not observed in the semi-insulating sample. The spectra are organized in order of doping concentration, with the highest at the top and the lowest at the bottom: 7.1×10^{18} cm^{-3}, 5.5×10^{18}, 2.6×10^{18}, 2.1×10^{18}, and semi-insulating. The spectra are normalized to the peak at 777 cm^{-1}.

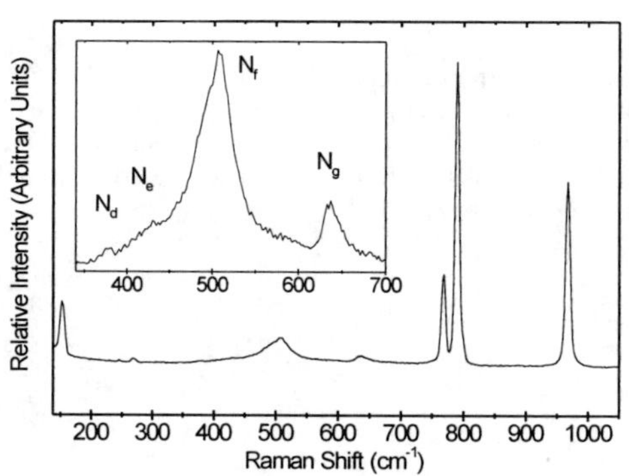

Figure 4: Raman spectrum from 6H-SiC with *n*-type doping concentration of 2.1×10^{18} cm^{-3}, taken at room temperature with excitation at 785 nm (1.58 eV). Insert is an enlargement of the region between 350 and 700 cm^{-1}. Note the appearance of peaks attributable to nitrogen labeled N$_d$, N$_e$, N$_f$, and N$_g$.

Figure 4 shows the Raman spectrum for a *n*-type nitrogen doped (2.1 x 10^{18} cm^{-3}) 6H-SiC sample taken at room temperature with 785 nm (1.58 eV) laser excitation. At least four additional peaks can be observed, which are labeled in the insert as N_d, N_e, N_f, and N_g. These peaks are not clearly visible in 6H-SiC at room temperature with 514 nm (2.41 eV) excitation (data not shown); in fact the electronic Raman scattering peak at 510 cm^{-1} is masked by the weakly scattering 6H-SiC phonons at 505 and 513 cm^{-1}.[3,4] We note that, for comparable nitrogen concentrations, it appears as if the electronic Raman scattering from nitrogen donor levels in 6H-SiC is much stronger than the 4H-SiC polytype.

The peaks observed in our Raman experiments are quite consistent with the values for electronic Raman scattering in *n*-type 6H-SiC established by Colwell and Klein.[7] The peak observed by Klein at 113 cm^{-1} in *n*-type 6H-SiC appears in our data as a tail in the Raman spectrum, as shown in Fig 5. Fano interference effects are observed in the peak at 204 cm^{-1} due to these tails, again similar to effects were observed by Klein.[7] The Raman peaks at 486, 505, 635, and 642 cm^{-1} measured by Klein[7] and Gorban[6] at low temperature are quite consistent with the asymmetric peaks we observe in 6H-SiC around 510 and 638 cm^{-1} at room temperature.[7] As pointed out by Gorban *et al.* the number of electronic Raman peaks is not equal to the number of inequivalent carbon sites for 6H-SiC.[6]

Nitrogen doped 4H- and 6H-SiC both have a greenish metallic color. In contrast, semi-insulating SiC is clear. The color is more intense in the heavily doped samples. The green color coincides with a broad electronic absorption in the near IR. This absorption band is the probable origin of the resonance enhancements observed in the electronic Raman scattering from the nitrogen doping levels. The near IR absorption seen in nitrogen doped SiC is most likely due to the formation of deep defects. It is reasonable to assume that nitrogen doping of SiC is not unrelated to the well-studied problem of nitrogen doping of diamond. Nitrogen forms a deep level in diamond associated with lattice relaxation and an observed UV absorption inside the gap.

We have also used time-resolved photoluminescence to examine n-type SiC wafers. In Figure 5 the time-resolved spectra clearly show three peaks at zero time delay. These can be attributed to nitrogen, aluminum and boron in the sample. Further work is underway in our laboratory to understand these spectra.

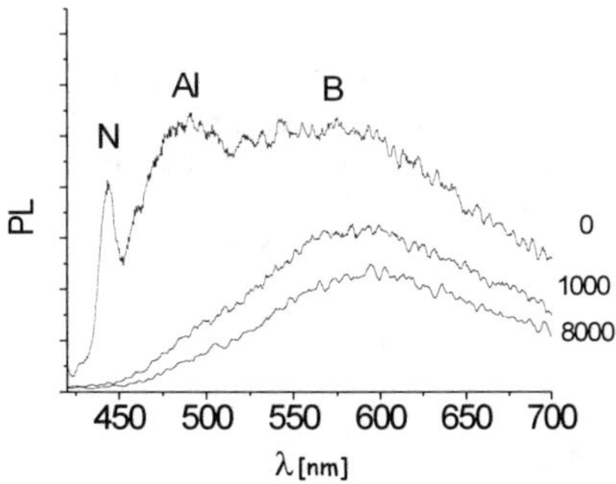

Figure 5: Time-resolved photoluminescence from n-type SiC at three time delays in ps.

CONCLUSIONS

We have examined the electronic and vibrational spectroscopy of 4H- and 6H- SiC. We have used both one- and two-phonon Raman spectroscopy to investigate the vibrational dynamics of SiC. The Raman spectra of SiC were found to be polytype dependent. The two-phonon Raman spectra are not simply twice the one-phonon Raman spectra, but reflect the full three-dimensional vibrational band structure of the SiC. We have also observed electronic Raman scattering from nitrogen donor levels in both 4H- and 6H-SiC. We have found that the electronic Raman scattering is enhanced with red or near IR laser excitation. This resonance is due to the near IR absorption, typical of nitrogen doped SiC, which has been attributed to deep defects in the material. The number of peaks observed in the electronic Raman scattering from nitrogen donor levels in both 4H- and 6H-SiC strongly suggests that the standard picture of nitrogen sitting in carbon vacancies is incomplete.

ACKNOWLEDGEMENTS

FHL would like to thank the Rutgers Research Council for financial support. Acknowledgement is also made to the donors of the Petroleum Research Fund, administered by the ACS, for the partial support of this research. Work at EMCORE was supported in part by ONR contract number N00014-97-C-0210 and monitored by Dr. Colin Wood.

REFERENCES

1   W. J. Choyke and G. Pensl, MRS Bulletin **22,** 25-29 (1997).
2   R. C. Glass, D. Henshall, V. F. Tsvetkov, and C. H. Carter, MRS Bulletin, 30-35 (1997).
3   J. Burton, L. Sun, M. Pophristic, S. Lukacs, F. H. Long, Z. C. Feng, I. T. Ferguson, J. Appl. Phys. **84** 6268-6274 (1999).
4   S. Nakashima and H. Harima, Physica Status Solidi A **162,** 37-63 (1997).
5   J. Burton, L. Sun, F. H. Long, Z. C. Feng, I. T. Ferguson, Phys. Rev B **59** 7282-7284 (1999).
6   I. S. Gorban, V. A. Gubanov, V. D. Kulakovskii, A. S. Skirda, and B. N. Shepel', Sov. Phys. Semicond. **30,** 928-930 (1988).
7   P. J. Colwell and M. V. Klein, Phys. Rev. B **6,** 498-515 (1972).

GROWTH OF SIC THIN FILMS ON (100) AND (111) SILICON BY PULSED LASER DEPOSITION COMBINED WITH A VACUUM ANNEALING PROCESS

Jipo Huang[*], Lianwei Wang[*], Jun Wen[**], Yuxia Wang[**], Chenglu Lin[*],
Carl-Mikael Zetterling[***], and Mikael Östling[***]

[*]State Key Laboratory of Functional Materials for Informatics, Shanghai Institute of Metallurgy,
Chinese Academy of Sciences, Shanghai, 200050, P. R.. China
[**]Department of Material Science and Engineering, University of Science and Technology of
China, Hefei, 230026, P. R.. China
[***]Royal Institute of Technology (KTH), Department of Electronics, Electrum 229,
SE-16440, Kista-Stockholm, Sweden

Keywords: SiC thin films; Pulsed laser deposition; Annealing

Abstract

Crystalline 3C-SiC thin films were successfully grown on (100) and (111) Si substrates by using ArF pulsed laser ablation from a SiC ceramic target combined with a vacuum annealing process. X-ray diffraction (XRD) and Fourier transform infrared spectroscopy (FTIR) were employed to study the effect of annealing on the structure of thin films deposited at 800°C. It was demonstrated that vacuum annealing could transform the amorphous SiC films into crystalline phase and that the crystallinity was strongly dependent on the annealing temperature. For the samples deposited on (100) and (111) Si, the optimum annealing temperatures were 980 and 920°C, respectively. Scanning electron microscope (SEM) micrographs exhibited different characteristic microstructure for the (100) and (111) Si cases, similar to that observed for the carbonization layer initially formed in chemical vapor deposition of SiC films on Si. This also showed the presence of the epitaxial relationship of 3C-SiC[100]//Si[100] and 3C-SiC[111]//Si[111] in the direction of growth.

1. Introduction

Silicon carbide's excellent physical and electrical properties, such as wide band- gap, high thermal conductivity, high breakdown electric field, high saturated electron drift velocity and resistance to chemical attack, provides a promising material for high-temperature, high-power and high-frequency electronic devices [1-2], as well as optoelectronic devices [3-4]. Cubic SiC (3C-SiC) is attractive owing to high electron mobility and high saturation velocity. As a result, there is an increasing interest in the growth of high quality 3C-SiC heteroepitaxial films on silicon. Since Matsunami and coworkers pioneered the single crystalline growth of 3C-SiC films on Si(100) by chemical vapor deposition(CVD)[5], CVD has become the most frequently used method for growth of SiC thin films [6-9]. In a typical CVD process, reactants consisting of a mixture of SiH_4 and C_3H_8 diluted in H_2 are flowed over a silicon substrate heated to 1300°C or above, to achieve epitaxial growth of SiC films. However, this method suffers from high hydrogen content and crystalline lattice defects in SiC films as a result of the very high substrate

Mat. Res. Soc. Symp. Proc. Vol. 572 © 1999 Materials Research Society

temperatures. Although epitaxial 3C-SiC films on Si have been reported at 750°C [10], the development of other lower temperature methods is desirable.

Recently it has been demonstrated that crystalline SiC films can be grown on Si substrates by pulsed laser deposition (PLD), a novel and powerful thin film growth technique at relatively low temperature. Rimai [11,12], Balloch [13], and Capano [14] have all reported preparation of 3C-SiC films on silicon. However, the former employed temperatures of 1000°C or higher, while the crystallinity of the SiC films reported by the latter two was poor. In this study, we report oriented SiC films on (100) and (111) Si substrates grown by pulsed laser deposition (PLD) at 800°C combined with a vacuum annealing process. The purpose is to investigate the effect of annealing on PLD-deposited SiC films, as well as the possibility of fabricating well-crystallized SiC thin films at relatively low temperature by using this method.

2. Experimental

The SiC thin films were deposited inside a stainless steel vacuum chamber evacuated by a turbomolecular pump to a base pressure of 2×10^{-7} Torr and a working pressure was maintained at $5\sim6\times10^{-6}$ Torr. Radiation from an ArF excimer laser was focused to a spot size of 0.1×0.4 cm^2 on a rotating sintered SiC target at an angle of 45°. The pulse repetition frequency was 3 Hz and a pulse energy of 150~200 mJ yielded an energy density of 2~4 J/cm^2 on the target. The plume of atoms and ions emerged in a cone normal to the target and impinged on the substrate with the area of 2×3 cm^2. P-type (100) and (111) Si wafers (20~35Ωcm, 350 μm thick) substrates cleaned with a standard silicon IC procedure were located about 3~5 cm away from and parallel to the surface of targets. The substrate was resistively heated up to 800°C, which was monitored by a thermocouple. The typical average growth rate of SiC films with the energy density of 2~4 J/cm^2 was about ~0.05 nm per pulse. After deposition, the samples were taken out and cut into several slices, and then annealed in a vacuum furnace with a pressure of 1×10^{-5} Torr and temperatures ranging from 900~1050°C for 30 min.

The crystal structure of the SiC films was characterized by X-ray diffraction (XRD) using Cu Kα radiation. Fourier transform infrared spectroscopy (FTIR), X-ray photoelectron spectroscopy (XPS) and scanning electron microscopy (SEM) were used to investigate composition and surface morphology.

3. Results and discussion

The XRD analysis of the SiC films deposited on Si(100) and Si(111) substrates at 800°C for 30 min could not detect any characteristic SiC diffraction peaks, indicating that the films were amorphous. The lack of crystallinity in the SiC films appeared to be related to the low growth temperature. It was found that the intensity of the refraction peaks varied as a function of the annealing temperature for both Si(100) and Si(111) cases, suggesting that there was an optimum annealing temperature in order to obtain well-crystallized films. The optimum temperatures in our experiment were 980°C and 920°C, respectively. The lower optimum temperature for (111)Si over (100) might be explained by the lower free surface energy for the (111) plane. Fig. 1 shows the XRD patterns of the SiC films annealed in a vacuum for 30 min, with the annealing temperature for Si(100) substrate being 980°C, and 920°C for the Si (111) . For films on Si(100), the lines corresponding to the 200 (2θ=41.5°) reflections are observed, whereas the 111 reflection (2θ=35.6°) is absent. The reverse is true for films on Si(111).

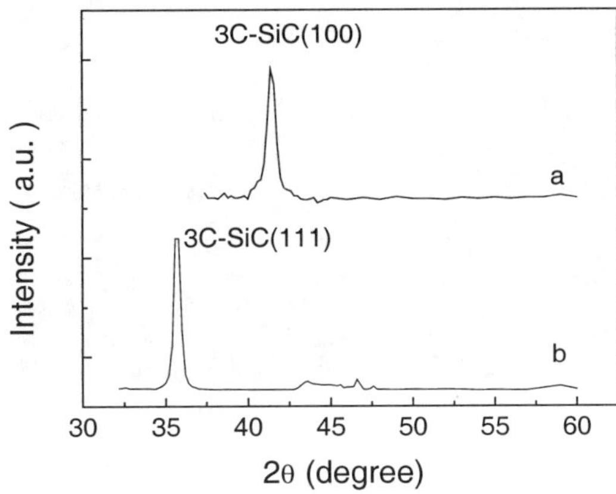

Fig. 1 XRD patterns of PLD deposited SiC films (a) on Si(100) annealed at 980°C
(b) on Si(111) annealed at 920°C.

The characteristics of infrared active modes in SiC were investigated by Fourier transform infrared spectroscopy. In the FTIR transmittance spectra of SiC films annealed in a vacuum on both (100) and (111) Si cases, the strong phonon absorption peak centered at 800 cm^{-1} is observed, which corresponds to the characteristic phonon mode of the Si-C bond [10]. These phonon modes clearly show that the films contain pure SiC phase.

Fig. 2(a) and 2(b) show SEM micrographs of the SiC films on Si(100) and Si(111) annealed in a vacuum for 30 min, respectively. The well-defined oriented arrangements, rectangular morphology for the (100) case and triangular morphology for the (111) case, could be observed, which was similar to that observed for the carbonization layer initially formed in chemical vapor deposition of SiC film on Si and was viewed as the indicative of the epitaxial growth of cubic SiC films and the presence of crystalline films [12,17]. Clearly, the epitaxial relationship was 3C-SiC[100]//Si[100] and 3C-SiC [111]//Si[111] in the direction of the growth. It is possible that these patterns originate in the underlying substrates and were associated with the SiC/Si interface. The pits on the corresponding substrates diffused into the upper films during the annealing, and the characteristic geometric patterns came into being gradually.

Fig. 3(a) and 3(b) present the XPS spectra of Si2p and C1s of the SiC film on Si(100), annealed at 980°C. As is evident from Fig. 3, the Si2p and C1s exhibit a single peak at 100.2 eV and 283.3 eV, respectively, but no peaks at 99.2 eV (corresponding to pure Si) and 282.6 eV (corresponding to graphite). This result also indicated the formation of crystalline SiC and the measured binding energies of the Si2p and C1s states agreed well with previous reports [15,16]. The atomic ratio of C/Si using the ratio of peak area and the empirical sensitivity factors [16] (0.87 for Si and 1.0 for C) was 1.16, i.e. nearly stoichiometric.

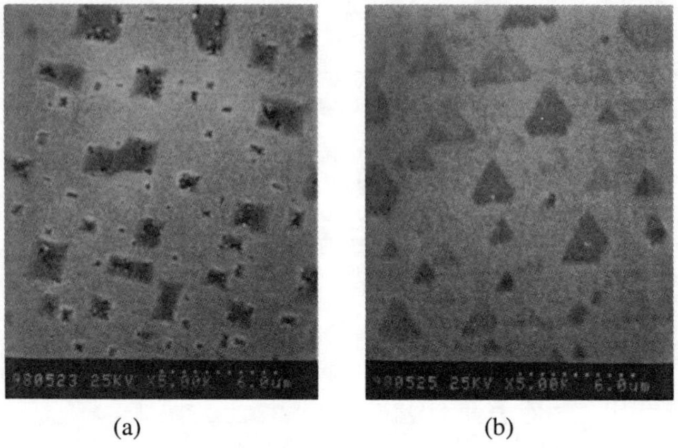

Fig. 2 SEM micrographs of the SiC films on (a) Si(100) annealed at 980°C and (b) Si(111) annealed at 920°C

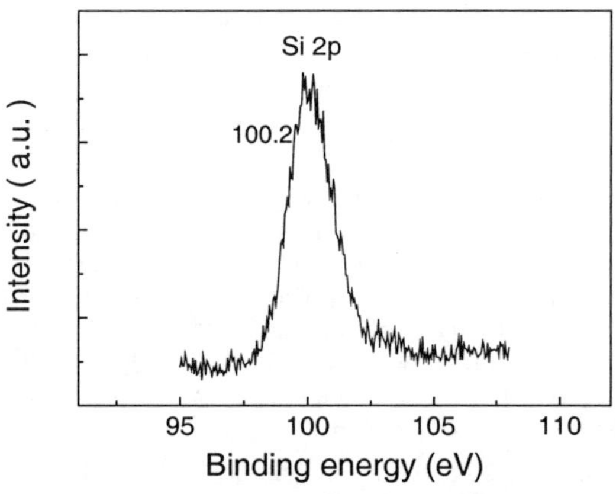

(a)

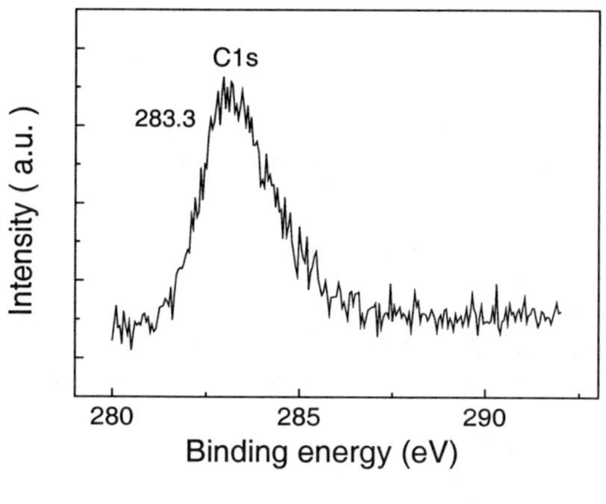

(b)

Fig. 3 XPS spectra of the SiC film deposited at 800°C on Si(100) and annealed in vacuum
at 980°C for 30 min (a) Si 2p (b) C 1s

4. Conclusion

Pulsed excimer laser deposition combined with a vacuum annealing process has been developed to prepare stoichiometric crystalline SiC thin films on Si(100) and Si(111) substrates. The amorphous SiC films deposited at 800°C could be transformed into crystalline phase after being vacuum annealed for 30 min. Both for the samples on (100) and (111)Si substrates, there were an optimum annealing temperature, 980 and 920°C, respectively. Different characteristic microstructure for (100) and (111) Si cases exhibited in SEM observation revealed the presence of the epitaxial growth relationship of 3C-SiC[100]//Si[100] and 3C-SiC[111]//Si[111].

Acknowledgments

This work is supported by National Natural Science Foundation of China under Grant No. 69776003 and the Shanghai Youth Foundation under Grant No. 97QD14034.

References

1. G. Müller, G. Krötz, and E. Niemann, Sensors and Actuators A **43**, 259 (1994).
2. D. M. Brown, E. Downey, M. Grezzo, J. Kretchmer, V. Krishnamurti, W. Hennessy and G. Michon, Solid State Electronics **59**, 1531 (1996).
3. J.W. Palmour, J. A. Edmond, H. S. Kong and C. H. Carter Jr., Physica B **185**, 461 (1993).
4. S. Sheng, M. G. Spencer, X. Tang, P. Zhou, K. Wongchotigul, C. Taylor and G. L. Harris, Mat. Sci. Eng. B **46**, 147 (1997).
5. H. Matsunami, S. Nishino, and H. Ono, IEEE. Trans. Electron Devices **28**, 1235 (1981).
6. S. Nishino, H. Suhara, H. Ono, and H. Matsunami, J. Appl. Phys. **61**, 4886 (1987).
7. Y. Gao, J. H. Edgar, J. Chaudhuri, S. N. Cheema, M. V. Sidorov, D. N. Braski, J. Crystal Growth **191**, 439 (1998).
8. W. Wesch, Nucl. Instr. and Meth. in Phys. Res. B **116**, 305 (1996).
9. J. A. Cooper, Jr., Mat. Sci. Eng. B **44**, 387 (1997).
10. I. Golecki, F. Reidinger and J. Marti, Appl. Phys. Lett., **60**, 1703 (1992).
11. L. Rimai, R. Ager, W.H. Weber and J. Hangas, Appl. Phys. Lett., **59**, 2266 (1991).
12. L. Rimai, R. Ager, W.H. Weber, J. Hangas, A. Samman and W. Zhu, J. Appl. Phys. **77**, 6601 (1995).
13. M. Balloch, R. J. Tench, W. J. Sielhause, M. J. Allen, A. L. Conor and D. R. Olander, Appl. Phys. Lett., **57**, 1540 (1990).
14. M. A. Capano, S. D. Walck and P. T. Murray, Appl. Phys. Lett. **64**, 3413 (1994).
15. Ming. Y. Chen and P. Terrence. Murray, J. Mat. Sci. **25**, 4929 (1990).
16. A. Watanabe, M. Mukaida, T. Tsunoda and Y. Imai, Thin Solid Films **300**, 95 (1997).
17. H.J. Kim and R.F. Davis, J. Electrochem. Soc. **134**, 2269 (1987).

ON THE ROLE OF FOREIGN ATOMS IN THE OPTIMIZATION OF 3C-SiC/Si HETEROINTERFACES

P. MASRI *, N. MOREAUD *, M. AVEROUS *, Th. STAUDEN **, T. WÖHNER **, J. PEZOLDT **
*Groupe d'Etude des Semiconducteurs, CNRS, UMR 5650, Université Montpellier 2, cc074, Place E. Bataillon, 34095 Montpellier cedex, France, masri@int1.univ-montp2.fr
**TU Ilmenau, Institut für Festkörperelektronik, PSF 100565, 98684 Ilmenau, Germany

ABSTRACT

3C-SiC/Si structures with Ge incorporation are elaborated by solid source molecular beam epitaxy (SSMBE). A comparison of the flatness of the SiC-surface and the interface between SiC and Si by comparing the deposition with and without Ge is made. The results are analyzed within the framework of a theoretical approach based on the theory of elasticity.

INTRODUCTON

Silicon carbide has garnered a particular attention because it has been guessed as a promising material for high-temperature and high-power electronic devices. Sensors based on SiC are indeed aimed because of the high thermal, chemical and mechanical stability of this material. The elaboration of high quality epilayers, which can be used as an efficient active part of a device, is strongly dependent on substrate surface quality. Among others, two approaches of SiC film epitaxial growth are currently considered. The investigation of the state of art of SiC wafer elaboration shows that wafer surface crystalline quality suffers from the existence of extended defects as dislocations and micropipes. This really represents a serious disadvantage of homoepitaxial growth as these structural defects can have dramatic consequences on device performance, especially because of their replication in the grown SiC active layer.

Another alternative is afforded by heteroepitaxial growth of SiC on Si substrates. The reasons for using Si is that large area of Si wafers, cheaper than SiC wafers, can be elaborated with a high surface crystalline quality. However, when using this solution, an important problem, related to the lowering of residual stress, must be solved. This methodology presents indeed a disadvantage due to the large mismatches of lattice parameters and thermal expansion coefficient respectively equal to 20 % and 8 %. Because of these mismatches and beyond layer critical thickness, extended defects, like for instance dislocations, can be favored from the formation energy point of view. Moreover, because of lattice parameter differences separating SiC and Si, stresses can not be fully relaxed by the dislocation network introduced at the SiC/Si interface during the carbonization of Si-substrate surfaces. As extended defects play a vital role in the quality of the final product, it is then worth seeking for a methodology which enables to smooth out their differences.

Among the different techniques which can smooth out the effect of large lattice mismatch, the buffer layer technique has been widely used to assist heteroepitaxial growth. The carbonization of clean Si surface using hydrocarbon radicals is generally performed at the early stage of the growth process. Gas source molecular beam epitaxy (GSMBE) experiments [1] have shown that when one uses $(CH_3)_2GeH_2$ (DMGe) as carbon source, the morphology of the carbonized surface layer improves when compared with the one without Ge (carbon is then provided by hydrocarbon radicals from cracked $-C_3H_8$). The thickness of the smooth carbonized layer is ~ 40 Å with a Ge concentration showing a saturation around 0.4 % , and the pur-

Mat. Res. Soc. Symp. Proc. Vol. 572 © 1999 Materials Research Society

pose of using DMGe is to introduce a large size atom at the heterointerface. Similar approaches were also carried out by [2, 3].

In this work, we investigate the effect of the incorporation of Ge atoms into the 3C-SiC/Si structure grown by SSMBE and we discuss the results within the framework of a theoretical approach based on the theory of elastic. This latter approach, which we have recently developed [4], as proved itself to be very useful in many aspects of the physics of epitaxy.

EXPERIMENT

The SiC layers were deposited on on-axis (111) Si substrate by using solid source molecular beam epitaxy. For the deposition of the 300 nm thick SiC layers on Si, we used an UMS 500 Balzers MBE system similar to the technique reported in [3]. The deposition procedure was similar to the method reported in [5] and consists of the following process steps: (1) silicon wafer cleaning, described in detail in [6], (2) 0-1 ML Ge deposition on the 7x7 reconstructed Si surface at 325 °C by electron beam evaporation, (3) 6 ML carbon deposition on the Si substrate, (4) heating up the Si wafer to 600 °C for 3 minutes, (5) gradually increase of the substrate temperature in periods of 2 minutes by 50 °C up to 1050 °C, (6) the SiC deposition started at 950 °C with a growth rate of 1nm/min. The final epitaxial temperature was 1050°C. The deposition process was monitored by in-situ RHEED and in-situ spectroscopic ellipsometry. Ex-situ, the films were investigated by atomic force microscopy (AFM). The interface width was determined with spectroscopic ellipsometry by using a three layer model consisting of substrate-interface-silicon carbide-surface.

EXPERIMENTAL RESULTS

For a good substitutional incorporation of Ge into SiC during the carbonization process, it is necessary to prevent the formation of Ge bonds in the Ge_xSi_{1-x} surface alloy. For this reason the (7x7)-Ge reconstruction was chosen. The (7x7)-Ge on (111)Si surface reconstruction is stabilized for Ge coverages below 1-1.5 ML [7] or for a Ge_xSi_{1-x} alloy with up to 0.24 according to [8]. For this structure, the Ge atoms randomly substitutes the sites of Si atoms [9]. Furthermore the Si dangling bonds are replaced in the (7x7)-Ge by those of Ge and that no Ge-Ge bonds are formed [10]. During the Ge deposition the (7x7)-Ge reconstruction remains unchanged and no three dimensional islands were observed. This assumption was supported by our experiments where a deterioration of the SiC nucleation behavior for Ge coverages above 2 ML was observed.

The deposition of carbon onto the silicon surface is an effective way to introduce large carbon concentrations into the near surface region [11]. For this reason 6 ML carbon were deposited onto the (7x7)-Ge (111)Si surface at 325°C. To relax the so formed compound system, it was annealed at the deposition temperature for several minutes. The following step wise annealing procedure had the task to form a thin initial SiC layer and to stimulate the Ge incorporation into the SiC lattice. During the subsequent epitaxy, no significant differences were found in the RHEED pattern for the different Ge coverages. All the 3C-SiC layers grown on these carbonized pseudo-substrates were single domain and show only two dimensional diffraction with a strong (3x3)-Si reconstruction without transmission features. For that reason, the in-situ spectroscopic ellipsometry was used to achieve information on the interface and ex-situ AFM investigations were used to characterize the morphology.

The experimental results of terrace width/grain diameter, surface roughness and interface width obtained after the growth of 300 nm SiC on Si are given in Table I for several Ge coverages ranging from 0 up to 1 ML, where ML represents the number of monolayers with respect to the (111) Si surface.

Table I Variation of geometric surface and interface features in function of Ge coverages

Ge coverage (ML)	Terrace width/grain diameter (nm)	Surface roughness, rms (nm)	Interface width (nm)
0	133	1.85	1.2
0.25	145	1.73	0.5
0.5	192	1.48	1.1
1	169	2.19	6.1

The interface width given in Table I was determined after the growth of a 100 nm thick layer during the deposition process. This was done to have information about the near starting conditions.

As can be seen, Ge incorporation into the Si surface leads to an improvement of the flatness of the SiC surface and to an increase of the grain size during conversion process. This might be an indication that the Ge incorporation into the Si leads to a decrease of the sum of the surface energy of the SiC nuclei and the interface energy. Additionally, the Ge incorporation improves the flatness of the SiC-Si interface. From the obtained results, it can be concluded that under our experimental conditions the largest positive effect can be expected for Ge coverages below 0.5 ML.

THEORY

The lattice misfit between two materials A (substrate Si) and B (overlayer SiCGe mixed layer or 3C-SiC) is related to the difference between their lattice parameters $\Delta = |\, a_A - a_B$. Furthermore, we assume that the corresponding periodicity of the interface unit cell must match the A and B lattices by a vernier process. If $a_A < a_B$, the mathematical expression of this matching is [12] :

$$(n_1+1)a_A = n_1 a_B \ (1.a), \text{ with } n_1 = a_A/(a_B - a_A) \ (1.b)$$

Eq. (1.b) holds when the B lattice is under compression and the A lattice under extension. It states that after n_1 jumps on the B lattice and (n_1+1) jumps on the A lattice, we may find in coincidence two interface sites belonging respectively to A and B. If the material A is under compression and B is under extension $(a_A > a_B)$, the geometric approach gives:

$$n_1 = a_B/(a_A - a_B) \tag{2}$$

In ref. [13], the authors have developed the idea of material phases stabilized by epitaxial strains. As the phase determining procedure is the minimization of lattice total energy as a function of substrate lattice parameter after growth-axis relaxation, one must emphasize the role of interface conditions. This means that now we have to deal with in-heterostructure material physics to take account of host material interactions. For cubic materials, the following effective elastic constants $f(C_{ij})$, for longitudinal phonon modes along directions [100], [110] and [111], are relevant to the relaxation energy [13]:

$$f(C_{ij})=C_{11} \ (a), \ f(C_{ij})=C_{11}+C_{12}+2C_{44} \ (b), \ f(C_{ij})=C_{11}+2C_{12}+4C_{44} \ (c) \tag{3}$$

As far as this in-heterostructure material physics is aimed, we must seek for relationships analogous to Eqs. 1 and 2, but which can (i) unify existing approaches on epitaxially constrained interfaces and (ii) provide way to interface optimization. In what follows, we present the methodolgy which enables to answer questions (i) and (ii).

To tackle these problems, a good starting point is to consider, for cubic crystals, the equations of the elasticity theory [13] which relate strain to lattice dynamics features:

$$\frac{\partial^2 u}{\partial t^2} = \frac{C_{11}}{\rho}\frac{\partial e_{xx}}{\partial x} + \frac{C_{12}}{\rho}\left(\frac{\partial e_{yy}}{\partial x} + \frac{\partial e_{zz}}{\partial x}\right) + \frac{C_{44}}{\rho}\left(\frac{\partial e_{xy}}{\partial y} + \frac{\partial e_{zx}}{\partial z}\right) \qquad (4)$$

where u is the x component of the displacement, ρ is the density, the C_{ij} are the elastic constants, and the $e_{\sigma\sigma'}$ are the strain components ($\sigma,\sigma'=x,y,z$). Similar equations for y and z directions can be deduced from Eq. 4. The left hand side of this equation is proportional to ω^2, the square of the angular frequency ω. Because of this feature, these equations may be considered as the signatures of a strain-dynamics correlation via the $S=f(C_{ij})/\rho$ factor.

In the case of periodic lattices, we can carry out a Bloch waves analysis of ω (g), where g represents the phonon wave vector in the growth plane, say g (g_x, g_y), where the g_x and g_y belong to the first two dimensional Brillouin zone (2DBZ) associated with the interface reciprocal lattice. For each interface configuration, i.e. for a specific growth plane, the dynamics equations involve effective elastic constants $f(C_{ij})$ which correspond to elastic waves propagating along the principal symmetry directions in cubic crystals. The expressions of $f(C_{ij})$ for the longitudinal and transverse modes in cubic crystals are given in ref. [14].

In what follows, we consider the case of a system constituted of two host materials A (B) representing the substrate (overlayer) with a lattice parameter a_A (a_B) and an S-factor equal to S_A (S_B). The extension of the 2DBZ associated with the surface/interface lattice is scaled by the wavevector components $g_x\sim\pi/ma$ and $g_y\sim\pi/na$ where a is the lattice parameter, and m and n are integers which scale the corresponding extension of the unit cell associated with a periodic lattice network (mxn). Then we have:

$$\omega^2 \propto SG^{-1}(a) \text{ where } G\equiv 1/g_x g_y \propto (m\times n)a^2 \text{ (b)} \qquad (5)$$

If we write down such a relationship for each host material A and B, we must introduce the following quantities:

$$S_{A,B}=[f(C_{ij})/\rho]_{A,B}(a) ; G_A \propto (m\times n)a^2_A(b) \text{ and } G_B \propto (p\times q)a^2_B \text{ (c)}. \qquad (6)$$

G_A and G_B represent the geometric factors associated with A and B, respectively. If we match up strain gradient components, we obtain the relationship :

$$G_A/S_A=G_B/S_B . \qquad (7)$$

If isotropic geometric conditions are assumed for the MIS (m=n and p=q) and if we apply a vernier procedure to both A and B lattices, we end up with the equation :

$$n_S=a_A(a_B/\sqrt{S} -a_A)^{-1} \qquad (8)$$

with $S=S_B/S_A$ and $a_B>a_A$ (A is under extension and B under compression). Eq. 8 means that after n_S jumps on lattice B and (n_S+1) jumps on lattice A, both lattices are in coincidence. If the material A is under compression and B is under extension ($S=S_B/S_A$, $a_A>a_B$), we obtain :

$$n_S=a_B/(a_A-a_B) \qquad (9)$$

Eqs. 8 and 9 are analogous to Eqs. 1, but with in addition a lattice parameter renormalizing factor ($\sqrt{S}$). If S=1, one can see that $n_S=n_1$: the condition S=1 implies that we have a matching of the elastic constant-density ratios. It is only in this case that the density of dislocations predicted by the S-theory is equal to that given by Eqs. 1. If the heterosystem is formed out of three host materials A (substrate), B (buffer layer) and C (overlayer), we have to consider two interfaces, namely B/A (interface 1; $S=S_1$) and C/B (interface 2 ; $S= S_2$). The basic equation of the interface optimization theory with respect to the S factors is the continuity of S_j throughout the heterosystem [4] :

$$S_1=S_2 \qquad (10)$$

APPLICATION TO THE 3C-SiC/Si INTERFACE

We have shown that during the Ge deposition, the (7x7)-Ge surface superstructure remains unchanged, corresponding to a stable interface symmetry. We can analyse this effect within the framework of our theory. To carry out this analysis, we use Eqs. 7 and 8 to calculate the interface superstructure (pxq) after film deposition, by assuming that the bare (111) Si surface is (7x7) reconstructed. We are then able to show that, in terms of Ge and C composition, there exist alloyed $Si_{1-y-x}C_yGe_x$ phases which can leave unchanged the (7x7) reconstruction. Moreover, we find that Ge plays a compensating role in respect to C: when the C composition of the alloy increases, we must increase Ge incorporation in order to stabilize the surface structure. These results, depicted on Fig. 1, show a linear variation of the Ge composition (x) in function of C composition (y).

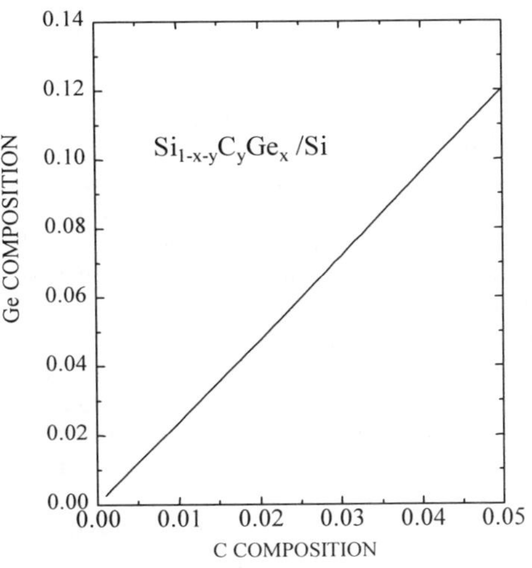

Fig.1 Variation of Ge composition vs y

Considering the use of buffer layer technique for 3C-SiC/Si heteroepitaxy, we concentrate on two approaches in SSMBE experiments. The first one is based on the elaboration of $Si_{1-y}C_y$ alloy layers deposited on Si substrates. The second one concerns the elaboration of $Si_{1-y-x}C_yGe_x$ as buffer layers aiming to improve the 3C-SiC/Si interface quality. We have applied our theory to the system 3C-SiC/$Si_{1-y-x}C_yGe_x$/Si (C/B/A) where $Si_{1-y-x}C_yGe_x$ is introduced as a buffer layer. In this case, interfaces 1 (S_1) and 2 (S_2) correspond to $Si_{1-y-x}C_yGe_x$/Si and 3C-SiC/$Si_{1-y-x}C_yGe_x$, respectively. The results shown on Fig. 2 demonstrate that the continuity condition on S_j can be fulfilled for Ge composition as low as 0.4 % in agreement with experimental results [1] while C composition corresponds to y≅ 5 %. This latter value agrees with that of ref. [15]. One must bear in mind that the C composition of the buffer layer must remain low (5 % is a reasonable value) because of the distortions introduced by C in the Si lattice. Then, a gradual increase of C composition must be attempted in order to meet the conditions of a pseudomorphic growth of 3C-SiC layer [7]. The incorporation of Ge atoms into the SiC heterosystem induces new properties which, in our approach, depend on lattice parameter, elastic constants and

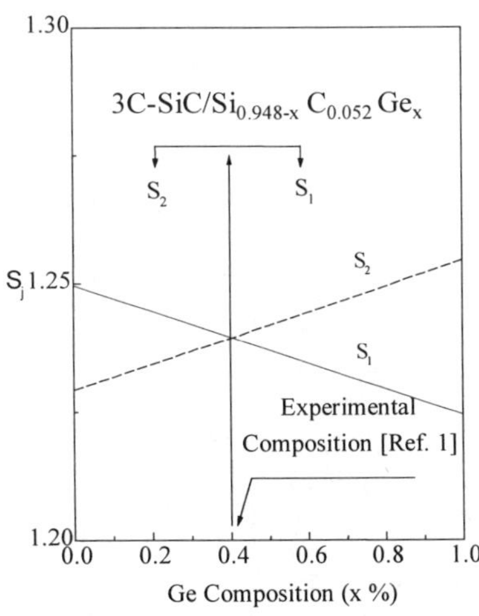

Fig. 2 Variation of S_j vs x

density. This enables to achieve a better matching of strain gradient factors (S_j) associated with the host materials involved in the heterostructure. The agreement between experimental and theoretical results demonstrate the relevance of S_j for heteroepitaxy optimization needs.

CONCLUSION

By using a theoretical approach based on the elasticity theory, namely the S-correlated theory of interface optimization and its continuity condition, we have determined the composition of a buffer layer formed out of Si, C and Ge species which is meant to improve the 3C-SiC/Si interface quality. The calculated composition, i.e., $Si_{0.944}C_{0.052}Ge_{0.004}$ is in agreement with the experimental results [1]. Ge is probably not the only host atom which can improve the 3C-SiC/Si heterointerface and several alternatives must be investigated. In our approach, the host layer thickness is not introduced as an explicit parameter. If we consider the parameters relevant to the S-theory, we can perform the following analysis. The density (ρ) is a material property and it should not depend on layer thickness. Lattice parameters depend on the considered host material phase. This means that the corresponding layer thickness must be at least that for which bulk structural features prevail. Eventually, elastic constants can be different for very thin and very thick layers, although in this work we have used bulk values.
The theoretical approach could be verified by experimental preparation of samples with different Ge content incorporated into the Si-SiC interface. Ge coverages of approximately 0.5 ML did influence the properties of the epitaxial layers positively. Investigations by AFM confirmed a decrease in surface roughness and an increase in grain size. The fit of in situ ellipsometric spectra using an optical model consisting of four layers (surface roughness, 3C-SiC layer, interface width and Si substrate) determines a reduction in interface width.

REFERENCES

1. T. Hatayama, N. Tanaka, T. Fuyuki, and H. Matsunami, J. Electronic Mater. 26, p. 160 (1997).
2. K. Zekentes and K. Tsagaraki, Mater. Sci. Eng. B56, (1999), in press
3. S. Mitchel, M. G. Spencer and K. Wongchotigul, Mater. Sci. Forum **264-268**, 231 (1998).
4. P. Masri, Phys. Rev. B**52**, 16627 (1995).
5. A. Fissel, K. Pfenninghaus, U. Kaiser, J. Kräußlich, H. Hobert, B. Schröter and W. Richter, Mater. Sci. Forum **264-268**, 255 (1998).
6. J. Pezoldt, V. Cimalla, Th. Stauden, G. Ecke, G. Eichhorn, F. Scharmann, D. Schipanski, Diam. Rel. Mater. **6**, 1311 (1997).
7 T. Ichikawa, S. Ino, Surf. Sci., **136**, 267 (1984)
8 P. Martensson, W.-X. Ni, and G.V. Hansson, Phys. Rev. B**36**, 5974 (1987).
9 B. Zhang, J.E. Northrup,, M.L. Cohen, Surf. Sci. **145**, L465 (1984).
10 S. Hasegawa, H.Iwasaki,S.-T. Li, S. Nakamura, Phys. Rev. B**32**, 6949 (1985).
11 S. Ruvimov, E. Bugiel, H.J. Osten, J. Appl. Phys. **78**, 2323 (1995).
12. See, for example, J. W. Matthews, in *Epitaxial growth*, Part B, Academic Press, New York, 505 (1975).
13. Sverre Froyen, Su-Huai Wei, Alex Zunger, Phys. Rev. B**38**, 10124 (1988).
14. C. Kittel, in *Introduction to Solid State Physics*, 3rd edn., Wiley, New York, ((1968) p. 119 Eq. (31).
15. K. Eberl, S. S. Iyer, J. C. Tsang, M. S. Goorsky and F. K. Legoues, J. Vac. Sci. Technol. **B10**, 934 (1992).

3C-SiC BUFFER LAYERS CONVERTED FROM Si AT A LOW TEMPERATURE

H. M. Liaw*, S. Q. Hong*, P. Fejes*, D. Werho*, H. Tompkins*, S. Zollner*, S. R.Wilson*, K.J. Linthicum**, and R. F. Davis**

*Motorola Inc., Semiconductor Products Sector, 2100 E. Elliot Road, Tempe, AZ 85284, rwd720@email.mot.com
**Department of Materials Science and Engineering, North Carolina State University, Raleigh, NC 27695

ABSTRACT

We have obtained single-crystal 3C-SiC films via conversion of the surface region of Si (111) and (100) wafers at 970 °C by reaction with C_2H_4 in an MBE reactor. The major defects in the films were clusters, voids, and misfit dislocations. Investigation by high resolution TEM images showed low lattice strains in the epitaxial layer due to the formation of 1 misfit dislocation for every 4 to 5 regular SiC planes that are bonded to Si at the interface. The clusters and voids often occurred in pairs. A model for forming the void-cluster pairs is suggested.

INTRODUCTION

Heteroepitaxial growth of 3C-SiC on Si has been difficult due to the large mismatch in lattice parameters. The difficulty has been partially overcome by use of a two-step growth process. The process involves the growth of a buffer layer followed by the growth of second epitaxial layer. The buffer layer growth is considered to be a critical step for the successful single-crystal growth of 3C-SiC on Si. The first and most common method for growing a buffer layer has been via the reaction of the Si substrate with propane as initially described by Nishino et al. [1, 2]. Propane was introduced into an epitaxial reactor, and the Si substrate was heated rapidly from room temperature to approximately 1360 °C and held at that temperature for a short time (< 5 minutes). Carbon was deposited on the Si substrate from the decomposition of propane. The Si top surface was converted to SiC by the reaction with the deposited C at elevated temperatures. The thin SiC films achieved by this process were approximately 20 nm thick and were reported as polycrystalline 3C-SiC. A single-crystal SiC buffer layer is preferable for achieving higher crystal perfection in the second epitaxial layer.

The growth of a single-crystal buffer layer has been pursued by Cheng et al. [3] using a lower temperature (in range of 1000 °C to 1170 °C) and a shorter carbonization time (in the range of 5 to 30 seconds). They observed that the films started to grow as thin circular islands. Lateral growth of the islands led them to coalesce and finally became a continuous film if the proper process parameters were applied. A continuous SiC film (approximately 18 nm thick) could only be obtained under a narrow process window that was at 1070 °C for 15 seconds. The converted films were single crystals but highly defective. High concentrations of defects were observed at the islands centers where nucleation initiated. The defect density decreased away from the island centers. However, defects such as stacking faults were still present. In addition, the films were embedded with high densities of etched pits which were created during the Si-to-SiC conversion process. The SiC films in the vicinity of the pits were misoriented and contained numerous stacking faults.

Mat. Res. Soc. Symp. Proc. Vol. 572 © 1999 Materials Research Society

Further decreases in carbonization temperatures were carried out by Yoshinobo et al. [4] with MBE. They used a temperature range from 720 °C to 830 °C with C_2H_2 as the carbon source. Single crystal 3C-SiC layers were obtained only in a narrow range of substrate temperatures near 790 °C. They used RHEED to confirm the growth of a single-crystal film. No further characterization of the defects in the films was conducted.

In this study we carried out the carbonization using gas-phase MBE similar to that of Yoshinobo et al. [4]. The single-crystal films were grown at 970 °C using C_2H_4. This paper reports the characterization of these films deposited under this condition. The mechanisms for low temperature growth of the converted SiC films will be proposed based on the characterization data.

EXPERIMENTAL

The conversions of Si (111) and (100) surfaces to SiC were achieved simultaneously using a custom MBE reactor designed, constructed and commissioned at NCSU for the growth of SiC thin films [5]. The size of each substrate was approximately 1cm by 1cm. They were (111) and (100) orientations with 3 degrees tilted toward the <110> flat. The native oxides on the Si substrates were removed by dipping the substrates in a buffered HF solution prior to loading into the reactor. After evacuation of the MBE chamber, the 1.8 sccm flow of C_2H_4 was introduced at room temperature. The temperature was ramped to 820 °C with a ramp rate of 6 °C/min and then ramped from 820 °C to 970 °C with a ramp rate of 3 °C/min. At 970 °C the substrates continued to be exposed to the 1.8 sccm flow of C_2H_4 for 60 minutes. This completed the conversion process.

Scanning electron microscopy (SEM) was used to evaluate the surface defects introduced by the carbonization. High resolution transmission electron microscopy (HREM) was used to characterize the microscopic defects. The film thickness was evaluated by spectroscopic ellipsometry. The film composition and carbon bondings were evaluated by X-ray photoelectron spectroscopy (XPS).

RESULTS

1. Film defects

The converted SiC films were continuous across the entire surface of each substrate. Investigation via SEM showed the presence of scattered clusters in the films. The clusters varied in size. The largest clusters were as large as several micrometers, although the largest size shown in Figure 1 is only hundreds of nm. The density of the clusters was estimated to be in the range of 4 x10^4/cm^2 to 10 x10^4/cm^2. The clusters contained voids between the film and the Si (111) substrate, as shown in Figure 2. Each cluster was the aggregate of several micro-clusters. The micro-clusters occurred preferentially along the edges of the voids. The edges of small voids were bounded by (110) crystallographic planes and indicate a preferentially etched pattern. The large voids became more rounded and were not crystallographically oriented.

Figure 3 is a plan view TEM micrograph showing a large cluster in the converted SiC film deposited on a Si (100) substrate. The gray white contrast regions are the voids underneath the film. The diameter of this cluster is approximately 5 µm. The voids along the edges of the cluster are significantly larger than those in the internal region of the cluster. It also shows that the edges of small voids are parallel to the {110} planes.

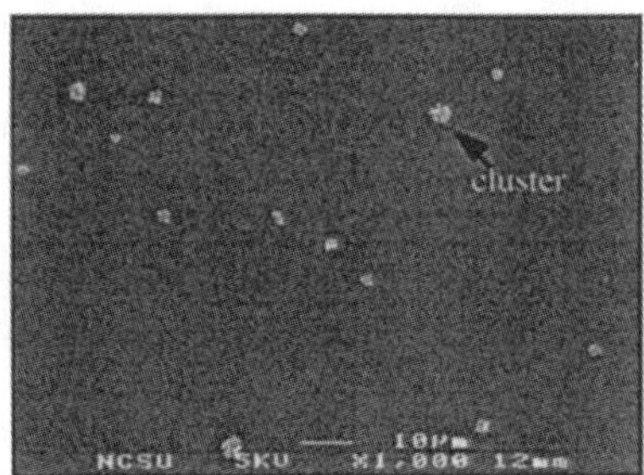

Figure 1. A SEM plan view of the converted SiC film on Si (111) showing presence of clusters

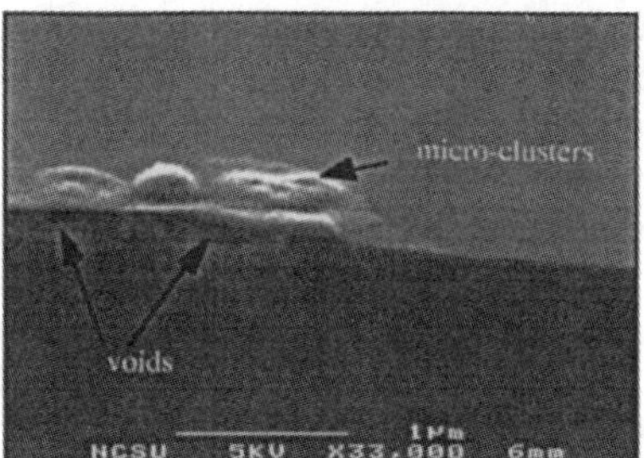

Figure 2. A cross-sectional SEM micrograph of a cluster containing voids and micro-clusters.

An electron diffraction pattern taken from the cluster is shown in Figure 4. The diffraction spots from the SiC film as well as from the Si substrate are observed. The major crystal planes are assigned and shown in Figure 4. It shows that the crystal planes of the SiC film align with the same Miller indices as the Si substrate. This confirms the epitaxial relationship and the monocrystalline character of the SiC film. Extra (or satellite) diffraction spots from the SiC planes of the same Miller indices are also present in the pattern. The satellite spots result from double or multiple diffractions but not from twinning in the film. The SiC diffraction spots are distorted from circular to elliptical shapes. This suggests that there are slight mis-orientations in some parts of the film. It is likely that the misorientations reside at or near micro-clusters.

High resolution electron micrographs (HREM) did not reveal polycrystalline regions and twinning in the converted SiC film. However, high densities of misfit dislocations were observed. Figure 5 shows the lattice image of the SiC film and the SiC/Si interface. The arrows in Figure 5 show the points of misfit dislocation initiation. These points are only several atomic spacings above the SiC/Si interface. This micrograph also shows that there is approximately one misfit dislocation (an extra SiC plane non-bonded to the Si substrate) for every 4 to 5 regular SiC planes that are bonded to Si at the interface.

Figure 5 also reveals that the boundary (or interface) between the Si substrate and the converted SiC is diffuse and wavy. The diffuse boundary may result from a tilt of the specimen

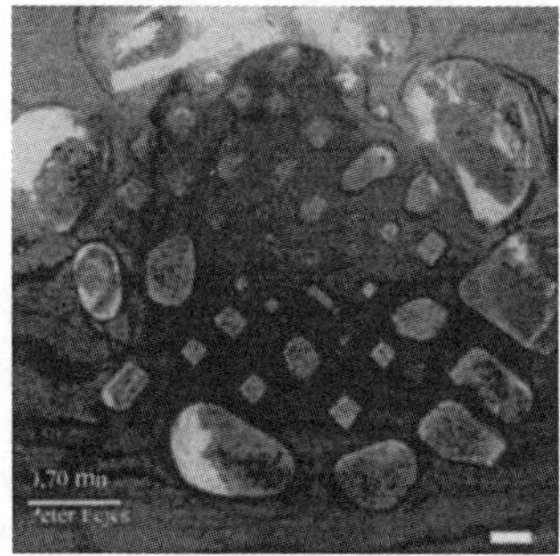

Fig. 3 A plane viewed TEM shows a cluster in the converted SiC (100) surface.

Figure 4. Electron diffraction pattern taken from the cluster shown in Figure 3.

surface away from perpendicular to the electron beam. Examination of several interface regions of the TEM specimen found few interfaces sharper than that shown in Figure 5. It suggests that this is the most common character of these interfaces. The width of undulation at the interface is in the range of several nm.

2. Lattice strains of the film and Si substrate

Lattice strains of the SiC films and the Si near the interface were calculated by measuring the change in lattice parameters from their theoretical values. The inter-planar distance of a given family of crystallographic planes was obtained by Fourier transform of the lattice image. The lattice parameters of the Si substrate further away from the interface were assumed to be the same as the bulk Si values and were used for calibration of the measurement accuracy taken at the vicinity of the interface. The measured results together with the calculated strains are listed in Tables 1 for a converted SiC film. The results show that the lattice strains of the SiC film are less than 1.2 %. We also evaluated the lattice strains of substrate in the vicinity of the interface and found it to be less than 0.5%. It suggests that stresses from a large lattice mismatch are totally alleviated by the misfit dislocation formation in the film. This agrees with the misfit dislocation density evaluated from HREM micrographs. In other words, approximately 20 % of the film lattices are dislocated in the misfit manner. This accounts for the disappearance of film strain supposedly caused by the 20% lattice mismatch between the film and substrate.

3. Film thickness and uniformity

The film thickness was measured using a spectroscopic ellipsometer. The measured optical spectra fitted fairly well with a layer model that assumed 25% voids in the films. The optical constants of the SiC films used in the model were assumed to be the same as those of bulk SiC. The thickness measured from the model was 48.8 Å and 48.1Å for the (100) and (111) films, respectively.

Figure 5. A cross-sectional HREM micrograph showing the initiation points of misfit dislocation generations.

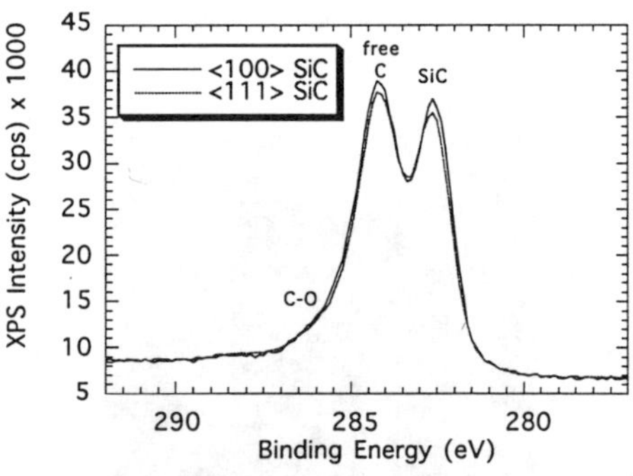

Figure 6. XPS spectra of the converted SiC films showing the bonded C peak at 282.5 eV and the free C peak at 284.2 eV.

Table 1. Measured lattice parameters of crystallographic planes and calculated lattice strains.

Atomic plane	SiC on (100)Si	Bulk SiC	Lattice strain	SiC on Si(111)	Lattice strain
film (220), vertical to the substrate	1.53 Å	1.54 Å	-0.0064935		
film (111), + slant	2.55 Å	2.52 Å	0.01190476	2.54 Å	0.0079365
film (111),- or zero slant	2.55 Å	2.52 Å	0.01190476	2.49 Å	-0.011905
film (200)	2.15 Å	2.18 Å	-0.0137615	2.19 Å	0.0045872

The thickness of the converted SiC measured by XTEM showed a large variation with location. The thickness varied in the range of 30 to 60 Å for the regions of films containing no voids. The thickness of the film located above the voids is considerable higher. This was determined by obtaining electron diffraction patterns from a plan view TEM sample. The diffraction pattern was visible when it was taken from a cluster. It was not visible when the pattern was taken from a smooth area because it was too thin. The cross-sectional SEM micrographs showed that the film was extremely rough at cluster regions. The film thickness in these regions was estimated to be from 1to 6 times of that at the smooth areas.

4. Film composition

The film composition was analyzed by XPS. A high resolution spectrum taken using C-1s was analyzed for free carbon vs. bonded carbon. Figure 6 shows the bonded C peak at 282.5 eV and the free C peak at 284.2 eV. A comparison of peak intensities suggests that about half of the C deposited is not yet converted to SiC. It also shows that the degree of the conversion is approximately the same for the SiC (111) and SiC (100). The Si 2p peak at 100.5 eV confirms little difference in the amount of conversion from (111) and (100) Si (not shown).

DISCUSSION

Single-crystal 3C-SiC(111) and 3C-SiC(100) films have been obtained via conversion of the near-surface region of Si (111) and Si (100) at 970 °C by reaction with C_2H_4 in a MBE reactor. The major defects in the films were clusters, voids, and misfit dislocations. The clusters and voids often occurred in pairs. The cluster-void pairs were also observed in 3C-SiC epitaxial layers formed by carbonization at higher temperatures [3] as well as by direct gas-phase MBE growth without any carbonization [6].

A model for formation of the cluster-void pairs has been proposed by Schmitt et. al. [6]. In this model the voids (silicon etch pits) were assumed to be formed first by Si evaporation. The clusters (referred to as islands in Ref. 6) subsequently nucleated and grew at corners of the voids. It was assumed in this model that most voids were unsealed and that the clusters were made of polycrystalline SiC.

Examination of our films showed no unsealed clusters and no polycrystalline SiC. Therefore, a more suitable model for forming the void-cluster pairs in our films is suggested as follows: (1) deposition of a thin carbon layer on the substrate surface, (2) conversion of the carbon in contact with Si into SiC, (3) diffusion of Si underneath the Si/SiC interface into the free carbon and the formation of SiC, and (4) continued deposition of free carbon on the top carbon layer and formation of voids by outdiffusion of Si from the void surfaces.

Cheng et al. [3] observed that their converted SiC films were grown by the coalescence of SiC islands. There is no evidence in our work that the films were formed by this mechanism. It suggests that a 2-dimensional growth mechanism is dominant over a 3-dimensional growth mechanism at 970 °C.

The undulations of the SiC/Si interface were likely caused by the inter-diffusion of C and Si across the interface. The large thickness variation in the films resulted from locally enhanced diffusion of Si. The Si can out-diffuse faster along the void surfaces than through the bulk of the films.

It is likely that the deposited carbon layers before the conversion were amorphous. During the conversion process, only the carbon atoms landing on top of Si atoms formed epitaxial SiC on Si. The remaining un-aligned carbon atoms, landed between the Si lattice planes, were also converted by inter-diffusions and formed extra planes or misfit dislocations. The misfit dislocations had released the lattice strain in the films.

The XPS analyses showed that the concentration of the unreacted C is fairly high. The non-bonded C can provide a driving force for the outdiffusion of Si. The excessive out-diffusion of Si can lead to the void formation and void growth.

The concentrations of misfit dislocations and the degree of lattice distortion in the converted SiC are difficult to improve, since the growth is not pseudomorphic. However, the voids and clusters may be reduced by optimizing the carbon layer thickness such that it is sufficiently thick to encapsulate the Si surface and prevent evaporation yet sufficiently thin to avoid excessive Si and C inter-diffusion. It is desirable to have a process condition so that the conversion does not start before complete carbon coverage of the substrate surface. Moreover, carbon deposition should not be continued after the conversion.

ACKNOWLEDGMENTS

The authors acknowledge the support of the Semiconductor Research Corporation and the Motorola Inc. R. Davis was supported in part by Kobe Steel, Ltd. University Professorship.

REFERENCES

1. S. Nishino, J. A. Powell, and H. A. Will, Appl. Phys. Lett., 42, p.460 (1982).
2. S. Nishino, in " Properties of SiC", p. 204, EMIS, Dataview Series, an Inspec Publication (1995).
3. T. T. Cheng, P. Pirouz, and J. A. Powell, Mat. Res. Soc. Symp. Proc. Vol. 148, p. 229 (1989)
4. T. Yoshinobu, T. Fuyuki and H. Matsunami, Jap. J. Appl. Phys. 30, p.L1086 (1991).
5. L. B. Lowlane, F. Tanaka, R. S. Kern and R. F. Davis, J. Mats. Res. 8, p.2310 (1993).
6. J. Schmitt, T. Troffer, K. Christiansen, R. Helbig, G. Pensl, and H. P. Strunk, Materials Science Forum vol.264-268, pt.1 p.247-50 Conference Title: Silicon Carbide, III-Nitrides and Related Materials. 7th International Conference

TIME RESOLVED PHOTOLUMINESCENCE OF CUBIC MG DOPED GAN

R. Seitz[*], C. Gaspar[*], T. Monteiro[*], E. Pereira[*], B. Schoettker[**], T. Frey[**], D.J. As[**], D. Schikora[**], K. Lischka[**]
*University of Aveiro, Department of Physics, Aveiro, PORTUGAL, seitz@fis.ua.pt
** University of Paderborn, Fachbereich Physik, Paderborn, GERMANY

Abstract

Mg doped cubic GaN layers were studied by steady state and time resolved photoluminescence. The blue emission due to Mg doping can be decomposed in three bands. The decay curves and the spectral shift with time delays indicates donor-acceptor pair behaviour. This can be confirmed by excitation density dependent measurements. Furthermore temperature dependent analysis shows that the three emissions have one impurity in common. We propose that this is an acceptor level related to the Mg incorporation and the three deep donor levels are due to compensation effects.

Introduction

GaN is found to exist in a thermodynamically stable hexagonal and in a metastable cubic phase [1]. Substrates for the cubic material are much easier available, cleavage is less complicated and due to the higher symmetry superior electronic properties are expected. In order to achieve optoelectronic devices based on p-type cubic GaN it is important to study its fundamental properties. As in hexagonal samples, the basic properties like recombination channels of Mg doped cubic GaN samples are still not completely understood. Lightly and moderately doped samples show besides the excitonic transitions at 3.27eV, the donor-acceptor pair DAP transition at 3.15eV and a Mg related DAP recombination at 3.04eV [2]. For heavily doped samples ([Mg]>10^{18}cm^{-3}) deep Mg related transitions are observed. In this work these deep blue emissions are studied by steady state (SS) and time resolved (TR) photoluminescence.

Experiment

The cubic GaN layers were grown by plasma-assisted molecular-beam epitaxy on semi-insulating (100) GaAs substrates at a substrate temperature of 720^{0}C. Details of the growth procedure were reported elsewhere [3]. For Mg doping, a commercially available effusion cell with an orifice of 3mm was used. The thickness of the GaN layers varies between 700 and 900nm. Mg concentration and depth distribution was determined by secondary ion mass spectroscopy. In this work samples with Mg concentration from $2.3*10^{18}$ to $5*10^{18}$cm^{-3} were analysed. Results are compared with an non-intentionally doped (nid) sample.

For steady state (SS) photoluminescence PL measurements we used the 325nm line of a HeCd laser. Excitation intensity was varied by neutral density (ND) filters. Time resolved (TR) measurements were carried out with a pulsed Xe lamp as an excitation source and a boxcar system for detection. The emission was dispersed and detected in both cases by a monochromator (1m, 1200/mm) and a photomultiplier, respectively. The samples were cooled down to 10K by a closed-cycle He cryostat. The temperature could

Mat. Res. Soc. Symp. Proc. Vol. 572 ©1999 Materials Research Society

be varied from 10K to room temperature by a heating resistance. The temperature was measured by a thermocouple.

Results and discussion

Figure 1 shows SSPL spectra of a heavily Mg doped and a nid sample. In the nid layer there is a strong emission at 3.27eV and 3.15eV ascribed to excitonic and pair emission, respectively. In the heavily Mg doped samples the PL spectra are dominated by a broad blue band composed of three emissions. The 3.27eV and 3.15eV emissions are hardly detectable in these layers.

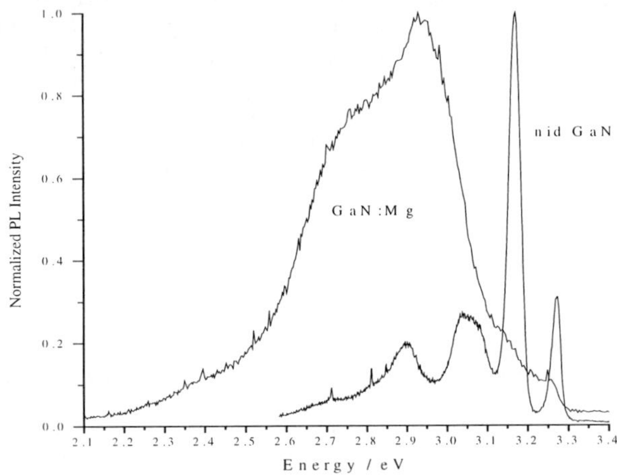

Fig. 1: Low temperature (10K) SSPL spectra of Mg doped and undoped GaN sample

The exciton recombination at 3.27eV and the DAP emission at 3.15eV are quenched and recombination takes place at lower energies. Doping cubic GaN layers with Mg introduces alternative recombination channels.

The comparison of SS and TR spectra shows a shift to lower energies in TR spectra (figure 2). Furthermore the low energy side of the emission is less quenched than the high energy side. This is an indication for an energy dependent transition probability. For a quantitative analysis the SS and TR spectra for various time delays were fitted to three Gaussian curves. In the SSPL spectrum the three Gaussians are peaked at 2.968eV (FWHM: 173meV), 2.789eV (FWHM: 239meV) and 2.512eV (FWHM: 318meV). The energetic distance between the three bands were maintained when fitting the spectra. The three bands shift to lower energies with increasing time delay (TD). The maximal shift (79meV) is reached at a TD of 10ms. Further increase of the TD does not result in a further shift. The shift to lower energies is accompanied by a very small narrowing of the bands. The energetic distance between the three bands could be maintained which means that all three bands show the same shift.

Decay curves were determined at various energies of the spectrum (figure 3). They all follow a power law $I(t) \sim t^{-a}$ rather than exponential decays. The exponent a is increasing from 1.00 to 1.43 with increasing energy position. This is a sign for slower emission at the low energy side. This behaviour is typical for distant DAP recombination.

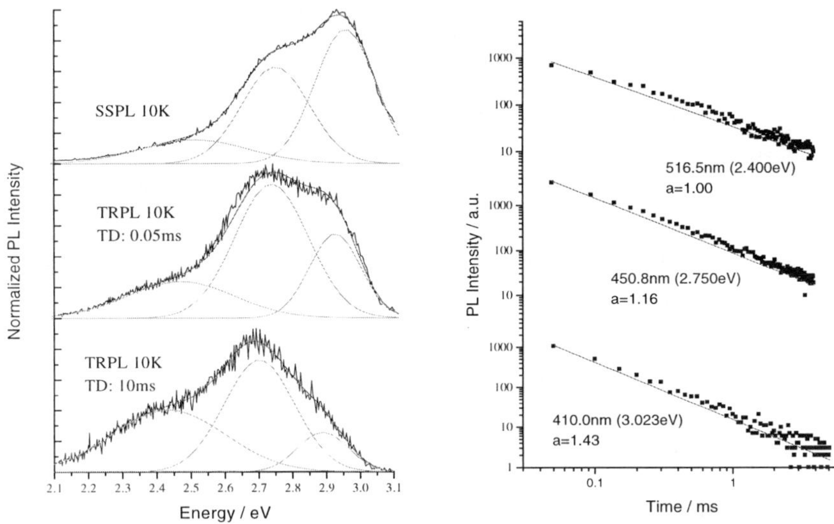

Fig. 2: SS and TR PL spectra of a heavily doped GaN:Mg sample

Fig. 3: Decay curves at various energies.

The recombination energy of a DAP is [4]:

$$h\nu = E_g - (E_A + E_D) + e^2/4\pi\varepsilon_0\varepsilon R \tag{1}$$

where E_D and E_A are the donor and acceptor ionisation energies and R is the distance between the donor and the acceptor. The recombination energy decreases with the distance between the impurities and so does the overlap of their wavefunctions. Therefore the recombination probability is lower for more distant pairs and thus the decay time is longer. Calculating the exact decay curves is rather complicated but it has been shown [5] that the decay follows approximately a power law with exponents between 1 and 2.

In figure 4 the spectra are shown for 3 different excitation intensities. With higher excitation intensity the whole band shifts to higher energies and the high energy side increases in intensity. The shift rate amounts approximately to 7meV per decade of excitation intensity. The increase of emission intensity with excitation density follows a power law:

$$I_{emission}=C*I^{b}{}_{excitation} \qquad (2)$$

with b showing values of 1.0 (2.95eV band), 0.9 (2.75eV band and 0.8 (2.5eV band), typical for recombinations at impurities (DAP or deep defects) [6]. Excitonic recombinations would behave superlinearily (b>1). The shift to higher energies of the peak positions with increasing excitation intensity can also be explained with recombinations at DAPs. With increasing excitation intensity, the very distant (low energy) pairs get saturated and the recombination takes place at closer (high energy) pairs.

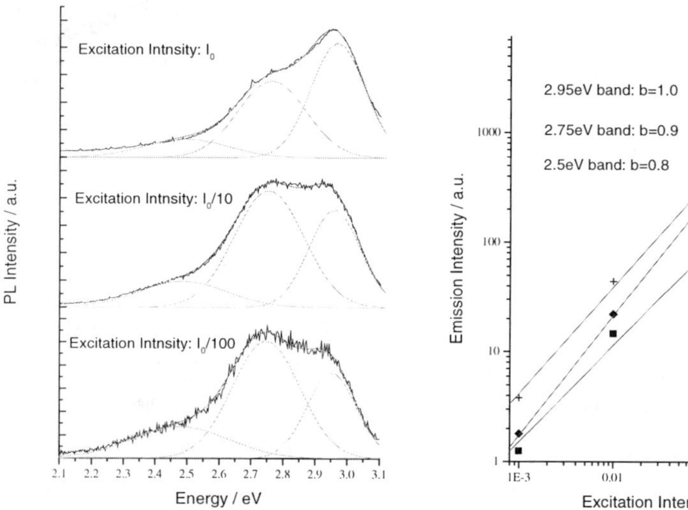

Fig. 4: PL spectra for various excitation densities

Fig. 5: Emission Intensity as a function of excitation intensity for the three bands

Temperature dependent spectra show that there is hardly any change in lineshape of the emission (figur 6). There is a shift of all three bands to lower energies with increasing temperature. The shift follows exactly the temperature dependence of the band gap of cubic GaN [7]. The energetic distance between the bands is maintained. The shift is 24meV between 10K and 165K. Above 165K the emission is very poor and a quantitive assessment is not possible anymore. DAP emissions are expected to behave like this because both donor and acceptor levels maintain their energetic distance to the corresponding band.

Usually the temperature dependence of DAP show exponential dependency of the following form:

$$I(T)=I_0/[1+C_1*\exp(-E_{a1}/k_BT)+C_2*\exp(-E_{a2}/k_BT)] \qquad (3)$$

C_1 and C_2 are two temperature independent constants and k_B is the Boltzmann constant.

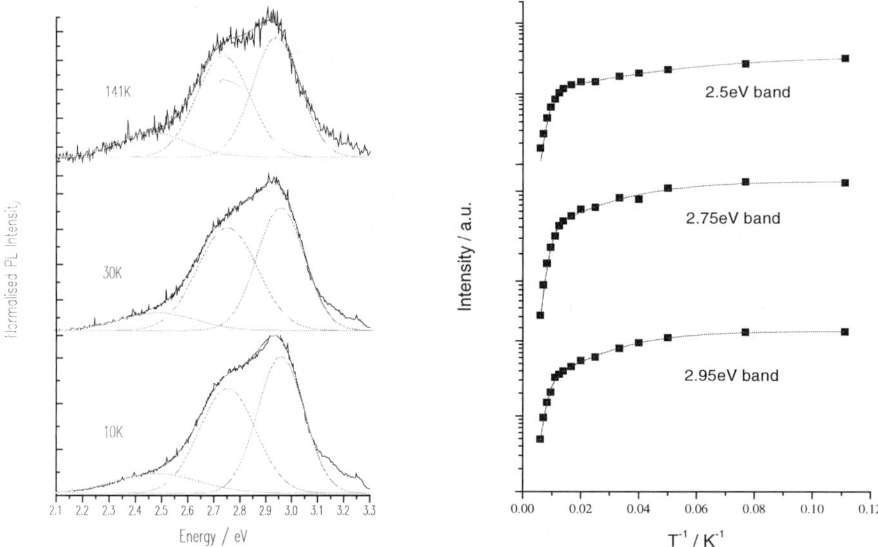

Fig. 6: SSPL spectra for various temperatures

Fig. 7: PL intensity of the three bands as a function of

All three bands show a similar behaviour with E_{a1} ~50meV for the 3 bands and E_{a2} varying between 4 and 6meV (figure 7). We could propose a recombination model with one common impurity for the 3 transitions. Mg on the Ga site is supposed to introduce an acceptor level in the band gap which however might be compensated by three donor levels. Such a compensation mechanism was already found in hexagonal GaN [10]. Evidence for additional 3 donor levels is given by As et al. [11]. However since the activation energy of 50meV cannot be directly related to any energy level of impurities mentioned in literature it is very likely that a more complicated quenching mechanism is present as proposed by Brandt [9]. No further information about the energy levels can be obtained from the data as the bands are broad and structureless and probably a superposition of unresolved single pair emissions and vibronic coupling.

Conclusions

We studied cubic GaN samples heavily doped with Mg. The results of SSPL and TRPL measurements depending on excitation density and temperature are all consistent with three DAP emissions. Since these emissions appear only in Mg doped samples we conclude that Mg doping introduces levels in the band gap which result in DAP recombination. At the moment we cannot determine the impurity levels with certainty.

References

1.   J W Orton, Semicond. Sci. Techonol. 10, 101 (1995).
2.   D.J. As, T. Simonsmeier, B. Schöttker, T. Frey, D. Schikora, W. Kriegseis, W. Burkhardt, B.K. Meyer, Appl. Phys. Lett. 73, 1835 (1998)
3.   D. Schikora, M. Hankeln, D.J. As, K. Lischka, T. Litz, A. Waag, T. Buhrow, and F. Henneberger, Phys. Rev. B 54, R8381 (1996)
4.   D. Thomas, J. Hopfield, W. Augustyniak, Phys. Rev.140 (1A), A202 (1965)
5.   P. Dean, Progress in Solid State Chem. 8, 1 (1973)
6.   T. Schmidt, K. Lischka, W. Zulehner, Phys. Rev. B 45, 8989 (1992)
7.   G. Ramírez-Flores, H. Navarro-Contreras, A. Lastras-Martínez, R.C. Powell, and J.E. Greene, Phys. Rev. B 50, 8433 (1994)
8.   U. Kaufmann, M. Kunzer, H. Obloh, M. Maier, Ch. Manz, A. Ramakrishnan, B. Santic, Phys. Rev. B 59, 5561 (1999)
9.   O. Brandt, Group III Nitride Semiconductor Compounds – Physics and Application, 11, 417-459, edited by B. Gil, Oxford Science Publications, Clarendon Press, Oxford (1998)
10.   L. Eckey, U. Gfug, J. Holst, A. Hoffmann, B. Schineller, K. Heime, M. Heuken, O. Schön, R. Beccard, J. Crys. Growth 189/190, 523 (1998)
11.   D. J. As, Phys. Stat. Sol. (b) 210, 445 (1998)

Acknowledgements

One of the authors (R. Seitz) gratefully acknowledges financial support by the *Fundação Para A Ciência E A Tecnologia*. D.J. As , T. Frey, D. Schikora, and K. Lischka acknowledges SIMS measurements by W. Kriegseis, W. Burkhardt and B.K. Meyer and financial support by *Deutsche Forschungsgemeinschaft*.

For further information, contact:
Roland Seitz
University of Aveiro, Department of Physics, 3800 Aveiro, Portugal
Fax: +351 34 424965   Tel: +351 34 370824   e-mail: seitz@fis.ua.pt

DIELECTRIC FUNCTION OF AlN GROWN ON Si (111) BY MBE

Stefan Zollner *, Atul Konkar *, R.B. Gregory *, S.R. Wilson *, S.A. Nikishin **, H. Temkin **
*Motorola Semiconductor Products Sector, Embedded Systems Technology Laboratories, MD
M360, 2200 West Broadway Road, Mesa, AZ 85202
**Texas Tech University, Dept. of Electrical Engineering, Box 43102, Lubbock, TX 79409

ABSTRACT

We measured the ellipsometric response from 0.7-5.4 eV of c-axis oriented AlN on Si (111)
grown by molecular beam epitaxy. We determine the film thicknesses and find that for our AlN
the refractive index is about 5-10% lower than in bulk AlN single crystals. Most likely, this
discrepancy is due to a low film density (compared to bulk AlN), based on measurements using
Rutherford backscattering. The films were also characterized using atomic force microscopy and
x-ray diffraction to study the growth morphology. We find that AlN can be grown on Si (111)
without buffer layers resulting in truely two-dimensional growth, low surface roughness, and
relatively narrow x-ray peak widths.

INTRODUCTION

AlN, like all group-III nitrides, has recently received much attention because of its potential
applications in high-power and high-frequency electronics, optoelectronics, electronic packaging,
and deep-UV lithography. The resulting flurry of activities in this area, however, does not mean
that AlN is a new material. Publications on the physical properties of AlN go back more than 35
years [1]. Much of the early results are based on pioneering work by Slack, who synthesized
high-purity AlN single-crystals and studied their properties, particularly the high thermal conductivity [2,3]. Recent interest in AlN has shifted to thin films prepared on different substrates
(Si, SiC, sapphire) using a variety of techniques, such as molecular beam epitaxy (MBE), chemical vapor deposition [4], and RF sputtering [5].

The optical properties of AlN, i.e., the refractive index n and extinction coefficient k, have
recently been reviewed by Loughin and French [3]. Below the band gap of about 6.1 eV, the
material is essentially transparent, allowing the determination of n using bulk crystals. Such measurements are usually carried out employing the minimum-deviation prism method with
extremely high accuracy, not achieveable on thin films. Bulk measurements are not affected
much by surfaces and interfaces, although the chemical composition is important. It is generally
assumed that the crystals grown by Slack contain a very little oxygen or other impurities [3].

Therefore, the purpose of our work is not to determine the refractive index of AlN, but rather
to compare the refractive index of AlN thin films grown by MBE with that of bulk crystals and to
draw conclusions about the properties of the film. Our work is similar to that of Jones *et al.* [5],
except that their films were prepared by RF sputtering. Our AlN films were grown using MBE on
low-resistivity Si (111), cut 3° off towards (110), at 860°C with ammonia as the nitrogen source,
therefore they might contain some hydrogen or silicon, which tend to reduce n. Because of the
high affinity of AlN for oxygen, contamination with oxygen is a possibility [4], which would also
reduce the value of n. Infrared transmission measurements of AlN on high-resistivity two-side
polished Si are planned to determine the impurity content. Two films were studied here, one
displaying predominantly 2-D growth (#1, nominal thickness 1500 Å), the other mixed 2-D and
3-D growth (#2, nominal thickness 2000 Å).

Mat. Res. Soc. Symp. Proc. Vol. 572 © 1999 Materials Research Society

AlN FILM PROPERTIES BY X-RAY DIFFRACTION, RUTHERFORD BACKSCATTERING, AND ATOMIC FORCE MICROSCOPY

Since the AlN films are grown on Si (111), the crystal structure template defined by the substrate encourages growth of wurtzite AlN with the hexagonal axis along the growth direction. This is confirmed by a ϑ-2ϑ x-ray diffraction scan using a D-MAXB single-axis goniometer with a Rigaku RU-200BH rotating-anode x-ray generator, see Fig. 2 (a). Because of the substrate miscut, which cannot be compensated by our experimental arrangement, the Si (111) peak is broad and weak. The AlN (0002) peak for film #1 is found at 2ϑ=36.05° corresponding to a lattice constant c=4.98 Å, in agreement with the literature for bulk AlN [1]. Therefore, the strain in the film due to the lattice mismatch between AlN and Si is completely relaxed by misfit dislocations. The width of the 2θ peak is 0.3-0.4°, slightly larger than the instrumental resolution of 0.2°. A rocking curve with the detector fixed at 2ϑ=36.05° shows a peak at 16.2°, indicating a tilt between the c-axis of the film and the surface normal because of the substrate miscut. The magnitude of the tilt cannot easily be determined with our goniometer. The width of the AlN (0002) rocking curve peak, see Fig. 2 (b), is slightly less than 1°, giving an estimate of the mosaic spread due to dislocations. The width of the AlN (0004) rocking curve peak is 0.86° (not shown). Film #2 shows similar x-ray diffraction results.

Atomic force microscopy (AFM) analysis of the AlN films were carried out on a Dimension 3000 instrument in tapping mode. The surface roughness of the two AlN films was substantially different. Whereas the film #2 with mixed 2-D and 3-D growth mode had a surface roughness of about 142 Å, film #1 with 2-D growth mode was much smoother and had surface roughness of about 10 Å. Shown in Fig. 1 is a perspective view AFM image of the AlN film #1 evidencing the high growth quality. The film exhibits a clear terrace and step structure that is typical of films with layer-by-layer 2-D growth on tilted substrates. The angle subtended by the terraces with the average film surface is about 2°. Assuming that the terraces are the AlN (0001) planes the miscut angle for AlN would be 2° which compares favorably with the intended miscut of the Si substrate of 3°. The average step height measured by AFM is greater than 30 Å. Since the thickness of 1 ML of AlN is 2.49 Å, the AFM results show that there is significant step bunching during AlN growth resulting in steps that are greater than 10 MLs high.

Rutherford backscattering (RBS) finds that the density of the AlN film #1 is 3.0 g/cm^3, about 10% lower than the published density of 3.26 g/cm^3 [1]. For film #2, the density is 2.9 g/cm^3, i.e., somewhat lower. Most likely, the differences in density are due to hydrogen (possibly saturating dangling bonds at structural defects) or other impurities (such as oxygen) propagating along dislocations or other growth defects (particularly in film #2). There is also a very small RBS yield due to a contaminant with a mass similar to that of Ga (less than 0.1% by composition), which is not expected to have a measureable influence on the optical properties.

ELLIPSOMETRY MEASUREMENTS

Ellipsometry measurements were carried out using a Woollam variable-angle ellipsometer with compensator in the 0.74 to 5.4 eV spectral range at three angles of incidence ϕ=65°, 70°, and 75°. We obtained the ellipsometric angles ψ and Δ, which are defined by r_p/r_s=tanψexp(iΔ), where r_p and r_s are the complex Fresnel reflection coefficients (amplitude and phase) for p- and s-polarized light. The finite bandwidth of our monochromator (about 1.5 nm) leads to a depolarization with an amplitude of no more than 6%. There is no additional depolarization due to thickness nonuniformities, which is also evident by the uniform coloration of the films across the wafer surface. The ellipsometric angles for film #1 are shown in Fig. 3 (dashed lines).

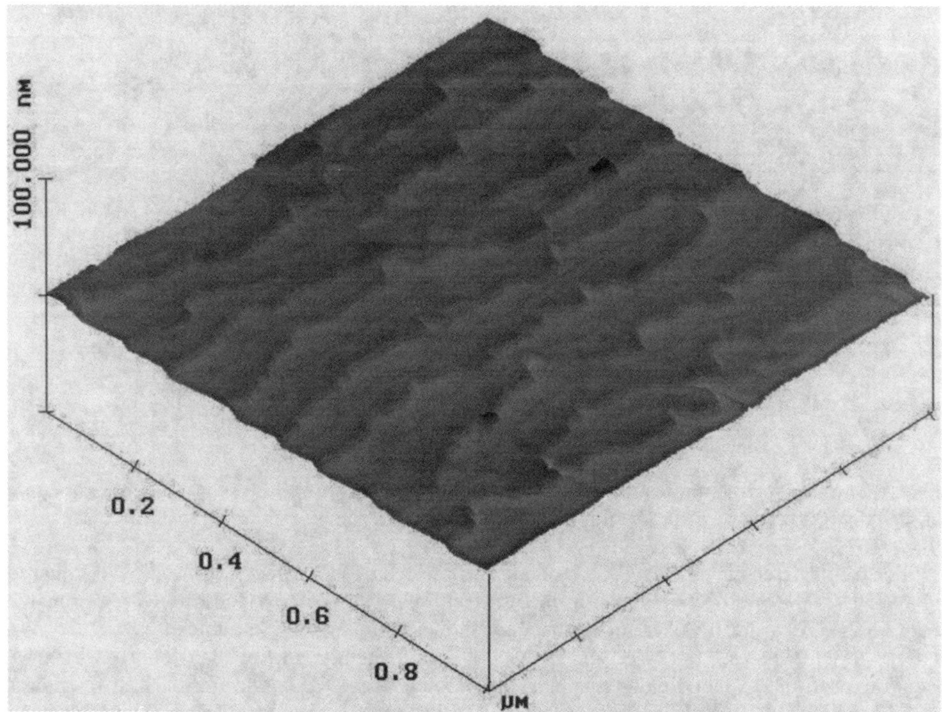

Figure 1: Perspective view atomic force microscopy image of AlN film #1 with 2-D growth by the terrace and step structure of the film.

We describe our data using a four-layer (ambient/surface/AlN film/substrate) model. An interface layer between the AlN film and the substrate does not improve the agreement between the data and the model. (In contrast to the work described in [4], no buffer was grown between the substrate and the AlN film.) We first ignore the birefringence of AlN and describe n with a Cauchy equation $n=A+B/\lambda+C/\lambda^2$, where λ is the wavelength. (Similar results are obtained with a Sellmeier equation or using a Lorentz oscillator.) The extinction coefficient k of the film is assumed to be given by an exponential Urbach tail. The absorption in the film is very small and might be located at the AlN/Si interface. The surface layer is modeled within the Bruggeman effective medium theory as a 50/50 mix of AlN and voids. (An oxide with properties similar to those of SiO_2 or Al_2O_3 gives a similar description.) Our model thus contains five parameters for the optical constants of the film plus the thicknesses of the film and the surface layer. These eight parameters are varied using the Marquardt-Levenberg algorithm to minimize the least-square deviation between the data and the model. The calculated ellipsometric angles, shown by the solid lines in Fig. 3, are in good agreement with the measured data. This agreement can be improved slightly by taking into account the small difference Δn between the ordinary and extraordinary refractive index of wurtzite AlN, also described with a Cauchy model. The anisotropy effects are small, on the order of 2% of n, see Ref. [3].

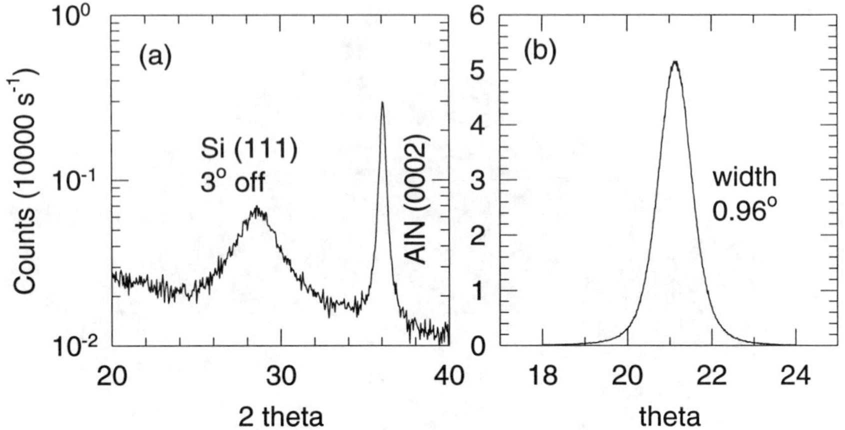

Figure 2: X-ray diffraction results of AlN on Si (111), film #1. (a) θ-2θ scan. (b) Rocking curve of the AlN (0002) peak. Data for film #2 are essentially the same.

For film #1 (see data in Fig. 3), we obtain thicknesses of the surface and AlN layers equal to 17 Å and 1500 Å, respectively, and a refractive index n=2.05 at 2 eV for AlN with a probable error on the order of 2%. AlN film #2 (data not shown) is found to have a thickness of 2026 Å, a surface layer thickness of 124 Å, and n=1.98 at 2 eV, slightly smaller than for #1, which we would expect due to the slightly smaller density found from RBS. The rms surface roughness found by AFM is 10 Å for film #1 and 142 Å for film #2, in reasonable agreement with the ellipsometry results.

DISCUSSION OF OPTICAL CONSTANTS

In Fig. 4, the ordinary refractive index determined from the Cauchy parameters and the extinction coefficient from the Urbach tail of the AlN film #1 are given by solid and the dash-double dotted lines, respectively. By fixing the thicknesses to their values determined in the model and solving the ellipsometric angles for n at each wavelength without any assumption about the dispersion model, we obtain n given by the dotted line, which is basically the same as in the Cauchy model. This confirms that the Cauchy description for n is adequate in our spectral range. The figure also shows the extraordinary refractive index found by the model (dashed line). We compare our results with the data of Ref. [3] for bulk AlN and of Refs. [6-7] for thin films, given by the symbols. The refractive index for the AlN film #2 is 3% lower (not shown).

DISCUSSION AND SUMMARY

It was shown that the refractive index of our AlN films grown on Si (111) by MBE is about 5-10% lower than that of bulk AlN single crystals [3]. Since RBS finds a density of our films that is about 10% lower than the density of bulk AlN [1], the difference in density is the most likely culprit for the low n. Impurities in the film (oxygen, hydrogen, or silicon) are also a possible

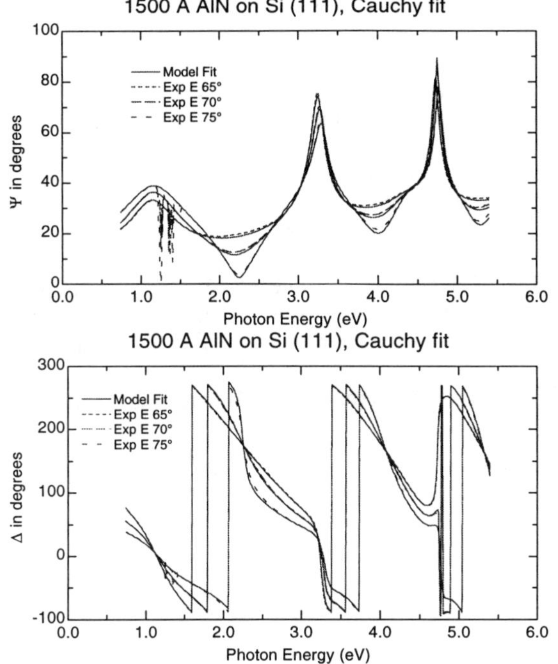

Figure 3: Ellipsometric angles for AlN on Si (111) grown by MBE (film #1). The dashed lines show the experimental data at three angles of incidence. The solid lines are the results of a calculation, based on (i) a Cauchy model with an Urbach tail for absorption to describe the AlN layer, (ii) a thin surface layer due to surface roughness or a native oxide, (iii) optical anisotropy due to the wurtzite crystal structure of GaN, (iv) finite bandwidth of the monochromator (1.5 nm).

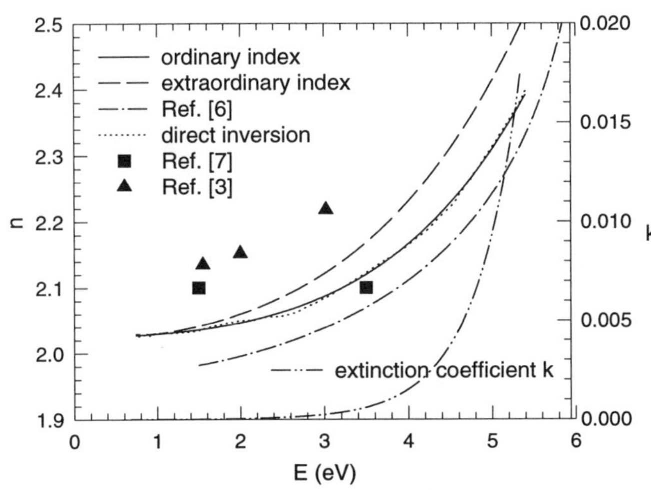

Figure 4: Ordinary (solid) and extraordinary (dashed) refractive index n and extinction coefficient k (double dot-dashed) of AlN film #1 based on the data and model calculation from Fig. 3. The dotted line shows n calculated at each wavelength from the ellipsometric angles, without assuming a specific model for the dispersion of AlN. Data from the literature are given for comparison.

explanation, although their presence has not been demonstrated. The different refractive index between two films is also due to density differences, as shown by RBS.

REFERENCES

1. O. Madelung, *Semiconductors - Basic Data*, 2nd ed. (Springer, Berlin, 1996), p. 69.
2. G.A. Slack, J. Phys. Chem. Solids **34**, 321 (1973).
3. S. Loughin and R.H. French, in *Handbook of Optical Properties of Solids*, edited by E.D. Palik, (Academic, Orlando, 1998), p. 373, and references therein.
4. S.Q. Hong, H.M. Liaw, K. Linthicum, R.F. Davis, P. Fejes, S. Zollner, M. Kottke, S.R. Wilson, this volume.
5. D.J. Jones, R.H. French, H. Müllejans, S. Loughin, A.D. Dorneich, P.F. Carcia, (in print).
6. O. Ambacher, M. Arzberger, D. Brunner, H. Angerer, F. Freudenberg, N. Esser, T. Wethkamp, K. Wilmers, W. Richter, M. Stutzmann, MRS Internet Journal of Nitride Semiconductor Research **2**, Article 22 (1997).
7. N.V. Edwards, M.D. Bremser, T.W. Weeks, R.S. Kern, R.F. Davis, and D.E. Aspnes, Appl. Phys. Lett. **69**, 2065 (1996).

**The comparative studies of chemical vapor deposition grown epitaxial
layers and of sublimation sandwich method grown 4H-SiC samples**

A. O. Evwaraye[a], S. R. Smith[b], and W. C. Mitchel
Materials Directorate, MLPO, Air Force Research Laboratory, Wright-Patterson Air Force
Base, Ohio, 45422-7077. U.S. A
[a] University of Dayton, Physics Department, 300 College Park, Dayton, Ohio 45469-2314
[b] University of Dayton Research Institute, 300 College Park, Dayton, Ohio 45469-0167

ABSTRACT

Thermal admittance spectroscopy was used to characterize the shallow dopants in chemical
vapor deposition (CVD) grown thin films and in sublimation sandwich method (SSM)
grown 4H-SiC layers. The values of the activation energy levels of $E_C - 0.054$ eV for
Nitrogen at the hexagonal site and of $E_C - 0.10$ eV for Nitrogen at the cubic site were indices
of comparison. The net carrier concentrations ($N_D - N_V$) of the films were determined by
capacitance-voltage measurements. The net carrier concentrations for the SSM films ranged
from 2×10^{17} to 7×10^{17} cm^{-3}. The two Nitrogen levels were observed in the CVD films.
Hopping conduction with an activation energy of $E_C- 0.0058$ eV was observed in one SSM
sample having $N_D - N_V = 7 \times 10^{17}$ cm^{-3}.

INTRODUCTION

It is well known that silicon carbide sublimates when heated above 1800^0C producing
Si, SiC_2, and Si_2C vapors. Sublimation of silicon carbide is the corner stone in the growth of
bulk silicon carbide single crystals. In the modified Lely technique or the physical transport
(PVT) method, these vapors are physically transported through the growth chamber onto a
cooler seed where they condense to form single crystals. The sublimation process has also
been successfully applied to the growth of thin epitaxial layers. In the so called " sandwich"
method (SSM), the source (polycrystalline) is separated from the substrate by a small gap
of about 1 mm. The source and the substrate are heated so that a temperature gradient of about
10^0 C exists. It is claimed that the conditions are near equilibrium conditions which allow the
growth to be carried out in wider ranges of temperature and pressure. Sublimation etching can
also be carried out before thin film deposition. This is accomplished under reverse
temperature gradient—this time the substrate is at a higher temperature. However, this leaves
the surface deficient of silicon and is carbon rich. Epitaxial layers of 1- 20 μm of 6H- and 4H-
SiC samples with residual impurity concentration of less than 10^{16} cm^{-3} and density of
dislocations less than 100 cm^{-2} have been reported[1,2]. Chemical vapor deposition (CVD) is
the more general method of growing thin films on substrates.

Nitrogen, which is incorporated into the crystalline lattice during growth, substitutes
for carbon atoms at the hexagonal (h) and cubic (k) sites in 4H- SiC. Nitrogen at the two
inequivalent sites leads to two nitrogen activation energy levels. These have been determined
to at $E_C - 0.054$ eV and at $E_C- 0.10$ eV for hexagonal and cubic sites respectively[3,4].

In this paper, we report the results of comparative studies of CVD and SSM grown
samples. The activation energy levels of nitrogen levels in 4H-SiC are the main indices of
comparison.

Mat. Res. Soc. Symp. Proc. Vol. 572 © 1999 Materials Research Society

EXPERIMENTAL DETAILS

The six samples used in this study are listed, with their net doping concentrations and their growth methods, in Table I.

Table I. List of samples with their net doping concentration and growth methods.

Sample #	$N_D - N_A$ (x 10^{17} cm^{-3})	Growth method
5500	0.0015	CVD
5635	0.0015	CVD
5722	2.0	SSM
5929	4.0	SSM
5930	7.0	SSM
5931	4.0	SSM

Samples 5500 and 5635 were grown by chemical vapor deposition and obtained from Cree Research Inc. and from Northrop-Grumman Science and Technology center respectively. The other four samples were SSM grown 4H-SiC and obtained from the Institute of Semiconductors, Kiev, Ukraine. After the usual cleaning in RCA solutions, Ni was sputtered onto the back side of the samples and annealed at 900^0 C to form the ohmic contact. Circular Schottky diodes were then fabricated on the thin films as described elsewhere[5]. Capacitance-voltage (C-V) measurements were made with a 1 MHz capacitance meter. The net doping concentrations listed in Table I were determined from the C-V data. The thermal admittance spectroscopy (TAS) is described in detail in reference 6. In this experiment, the conductance and the capacitance are measured as functions of frequency and temperature. TAS measurements were made at frequencies ranging from 100 Hz to 1 MHz, and a temperature range of 5 K to 300 K.

RESULTS AND DISCUSSION

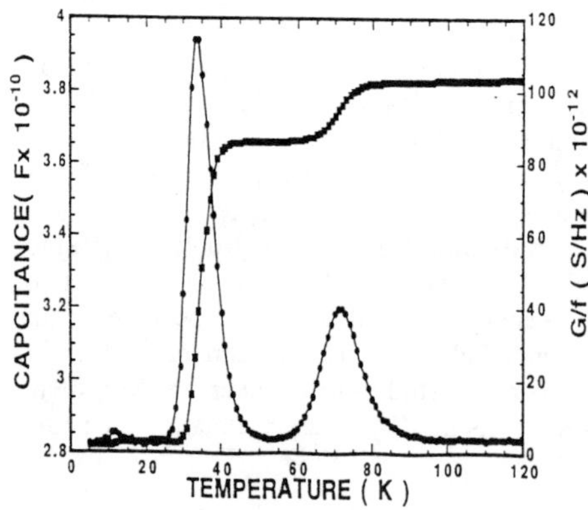

Figure 1. The two conductance peaks are due to excitation from N at hexagonal and cubic sites.

The thermal admittance spectrum of sample 5635 is shown in figure 1. The two conductance maxima or the two inflections in the capacitance curve correspond to emissions from nitrogen at the hexagonal and cubic sites respectively. The activation energies were obtained from the plots of $\ln(e_n/T^2)$ versus 1/kT. The activation energies so determined are E_C-0.054 eV and E_C-0.101 eV.

Figure 2 shows TAS of sample 5929. The device was first connected to the grid (planar) and TAS data taken at 20 kHz. However, an identical spectrum was obtained when the device was connected to the ground (front to back) thereby involving the substrate in the measurements.

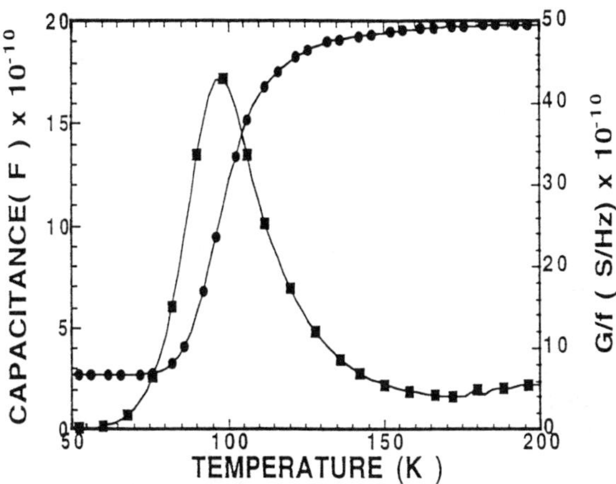

Figure 2. Thermal admittance spectroscopy of sample 5929. Only one conductance peak is observed.

Figure 3 is the plot of ln (e_n/T^2) versus $1/kT$; the slope of the straight line through the data points gives the activation energy of thedefect. The value is E_C-0.073 eV. The emission of nitrogen from the hexagonal site was not observed, due to the position of the Fermi level

Figure 4 shows the TAS of sample 5930 (N_D-N_A = 7 x 10^{17} cm^{-3}) taken at two frequencies (1 MHz and 80 kHz). The conductance maximum occurs at much lower temperatures than those in

figures 1 and 2. The Arrhenius plot of the TAS data yielded an activation energy of E_C- 0.0058 eV. This is clearly an activation energy for hopping conduction.[7] Normally hopping conduction in silicon carbide does not commence til the background doping is in the range of 4 x 10^{18} cm^{-3}. As expected emission from N into the conduction band was not observed.

The devices of sample 5931 were first connected to the grid so the measurements were planar, not involving the substrate. Figure 5 shows the TAS of sample 5931 at the measuring frequency of 50 kHz. The activation energy determined from the plot of ln(e_n/T^2) versus $1/kT$

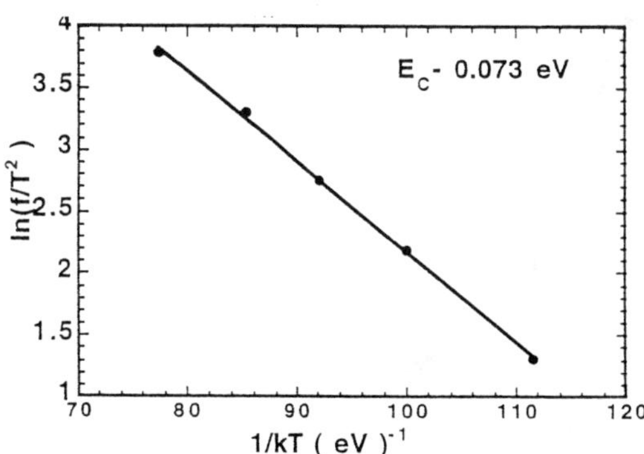

Figure 3. Arrhenius plot of TAS for sample 5929. The slope of the line fitted to the data yield the thermal activation energy.

is E_C –0.055 eV. This is the N level at the hexagonal site. However, when the devices were connected to the ground to involve the substrate in the measurements, an additional peak appeared in the spectrum. This is shown in figure 6. The additional conductance peak at 130 K must be due to the interface between the SSM film and the substrate. A SIMS analysis of a a silicon carbide specimen annealed at 1900⁰ C showed that the surface was carbon rich. So the secondary peak in the spectrum may due a carbon complex at the interface. The devices in

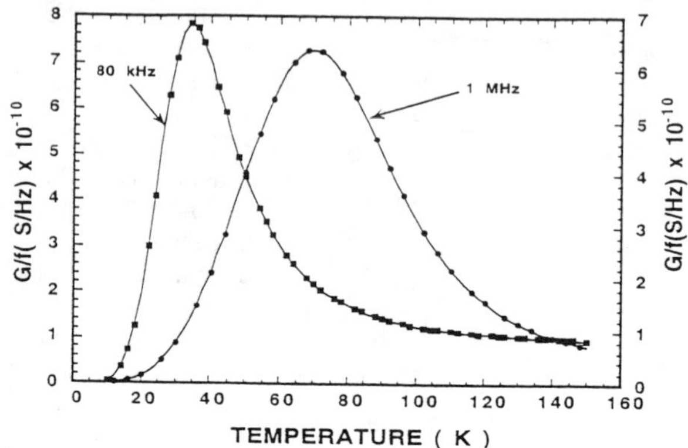

the SSM film in sample 5931 were stripped off and after thorough cleaning, a new set of diodes were fabricated on the substrate. The TAS obtained was identical to that in figure 5. It was not possible to deplete the substrate to the interface in order to observe the second conductance maximum of figure 6.

Figure 4. The TAS of sample 5930 at two different frequencies of 20 kHz and 1 MHz. The activation energy for hopping conduction is 58 meV.

The TAS of sample 5722 ($N_D - N_V = 2 \times 10^{17}$ cm⁻³) is identical to that of sample 5635 shown in figure 1. However, the activation energies were found to be E_C-0.033 eV and E_C- 0.080 eV respectively

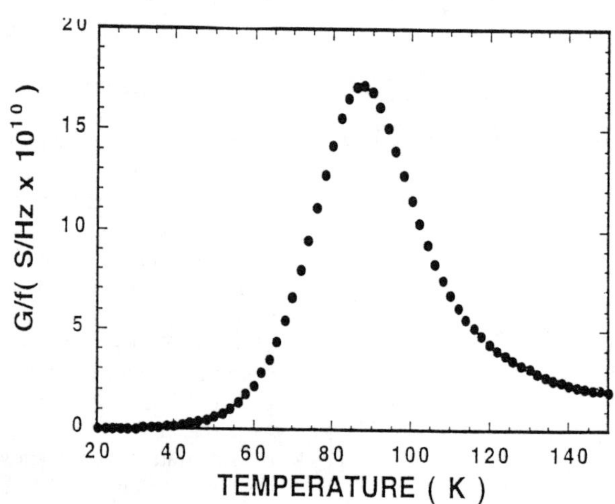

Figure 5. TAS of sample 5931; the devices were connected to the grid without involving the substrate in the measurements. The activation energy is $E_C - 0.055$ eV.

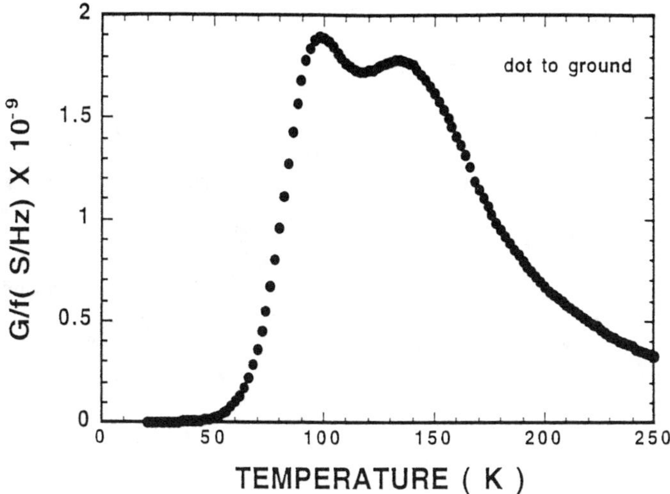

Figure 6. The TAS of sample 5931 when the device was wired to ground through the substance. The second peak is defect at the interface between the SSM film and the substrate.

CONCLUSION

Thermal admittance spectroscopy has been used to determine the nitrogen activation energy in two CVD grown samples and four SSM grown samples. The two N levels were observed in samples 5500   (CVD), 5635 (CVD) and 5722 (SSM). Two SSM samples 5929 and 5931, even though they have the same net doping concentration (4×10^{17} cm^{-3}) behaved quite differently. Nitrogen at the hexagonal site (E_c- 0.054 eV) was observed in sample 5931 while a deeper level E_c- 0.073 eV was observed in sample 5929. The degree of compensation is the difference between these samples. The fact that hopping conduction took place in sample 5930 is another indication that the SSM samples are compensated. The SSM samples behaved more like bulk silicon carbide than the traditional epitaxial layers grown by CVD.

ACKNOWLEDGEMENTS

We would like to acknowledge the technical contributions of Mr. Gerry Landis, Mr Robert Bertke, and Mr. Robert Leese for the preparation of specimens.

REFERENCES

1. M.M. Anikin, V. A. Dmitriev, and V. E. Chelnokov,176 Meeting of Electrochem. Society
    (Hollywood FL)
2. P.A. Ivanov and V. E. Chelnokov, Semicond. Sci. Technol. 7, 863, (1992)
3. A. O. Evwaraye, S. R. Smith, and W. C. Mitchel,, J. Appl. Phys. **79**,7726, (1996)
4. W. Gotz, A. Schoner, G. Pensl, W. Suttrop, W. J. Choyke, R. Stein, and S Leibenzeder, J.
    Appl. Phys. **73,** 3332 (1993)
5. A. O. Evwaraye, S. R. Smith, M. Skowronski, and W. C. Mitchel, J. Appl. Phys. **75,** 3472
    (1994)
6. D. L. Losee, J. Appl. Phys. **46,** 2204 (1975) see also A. O. Evwaraye, S. R. Smith, and
    W. C. Mitchel, J. Appl. Phys. **75,** 3472 (1994)
7. W. C. Mitchel, A. O. Evwaraye, S. R. Smith, and M. D. Roth, J. Electron. Mater. **26,**113 (
    1997)

Part III

SiC Bulk Growth and Characterization

IMPURITY EFFECTS IN THE GROWTH OF 4H-SiC CRYSTALS BY PHYSICAL VAPOR TRANSPORT

V. BALAKRISHNA, G. AUGUSTINE, and R. H. HOPKINS
Northrop Grumman Science & Technology Center, 1350 Beulah Road, Pittsburgh, PA 15235

ABSTRACT

SiC is an important wide bandgap semiconductor material for high temperature and high power electronic device applications. Purity improvements in the growth environment has resulted in a two-fold benefit during growth: (a) minimized inconsistencies in the background doping resulting in high resistivity (>5000 ohm-cm) wafer yield increase from 10-15% to 70-85%, and (b) decrease in micropipe formation. Growth parameters play an important role in determining the perfection and properties of the SiC crystals, and are extremely critical in the growth of large diameter crystals. Several aspects of growth are vital in obtaining highly perfect, large diameter crystals, such as: (i) optimized furnace design, (ii) high purity growth environment, and (iii) carefully controlled growth conditions. Although significant reduction in micropipe density has been achieved by improvements in the growth process, more stringent device requirements mandate further reduction in the defect density. In-depth understanding of the mechanisms of micropipe formation is essential in order to devise approaches to eliminate them. Experiments have been performed to understand the role of growth conditions and ambient purity on crystal perfection by intentionally introducing arrays of impurity sites on one half of the growth surface. Results clearly suggest that presence of impurities or second phase inclusions during start or during growth can result in the nucleation of micropipes. Insights obtained from these studies were instrumental in the growth of ultra-low micropipe density (less than 2 micropipes cm^{-2}) in 1.5 inch diameter boules.

INTRODUCTION

To effectively compete as a viable semiconductor material suitable for commercial and military technologies, SiC crystal perfection and diameter of SiC substrates must approach at least GaAs-like specifications. Silicon Carbide (SiC) crystals are of special interest compared to other contemporary high frequency semiconductors, such as Si and GaAs, because of its unique properties for high power density microwave generation, for high power switching, and for high temperature and radiation tolerant operation. As shown in Table I, SiC exhibits a higher thermal conductivity (k_T), a higher critical breakdown field (E_b), and a saturated velocity (V_{sat}) equal to

Table I -- Comparison of fundamental properties of Si, GaAs, and SiC for device applications

	E_g(eV)	E_b(MV/cm)	V_{sat}(10^7cm/s)	K_T(W/cmK)	ε
Si	1.12	0.6	1.0	1.5	11.8
GaAs	1.42	0.6	2.0	0.5	11.8
6H-SiC	3.02	3.2	2.0	3.0*	10.0
4H-SiC	3.26	3.0	2.0	3.0*	9.7

Mat. Res. Soc. Symp. Proc. Vol. 572 ©1999 Materials Research Society

GaAs at high fields desirable for high power devices. As a consequence, solid-state SiC electronics are having a pervasive impact on advanced DOD systems, as well as on commercial applications, ranging from passively cooled high power and high voltage switches, satellite communications, and cellular phones, to nuclear instrumentation, where high temperature and radiation tolerant SiC properties offer major advantages over silicon-based circuits. However, the success of the various commercial and military applications depend on the availability of large diameter, low defect, high purity SiC substrates at affordable prices. Contemporary SiC crystals are limited in the purity, and have increasing residual stresses with an increase in diameter. In addition, the physical vapor transport process is relatively an inefficient growth process resulting in low yield and high cost of the wafers. Increase in wafer diameter (up to 4") and reduction in microstructural defect density is necessary for widespread use of this unique semiconductor material.

PVT SiC CRYSTAL GROWTH

Bulk SiC crystals are mainly grown by the physical vapor transport technique (or the modified-Lely approach) [1-4]. Growth is achieved by the sublimation of the SiC source onto a SiC seed. A schematic of the PVT growth technique is shown in Figure 1. By appropriately controlling the temperature gradient between the source and the seed, the gas pressure inside the growth chamber, and the orientation and quality of the seed crystal, the growth rate and the polytype of the of the SiC crystal is defined. While, the baseline process at Northrop Grumman is the growth of 35 mm diameter SiC boules, with less than 100 micropipes cm^{-2}, single polytype crystals, we have previously demonstrated two- and three-inch diameter growth. The current focus is on the growth of 50 mm diameter (2 inch) whose properties are similar to the baseline crystals, and on the development of three inch diameter crystals. Emphasis is to increase wafer yield by accurate control over their electrical properties and minimize defect density, and towards this several aspects are being investigated which include:

(1) SiC Purity: The limitation to SiC purity arises from the lack of commercially available high purity SiC sources. As clearly evident in Table II, glow discharge mass spectroscopy (GDMS) and calibrated SIMS analysis indicates that most sources are high in metallics, transitional metals, or in nitrogen. Most of these impurities are electrically active and result in a high degree of variability in the electrical properties of the crystal [5-6]. For example, boron and aluminum are common shallow acceptor impurities in SiC, while nitrogen and vanadium are shallow and deep donors respectively. Hence, depending on the type of dominant impurity, the growth could result in p-type or n-

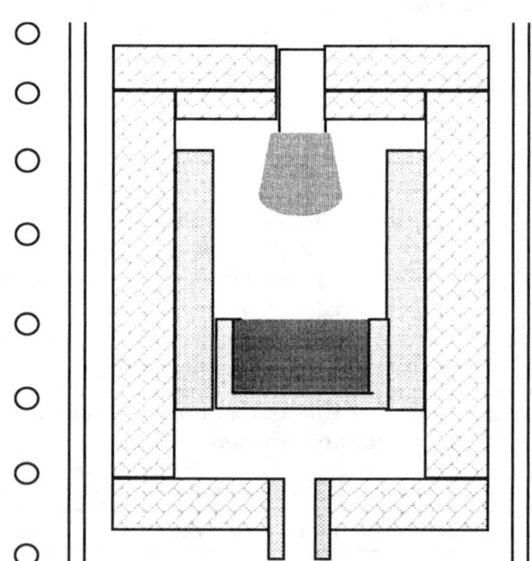

Figure 1 – Schematic of conventional PVT SiC furnace

type doping respectively. In addition, this work has clearly shown that presence of impurities can also result in the nucleation of micropipes, and significantly affect device yield.

(2) Microstructural defects: The presence of microstructural defects such as micropipes (which are almost unique to SiC), edge boundaries and cracks, and high dislocation densities, has limited the rapid commercialization of SiC technology. Micropipes are micron or submicron sized , tube-like voids present in PVT-grown SiC, which degrade epitaxial layer morphology and are detrimental to device performance and yield. Significant strides in the reduction of microstructural defects (to <30 micropipes cm^{-2}) and increase in wafer diameter (up to 2" diameter) has resulted in the fabrication of practical SiC devices such as MESFETs [7], static induction transistors (SITs) [8], and MOSFETs [9], opening the door for the economic exploitation of this semiconductor. Currently, the wafers fabricated in our laboratory are low in micropipe densities (ranging on average from <30 to about 300 micropipes cm^{-2}, with a best effort of <2 cm^{-2}), and the dislocation densities are in the range of 10^{3} to 10^{5} cm^{-2}. However, to benefit power device fabrication (for example, devices with 25mm^{2} footprint size or greater), and to provide a viable alternative for direct fabrication of devices on the substrate, further reduction in micropipe and dislocation densities are needed.

There are several factors or combination of factors that can result in the formation of micropipes [10]. For example, initial growth conditions favoring inhomogeneous nucleation and growth of the crystals, constitutional supercooling, poor seeding technique, impurities in the growth ambient, and lack of process control can result in the formation of micropipes or other microstructural defects such as second phase inclusions, boundaries, and dislocations.

SiC PURITY

The focus of this study was (1) to improve the background crystal purity by raising the source purity and the purity of the growth environment and thereby improve crystal yield, and (2) optimize growth conditions to fabricate large diameter SiC crystals, while simultaneously striving to reduce the defect density. Towards improving the crystal purity, effort was focussed on evaluating and improving the source purity, and purity of the growth ambient

Source Purity

Various commercially available and in-house sources were analyzed to define the

Table II- GDMS analysis of SiC source used for PVT crystal growths. Impurities in ppm by weight, nitrogen content measured by calibrated SIMS.

Source	N	B	Al	V	Ti
Reacted Powder-1	-	5	10	130	80
Reacted Powder-2	-	3.9	19	65	42
CVD Source-1	-	8	2	2	5
CVD Source-2	-	8	2	3	8
CVD Source-3	1.0×10^{18}	1.5	0.1	0.001	0.013

concentrations of various (known) electrically active impurities. Subsequently, N+ and semi-insulating 4H-SiC crystals grown from these sources were also analyzed to obtain insights into the segregation behavior of the impurities. Earlier work clearly reveals the detrimental role of boron and nitrogen in the growth of high resistivity 6H-SiC crystals [6], this study considers the growth of N+ and semi-insulating 4H-SiC crystals, and their potential role in the nucleation of micropipes. Glow discharge mass spectroscopy (GDMS) and secondary ion mass spectroscopy (SIMS) measurements conducted on a variety of starting source materials, see Table II, reveals the presence of acceptor impurities such as boron and aluminum in addition to a host of other metallics such as titanium and vanadium. However, CVD sources (as seen in Table II) have a significantly lower impurity concentrations and hence more suitable for SiC crystal growth. In addition to the impurity incorporation from the starting source material, the carbon components inside the growth chamber are a source of high contamination as shown in Table III. Fortunately, as is evident, stringent high temperature halogen cleaning procedures can reduce the impurity concentrations by over an order of magnitude.

Crystal Purity

High power switching devices, such as thyristors, operating at high temperatures require high purity SiC substrates, while, MESFET devices operating at microwave frequencies require a highly resistive (or semi-insulating) substrate. Sources with little or no boron, nitrogen, or aluminum is critical to consistently grow highly resistive substrates; however, Tables II and III clearly indicate the presence of these impurities in the source and in the carbon components used in the growth chamber. It is hypothesized that similar to 6H-SiC crystals, nitrogen is a shallow donor in 4H-SiC and hence compensates any residual acceptors that might be present in the crystal such as boron and aluminum. Consequently, using ultra-pure CVD sources of SiC with low background impurities in a high purity growth environment, crystals were successfully grown with greater than 5000 ohm-cm, p-type resistivity.

The ability to have a consistently pure starting material and a high purity growth environment has significant impact in the growth of low defect, semi-insulating 4H-SiC crystals. Using a baseline approach similar to that used in the development of 6H-SiC crystals, semi-insulating completely 4H-SiC crystals with room temperature resistivities in the 10^{15} ohm-cm range have been grown. To consistently grow semi-insulating 4H-SiC, vanadium is used to compensate for any additional acceptor level impurities such as boron present in the growth ambient either in the source or in the carbon components. Since vanadium, as an impurity, forms a deep level close to the mid bandgap of 4H-SiC similar to 6H-SiC, it compensates for any

Table III – High temperature purification of carbon parts significantly reduces impurity concentrations

	B	AL	V	Ti	Fe	Ni	Zn	S
Untreated Carbon 1	1.7	<2	6.1	20	80	19	35	160
Treated Carbon 1	0.3	<2	0.07	6	<2	8	5	120
Untreated Carbon 2	7.6	10	380	21	450	580	87	8900
Treated Carbon 1	3	<2	0.1	18	<2	10	9	11

additional acceptor impurities that might be present.

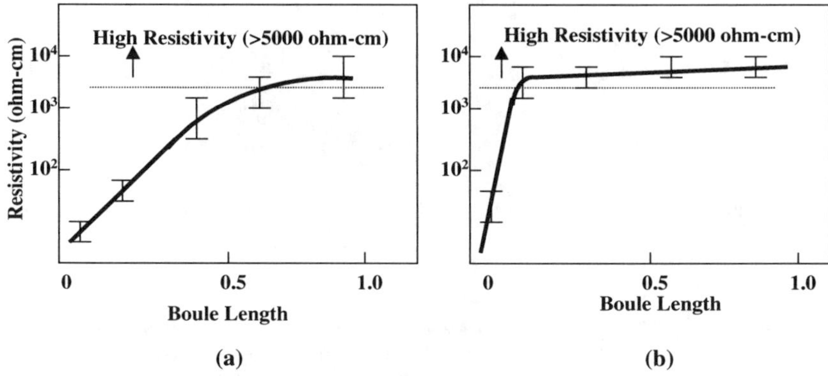

Figure 2-- Variation of resistivity along length of crystal boule in vanadium doped crystal; (a) non optimized growth, and (b) optimized growth in high purity environment

The variability in the resistivity of the grown crystals is attributed to the variability in the contamination levels of the various impurities, and has been one of the main aspects for poor yield of semi-insulating wafers. As shown in Figure 2, the yield of semi-insulating wafers increased from about 25%, as shown in Figure 2(a), to about 85%, as seen in Figure 2(b) due to improvements in source and growth environment purity, and due to the improved growth conditions.

MICROSTRUCTURAL DEFECTS

To understand the origin and mechanisms of micropipe formation various aspects of nucleation and growth have been carefully evaluated. Significant insights have been obtained in understanding the mechanisms of micropipe formation; recent results strongly suggest the dominant role of impurities, initial crystal nucleation, and non-stoichiometric growth conditions as the potential causes for micropipe formation. Result from the crystal nucleation experiments and effect of impurities on micropipe formation are presented.

Nucleation Effect

It is clear that initial nucleation conditions are critical in the growth of highly perfect crystals; as shown in Figure 3, a high quality seed is a necessary but not sufficient condition for the growth of low defect boules. A transmission optical micrograph of a longitudinal section of the boule clearly reveals the seed-crystal interface. Although the seed had a very low micropipe density (approx. 20 micropipes cm^{-2}), the ensuing crystal resulted in the formation of over 1000 micropipes cm^{-2}. It is hypothesized that perturbations at the growth interface, droplet formation, and/or presence of second phase inclusions can result in the formation of micropipes; dark second phase inclusions are clearly seen in the micrograph.

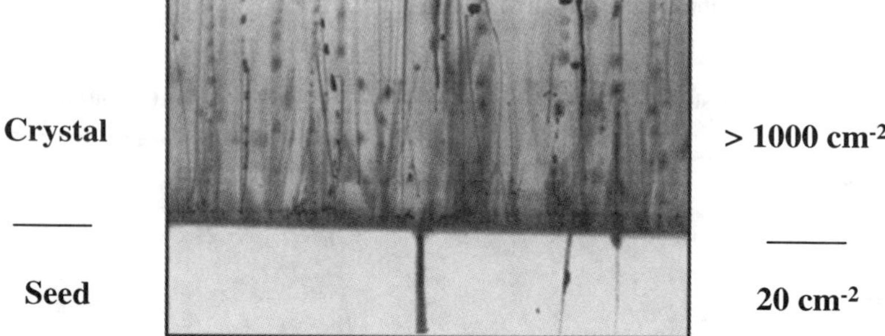

Figure 3—Presence of droplets and/or second phase inclusions can result in the formation of micropipes as clearly seen in the transmission optical micrograph of a longitudinal section near the seed-crystal interface

Impurity Effect

Impurities or second phase inclusions can also result in the nucleation of micropipes. For example, one factor limiting the yield of semi-insulating wafers is the presence of high micropipe densities. It is rationalized that the increased density of micropipes is a result of vanadium addition, which if not added in precise concentrations to compensate for the acceptor impurities can result in the formation of precipitates or second phase inclusions resulting in the nucleation of micropipes. Transmission optical micrograph (see Figure 4), verified by energy dispersive x-ray analysis, clearly reveals the presence of vanadium silicide and vanadium carbide precipitates formed in the crystals during growth, as shown in Figure 4(a). The second phase inclusions act as nucleation centers for the micropipes as evident in Figure 4(b).

To verify the hypothesis that droplet formation or presence of second phase particles during growth results in the formation of micropipes, controlled sites for nucleation of micropipes were created by patterning a series of impurity 'dots' (1000 angstroms thick) on one half of the seed face, see Figure 5. The dots varied in size from 2 to 50 μm in diameter, so as to

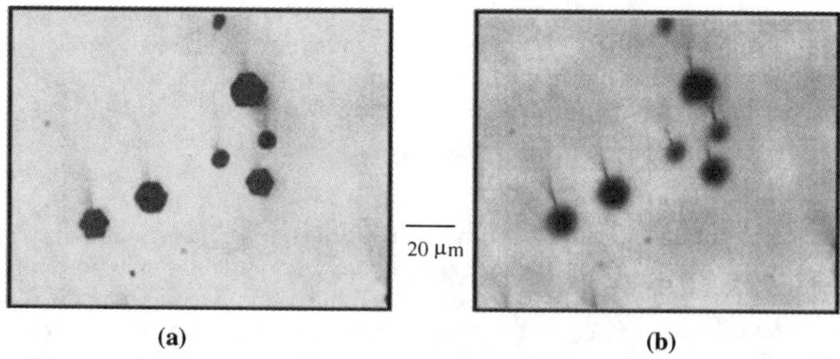

(a)　　　　　　(b)

Figure 4—Presence of vanadium silicide and vanadium carbide precipitates/inclusions can cause the nucleation of micropipes; (a) cluster of precipitates visible, and (b) nucleation of micropipes from the inclusions clearly visible

encompass a wide range of nucleation sites. The transmission optical micrograph of the ensuing crystal revealed the nucleation of a set of micropipes precisely at the location of the initial dots or impurity locations. Although these results are not complete, it clearly demonstrates a strong correlation between the presence of second phase inclusions on the growth surface and the nucleation of micropipes.

Considering various other models for micropipe formation mechanisms, and our own

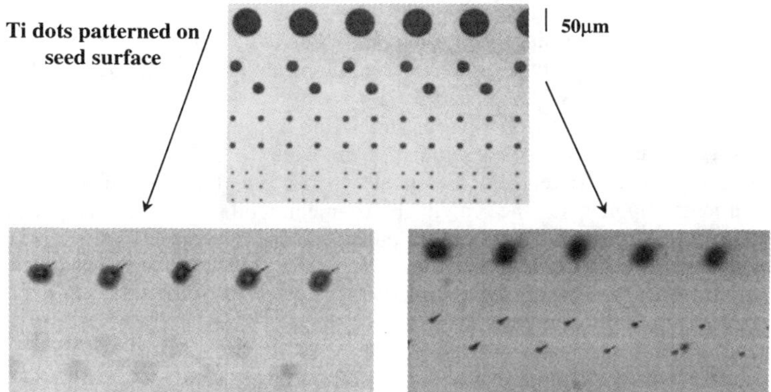

Figure 5—Micropipes emanating from each of the artificial nucleation site created on the seed surface

experimental results, we have steadily reduced the micropipe density in the 4H-SiC crystals.

Large diameter Crystals

The current focus of our work is to develop 2" diameter completely 4H polytype wafers as the baseline with less than 30 micropipes cm^{-2}, in addition to demonstrating three-inch

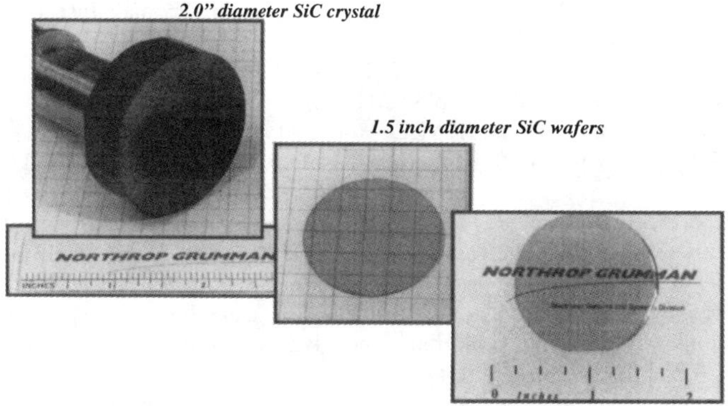

Figure 6—Highly transparent 1.5" diameter wafers and 2" boules grown under optimized growth conditions and in a high purity environment are representative of the current progress

diameter growth. Simultaneously, finite element analysis modeling is providing insights to minimize stresses in the crystals so as to reduce dislocation density and eliminate cracking. Highly transparent 1.5" diameter wafers and 2" diameter boules, grown in our laboratories, are clearly seen in Figure 6.

CONCLUSIONS

Large diameter (up to 50 mm diameter) 4H-SiC crystals grown by the PVT technique today exhibit average micropipe densities near 100 cm^{-2}, with the best being <2 cm^{-2}. Focus is on the development of highly perfect two- and three-inch diameter wafers. The results obtained thus far can be rationalized as follows:

1. Residual impurities play a critical role in improving the yield of high resistivity SiC crystals. Results suggest that sources of contamination are from the SiC source powder and from the growth environment. Therefore, it is critical to perform high temperature halogen treatment of the carbon components in the growth environment, while improving the purity of the SiC source powder. Further improvements in purity are still necessary.
2. Improvements in purity of the growth environment and in the growth process has resulted in yield improvements from 10-25% to 70-85% of high resistivity (> 5000 ohm-cm) 4H-SiC crystal and is attributed to the improvements in the source and growth environment purity.
3. Impurities also aid in the nucleation of micropipes. Our studies suggest that the presence of second phase inclusions, precipitates, or droplets during growth can result in the nucleation of micropipes. Addition of vanadium as a deep donor for semi-insulating growth can result in the formation of precipitates which can nucleate micropipes.

REFERENCES

1. Y. M. Tairov and V. F. Tsvetkov, Journal Cryst. Growth, **43**, 209 (1978).
2. G. Ziegler, P. Lanig, D. Theis, and C. Weyrich, IEEE Transactions on Electron Devices, ED-**30**, 277 (1983).
3. H. McD. Hobgood, D. L. Barrett, J. P. McHugh, R. C. Clarke, S. Sriram, A. A. Burk, J. Greggi, C. D. Brandt, R. H. Hopkins, and W. J. Choyke, Journal of Crystal Growth, **137**, 181 (1994).
4. V. Balakrishna. R. H. Hopkins, G. Augustine, G. T. Dunne, and R. N. Thomas, Inst. Phys. Ser. No **60**, DRIP-VII, Berlin, pp. 321-330 (1997)
5. J. R. Jenny, M. Skowronski, W. C. Mitchel, H. McD. Hobgood, R. C. Glass, G. Augustine, and R. H. Hopkins, Appl. Phys. Lett., **68** (14), p1963 (1996).
6. R. C Glass, , G. Augustine, V. Balakrishna, H. McD. Hobgood, R. H. Hopkins, J. Jenny, M. Skowronski, and W. J. Choyke, SiC and Related Materials, Inst. Of Phys. Ser. **142**, p 37-40 (1995)
7. S. Sriram, G. Augustine, A. A. Burk Jr., R. C. Glass, H. McD. Hobgood, P. A. Orphanos, L. B. Rowland, T. J. Smith, C. D. Brandt, M. C. Driver, and R. H. Hopkins, IEEE Electron Device Letters, **17**, 369 (1996).
8. R. C. Clarke, R. R. Siergiej, A. K. Agarwal, C. D. Brandt, A. A. Burk Jr., A. Morse, and P. A. Orphanos, Proceedings IEEE/Cornell Conference on Advanced Concepts in High Speed Semiconductor Devices and Circuits, **47** (1995).
9. J. W. Palmour, S. T. Allen, R. Singh, L. A. Lipkin, and D. G. Waltz, Technical Digest of the International Conference on Silicon Carbide and Related Materials, Kyoto Japan, 813 (1995).
10. V. F. Tsvetkov, D. N. Henshall, M. F, Brady, R. C. Glass, and C. H. Carter, Jr., Mat. Res. Soc. Sym. Proc. Vol. **512**, p89 (1998)

CHARACTERIZATION OF VANADIUM-DOPED 4H-SiC USING OPTICAL ADMITTANCE SPECTROSCOPY

S.R. Smith[a], A.O. Evwaraye[b], W.C. Mitchel, J.S. Solomon[c], and J. Goldstein, Materials Directorate, MLPO, Air Force Research Laboratory, Wright-Patterson Air Force Base, Ohio, 45422-7077. USA
[a]University of Dayton Research Institute, 300 College Park, Dayton, Ohio 45469-0178.
[B]University of Dayton, Physics Department, 300 College Park, Dayton, Ohio 45469-2314
[c]University of Dayton Research Institute, 300 College Park, Dayton, Ohio 45469-0167

ABSTRACT

Vanadium is an important dopant in SiC because it gives rise to donor levels near the middle of the bandgap which can be used to make the material semi-insulating, and semi-insulating material has many applications as a substrate material for high-power electronics. However, conventional means of characterizing electronic levels in the bandgap of the material require very high temperatures, in the neighborhood of 650-800 °C, in order to move the Fermi level to midgap and cause ionization of the V donors. The technique of Optical Admittance Spectroscopy permits the ionization of the midgap donors using light of the appropriate energy, and thus avoids the need for high temperatures. Using this technique we have examined several specimens of V-doped and high-resistivity 4H-SiC. We have identified levels previously associated with V, and new levels we attribute to Ti. Pinning of the Fermi level in some specimens was verified by high-temperature Hall effect measurements. SIMS measurements were used to determine impurity concentrations. IR absorption measurements were correlated with the Ti, V, and Cr concentrations determined by SIMS.

INTRODUCTION

High-power , high-temperature applications of semiconductors require a stable platform or substrate. This substrate should be able to conduct heat very well and conduct electricity very poorly. Semi-insulating 4H-SiC is such a material. SiC can be made semi-insulating by very close compensation of residual impurities, or by intentionally adding elements that give rise to electrical levels near the middle of the bandgap in quantities sufficient to pin the Fermi level there. One of the most popular elements for this purpose is vanadium. However, it is almost impossible to add just V because there are usually impurities present in the graphite parts of the growth chamber comprised of other transition metals (e.g. Ti, Cr, Zr). If their concentration is small enough, they are of little consequence, but if their concentration becomes large, it may be the impurities that affect the electrical characteristics of the semiconductor as much as the intentional dopant. For the most part these impurities have been assumed to be electrically insignificant, even though Ti has been determined to give rise to an isoelectronic center in SiC, as well as forming Ti-N pairs.[1,2]

253

It is difficult to determine the position in the bandgap of energy levels that are near midgap because of the extremely high temperatures that are required to ionize these elements in conventional thermal characterization experiments. What is needed is a technique that does not require high temperatures. One such technique is Optical Admittance Spectroscopy (OAS). This technique utilizes the response of a Schottky diode to the excess charge created when electrons are excited from either the valence band to an acceptor, or from a donor to the conduction band. Other transitions are possible depending on the charge state of the impurity, but the technique requires a free carrier to change the conductance of the Schottky diode. The technique has been described in detail elsewhere.[3,4]

In this paper, we describe the results of using OAS measurements to characterize SiC. The OAS spectra of several specimens of 4H-SiC were compared in the range from 200 nm (6.2 eV) to 1300 nm (0.95 eV). The specimens were also characterized by SIMS measurements to determine the concentration of transition metal impurities, principally Ti, V, and Cr. High-temperature Hall effect measurements were used to determine the position of the Fermi level. IR absorption data were correlated with the SIMS data to determine the effect of varying Ti, V, and Cr concentrations on the amplitudes of peaks at 9682 and 6380 nm which are normally attributed to vanadium.

An attempt was made to determine if there is a correlation between the magnitude of vanadium optical absorption lines and the concentration of vanadium. Optical absorption spectra were obtained using a Bomem DA-3 spectrometer fitted with an InSb detector. The specimen was cooled to 10 K in a Cryo Industries continuous flow cryostat. The temperature stability was better than ±0.03 K. It has been reported that the V absorption lines exhibit an increase in absorption following exposure to intense broad-band illumination from a quartz-halogen source;[5,6] therefore, a globar source was used to illuminate the specimens.

The V concentration, as well as the concentrations of Ti and Cr, were measured by SIMS. A total of 16 different specimens were examined. Some of the specimens exhibited very high resistivity and accumulated charge during the SIMS measurements, while others did not. The charging of the specimen in the course of the measurement probably affected the value of the concentration in a way for which we can not correct. 4H- and 6H specimens were used, and some mixed polytype specimens as well, were measured.

RESULTS and DISCUSSION

Table I lists the specimens in ascending order of V concentration as determined by SIMS. Also listed are the concentrations of Ti, Zr, and Cr. It is the purpose of this paper to show that Ti, at least, has an electrical transition that is detectable using OAS.

Table I. Specimens used in this study sorted by V concentration. Concentration was determined by SIMS.

Specimen	N (V)	N (Ti)	N (Zr)	N (Cr)
6147	9.00E+14	1.50E+16	ND	ND
5724	5.00E+15	1.50E+16	>1e17	>1e16
6129	2.30E+16	1.90E+16	3.90E+15	2.30E+17
5723	3.00E+16	6.10E+16	3.00E+16	>3.0e16
5633	6.00E+16	ND	ND	ND
5631	4.30E+17	6.00E+17	ND	ND
5713	4.30E+17	6.30E+17	ND	ND
5630	ND	7.00E+17	1.40E+17	3.80E+17
6130	1.10E+17	ND	ND	7.40E+17

ND = Not Detected

We determined where the Fermi level was pinned in the high-resistivity specimens by high-temperature Hall effect measurements. This was critical in order to determine the nature of the compensation. OAS can detect transitions from ionized acceptors in *n*-type material as well as from neutral donors. Furthermore, in *p*-type material transitions from the valence band to neutral acceptors, and ionized donors can be detected. Which center is giving rise to the transition can not be determined unless the energy coincides with thermal energy differences, or compensation rules out one or the other. Therefore it was important to know whether the specimen was *n*- or *p*-type at high temperature, though semi-insulating at room temperature.

Figure 1 is an optical admittance spectrum for specimen 5723 which, from Table I, it can be seen, had a V concentration of 3.0×10^{16} cm^{-3}, and a Ti concentration of at least 6×10^{16} cm^{-3}. Also present in non-negligible amounts were Zr and Cr. The unusual feature of this spectrum is the large triplet peak in the middle of the spectrum at 560 nm (2.21 eV). The peak at 800 nm (1.54 eV) is known from previous experiments to be attributable to the V donor. The abrupt edge at 850 nm corresponds to a photon energy of 1.45 eV

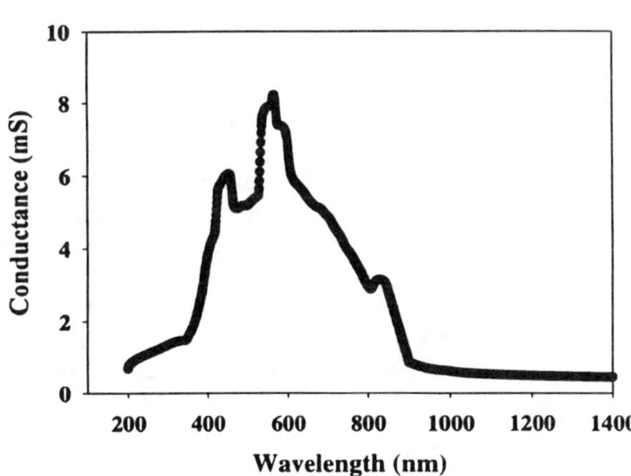

Figure 1. OAS spectrum of specimen 5723 containing 3×10^{16} cm^{-3} V, and containing 6×10^{16} cm^{-3} Ti.

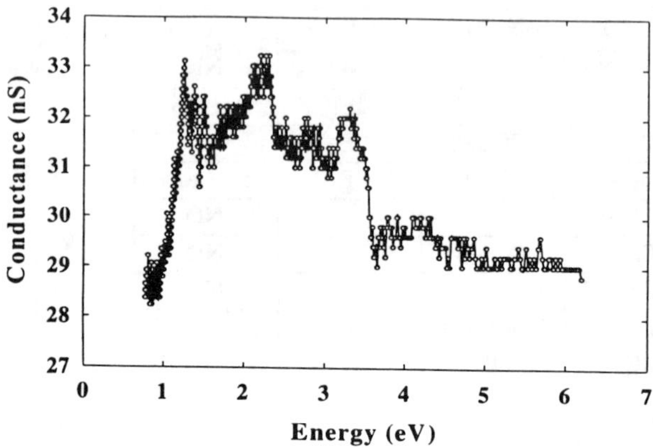

Figure 2. Optical admittance spectrum of 4H-SiC in which the Fermi level was shown to be pinned at 1.1 eV below the Conduction band.

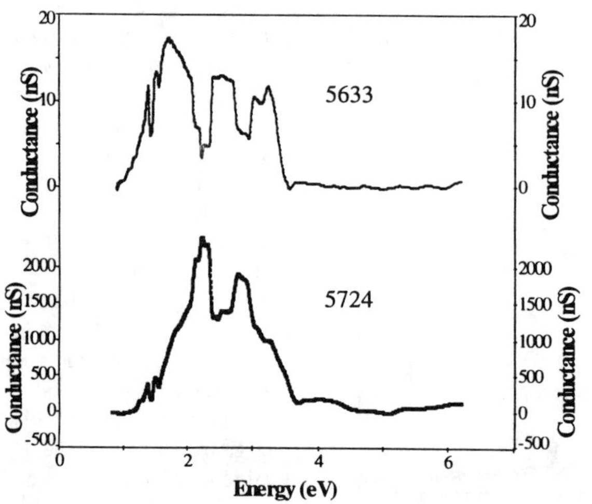

Figure 3. Comparison of OAS spectra of specimens 5633 and 5724. Both are 4H-SiC considered to be semi-insulating, but only 5724 had detectable Ti impurities.

which agrees with the high-temperature Hall effect determination of the Fermi level position. Another specimen (5713) indicated a Fermi level position of 1.1 eV; this was also verified by the abrupt edge in the OAS spectrum. This result is shown in figure 2.

We were the first to report the presence of a deep level at 1.1 eV below the conduction band in 1993.[refs. 3,4] At that time we suggested that this peak seen in the OAS spectrum was related to a complex defect involving vanadium or titanium. The defect had not been observed by any other techniques. The specimens cited above are the first incidence of the 1.1 eV defect being observed in SiC by Hall effect measurements. Very recent SIMS measurements have revealed the presence of oxygen in the specimens of SiC which had a Hall effect energy of 1.1 eV, however, the carrier concentration of the 1.1 eV level did not scale with the oxygen concentration, thus ruling out oxygen or O complexes as the source of this level. From Table I it is clear that we can rule out Cr and Zr as sources for the prominent peaks, as they were not detected in specimen 5713, but Ti was. The fact that the pinning of the Fermi level is seen in the 'edge' of the OAS response confirms the reality of the defect and its donor nature. However, both of these specimens these specimens contain V <u>and</u> Ti according to SIMS measurements making identification with either species difficult.

The OAS spectrum of specimen 5633 is shown in figure 3 compared to the spectrum of another SI 4H-SiC specimen (5724). The most striking difference is the absence of a multi-component peak at about 2.2 eV. Fitting the spectrum of specimen 5633 in the region of 2.2 eV results in a three-component peak having component energies of 2.09 eV, 2.20 eV, and 2.30 eV. Since we know from Hall effect data that the Fermi level was pinned at the midgap V donor, the defect from which these peaks arise must have electrons residing on the center to be excited to the conduction band by the incident light. Subtracting the excitation energy from the bandgap energy of 3.026 eV yields an energy difference of $E_V + 0.94$ eV, $E_V + 0.83$ eV, and $E_V + 0.73$ eV respectively. This means that the defect responsible for the triplet of peaks is either a very deep donor with component energies approximately two eV below the conduction band, or a compensated acceptor with component energies about 0.8 eV above the valence band. Further scrutiny of the spectra in figure 3 reveals that another peak centered near 2.9 eV is also inverted. Virtually all other features of these spectra are the same. As can be seen from Table I, the only difference between these specimens is the presence of Ti in one, and its absence in the other.

Dalibor, *et al.*, [ref. 1] have attributed deep acceptors with energies of EC - 0.117 eV and EC - 0.160 eV to Ti in 4H-SiC. They assigned these levels to the ionized Ti acceptor residing at the hexagonal and cubic sites, respectively. In the *n*-type specimens into which the transition metal impurities were implanted, Dalibor could not have detected acceptors close to the valence band via the DLTS measurements that they performed, nor would TAS measurements have revealed acceptors close to the valence band. We suggest that there may be Ti-related energy levels in the SiC bandgap near the valence band. Such a level has been seen by us in *p*-type SiC in Optical Admittance Spectroscopy experiments.[7] We reported a deep level at $E_V + 1.1$ eV in 1994. At that time, we could not assign this level to any known impurity. However, it appears that Ti, or a Ti-related defect may have been responsible for this line in the OAS spectra. Further work with SIMS and Hall effect will be used to examine those early specimens for Ti.

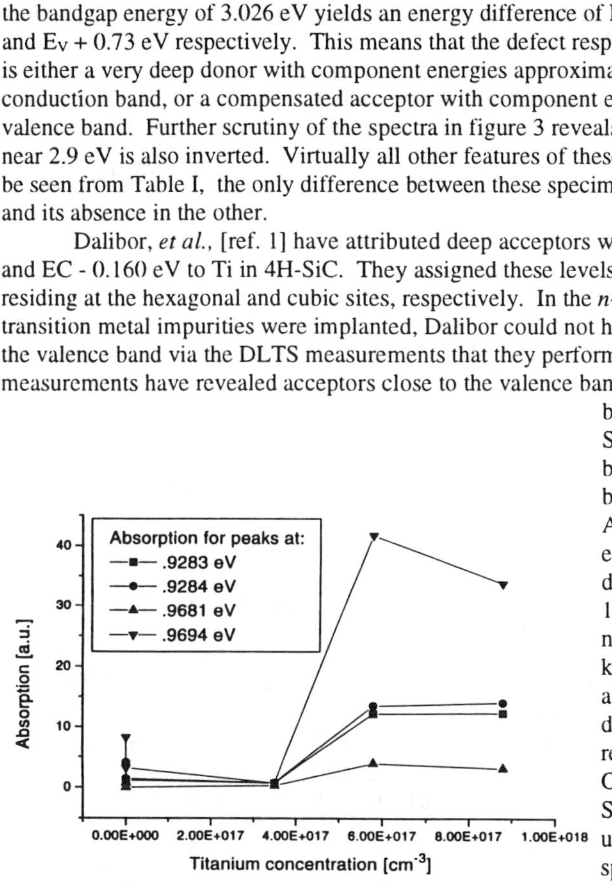

For the Cr ion, no correlation was seen with concentration for any of the absorption lines for any of the specimens. For the V ion, a clear trend was observed in the four specimens which had both sides polished and in which no charging was observed. The peaks examined were those at 0.9200 eV, 0.9196 eV, 0.9460 eV, 0.9474 eV, and 0.9481 eV. The trend is monotonically increasing to apparent saturation at the maximum V concentration. For

Figure 4. Correlation of Optical Absorption peak intensity with Ti concentration.

Ti, all 4H specimens exhibit a step-wise increase in absorption at a threshold concentration of 5×10^{17} cm^{-3}. The maximum Ti concentration available was 5×10^{18} cm^{-3}.

CONCLUSIONS

We have concluded that a peak in the OAS spectra of 4H-SiC, comprised of at least three components, is related to the presence of Ti in the SiC lattice that produces an electrically active site(s). Ti is present in most specimens that have been doped with transition-metal species in order to produce semi-insulating material, however, in at least one case, Ti was not added with the other impurities. In this specimen, the above mentioned peak was absent. The source of the metal impurities was not determined, however, the specimens were intentionally doped with V. Correlation of optical absorption peaks with V and Ti concentration was observed in all specimens. The correlation of the V concentration was monotonically increasing, whereas the correlation with the Ti concentration was step-wise with a threshold at 5×10^{17} cm^{-3}.

ACKNOWLEDGEMENTS

We would like to acknowledge the technical contributions of Mr. Gerry Landis, Mr. Robert Bertke, and Mr. Robert Leese for the preparation of specimens. One of us, SRS, was supported on Air Force Contract No. F615-95-C-5445.

REFERENCES

1. Thomas Dalibor, Gerhard Pensl, Nils Nordell, and Adolf Schöner, Phys. Rev. B **55**, p 13 618 (1997)
2. T. Dalibor, G. Pensl, H. Matsunami, T. Kimoto, W.J. Choyke, A. Schöner, and N. Nordell, Phys. Stat. Sol. (a) **162**, 199 (1997)
3. A.O. Evwaraye, S.R. Smith, and W.C. Mitchel, Mater. Res. Soc. Proc. **325**, 353, (1994)
4. A.O. Evwaraye, S.R. Smith, and W.C. Mitchel, J. Appl. Phys. **77**, 4477 (1995)
5. Jürgen Schneider and Karin Maier, Physica B **185**, 199 (1993)
6. J. Schneider, H.D. Müller, K. Maier, W. Wilkening, F. Fuchs, A. Dörnen, S. Leibenzeder, and R. Stein, Appl. Phys. Lett. **56**, 1184 (1990)
7. S.R. Smith, A.O. Evwaraye, and W.C. Mitchel, Phys. Stat. Sol. (a) **162**, 227 (1997)

ONLINE MONITORING OF PVT SiC BULK CRYSTAL GROWTH USING DIGITAL X-RAY IMAGING

P.J. WELLMANN, M. BICKERMANN, M. GRAU, D. HOFMANN, T.L. STRAUBINGER
AND A. WINNACKER
Materials Department VI, University of Erlangen, Martensstrasse 7, 91058 Erlangen,
GERMANY, Email: peter.wellmann@ww.uni-erlangen.de

ABSTRACT

An advanced method based on x-ray imaging is presented which allows us to visualize the
ongoing processes during physical vapor transport (PVT) growth of SiC. Using a high resolution
and high speed x-ray imaging detector based on image plates and digital recording we are able to
follow the SiC bulk single crystal growth as well as the evolution of the SiC powder source
inside the inductively heated graphite crucible on-line and quasi-continuously.

INTRODUCTION

SiC bulk crystals used to fabricate substrates for commercial device applications are presently
prepared by the physical vapor transport (PVT) method (so called modified Lely process) [1-4]
at high temperatures (T=2100°C ... 2400°C). This growth technique and the extreme thermal
conditions impose up to now intrinsic limitations to the visual control of SiC growth taking place
inside a graphite crucible. In order to improve the understanding of the PVT SiC growth process
and hence the crystal quality it would be highly advantageous to monitor the evolution of both,
the SiC crystal and the SiC source material during the whole process time. We have developed
an advanced method based on x-ray imaging which allowed us to visualize the ongoing
processes during PVT growth of SiC. Using a high resolution and high speed x-ray imaging
detector based on image plates and digital recording [5][6] we were able to follow the crystal
growth inside the inductively heated graphite crucible quasi-continuously. Analyzing the
digitized x-ray images of the graphite crucible taken during the growth we have (i) monitored the
evolution of the SiC crystal and (ii) investigated the degradation of the SiC source powder (i.e.
graphitization) with increasing process time. In this paper we will introduce our new on-line x-
ray imaging method of SiC growth and we will discuss its feasibility by showing several shots of
the high temperature SiC vapor growth process.

EXPERIMENT

6H and 4H SiC bulk single crystals have been prepared by the PVT technique in an
inductively heated graphite crucible at elevated temperatures of T = 2100°C ... 2300°C (figure
1). The temperature has been monitored at the seed end and at the source end of the crucible
using optical pyrometers. Argon has been used as carrier gas at a system pressure of p = 5mbar ...
40mbar. As a specialty of our setup, the induction coil can be moved electrically in vertical
direction giving rise to a wide range of geometrical configurations and the possibility of an in-
stationary growth process with increasing growth time. As seeds we have used (0001) oriented
30mm ... 40mm SiC crystals grown by PVT in our laboratory. The micropipe density varied
between 200cm^{-2} ... 500cm^{-2}. At defined times nitrogen was added for 10min to the argon carrier
gas in order to introduce doping striations revealing the shape of the growth interface [3]. The
source material (SiC powder) was synthesized from elemental Si and C. A typical growth run
can be divided in three parts: (i) Heating up is performed under an

Mat. Res. Soc. Symp. Proc. Vol. 572 © 1999 Materials Research Society

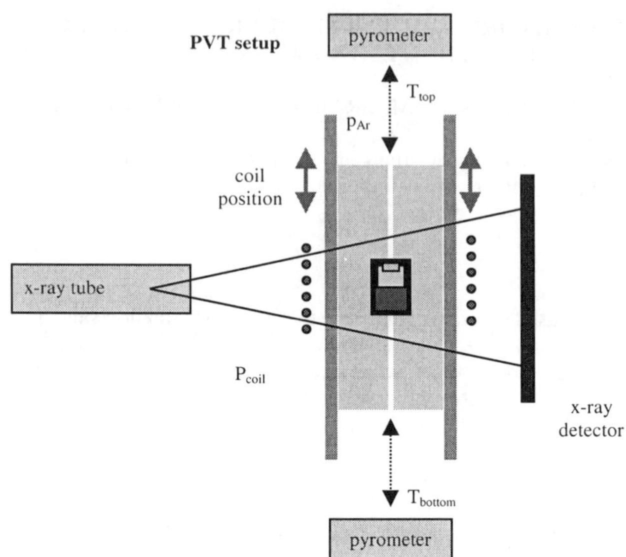

Figure 1. PVT setup for the growth of SiC bulk crystals. The inductively heated graphite crucible (loaded with a SiC seed crystal (top) and a SiC powder source (bottom)) is imbedded in a graphite like insulation inside a quartz ampoule. Using a x-ray tube and an image plate based detector and scanner, shots of the ongoing processes inside the graphite crucible were taken online during the growth process.

argon atmosphere (p = 800mbar). (ii) At the beginning of the growth (defined as 0h) the argon pressure is lowered to 5mbar ... 40 mbar giving rise to a physical vapor transport of various SiC species from the SiC source to the colder SiC seed. (iii) At the end of the growth run (typically after 72h) the inductive heating is switched off.

In order to visualize the ongoing processes inside the graphite crucible during growth we have applied a x-ray imaging system (x-ray tube, image plate detector and scanner) to the PVT setup (figure 1). During x-ray exposure by the x-ray tube an image of the graphite crucible interior (SiC seed/crystal and SiC source material) is transferred to the x-ray detector. While the x-ray source is a conventional x-ray tube for medical care (γ-energy = 60keV, I=7mA), a high speed, high resolution and high dynamic range x-ray detector based on image plates and digital recording was used.

An overview on the physics of image plate detectors and technical data of the image plate scanner FASTSCAN can be found elsewhere [5][6]. In short: x-rays produce radiation defect centers in x-ray storage phosphors (in our case a BaFBr:Eu based Fuji HR-V image plate) which give rise to fluorescence under light excitation of proper wavelength (in our case λ = 635nm). The intensity of the so called "photo-stimulated luminescence" is proportional to the x-ray dose. Thus, by scanning the image plate with a focused readout light beam one can extract the x-ray image by monitoring the photo-stimulated luminescence point by point. The image plate is recycled by exposure to intense light. Readout and recycling is done using the image plate scanner FASTSCAN [6]. FASTSCAN has a maximum scanning size of 30cm x 20cm at a resolution of up to 850dpi. The image readout time is 40sec for an image of 3500 x 5000 pixels. In practice the resolution is limited by the image plate and is about 100μm x 100μm.

There are several advantages of the digital x-ray detector over conventional film techniques: The fast digital image capture (t = 40sec) without time consuming developing stage allows an

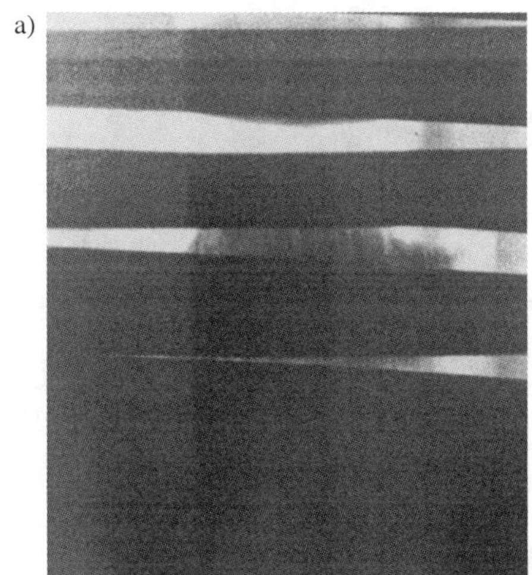 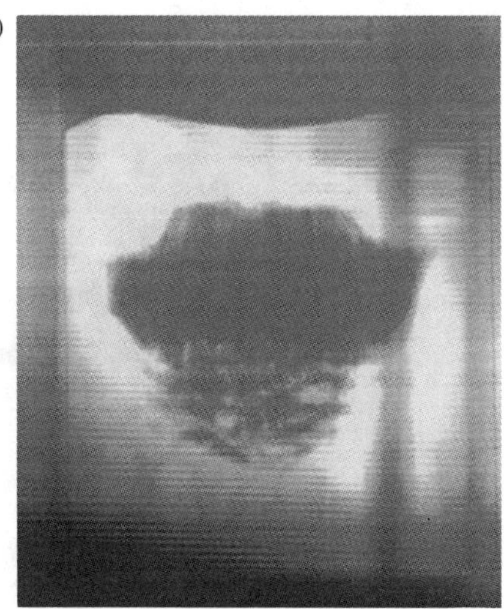

Figure 2. x-ray images of the hot graphite crucible (T=2270°C) (a) with a fixed induction coil during x-ray exposure and (b) with a moving induction coil during x-ray exposure.

online monitoring of the crystal growth process with the option of online process control. Due to the large dynamic range (10^6 for image plates versus $10^2 \ldots 10^3$ for conventional x-ray films) and high sensitivity the choice of a matching x-ray dose is uncritical. By computer processing of the digitized image a large contrast of the graphite crucible interior can be obtained.

X-ray and γ-ray techniques have been used in the past to study the solid-liquid interface during crystal growth from the melt (see for example [7-9]). However, the experimental challenge in the case of an inductively heated crucible as in our case is the large x-ray absorption coefficient of the induction coil: The absorption of the x-ray beam is about 10^4 times larger in the copper coil as compared to the graphite parts of the crucible, the SiC crystal and the SiC source. Dark stripes are introduced on the x-ray detector by the induction coil (figure 2a). Although digital image processing could partly uncover hidden parts of the digitized image of the crucible interior, we have developed a hardware solution to get rid of the unwanted perturbation by the copper coil. While moving the induction coil in vertical direction for at least 1/2 period (period is defined as the distance between two turns of the coil, in our case 2cm) we expose the x-ray detector continuously (typical time period = 3s). Hereby every part of the crucible interior will be captured by the image plate once, resulting in a stripe free image. Instead of a continuous exposure also a stepwise exposure (for example 16 steps) has been applied (figure 2b). After finishing the exposure the coil will be moved back to the starting position. The whole process takes about 10s to a few minutes. The temperature displays of the top and bottom pyrometers indicate that the thermal conditions inside the graphite crucible do not change during coil movement. Temperature changes of 1°C ... 2°C were observed after about 10 minutes if the coil was not moved back to the starting position after x-ray illumination.

RESULTS

Figure 3 shows a series of x-ray transmission shots of the hot graphite crucible (T=2270°C) at the beginning (0h, figure 3a), 8h (figure 3b) and 22h after the beginning (figure 3c) of the

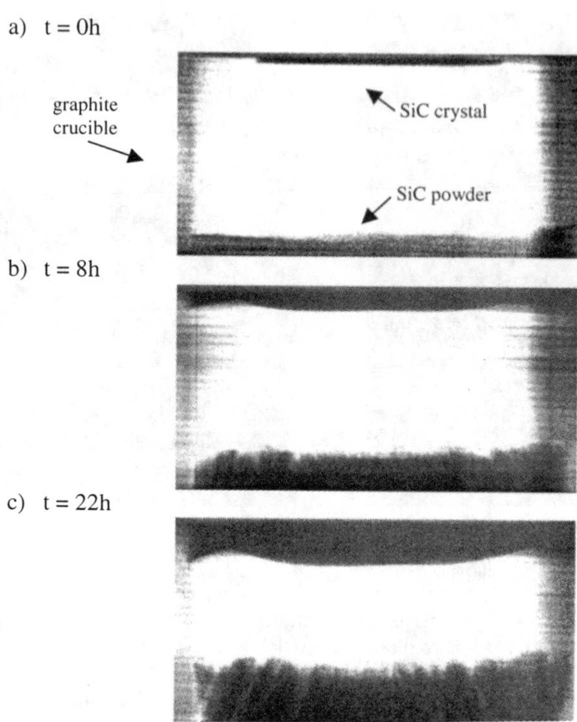

Figure 3. x-ray images of the hot graphite crucible (T=2270°C) (a) at the beginning of the growth process (0h), (b) 8h and (c) 22h after the beginning of the growth process.

growth. One can clearly distinguish the SiC crystal, the SiC source material and parts of the graphite crucible. The image resolution is about 100µm.

At the beginning of the growth (0h, figure 3a) the seed SiC crystal is visible as a dark stripe in the x-ray image. The flat surface of the seed and the thickness of 1mm is well reproduced indicating that the x-ray imaging system (x-ray source, detector and scanner) creates only minor image distortions. After 8h and 22h of growth the SiC crystal has reached a length of 4.6mm and 8.4mm (figure 3b and 3c), respectively. The shape of the growth interface is well resolved in the x-ray image. The growth interface has a convex shape in the central part (d<35mm); single crystal growth of high quality and low micropipe density has been observed. In the outer part (d>40mm) – at a diameter larger than the original seed crystal – a concave growth interface occurs. In this area polycrystalline growth and the formation of various polytypes have been observed. The SiC source material (figure 4) gives rise to a homogeneous contrast in the x-ray image at the beginning of the growth. However, after 8h (figure 4b) and 22h (figure 4c) of growth graphitization of the source material is observed in the vicinity of the hot graphite crucible. Due to its low density the residual graphite powder of the SiC source material is almost transparent for x-rays and gives rise to a light contrast in the x-ray image. The contrast of the core part has become darker indicating a compression of the SiC source powder due to sublimation in the hotter outer parts (close to the hot graphite crucible) and recrystallisation in the colder central part.

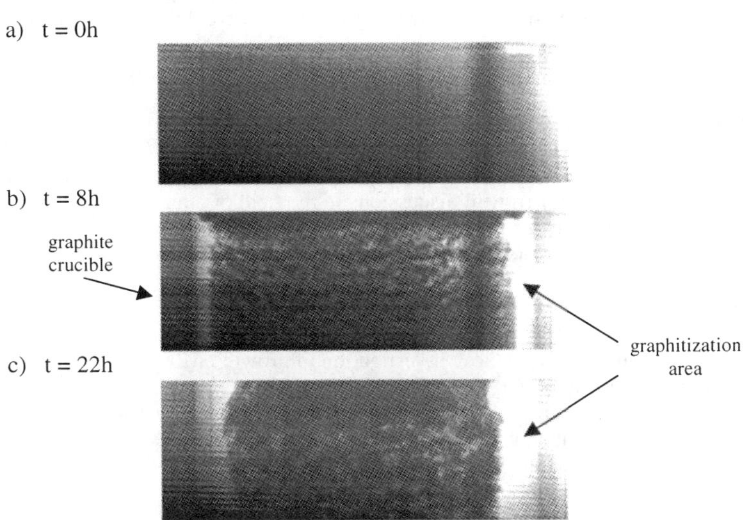

Figure 4. x-ray images of the SiC source material inside the hot graphite crucible (T=2270°C) (a) at the beginning of the growth process (0h), (b) 8h and (c) 22h after the beginning of the growth process.

CONCLUSIONS

Digital x-ray imaging turns out to be a useful tool for a routine control of SiC bulk crystal growth. The length of the SiC crystal and the status of the SiC source material can be monitored online. The clear distinction between available and consumed (graphitization area) SiC source material (figure 4) opens a way to optimize the yield (number of SiC wafers per crystal) for each growth run. For a continuous process control the growth rate, the growth interface as well as the consumption of the SiC source could be monitored online using automated digital image processing. Through a feedback loop the whole growth could be supervised by a computer controlling the inductive heating power (P_{coil}), the coil position and/or the argon pressure, respectively.

Digital x-ray imaging is as well a useful tool for understanding the physical vapor transport process of SiC from a research and development point of view. The growth rate and the shape of the growth interface can be studied without using demarcation doping striations by intermittently introducing nitrogen gas during the growth [3]. The shape of the SiC crystal growth interface can be determined with an accuracy better than 100μm (see figure 2 and 3). The study of the time evolution of the SiC source material is another crucial point for understanding the thermal conditions and underlying growth process. The graphitization which reflects the consumption of the source material the process of SiC powder compression due to sublimation and recrystallisation play an important role. Especially at the beginning of the growth (first 10 hours) the SiC powder changes its properties giving rise to a change of growth velocity and change of incorporated defects into the growing crystal. A detailed discussion of the time evolution of the SiC source and its impact on the growth process will be given in a forthcoming paper.

In summary we have for the first time presented a digital x-ray imaging technique which enable us to follow the SiC PVT process online during growth. We have introduced the underlying measurement procedure and shown various x-ray images of the growing SiC crystal and of the SiC source material inside the hot graphite crucible. Digital x-ray imaging is a powerful tool to study the processes going on during PVT of SiC. In addition digital x-ray imaging can be used for process control in the industrial growth of SiC.

ACKNOWLEDGMENTS

We would like to thank L. Kadinski and M. Selder (both Department of Fluid Dynamics, University of Erlangen) for fruitful discussions in the field of numerical modeling and heat transfer. This work has been supported by the Bayerische Forschungsstiftung (contract No. 176/96) and the Deutsche Forschungsgemeinschaft (contract No. Wi393/9).

REFERENCIES

1. Y.M. Tairov and V.F. Tsvetkov, Investigation of growth processes of ingots of silicon carbide single crystals, J.Cryst.Growth 43, 209, 1978.
2. G. Ziegler, P. Lanig, D. Theis and C. Weyerich, Single crystal growth of SiC substrate material for blue light emitting diodes, IEEE Trans. Electron. Devices 30, 277, 1983.
3. R. Eckstein, D. Hofmann, Y. Makarov, St.G. Müller, G. Pensl, E. Schmitt and A. Winnacker, Analysis of the sublimation growth process of silicon carbide bulk crystals, Mat. Res. Soc. Symp. Proc. 423, 215-220, 1996.
4. D. Hofmann, R. Eckstein, M. Kölbl, Y. Makarov, St.G. Müller, E. Schmitt, A. Winnacker, R. Rupp, R. Stein and J. Völkl, SiC-bulk growth by physical vapor transport and its global modeling, J.Cryst.Growth 174, 669-674, 1997.
5. A. Winnacker, x-ray imaging with photostimulable storage phosphors and future trends, Physica Medica IX 2-3, 95-101, 1993.
6. M. Thoms, H. Burzlaff, A. Kinne, J. Lange, H. von Seggern, R. Spengler and A. Winnacker, An improved x-ray image plate detector for diffractometry, Mater.Sci.Forum 107 (1), 228-231, 1995.
7. P.G. Barber, R.K. Crouch, A.L. Fripp, W.J. Debnam, R.F. Berry and R. Simchick, A procedure to visualize the melt-solid interface in Bridgeman grown germanium and lead tin telluride, J.Cryst.Growth 74, 228-230, 1986.
8. K. Kakimoto, M. Eguchi, H. Watanaba and T. Hibiya, In-situ observation of impurity diffusion boundary layer in silicon Czrochalski growth, J.Cryst.Growth 99, 665-669, 1990.
9. T.A. Campbell and J.N. Koster, Visualization of liquid-solid interface morphologies in gallium subject to natural convection, J.Cryst.Growth 140, 414-425, 1994.

POLYTYPE STABILITY AND DEFECT REDUCTION IN 4H-SiC CRYSTALS GROWN VIA SUBLIMATION TECHNIQUE

R.Yakimova[a,b], T.Iakimov[b], M.Syväjärvi[a], H.Jacobsson[a], P.Råback[c], A.Vehanen[d], E.Janzén[a]

[a]Dept of Physics and Measurement Technology, Linköping University, S-581 83 Linköping, Sweden
[b]Okmetic AB, Box 255, 17824 Ekerö, Sweden
[c]Center for Scientific Computing, P.O. Box 405, FIN-02101 Espoo, Finland
[d]Okmetic Ltd., PO Box 44, FIN-01301 Vantaa, Finland

ABSTRACT

Reproducible growth of 4H-SiC with good crystalline quality has been obtained in a temperature interval around 2350°C and on 4H-SiC C-face seeds. It has been observed that morphological instability may appear at the initial stage of growth, causing formation of defects. Experimental evidence has been found that supersaturation and surface kinetics are responsible for the polytype stability, while growth front and growth mode address defect reduction. An explanation of the findings has been suggested. It has been shown that starting the growth with a relatively low growth rate (≈ 100 μm/h) can be beneficial for the crystal quality.

INTRODUCTION

Silicon Carbide (SiC) is a material of expectation for high temperature power switching and high frequency power generation. While SiC may offer an exclusive combination of physical and electronic properties for many applications, the high temperature and chemical stability of this material, as well as the variety of stacking sequence along the c-direction in the close-packed structure of SiC, cause difficulties for growth of device quality crystals, especially of large single crystals.

Among different crystallographic modifications of SiC, 4H polytype is the most interesting for power device applications. However due to the low stacking fault energy it is difficult to restrict syntaxy (parasitic polytype formation) during bulk crystal growth and thus to grow a single polytype material. Another well known problem is the large number of structural defects such as micropipes, mosaicity and dislocations. Moreover, these problems are interrelated to a large extent; defects easily result in polytype disturbances [1], while polytype inclusions may lead to defect formation [2]. However, little is known about the kinetics and thermodynamics of polytype formation, growth stability, and also the mechanism that produces the periodic sequences. As discussed in Reference [3] 3C-SiC may be the initial polytype that forms at virtually all growth temperatures and thus acts as a necessary precursor for the phase transformation to other polytypes. Several growth parameters, such as the growth temperature [4,5], supersaturation [3,4], vapor phase stoichiometry and impurities [3,6] and polarity of seed surface [7] have been discussed to influence the polytype stability.

Seeded sublimation growth has been the most successful method to date for growth of large 4H-SiC boules that can be sliced into wafers. 4H wafers of 50 mm diameter are commercially available and 35 mm wafers with 7 micropipes (0.7 cm^{-2}) have been reported [8]. Generally, it is more difficult to grow 4H polytype in comparison with 6H-SiC, considering the size of crystals and the yield. On a large scale, the problem with micropipes, dislocation networks and stress in the crystals still remains [9]. Although significant progress has been made in the polytype control and defect reduction in SiC crystals, all mechanisms governing these processes have not been completely understood.

This paper describes a study of the influence of the early stage of 4H-SiC crystal formation on the polytype uniformity and defect occurrence in the subsequently grown crystals. An attempt has been made to reveal the role of growth characteristics such as seed surface polarity, supersaturation distribution and temperature for the morphological stability.

Mat. Res. Soc. Symp. Proc. Vol. 572 © 1999 Materials Research Society

EXPERIMENTAL

The growth experiments were performed with seeded sublimation technique in an Epigress SB50 sublimation system [10] with inductively heated graphite crucible and movable RF-coil. Rigid graphite insulation was used for thermal shielding and the reactor outer walls were air-cooled. The temperature was monitored both at the top and the bottom of the crucible with two-color pyrometers. The growth was performed in the temperature range of 2300-2450°C measured at the bottom of the crucible (T_b) and maintained constant during the growth run. The top temperature was used only as a reference when different runs were compared. Growth took place at a reduced Ar pressure ranging 5-30 mbar depending on the growth temperature. The SiC source powder was purified and sintered before growth to prevent contamination of the seed crystal at the initial stage of the growth process. As seeds we employed 4H polytype crystals produced via sublimation technique, (0001) well oriented or misoriented to the [11$\overline{2}$0] direction. Growth was performed on both Si- and C-terminated face. This variety of seeds was particularly necessary in order to investigate the seed influence on the polytype uniformity and structural quality. When the growth temperature was reached under nearly an atmospheric pressure the growth was initiated by applying a controlled pressure reduction. In order to vary the starting supersaturation we used two different pressure reduction schemes following an exponential decay of 200 sec and 15 min. The supersaturation was changed also by changing the temperature gradient when keeping constant T_b.

We examined two groups of samples. First, we studied as-grown surfaces after 6 hours of growth, referred to as early stage of growth. This allowed observation of the growth morphology after nucleation had been completed and the crystal habit had been founded. The second group of samples comprised wafers cut from boules and properly processed to permit structural and polytypic uniformity evaluation. The grown material was investigated using an optical microscope with Nomarski interference contrast and crossed polarizers, as well as by high resolution X-ray diffraction (HRXRD). Etching in molten KOH at 500°C for different times was also performed. The optical investigations are suitable to observe micropipes and domain boundaries on bare or KOH etched surfaces, while HRXRD was used to assess domain misorientation and strain in the crystal by recording ω rocking curves and $2\theta/\omega$ diffraction curves. X-ray diffraction measurements were utilized to determine the polytype. Computer simulation was used to better understand temperature and supersaturation distributions near the growth interface.

RESULTS

Growth results showing conditions for 4H crystal polytype occurrence and stability are summarized in Table I. Stable 4H growth with good structural quality is achieved within a narrow temperature range, 2350°C-2375°C and only on C-face. The pressure of the process gas is 5 mbar but we did not observe an effect of the pressure and pressure reduction rate on the polytype formation. Similarly the type of source material did not affect the polytype. The growth rate is about 100 μm/h in the beginning and it reaches 0.5 mm/h approximately after 1 mm of growth. Depending on crucible geometry the crystal shape is either cylindrical or slightly conical while the growth front is nearly flat or slightly convex. The largest diameter of the crystals is 38 mm. The thickness of the polycrystal ring around the monocrystal area does not exceed 5 mm.

Fig. 1. A typical pattern on as-grown surface of 4H SiC, C-face.

Table I. 4H polytype occurrence at different temperatures and 4H seed orientations.

Growth temperature [°C]	Surface orientation	Seed surface polarity	Grown crystal polytype	Remarks
2350-2375	on-axis {	Si-face C-face	6H, 15R inclusions 4H (100%)	} spiral growth competition
	off-axis {	C-face Si-face	4H (100%) 6H, 15R inclusions	} single spiral with preferred lateral growth in [11$\bar{2}$0]
<2350	low crystal quality			
>2375	polytype conversion 4H $\Rightarrow$ 6H			

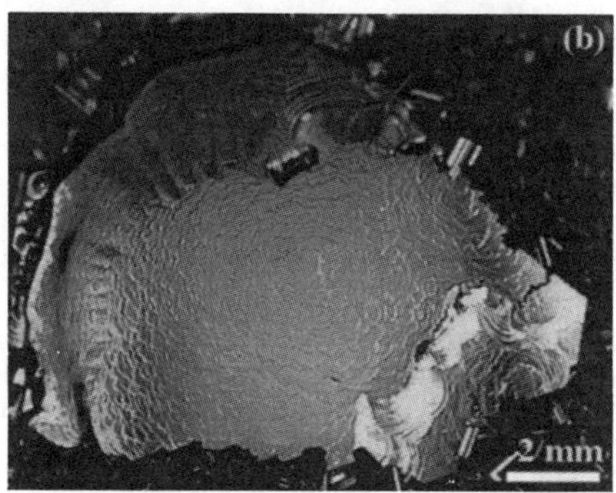

Fig. 2 (a) Non-uniform growth morphology manifesting growth instability; (b) stable growth morphology with one growth center.

Morphologies of as-grown surfaces on C-terminated faces of 4H SiC seeds at early stage of growth showed patterns (Fig. 1), which have been previously reported [11]. Typically, C-face exhibits a small pseudo-flat area (mesa) with six ridges emerging in the six equivalent directions <11$\bar{2}$0>. Often polygonized spiral can be distinguished on the mesa. Microsteps are observed in between the ridges, running along <1$\bar{1}$00> directions and thus the pattern reflects the six-fold crystal symmetry. Morphological stability at early stage of growth is an important characteristique, which was found to affect 4H polytype stability and defect formation. Fig. 2 displays uneven growth morphology (2a) and regular morphology (2b) in case of two different supersaturations over the growing surface when the other process parameters are the same. In the first case one can observe several misoriented growth centers with an enhanced growth at the edges. After several millimeters of growth the growth front acquired a concave shape, resulting in formation of domain boundaries, micropipes and dislocations. At some defective areas 15R inclusions were found. Fig. 3 depicts calculated temperature at the axis and inner crucible walls at different coil positions, provided T_b is fixed. From this and our temperature measurements it was estimated that temperature gradient can be tuned from 4-5°C/cm to 15-20 °C/cm. Large values of the temperature gradient resulted in morphological instabilities (Fig. 2a). To avoid this we used a low temperature gradient (4-5°C/cm) corresponding to a growth rate of approximately 100 μm/h. When growth conditions are properly selected there is only one growth promoting center (Fig. 2b) from which steps spread out over the whole growing surface in the course of the crystal growth. We examined the as-grown surface of the top wafer of a 4H SiC boule. An optical micrograph taken under crossed polarizers is shown in Fig. 4. There is strain-associated contrast located at the growth center. A wafer from the region just below the top surface was polished and subjected to structural investigations. Fig. 5 represents images taken in reflection light (5a) and crossed polarizers (5b) modes from 1 cm^2 area of the wafer after KOH etching. It was possible to trace the growth

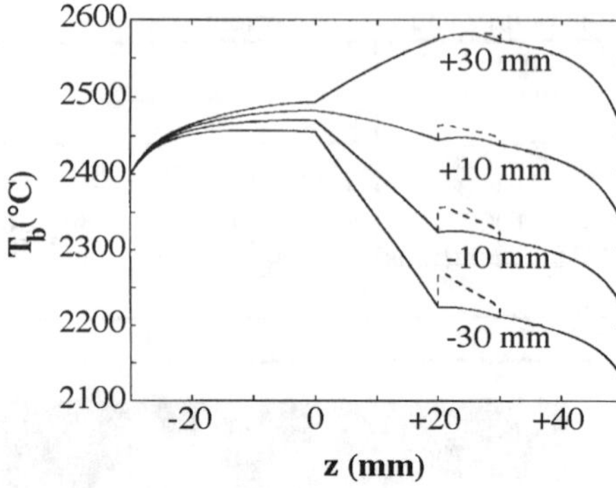

Fig. 3. Calculated temperature at the axis and inner crucible walls (dashed line) at four different RF-coil positions when the bottom temperature is constant.

Fig. 4. Optical micrograph under crossed polarizers of the as-grown surface of a 4H-boule.

promoting center and observe rows of dislocations along the six symmetrical ridges. Strains are still seen, however no micropipes are visible exactly in the spiral center. The two large hexagonal etch pits outside the spiral center do not appear to be micropipes. The sample was "mapped" by utilizing ω rocking curves. Fig. 6 gives two representative ω-scans from two areas of the sample, A and B, with a spot size of 2x14 mm. The peaks over a large part (A) of the sample (not illustrated) are sharp and symmetric with high intensity and FWHM value of 14". On the area with defects (B), corresponding to Fig. 5, the rocking curves show peak broadening, asymmetry and lower intensity. The 2θ/ω-scan, taken with a large spot of 10x14 mm, shows a sharp peak with FWHM of 17" and an intensity of 40 000 counts per second.

DISCUSSION

Polytype stability

From our results it follows that the occurrence and stability of 4H polytype depend on the temperature and the type of seed. Similar findings have been reported in Ref. [4,5] and [7], respectively. The temperature range of 4H stability is different in different studies. Commonly the temperatures are lower than for the growth of 6H-SiC. The upper limit in our experiments is 2375°C above which 4H polytype tends to convert to 6H, while below 2350°C the crystal quality is the limiting factor (Table I). The authors of Reference [12] suggest that SiC polytypes are kinetically determined metastable phases rather than true thermodynamic phases and therefore a well defined temperature stability range for each polytype can not be expected. Note that in our experiments the pressure is not effective in determining the polytype. As to the effect of the face polarity of the seed, one can speculate that the low surface free energy of the C-face facilitates the nucleation of 4H polytype which otherwise is more difficult to form due to the higher formation enthalpy than that of 6H-SiC. Consequently, other polytype inclusions can occur when growing on the Si-face. 4H polytype maintains stable growth in two modes depending on the seed orientation. On on-axis seeds the growth takes place via spiral competition ending with a dominant center in the middle of the crystal. The growth proceeds with layer-spiral mechanism. On off-axis seeds the growth center is at the seed edge and the lateral step growth is more pronounced in the tilt direction. Both situations allow growth of single polytype material.

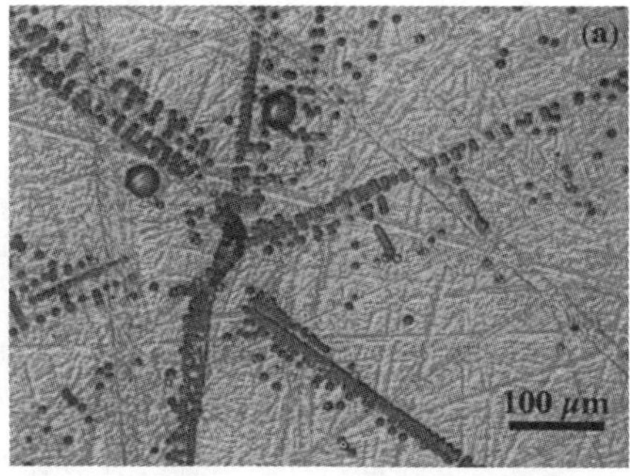
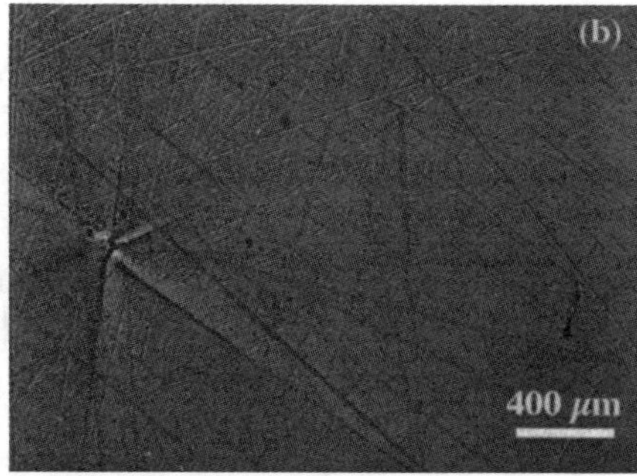

Fig. 5. Optical micrographs taken from an etched (0001) 4H sample with defects under (a) reflection light and (b) crossed polarizers; random lines are polishing scratches.

Morphological stability

Non-uniform growth illustrated in Fig. 2a can be discussed in terms of a morphological instability due to an enhanced nucleation at the edges of the seed. This is known to appear in case of polyhedral crystal growth when the supersaturation exceeds some critical value [13]. The solution of the diffusion equation yields a non-uniform supersaturation over the growing face, being largest at the corners and smallest at the centers of faces [14]. Therefore growth in diffusion limited sublimation growth is favored at the seed edges. Two consequences may be expected then: (i) many growth centers with different orientations, respectively domains forming grain boundaries with defects [2] and (ii) non-uniform growth rate leading to non-uniform material properties, e.g. inclusions. On the other hand, surface kinetics may smooth the growing surface by adjusting the density of growth steps. This can happen if the surface kinetics is fast enough and supersaturation is not too high. We have shown that by decreasing the supersaturation (low growth rate) and increasing the seed temperature, growth can start with a stable morphology (Fig. 2b). The described instabilities are more pronounced in 4H growth due to a lower surface diffusion coefficient and higher supersaturation over the growing surface in comparison with 6H growth. To confirm that growth favored at the edges of the seed is not due to an incorrect crucible design, we performed growth on two adjacent seeds. The results have clearly indicated that this is not the case. These experiments are described in Reference [14]. There could be several other solutions to the discussed problem. One is to use a large area seed. Our tests indicated that growth can proceed in an uniform way once the tendency of many center growth is suppressed. Similarly, morphologically stable growth can be achieved if diffusion mass transport is replaced by Stefan flow mass transfer. Finally, we simulated a particular temperature profile over the seed, which may lead to an even growth rate at each point of the seed [15].

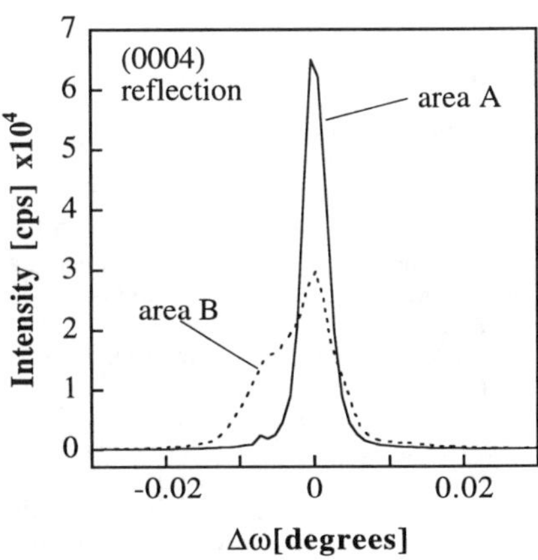

Fig. 6. Two representative ω-scans from area B shown in Fig. 5 and a good quality area A.

Defects

The probability of defect formation during growth increases with increasing the number of growth centers. This motivates the importance of starting boule growth with only one growth center. According to our results a large number of dislocations penetrating the growing surface may be formed in the vicinity of the growth promoting center. These are not dislocations associated with the spiral growth mechanism since we do not observe spirals emerging from them. They are found at the ridges spreading in six equivalent $<11\bar{2}0>$ directions, Fig. 5a. We suggest an explanation of the defect origin based on the anisotropy of the step distribution from the growth center e.g. the step velocity is different in the $<11\bar{2}0>$ and $<1\bar{1}00>$ directions. Furthermore, because of six-fold symmetry of (0001) SiC face the steps undergo transitions at the $<11\bar{2}0>$ directions to the adjacent facet. This may lead to strain accumulation, which later relaxes in dislocation formation. The shape of the rocking curves indicates that in the defective area there are domains that are slightly misoriented only but do not form a mosaic structure. In addition, by optical examination of the etched sample we did not observe imperfect domain boundaries. It is worth noting that the average micropipe density in this sample is 170 cm^{-2} which corresponds to the micropipe density in the seed crystal. It seems that our growth regimes do not provide conditions for creating new micropipes by opening hollow cores of giant screw dislocations. The average density of randomly distributed dislocations is about 15 000 cm^{-2}.

CONCLUSION

4H polytype crystals have been grown having a single polytype structure and micropipe density comparable with the seed. By controlling morphological stability, growth defects, such as micropipes and mosaicity, can be reduced. Concerning 4H polytype growth stability, surface kinetics plays an important role most probably due to the lower surface diffusion coefficient and lower equilibrium vapor pressure compared with 6H. We propose to start the growth with a relatively low growth rate compared with the following boule growth, which limits defect formation and ensures polytype uniformity.

ACKNOWLEDGMENT

The SSF SiCEP Program, NUTEK and Okmetic Ltd. are gratefully acknowledged for support.

REFERENCES

1. M. Syväjärvi, R. Yakimova, P-A. Glans, A. Henry, M.F. MacMillan, L.I. Johansson, and E. Janzén, J. Cryst. Growth **198-199** 1019-1025 (1999).
2. M. Tuominen, R. Yakimova, E. Prieur, A. Ellison, T. Tuomi, A. Vehanen and E. Janzén, Diamond and Related Materials **6** 1272-1275 (1997).
3. Yu.M. Tairov and V.F. Tsvetkov, Progr. Crystal Growth Characterization **4** 111-162 (1982).
4. M. Kanaya, J. Takahashi, Y. Fujiwara, and A. Moritani, Appl. Phys. Lett. **58** 56-58 (1991).
5. G. Augustine, H. McD. Hobgood, V. Balakrishna, G. Dunne, R.H. Hopkins, Phys. Stat. Sol. (b) **202** 137-148 (1997).
6. Yu.A. Vodakov, E.N. Mokhov, A.D. Roenkov, M.M Anikin, Sov. Tech. Phys. Lett. 5 147-148 (1979).
7. R.A. Stein and P. Lanig, J. Cryst. Growth **131** 71-74 (1993).
8. C.H. Carter, Jr., V.F. Tsvetkov, D. Henshall, O. Kordina, K. Irvine, R. Singh, S.T. Allen and J.W. Palmour, 2nd ECSCRM'98, Sept. 2-4, 1998, Montpellier, France, Abstracts pp.1-2 (1998).
9. P.G. Neudeck, W. Huang, and M. Dudley, Mat. Res. Soc. Symp. Proc. **483** 285-294 (1998)
10. Epigress product information bulletin, 98.11.04
11. A. Okamoto, N. Sugiyama, T. Tani and N. Kamiya, Mat. Sci. Forum **264-268** 21-24 (1998).
12. S. Limpijumnong and W.R.L. Lambrecht, Phys. Rev. B **57** 12017-12022 (1998).
13. T. Kuroda, T. Irisawa, A. Ookawa, J. Crystal Growth **42** 41-46 (1977).
14. R. Yakimova, M. Syväjärvi, M. Tuominen, T. Iakimov, P. Råback, A. Vehanen, and E. Janzén, Mater. Sci. Eng. B (1999) in press
15. P. Råback, Ph.D. Thesis, Helsinki University of Technology, Finland (1999).

Growth and Characterization of 2" 6H-Silicon Carbide

Erwin Schmitt, Robert Eckstein, Martin Kölbl, Arnd-Dietrich Weber

SiCrystal AG, Heinrich-Hertz-Platz 2, 92275 Eschenfelden, GERMANY
Email: e.schmitt@sicrystal.com

ABSTRACT

For the growth of 2" 6H-SiC a sublimation growth process was developed. By different means of characterization crystal quality was evaluated. Higher defect densities, mainly in the periphery of the crystals were found to be correlated to unfavourable process conditions. Improvement of thermal boundary conditions lead to a decreased defect density and better homogeneity over the wafer area.

INTRODUCTION

During the last years silicon carbide has attracted more and more attention in many fields of semiconductor research. A rising number of companies choose silicon carbide as a substrate for device processing. The production of the blue Light Emitting Diodes based on 6H-SiC has already started. Mass production is first of all a question of yield. Thus the prerequisites for silicon carbide are size and quality of the wafers. In the case of 6H silicon carbide the minimum diameter desired is 2". It is inevitable, that wafers have to fullfill the quality requirements not only in a selected, but on a large fraction of their area. Achieving homogeneous material properties is one of the major tasks at the present stage of SiC crystal growth.

CRYSTAL GROWTH AND CHARACTERIZATION

Growth of semiconductor grade silicon carbide by sublimation was first reported by Lely [1] in 1955. Another breakthrough in bulk growth was marked by Tairov and Tsvetkov [2], who used single crystalline material as seeds. During the following years the sublimation technique for silicon carbide has been further improved [3, 4]. Our growth experiments for 6H-SiC are based on the principles of the above mentioned technique. Evaluation of material properties was performed by different means of characterization, as described in the following: Optical microscopy of grown surfaces and wafers was carried out. The utilization of crossed polarizers enables stress birefringency images of the area under observation. Micropipe densities (MPD) were determined by focussing through the complete sample thickness. For a selected amount of samples also KOH etching (10 minutes at 600 °C) and scanning electron microscope investigations were conducted. No additional micropipes could be detected by these methods, but other defects like grain boundaries, cracks, dislocations and inclusions could be revealed. To determine the composition of inclusions, scanning auger spectroscopy was applied. From rocking curves (XRD-analysis) the full width at half maximum (FWHM) value of the (0006) reflexion was determined. The broadening of the FWHM-value by a superposition of indivual peaks could be interpreted as existence of slightly misoriented grains [5]. Measuring the sheet resistance (induction of eddy currents), which is a comfortable and non-distructive technique, allows the calculation of the specific electrical resitivity. A combination of resistivity and Hall-measurements gives the correlation between specific electrical resistance and net doping concentration N_D-N_A.

Mat. Res. Soc. Symp. Proc. Vol. 572 © 1999 Materials Research Society

RESULTS

The micropipe density is an important value to determine the material quality of the grown crystals. After upscaling of boule diameter we found out, that there was a strong variation of the MPD over the 2"-wafer. In all cases the MPD at the periphery was significantly higher than in the center of the wafer. Figure 1 illustrates, how average micropipe densities were determined, taking into account different diameters for the area under observation. Applying this method to our 2"-wafers we found, that for a "virtual" diameter of up to 1,625" the average MPD was below 200 cm^{-2} (see Figure 2, series I). For the whole wafer ("real" diameter of 2") the average MPD was up to 380 cm^{-2}. These growth experiments of series I also showed a poor crystalline quality according to the rocking curves of X-ray diffraction. Additionally stress birefringency images of wafers cut from these boules were made, to reveal the lateral variation of stress contrast. One typical image is illustrated in Fig. 3a. Areas with low stress contrast are restricted to a small central region of the wafer. Most of the wafer area has a high contrast in the stress birefringency. Both, micropipe-distribution and stress birefringency show, that the desired aim of homogeneity of crystal quality over the whole wafer was not achieved so far.

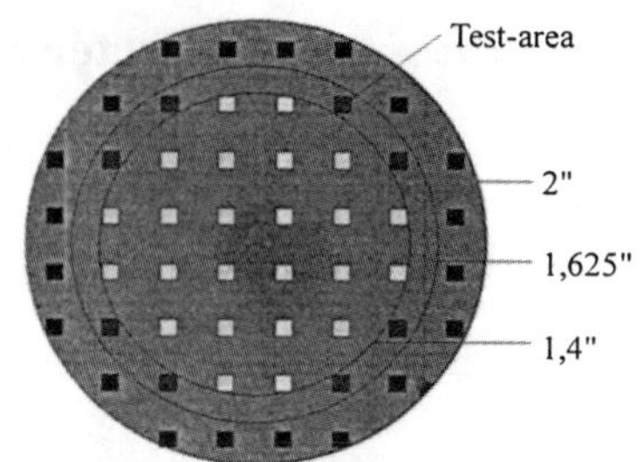

Figure 1: Determination of average micropipe densities (MPD)

■+■+□ = 2 " ;   ■+□ = 1,625 " ;   □ = 1,4 "

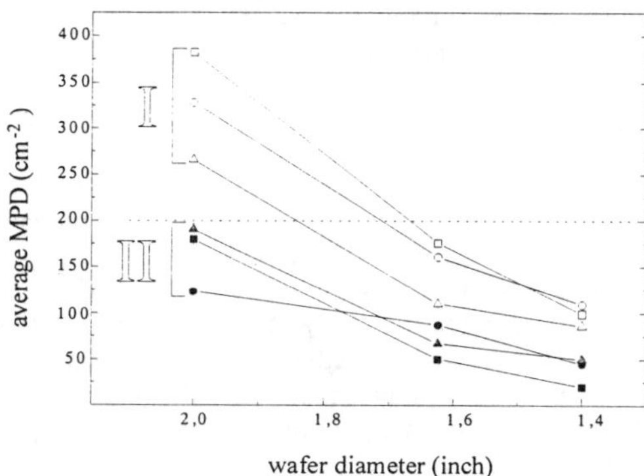

Figure 2: Dependence of average micropipe density (MPD) on wafer area under observation for 2" 6H-SiC wafers

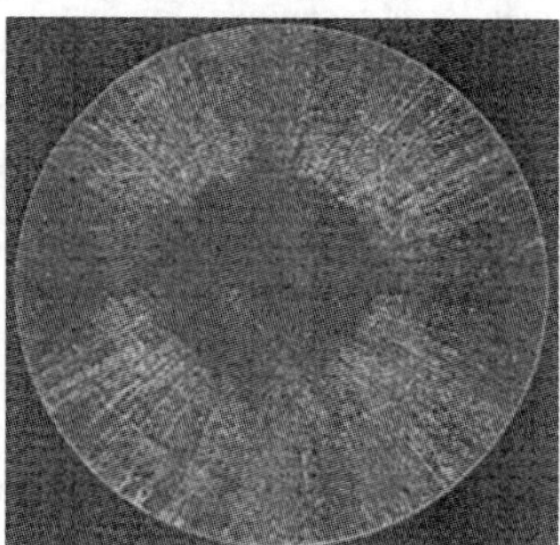

Figure 3a: Stress birefringency image of a 2" 6H-SiC wafer with a high stress contrast over a large area.

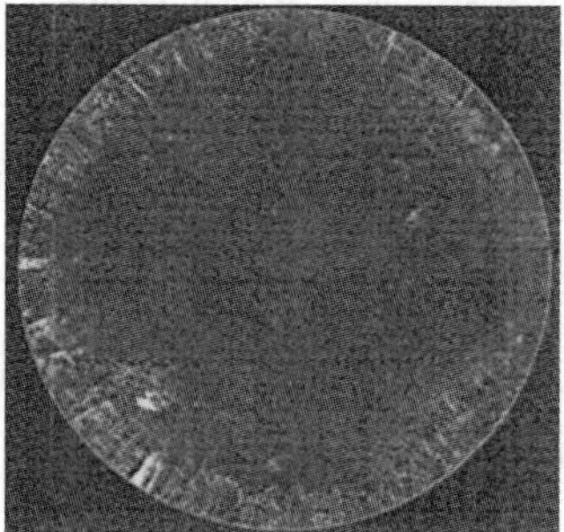

Figure 3b: Stress birefringency image of a 2" 6H-SiC wafer with a nearly contrast-free central region.

For the investigation of the influence of different process conditions on the crystal quality, two comparable wafers from the same boule were selected. Therefore the influence of seed quality can be neglected. These samples showed nearly identical micropipe densities and micropipe distribution, stress birefringency images and FWHM-values. From these two seeds two 6H-SiC crystals were grown under different thermal conditions. The results can be seen in Figure 4. High temperature gradients (indicated by ↑) lead to a poorer crystalline quality (A1), while more moderate thermal boundary conditions (indicated by ↓) resulted in smaller FWHM-values (B1). By using wafers from the boules A1 and B1 as seeds for the subsequent growth-runs, the two experimental series A and B were carried out. In all experiments the achieved crystalline quality could be correlated with the applied thermal conditions. Based on these results a significant improvement of wafer quality became possible. From Figure 2 (series II) one can see a reduction of average micropipe densities below the level of 200 cm^{-2}, over the whole wafer area. A comparision between Figures 3a and 3b shows an obvious stress reduction.

In addition inclusions in silicon carbide were examined with auger electron spectroscopy. These inclusions could generate micropipes and are so far ascribed to graphite [7, 8].

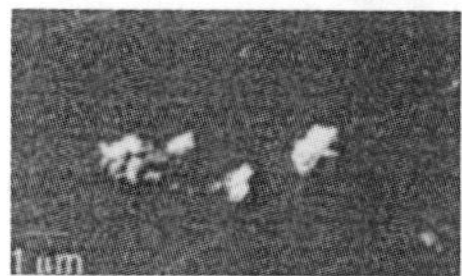

Figure 5a: SEM-image of an inclusion, exposed after successive sputtering.

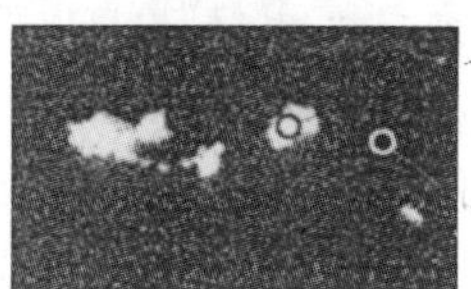
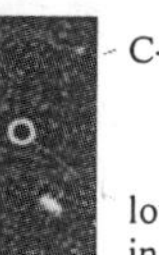

Figure 5b: AES-image for Carbon-Mapping

Figure 5c: AES-image for Silicon-Mapping

Fig. 5a shows inclusions, that were exposed on the surface of silicon carbide samples after successive sputtering with a scanning auger microscope. Silicon- and carbon-sensitive auger-mappings of this surface showed that these inclusions have a high carbon and a low silicon content, relative to the surrounding silicon carbide matrix (Fig 5b, 5c). Carbonization due to unfavourable process conditions could be the origin of these inclusions. The reduction of the inclusion content could contribute to a further lowering of MPD.

Besides crystal quality and its lateral distribution, the homogeneity of electrical properties of wafers is also of great importance. The results of a typical specific electrical resistivity mapping is shown in Fig. 6. The resistivity lies between 0.0382 Ωcm and 0.0496 Ωcm, with an average value of 0.0451 Ωcm and a standard deviation of 0.0028 Ωcm. The average value for the resitivity corresponds to a net doping concentration N_D-N_A of $3,3*10^{18}$ cm^{-3} .

SUMMARY AND OUTLOOK

We have shown that the enlargement of crystal diameter can lead to a decreased crystal quality, especially in the periphery of the crystals. By improving process conditions we succeeded in growing boules with significantly reduced defect densities and improved homogeneity. For further investigations, concerning the connection between process conditions and crystal quality, additional quality criteria like dislocation density, grain

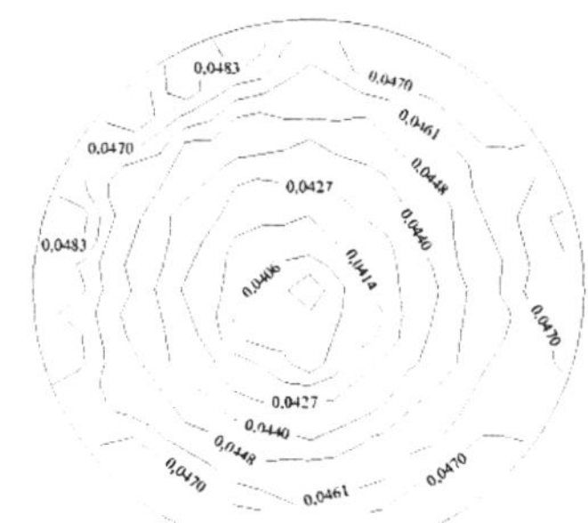

Figure 6: Mapping of specific electrical resistivity of a nitrogen-doped 2" 6H-SiC wafer.

boundaries, etc. have to be taken into account. The calculation of the temperature distribution inside the growth reactor by means of numerical modelling will also contribute to a better understanding of the interactions of the growth process.

ACKNOWLEDGEMENTS

This work is supported by the Bavarian Research Foundation under contract No. 176/96. The authors wish to thank the partners of the above mentioned cooperation project for their contributions. That are: Inst. of Materials Science 6, Dept. for Appl. Physics and Dept. of Fluid Mechanics (all University of Erlangen-Nürnberg). Special thanks to Inst. of Materials Science 4 (University of Erlangen-Nürnberg) for auger scanning microscopy.

REFERENCES

[1]   J. A. Lely, Ber. Deut. Keram. Ges., **32**, 229 (1955)

[2]   Yu.M. Tairov, V.F. Tsvetkov, J. Cryst. Growth **43**, 209 (1978)

[3]   G. Ziegler, P. Lanig, D. Theis, C. Weyrich, IEEE TRANSACTIONS ON ELECTRON DEVICES; VOL. **ED-30**, No. 4, 277 (1983)

[4]   D.L. Barret, J.P. McHugh, H.M. Hobgood, R.H. Hopkins, P.G. McMullin, R.C. Clarke, J. Cryst. Growth **128**, 358 (1993)

[5]   M. Tuominen, R. Yakimova, R.C. Glass, T. Tuomi, E. Janzen, J. Cryst. Growth **144**, 267 (1994)

[6]   S. Milita, R. Madar, J. Baruchel, A. Mazuelas, Materials Science Forum Vols. **264-268**, 29 (1998)

[7]   T. Tsvetkov, R. Glass, D. Henshall, D. Asbury, C.H. Carter, Jr., Materials Science Forum Vols. **264-268**, 3 (1998)

[8]   D. Hofmann, M. Bickermann, R. Eckstein, M. Kölbl, E. Schmitt, A. Weber, A. Winnacker, J. Cryst. Growth (1999), in press

EXPERIMENTAL AND THEORETICAL ANALYSIS OF THE HALL-MOBILITY IN N-TYPE BULK 6H- AND 4H-SIC

ST.G. MÜLLER, D. HOFMANN, A. WINNACKER
Materials Science Institute VI, University Erlangen-Nürnberg,
Martensstr. 7, D-91058 Erlangen Germany

ABSTRACT

The electrical properties of nitrogen doped n-type 6H- and 4H-SiC bulk crystals grown by the Lely- or modified Lely-method have been characterized by Hall-measurements. The doping densities were determined by a fit of the neutrality equation to the experimental data, accounting for in-equivalent lattice sites and the temperature dependence of the effective density-of- states-mass extracted from recent results of ab-initio-calculations of the 6H- and 4H-SiC bandstructure [1]. The theoretical analysis of the Hall-mobility is based on an extended form of the Rode-Nag iteration algorithm [2]. The calculation scheme considers all relevant elastic and inelastic scattering mechanisms, the anisotropy of the crystal modifications and the possible effect of spatial inhomogeneities in the distribution of donors, acceptors or in the related electron system. Within these concepts it is possible to achieve a quantitative agreement between theoretical and experimental mobility data in 4H- and 6H-SiC over the whole temperature range of band conduction. New values for the acoustic deformation potentials E_{ac} [$15.0\pm0.5\,eV$ (6H), $14.8\pm0.5\,eV$ (4H)] and the coupling constants for intervalley phonon scattering D_{int} [$2.3 \pm 0.1 \times 10^9\,eV/cm$ (6H), $2.6 \pm 0.1 \times 10^9\,eV/cm$ (4H)] are given.

INTRODUCTION

For the understanding of the limiting factors of the electron mobility in SiC a detailed analysis of the scattering mechanisms is essential. A principal problem for early scientific approaches to this subject [3, 4, 5] was the lack of critical material parameters of SiC at that time. Recent theoretical results of ab-initio-calculations of the 6H- and 4H-SiC bandstructure [1] and experimental data for the effective electron masses [6, 7, 8] provide the basis for a quantitative, theoretical mobility-analysis of these polytypes. Until now a detailed calculation of the electronic mobility of SiC without the relaxation time approximation was only performed for the cubic modification (3C-SiC) [9]. Similar to this work the mobility-analysis of n-type 6H- and 4H-SiC bulk crystals grown by the Lely- or modified Lely-method presented in this study is based an extended version of the Rode-Nag iteration method [2], but modified to account for the specific conductivity-anisotropy [10] of these hexagonal polytypes.

EXPERIMENT

The investigated nitrogen doped n-type 6H- and 4H-SiC samples (typical dimensions: $5 \times 5 \times 0.5\,mm^3$ with the orientation of the basal plane perpendicular to the cristallographic c-axis) were cut from crystals grown by the Lely- (L) or modified Lely-method (ML). Further details of the SiC sublimation growth at the Materials Science Institute VI (Erlangen, Germany) can be found in [11]. Ohmic contacts at the four corners of the samples were prepared by vaporization of titanium, followed by an annealing step at $900°C$ for 10 min in vacuum. For the measurement of the electrical conductivity σ and the Hall-constant R_H at an magnetic field B of $0.42\,T$ the *van der Pauw* [12] method was used. The electron density $n = r_H/(e|R_H|)$ and the Hall-mobility $\mu_H = \sigma(B = 0)|R_H|$ were calculated with the simplified assumption of the Hall-scattering factor $r_H = 1$. Based on the number of hexagonal and cubic lattice sites the donor concentration ratios $N_D(h) : N_D(k)$ for 4H- and 6H-SiC are 1 : 1 and 1 : 2, respectively. Hereby the energetic difference of the two cubic sites (k_1, k_2) in 6H-SiC is too small to be seperated by Hall-measurement. The donor concentration $N_D = N_D(h) + N_D(k)$, the compensating acceptor concentration N_{comp} and the ionization energies $(\Delta E(h), \Delta E(k))$ were determined by direct numerical fitting of the neutrality equation to the $n(T)$ data. In the neutrality equation the temperature dependence of the effective density of states mass [1] was considered and the valley-orbit splitting of the

Mat. Res. Soc. Symp. Proc. Vol. 572 © 1999 Materials Research Society

nitrogen ground-state [13, 14, 15] was taken into account by modifying the spin degeneracy factor accordingly [16]. The numerical fitting results for the samples of Fig.1 and Fig.2 are summarized in the following table:

polytype	*sample*	N_D $[cm^{-3}]$	N_{comp} $[cm^{-3}]$	$\Delta E(h)$ $[meV]$	$\Delta E(k)$ $[meV]$
6H	ML1	2.3×10^{17}	2.9×10^{16}	94	119
	L1	3.2×10^{18}	4.0×10^{17}	77	112
	L2	7.8×10^{18}	1.4×10^{18}	72	101
4H	ML4	1.5×10^{17}	6.7×10^{16}	51	109
	ML5	3.8×10^{18}	1.1×10^{18}	53	98

The dependency of the activation energies from the doping concentrations can be understood within the theory of the formation and broadening of a donor impurity band [17]. Due to the comparatively high donor binding energies in SiC the mechanism of *Hopping-conductivity* [17] becomes important at relatively high temperatures of 40-75K, depending on doping and compensation. Connected to this is the high temperature dependence of μ_H (Fig.1 and Fig.2) in this temperature regime [18], where the theoretical interpretation within the formalism of the *Boltzmann*-equation used in the following chapter is not valid.

From the Hall-analysis of a series of n-type bulk 4H- and 6H-SiC samples, typically an increase of N_{comp} was found with increasing N_D, despite the fact, that in the low temperature photoluminescence of the ML-samples no donor-acceptor pair spectrum was visible in the blue spectral region. This is to be expected for the presence of flat acceptors like *Al* [19]. The photoluminescence spectrum is dominated by a broad emmission at $1.8\,eV$ (6H) and $2.1\,eV$ (4H), which can be correlated to the presence of deep intrinsic acceptors [20] and the possible mechanism of self-compensation [21].

THEORY

The presented, theoretical analysis of the Hall-mobility is based on the linearized *Boltzmann*-equation. In the underlying coordinate system the z-axis is by definition parallel to the c-axis of the hexagonal unit cell of SiC, while the y-axis is parallel to one of primitive basis-vectors a_1 of the hexagon and the x-axis perpendicular to the corresponding face. For the Hall-geometry under investigation with the electrical current density $j \perp c$-axis and the magnetic field $B = (0, 0, B_z) \parallel c$-axis the electrical field is without loss of generality defined by $\mathcal{E} = (\mathcal{E}_x, 0, 0)$. The results of bandstructure calculations show, that due to the low symmetry (C_{2v}) of the k-vector in the $M - L$ direction of the *Brillouin-zone*, the tensor of the effective electron mass for 4H- und 6H-SiC has 3 independent components [1]. In the parabolic approximation for 6H- and 4H-SiC there are respectively 6 and 3 equivalent conduction band minima with ellipsoidal surfaces of constant energy described by the effective masses parallel to the principal axis (m_1 ($\parallel M - \Gamma$), m_2 ($\parallel M - K$), m_3 ($M - L$)) have to be considered. Using the *Herring-Vogt Transformation* [2] the ellipsoidal energy surface in k-space can be transformed to a spherical surface in the new w-space with the energy dispersion $E_w = \hbar^2 w^2 / 2m_\sigma$ (m_σ: arbitary normalization factor). The definition of effective field intensities finally leads to an equivalent description of the carrier transport by the *Boltzmann*-equation as in the case of an isotropic effective electron mass. After linearization the coupled equations for the calculation of the electronic relaxation functions ϕ_x and ϕ_y (not to be confused up with relaxation times) are given by [2]

$$L_C \phi_x - \omega_z \phi_y = 1 \tag{1}$$
$$\omega_z \phi_x + L_C \phi_y = 0 \tag{2}$$

with $\omega_z = eB_z/(m_1 m_2)^{1/2}$ and the collision operator

$$L_C \phi_i = \frac{V_C}{(2\pi)^3} \frac{(m_1 m_2 m_3)^{1/2}}{m_\sigma^{3/2}} \int dw' \frac{1 - f_0(E')}{1 - f_0(E)} T_{w,w'} \left[\phi_i(E) - \frac{w_i'}{w_i} \phi_i(E') \right]. \tag{3}$$

(f_0: Fermi-Dirac distribution, $T_{\boldsymbol{w},\boldsymbol{w}'}$: transition probability, $E \equiv E_{\boldsymbol{w}}$, $E' \equiv E_{\boldsymbol{w}'}$). After the back-transformation from $\boldsymbol{w}$-space to $\boldsymbol{k}$-space, taking into account the configuration of the equivalent conduction band minima, the conductivity tensor for 4H- and 6H-SiC for the *special* configuration $\boldsymbol{B} \parallel \boldsymbol{c}$-axis is:

$$\tilde{\sigma} = \begin{pmatrix} \sigma_{11} & \sigma_{12} & 0 \\ -\sigma_{12} & \sigma_{11} & 0 \\ 0 & 0 & \sigma_{33} \end{pmatrix} \tag{4}$$

with

$$\sigma_{11} = \frac{ne^2}{2} \left(\frac{\langle \phi_x \rangle}{m_1} + \frac{\langle \phi_x \rangle}{m_2} \right) , \ \sigma_{12} = \frac{ne^2}{(m_1 m_2)^{1/2}} \langle \phi_y \rangle , \ \sigma_{33} = \frac{ne^2}{m_3} \langle \phi_x(\boldsymbol{B} = 0) \rangle \tag{5}$$

and

$$\langle \phi_i \rangle \equiv -\frac{2}{3} \int_0^\infty dE \, \phi_i \, E^{\frac{3}{2}} \ / \int_0^\infty dE \, f_0 \, E^{\frac{1}{2}} \tag{6}$$

In this case the Hall-constant and Hall-mobility are connected to the components of the electrical conductivity tensor by:

$$R_H = \sigma_{12} \left[B_z \left(\sigma_{11}^2 + \sigma_{12}^2 \right) \right]^{-1} , \ \mu_H = \sigma_{11}(0) \, |R_H| \tag{7}$$

Within the existing accuracy of the mobility measurements and the lack of accurate data for electron-phonon coupling constants in 4H- and 6H-SiC the extensive calculations involved in an exact evaluation of (3) cannot be justified. Therefore in this work the transition probability $T_{\boldsymbol{w},\boldsymbol{w}'}$ is approximated as an isotropic process with the effective density of states mass $m_d \equiv (m_1 m_2 m_3)^{1/3}$, which is exact for *randomising scattering*. The anisotropy of the conductivity is then primarily connected to the superposition of the different contributions to the conductivity expressed in (5). All collision integrals in (3) for the elastic and inelastic scattering processes [2, 22] were calculated by numerical integration. No approximations (like e.g. the *Brooks-Herring* formula to describe impurity scattering) were used, as this turned out to be critical for the accuracy of the results. It has also to be stressed, that due to the relatively high LA- (6H: $76.5\,meV$, 4H: $76.6\,meV$ [23]) and LO- (6H: $119.98\,meV$, 4H: $119.73\,meV$ [7]) phonon energies involved in the inelastic electronic scattering processes in SiC (inter-valley scattering: LA, non-polar and polar optical phonon scattering: LO) theoretical results based on the relaxation time approximation [24] are strictly speaking not valid. Therefore in this work the calculations are based the *Rode-Nag* iteration method [2], which was extended to account for several inelastic processes at one time. It was shown by *Tsukioka* [9], that for the explanation of mobility data of 3C-SiC in addition to the scattering by isolated ionized impurities it was necessary to introduce the concept of dipole-scattering [22] to the theoretical mobility analysis, accounting for inhomogeneities in the impurity or related electron configuration. Also for a consistent mobility-analysis of n-type bulk 4H- and 6H-SiC it turned out to be necessary to split the concentration of ionized impurities into

$$N_{dipole} = x_{dipole} \cdot N_A^- \approx x_{dipole} \cdot N_A \tag{8}$$
$$N_{isolated}^{+/-} = N_D^+ + N_A^- - 2N_{dipole} \approx n + 2N_A(1 - x_{dipole}) \tag{9}$$

with the fraction x_{dipole} of the acceptors (density: N_A) forming dipoles of a length d_{dipole} with donors (density: N_D). Possible reasons for this may be attributed to the already mentioned self-compensation mechanisms or the preferencial formation of so called "1-complexes" in a impurity band [17]. The considered elastic scattering processes additionally included acoustical deformation potential (E_{ac}) scattering, while piezoelectric- and neutral impurity scattering were ineffective for the investigated samples in the temperature regime of band conduction. E_{ac}, x_{dipole}, d_{dipole} and D_{int} (coupling constant for intervalley-scattering) were used as fitting parameters for the calculations. The coupling constant D_{npo} of the negligible non-polar optical phonon scattering ($\hbar\omega_{LO} > \hbar\omega_{LA}$!) cannot be extracted by the fitting process. Thus $D_{npo} \equiv D_{int}$ was defined. Within these concepts, for the first time a consistent, quantitative agreement between theory and experiment (Fig.1 and Fig.2) could be achieved within the entire range of band conduction. An increase of x_{dipole} is found with increasing donor concentrations. The fitting results correspond well to the parameters found by *Tsukioka* [9] for 3C-SiC:

polytype	E_{ac} [eV]	D_{int} [eV/cm]	d_{dipole} [nm]	x_{dipole}
3C [9]	14	2.2×10^9	0.74	100% (postulate)
6H	15 ± 0.5	$2.3 \pm 0.1 \times 10^9$	0.7 ± 0.05	variable
4H	14.8 ± 0.5	$2.6 \pm 0.1 \times 10^9$	0.7 ± 0.05	variable

The temperature dependence of the Hall-factor $r_H(T)$ also resembles the principal dependence, recently measured for 4H-SiC samples [25]. By correcting from some examples the free electron concentration $n(T)$ with the theoretical results of $r_H(T) \neq 1$ the following values are found after a new fit of the neutrality equation:

polytype	*sample*	N_D [cm^{-3}]	N_{comp} [cm^{-3}]	$\Delta E(h)$ [meV]	$\Delta E(k)$ [meV]
6H	ML1	2.2×10^{17}	2.6×10^{16}	96	135
6H	L2	7.8×10^{18}	2.0×10^{18}	61	98
4H	ML4	1.7×10^{17}	6.3×10^{16}	53	116
4H	ML5	4.1×10^{18}	1.3×10^{18}	47	97

The deviations from the previous fitting results with $r_H = 1$ are within the limits given by the accuracy of the underlying experimental data and the parameters used for the fitting procedure. It is therefore important to note, that the given analysis justifies the use of $r_H = 1$ for the evaluation of SiC-Hall-data for practical purposes. Only by this fact it can be understood, that it was possible to achieve a quantitative aggreement between the described theory and experimental data at all. Otherwise a far more complex iteration procedure between the mobility calculation and correcting the experimental data with the theoretical $r_H(T)$ values would have been necessary.

CONCLUSIONS

The doping densities and ionization energies of n-type bulk 6H- and 4H-SiC samples were extracted from Hall-measurements considering recent results of ab-initio bandstructure calculations [1]. The theoretical analysis of the Hall-mobility is based on an extended version of the Rode-Nag iteration algorithm [2] taking into account all relevant elastic and inelastic scattering mechanisms. In order to achieve a consistent, quantitative agreement to experimental data the concept of dipole scattering has to be introduced into the analysis to account for possible spatial inhomogeneities in the impurity- or related electron-system. The given fitting parameters for the acoustic deformation potentials and the inter-valley coupling constants are close to recently published values extracted from high-field transport Monte Carlo simulations for 4H- and 6H-SiC [26]. The results also justify the use of $r_H = 1$ for practical purposes of Hall-data evaluation. By transfering the presented theoretical concept of dipole-scattering for the interpretation of Hall-data of SiC epi-layers one has to keep in mind, that the Hall measurements in this case may be strongly influenced by the conductive substrate and as the growth conditions for epi-layers are typically quite different from SiC bulk growth, this may particulary influence selfcompensation effects and the spatial distribution of donors and acceptors.

ACKNOWLEDGEMENTS

The author wants to thank A. Schöner (Industrial Microelectronics Center, Kista, Sweden) for providing Hall-data of low doped 4H-SiC samples. This work was financially supported by the Bundesministerium für Bildung und Wissenschaft (BMBF) (FKZ: 03 M 2746)

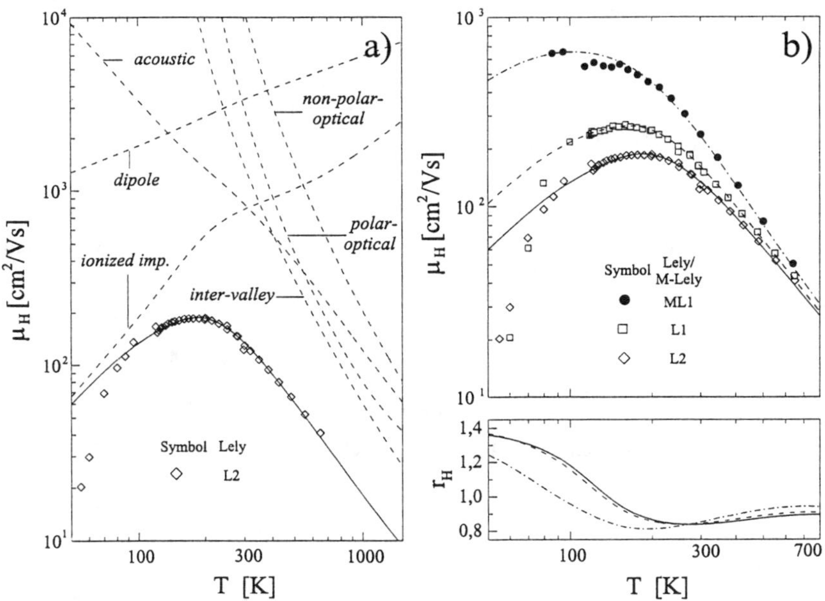

Figure 1. Theoretical and experimental Hall-mobility in 6H-SiC: a) scattering mechanisms, b) Hall-mobility μ_H and Hall-factor r_H for different samples [x_{dipole}: ML1(0%), L1(40%), L2(70%)]

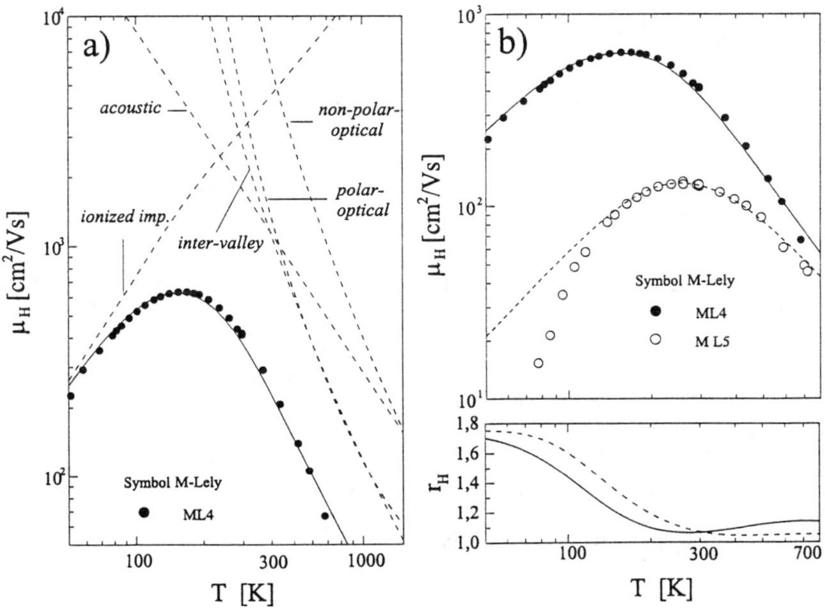

Figure 2. Theoretical and experimental Hall-mobility in 4H-SiC: a) scattering mechanisms, b) Hall-mobility μ_H and Hall-factor r_H for different samples [x_{dipole}: ML4(0%), ML5(15%)]

REFERENCES

[1] G. Wellenhofer, U. Rössler, phys. stat. sol. (b) **202**, 107 (1997)

[2] B.R. Nag,*Electron Transport in Compound Semiconductors*, vol. 11 of *Springer Series in Solid-State Sciences* (Springer-Verlag, Berlin, 1980)

[3] L. Pattrick, J. Appl. Phys. **38**, 50 (1967)

[4] J.J. Daal, Philips Res. Rept. Suppl.3 **70**, 1 (1965)

[5] B.W. Wessels, H.C. Gatos, J. Phys. Chem. Solids **38**, 345 (1977)

[6] N.T. Son, O. Kordina, A.O. Konstantinov, W.M. Chen, E. Sörman, B. Monemar, E. Janzén, Appl. Phys. Lett. **65**, 3209 (1994)

[7] H. Harmia, S. Nakashima, T. Uemura, J. Appl. Phys. **78**, 1996 (1995)

[8] D. Volm, B.K. Meyer, D.M. Hofmann, W.M. Chen, N.T. Son, C. Persson, U. Lindefelt, O. Kordina, E. Sörmann, A.O. Konstantinov, B. Monemar, Phys. Rev. B **53**, 15409 (1996)

[9] K. Tsukioka, Inst. Phys. Conf. Ser. **142**, 397 (1996)

[10] M. Schadt, G. Pensl, R.P. Devaty, W.J. Choyke, R. Stein, D. Stephani, Appl. Phys. Lett. **65**, 3120 (1994)

[11] St.G. Müller, R. Eckstein, W. Hartung, D. Hofmann, M. Kölbl, G. Pensl, E. Schmitt, E.J. Schmitt, A.-D. Weber, A. Winnacker, Mat. Sci. For. **264-268**, 33 (1998)

[12] L.J. van der Pauw, Philips Res. Rep. **13**, 1 (1958)

[13] W. Suttrop, G. Pensl, W.J. Choyke, R. Stein, S. Leibenzeder, J. Appl. Phys. **72**, 3708 (1992)

[14] W. Götz, A. Schöner, G. Pensl, W. Suttrop, W.J. Choyke, R. Stein, S. Leibenzeder **73**, 3332 (1993)

[15] P.A. Colwell, M.V. Klein, Phys. Rev. B **6**, 498 (1972)

[16] J.S. Blakemore, *semiconductor statistics*, (Pergamon Press, Oxford, 1962)

[17] B.I. Shklovskii, A.L. Efros, *Electronic Properties of Doped Semiconducters*, vol. 45 of *Springer Series in Solid-State Sciences* (Springer Verlag, Berlin, 1984)

[18] B. Molnar, J. Mat. Res. **7**, 2465 (1992)

[19] M. Ikeda, H. Matsunami, T. Tanaka, Phs. Rev. B **22**, 2842 (1980)

[20] A. Wysmolek, P. Mroziński, R. Dwiliński, S. Vlaskina, M. Kamińska, Acta Physica Polnica A **87**, 437 (1995)

[21] S.I. Vlaskina, Y.P. Lee, V.E. Rodionov, M. Kamińska, Mat. Sci. For. **264-268**, 577 (1998)

[22] R. Stratton, J. Phys. Chem. Solids **23**, 1011 (1962)

[23] W.J. Choyke, R.P. Devaty, L.L. Clemen, M.F. MacMillan, M. Yoganathan, G. Pensl, Inst. Phys. Conf. Ser. **142**, 257 (1996)

[24] T. Kinoshita, K.M. Itoh, J. Muto, M. Schadt, G. Pensl, K. Takeda, Mat. Sci. For. **264-268**, 295 (1998)

[25] G. Rutsch, R.P. Devaty, D.W. Langer, L.B. Rowland, W.J. Choycke, Mat. Sci. For. **264-268**, 517 (1998)

[26] R. Mickevičius, J.H. Zhao, Mat. Sci. For. **264-268**, 291 (1998)

MID INFRARED PHOTOCONDUCTIVITY SPECTRA OF DONOR IMPURITIES IN HEXAGONAL SILICON CARBIDE

R. J. LINVILLE, G. J. BROWN, W. C. MITCHEL, A. SAXLER AND R. PERRIN
Air Force Research Laboratory, Materials & Manufacturing Directorate (AFRL/MLPO)
3005 P ST, Wright-Patterson AFB, OH 4533-7707

ABSTRACT

Mid-infrared photoconductivity (PC) is a useful technique for identifying and investigating donor and acceptor centers in many semiconductors. This is especially true when the PC results are combined with other measurements such as Hall Effect and DLTS. We report on the first Fourier Transform Infrared (FTIR) photoconductivity spectra for n-type 6H and 4H-SiC. The samples studied had temperature dependent Hall activation energies around 45 meV and 85 meV in the 4H samples, and a single activation energy of 106 meV in the 6H. For the 4H samples, the PC spectra showed an increase in photoresponse between 40 and 47 meV, with another sharp increase at 120 meV. In the 6H-SiC, the photoresponse also had a rapid increase at 120 meV, and at 77 meV in one sample. The photoresponse spectra of the n-type 4H and 6H-SiC samples were distinctly different in the mid-infrared.

INTRODUCTION

SiC has, in comparison with Si, superior properties regarding high-power, high-frequency and high-temperature electronics. The material has high thermal conductivity, can withstand high electric fields before breakdown and also high current densities. The wide bandgap results in a low leakage current even at high temperatures. The bandgap of SiC depends on the polytype and ranges from 2.4eV (3C-SiC, T=4.2K) to 3.3eV (2H-SiC, T=4.2K).[1] Nitrogen is the main donor impurity in all the polytypes of SiC[2] and dominates the electrical properties of n-type SiC since it has shallow levels in the upper half of the forbidden gap. Nitrogen substitutes for carbon in the SiC lattice.[2] The activation energy of the nitrogen levels depends on the polytype and the substitutional lattice site involved. In recent years, there have been several studies on identifying the nitrogen levels in 6H and 4H silicon carbide.

One technique that has not been used previously to study these energy levels is mid-infrared photoconductivity. The only reported photoconductivity spectra for SiC are for deep levels observed at energies higher than 0.5 eV. Mid-infrared photoconductivity (0.025 to 0.5 eV) can be a sensitive technique for identifying both defect levels and impurities in semiconductors.[3,4] Typically the optical activation energies identified in photoconductivity spectra agree well with the thermal activation energies measured by temperature dependent Hall effect for samples from the same wafer. In this study we compare the results from Hall effect and Fourier transform infrared spectral photoconductivity for several samples of 6H and 4H n-type SiC.

EXPERIMENTAL

The 4H-SiC and 6H-SiC samples have been grown by physical vapor transport. All of the samples were found to be of the n-type conductivity by Hall effect measurements. The

Mat. Res. Soc. Symp. Proc. Vol. 572 © 1999 Materials Research Society

investigated samples were cut from (0001) wafers of SiC crystals which were either vicinal or on-axis. For the Hall effect measurements we used samples ~1 cm^2 dimensions. The ohmic contacts were fabricated in the four corners of the samples (Van der Pauw geometry) by evaporating Ni/Au contacts (n-type samples) and by subsequent annealing in forming gas at 925° C for 5 min. The resistivity and Hall constant at a magnetic field of 5-8 Kilogauss were measured in the temperature range of 10-1000 K.

The photoresponse measurements were made using a Bio-Rad FTS-6000 Fourier Transform spectrometer with a wavelength range from 2 to 50 microns. The samples was mounted on the coldfinger of a closed cycle refrigerator allowing the sample temperature to be varied from 10 K to 290 K. The sample was mounted in series with a load resistor and voltage biased in a range from 10 to 1000 volts, depending upon signal-to-noise requirements. The spectral resolution was varied between 4 cm^{-1} to 1 cm^{-1} with the higher resolution reserved for samples with initially high signal-to-noise ratios. For the photoresponse studies, the samples were 8mm x 6mm x 0.5mm dimensions with two Ni/Au strip contacts covering all but a 1 mm strip down the center of the sample's top surface. Gold wires were attached to the contacts with indium solder.

6H-SiC RESULTS

Figure 1 shows the measured free-electron carrier concentration (n) versus inverse temperature for a n-type 6H-SiC sample in the concentration range from 10^{10} to 10^{17} cm^{-3} as a function of the reciprocal temperature. The free-electron concentration was determined from the experimental Hall constant R$_h$ and given by:

$$n = 1/eR_H. \tag{1}$$

The carrier concentration versus inverse temperature data are fit by the least squares method to the multi-donor charge balance equation:

$$n + N_a = \sum_i \frac{N_{d_i}}{1 + \dfrac{n g_i}{N_C} e^{E_{a_i}/kT}} \tag{2}$$

Here n is the carrier concentration, N$_{di}$ is the concentration of the ith donor level with activation energy E$_{ai}$ and degeneracy factor g$_i$, N$_a$ is the concentration of compensating acceptors, and the density of states in the conduction band is:

$$N_C = 2M_C(2\pi m^* kT/h^2)^{3/2} \tag{3}$$

where M$_C$ is the number of conduction band minima and m* is the electron effective mass. For 6H-SiC M$_C$ = 6 and m* = 0.588. For 4H-SiC, M$_C$ = 3 and m* = 0.390.

The fitting results for both 6H and 4H-SiC samples are given in Table I. This table shows donor concentration, Hall activation energies, as well as the standard accepted energy levels, based on IR absorption data, for the nitrogen donor in both SiC polytypes. The fits for the sample in Fig. 1 indicated only one donor level with an activation energy of 106.5 meV. The other 6H sample was highly resistive and the Hall data indicated a poorly defined deep level.

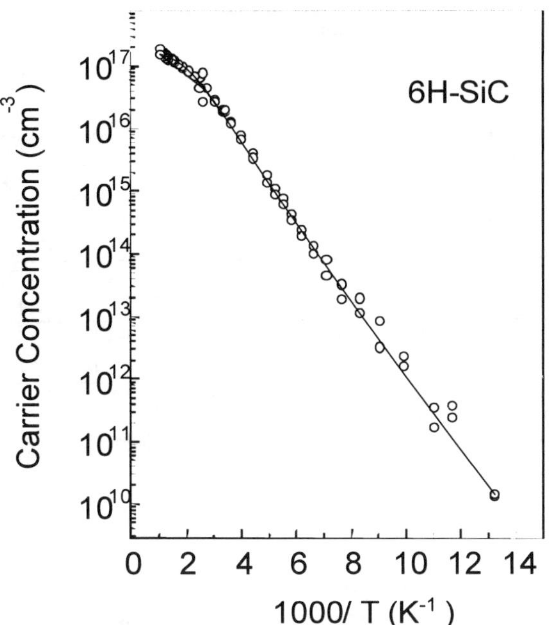

FIG. 1. Measured electron carrier concentration versus reciprocal temperature as obtained form Hall Effect measurement of n-type 6H-SiC.

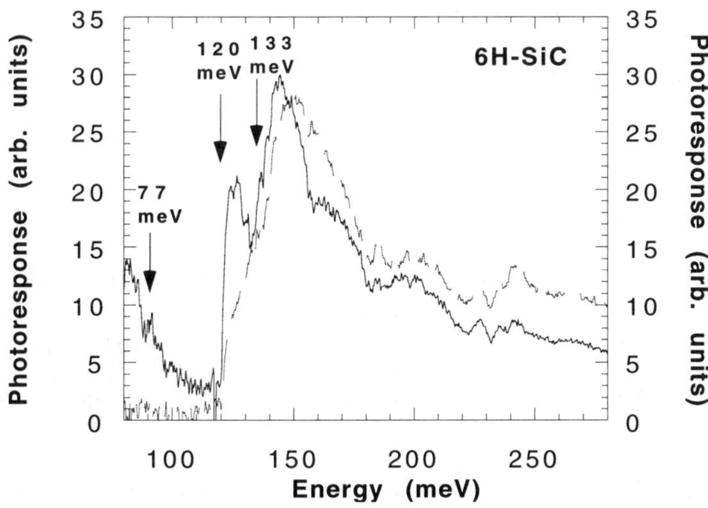

FIG. 2. Fourier Transform Infrared photoresponse of n-type 6H-SiC samples.   Sample 4 (dashed line) is the same as shown in the Hall data. Sample 3 (solid Line) shows response from the shallower nitrogen level of 6H-SiC

TABLE I. SiC Donor Activation Energies

Sample	N_d (atoms/cm^3)	E_{act} (IR Abs.) (meV)	E_{act} (Hall) (meV)	E_{act} (PC) (meV)
1 (4H)	1.2×10^{17}	52[a]	46.8	47.1
	7.7×10^{16}	92[a]	87.0	120
2 (4H)	1.9×10^{17}	52[a]	44.6	40.9
	1.8×10^{17}	92[a]	83.8	120
3 (6H)	--	81[b]	--	77
		138,142[b]		120
4 (6H)	8.4×10^{17}	81[b]		--
		138,142[b]	106.5	120

a) W. Gotz, A. Schoner, G. Pensl, W. Suttrop, W. J. Choyke, R. Stein and S. Leibenzeder, J. Appl. Phys. **73**, 3332 (1993).
b) W. Suttrop, G. Pensl, W. J. Choyke, R. Stein and S. Leibenzeder, J. Appl. Phys. **72**, 3708 (1992).

Figure 2 shows two FTIR photoresponse spectra of n-type 6H-SiC material. The spectrum for sample 4, in the wavelength range 50-300 meV, shows a rapid increase in the photoresponse at 120 meV with no lower energy states observed, probably due to the high compensation level in this sample. A rapid increase in the photoconductivity occurs when the photon energy is greater than the energy required to optically excite a donor electron from its ground state to the conduction band. The energies determined from the photoresponse onset are generally accurate measures of the optical activation energy of the impurity or defect levels in the material. The 120 meV activation energy is between the value from the Hall fitting and the expected standard value. The sample 3 spectrum shows two distinct photoresponse onsets at 77 and 120 meV.

The mobilities for each of the above are less than 10 cm^2/Vs at low temperature. These low mobilities required high electric fields to be applied between the contacts and a very narrow optical area between the contacts. The 6H samples had an applied bias voltages from 45.1 volts to 1000 volts. To obtain measurable photoresponse, sample 3 had an applied voltage of 45.1 at 9.0K and sample 4 had an applied voltage from 500 volts to 1000 volts at 9.0K.

4H-SiC RESULTS

Hall effect results for one of the 4H samples is shown in Fig. 3. The data for both samples were fit to eq. 2 with the 4H parameters. Both samples had two donors levels near the accepted values for the hexagonal and cubic sites of nitrogen and the results are given in Table I. The 4H-SiC samples had lower concentration levels approximately 10^{17} cm^{-3} and higher mobilities on the order of 150 cm^2/Vs at low temperature (~30 K). For both 4H-SiC crystals a maximum value of 1000 cm^2/Vs at T=100 K was obtained.

The low temperature photoconductivity spectra for two n-type 4H SiC samples are compared in Fig. 4. The two spectra are nearly identical in shape throughout the mid-infrared. Both spectra show strong increases in the photoresponse around 40 meV and at 120 meV. The two samples show two slightly different energies for the onset of the shallow nitrogen level, 40 and 47 meV. As previously reported in DLTS studies, the activation energy of nitrogen at the

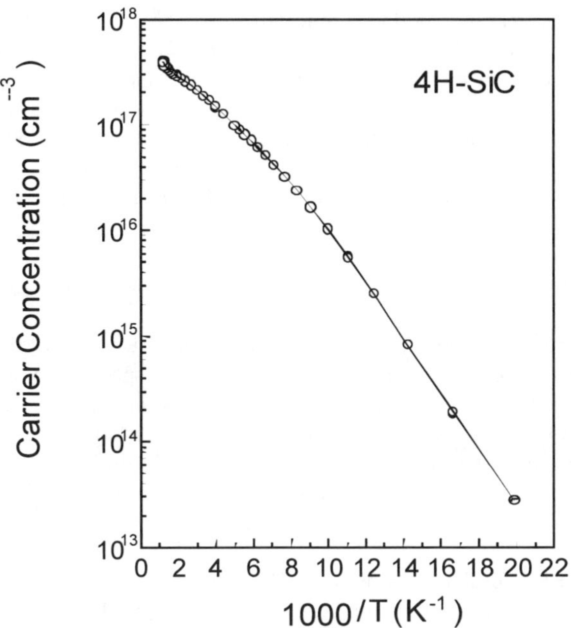

FIG. 3. Measured electron carrier concentration versus inverse temperature as obtained from the Hall measurement of n-type 4H-SiC sample 1.

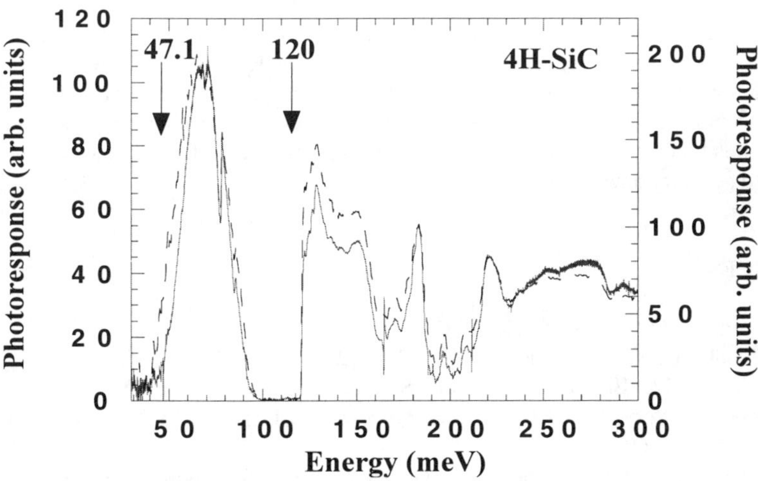

FIG. 4. Photoresponse spectra of n-type 4H-SiC. Sample 1 (solid line) is the same as shown in the Hall data. Sample 2 (dashed line) has a higher donor sample concentration than sample 1.

hexagonal site decreases with donor concentration.[5] The 120 meV level is higher in energy than expected. As shown in Table I, the theoretical energy for nitrogen on a cubic site is 92 meV.

The other influence on the SiC photoresponse spectrum is the strong absorption bands due to the excitation of lattice vibrational modes. This process generates no electrons and hence no additional photoresponse signal. Instead, these phonon modes compete with any photoexcitation processes for the available photons at selected wavelengths and create drops in the background photoresponse. The influence of the SiC lattice absorption is clearly seen in the 4H spectra in the region from 95 mev to 220 meV.

CONCLUSIONS

Another powerful optical technique for examining impurity states in SiC has been demonstrated. The mid-infrared photoconductivity spectra revealed the optically active impurity states in n-type 6H and 4H-SiC. These photoionized impurity transitions are most likely related to the nitrogen states known to be common in these materials. The photoresponse spectra from 6H and 4H-SiC are observed to be distinctly different in many respects. However, all the spectra show very strong photoresponse onsets at 120 meV. In principle, the 6H and 4H polytypes should have different activation energies for substitutional nitrogen on the cubic lattice sites.

If only the 6H results are considered, the agreement between the energies determined from photoresponse, 77 and 120 meV, are in good agreement with Hall results in the literature, 85 and 125 meV. When the 4H spectra are considered however, the assignment of the 120 meV onset to an optical activation energy may be in doubt. In 4H the nitrogen energy for the cubic sites should be closer to 100 meV. One factor maybe the strong optical phonons known to occur between 98.7 meV (E_1 TO) and 119.9 mev (A_1 LO). Further work will be done on thinned SiC samples to reduce the intensity of these lattice absorptions to see if any photoresponse is revealed below the 120 meV limit.

The agreement for the activation energies of substitutional nitrogen on the hexagonal sites in both polytypes determined from the photoresponse was excellent. The 4H-SiC spectra revealed a shallow nitrogen level at 47 meV for N_d= 1.2E17 cm^{-3} and 41 meV for N_d=1.9E17 cm^{-3}. This shift with donor concentration has been observed by other researchers. There are sharper features in the photoresponse spectra of 4H-SiC related to nitrogen donor excited states that will be published at a later date.

REFERENCES

1. W. J. Choyke and G. Pensl, Physica B **47**, 212 (1991).
2. H. H. Woodbury and G. W. Ludwig, Phys. Rev. **124**, 1083 (1961).
3. W. C. Mitchel, G. J. Brown, L. S. Rea and S. Smith, J.Appl. Phys. **71**, 246 (1992).
4. J. J. Rome, R. J. Spry, T. C. Chandler, G. J. Brown, R. J. Harris and B. C. Covington, Phys. Rev. **B25**, 3615-3618 (1982).
5. A. O. Evwaraye, S. R. Smith and W. C. Mitchel, Mat. Res. Soc. Symp. Proc. On <u>Defect and Impurity Engineered Semiconductors II</u>, **510**, 187 (1998).

Part IV

GaN Growth and Characterization

The influence of the sapphire substrate on the temperature dependence of the GaN bandgap

Joachim Krüger, Noad Shapiro, Sudhir Subramanya, Yihwan Kim, Henrik Siegle, Piotr Perlin, and Eicke R. Weber
Department of Materials Science, University of California at Berkeley
and
Materials Science Division, Lawrence Berkeley National Laboratory,
Berkeley, California 94720, USA

William S. Wong and Timothy Sands
Department of Materials Science, University of California at Berkeley
Berkeley, California 94720, USA

Nathan W. Cheung
Department of Electrical Engineering and Computer Science, University of California,
Berkeley, California 94720, USA

Richard J. Molnar
Lincoln Laboratory, Massachusetts Institute of Technology,
Lexington, Massachusetts 02173, USA

ABSTRACT

This paper analyses the influence of the sapphire substrate on stress in GaN epilayers in the temperature range between 4K and 600K. Removal of the substrate by a laser assisted liftoff technique allows, for the first time, to distinguish between stress and other material specific temperature dependencies. In contrast to the prevailing assumption in the literature, that the difference in the thermal expansion coefficients is the main cause for stress it is found that the substrate has a rather small influence in the examined temperature range. The measured temperature dependence of stress is in contradiction to the published values for the thermal expansion coefficients for sapphire and GaN.

INTRODUCTION

In recent years, GaN and related compounds have attracted a lot of academic as well as commercial interest. This is due to the potential applications for UV-based opto-electronic applications as well as high-temperature electronics [Kah1]. Very bright blue and green InGaN single quantum well diodes light-emitting diodes have been developed and commercialized [Nak1], and a laser diode consisting of 4 InGaN multi quantum wells has been reported to have a room temperature cw-operational lifetime of more than 10.000 hrs [Nak2].

Since large-scale GaN substrates are not available, epitaxial layers of GaN are deposited for the most part on foreign substrate materials like sapphire and SiC. These materials are known to result in stress in the GaN main layer which can reach values of up to 1.2 GPa [Kru1], either compressive or tensile. It is commonly argued that the lattice mismatch between layer and substrate and the difference in thermal expansion coefficient (TEC) are the main causes of stress in the GaN layer at room temperature. Consequently, considerable efforts have been spent on the exploration of alternative substrate materials. These have been either focused on matching the lattice constant

Mat. Res. Soc. Symp. Proc. Vol. 572 © 1999 Materials Research Society

(Si, GaAs, ZnO, LiAlO₃) or matching the thermal expansion coefficient (Ge [Sie1]). As to date, however, best results for GaN are still achieved for growth on sapphire. Though methodologies have been empirically developed to control stress to some extent, very little fundamental understanding of the exact causes of stress has been gained.

Based upon the linear thermal expansion coefficients for GaN (α = 5.59 x 10⁻⁶/K), for sapphire (α = 7.5 x 10⁻⁶/K), and for SiC (α = 4.2 x 10⁻⁶/K), all values are measured at room temperature [Lan1], GaN crystals grown on sapphire should be compressively stressed, whereas crystals grown on SiC should be under tensile stress. This statement assumes that stress caused by the mismatch of the lattice constants is completely relaxed. For most reported characterization studies, this assumption certainly holds, though it has been recently shown that an adequate buffer layer design allows to grow GaN in tension on sapphire [Kru1] and compressively stressed GaN on SiC [Edw1]. Up to this point, however, no publication is able to *quantitatively* explain the influence of the substrate on the amount of stress in the GaN main layer. Due to experimental constraints, it is unfortunately difficult to determine stress *in-situ* during growth and subsequent cooling down. Therefore, in most cases, post-growth attempts to monitor the temperature dependence of stress are limited to the range between 4 K and 800 K. Extrapolation of stress data taken by Raman spectroscopy between 300 K and 750 K [Age1] reveal a discrepancy between the expected absolute amount of stress and actual measured stress values for MOCVD and MBE grown samples. This effect was explained by the assumption of onset of plastic flow releasing stress during cool-down to an unknown amount, [Kie1]. On the contrary, S.Hearne *et.al.* [Hea1] have recently shown by *in-situ* stress measurements that GaN tends to grow under tensile stress and that plastic relaxation during cool-down is negligible.

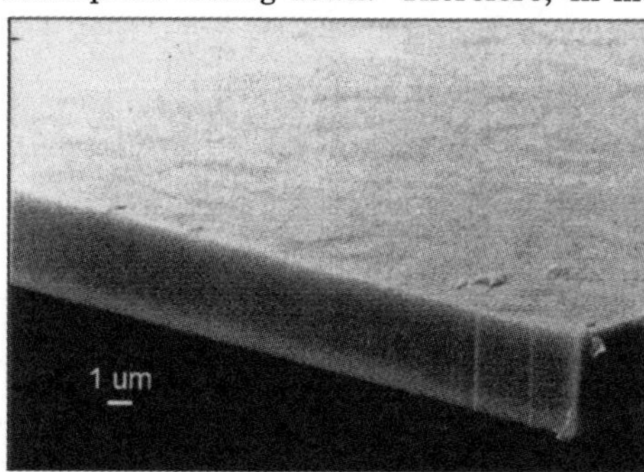

Figure 1: *SEM image of a free-standing GaN membrane lifted off from sapphire.*

Stress can be most conveniently determined by photoluminescence at cryogenic temperatures. It is known to shift the PL transition of the donor bound exciton (DX) spectrum by 27meV/GPa with the stress-free position located at 3.467 eV, [Kie1]. Besides being subject to built-in stress at a given temperature, near band gap transitions are also temperature dependent, as they follow the temperature dependence of the bandgap. For the last 25 years, numerous publications have attempted to determine and explain the bandgap temperature dependence of hetero-epitaxially grown GaN. The huge differences in the observed dependencies have been ascribed to various material and growth parameters, such as the growth method, post-growth cooling down rate, sample thickness, to name a few. None of these publications, however, is in the position to sort out one particular parameter and to describe or even explain its influence on the GaN bandgap temperature dependence.

This study now takes an entirely new approach to the described problems. Removal of the sapphire substrate via a laser-assisted liftoff technique [Won1] provides a reference sample which allows studying *exclusively* the temperature dependence of stress. For the first time, it is therefore possible to distinguish between intrinsic and material specific effects of the band gap temperature dependence on the one hand and stress effects caused by the thermal mismatch between sapphire substrate and GaN layer on the other hand. To prove the general applicability of the described findings, HVPE and MOCVD grown samples are analyzed.

EXPERIMENTAL

All GaN samples used in this study were grown on sapphire. The samples grown by Metal-Organic Chemical Vapor Deposition (MOCVD) were supplied by Cree Research and represent commercially available state-of-the-art material. The n-type carrier density was specified as in the low 10^{17}cm^{-3}; the sample thickness was 2 μm. The Hydride Vapor Phase Epitaxy (HVPE) grown samples were about 25 μm thick. Growth details are described in [Mol1].

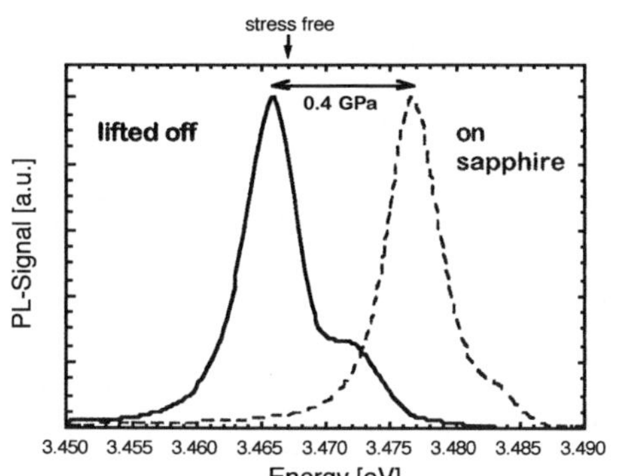

Figure 2: *Low-temperature PL spectra of a free-standing and a GaN film still attached to its sapphire substrate.*

For each set of samples, a reference sample was produced by removing its sapphire substrate by a recently developed laser-assisted lift-off technique [Won1]. This method is known to preserve the optical and structural quality of the GaN film, see figure 2.

Photoluminescence was excited by a 50mW HeCd laser, diffracted by a 0.85m double grating monochromator and detected by an UV-sensitive photo-multiplier. Special attention was spent to keep the excitation density constant for all samples. Further, with neutral density filters the excitation density was sufficiently reduced to exclude any warming effects. For temperatures between 15 K and 320 K a closed cycle refrigerator was used, temperatures between 300 K and 600 K were achieved by using a hot plate. Sample mounting with vacuum grease warranted a stress-free sample fixture and sufficient thermal contact with the copper holder.

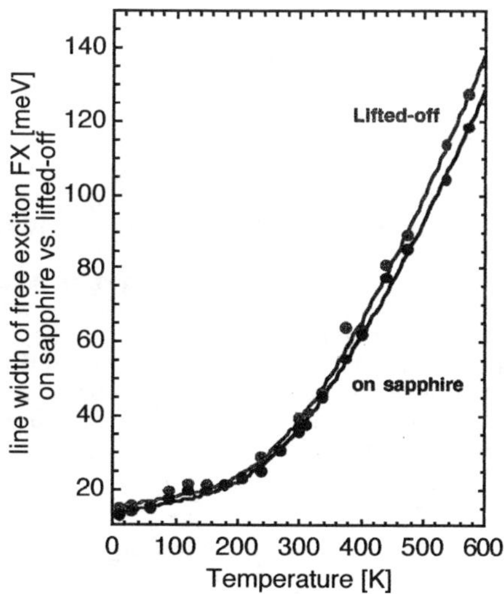

Figure 3: *Temperature dependence of free exciton line full width at half maximum.*

RESULTS

Comparison between the low-temperature PL spectra of the MOCVD grown GaN layer still attached to its substrate and the free-standing layer shows a red shift of the donor bound exciton transition, figure 2. The energetic difference indicates a stress release of 0.4 GPa, according to the calibration provided by [Kie1]. In particular, it is found that the stress gradient across the cross section of the HVPE grown layer which can be as much as 0.8 GPa, [Sie2], has been completely relaxed, too. The energetic positions of the PL spectra excited from the top and the former interface side are identical.

The near band gap PL signal at cryogenic temperatures is comprised of three transitions: the donor bound exciton DX = 3.466eV, the free exciton A FX(A) = 3.472eV, and the free exciton B FX(B) = 3.479eV; energetic positions are given for the free-standing sample. For the whole temperature range between 4 K and 600 K the entire near band gap spectrum could be perfectly simulated by combination of these lines under the assumption of a Lorentzian line type. This finding proves that the nature of the luminescing transition does not change over the whole range. In particular it should be noted that the free excitonic transition is traceable for the whole temperature range and no contribution from band-to-band transitions has been noticed, in agreement with [Her1]. For most samples, the donor bound exciton has been visible for temperatures up to 150 K. The free exciton B is clearly visible up to 260 K. Due to the rapidly increasing line width (fig.4) and small energetic distance to the free exciton A, it can not be resolved anymore for higher temperatures. As expected, the energetic spacing between these transitions are temperature-independent, the found energetic differences are: (FX(A) - DX = 5.5meV, FX(B) - FX(A) = 6.5 meV)

$$\Gamma = \Gamma_0 + \gamma_{ph}T + \frac{\Gamma_{LO}}{\exp[h\nu_{LO}/k_BT] - 1}$$

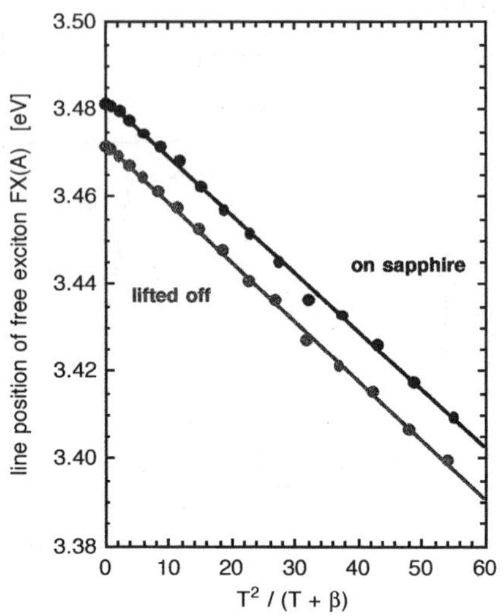

Figure 4: *Varshni plot of the energetic position of the free exciton A for both the free-standing and the GaN film on sapphire.*

The temperature dependence of the line width can be broken down into three major regimes, figure 3. At cryogenic temperatures, the line width is dominated by interface roughness, the exciton-exciton interaction, and exciton scattering by impurities, [Vis1]. The intermediate region is given by the coupling strength of the interaction between excitons and acoustical phonons. For temperatures above 240K, the line width is entirely dominated by the exciton interaction with longitudinal optical phonons. We find the following parameters: Γ_0 = 13 meV, γ_{ph} = 31±2 µeV, and Γ_{LO} = 500±25 µeV. These values are in agreement with a previous publication [Vis1].

Figure 4 shows the temperature dependence of the free exciton transition energy FX(A) for both the layer on sapphire and its free-standing reference sample. It should be noted, that the GaN band gap energy is by 27 meV [Buy1] larger than the free exciton energy; the difference accounts for the free exciton binding energy. Since this paper is only concerned about relative changes in the band gap energy, the given values are not corrected for this energy.

For most semiconductors the variation of the band gap with temperature can be described by the semi-empirically found Varshni formula [Var1]:

$$E(T) = E(T{=}0) - \frac{\alpha\, T^2}{\beta + T}$$

For both the free-standing and the layer still attached to its sapphire substrate, we find the same parameters within the experimental error, table 1. This is very remarkable, since the values

Experimental technique	α [10^4 eV/K]	β [K]	Reference
Photoluminescence of FX (A)	5.0	400	[Vis1]
FX (B)	5.2	450	"
PL excitation	- 5.08	- 996	[Mon1]
Photoabsorption	9.39	772	"
Photoreflection of FX (A)	8.32	835.6	[Sha1]
FX (B)	10.9	1194	"
Absorption of MOCVD grown GaN	5.66	737.9	[Man1]
Absorption of MBE grown GaN	11.56	1187.4	"
PL of FX (A) MOCVD, on sapphire	13.25	1539	this study
PL of FX (A) MOCVD, lifted off	13.56	1570	"

Table 1: *Comparison of Varshi parameters determined for various GaN materials.*

reported in the literature scatter considerably; for instance, the parameter β which is proportional to the specific heat Debye temperature [Man1] has been found to vary between β = -996 K [Mon1] and β = 3690 K [Sha1]. This clearly proves that the parameters α and β are strongly sample dependent and comparison of results taken from different authors on different samples is less meaningful. We argue that this is due to the differing amount of point and extended defects which certainly influences the Debye temperature. As the results of this study prove, stress caused by thermal mismatch with the substrate plays only a minor role at this scale.

It is more illustrative to analyze the temperature dependence of the difference in the FX(A) position between the sample on sapphire and its respective free-standing reference sample, figure 5. This should follow the temperature dependence of stress in the GaN main layer. Apparently, with decreasing temperature also the difference decreases indicating that the GaN layer is *less compressively* stressed. This finding is surprising and in contradiction to the prevailing understanding in the literature. Since the TEC of sapphire is higher than the TEC of GaN, the GaN main layer should be *increasingly compressively* stressed with decreasing temperature – provided that the TECs are temperature-independent. Consequently, the energetic difference between the PL spectrum of the free-standing layer and the layer still attached to the substrate should increase. Strikingly, the opposite tendency is experimentally observed.

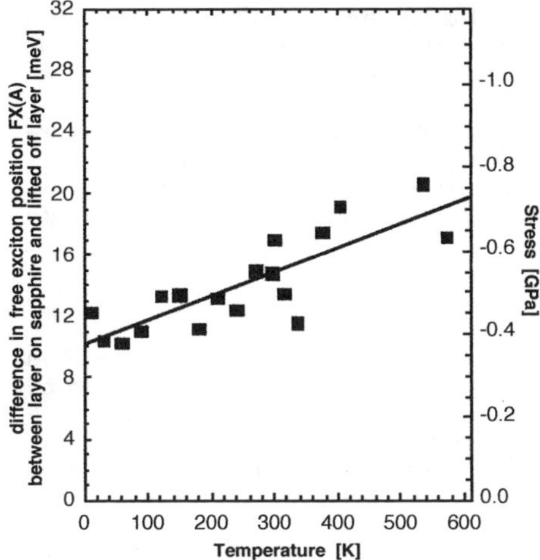

Figure 5: *Difference in energetic position of the free exciton A between GaN film on sapphire and the free-standing reference sample.*

DISCUSSION

The unexpected temperature behavior of stress in GaN hetero-epitaxially grown on sapphire is the most interesting aspect of this study. Our results clearly indicate that the influence of the thermal mismatch between sapphire and GaN epilayer on the GaN band gap temperature dependence between 4 and 600 K is much less than generally assumed in the literature. Monemar *et.al.* [Mon2], for instance, have found that the PL spectra of GaN grown on sapphire and on SiC differ in their energetic positions by more than 30meV when cooled down to helium temperatures. If this difference was solely due to stress, the two different substrates would result in a difference of thermal stress of more than 1 GPa!

Assuming a (temperature-independent) Young modulus of E = 200 GPa [Kie1] and a Poisson ratio of ν = 0.27 [Kie1], one estimates for a change of temperature from 300K to 4 K the stress for GaN on sapphire to be σ = -0.16 GPa. This difference should result in a blue shift of the PL spectrum by 4.3 meV. This should be clearly visible but is experimentally not observed.

Two possible explanations for the findings of this study should be discussed: First, the variation of stress in GaN grown on sapphire in the respective temperature range is not governed by the thermal mismatch between GaN and sapphire. The second and more likely cause is given by the uncertainty of published thermal expansion coefficients for both sapphire and GaN, [Lan1, Les1]. The reported values differ by a factor of two, even when only values taken at room temperature are compared. The TEC temperature dependence is even less known so that one may speculate that an

293

unanticipated temperature behavior of these parameters might be responsible for the here observed stress dependence.

On the other hand it should be noted that stress is usually determined by photoluminescence at *cryogenic* temperatures, whereas strain is almost always assessed by X-ray diffraction at *room* temperature [Kie1]. Our results prove that the difference in measurement temperature does not translate into an inconsistency between the two methods, the maximum deviation is about 0.1 GPa which falls into the range of experimental error of most methods.

The results of this work, however, strongly suggest to measure the temperature dependence of the a and c lattice parameter by X-ray diffraction. Preliminary studies are already available [Hei1, Les1], but their results are not conclusive yet. Measurements of a lifted-off GaN film will provide the best basis for a solid data collection.

SUMMARY

This paper analyses, for the first time, the temperature dependence of stress in MOCVD and HVPE grown GaN films on sapphire in respect to a free-standing reference GaN film which has been lifted off from its substrate. The measured temperature dependence of stress in the temperature range between 4 and 600 K is in contradiction to the published values for the thermal expansion coefficients for sapphire and GaN.

ACKNOWLEDGMENT

This work was supported by the Office of Energy Research, Office of Basic Energy Sciences, Division of Materials Sciences of the U.S. Department of Energy under Contract No. DE-AC03-76SF00098.

REFERENCES

[Age1] J.W.Ager, *et.al.*; Mat.Res.Soc.Symp.Proc. **449**, 775 (1997)

[Buy1] I.A.Buyanova, *et.al.*; Appl.Phys.Lett. **69**, 1255 (1996)

[Cod1] G.D.Cody; in *Hydrogenated Amorphous Silicon*; ed. By J.Pankove, Semiconductors and Semimetals Vol. **21**, Part b, Ch. 2, p. 11 (Academic, New York 1984)

[Edw1] N.V.Edwards, *et.al.*; Appl.Phys.Lett. **73**, 2808 (1998)

[Hei1] H. Heinke, *et.al.*; J. of Crystal Growth **189/190**, 375 (1998)

[Her1] H.Herr, V.Alex, and J.Weber; Mater. Res. Soc. Symp. Proc. **482**, 719 (1998).

[Kie1] C. Kisielowski in *Gallium Nitride II*; Semiconductors and Semimetals Vol. **57**, Ch. 7, p. 275(Academic, New York 1999)

[Kru1] J.Krüger *et.al.*; to be published

[Les1] M. Leszczynski, *et.al.*; J.Appl.Phys. **76**, 4909 (1994)

[Hea1] S. Hearne, *et.al.*; Appl.Phys.Lett. **74**, 356 (1999)

[Lan1] *Landolt–Bornstein: Numerical Data and Functional Relationships in Science and Technology*, Springer, Berlin, 1982, Vol. **17b**.

[Man1] M.O.Manasreh; Phys.Rev.B. **53**, 16425 (1996)

[Mol1] R.J. Molnar, *et.al.*; J.Crystal Growth **178**, 147 (1997)

[Mon1] B.Monemar; Phys.Rev.B. **10**, 676 (1974)

[Mon2] B.Monemar, *et.al.*; MRS Internet J. Nitride Semicond. Res. 1, 2(1996)

[Sie1] H.Siegle, *et.al.*; MRS '99 Spring meeting; these proceedings

[Sie2] H. Siegle, *et.al.*; Appl.Phys.Lett. **71**, 2490 (1997)

[Sha1] W.Shan, *et.al.*; Appl.Phys.Lett. **66**, 985 (1995)

[Var1] Y.P.Varshni, Physica **34**, 149 (1967)

[Vis1] A.K.Viswanath, *et.al.*; J.Appl.Phys. **84**, 3848 (1998)

[Won1] W. S. Wong, *et.al.*; Appl.Phys.Lett. **72**, 599 (1998)

EFFECT OF N/Ga FLUX RATIO IN GaN BUFFER LAYER GROWTH BY MBE ON (0001) SAPPHIRE ON DEFECT FORMATION IN THE GaN MAIN LAYER

S. Ruvimov,* Z. Liliental-Weber, J. Washburn,
Lawrence Berkeley National Laboratory, MS 63-203, Berkeley, CA 94720,
* present address: Mitsubishi Silicon America, 1351 Tandem Av., Salem, Oregon, 97303

Y. Kim, G. S. Sudhir, J. Krueger, and E. R. Weber
Department of Materials Science and Mineral Engineering, University of California at Berkeley, Berkeley, California, 94720

Transmission electron microscopy was employed to study the effect of N/Ga flux ratio in the growth of GaN buffer layers on the structure of GaN epitaxial layers grown by molecular-beam-epitaxy (MBE) on sapphire. The dislocation density in GaN layers was found to increase from $1^{\times}10^{10}$ to $6^{\times}10^{10}$ cm^{-2} with increase of the nitrogen flux from 5 to 35 sccm during the growth of the GaN buffer layer with otherwise the same growth conditions. All GaN layers were found to contain inversion domain boundaries (IDBs) originated at the interface with sapphire and propagated up to the layer surface. Formation of IDBs was often associated with specific defects at the interface with the substrate. Dislocation generation and annihilation were shown to be mainly growth-related processes and, hence, can be controlled by the growth conditions, especially during the first growth stages. The decrease of electron Hall mobility and the simultaneous increase of the intensity of "green" luminescence with increasing dislocation density suggest that dislocation-related deep levels are created in the bandgap.

1. INTRODUCTION.

Epitaxial GaN is a promising material for electronic applications such as visible light-emitting diodes (LEDs) [1-4], blue lasers [3,4] and metal-semiconductor field-effect transistors [5]. Device quality GaN on sapphire has been successfully achieved by metal-organic vapor phase epitaxy using low temperature (LT) GaN or AlN buffer layers [6]. The LT buffer layer provides a high density of nuclei for growth of the main GaN layer at high temperature and promotes lateral growth of GaN [6]. The growth and subsequent coalescence of GaN islands finally leads to quasi-two-dimensional (2D) growth [7]. The structural quality of the GaN layer appears to depend on growth evolution. Because dislocations in the GaN have a low mobility [8] there is a low probability for their interaction and annihilation compared to other III-V materials such as GaAs. The defect density in GaN layers can be significantly reduced by optimization of growth conditions, especially at initial growth stages. Thus, the structure of a LT buffer may significantly effect the dislocation density in the GaN epilayer.

The LT buffer layer was found to transform during the temperature ramp/anneal with an increase of its average grain size and root-mean-square (rms) roughness [9-11]. The growth evolution can be affected by a number of parameters including the thickness of the LT buffer, the temperatures for LT and HT growth, growth rate, Ga/N flux ratio, gas ambient, etc [12-19]. However, the effect of the Ga/N flux ratio in the growth of the LT buffer layer has not yet been studied in detail. Here we report the results of a TEM study of the effect of the N/Ga flux ratio during growth of GaN buffer layers on defect formation in epitaxial GaN layers grown by MBE on (0001) sapphire.

Mat. Res. Soc. Symp. Proc. Vol. 572 © 1999 Materials Research Society

2. EXPERIMENT.

GaN epitaxial layers were grown by molecular beam epitaxy (MBE) on (0001) sapphire substrates using a Riber 1000 system. Growth details are described elsewhere [20]. After cleaning and nitridation of the sapphire substrate (700 $^{\circ}$C, 10 min), a LT GaN buffer layer was deposited (500 $^{\circ}$C, 5 min) followed by the growth of a 2 μm-thick GaN epilayer (725 $^{\circ}$C, 4 h). A set of 4 GaN layers were grown under various nitrogen flow rates (5, 15, 25 and 35 sccm) during the buffer deposition with a fixed N flux rate of 35 sccm for the main layer. The Ga cell temperature was kept constant at 880 $^{\circ}$C during the growth. TEM studies were carried out on Topcon 002B and JEOL 200CX microscopes operated at 200 kV and on an ARM microscope operated at 800 kV. Cross-sectional specimens were prepared for TEM study by dimpling followed by ion milling.

Atomic force microscopy (AFM) was employed to determine the surface morphology of GaN layers as well as LT buffer layers with and without annealing at high temperature. Electrical properties of GaN layers were evaluated by Hall effect measurements at room temperature. X-ray diffraction was performed on a Siemens D-5000 diffractometer to determine the crystalline quality of GaN layers using both symmetric and asymmetric Bragg reflection.

3. RESULTS AND DISCUSSION.

TEM indicated that all layers contained a high density of threading dislocations and inversion domain boundaries (IDBs). Growth parameters and structural characteristics of the GaN layers are shown in Table 1.

Table 1. Growth and structural parameters of GaN layers.

Sample	N flux (buffer), sccm	N flux (epi-layer), sccm	Buffer roughness, nm	Dislocation density, cm^{-2}	FWHM of symmetric (0002) reflection, arcmin	FWHM of asymmetric (0002) reflection, arcmin
A	5	35	1.1	$1.8^{x}10^{10}$	2.3	16
B	15	35	1.1	$0.5^{x}10^{10}$	3.5	15
C	25	35	-	$1.0^{x}10^{10}$	7.0	17
D	35	35	1.3	$5.6^{x}10^{10}$	8.5	34

Figure 1 shows two plan view TEM images of the same area taken under different diffraction conditions where either dislocations (a) or IBDs (b) were visualized. Dislocations were often lying within IDBs which form closed domains of 15-30 nm in diameter [Fig. 1 (b)].

Threading dislocations were usually arranged in small angle boundaries with both tilt and twist components dividing the epitaxial layer into columnar grains. Diffraction analysis shows that a majority of the threading dislocations are edge dislocations with b=a/3$<11\bar{2}0>$ lying in the c-plane. Therefore, grains were mainly misoriented around the c-axis being almost perfectly oriented in the c-plane. As a result, the full width at half maximum (FWHM) for a symmetric (0002) x-ray rocking curve from the GaN layer is typically much smaller than that of an asymmetric x-ray rocking curve (see Table 1).

Many dislocations and all IDBs originated at the interface with the sapphire and propagated up to the layer surface. Some dislocations form half loops near the interface and don't propagate further into the GaN layer.

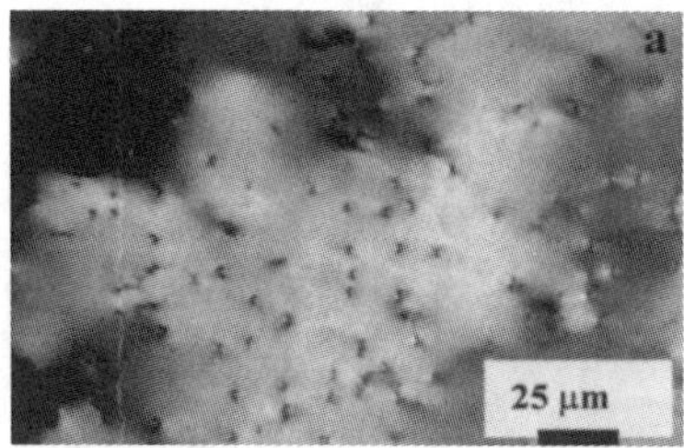
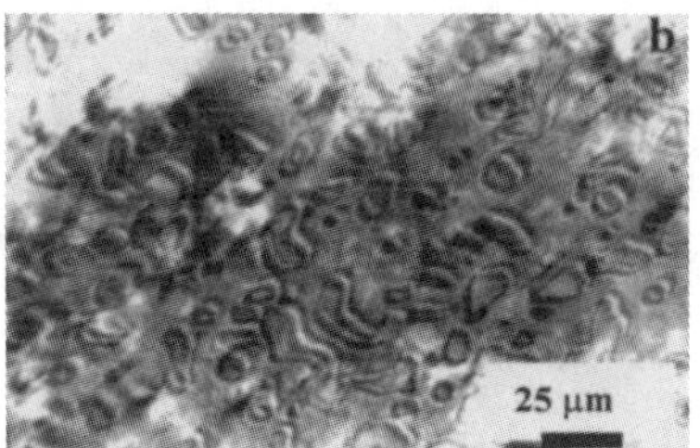

Fig. 1, a-b. Plan-view TEM images of GaN layer (the same area) taken at different **g**.

The GaN/sapphire interface is almost atomically abrupt, but contains steps of about 1-2 monolayer height. Formation of IDBs was often associated with specific defects at the interface with a substrate (Fig.2). Fig 2 (b) shows an IDB associated with an atomic step at the interface. The white contrast at the interface might indicate the presence of a thin AlN layer between the GaN and sapphire. A one to two monolayer thick AlN layer is expected to form during the nitridation of the sapphire. This AlN layer may reduce the possible deterioration of sapphire and, hence, the interface roughening. As a result, it leads to a reduction of threading dislocation density and dislocation rearrangement in both GaN buffer and epi layer.

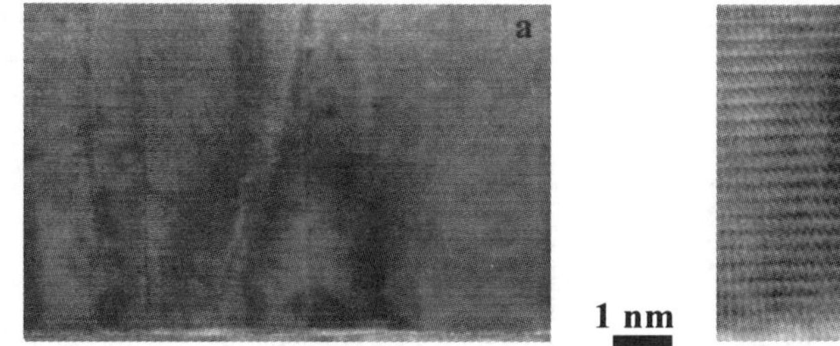
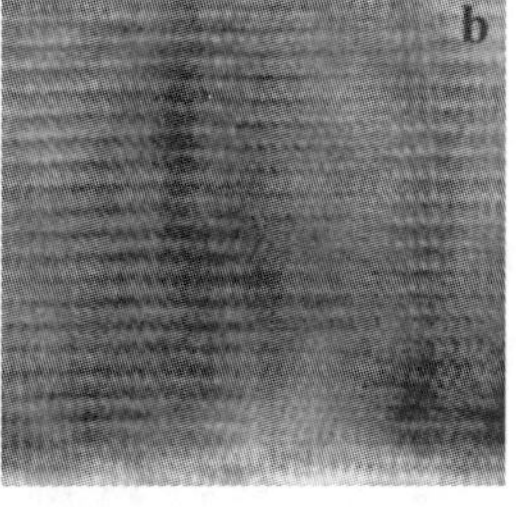

Fig. 2. Cross-sectional TEM (a) and HREM (b) images of GaN epitaxial layer

A white contrast often appears under the IDBs (see e.g. Fig. 2.) suggesting that AlN might locally change the polarity and give rise for the formation of IDBs. Therefore, IDBs are expected to originate at the earliest growth stage, during the buffer layer deposition. IDBs were observed in the GaN buffer layer before and after the high temperature ramp/anneal. Fig. 3 shows a typical HREM image of a LT GaN buffer layer after the high temperature anneal to 725 °C. The buffer layer has a hexagonal structure and contains a high density of IDBs originating at the interface. Before annealing the buffer layer often contained a fraction of the cubic phase (not shown) that transforms into the hexagonal structure during the high temperature ramp/anneal. LT GaN in Fig. 3 has relatively rough surface and smaller grains compared to the AFM data in Table 1. The small grains (< 50 nm in diameter) were not visible on AFM images because they were buried under a thin amorphous-like layer (not shown). The nature of this layer is unknown.

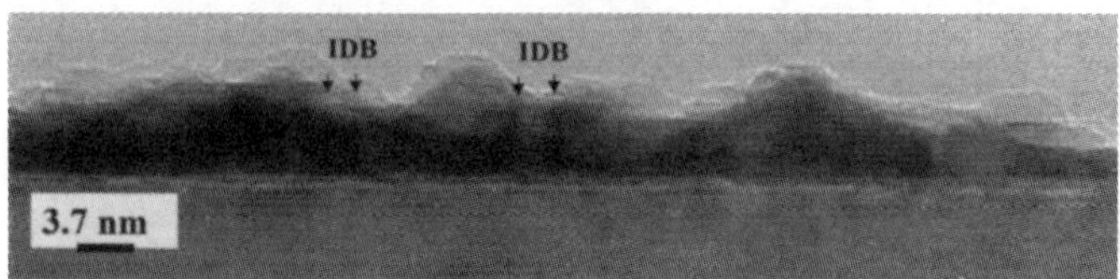

Fig. 3. Cross-sectional HREM image of LT GaN buffer layer after thermal anneal at 725 °C.

Initial growth of the GaN layer proceeds in three dimensional (3D) fashion due to a high mismatch in lattice parameters between GaN and sapphire. The GaN/sapphire interface contains a high density of misfit dislocations which accommodate almost all the misfit between their crystalline lattices at growth temperature during the first growth stages. The orientation relationship between GaN and sapphire, $(0001)_{GaN}//(0001)_{Al2O3}$, $[1\bar{1}00]_{GaN}//[11\bar{2}0]_{Al2O3}$, provides the best match between their crystalline lattices, but mismatch still remains high. The orientation relationship results in a 6 to 7 "magic" ratio between crystalline lattices of GaN and sapphire: every 6 planes ($[1\bar{1}00]$ or $[11\bar{2}0]$) of GaN fits to 7 planes ($[11\bar{2}0]$ or $[1\bar{1}00]$, respectively) of sapphire with an error of about 2 % (Fig. 4). The period of 1.7 nm for the contrast oscillations observed at the GaN/sapphire interface fits well to this "magic" value. The atomic structure of this GaN/sapphire interface suggests formation of a high density of small coherent hexagonal nucleai as the first stage of the growth and, then, their coalesce into larger incoherent GaN islands with misfit dislocations at the interface. Threading dislocations in GaN appear at points where the misfit dislocation network is disturbed, for example, by the presence of atomic steps at the interface or at merging points of adjacent islands.

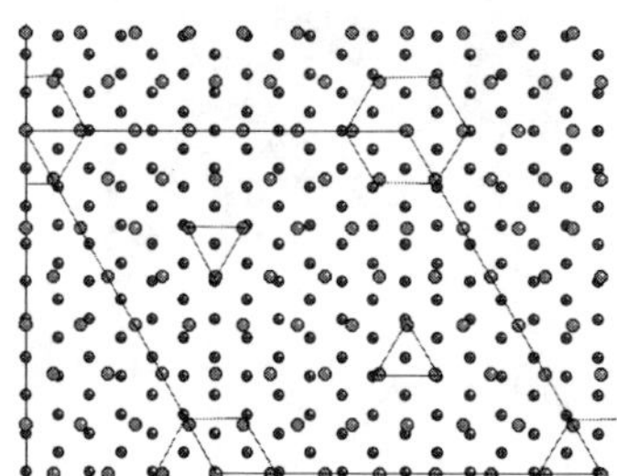

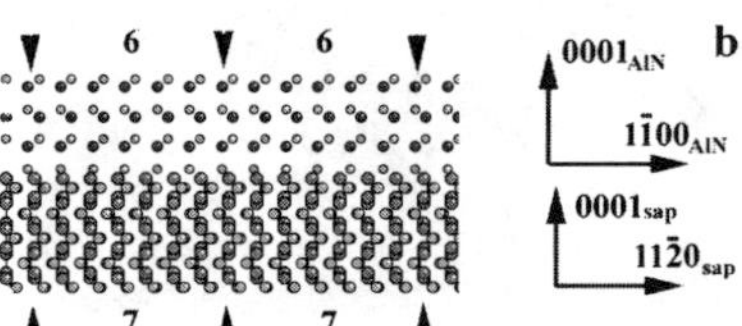

Fig. 4. Coincidence site lattice (a) for adjacent Ga (or Al) layers (two-dimensional lattices of Ga (or Al) atoms) in crystalline lattices of GaN (or AlN) and sapphire, and GaN (AlN)/sapphire interface (b).

The dislocation density is high in the GaN buffer layer and then drastically decreases over 0.2 µm toward the GaN top surface (Fig.2). This dislocation distribution being typical for many GaN samples indicates that most of threading dislocations originate and annihilate during the early stages of growth. Because of low dislocation mobility at the growth temperature [8], the dislocation distribution in the GaN layer is, to a large extent, frozen after the layer growth and, hence, reflects the growth process itself. Dislocation generation and annihilation are mainly growth-related processes and, hence, can be controlled by the growth conditions, especially during the first growth stages.

Formation of dislocation half loops near the interface can be caused by lateral overgrowth of GaN islands at high temperature. The growth conditions for various grains are locally different because the islands differ in their initial size and strain distribution within each island. Therefore, larger islands with lower strain will grow faster than others. As a result, these islands will laterally overgrow the others leaving them buried near the interface.

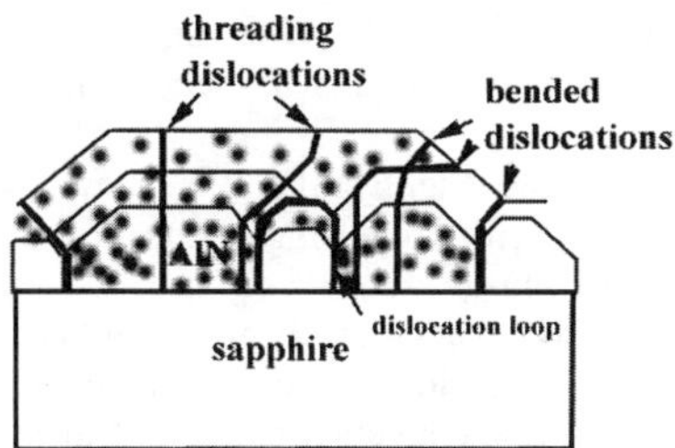

Threading dislocations that accommodate the misfit forced by this lateral overgrowth to bend into basal planes, interact with other dislocations and often annihilate. Schematically it is shown in Fig. 5. This process results in the formation of many half loops in the buffer and in the first 0.2 μm of the GaN and in the decrease of overall dislocation density in the growing GaN layer.

Fig. 5. Lateral overgrowth of islands during the early growth stages.

The efficiency of this process depends on both thickness and structure of the buffer layer, and on the growth conditions. This suggests that the structure of the buffer layer may effect the defect formation and annihilation in the GaN epilayer. The roughness of LT buffer layer is highest at N flux of 35 sccm (Table 1) while the average grain size of about 60 nm (measured by AFM) doesn't change with the N flux. Accordingly the dislocation density in GaN layers also increases up to $6^{x}10^{10}$ cm^{-2} for N flux of 35 sccm during the growth of the GaN buffer layer. The increase of the dislocation density from 1.0×10^{10} to 6.0×10^{10} cm^{-2} corresponds to a decrease in a grain size of the main GaN layer from 100 nm to 40 nm. This suggests that the lateral overgrowth during the deposition of the main GaN layer may be affected by a "micro-roughness" of the LT buffer layer.

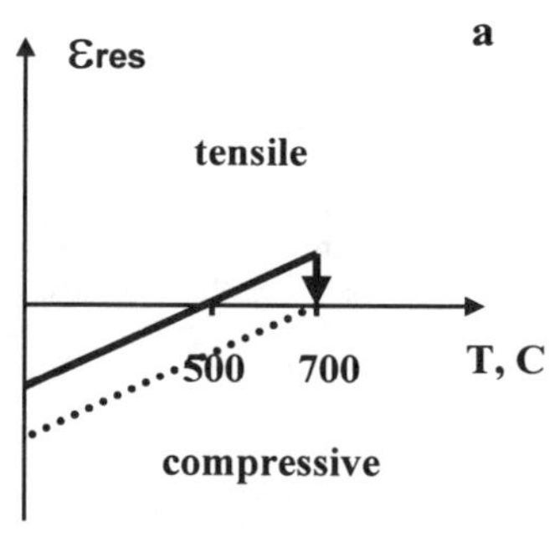

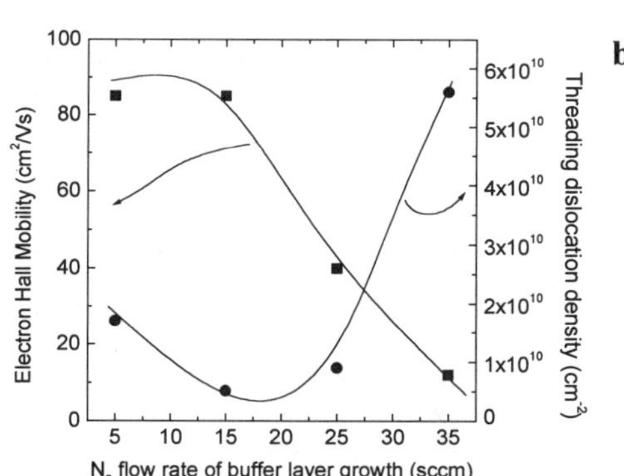

Fig. 6. (a) An increase of the residual stress at room temperature due to partial stress relaxation of the LT buffer at high temperature. (b) Electron Hall mobility and threading dislocation density of GaN epitaxial layers as a function of nitrogen flow rate during the buffer layer growth.

The changes in the dislocation density with N flux are similar to those for the FWHM of asymmetric x-ray rocking curves (see Table 1). This agrees well with the result of diffraction analysis showing that the majority of threading dislocations is of edge type. FWHM of a symmetric (0002) x-ray rocking curve is more sensitive to misorientations in c-plane so that its increase with N flux during the buffer layer growth may indicate on the increasing fraction of mixed dislocation in the GaN layer. Residual compressive stress in the GaN layer decreases with N flux during the buffer layer growth according to the absorption spectroscopy measurements. A shift in the absorption edge has been observed in absorption spectra due to the residual stress in the GaN layer. An increase of the residual stress at room temperature for the layer grown on a "Ga-rich" buffer can be associated with partial stress relaxation of the LT buffer during the high temperature anneal. Schematically it is shown in Fig. 6 (a).

The decrease of electron Hall mobility from 85 to 12 cm^2/Vs [Fig. 6 (a)] and the simultaneous increase of the intensity of "green" luminescence (not shown) with increasing dislocation density suggests that dislocation-related deep levels are created in the band-gap.

In conclusion, a decrease of the dislocation density in GaN layers was observed for "Ga-rich" LT GaN buffer layers when the nitrogen flux was below 25 sccm during the growth of the GaN buffer layer under otherwise the same growth conditions. All GaN layers were found to contain IDBs originating at the interface with the sapphire and propagating up to the layer surface. Dislocation generation and annihilation were shown to be mainly growth-related processes and, hence, can be controlled by the growth conditions, especially during the early growth stages. The decrease of electron Hall mobility and the simultaneous increase of the intensity of "green" luminescence with increasing dislocation density suggest that dislocation-related deep levels were created in the band-gap.

This study was supported by the Director, Office of Energy Research, U.S. Department of Energy under Contract No. DE-AC03-76SF00098. The use of the facilities at National Center for Electron Microscopy and assistance of W. Swider for sample preparation are appreciated.

1. H. Amano, M. Kito, X. Hiramatsu, and I. Akasaki, Jpn. J. Appl. Phys. **28**, L2112 (1989)

2. S. Nakamura, T. Mukai, and M. Senoh, Jpn. J. Appl. Phys. **30**, L1998 (1991)

3. S. Nakamura, M. Senoh, S. Nagahama, et al, Appl. Phys. **69**, 4056 (1996)

4. F.A. Ponce, D.P. Bour, Nature **386**, 351 (1997)

5. S.N. Mohammad, A. Salvador, and H. Morkoç, Proc. IEEE **83**, 1306 (1995)

6. H. Amano, N. Sawasaki, I. Akasaki, and Y. Toyoda, Appl. Phys. Lett. **48**, 353 (1986)

7. H. Amano, I. Akasaki, K. Hiramatsu, N. Koide, and N. Sawasaki, Thin Solid Films **163**, 415 (1988)

8. L. Sugiura, J. Appl. Phys. 81, 1633 (1997)

9. A.E. Wickenden, D.K. Wickenden, and T.J. Kistenmacher, J. Appl. Phys. **75**, 5367 (1994)

10. T. George, W.T. Pike, M.A. Khan, J.N. Kuznia, and P. Chang-Chien, J. Electron.Mat. **24**, 241 (1995)

11. X.H. Wu, D. Kapolnek, E.J. Tarsa, B. Heying, S. Keller, B.P. Keller, U. Mishra, S. P. DenBaars, and J.S. Speck, Appl.Phys.Lett. **68**, 1371 (1996)

12. S. Nakamura, Jpn. J. Appl. Phys. **30**, L1705 (1991)

13. S. Ruvimov et al., *unpublished*

14. Y. Kim et al Mat.Res. Soc. Symp. **449**, 227 (1997)

15. J. Han, T.-B. Ng, R.M. Biefeld, M.H. Crawford, and D.M. Follstaed, Appl. Phys. Lett. **71**, 3114 (1997)

16. D. Kapolnek, X.H. Wu, B. Heying, S. Keller, B.P. Keller, U.K. Mishra, S.P. DenBaars, and J.S. Speck, Appl. Phys. Lett. **67**, 1541 (1995)

17. E.H. Tarsa, B. Heying, X.H. Wu, P. Fini, S.P. DenBaars, J.S. Speck, J. Appl. Phys. **82**, 5472 (1997)

18. P. Hacke, G. Feuillet, H. Okamura, and S. Yoshida, Appl. Phys. Lett. **69**, 2507 (1996)

19. Z. Yu, S.L. Buczkowski, N.C. Giles, T.H. Mayers, and M.R. Richards-Babb, Appl. Phys. Lett. **69**, 2731 (1996)

20. Y. Kim et al, *unpublished*

ENHANCED OPTICAL EMISSION FROM GaN FILM GROWN ON COMPOSITE INTERMEDIATE LAYERS

Xiong Zhang, Soo-Jin Chua, Peng Li , and Kok-Boon Chong
Center for Optoelectronics, Department of Electrical Engineering
National University of Singapore
10 Kent Ridge Crescent, Singapore 119260, elezx@nus.edu.sg

ABSTRACT

GaN films have been grown on silicon-(001) substrate with specially designed composite intermediate layers consisting of an ultra-thin amorphous silicon layer and a $GaN/Al_xGa_{1-x}N$ (x=0.2) multilayered buffer by metal-organic chemical vapor deposition and characterized by photoluminescence and x-ray diffraction spectroscopy. It was found that the GaN films grown on the composite intermediate layers gave comparable or slightly stronger optical emission than those grown on sapphire substrate under identical reactor configuration. Moreover, the full width at half maximum for the GaN band-edge-related emission is 40 meV at room temperature. This fact indicates that, by using the proposed composite intermediate layers, the crystalline quality of GaN-based nitride grown on a silicon substrate can be significantly improved.

INTRODUCTION

Recently there has been successful demonstration of commercially available blue and green light-emitting diodes (LEDs)[1] and long life-time laser diodes (LDs)[2] fabricated from group-III nitrides which are generally grown on sapphire substrate. However, besides the large difference in lattice constant and thermal expansion coefficient between the group-III nitride and sapphire substrate, sapphire is an insulating material and extremely rigid. Therefore, it is not easy to fabricate a group-III nitride-based semiconductor device on a sapphire substrate. Silicon is one of the proposed substrate materials to overcome this shortcoming because of its high crystal quality, large area size, low manufacturing cost, and the potential application in integrated opto-electronic devices. However, due to the even larger differences in lattice constant and thermal expansion coefficient between the group-III nitride and silicon as compared with sapphire, it is really difficult to grow high quality epitaxial layer of group-III nitride on a silicon substrate. Attempts have been made to grow group-III nitrides on silicon substrates in the past decade using various kind of materials as the intermediate layer between group-III nitride and silicon substrate. These include AlN,[3, 4] carbonized silicon,[5, 6] nitridized GaAs,[7] oxidized AlAs,[8] and γ-Al_2O_3.[9] In particular, by using AlN thin film as the intermediate layer, ultraviolet and violet-light emitting diodes of group-III nitride have been fabricated on a silicon substrate by molecular beam epitaxy (MBE) recently.[3] However, the turn-on voltages as well as the brightness of these diodes do not approach the performance levels of the corresponding devices grown on a sapphire substrate by metal-organic vapor phase deposition (MOCVD). Therefore, both the constitution of the intermediate layer and the growth method for it need to be further optimized in order to enhance the crystallinity of the group-III nitrides. In this letter, we report the growth of high quality GaN epitaxial layers on a silicon substrate by MOCVD using the specially-designed composite intermediate layers (CILs).

Mat. Res. Soc. Symp. Proc. Vol. 572 © 1999 Materials Research Society

EXPERIMENT

GaN epitaxial layers were grown on silicon-(001) substrates by low pressure (100±2 Torr) MOCVD in an EMCORE D125 vertical reactor. Trimethylgallium (TMGa), Trimethylaluminium (TMAl), and high purity ammonia (NH$_3$) were used as Ga, Al, and N precursors, respectively, and hydrogen-diluted silane (SiH$_4$) was employed for depositing thin amorphous silicon film. As shown in Fig. 1, after a chemical cleaning process, the silicon-(001) substrate was heated to 1,030 $^{\circ}$C under hydrogen ambient for 10 min in order to produce a clean, oxide-free surface. The silicon substrate was then cooled down to 525 $^{\circ}$C, and an ultra-thin (less than 5 nm) amorphous silicon film was deposited onto the surface of the silicon-(001) substrate by flowing the hydrogen-diluted silane to form a "soft" buffer. Subsequently, a three period GaN/Al$_x$Ga$_{1-x}$N (x=0.2) multilayered buffer (MLB) was grown on the top of the formed ultra-thin amorphous silicon film to constitute the CILs. The details for the growth of GaN/Al$_x$Ga$_{1-x}$N (x=0.2) MLB has been described elsewhere.[10, 11] Finally, the temperature was raised to 1,000 $^{\circ}$C and a 1 μm-thick unintentionally doped GaN epitaxial layer was grown on the surface of the formed CILs consisting of the ultra-thin amorphous silicon film and the MLB. No cap layer was deposited on the top of GaN epitaxial layers. For the purpose of comparison, undoped GaN films have also been grown over a conventional 25 nm-thick GaN single layer buffer on sapphire substrates.

After the growth, room-temperature photoluminescence (PL) and x-ray diffraction (XRD) measurements were carried out in order to characterize the crystalline quality of the grown undoped GaN epitaxial layers. The optical properties, more specifically, the intensity and linewidth of the PL emission and XRD peaks of the undoped GaN samples grown on silicon substrates using the newly-developed CILs were compared with those grown on sapphire substrate using the conventional GaN single layer buffer.

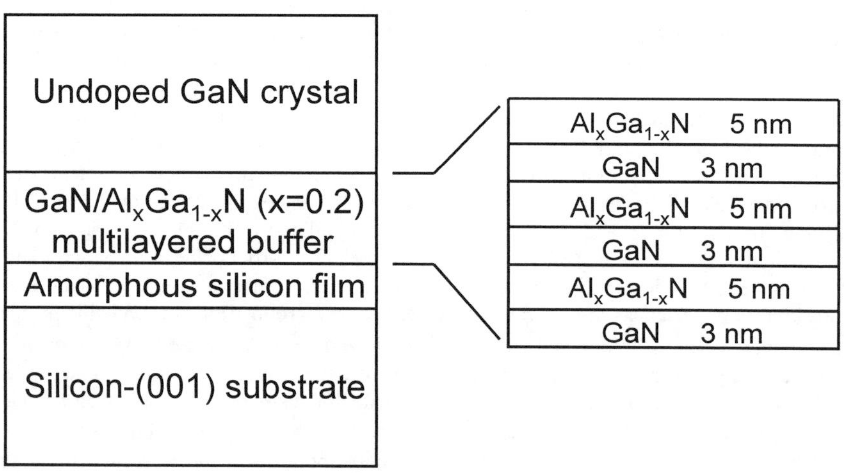

FIG. 1. Schematic sectional view showing an undoped GaN crystal grown over the composite intermediate layers (CILs) consisting of an ultra-thin (less than 5 nm) amorphous silicon film and a three period GaN/Al$_x$Ga$_{1-x}$N (x=0.2) multilayered buffer on a silicon-(001) substrate.

RESULTS AND DISCUSSION

As shown in Fig. 2, the PL intensity of the dominant emission peak around 3.4 eV which is attributed to the band-edge-related transition in wurtzite GaN epitaxial layer grown over the CILs on a silicon substrate (in solid line), is comparable or even slightly stronger than that for the undoped GaN sample grown over a conventional GaN single layer buffer on a sapphire substrate (in dotted line) under identical experimental configuration. The defect-related yellow-band emission centered at 2.35 eV is much weaker as compared with the dominant GaN band-edge emission. Although it is impossible for us to compare the PL intensity with the results reported by other research groups of the GaN samples grown over different intermediate layers on silicon substrates, [3-9] we can still evaluate some conclusions by focusing on the PL linewidth at room temperature, which is known to be nearly independent on the measurement conditions, such as the power density of the excitation laser beam. The full width at half maximum (FWHM) of the dominant GaN band-edge-related emission peak for the GaN/Si(001) sample shown in Fig. 2, for example, was measured to be approximately 40 meV. This value is 38 % narrower than the best result achieved so far for GaN film grown on a silicon substrate, 65 meV which was recently reported by Osinsky et al.[4], and quite comparable to that for the low-doped GaN film grown on a sapphire substrate.[12] This fact indicates that the crystal quality of GaN-based nitrides grown on silicon substrate can be significantly improved by using the CILs technique.

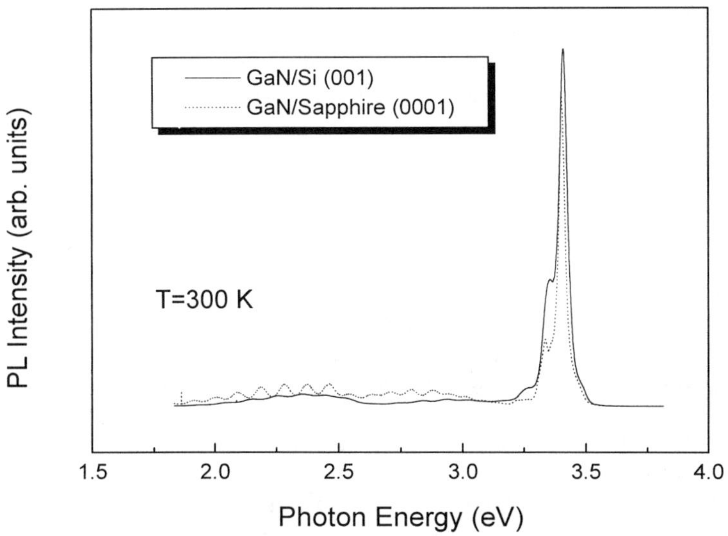

FIG. 2. The 300 K PL spectra for undoped GaN film grown over the CILs on a silicon substrate (solid line) and that grown over a conventional GaN buffer on a sapphire substrate (dotted line) under identical experimental configuration. The full width at half maximum (FWHM) of the band-edge-related emission peak at 3.4 eV is 40 meV which is 38 % narrower than the narrowest value reported, 65 meV, for an undoped GaN film grown on a silicon substrate.

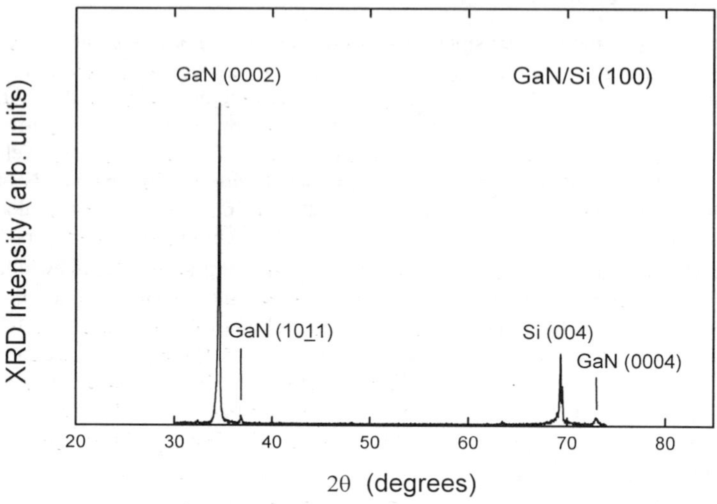

FIG. 3. X-ray diffraction profile and its rocking curve of the (0002) reflection for an undoped GaN film grown over the CILs on Si(001) substrate. The dominant diffraction peak at 34.6 arc-degrees with a FWHM of 40 arc-minutes is attributed to the (0002) diffraction of the wurtzite GaN, and is much stronger than the signal from the silicon substrate.

On the other hand, the XRD measurement result is illustrated in Fig. 3, showing a x-ray diffraction profile and its rocking curve of the (0002) reflection for an undoped GaN epitaxial layer grown over the CILs on a Si(001) substrate. Compared with the diffraction peak from the silicon substrate at 69.3 arc-degrees, Si(004), the dominant diffraction peak at 34.6 arc-degrees is much more intense, and identified as the (0002) diffraction from the wurtzite GaN crystal. The weak diffraction peak near 73 arc-degrees is attributed to the (0004) diffraction of the wurtzite GaN, and no diffraction peak from the zinc-blende (cubic) GaN is observed. The FWHM for the dominant GaN (0002) diffraction peak was measured to be as narrow as 11 arc-minutes which is much better than the corresponding value (108 arc-minutes) recently reported by Wang *et al.*[9] who employed a thin γ-Al$_2$O$_3$ film as the intermediate layer. These facts reveal once again that the crystal quality of GaN-based nitrides can be greatly improved by using the proposed CILs.

It is clear that as the temperature is raised from a relatively low temperature of 525 °C to a high temperature of 1,000 °C, both the ultra-thin amorphous silicon film and the amorphous or poly-crystalline GaN/Al$_x$Ga$_{1-x}$N MLB formed at 525 °C will completely or partially change to mono-crystalline or poly-crystalline due to the recrystallization. In fact, they serve as the seed crystals for the subsequent growth of the GaN-based nitrides. Compared with the conventional low-temperature-grown AlN intermediate layer, the CILs in this study turned out to be able to accommodate the strain arising from the lattice mismatch between the group-III nitrides and the silicon substrate, and to form the seed crystals more effectively. In other words, because the strain-accommodating and recrystallizing effects are of crucial importance in improving the crystal quality of the group-III nitrides, and they seem to be more profound in the CILs than in the conventional intermediate layers, the crystal quality of the group-III nitrides has been significantly improved by utilizing the CILs, as confirmed by the strong and sharp GaN-related PL and XRD signals described above.

Note that the optimal values in total layer thickness and solid composition for the CILs apparently depend on the selection of the constituent materials in the CILS as well as the subsequently grown group-III nitrides. At the present time, however, since the physical origin of the CILs is unfortunately not very clear, it is truly difficult to theoretically determine or predict the optimal layer thickness of CILs for a specific material combination. In other words, the optimal value for a specific material combination can now only be determined by experiment. However, the existence of the optimal layer thickness for the CILs can be interpreted qualitatively as follows. Generally an intermediate layer grown at a low temperature provides seed crystals which act as nucleation sites with low orientational fluctuation to promote the lateral growth of the group-III nitrides. The CILs in this study consisting of an ultra-thin amorphous silicon film and a MLB provide more seed crystals and interfaces than a conventional single intermediate or buffer layer to accommodate the strain and to terminate the misfit dislocations. However, if the CILs are too thin, they may neither provide sufficient amount of seed crystals necessary for the subsequent growth of the group-III nitrides nor effectively accommodate the strain and terminate the misfit dislocations. On the other hand, if the CILs are too thick, they tend to bring about excessive amount of the seed crystals with high orientational fluctuation. Therefore, there should be an optimal value for the total layer thickness of the CILs.

CONCLUSIONS

We have grown high quality GaN films on silicon substrates using a novel CILs technique by low-pressure MOCVD. The CILs consist of an ultra-thin (<5 nm) amorphous silicon film and a three-period $GaN/Al_xGa_{1-x}N$ (x=0.2) superlattice-like MLB. Based on the detailed study of PL and XRD spectroscopy, the GaN films grown over CILs on the Si(001) substrates were verified to be nearly pure wurtzite crystals, which gave comparable or even slightly stronger PL emission than those grown over a conventional GaN single layer buffer on a sapphire substrate under identical reactor configuration. Moreover the FWHM for the GaN band-edge-related PL peak is measured to be as narrow as 40 meV at room temperature, which is the narrowest value ever reported. This fact indicates that by using the proposed CILs technique, the crystalline quality of GaN-based nitride grown on a silicon substrate can be significantly improved.

REFERENCES

[1] S. Nakamura, M. Senoh, N. Iwasa, and S. Nagahama, Jpn. J. Appl. Phys. **34**, L797(1995).

[2] S. Nakamura, M. Senoh, S. Nagahama, N. Iwasa, T. Yamada, T. Matsushita, H. Kiyoku, Y. Sugimoto, T. Kozaki, H. Umemoto, M. Sano, and K. Chocho, Jpn. J. Appl. Phys. **37**, L309(1998).

[3] S. Guha and N. Bojarczuk, Appl. Phys. Lett. Vol. **72**, 415(1998).

[4] A. Osinsky, S. Gangopadhyay, J. W. Yang, R. Gaska, D. Kuksenkov, H. Temkin, I. K. Shmagin, Y. C. Chang, J. F. Muth, and R. M. Kolbas, Appl. Phys. Lett. **72**, 551(1998).

[5] A. J. Steckl, J. Devrajan, C. Tran, and R. A. Stall, Appl. Phys. Lett. **69**, 2264(1996).

[6] Y. Hiroyama and M. Tamura, Jpn. J. Appl. Phys. **37**, 630(1998).

[7] J. W. Yang, C. J. Sun, Q. Chen, M. Z. Anwar, M. A. Khan, S. A. Nikishin, G. A. Seryogin, A. V. Osinsky, L. Chernyak, H. Temkin, C. Hu, and S. Mahajan, Appl. Phys. Lett. **69**, 3566(1996).

[8] N. P. Kobayashi, J. T. Kobayashi, P. D. Dapkus, W. J. Choi, A. E. Bond, X. Zhang, and D. H. Rich, Appl. Phys. Lett. **71**, 3569 (1997).

[9] L. Wang, X. Liu, Y. Zan, J. Wang, D. Wang, D. Lu, and Z. Wang, Appl. Phys. Lett. **72**, 109(1998).

[10] X. Zhang, S. J. Chua, P. Li, K. B. Chong, and W. Wang, Appl. Phys. Lett. . **73**, 1772(1998).

[11] X. Zhang and S. J. Chua, Singapore Patent, 9801054-9 (1998).

[12] E. F. Schubert, I. D. Goepfert, W. Grieshaber, and J. M. Redwing, Appl. Phys. Lett. **71**, 921(1997).

PENDEO-EPITAXIAL GROWTH OF GAN ON SIC AND SILICON SUBSTRATES VIA METALORGANIC CHEMICAL VAPOR DEPOSITION

K.J. Linthicum*, T. Gehrke*, D. Thomson*, C.Ronning*, E.P. Carlson*, C.A. Zorman**, M. Mehregany**, and R.F.Davis*

*Department of Materials Science and Engineering, North Carolina State University, Raleigh, NC 27965
**Department of Electrical, Systems and Computer Engineering and Science, Case Western Reserve University, Cleveland, OH 44106

ABSTRACT

Pendeo-epitaxial lateral growth (PE) of GaN epilayers on (0001) 6H-silicon carbide and (111) Si substrates has been achieved. Growth on the latter substrate was accomplished through the use of a 3C-SiC transition layer. The coalesced PE GaN epilayers were characterized using scanning electron diffraction, x-ray diffraction and photoluminescence spectroscopy. The regions of lateral growth exhibited ~0.2° crystallographic tilt relative to the seed layer. The GaN seed and PE epilayers grown on the 3C-SiC/Si substrates exhibited comparable optical characteristics to the GaN seed and PE grown on 6H-SiC substrates. The near band-edge emission of the GaN/3C-SiC/Si seed was 3.450 eV (FWHM ~ 19 meV) and the GaN/6H-SiC seed was 3.466 eV (FWHM ~ 4 meV).

INTRODUCTION

Prior to the advent of the lateral growth techniques, heteroepitaxial gallium nitride films typically contained threading defect densities that exceeded $10^9/cm^2$. These defects seriously compromised the properties of subsequently fabricated devices. New growth techniques leading to low defect-density nitride films were required to achieve commercialization of viable optoelectronic and microelectronic devices.

The advancement of lateral epitaxial overgrowth (LEO) as a technique capable of producing GaN epilayers with defect densities reduced to $10^5/cm^2$ has recently been demonstrated by several research groups.[1,2,3,4] The significance of lateral overgrowth was immediate as Nichia Chemical reported in 1997 that use of LEO aided in increasing the lifetime of their blue laser diode from a few hundred hours to over 10,000 hrs,[5] and then subsequently introduced the first commercially available GaN blue laser diode.[6] However, to benefit from this reduction in defects, the placement of devices incorporating LEO technology is confined to regions on the final GaN device layer that are located over the masked regions and not over the window regions of the GaN seed layer. Figure 1 shows a scanning electron microscopy (SEM) cross-sectional image of GaN grown using the LEO technique.

The authors are currently investigating another lateral growth technique for GaN, namely, pendeo-epitaxy,[7,8,9,10,11] as a process route to resolve the aforementioned alignment problem. Pendeo-epitaxy (from the Latin: *pendeo*—to hang, or to be suspended) incorporates mechanisms of growth used by the conventional LEO process by using masks to prevent vertical propagation of threading defects, *and* extends the phenomenon to employ the substrate as a *pseudo-mask*. Pendeo-epitaxy (PE) differs from conventional LEO in that growth does not initiate through

Mat. Res. Soc. Symp. Proc. Vol. 572 ©1999 Materials Research Society

open windows but begins on sidewalls etched into the GaN seed layer. As the lateral growth from the sidewalls continues, vertical growth of GaN begins and results in the eventual lateral overgrowth of the masked seed form. Pendeo-epitaxial growth ultimately results in coalescence over and between seed forms, producing a continuous layer of GaN, as shown in Figure 2.

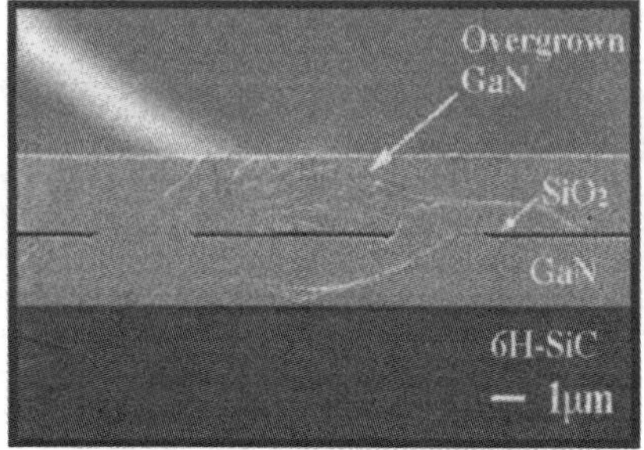

<table>
<tr><td>

Figure 1. Cross-sectional SEM micrograph of LEO GaN on SiC.

</td><td>

Figure 2. Cross-sectional SEM micrograph of PE GaN on SiC.

</td></tr>
</table>

Although both LEO and PE research have led to low defect-density GaN material on two-inch SiC and sapphire substrates, the use of this substrate does not resolve the problem of achieving low-cost, large area GaN films necessary for commercialization of microelectronic devices. Meeting this objective has renewed interest in using Si(111) substrates as an alternative to SiC and sapphire. Several recent demonstrations of lateral growth process routes achieving growth of (0001)GaN on (111)Si have been reported.[8,12,13] In this paper we report on the process steps for and characterization of PE GaN epilayers grown on both (0001) 6H-SiC and (111) Si substrates.

EXPERIMENTAL PROCEDURES

GaN Seed Layers

(0001) 6H-SiC on-axis substrates: The initial 500nm thick GaN seed layers were grown on 100nm thick high-temperature AlN buffer layers previously deposited on the substrates via metalorganic vapor phase epitaxy (MOVPE), as detailed in Ref. 14.

(111)Silicon substrates: 500nm thick (111)3C-SiC transition layers were initially grown on very thin (111)3C-SiC layers produced by conversion of Si substrate surfaces via reaction with C_3H_8 entrained in H_2. Both the conversion step and SiC film deposition were achieved using atmospheric pressure chemical vapor deposition (APCVD), as detailed in Ref. 15. The 500nm thick GaN seed layers were subsequently deposited on 100 nm thick AlN buffer layers in the manner used for the 6H-SiC substrates as noted above.

Pendeo-epitaxial growth of GaN

A 100 nm silicon nitride growth mask was deposited on the seed layers via plasma enhanced CVD. A 150 nm nickel etch mask was subsequently deposited using e-beam evaporation. Patterning of the nickel mask layer was achieved using standard photolithography techniques and Ar-plasma sputtering. The final, tailored microstructure consisting of silicon nitride masked GaN seed forms was fabricated via inductively coupled plasma (ICP) etching of portions of the silicon nitride growth mask, the GaN seed layer and the AlN buffer layer. The seed-forms used for this study were raised rectangular stripes oriented along the $<1\overline{1}00>$ direction. Various seed form widths and separation distances were employed. Pendeo-epitaxial growth of GaN was achieved within the temperature range of 1050-1100°C and a total pressure of 45 Torr. The precursors (flow rates) of triethylgallium (26.1 μmol/min) and NH_3 (1500 sccm) were used in combination with a H_2 diluent (3000 sccm). Additional experimental details regarding the pendeo-epitaxial growth of GaN and $Al_xGa_{1-x}N$ layers employing 6H-SiC substrates are given in Refs. 7-11.

The morphology and defect microstructures have been investigated using scanning electron microscopy (SEM) (JEOL 6400 FE), transmission electron microscopy (TEM) (TOPCON 0002B, 200KV) and X-ray diffraction (XRD) (Philips X'Pert MRD X-ray diffractometer). Optical characterization was performed using a He-Cd laser (λ=325 nm).

RESULTS AND DISCUSSION

GaN Seed Layers

The pendeo-epitaxial phenomenon is made possible by using growth and mask mechanisms similar to conventional LEO techniques and by using the substrate itself as a *pseudo*-mask, i.e. the GaN does not nucleate on the exposed SiC surface when higher growth temperatures are employed to enhance lateral growth. The Ga- and N-containing species more likely either diffuse along the surface or evaporate (rather than having sufficient time to form GaN nuclei) from this substrate. Since the newly deposited laterally grown GaN is suspended above the SiC substrate, there are no defects associated with the mismatches in lattice parameters between the PE GaN and the SiC substrate.

Pendeo-epitaxial techniques can be applied in general to other substrates as long as they, or a transition layer deposited on the substrate, act similarly as a *pseudo*-mask. As demonstrated above, silicon carbide meets this requirement. Therefore, by using a transition layer of 3C-SiC, silicon can be successfully used as a substrate for PE growth. In addition to acting as a *pseudo*-mask, the 3C-SiC performs two other key functions. Firstly, deposition on (111)Si results in the growth of (111)3C-SiC. The atomic arrangement of the (111) plane is equivalent to the (0001) plane of 6H-SiC; thus it facilitates the sequential deposition of a high temperature (0001) 2H-AlN buffer layer of sufficient quality needed for the GaN seed layer. Secondly, since the AlN buffer does not act as a *pseudo*-mask, it must also be removed from the areas between the GaN seed forms to prevent the undesired nucleation of 'defective' GaN in these areas. Without the presence of an AlN buffer or other transition layer, the silicon substrate is exposed to the growth environment and consequent effects resulting from the reaction of Si atoms with the Ga and N species. Therefore the 3C-SiC transition layer acts as a reaction/diffusion barrier between the Si substrate and the GaN.

Figure 3 shows a SEM micrograph of a 500 nm (111) 3C-SiC transition layer deposited on an on-axis (111)Si substrate. A typical high resolution DCXRD scan is shown in Figure 4. Although the quality of the 3C transition layer is less than optimal, as indicated from the relatively rough morphology visible in Figure 3, and typical XRD FWHM values of 0.5°, it is of sufficient quality for the deposition of 2H-AlN and 2H-GaN seed layers.

The surface morphology of the GaN seed layers deposited on the (111) 3C-SiC transition layers were comparable to GaN seed layers deposited on (0001) 6H-SiC substrates, and had very smooth surfaces. Low temperature (14K) photoluminescence analysis of GaN seed layers grown on both 6H-SiC and 3C-SiC/Si substrates are shown in Figure 5.

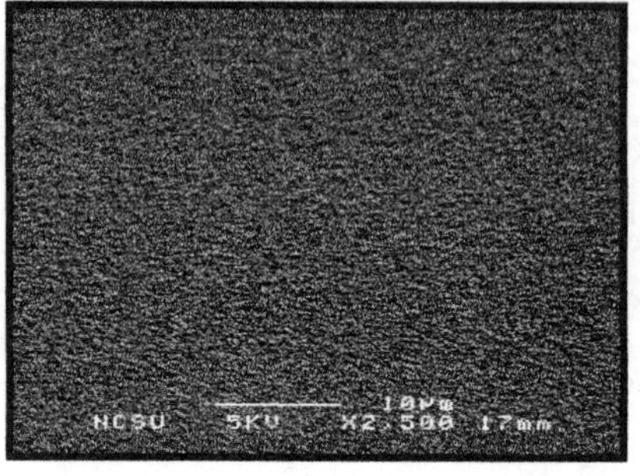

<table>
<tr><td>

Figure 3. Plane-view SEM
of a 0.5 micron thick 3C-SiC
transition layer on (111)Si.

</td><td>

Figure 4. X-ray rocking curve of
the 111 diffraction of a 0.5 micron
thick 3C-SiC transition layer.

</td></tr>
</table>

The near band-edge emission was 357.7 nm (3.466 eV, FWHM ~ 4meV) and 359.4 nm (3.450 eV, FWHM ~ 19 meV), respectively, and has been attributed to an exciton bound to a neutral donor ($X\text{-}D^0$). These results show that the quality of the GaN seed layers deposited on 3C-SiC/Si substrates is approaching the optical quality of GaN grown on 6H-SiC substrates.

Pendeo-Epitaxial Growth:
Figure 2 shows an SEM micrograph of a PE grown GaN using a 6H-SiC substrate with a 1 micron thick GaN seed layer and a 100nm thick silicon nitride mask. The seed microstructure was comprised of 3 µm wide posts and 1.5 µm wide trenches oriented in the $<1\bar{1}00>$ direction. Analysis of Figure 2 reveals coalescence between and above the seed structures resulted forming a continuous epilayer of GaN.

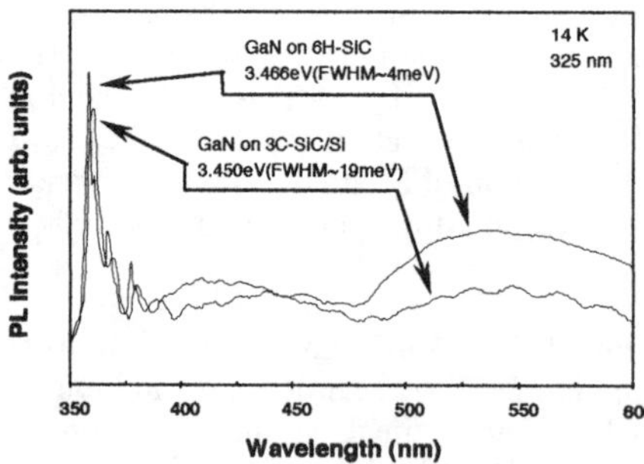

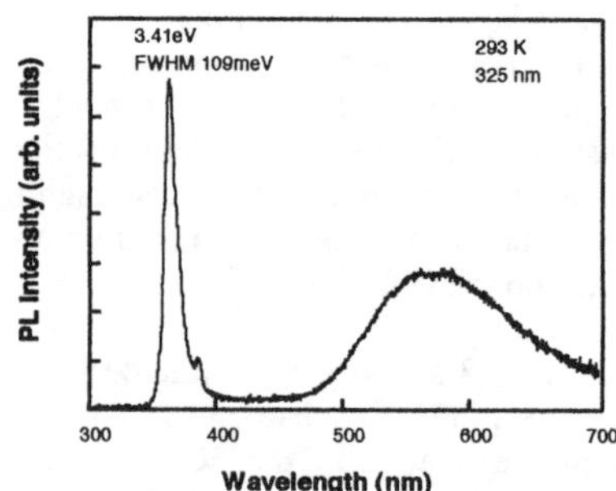

<table>
<tr><td>

Figure 5. Low temperature (14 K)
PL spectra of GaN seed layers grown
on 6H-SiC and 3C-SiC/Si substrates.

</td><td>

Figure 6. Room temperature
PL spectrum of coalesced PE
GaN on a 3C-SiC/Si substrate.

</td></tr>
</table>

Figure 6 shows the room temperature PL spectrum for a coalesced GaN PE epilayer grown on a 3C-SiC/Si substrate. A band-edge emission of 363.6 nm (3.41 eV) was observed and indicates the PE GaN on silicon is under tensile stress.

Figures 7 and 8 show SEM micrographs of PE grown GaN using 3C-SiC/Si substrates. The seed microstructures were 2 μm wide posts and 3 μm wide trenches oriented in the $<1\bar{1}00>$ direction. The sample shown in Figure 7 employed a 200 nm silicon nitride mask on top of the seed forms. Analysis of this sample revealed void formation and poor coalescence over the 'thick' seed mask. Void formation in the trench regions is also visible; however this does not appear to effect the quality of the coalesced GaN above the trench. This is typical of PE GaN grown under nonoptimized conditions. The sample in Figure 8 is a maskless PE GaN epilayer. PE GaN grown using this seed microstructure configuration is the equivalent of LEO techniques, such that vertical propagation of threading defects from the original GaN seed is not prevented. Analysis of Figure 8 reveals the same void formation at the coalescence point in the trench regions.

Although the cause of these voids is not yet fully understood, it is believed that optimization of the initial seed microstructure, including post-trench ratio, seed thickness, etch quality and mask thickness, result in their elimination. Preliminary results suggest that minimization of the trench width reduces the void formation in the trench region. Also, minimizing the seed mask thickness helps to eliminate void formation in and poor coalescence of GaN grown over the mask. Research optimizing all of the process and fabrication parameters is ongoing.

Figure 9 shows a typical X-ray rocking curve for a PE GaN sample grown on a 3C-SiC/Si substrate. Similar to the case of GaN grown using LEO techniques, there is ~ 0.2° tilting of the laterally grown GaN compared to that of the seed layer, in the $<11\bar{2}0>$ direction (i.e. perpendicular to the seed form orientation). However, unlike the LEO technique, it is difficult to determine if the PE material, the LEO material, or both are tilted since there is both pendeo-epitaxial growth between and LEO growth above the original seed forms. Work is in progress to make this determination.

Figure 7. 45° SEM view of PE GaN grown on a 3C-SiC/Si substrate.

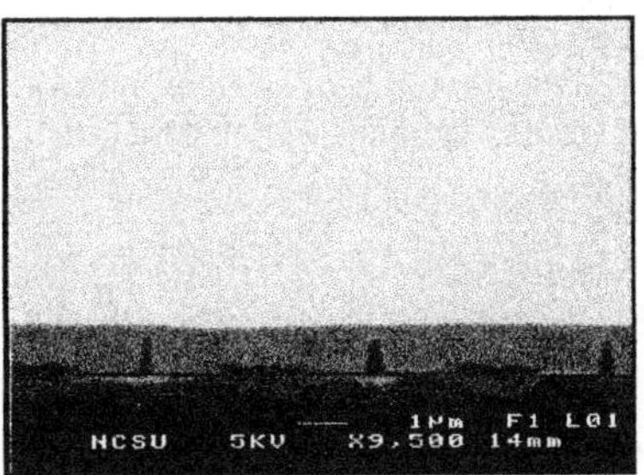

Figure 8. 45° SEM view of maskless PE GaN grown on a 3C-SiC/Si substrate.

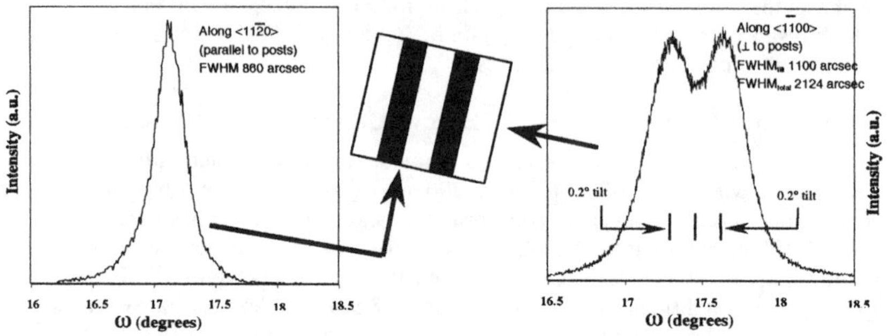

Figure 9. DCXRD rocking curves of the 0002 diffraction of PE GaN grown on 3C-SiC/Si substrates.

CONCLUSIONS

The pendeo-epitaxy process route has been developed as an alternative to the conventional GaN LEO technique and as a means of confining all the vertically threading defects, stemming from the GaN/AlN and AlN/SiC interfaces, within the seed forms. This results in the growth of a more uniform low defect-density GaN epilayer. Incorporation of silicon nitride masks and SiC *pseudo*-masks (either as the 6H-SiC substrate or a 3C-SiC transition layer) combined with etched sidewalls of GaN seed forms has allowed the achievement of PE GaN films with low dislocation densities over the entire GaN epilayer surface. The quality of GaN seed layers grown on 3C-SiC/Si substrates was shown to be comparable to GaN layers grown on 6H-SiC. Investigations regarding the optimization of the PE growth technique, including determination of the ideal GaN seed thickness, ideal seed microstructure geometry (e.g. post width, trench width, etc.) and ideal seed mask material, is ongoing. Additionally, optimization of fabrication steps for the PE GaN seed forms (e.g., photolithography, mask alignment, ICP etching process, etc.) is underway.

ACKNOWLEDGEMENTS

The authors acknowledge Cree Research, Inc. and Motorola for the SiC and the silicon wafers respectively. This work was supported by the Office of Naval Research under contracts N00014-96-1-0765 (Colin Wood, monitor) and N00014-98-1-0654 (John Zolper, monitor).

REFERENCES

[1] O. Nam, T. Zheleva, M. Bremser, and R. Davis, Appl. Phys. Lett., **71**, 2638 (1997).

[2] A. Sakai, H. Sunakawa, and A. Usui, Appl. Phys. Lett., **73**, 481 (1998).

[3] H. Marchand, X. Wu, J. Ibbetson, P. Fini, P. Kozodoy, S. Keller, J. Speck, S. Denbaars, and U. Mishra, Appl. Phys. Lett., **73**, 747 (1998).

[4] H. Zhong, M. Johnson, T. McNulty, J. Brown, J. Cook Jr., J. Schezina, *Materials Internet Journal, Nitride Semiconductor Research, 3*, 6, (1998).

[5] S. Nakamura, M.Senoh, S. Nagahama, N. Iwasa, T. Yamanda, T. Matsushita, H. Kiyoku, Y. Sugimoto, T. Kozaki, H. Umemoto, M. Sano, and K. Chocho, Proc. of the 2nd Int. Conf. On Nitride Semicond., Tokushima, Japan, October, 1997

[6] MRS Internet Journal of Nitride Semiconductor Research, January 13, 1999

[7] K. J. Linthicum, T.Gehrke, D.B. Thomson, E.P. Carlson, P. Rajagopal, T. Smith, and R.F.Davis, (submitted to Applied Physics Letters)

[8] K. J. Linthicum, T. Gehrke, D.B. Thomson, K.M. Tracy, E.P. Carlson, T,P.Smith, T.S. Smith, T.S. Zheleva, C.A. Zorman, M. Mehregany, and R.F.Davis, *MRS Internet J. Nitride Semicond. Res 4S1*, G4.9 (1999).

[9] T. Gehrke, K. J. Linthicum, D.B.Thomson, P. Rajagopal, A. D. Batchelor and R. F. Davis, *MRS Internet J. Nitride Semicond. Res 4S1*, G3.2 (1999).

[10] D.B.Thomson, T. Gehrke, K. J. Linthicum, P. Rajagopal, P. Hartlieb, T. S. Zheleva and R. F. Davis, *MRS Internet J. Nitride Semicond. Res 4S1*, G3.37 (1999).

[11] T. S. Zheleva, D.B.Thomson, S. Smith, P. Rajagopal, K. J. Linthicum, T. Gehrke, and R. F. Davis, *MRS Internet J. Nitride Semicond. Res 4S1*, G3.38 (1999).

[12] H.Marchland, N. Zang, L. Zhao, Y. Golan, S.J. Rosner, G. Girolami, P.T. Fini, J.P. Ibbetson, S. Keller, S.Denbaars, J. Speck, and U.K. Mishra, *MRS Internet J. Nitride Semicond. Res 4*, 2 (1999).

[13] P. Kung, D. Walker, M. Hamilton, J. Diaz, and M. Razeghi, Appl. Phys. Lett., **74**, 570 (1998).

[14] T. Weeks, M. Bremser, K. Ailey, E. Carlson, W. Perry, and R. Davis, Appl. Phys. Lett., **67**, 401 (1995).

[15] C. Zorman, A. Fleischman, A. Dewa, M. Mehregany, C. Jacob, S. Nishino, and P. Pirouz, J. Appl. Phys., **78**, 5136 (1995).

MASKLESS LATERAL EPITAXIAL OVERGROWTH
OF GaN ON SAPPHIRE

P. FINI*, H. MARCHAND**, J.P. IBBETSON**, B. MORAN*, L. ZHAO*,
S.P. DENBAARS*, J.S. SPECK*, U.K. MISHRA**
*Materials Dept., Univ. of California, Santa Barbara; Santa Barbara, CA 93106
**ECE Dept., Univ. of California, Santa Barbara; Santa Barbara, CA 93106

ABSTRACT

We demonstrate a technique of lateral epitaxial overgrowth (LEO) of GaN, termed 'maskless' LEO, in which no mask is deposited prior to LEO regrowth. Instead, a bulk (> 2 μm) GaN layer on sapphire is selectively dry etched, leaving ~5μm-wide stripe mesas oriented in the $<10\bar{1}0>_{GaN}$ direction, with a 20 μm period. These stripes serve as seeds for LEO GaN growth, which proceeds from the tops of the stripes and expands laterally, resulting in a 'T', or overhang, morphology. As for LEO over an SiO_2 mask, significant defect reduction (from ~10^9 cm^{-2} to ~10^6 cm^{-2}) is observed in cross-sectional transmission electron microscopy (TEM). Atomic force microscopy of the top surface of the LEO GaN reveals that no threading dislocations with screw component terminate at the surfaces of laterally overgrown regions. X-ray diffraction measurements reveal that the wings exhibit a crystallographic tilt away from the seed regions in an azimuth perpendicular to the stripe direction; the tilt angle (~0.4 − 0.5°) is relatively independent of growth temperature and wing aspect ratio.

INTRODUCTION

The technique of lateral epitaxial overgrowth (LEO) involves regrowth on a selectively masked epilayer, such that the regrown material emerges from unmasked areas and overgrows the mask. Through blocking and/or redirection of threading dislocations, the overgrown material has a lower residual defect density than the seed material. Significant extended defect density reduction has been demonstrated in GaN films grown by metalorganic chemical vapor deposition (MOCVD) and hydride vapor phase epitaxy (HVPE) on sapphire,[1-8] SiC,[9-11] and Si(111).[5] Reductions in threading dislocation density (TDD) have led to direct benefits in device performance such as long-lifetime C-W blue lasers,[12-14] low reverse-bias leakage current LED's[15,16] and p-n junctions,[17] low gate leakage current $Al_xGa_{1-x}N/GaN$ FET's,[18] and low dark current, sharp cutoff $Al_xGa_{1-x}N$-based solar-blind photodetectors.[19]

Much of the recent work in the LEO of GaN has utilized dielectric masks such as SiO_2 and SiN_x, patterned with periodic arrays of stripes or holes. The use of SiO_2 in particular presents difficulties in controlling the background carrier concentration of the LEO GaN, since outdiffusion of Si and O may occur at the elevated temperatures of regrowth (*e.g.*, 1050°C). Such autodoping of the LEO GaN limits its usefulness for devices which require an insulating base layer, such as field effect transistors. In the effort to obtain semi-insulating LEO GaN, one approach is to eliminate the mask altogether, while still benefitting from extended defect reduction. The present technique, termed 'maskless' LEO, has the potential of achieving this by regrowing on GaN stripes that are separated by the sapphire upon which the seed layer was grown. It is expected that the chemical and physical stability of the sapphire as well as the regrowth morphology will lead to LEO GaN with lower unintentional impurity incorporation. Thus maskless LEO has potential advantages over related techniques that yield similar morphologies but use a dielectric mask.[20] In this paper we present the structural and morphological characterization of maskless LEO GaN grown under various conditions.

Mat. Res. Soc. Symp. Proc. Vol. 572 © 1999 Materials Research Society

EXPERIMENT

Bulk (>2µm) GaN layers were grown on sapphire using a standard two-step process.[21] These layers were coated with photoresist (PR) and patterned using standard UV photolithography, such that 5 µm wide PR stripes were formed, with 20 µm period and oriented in the $<10\bar{1}0>_{GaN}$ direction. Samples were etched using Cl_2 reactive ion etching, resulting in all GaN being removed except the material beneath the PR stripes. Etching in this manner resulted in stripes with some etch damage on the upper surfaces, and occasionally areas between stripes where material ('grass') remained on the sapphire. Examples of both of these features are shown in Fig. 1. It has been found in the present study that the selectivity of the LEO growth is not affected by the features in Fig. 1(b).

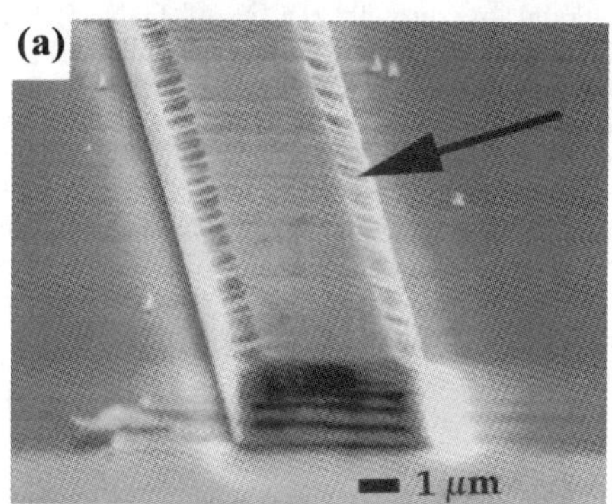

Figure 1. Scanning electron micrographs of as-processed GaN stripes, showing (a) etch damage on upper corners of stripe; (b) 'grass', or post-etch residual material.

The LEO GaN was deposited at 76 Torr in a H_2 carrier, with an NH_3 flow fixed at 1.77 slpm. Two trimethylgallium (TMG) flows (52 and 104 µmol/min) and three surface temperatures (1015°C, 1060°C, and 1100°C) were studied. The approximate surface temperature was calibrated using an optical pyrometer. Growth durations of 15 min and 30 min were used for f_{TMG} = 104 µmol/min and 52 µmol/min, respectively, to supply equimolar amounts of Ga to both series of samples. Uncoated samples were characterized in cross section by scanning electron microscopy (SEM) using a JEOL 6300F field emission microscope operating at 15 kV. Specimens for transmission electron microscopy (TEM) were prepared by wedge polishing followed by standard Ar^+ ion milling and images were recorded on a JEOL 2000FX microscope operated at 200 kV. Atomic force microscopy (AFM) of LEO GaN surface morphology was carried out on a Digital Instruments D3000 scanning probe microscope. X-ray diffraction was performed using four-bounce Ge(220)-monochromated Cu-Kα radiation in a four-circle diffractometer operating in receiving slit mode, with a 1.0 mm slit on the detector arm. Specifically, rocking curves of the GaN 0002 peak were measured, with the scattering plane perpendicular to the stripe direction. In this orientation, the rotation (rocking) axis is parallel to the stripe direction.

RESULTS

The maskless LEO stripe morphologies that result from 15 min of growth at a TMG flow of 104 µmol/min (nominal V/III = 750) are shown in Fig. 2. At the lowest temperature, the LEO GaN is bound by the (0001) basal plane and both inclined $\{11\bar{2}n\}$ (where n ≈ 2) and vertical $\{11\bar{2}0\}$ sidewalls. As the growth temperature is increased, the $\{11\bar{2}0\}$ sidewalls dominate and the lateral growth rate increases, which has been observed previously on LEO over an SiO_2

mask.[7] At all three temperatures studied, the LEO GaN grows vertically and laterally from the upper part of the 'seed' GaN stripe, and not from the base of the $\{11\bar{2}0\}$ seed sidewalls. This is believed to be due to etch damage such as shown in Fig 1(a), which would cause the LEO GaN to start growing on the top surfaces of the stripes, and then expand laterally. Subsequently, downward expansion of the LEO 'wings' also occurs, the extent of which increases with temperature. This implies that the growth rate of not only the $\{11\bar{2}0\}$ sidewalls but also the $(000\bar{1})$ plane increases with temperature.

Figure 2. SEM micrographs of as-cleaved LEO stripe cross sections, grown for 15 min with 104 μmol/min TMG at (a) 1015°C, (b) 1060°C, and (c) 1100°C. Bar length is 1 μm.

When the TMG flow is set at 52 μmol/min (nominal V/III = 1500), the LEO stripe morphologies shown in Fig. 3 result. The same temperature dependence of facet formation as seen above is observed, with the inclined $\{11\bar{2}n\}$ facets disappearing at higher temperatures. However, since the V/III ratio is higher in this case, the extent of lateral growth is also higher, as observed for LEO over an SiO_2 mask.[22] At the highest temperature, the LEO wings have expanded downward to such an extent that they are in contact with the sapphire substrate. This morphology is facilitated by the enhanced growth rates of the $(000\bar{1})$ and $\{11\bar{2}0\}$ facets, which could be directly dependent on the NH_3 partial pressure. If it assumed that the decomposition of NH_3 is incomplete in the temperature range studied and the active nitrogen species have a short residence time on the growing facets, then the 'local' V/III ratio in the immediate vicinity of the stripe increases with increasing temperature. The relation between the ideal surface atomic structures of the facets and their stability is discussed in more detail elsewhere.[22]

Figure 3. SEM micrographs of as-cleaved LEO stripe cross sections, grown for 30 min with 52 μmol/min TMG at (a) 1015°C, (b) 1060°C, and (c) 1100°C. Bar length is 1 μm.

Although the LEO stripes shown in Figs. 2 and 3 have (initially) suspended wings, it has also been observed in the present study that the wings may be contact with the sapphire from the onset of regrowth. This growth mode affects the dislocation density in the wings, as discussed below. We believe that variations in the degree of etch damage on the seed sidewalls may alter the manner in which LEO GaN initially grows on the seed.

Cross-sectional TEM micrographs of the stripes in Figures 2(a) and (c) are shown in Fig. 4. The bright-field image in Fig. 4(a) reveals that some dislocations with edge component have been redirected into the wing, which is expected when $\{11\bar{2}n\}$ facets are present. The weak-beam images in Figures 4(b) and (c) reveal that the dislocations with screw and edge component, respectively, have significantly lower density in the LEO wings than the seed regions. However, the small wing volume in these samples precludes accurate estimation of the TDD at this time.

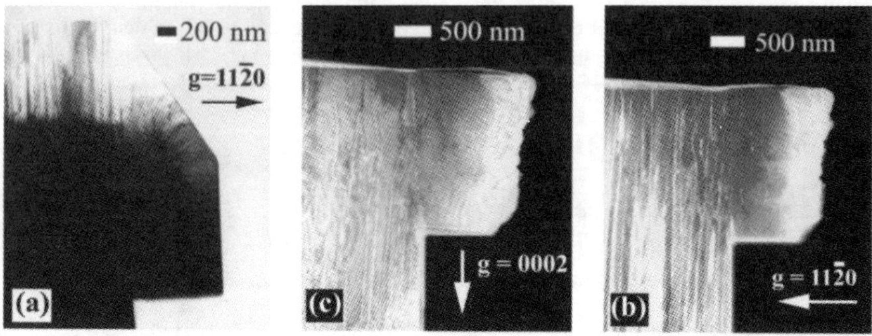

Figure 4. TEM micrographs of LEO stripes grown with 104μmol/min TMG: (a) bright-field cross-section; (b),(c) weak-beam images with two different diffraction conditions.

When the LEO wings are in contact with the sapphire while expanding laterally, an overall reduction in defect density to ~10^6 cm^{-2} is observed, as shown in Fig. 5(a). However, vertical arrays of edge dislocations with line directions parallel to the stripe and Burgers vector $\mathbf{b}$ = 1/3<11$\bar{2}$0> are visible at various locations in the wings. These vertical arrays, highlighted by arrows, would be expected to give rise to crystallographic tilt, since they are essentially low-angle tilt boundaries. Additionally, in Fig. 5(b) mixed-character threading dislocations generated at the interface between the wing and the sapphire are highlighted with arrows.

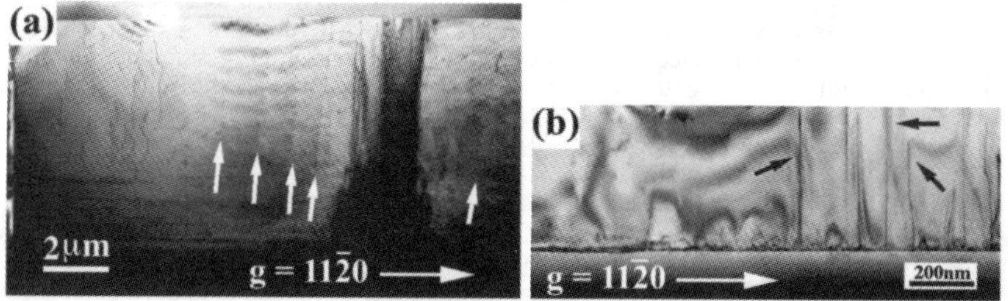

Figure 5. Bright field TEM micrographs of (a) seed region and LEO wing in contact with sapphire (arrows highlight vertical edge dislocation arrays), (b) close-up of LEO wing / sapphire interface, with TDs highlighted by arrows.

The threading dislocation density reduction observed in cross-sectional TEM is also evident in atomic force microscopy (AFM) micrographs of the top surfaces of the stripes, an example of which is shown in Fig. 6. The right end of the micrograph is the LEO GaN located directly above the seed, whereas the remainder is the wing region. The absence of step terminations in this area indicates that there are no threading dislocations with a screw component emerging at the surface, since their intersection with a free surface necessitates termination of two steps or a single paired step.[4] Step pairing as well as preferential orientation step orientation (dictated by the in-plane symmetry of the wurtzitic crystal structure) is seen in the wing regions, as they are unconstrained by step terminations.

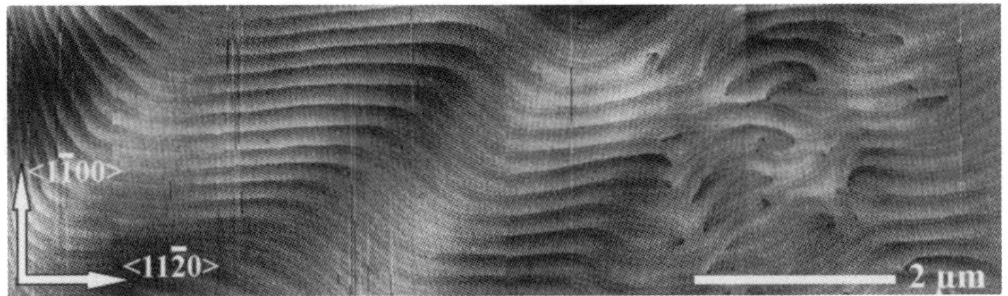

Figure 6. Montage of AFM images for a stripe in contact with sapphire, showing 'wing' (overgrown) region on left, and seed region on right. The total z-range is 2 nm. Note that vertical lines are scan artifacts.

The x-ray diffraction rocking curves of the GaN 0002 peak in Fig. 7(a) reveal that for the f(TMG) = 104 μmol/min series, the LEO wings have a crystallographic tilt with respect to the seed regions. This tilt, ~0.4-0.5° for all three growth temperatures, is only observed in an azimuth perpendicular to the stripe direction; no tilt is discernable along the $<10\bar{1}0>_{GaN}$ stripe direction. The same is true for LEO GaN stripes grown at f(TMG) = 52 μmol/min, which have tilts in the same range, as shown in Fig. 7(b). The fact that tilt is independent of aspect ratio (which depends on temperature and V/III ratio) for these suspended morphologies may indicate that it emerges at the onset of LEO. In contrast, LEO wings which grow in contact with an SiO_2 mask exhibit tilts that increase progressively with extent of lateral overgrowth.[23]

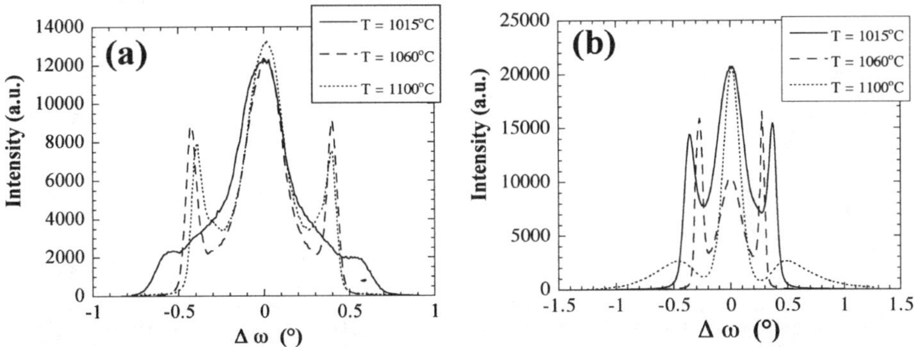

Figure 7. X-ray diffraction rocking curves of LEO stripes grown with (a) 104 μmol/min for 15 min and (b) 52 μmol/min for 30 min.

CONCLUSIONS

We have demonstrated a novel GaN lateral overgrowth technique, termed maskless LEO, in which selective regrowth is achieved without the need for mask deposition. For the range of growth conditions studied, morphologies with suspended wing (LEO GaN) regions largely result. It is believed that RIE etch damage on the seed GaN stripes is the cause of this 'T' morphology. TDD reduction has been observed in both TEM and AFM of LEO wing regions. The wings exhibit crystallographic tilts away from the seed region, which are relatively independent of wing aspect ratio. The origin of this tilt is the subject of ongoing study, as it is important in determining the nature of the merge front when two wings coalesce.

ACKNOWLEDGMENTS

The authors would like to acknowledge funding from the Office of Naval Research (ONR), monitored by C. Wood, and the Air Force Office of Scientfic Research (AFOSR), monitored by G. Witt. PF acknowledges support from a NDSEG Fellowship, sponsored in part by the Office of Naval Research. HM acknowledges support from the NSERC (Canada) and a Raychem Fellowship.

REFERENCES

1. A. Usui, H. Sunakawa, A. Sakai, and A. A. Yamaguchi, Jpn. J. Appl. Phys., Pt. 2 **36** (7B), L899-902 (1997).
2. A. Sakai, H. Sunakawa, and A. Usui, Appl. Phys. Lett. **71** (16), 2259-61 (1997).
3. A. Kimura, C. Sasaoka, A. Sakai, and A. Usui, presented at MRS Nitride Semicond. Symp., Boston, MA, USA, 1997 (unpublished).
4. H. Marchand, J.P. Ibbetson, P. Fini, P. Kozodoy, S. Keller, J.S. Speck, S.P. DenBaars, and U.K. Mishra, MRS Internet J. Nitride Semicond. Res. **3** (3) (1998).
5. H. Marchand, N. Zhang, L. Zhao, Y. Golan, P. Fini, J.P. Ibbetson, S. Keller, S.P. DenBaars, J.S. Speck, and U.K. Mishra, presented at 25th Int. Symp. Comp. Semicond., Nara, Japan, 1998 (unpublished).
6. A. Sakai, H. Sunakawa, and A. Usui, Appl. Phys. Lett. **73** (4), 481-3 (1998).
7. H. Marchand, J.P. Ibbetson, P. Fini, S. Chichibu, S.J. Rosner, S. Keller, S.P. DenBaars, J.S. Speck, and U.K. Mishra, presented at 25th Int. Symp. Comp. Semicond., Nara, Japan, 1998 (unpublished).
8. H. Marchand, J.P. Ibbetson, P. Fini, X.H. Wu, S. Keller, S.P. DenBaars, J.S. Speck, and U.K. Mishra, MRS Internet J. Nitride Semicond. Res. **G 4.5** (G4.5) (1998).
9. T. S. Zheleva, Nam Ok-Hyun, M. D. Bremser, and R. F. Davis, Appl. Phys. Lett. **71** (17), 2472-4 (1997).
10. Ok-Hyun Nam, M. D. Bremser, T. S. Zheleva, and R. F. Davis, Applied Physics Letters **71** (18), 2638-40 (1997).
11. J. A. Freitas, Jr., O.-H. Nam, T. S. Zheleva, and R. F. Davis, J. Cryst. Growth **189-190**, 92-6 (1998).
12. S. Nakamura, M. Senoh, S. I. Nagahama, N. Iwasa, T. Yamada, T. Matsushita, H. Kiyoku, Y. Sugimoto, T. Kozaki, H. Umemoto, M. Sano, and K. Chocho, J. Cryst. Growth **189/190**, 820-5 (1998).
13. S. Nakamura, M. Senoh, S. Nagahama, N. Iwasa, T. Yamada, T. Matsushita, H. Kiyoku, Y. Sugimoto, T. Kozaki, H. Uimemoto, M. Sang, and K. Chocho, Japn. J. Appl. Phys. Lett. **37** (3B), L309-12 (1998).
14. S. Nakamura, M. Senoh, S. I. Nagahama, N. Iwasa, T. Yamada, T. Matsushita, H. Kiyoku, Y. Sugimoto, T. Kozaki, H. Umemoto, M. Sano, and K. Chocho, Appl. Phys. Lett. **72** (2), 211-13 (1998).
15. T. Mukai, K. Takekawa, and S. Nakamura, Jpn. J. Appl. Phys. **37** (7B), L839-41 (1998).
16. C. Sasaoka, H. Sunakawa, A. Kimura, M. Nido, A. Usui, and A. Sakai, J. Cryst. Growth **189/190**, 61-6 (1998).
17. P. Kozodoy, J.P. Ibbetson, H. Marchand, P.T. Fini, S. Keller, J.S. Speck, S.P. DenBaars, and U.K. Mishra, Appl. Phys. Lett. **73** (7), 975-7 (1998).
18. R. Vetury, H. Marchand, J.P. Ibbetson, P. Fini, S. Keller, J.S. Speck, S.P. Denbaars, and U.K. Mishra, presented at 25th Int. Symp. on Compound Semicond., Nara, Japan, 1998 (unpublished).
19. G. Parish, S. Keller, P. Kozodoy, J.P. Ibbetson, H. Marchand, P. Fini, S.B. Fleischer, S.P. DenBaars, and U.K. Mishra, presented at 1998 Conference on Optoelectronic and Microelectronic Materials And Devices, University of Western Australia, Perth, Australia, 1998 (unpublished).
20. T. Gehrke, K.J. Linthicum, D.B. Thomson, P. Rajagopal, A.D. Batchelor, and R.F. Davis, MRS Internet J. Nitride Semicond. Res. **4S1**, G3.2 (1999).
21. X. H. Wu, P. Fini, S. Keller, E. J. Tarsa, B. Heying, U. K. Mishra, S. P. DenBaars, and J. S. Speck, *Japn. J. Appl. Phys.* **35** (12B), L1648 (1996).
22. H. Marchand, J.P. Ibbetson, P.T. Fini, S. Keller, S.P. DenBaars, J.S. Speck, and U.K. Mishra, J. Cryst. Growth **195**, 328-332 (1998).
23. P. Fini, J.P. Ibbetson, H. Marchand, L. Zhao, S.P. DenBaars, and J.S. Speck, unpublished (1999).

REPRODUCIBILITY AND UNIFORMITY OF MOVPE PLANETARY REACTORS® FOR THE GROWTH OF GaN BASED MATERIALS

M. Heuken*, H. Protzmann*, O. Schoen*, M. Luenenbuerger*, H. Juergensen*, M. Bremser**, E. Woelk**

*AIXTRON AG, Kackertstr. 15 - 17, D-52072 Aachen, Germany, mail@aixtron.com

**AIXTRON Inc., 1670 Barclay Blvd., Buffalo Grove, IL 60089 USA, woe@aixtron.com

Production scale MOVPE reactors such as the AIXTRON 2000HT Planetary Reactor® offer unique possibilities to fabricate highly efficient GaN based devices at a low cost of ownership. The scope of this investigation is to understand the dependence of wavelength, thickness and doping uniformity on parameters such as total gas flow, temperature distribution in the reactor and purity of the precursors. Wafer to wafer uniformity in the 7x2" wafer configuration as well as run to run reproducibility will be discussed. We obtained a wafer to wafer standard deviation of 2.7% for the sheet resistance of Si-doped GaN/InGaN/GaN double heterostructures. The wafer to wafer standard deviation of the main PL emission wavelength at 412.3 nm is 1.8 nm. The run to run reproducibility of the main emission wavelength is <3 nm. We obtained reproducible resistivities of GaN:Mg layers of less than 1 Ωcm which corresponds to $5\text{-}10\times10^{17}$cm^{-3}. Statistical data of p-type doping taking 20 runs into account gave an average hole concentration of 5.5×10^{17}cm^{-3}. Together with the wafer to wafer thickness uniformity of <1% the most sensitive layer properties are well controlled to allow a cost-effective mass production process. Structures such as SQW and MQW structures were grown to understand the performance of a production system with respect to interface properties.

1. Introduction

The market potential of blue, green, amber, red and even white LEDs based on GaN heterostructures will increase dramatically. Since the profit margin of a LED is small, cost saving production processes with a low consumption of expensive precursors and a high device yield is essential [1, 2]. Therefore we developed the AIXTRON multiwafer Planetary Reactor® for the growth of GaN-based material. This paper discusses reproducibility and uniformity issues under production aspects.

2. Experimental results

Undoped GaN layers were grown at T_D = 1160 - 1180°C using a total pressure between 100 - 500 mbar. A typical V/III ratio for undoped GaN is ~1000. These parameters result in a GaN growth rate of about 2 μm/h and an efficiency of the Ga precursor of 20%. Higher growth rates of 3 μm/h and Ga precursor efficiencies up to 25% were also achieved. The typical background carrier concentration of undoped GaN is 1×10^{16} cm^{-3}. Depending on the ammonia purity and on layer thickness background carrier concentrations as low as 5×10^{15} cm^{-3} and mobilities as high as 600 cm^2/Vs were achieved. The thickness uniformity on one single wafer without edge exclusion measured by white light interference can be tuned to ~1%. The wafer to wafer thickness uniformity

Mat. Res. Soc. Symp. Proc. Vol. 572 © 1999 Materials Research Society

in one growth run is well below 1%. Silicon doping of GaN was achieved by using SiH$_4$. Fig. 1 shows the sheet resistivity maps of four GaN layers grown in one run. The average value of the wafer is 76.4 $\Omega/\square$. The standard deviation on each wafer is ~1%. The wafer to wafer standard deviation is <3%. Lines of constant sheet resistance are concentric circles on the wafer. This indicates a linear depletion of the gas phase across the wafer together with a very uniform temperature profile. Single and multiple quantum well structures of InGaN/GaN were grown and investigated. The interface switching sequence, the growth temperature, total pressure for the InGaN growth and the carrier gas for InGaN play an important role. We used in-situ reflectometry measurements to optimize the growth step at the interfaces.

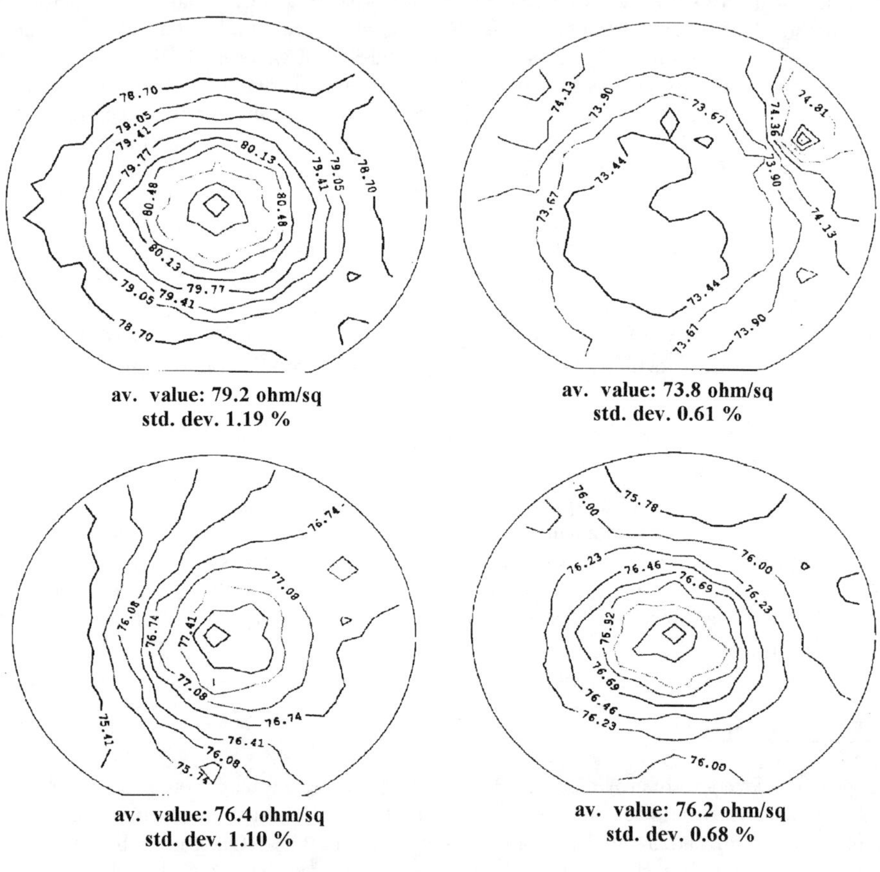

av. value: 79.2 ohm/sq
std. dev. 1.19 %

av. value: 73.8 ohm/sq
std. dev. 0.61 %

av. value: 76.4 ohm/sq
std. dev. 1.10 %

av. value: 76.2 ohm/sq
std. dev. 0.68 %

wafer-to-wafer standard deviation: 2.7%

Fig. 1: Wafer-to-wafer uniformity of Si-doped GaN/InGaN/GaN-DHS

Switching the precursor, ramping the pressure to the appropriate elevated pressure for InGaN and the transition from H_2 to N_2 carrier gas was done in that way that the reflectivity of the grown layer did not drastically decrease and that additional high quality GaN (as indicated by oscillation of the reflectometry) can be grown on top of the quantum well. This growth technology was developed to enable growth of high-quality LED structures.[3, 4]. Due to GaN specific design of the temperature control system, the low thermal mass of the heated parts and the RF heating system, fast heating and cooling cycles are possible. Fig. 2 shows the normalized emission intensity as a function of emission wavelength at room temperature for seven satellite positions of an InGaN/GaN multiple quantum well.

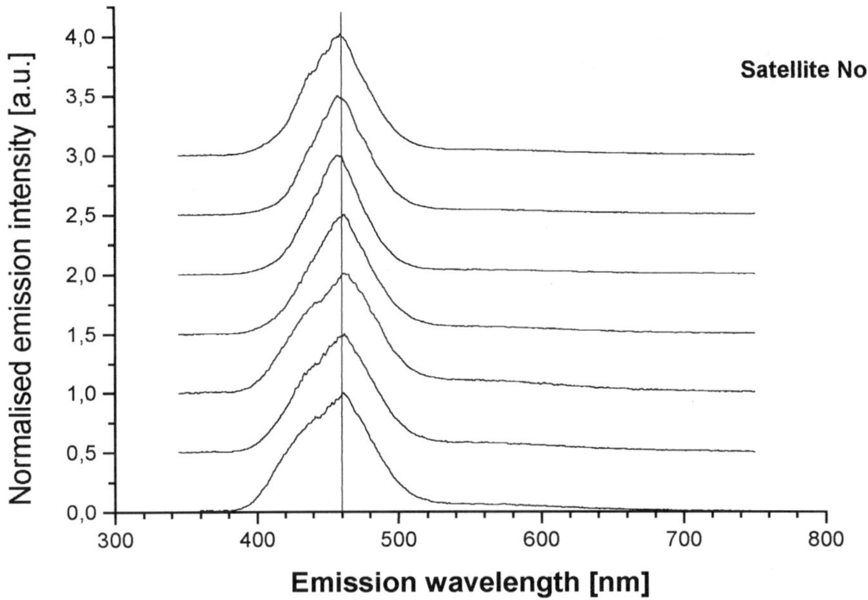

Fig. 2: Normalized emission intensity as function of emission
wavelength for seven satellite positions

The emission wavelength of all layers is at 460 nm as indicated. Photoluminescence spectra measured with a higher resolution show that the wafer to wafer difference is less than ± 1 nm. Sometimes Fabry-Perot oscillations are superimposed on the PL making a precise evaluation of the peak wavelength more difficult. Fig. 3 shows a PL wavelength mapping of one single wafer. The average wavelength is 462.7 nm and the standard deviation is 1.9 nm. This result is in good agreement with the conclusions of the measurements shown in Fig. 1 and in Fig. 2. To achieve these data, an extremely uniform temperature profile across the reactor is required. Similar structures were grown emitting at lower wavelength e.g. 394 nm. The standard deviation was 1.2 nm across one two inch wafer. Wafer to wafer standard deviations of less than 2 nm were already achieved for structures emitting at 412 nm. To achieve InGaN quantum well structures emitting at longer wavelength, tuning of the process parameters was necessary. Emission wavelengths up to 520 nm were achieved.

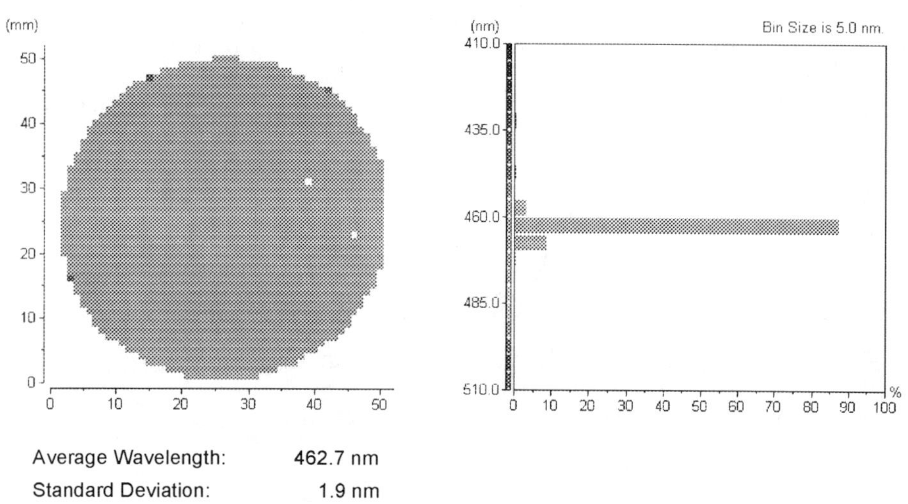

Fig. 3: InGaN MQW structure grown on AIX 2000HT

In addition high temperature lasing experiments were carried out using these InGaN/GaN MQWs under photo-excitation. Laser action was achieved up to 585K. The peak wavelength and the FWHM of the lasing spectrum at 300 K were 423.2 nm and 9 meV, respectively. The laser oscillation appeared near one of the peaks of the stimulated emission with increasing excitation intensity. Lasing thresholds at T = 78 K, 300 K, 585 K were 24 kW/cm^2, 95 kW/cm^2, 555 kW/cm^2, respectively. A characteristic temperature of 164K was derived from the temperature dependence of the lasing threshold. This is in agreement with the value of 162 K reported by others[5, 6]. Further details will be presented elsewhere [7].

3. **Conclusion**

Optimized growth conditions for a GaN-based structures containing InGaN/GaN MQWs require the use of different carriers gases, growth temperature and growth pressure. Our specially designed MOCVD systems were able to achieve the results discussed here by precise control of thermal and flow conditions. The high group III precursor efficiency, low V/III ratio and excellent uniformity obtained will enable the low-cost production of GaN-based layers for LED fabrication.

4. **References**

1. Y. Park, B.J. Kim, J. W. Lee, O. H. Nam, C. Sone, H. Park, Eunsoon Oh, H. Shin, S.Chae, J. Cho, Ig-Hyeon Kim, J. S. Khim, S. Cho, T. I. Kim, MIJ_NSR, Vol. **4**, Article No. 1.

2. O. Schoen, M. Schwambera, B. Schineller, D. Schmitz, M. Heuken, H. Juergensen Journal of Crystal Growth **195** (1998) 297-303.

3. Y. Kawaguchi, M. Yamaguchi, N. Sawaki, K. Hiramatsu, W. Taki, N. Kuwano, K. Oki, T. Zheleva and R. F. Davis, Record of the 16th Electronic Materials Symposium, L. Minoo, July 9 - 11, 1997.

4. O. Schoen, B. Schineller, M. Heuken, R. Beccard Journal of Crystal Growth **189/190** (1998) 335-339.

5. S. Bydnik, T. J. Shmidt, Y. H. Cho, G. H. Gainer and J. J. Song, S. Keller, U. K. Mishra and S. P. DenBaars., J. Appl. Phys. **72**, 13, (1998) p. 1623.

6. S. Bydnik, Y. H. Cho, T. J. Schmidt, J. Krasinski, J. J. Song, S. Keller, U. K. Mishra, S. P. DenBaars, Mat. Res. Soc. Symp. Proc. Vol. **512** (1998).

7. I. P. Marko, E. V. Lutsenko, V. N. Pavlovskii and G. P. Yablonskii, O. Schoen, H. Protzmann, M. Luenenbuerger, M. Heuken, B. Schineller and K. Heime, to be presented at the ICNS Montpellier 1999.

SYNCHROTRON X-RAY TOPOGRAPHY STUDIES OF EPITAXIAL LATERAL OVERGROWTH OF GaN ON SAPPHIRE

PATRICK J. MCNALLY*, T. TUOMI**, R. RANTAMÄKI**, K. JACOBS***, L. CONSIDINE****, M. O'HARE *, D. LOWNEY*, A.N. DANILEWSKY*****
*Microelectronics Research Laboratory, Dublin City University, Dublin 9, Ireland.
**Optoelectronics Laboratory, Helsinki University of Technology, 02015 TKK, Finland.
***Dept of Information Technology (INTEC), Univ. of Gent, B-9000, Belgium.
****Thomas Swan & Co. Ltd., Harston, Cambridge CB2 5NX, U.K.
*****D-79108 Freiburg, Germany.

ABSTRACT

Synchrotron white beam x-ray topography techniques, in section and large-area transmission modes, have been applied to the evaluation of ELOG GaN on Al_2O_3. Using the openings in 100 nm thick SiO_2 windows, a new GaN growth took place, which resulted in typical overgrowth thicknesses of 6.8 μm. Measurements on the recorded Laue patterns indicate that the misorientation of GaN with respect to the sapphire substrate (excluding a $30°$ rotation between them) varies considerably along various crystalline directions, reaching a maximum of a ~$0.66°$ rotation of the (0001) plane about the [01•1] axis. This is ~3% smaller than the misorientation measured in the non-ELOG reference, which reached a maximum of $0.68°$. This misorientation varies measurably as the stripe or window dimensions are changed. The quality of the ELOG epilayers is improved when compared to the non-ELOG samples, though some local deviations from lattice coherence were observed. Long range and large-scale (order of 100 μm long) strain structures were observed in all multi quantum well epilayers.

INTRODUCTION

Epitaxial lateral overgrowth (ELOG) holds out the potential for significant reductions in threading dislocation densities for mismatched hexagonal-GaN on sapphire epitaxy [1-3]. Using openings in a relatively thick SiO_2 mask a new MOVPE GaN growth is carried out. After an initial phase of vertical growth upward through the mask window, the growth then proceeds laterally over the mask itself. It is thought that a significant reduction in threading dislocation densities can be achieved via mask blocking of vertically propagating dislocations and via a redirection of the propagation of some dislocations at the growth front [4-5]. Studies have shown that the ELOG technique can result in dislocation densities almost three orders of magnitude lower than in the non-ELOG case, wherein typical densities of ~1 x 10^{10} cm^{-2} are often observed [3]. GaN-based opto- and electronic devices are expected to benefit from this reduction in dislocation density and this technique has recently been applied to GaN blue laser production [6-7]. However, an understanding of the processes active during the ELOG procedure, and their impact on strain and dislocation generation, is still far from complete. In this study, synchrotron white beam x-ray topography, in section and large-area transmission [8-9] modes, have been applied to the evaluation of ELOG GaN on Al_2O_3.

Mat. Res. Soc. Symp. Proc. Vol. 572 © 1999 Materials Research Society

EXPERIMENTAL

The measurements were performed at the Hamburger Synchrotronstrahlungslabor at the Deutsches Elektronen-Synchrotron (HASYLAB am DESY, beamline F1), Hamburg, Germany, utilising the continuous spectrum of synchrotron radiation from the DORIS III storage ring bending magnet. The ring operated at a positron momentum of 4.45GeV/c and at typical currents of 80-150mA. The Laue patterns of topographs were recorded either on Kodak High-Resolution SO-343 or SO-181 Professional X-ray film having an emulsion grain size of about 0.05 μm. Transmission topographic techniques were employed in producing these topographs [8-9]. For transmission section topography, the incoming beam is collimated into a narrow ribbon by a slit typically 15-20 μm in width. When a single crystal is immersed in a white X-ray beam a number of lattice planes (hkl) select out of the continuous spectrum the proper wavelengths to be reflected according to Bragg's Law. Due to the low divergence of the synchrotron radiation beam (nominally 0.06 mrad FWHM), each spot of this particular Laue pattern is itself a high-resolution topograph Thus, a set of Laue case section topograph images of sample cross-sections are produced and, provided that the Bragg angle is not too small, the image gives detailed information about the energy flow within the crystal and direct depth information on the defects present in a particular crystal slice. For large-area transmission topographs the collimating slits are opened to provide an incoming beam whose dimensions are ~ 3 mm x 4mm.

The lattice mismatch between hexagonal wurtzite α-phase GaN and the underlying rhombohedral (hexagonal) α-Al_2O_3 can achieve values as high as 13.9% [10] and standard epitaxial deposition of GaN on sapphire leads to very high threading dislocation (TD) densities. However, as stated earlier, the implementation of ELOG leads to significant reductions in TD densities from ~10^{10} cm^{-2} to ~10^7 cm^{-2}. The epitaxial lateral overgrowth of GaN was carried out in a vertical rotating disk MOVPE (Metal Organic Vapor Phase Epitaxy) reactor manufactured by Thomas Swan & Co., operated at a pressure of 100 Torr. Trimethylgallium (TMG), trimethylindium (TMI) and ammonia (NH_3) were used as Ga, In and N precursors respectively. At first, a 0.85-1.2 μm GaN epilayer was grown on top of a low temperature thin (33-50 nm) GaN buffer layer on a c-plane sapphire substrate. Growth conditions of these layers are described elsewhere [11]. Afterwards 100-150 nm thick SiO_2 stripes were deposited using plasma enhanced chemical vapor deposition (PECVD) followed by conventional photolithography and dry etching. Stripe widths (L) of 2 and 3 μm were applied, whereas the window openings (W) varied between 3 and 5 μm. These stripes were oriented in the $<1\bar{1}00>$ direction in order to obtain large lateral growth rates. The resulting structure was then re-entered into the reactor after a 5 sec dip into a HF:H_2O solution (1:4) to remove surface oxide. The actual ELOG growth took place at temperatures ranging from 1060°C to 1080°C, using H_2 as the carrier gas. The thickness of the ELOG layer varied between 4.8 and 6.8 μm. Typical flow rates for TMG and NH_3 were 23.7 μmol/min and 5 l/min respectively. When Multi-Quantum Well (MQW) structures were implemented they were grown in N_2 at 865°C and contained 5 periods of 3 nm thick $In_{0.2}Ga_{0.8}N$ wells and 6 nm thick GaN barriers. Flow rates in this case consisted of 3 μmol/min for TMG and 9.6 μmol/min for TMI. A diagram of the ELOG scheme used in this study is shown in Figure 1. The wafers were divided into four regions, each with a differing fill factor, i.e. the ratio of W to the total mask period, W+L. This information is given in Table 1.

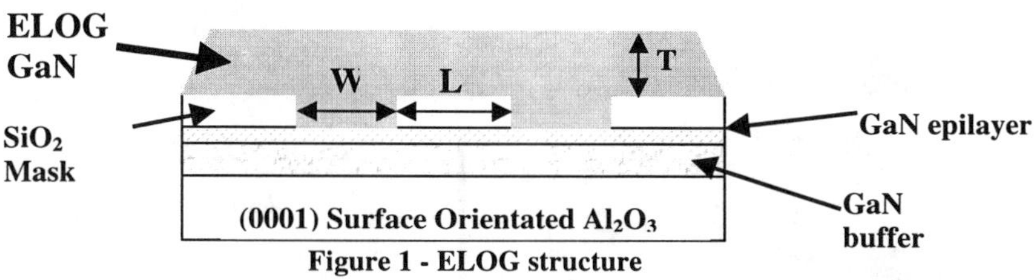

Figure 1 - ELOG structure

Sample Region	Window Opening W (μm)	SiO₂ Stripe Width L (μm)	Fill Factor W/(W+L)
a	5	3	0.625
b	4	3	0.571
c	4	2	0.667
d	3	3	0.500

Table 1 - ELOG Mask Dimensions

RESULTS

Figure 2(a) shows the Laue pattern taken from a complete film of Region a of the ELOG GaN on Al₂O₃ material. The sample to film distance (D) was 50 mm. The major flat of the ELOG GaN wafer is oriented in the [11$\bar{2}$0] direction of the GaN, so that means that the stripes on the sample run along the [1$\bar{1}$00] direction, which is perpendicular to the flat. Concerning the non-ELOG sample, the shortest straight edge has the [1$\bar{1}$00] orientation, the other one is oriented in the [11$\bar{2}$0] direction. Please note that all the aforementioned directions are for the GaN. There exists a 30° rotation between the crystallographic orientations of the GaN and the sapphire substrate, which means that the major flat of the wafer is one of the <1$\bar{1}$00> directions for the sapphire. The direction normal to the surface is the [0001] direction, so the surface is the so-called c-plane, or the (0001) plane.

The images of the substrate and epilayer are separated from each other in certain diffraction directions, e.g. the 11$\bullet\bar{3}$ Al₂O₃ and the 01$\bullet\bar{1}$ GaN reflections. A small separation between certain reflections of the substrate and epilayer is expected, since there exists a 30° rotation between the substrate and epilayer. A magnified image of the highlighted reflections is given in Figure 2(b) and the two stripe images are the section transmission topographs for the substrate (uppermost) and the ELOG GaN (lower image). The expected separation of the two reflections, based solely on the designed 30° mutual rotation, is indicated by 2$\Delta\phi$. However, the measured separation, 2$\Delta\theta$, is much greater. Therefore it must be concluded that the observed misorientation between the substrate and ELOG GaN is mainly due to lattice dilatation or rotation. In the case of strained layer GaN epitaxy, the perpendicular component of the lattice parameter will change while the parallel components remain unchanged. This results in a distortion of the relative positions of the lattice planes in the epilayer with respect to the substrate. Thus, along directions where the diffraction vector (**g**) and distortion vector running normal to the distortion (**h**) are mostly parallel (i.e. any region where **g.h** $\neq$ 0) there will be an observed shift in the position of the diffracted image of the epilayer with respect to the remainder of the crystal, i.e. the substrate.

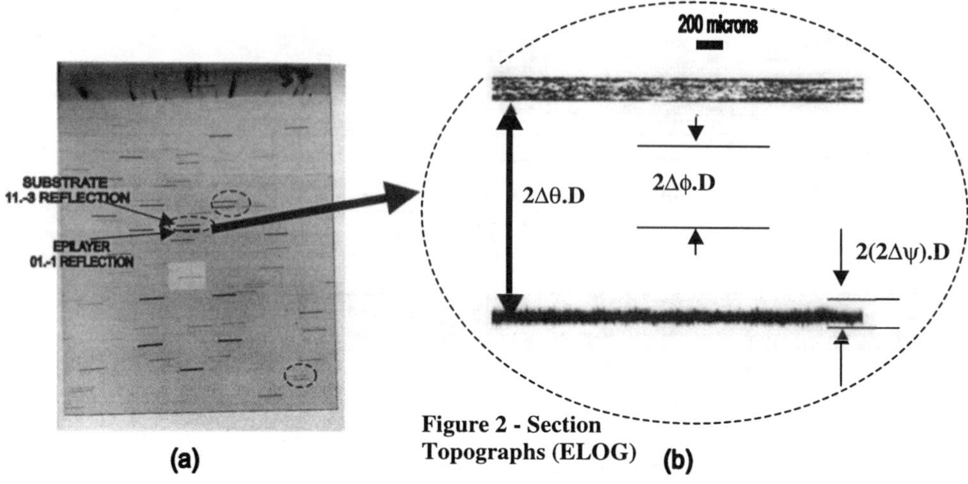

Figure 2 - Section Topographs (ELOG)

(a)   **(b)**

It was found that the $\Delta\theta_{max}$ for non-ELOG GaN epilayers was ~0.68°, while the $\Delta\theta_{max}$ for the ELOG GaN was ~0.66°, representing a combination of rotation of the (0001) plane about a [01•1] axis or lattice dilatation. Further experimentation will be required to determine which, if either, mechanism is dominant. In particular SXRT may be sensitive to tilting of the c-planes in the ELOG GaN with respect to the seed material. The quality of the sapphire substrate appears to be rather good. However, one can see that the image of the epilayer is severely broadened and appears to be almost as thick as the substrate itself. This effect is most likely due to the fact that the quality of the epilayer is far from perfect and this manifests itself via local deviations from lattice coherence throughout the epilayer. Various regions in the epilayer, each slightly misorientated with respect to its neighbour, though still generally macroscopically aligned, will each contribute to a topographic image. However, each of these regions will produce images at slightly different locations on the film. If it is assumed that these deviations are symmetrically distributed around the nominal crystallographic plane positions, the measured broadening of the epilayer image in the section topograph is given by $2\Delta\psi$. For the example shown here we find these local lattice misorientational deviations to give values of $\Delta\psi=\pm0.07° \approx \pm250$ arcsec across an 8.5 mm length of epilayer. Significant differences between this non-ELOG sample and the ELOG wafers are observed.

For the non-ELOG wafer the overall dilatational misorientation ($\Delta\theta_{max}$) reaches a larger maximum of a 0.68° rotation of the (0001) plane about the [01•1] axis. This is >3% larger than previously recorded for the ELOG wafers. The local variations from coherence within the epilayer are much larger and values of $\Delta\psi$ are as large as $\Delta\psi=\pm0.12°$ across an 8.5 mm length of epilayer. This is principally due to the apparent growth of highly misorientated features within the epilayer (arrows B and C in Figure 3). These features correspond to highly misorientated features on the substrate. The epitaxial replication of defects in the substrate **is not** observed for the ELOG samples.

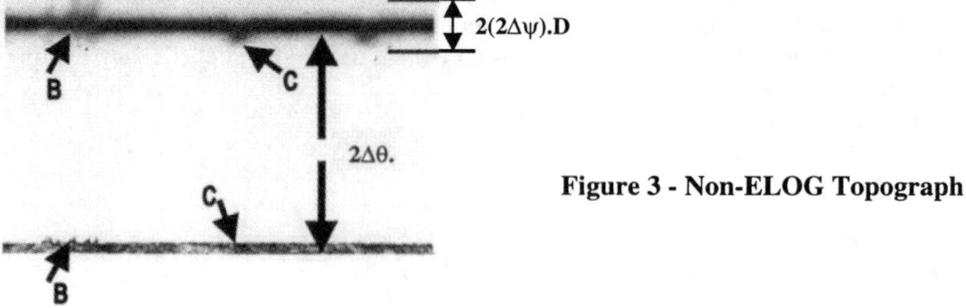

Figure 3 - Non-ELOG Topograph

It was also observed that the misorientation varies as a function of the window/stripe separation. This was especially apparent when comparing two regions on the ELOG mask with greatly differing fill factors. An example of this is shown in Figure 4, which is a $01\bullet\bar{1}$ transmission section topograph as before. The image is left in its original orientation on the recording film, i.e. the substrate image would appear above the epilayer image.

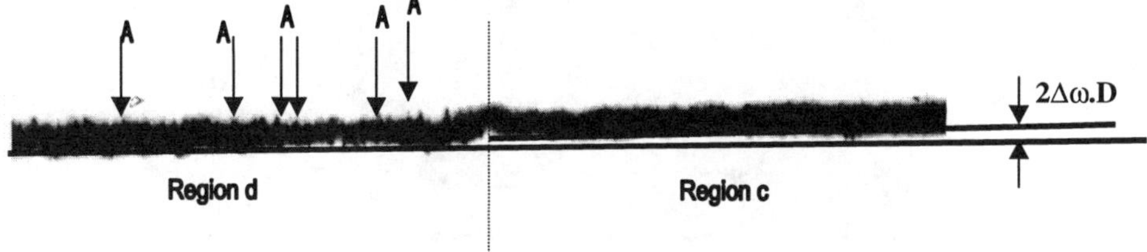

Figure 4 - Misorientation across two regions of the ELOG mask

In this case the beam scanned two regions simultaneously, i.e. Regions c and d as indicated on the figure. In region d large spikes are observed - indicated by arrows A- (similar features are also seen in Figure 2(b)) and could be caused by a build-up of strain/misorientation at the SiO_2 stripe edges. Since the stripes in this region are close to the threshold of resolution for SXRT (ca. 1.7 μm at HASYLAB), it is difficult to discern individual windows. However, this argument is supported by the fact that these features are not easily seen for Region c, wherein the stripe widths are reduced by 33%. There is a clear shift in the separation of the two regions of the epilayer with respect to the substrate. Region d displays a slightly greater misorientation than Region c. This difference is indicated on Figure 4 by the angle Δω, and measurements yield a value of Δω=0.015°. Figures 2, 3 and 4 indicate that the epilayer signals from the seed GaN and the ELOG and non-ELOG GaN epilayers cannot be easily distinguished. This is most likely due to the fact that the seed layer is much thinner than subsequent layers and cannot be easily separated on the topographic images. Nevertheless, the clearly distinguishable "kink" in the epilayer image of Figure 4 suggests that an, as yet unconfirmed, strain/relaxation effect impacts on the observed relative misorientation.

Figure 5 is a $11\bullet\bar{3}$ (substrate)/$01\bullet\bar{1}$ (epilayer) large-area transmission topograph (LAT) of a sample upon which MQW structures have been fabricated. The LAT image covered two regions, namely MQWs on ELOG GaN epilayer and MQWs on a non-ELOG GaN epilayer. Again, due to the inherent misorientation of the sapphire substrate with respect to the GaN-based materials, the images of the substrate and epilayers are conveniently separated. There is a slight overlap where the substrate and epilayer images meet. Thus, the topographic images of the epilayers are unambiguously not contaminated by substrate contributions. The topographic images of the sapphire substrate are identical for both cases. However, clear differences emerge in the topographs of the epilayers. The MQW structures fabricated on the ELOG side of the wafer contains strain features, which are not observed on the non-ELOG side. These features manifest themselves as an extra "orange peel" roughening of the image as seen on the bottom left hand side (LHS) of Figure 5. These features are of the order of 100 μm in size and their origins are uncertain and were only observed for MQW fabrication on ELOG material.

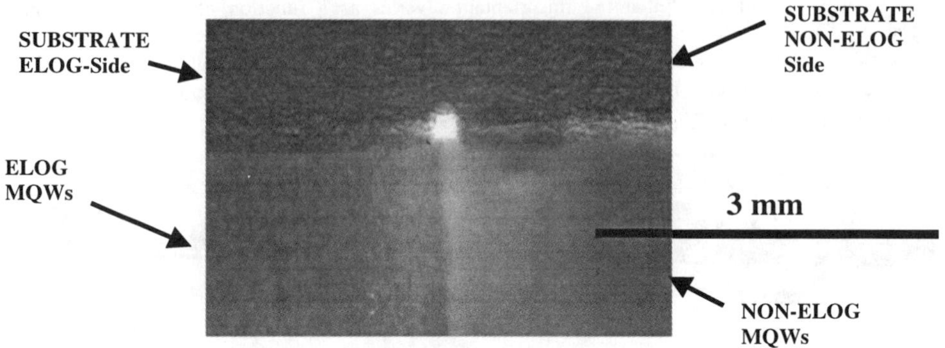

Figure 5 - Large area transmission topograph of MQW structures

CONCLUSION

SXRT was used to monitor the improvement in the quality of ELOG GaN on sapphire when compared to non-ELOG material. Some local deviations from lattice coherence were observed. In the best ELOG epilayers, section topographic measurements yielded misorientational deviations of up to ±0.07° across an 8.5mm length of epilayer. Topographic measurements also revealed variations in ELOG epilayer quality as stripe/window dimensions changed, though the non-ELOG layers were invariably worse. The fabrication of MQW structures on ELOG GaN produces an "orange peel" strain structure in the epilayers, whose dimensions are of the order of 100 μm.

ACKNOWLEDGEMENTS

The support of T. Wroblewski at HASYLAB, Hamburg, Germany is greatly appreciated. This project was supported by the TMR-Contract ERBFMGECT950059 of the European Union and the Irish Forbairt International Collaboration Programme.

REFERENCES

1. K. Kato, S. Kitamura and N. Sawaki, J. Crystal Growth, **144**, 133 (1994).
2. D. Kapolnek, S. Keller, R. Vetury, R.D. Underwood, P. Kozodoy, S.P. DenBaars and U.K. Mishra, Appl. Phys Lett., **71**, 1204-1206 (1997).
3. A. Usui, H. Sunakawa, A. Sakai and A.A. Yamaguchi, Jpn. J. Appl. Phys., **36**, L899 (1997).
4. T.S. Zheleva, O.K. Nam, M.D. Bremser and R.F. Davies, Appl. Phys. Lett., **71**, 2472-2474 (1997).
5. A. Sakai, H. Sunakawa and A. Usui, Appl. Phys. Lett., **71**, 2259-2261 (1997).
6. S. Nakamura, M. Senoh, S. Nagahama, N. Iwasa, T. Yamada, T. Matsushita, H. Hiyoku, Y. Sugimoto, T. Kozaki, H. Umemoto, M. Sano and Chocho, Appl. Phys. Lett., **72**, 211-213 (1998).
7. S. Nakamura, M. Senoh, S. Nagahama, N. Iwasa, T. Yamada, T. Matsushita, H. Hiyoku, Y. Sugimoto, T. Kozaki, H. Umemoto, M. Sano and Chocho, Jpn. J. Appl. Phys., **36**, L1568 (1997).
8. M. Sauvage-Simkin in <u>Synchrotron Radiation Research</u>, eds. H. Winick and S. Doniach, pp. 179-204, Plenum (1982).
9. T. Tuomi, K. Naukkarinen and P. Rabe, phys. stat. sol. (a), **25**, 93-106 (1974).
10. O Ambacher, J. Phys. D: Appl. Phys, 31, 2653-2710 (1998).
11. L. Considine, E.J. Thrush, J.A. Crawley, K. Jacobs, W. Van der Stricht, I. Moerman, P. Demeester, G.H. Park, S.J. Hwang, J.J. Song, J. Crystal Growth **195**, 192-198 (1998).

CONDUCTING (Si-DOPED) ALUMINUM NITRIDE EPITAXIAL FILMS GROWN BY MOLECULAR BEAM EPITAXY

J.G. Kim, Madhu Moorthy and R.M. Park,
Department of Materials Science and Engineering, University of Florida,
Gainesville, FL 32611
Jgkim@solid.ssd.ornl.gov

ABSTRACT

As a member of the III-V nitride semiconductor family, AlN, which has a direct energy-gap of 6.2eV, has received much attention as a promising material for many applications. However, despite the promising attributes of AlN for various semiconductor devices, research on AlN has been limited and n-type conducting AlN has not been reported. The objective of this research was to understand the factors impacting the conductivity of AlN and to control the conductivity of this material through intentional doping. Prior to the intentional doping study, growth of undoped AlN epilayers was investigated. Through careful selection of substrate preparation methods and growth parameters, relatively low-temperature molecular beam epitaxial growth of AlN films was established which resulted in insulating material. Intentional Si doping during epilayer growth was found to result in conducting films under specific growth conditions. Above a growth temperature of 900°C, AlN films were insulating, however, below a growth temperature of 900°C, the AlN films were conducting. The magnitude of the conductivity and the growth temperature range over which conducting AlN films could be grown were strongly influenced by the presence of a Ga flux during growth. For instance, conducting, Si-doped, AlN films were grown at a growth temperature of 940°C in the presence of a Ga flux while the films were insulating when grown in the absence of a Ga flux at this particular growth temperature. Also, by appropriate selection of the growth parameters, epilayers with n-type conductivity values as large as 0.2 $\Omega^{-1}cm^{-1}$ for AlN and 17 $\Omega^{-1}cm^{-1}$ for $Al_{0.75}Ga_{0.25}N$ were grown in this work for the first time.

INTRODUCTION

Research on the III-V nitride materials system (AlN, GaN, InN, and their ternaries) has been one of the hottest issues in current materials research. Ever since the successful fabrication of the first highly efficient blue-light-emitting diodes [1] and blue-diode lasers [2], enormous attention has been given to the III-V nitrides. The III-V nitride materials are not only good candidates for optoelectronic devices, but also promising materials for high temperature, high frequency, and high power electronic device applications [3]-[6].

Among the III-V nitrides, AlN has recently drawn attention due to its potential for use in many device application areas. AlN displays high thermal conductivity, high temperature stability and a large direct band-gap of 6.2eV and has attractive piezoelectric properties, which render it suitable for surface acoustic wave device applications [7]. AlN (and high Al content AlGaN) is also of particular interest due to its negative electron affinity [8], which can be exploited for cold cathode applications in high power vacuum electronics and flat panel displays.

Despite these promising attributes, however, progress in the development of AlN-based devices has been limited due to difficulties encountered in doping the material. Attempts to dope AlN have been made by several researchers [9]-[11], however, all of these attempts have resulted in the production of highly resistive material ($\rho > 1000$ Ω-cm).

Mat. Res. Soc. Symp. Proc. Vol. 572 © 1999 Materials Research Society

The work described in this paper was motivated by the need to develop methods of producing low resistivity AlN and high Al-content AlGaN, aided by an understanding of the mechanisms that have prevented such an achievement thus far. The objectives of this work were to develop ways to obtain conducting intentionally doped n-type AlN and to understand how it is possible to produce such material.

EXPERIMENT

The samples in this work were grown by molecular beam epitaxy employing the previously reported rf plasma discharge, nitrogen free-radical source [12]-[14]. In order to minimize hydrogen effect during intentional doping of silicon, N_2 gas was introduced to the free-radical source instead of NH_3 gas. Also, standard effusion cells were used for gallium, aluminum and silicon. First, sapphire (0001) surfaces were etched in H_2SO_4 : H_3PO_4 (3:1) solution at 160°C for 90 minutes, after which gives smooth (5µm x 5µm AFM RMS value of 1.0Å) and featureless (free of polishing damage) surface. The etched sapphire samples were further cleaned in the MBE growth chamber by heating up to 940°C for 60 minutes followed by 2-hour nitridation. The rf power used for nitridation was 250W and a nitrogen flux was $1x10^{-5}$ torr.

Three types of buffer layers were attempted, namely, GaN, low temperature AlN ($T_{substrate}$ = 750°C), and high temperature AlN ($T_{substrate}$ = 900°C). A 1µm-thick GaN epilayer was grown at 750°C after each type of buffer layer was grown. From the room temperature Hall-effect measurements of the GaN epilayers, the high temperature AlN buffer layer was chosen to perform the growth and Si-doping of AlN epilayers since only the GaN layer grown on high temperature AlN buffer was insulating. Furthermore, intentionally Si-doped GaN layers grown on high temperature AlN buffer layer exhibits higher room temperature free-electron mobility (~200 $cm^2V^{-1}s^{-1}$ at $10^{16} < n < 10^{17}$ cm^{-3}) than the values of the layers grown on the other buffer layers.

Three sets of 0.4~1.0µm thick undoped and Si-doped AlN epilayers were prepared on previously mentioned high temperature AlN buffer layers. First, a series of undoped and Si-doped AlN films was deposited at various substrate temperatures (850, 880, and 940°C) under a fixed III/V ratio to 2.7 x 10^{-3}. The second set of samples was grown at the same series of conditions with additional 0.1µm thick undoped GaN buffer layers. Finally, the conditions used in second series of samples were used except an additional Ga flux of 7.0 x 10^{-8} Torr was included during growth. In all cases, the Si effusion cell temperature was 1280°C, which resulted in a Si doping concentration of mid 10^{19} cm^{-3} in GaN grown using a similar growth rate (~0.1 µm/hr). The samples were analyzed using several analytical techniques, namely, reflection high electron energy diffraction (RHEED), Hall-effect measurement, secondary ion mass spectrometry (SIMS), AFM, cross-sectional secondary electron microscopy (SEM), and electron microprobe.

RESULTS AND DISCUSSION

Figure 1a ~ 1c is the AFM images of sapphire (0001) surface change before and after various etching times. As the etching progressed, polishing damage was observed (Figure 1b) which was removed after 90 minutes (Figure 1c).

All undoped AlN and high Al content AlGaN epilayers were insulating. RHEED pattern during growth showed that most of the samples were weakly (2x2) reconstructed. At a lower growth temperature, it was more difficult to see or maintain the reconstruction during growth. The first set of Si-doped samples, directly grown on high temperature AlN buffer layer, was all insulating while the second set of Si-doped samples, grown on additional GaN buffer layer, showed a temperature dependence of resistivity (Figure 2).

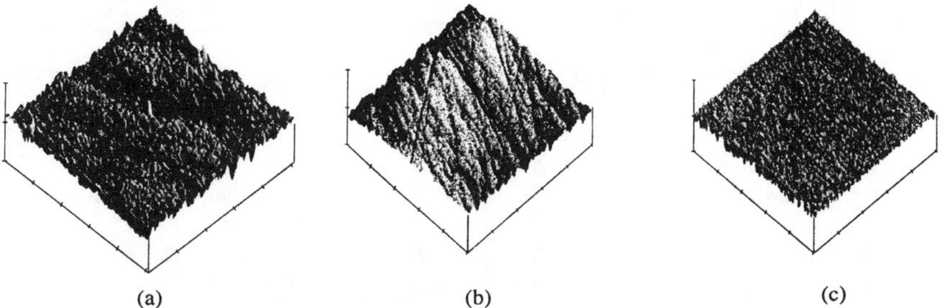

Figure 1. 5 x 5 µm AFM images of (a) as-received sapphire (0001) surface (RMS = 3.32Å),
(b) after 30-minute etching (RMS = 2.34Å) and
(c) after 90-minute etching (RMS = 1.15Å) in H_2SO_4 : H_3PO_4 (3:1) solution at 160°C.

The figure indicates that above the growth temperature of 900°C, AlN films were insulating, however, below the growth temperature of 900°C, the AlN films were conducting.

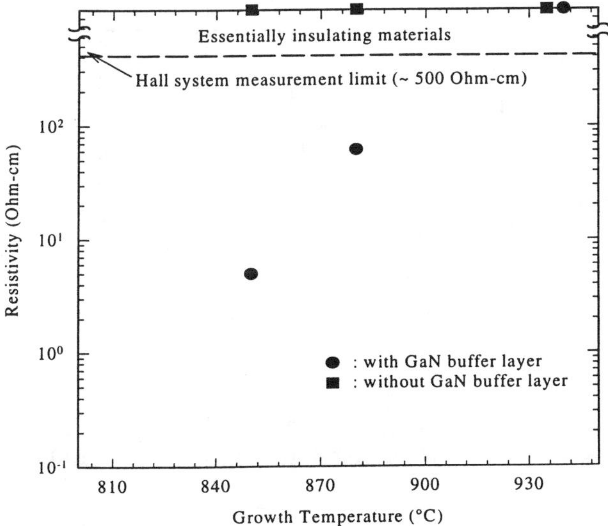

Figure 2. Resistivity data obtained from Si-doped AlN films grown
at various substrate temperatures with and without a GaN
buffer layer.

Even though n-type conduction was observed from Hall-effect measurement, the Hall voltage was fluctuating greatly making it difficult to determine either carrier mobility or concentration. Figure 3a ~ 3b is the SIMS result of the Si-doped samples. Due to the interference of N_2 or CO signal, isotope 29Si signal was used to compare the Si dose. The result indicates that regardless of the growth temperature the amount of Si in the film was about the same. Therefore, it seems that the compensation, not Si incorporation, is the reason for these resistivity changes.

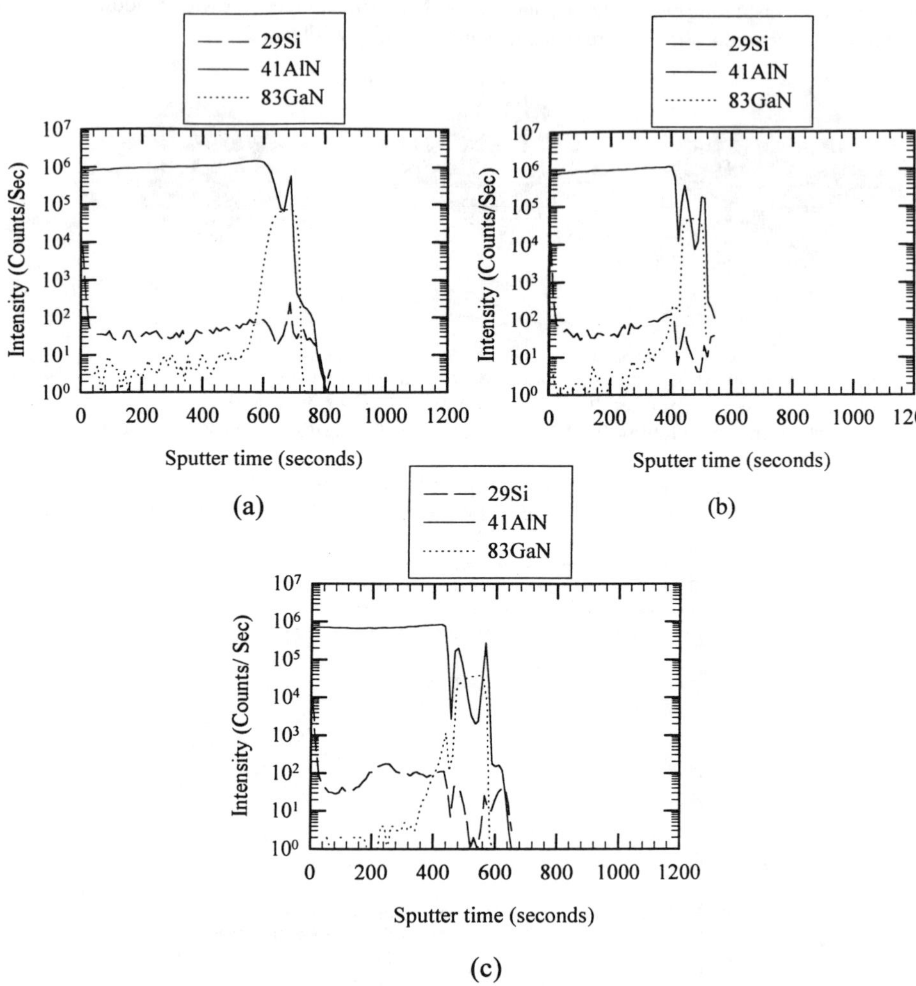

Figure 3. SIMS depth profile of a Si-doped AlN film grown at (a) 850°C, (b) 880°C and (c) 940°C

Further reduction in resistivity was achieved from the samples grown with a fixed additional Ga flux of 7.0 x 10^{-8} Torr (Figure 4). Fraction of the Ga flux was incorporated after electron microprobe analysis for the samples grown at 880 and 850°C. However, no evidence of Ga incorporation was detected on the sample grown at 940°C even after SIMS analysis. The SIMS results of the samples grown at 940°C with Ga flux and without Ga flux were compared

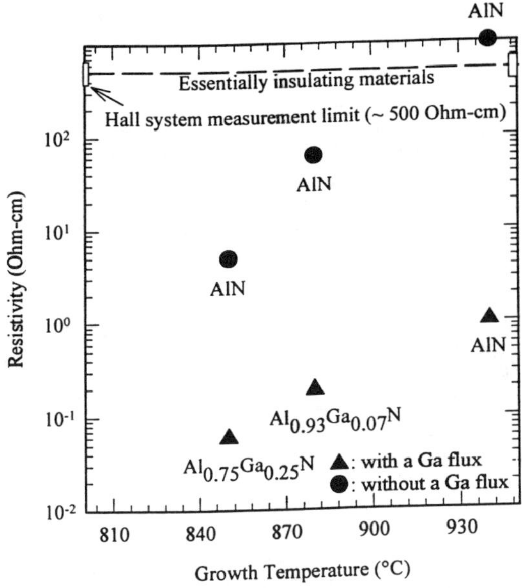

Figure 4. Resistivity data obtained from Al$_x$Ga$_{1-x}$N films grown at various substrate temperatures with and without the presence of a Ga flux. Ternary compositions were determined by microprobe analysis.

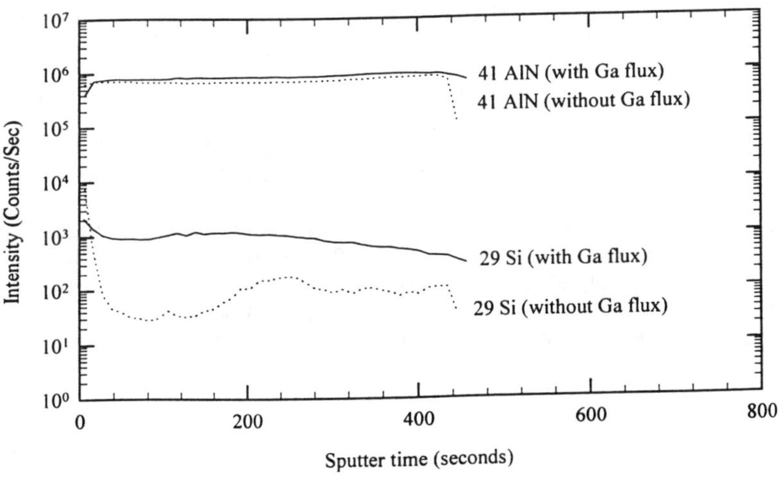

Figure 5. Si profiles obtained by SIMS analysis in Si-doped AlN films grown with and without the presence of a Ga flux at a substrate temperature of 940°C.

in Figure 5, and an order of magnitude increase in Si dose was observed from the sample with Ga flux.

CONCLUSIONS

In this study, n-type conducting (conductivities as high as 17 Ω^{-1}-cm^{-1}) $Al_xGa_{1-x}N$ (x $\geq$ 0.75) epilayers were produced for the first time by Si-doping. The influence of growth parameters, such as growth temperature, and the addition of a Ga flux, on the conductivity was studied. Conductivity values increased with decreasing growth temperature and also with the addition of a Ga flux.

ACKNOWLEDGMENTS

The author would like to thank Mr. Lynn Calhoun for his valuable assistance and discussions regarding this research and Dr. Margaret Puga, who performed the SIMS measurements.

REFERENCES

1. S. Nakamura, T. Mukai and M. Senoh, Appl. Phys. Lett. **64**, 1687 (1994).
2. S. Nakamura, Proc. of the SPIE **V2693**, 43 (1996)
3. M.A. Khan, Q. Chen, M.S. Shur, B.T. Dermott, J.A. Higgins, J. Burm, W. Schaff and L.F. Eastman, Electronic Letters **32**, 357 (1996).
4. Y.F. Wu, B.P. Keller, S. Keller, D. Kapolnek, P. Kozodoy, S.P. Denbaars and U.K. Mishra, Appl. Phys. Lett. **69**, 1438 (1996).
5. S.C. Binari, J.M. Redwing, G. Kelner and W. Kruppa, Electronic Letters **33**, 242 (1997).
6. Q. Chen, R. Gaska, M. Asif Khan, M.S. Shur, A. Ping, I. Adesida, J. Burm, W.J. Schaff and L.F. Eastman, Electronic Letters **33**, 637 (1997).
7. G.R. Kline and K.M. Lakin, Appl. Phys. Lett. **43**, 750 (1983).
8. M.C. Benjamin, C. Wang, R.F. Davis and R. J. Nemanich, Appl. Phys. Lett. **64**, 3288 (1994).
9. T.L. Chu, D.W. Ing and A.J. Noreika, Solid State Electron. **10**, 1023 (1967).
10. X. Zhang, P. Kung, A. Saxler, D. Walker, T.C. Wang and M. Razeghi, Appl. Phys. Lett. **67**, 1745 (1995).
11. M.D. Bremser, W.G. Perry, T. Zheleva, N.V. Edwards, O.H. Nam, N. Parikh, D.E. Aspnes and R.F. Davis, MRS Internet Journal **Vol. 1**, Article 8 (1996).
12. H. Liu, A.C. Frenkel, J.G. Kim and R.M. Park, J. Appl. Phys. **74**, 6124 (1993).
13. J.G. Kim, A.C. Frenkel, H. Liu and R.M. Park, Appl. Phys. Lett. **65**, 91 (1994).
14. H. Liu, J.G. Kim, H. Ludwig and R.M. Park, Appl. Phys. Lett. **71**, 347 (1997).

INVESTIGATION OF THE MORPHOLOGY OF AlN FILMS GROWN ON SAPPHIRE BY MOCVD USING TRANSMISSION ELECTRON MICROSCOPY

W.L. Sarney[1], L. Salamanca-Riba[1], P. Zhou[2], S. Wilson[2], M.G. Spencer[2], and K.A. Jones[3]

[1]Dept. of Materials & Nuclear Engineering, University of Maryland, College Park, MD
[2]Materials Science Research Center of Excellence, Howard University, Washington, D.C.
[3]U.S. Army Research Laboratory, Adelphi, MD

ABSTRACT

To determine the effect of growth conditions on AlN film morphology, we investigated several AlN films grown on sapphire by MOCVD with various V/III ratios. Transmission electron microscopy was used to characterize the film's crystalline quality and defect morphology. TEM results show that the resulting film morphology depends on the V/III ratio. Films grown with NH_3 flow rates below 170 sccm have high crystalline quality. In contrast, we observe columnar growth and secondary interfaces in films grown with NH_3 flow rates at or above 170 sccm. The secondary interfaces are likely to be inversion domain boundaries (IDBs) and may be associated with strain relaxation. We discuss the V/III ratio's effect on crystalline quality, surface roughness, and IDB and columnar structure formation.

INTRODUCTION

Because of its large energy gap, AlN is a good candidate for the fabrication of high temperature semiconductor devices and as the dielectric material in high power, high frequency devices operating at high temperatures. However, the quality of AlN films still needs to improve before this material can be used in practical applications. Due to the large lattice mismatch, AlN films grown on sapphire tend to grow three dimensionally, leading to a high defect density and in many instances, a columnar structure. In order to reduce the lattice mismatch from 35% to an effective mismatch of 13%, AlN films grown on sapphire have a 30° in plane rotation with respect to the substrate. Three dimensional growth negatively affects the optical and electrical properties of AlN, therefore inhibiting its potential for high power, high frequency, and high temperature applications. The goal of this experiment was to determine which growth parameters result in the highest quality films. In a previous study, we have obtained the optimum growth temperature range (1110°C - 1150°C) for high quality AlN films. In this study, we obtain the optimum V/III ratio for this temperature range.

EXPERIMENT

Several AlN films were grown on (0001) sapphire in a low pressure MOCVD reactor. All of the films were grown for 15 minutes and at a pressure of 10 Torr. Samples were grown at four different temperatures ranging from 1110° C - 1160° C. The growth precursors, trimethylaluminum (TMA), and ammonia, the hydrogen flow rate and the temperature were held constant during the growth of each individual sample. Various NH_3 flow rates, ranging from 130 - 190 sccm, were used for different samples. Since the TMA flow rate was 15 sccm for each sample, an increase in the ammonia flow results in an increased V/III ratio.

Mat. Res. Soc. Symp. Proc. Vol. 572 © 1999 Materials Research Society

Cross-sectional TEM samples were prepared using a tripod polishing and ion milling, and were examined in a JEOL 4000FX TEM operated at 300 KV. The film's surface was examined with atomic force microscopy (AFM) and initial crystallinity was assessed with in-situ RHEED at the end of the growth.

RESULTS

Sample A is typical of the high quality AlN films we have grown and is shown in Figs. 1-2. Figure 1 shows that the film is single crystalline and has a relatively low density of threading dislocations. High resolution images (Fig. 2) show that the interface in this sample is sharp and that the film is of high crystalline quality. The film does not show the columnar structure that is often observed in AlN/Sapphire films. The white and dark bands seen near the interface are due to strain contrast, as typically observed in highly mismatched systems. Sample A was grown with a low ammonia flow of 150 sccm. AFM results show that sample A has a very smooth surface.

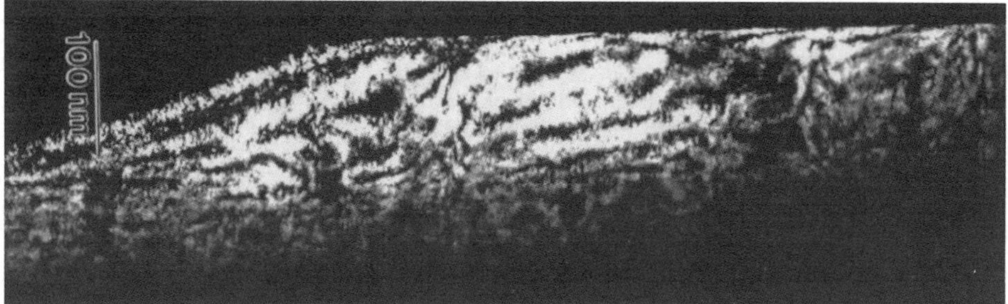

Fig. 1 (0002) Dark field image (DF) of sample A

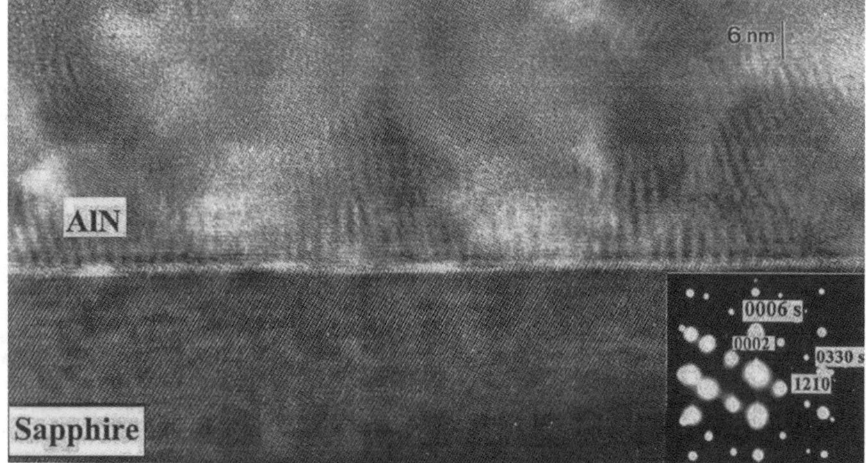

Fig. 2. High resolution image of sample A. The s subscript in the diffraction pattern indices denote the substrate spots. The zone axes are $(1\bar{2}10)_s/(10\bar{1}0)_f$.

Several of the AlN samples have an interface within the film, which lies 18 – 20 nm above the AlN/Sapphire interface. These films were grown with NH_3 flow rates at or exceeding 170 sccm. One such film is shown in Figs. 3 and 4, and will be referred to as sample B in this paper. The upper interface was not expected since the growth was carried out under constant conditions and with no interruptions. The (0002) dark field image (DF) shown in Fig. 3 clearly shows the second interface. High resolution images and diffraction patterns (Fig. 4) indicate that the film consists of the 2H polytype on both sides of this interface. The interface may be an inversion domain boundary, which is caused by Al-Al or N-N atomic bonding. IDBs have been observed by TEM in AlN films in previous experiments[1-2]. The lower AlN layer is single crystalline, as seen in Fig 4. The upper AlN layer is crystalline near the interface, but becomes polycrystalline as the film thickness increases. The diffraction pattern, shown as inset to Fig 4, was taken from the film and substrate using a large selected area aperture. The pattern shows the superposition of the $(0\bar{1}10)$ pattern for the sapphire substrate, the $(\bar{1}2\bar{1}0)$ pattern from the single crystalline AlN layer, and several spots in a ring pattern representing the polycrystalline portion of the film. Fig 3. shows that the film, especially the upper layer, is highly defective. The upper layer is columnar and contains many regions of Moiré fringes, which arise due to the film's polycrystallinity. AFM results show that sample B has a rougher surface than sample A, and contains several small grains.

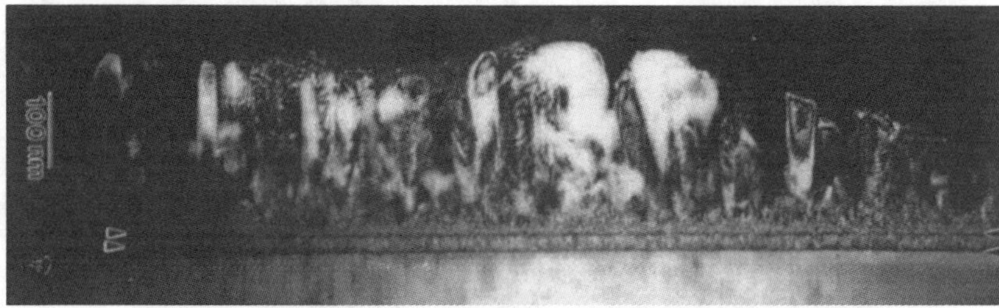

Figure 3. (0002) DF image of sample B

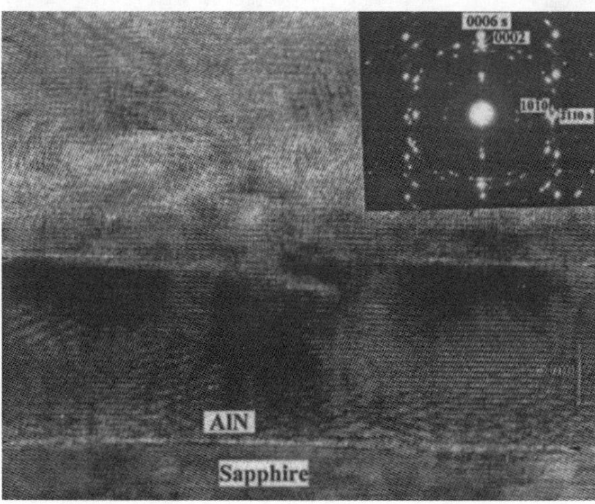

Figure 4. High resolution image of sample B. The zone axes are $(0\bar{1}10)_s / (\bar{1}2\bar{1}0)_f$.

Another example of a sample showing a second interface is shown in Figs. 5 – 6, and will be referred to as sample C. This sample was grown with a high ammonia flow of 190 sccm. The second interface is not as sharp as seen in sample B, and appears to consist of several dislocations and stacking faults. The film is crystalline 2H AlN on both sides of the interface and throughout the entire film. The (0002) DF image in Fig. 5. shows that this film is columnar, but does not contain Moiré fringes as seen in sample B. AFM results show that sample C has small grains and is slightly rougher than sample B.

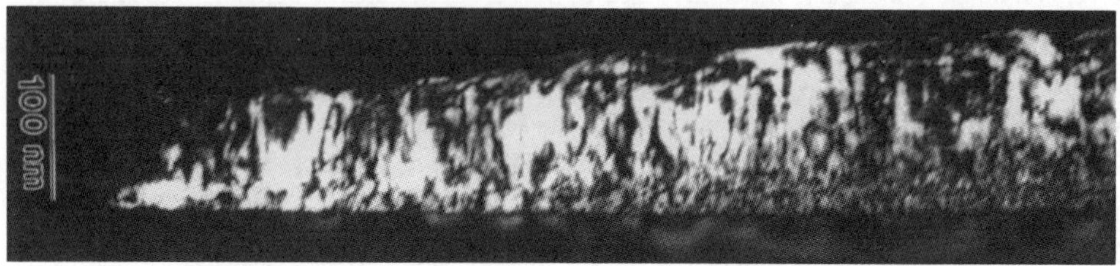

Figure 5. (0002) DF image of sample C

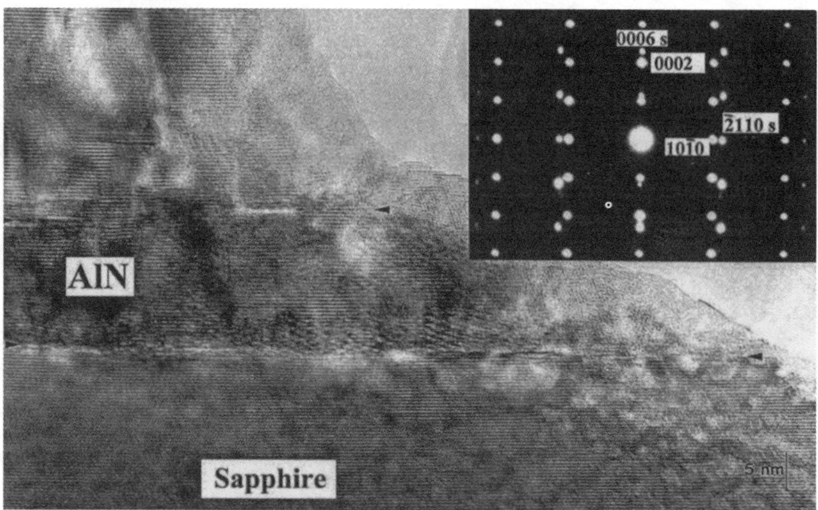

Figure 6. High resolution TEM image of sample C. The zone axes are $(0\bar{1}10)_s/(\bar{1}2\bar{1}0)_f$.

Fig. 7 is a high resolution image of sample D, which contains a second and third interface. This film was grown with an ammonia flow rate below 170 sccm, but at a higher temperature than all of the other films grown in this experiment. The secondary interface is located 18 nm from the film/substrate interface, similar to samples B and C. High resolution images show that the film below the interface, labeled region 1 in Fig. 7, consists of the 2H AlN polytype. The region above the interface (region 2) shows a markedly different atomic arrangement than the one seen in the film below the interface. This region may consist of 6H

AlN. The diffraction pattern (inset to Fig. 7), shows the superposition of the $(0\bar{1}10)$ pattern for the substrate, the $(\bar{1}2\bar{1}0)$ pattern for the film, and several extra spots. Microdiffraction confirms that these extra spots come from the region between the second and third interfaces. The film above the third interface (region 3) consists of the 2H polytype. We do not currently know why our growth conditions would lead to the film morphology seen in this sample. Further investigation, including photoluminescence and x-ray analysis, are needed to identify the structure seen in region 2. AFM images show that the surface of sample D consists of several large grains and is approximately 3 times rougher than sample B or sample C.

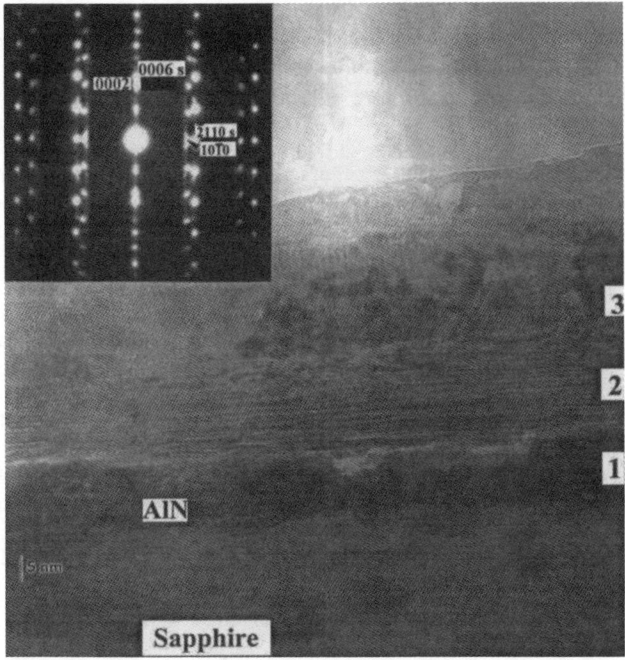

Fig. 7. High resolution TEM image of sample D. The zone axes are $(0\bar{1}10)_s/(\bar{1}2\bar{1}0)_f$.

DISCUSSION

Secondary interfaces, which we believe may be inversion domain boundaries, were seen in several samples, such as those represented by samples B and C. The IDBs were consistently located between 18 and 20 nm from the substrate. In a similar experiment with AlN/6H-SiC films, we observe IDBs consistently located at 100 nm from the film/substrate interface. It is possible that the location of the IDB is related to the degree of film/substrate lattice mismatch, which is in turn related to the threading dislocation density. The DF images of samples containing IDBs indicate that there is a higher density of threading dislocations below the IDB than above it. Also, the majority of the dislocations are not continuous across the IDB. Since dislocations cannot end inside a crystal, the threading dislocations must bend on the plane of the IDB. Despite the reduction in dislocation density above the IDB, the film quality generally degrades above the IDB as the film becomes columnar.

All of the higher quality films we have observed, such as sample A, were grown with ammonia flow rates below 170 sccm and at temperatures ranging from 1110° to 1150° C. These films do not have the columnar structure, IDBs, or Moiré fringes commonly observed in AlN/Sapphire films, such as samples B and C.

Sample D, which was grown with an ammonia flow rate of 130 sccm but at a higher temperature of 1160° C showed regions of mixed polytype. We do not currently now why these growth parameters would induce a polytype transformation. Further experiments are planned to determine which MOCVD reactor conditions would cause the structure seen in sample D to arise. Other than sample D, which was grown at the higher temperature than all of the other samples in this study, we did not find a correlation between small temperature deviations and the film morphology.

Films grown with ammonia flows above 170 sccm and within the temperature range 1110°C - 1150°C such as samples B and C, contain a high density of defects and secondary interfaces. Inversion domain boundaries are seen in all samples grown with ammonia flows at or above 170 sccm. We suggest that the excess ammonia may have caused N-N atomic bonding, causing the secondary interface to arise.

CONCLUSION

We have found the optimum TMA/NH_3 flow rate ratio range for AlN films grown on sapphire by MOCVD. We find that samples grown within the temperature range 1110° – 1150° C with NH_3 flow rates between 130 and 170 sccm and TMA rates of 15 sccm are of high crystalline quality, have a smooth surface, and do not have inversion domain boundaries or a columnar structure. In contrast, samples grown with higher NH_3 flow rates consist of a columnar structure, have a rougher surface and contain inversion domain boundaries at 18-20 nm from the film/substrate interface. These IDBs seem to be related to threading dislocations in the film. Above the IDBs the film's quality is degraded and in some instances the film becomes polycrystalline. Further work is necessary to understand the formation of IDBs and their dependence on the growth conditions.

ACKNOWLEDGEMENTS

This work was supported by MRCP Army Grant No. DAAL 019523530.

REFERENCES

1. J.P. Michel, I. Masson, S. Choux, and A. George, Phys. Stat. Sol, **146**, 97 (1994).
2. A. Westwood, M. Notis, J. Am. Ceram. Soc. **74**, 1226 (1991).

TEMPERATURE DEPENDENT MORPHOLOGY TRANSITION OF GaN FILMS

A.R.A. ZAUNER, F.K. DE THEIJE, P.R. HAGEMAN, W.J.P. VAN ENCKEVORT, J.J. SCHERMER, AND P.K. LARSEN
Research Institute for Materials, University of Nijmegen, Toernooiveld, 6525 ED Nijmegen, The Netherlands

ABSTRACT

The temperature dependence of the surface morphology of GaN epilayers was studied with AFM. The layers were grown by low pressure MOCVD on (0001) sapphire substrates in the temperature range of 980°C-1085°C. In this range the (0001) and $\{1\bar{1}01\}$ faces completely determine the morphology of 1.5 μm thick Ga-faced GaN films. For specimens grown at 20 mbar and temperatures below 1035°C the $\{1\bar{1}01\}$ faces dominate the surface, which results in matt-white layers. At higher growth temperatures the morphology is completely determined by (0001) faces, which lead to smooth and transparent samples. For growth at 50 mbar, this transition takes place between 1000°C and 1015°C. It is shown that the morphology of the films can be described using a parameter α_{GaN}, which is proportional to the relative growth rates of the (0001) and the $\{1\bar{1}01\}$ faces.

INTRODUCTION

Due to its material properties gallium nitride (GaN) has attracted enormous attention in recent years, certainly after the success of GaN-based blue light emitting diodes. The lack of lattice matched substrates for epitaxial growth of GaN films has led to the application of a variety of substrates of which sapphire is most frequently used [1].

Despite the large mismatch in lattice constants between GaN and sapphire, device quality layers can be obtained using metalorganic chemical vapour deposition (MOCVD) and a two step growth procedure. On top of an initial buffer layer [2] a GaN film is deposited at relatively high temperatures. In this paper the influence of the deposition temperature on the surface morphology of the layers is investigated using atomic force microscopy (AFM) and scanning electron microscopy (SEM).

EXPERIMENTAL

The GaN layers were grown in a horizontal MOCVD reactor equipped with a SiC coated graphite susceptor with a hydrogen gas driven rotating disc to obtain optimum uniformity during the growth process. The growth temperatures calibrated for the centre of the disc are determined by a thermocouple. The discharge of the hydrogen flow causes a small temperature decrease towards the periphery of the disc.

Immediately before growth, the two-inch sapphire (0001) substrates were cleaned in organic solvents, etched in a solution of $HCl:HNO_3 = 3:1$, and finally rinsed in de-ionised water. All deposition runs started with nitridation of the substrate surface, carried out in a nitrogen/ammonia (N_2/NH_3) gas stream at 1030°C, followed by deposition of a 20 nm thick GaN buffer layer at 500°C. On this buffer layer, the GaN films were grown.

Using trimethylgallium (TMG) and NH_3 as precursors, and hydrogen (H_2) as carrier gas, two series of GaN layers, one at 20 mbar and the other at 50 mbar total reactor pressure, were grown at temperatures between 980°C and 1085°C. Growth was performed with a TMG flow of 63 μmol/min and a NH_3 flow of 2.5 standard litre per minute (slm), diluted with H_2 to a total

345

Mat. Res. Soc. Symp. Proc. Vol. 572 © 1999 Materials Research Society

flow of about 5 slm. All samples were grown for 1 hour which resulted in a GaN layer thickness of about 1.5 μm. AFM and SEM were applied to study the surface morphology of the samples.

RESULTS AND DISCUSSION

For the naked eye, lower deposition temperatures result in matt-white samples, whereas at higher deposition temperatures the layers look colourless and mirror-like. In figure 1 the specimen appearance is given as function of growth temperature and pressure. It shows that the transition from matt-white to mirror-smooth appearance occurs between 1000°C and 1015°C for samples grown at 50 mbar, and at 1035°C for those grown at 20 mbar. To ensure that changes in polarity are not the cause for the observed morphology transition, the polarity of the deposited GaN layers is determined by anisotropic etching [3]. For all samples it is found that growth occurred in the [0001] direction (Ga-face).

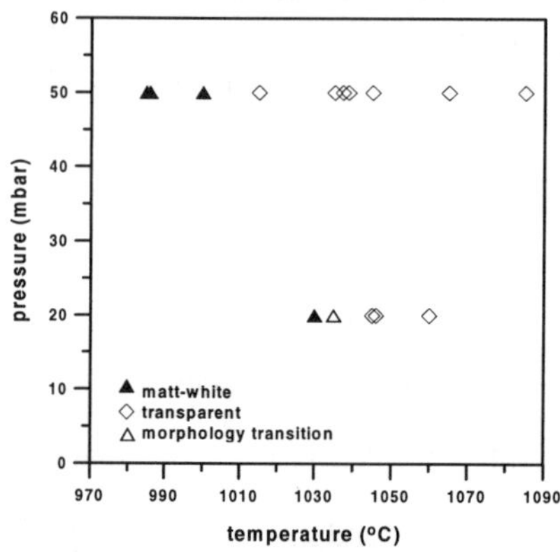

Figure 1. The appearance of about 1.5 μm thick GaN layers versus growth temperature for two different pressures

One sample, grown at 20 mbar and 1035°C, forms an excellent specimen to study the morphology transition, since it exhibits a mixture of both appearances. This sample, which will be referred to as sample 'B', has a matt-white appearance at the edges, where the deposition temperature has been slightly lower than at the central region, and is mirror-like and transparent at the centre. To determine the temperature difference during growth along sample B, the centre of a complete matt-white sample grown at 1030°C and 20 mbar and slightly thicker as sample B is used as a reference. AFM examination of this sample shows a morphology comparable to that on sample B at a position halfway between the periphery and centre. This indicates, taking the thickness difference into account, that the growth temperature near the edges of sample B has been 1027 ± 3 °C. For which can be concluded that a temperature gradient of only about 8°C can cause a dramatic change in morphology.

In figure 2 an AFM image from the periphery of sample B is shown. The main characteristic of the surface is formed by large and irregularly shaped pits of about 300-600 nm wide, with crystallographically oriented side faces and with a density of about 1.2 x 10^9 cm^{-2}. The depth of the pits is at least 250 nm, however, the sides are too steep to give reliable depth measurements using AFM. SEM observations showed that the angle between the pit walls and the (0001) face is about 60°.

From the crystal structure of wurtzite-GaN and its lattice constants [4] the theoretical crystal morphology of GaN can be determined using the 'connected net' theory [5,6]. It states that {hklm} faces parallel to a network of atoms interconnected by strong bonds, within a slice thickness d_{hklm}, determine the morphology of a crystal. Three faces, {0001}, {1$\bar{1}$01} and

$\{1\bar{1}00\}$, are found to be parallel with such a connected net. The occurrence of these faces was indeed reported in literature [7,8].

The intersection lines and the inclination of the previously mentioned pit walls with the (0001) surface indicate that these side walls are in fact $\{1\bar{1}01\}$ faces, for which an angle of 61.9° with respect to the (0001) face can be calculated. Since growth is performed on (0001) sapphire substrates and no etching of the GaN surface is expected to occur, it can be concluded that the pits are induced by the coalescence of islands bounded by $\{1\bar{1}01\}$ facets. The same conclusion was drawn in references [9,10] on the basis of surface topographs showing the nucleation, lateral expansion, and coalescence of islands for GaN grown on c-plane sapphire by MOCVD.

For the surface region located in the intermediate zone between the edge and centre of sample B, an AFM image is given in figure 3. It shows the transition stage between the lateral expansion of the islands and the dislocation induced step flow growth of the planar (0001) surface. The density of pits is still about 1.2×10^9 cm^{-2}, but most of them are partially closed to form about 7 nm wide grooves.

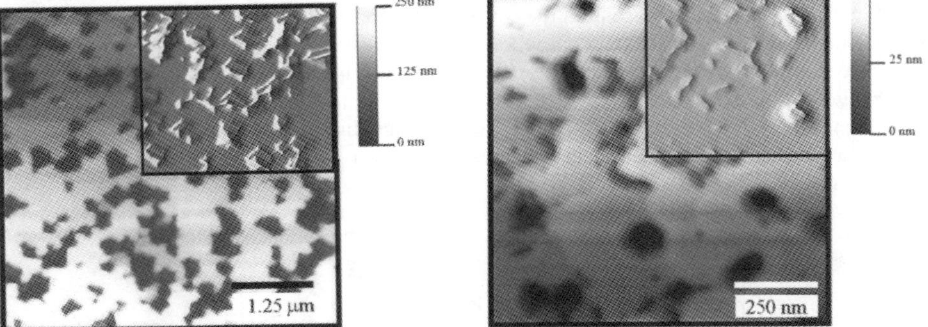

Figure 2. Height and deflection (inset) AFM image of a surface area near the periphery of sample B, grown at 1035°C and 20 mbar. Pits can be recognised which are bounded by $\{1\bar{1}01\}$ faces.

Figure 3. Height and deflection (inset) AFM image of a surface area between the periphery and the centre of sample B. Pits and grooves can be recognised.

At the centre of sample B (figure 4) the same kind of cavities is found as shown in figures 2 and 3. However, their average sizes and density are considerably less, and are found to be, respectively, 100 to 300 nm and about 4.0×10^8 cm^{-2}, indicating that the coalescence of the islands is almost complete.

From the results discussed above, it is clear that the deposition temperature largely influences the surface morphology of the GaN layers. It can be argued that a lower density of nucleation islands at the periphery induces the larger size of pits near the edge of the sample. However, it is known that a lower temperature is expected to result in higher chemical driving force for GaN growth, therefore, the nucleation rate will certainly not decrease [9]. To determine the nucleation density along the samples, a specimen is grown for 20 minutes to a thickness of 0.46 μm (instead of ≈ 1.5 μm) under simular conditions.

On this thinner sample, indicated as sample T, the separate islands can be distinguished. The island density proved to be constant along the entire sample, therefore the different morphologies cannot be explained by a different nucleation density.

At lower temperatures, the growth velocity in the $\langle 1\bar{1}01 \rangle$ directions is low compared to that in the [0001] direction, causing slow lateral growth of the islands, and the morphology of the 1.5 µm thick layers is dominated by $\{1\bar{1}01\}$ faces. For increasing temperatures, growth velocity in the $\langle 1\bar{1}01 \rangle$ direction increases compared to the growth velocity in the [0001] direction and therefore, the slow growth in the [0001] direction dominates the surface morphology.

Analogous to the CVD growth of diamond crystallites [11,12], a growth rate factor is defined to describe the morphology of the Ga faced GaN films. The growth rate factor α_{GaN} equals $(v_{[0001]} / v_{[1\bar{1}01]})\cos\varphi$, where $v_{[0001]}$ and $v_{[1\bar{1}01]}$ are the growth rates in the [0001] and $[1\bar{1}01]$ direction, respectively, and φ (=61.9°) is the angle between the (0001) and $\{1\bar{1}01\}$ faces. For $\alpha_{GaN} = 0$ no $\{1\bar{1}01\}$ faces develop and a completely closed (0001) layer is formed, for $\alpha_{GaN} \geq 1$ only $\{1\bar{1}01\}$ facets are present on the surface. The present work shows that besides a strong dependence on the deposition temperature, α_{GaN} is also pressure dependent.

Assuming that pits and islands are hexagonal in shape and (0001) faces top the islands, the values of α_{GaN} can be determined (figure 5). From the average (0001) surface area and the nucleation density of the islands, the smallest centre-edge distance x of the hexagonal island can be calculated. Using this value of x and the layer thickness t, α_{GaN} is given by

$\alpha_{GaN} = t / (t + x \tan\varphi)$.

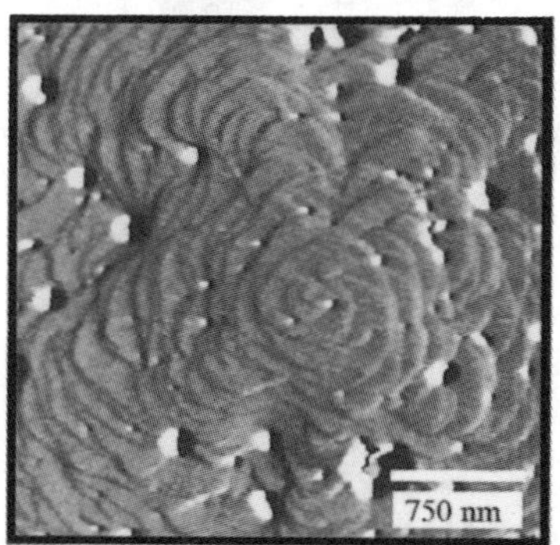

Figure 4. Deflection AFM image of the surface at the centre of sample B.

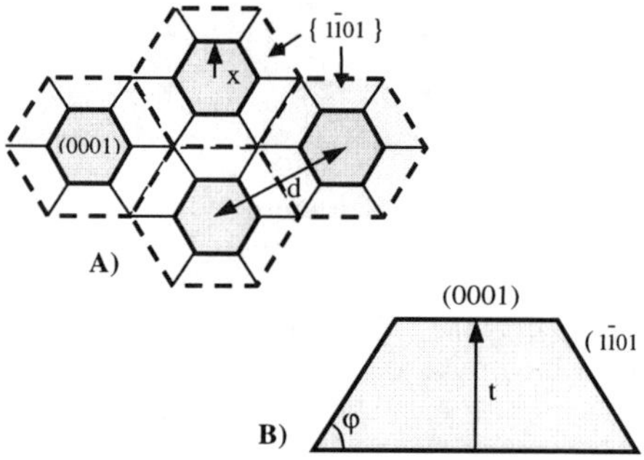

Figure 5. Top view (A) and cross section (B) of hexagonal islands.

The minimum layer thickness t_{min} necessary to obtain a smooth film depends on α_{GaN} and the distance d between the nuclei, and is given by

$$t_{min} = \frac{d\,\alpha_{GaN}\,\tan\varphi}{2 - 2\,\alpha_{GaN}} \qquad (1)$$

The surface area covered by the (0001) faces is estimated from AFM images along the surface of sample B. The nucleation density of the islands could be directly determined from sample T and was found to be 2.1×10^9 cm^{-2}. As already stated, the island density proved to be constant along sample T, it does not change as function of temperature and is assumed to be the same for sample B. SEM cross section observations on sample B showed that the growth rate in the [0001] direction was also constant, resulting in a constant layer thickness, t, of 1.48 ± 0.03 µm. For sample B a film with completely coalesced islands, i.e. the islands just touching each other with their (0001) faces, is obtained for α_{GaN} values ≤ 0.87. In table I it can be seen that the small temperature gradient across the substrate causes an increase of α_{GaN} values from 0.87 to 0.94. Since the layer thickness of sample B is constant, the change in α_{GaN} values is only caused by a difference in $v_{[1\bar{1}01]}$.

Table I. Growth rate factor α_{GaN} across sample B

distance from centre (10^{-2} m)	α_{GaN} values
0.00	0.87
0.25	0.88
0.50	0.89
0.75	0.91
1.00	0.93
1.25	0.94

More detailed AFM measurements on the centre of sample B (see figure 6) show a high density of interacting growth spirals, which emerge from dislocations with a screw component. Theoretical calculations [13] revealed that the GaN surface during growth has a surface energy comparable with a surface free of adsorbates. This implies high step and kink energies, which predict strongly polygonized growth spirals [14]. However, all spirals are found to be circular. This can be explained by assuming that crystal growth is limited by surface diffusion rather than by the integration of growth units at the step sites. The spirals consist of monoatomic (2.5 ± 0.3 Å) or double steps (5.0 ± 0.4 Å), favourably comparing with the d_{0002} distance of 2.59 Å. Steps emerging from dislocations have in general a height of 2 d_{0002}, indicating a screw component [0001].

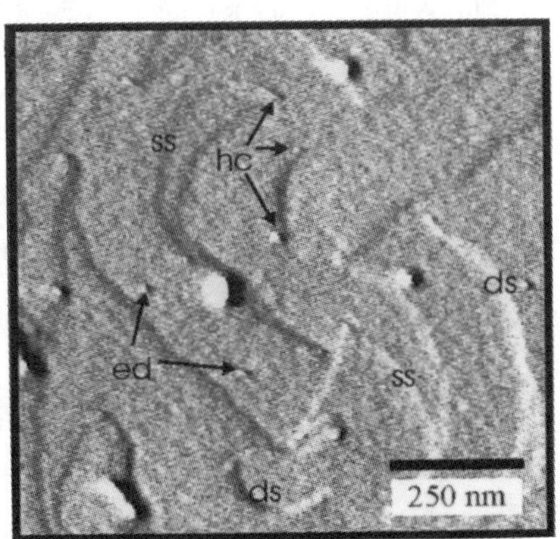

Figure 6. Deflection AFM image of the surface in the centre of sample B at high magnification. Monoatomic (ss) and double (ds) steps are shown emerging from hollow cores (hc) at dislocations with a screw component. Hollow cores related to edge dislocations (ed) can be recognised.

On these GaN layers also a number of small holes is observed with a density of 1.5×10^9 cm^{-2}. The holes often coincide with dislocations containing a screw component and have a narrow diameter distribution; 48 ± 9 nm. The diameter of these hollow cores is much wider than can be expected from standard stress theory of dislocations [15]. The large diameter can possible be explained by the precipitation of vacancies to form voids along the dislocation lines[16].

CONCLUSIONS

The morphology of ≈ 1.5 μm thick Ga-faced GaN epilayers, grown by MOCVD on (0001) sapphire substrates, is strongly temperature dependent. For the investigated temperature range of 980-1085°C the growth morphology is determined by (0001) and $\{1\bar{1}01\}$ faces. To describe the observed morphology a growth rate factor α_{GaN} is introduced. At higher growth temperatures the morphology is governed by growth in [0001] direction, whereas at lower growth temperatures, a slower lateral growth in the $\langle 1\bar{1}01 \rangle$ directions determines the morphology for a longer time. On the mirror-like surfaces single and double steps can be seen emerging from the hollow cores of dislocations with a screw component.

ACKNOWLEDGEMENTS

This work has been financially supported by the Dutch Technology Foundation (STW) and the Dutch Technology Foundation for Chemical Research (SON).

REFERENCES

[1] *III-V Nitrides*, edited by F.A. Ponce, T.D. Moustakas, I. Akasaki, and B.A. Monemar (Mater. Res. Soc. Proc. **449**, Pittsburgh, Pennsylvania, 1996)
[2] H. Amano, N. Sawaki, I. Akasaki, Y. Toyoda, Appl. Phys. Lett. **48**, 353 (1986)
[3] J.L. Weyher, S. Müller, I. Grzegory, S. Porowski, J. Crystal Growth **182**, 17 (1997)
[4] F.A. Pearton, in *Optoelectronic Properties of Semiconductores and Superlattices, volume 2:, GaN and Related Materials*, edited by S.J. Pearton (Gordon and Breach Science Publishers, Amsterdam, 1997)
[5] P. Bennema, in *Handbook of Crystal Growth, Volume 1*, edited by D.T.J. Hurle (North-Holland, Amsterdam, 1993) 477
[6] R.F.P. Grimbergen, H. Meekes, P. Bennema, C.S. Strom, L.J.P. Vogels, Acta Crystallogr. A **54**, 491 (1998)
[7] T. Akasaka, Y. Kobayashi, S. Ando, N. Kaboyashi, Appl. Phys. Lett. **71**, 2196 (1997)
[8] T. Kozawa, M. Suzuki, Y. Taga, Y. Gotoh, J. Ishikawa, J. Vac. Sci. Technol. B **16** (1998) 833
[9] X.H. Wu, P. Fini, S. Keller, E.J. Tarsa, B. Heying, U.K. Mishra, S.P. DenBaars, S.J. Speck, Jpn. J. Appl. Phys. **35**, L1648 (1996)
[10] P. Vennéguès, B. Beaumont, S. Haffouz, M. Vaille, P. Gibart, J. Crystal Growth **187**, 167 (1998)
[11] C. Wild, P. Koidl, W. Müller-Sebert, R. Kohl, N. Herres, R. Locher, R. Samlenski, R. Brenn, Diamond Relat. Mater. **2**, 158 (1993)
[12] R.E. Clausing, L. Heatherly, L.L. Horton, E.D. Specht, G.M. Begun, Z.L. Wang, Diamond Relat. Mater. **1**, 411 (1992)
[13] J.E. Northrup, R. Di Felice, Phys. Rev. B **56** (1997) R4325
[14] W.J.P. van Enckevort, in *Facets of Fourty Years of Crystal Growth*, edited by W.J.P. van Enckevort, H.L.M. Meekes, and J.W.M. van Kessel (University of Nijmegen, 1997, internal publication), 62
[15] F.C. Frank, Acta Crystallogr. **4** (1951) 327.
[16] P. Coulomb, J. Friedel, in *Dislocations and Mechanical Properties of Crystals,* edited by J.C. Fisher, W.G. Johnston, R. Thomson and T. Vreeland (Wiley, New York, 1957) 555.

COMPARATIVE STUDY OF EMISSION FROM HIGHLY EXCITED (In, Al) GaN THIN FILMS AND HETEROSTRUCTURES

B.D. Little*, S. Bidnyk*, T.J. Schmidt*, J.B. Lam*, Y.H. Kwon*, J.J. Song*, S. Keller**, U.K. Mishra**, S.P. DenBaars**, W. Yang***

*Center for Laser and Photonics Research and Department of Physics, Oklahoma State University, Stillwater, OK 74078
**Computer Engineering and Materials Departments, University of California, Santa Barbara, CA 93106
***Honeywell Technology Center, Plymouth, MN 55441

ABSTRACT

The optical properties of (In, Al) GaN thin films and heterostructures have been compared under the conditions of strong nanosecond excitation. The stimulated emission (SE) threshold from AlGaN epilayers was found to increase with increasing Al content compared to GaN, in contrast to InGaN epilayers, where an order of magnitude decrease is observed. Optically pumped SE has been observed from AlGaN films with aluminum concentrations as high as 26%. Room temperature SE at wavelengths as low as 327 nm has been achieved. In contrast to the increase of SE threshold seen for AlGaN films, we found that AlGaN/GaN heterostructures which utilize carrier confinement and optical waveguiding drastically enhance the lasing characteristics. We demonstrate that AlGaN/GaN heterostructures are suitable for the development of deep ultraviolet laser diodes.

INTRODUCTION

(Al, In) GaN epilayers and heterostructures have drawn much attention in recent years due to the potential for blue/UV light emitters and detectors for use in high density data storage, high temperature electronics, solar-blind detectors, atmospheric sensing, and medicine [1]. Recently, nearly a dozen research groups have demonstrated lasing in InGaN-based heterostructures, with the lowest reported emission wavelength being 376 nm [2]. However, to obtain shorter wavelength laser diodes, it is necessary to use AlGaN-based structures. Preliminary studies have shown that the incorporation of Al into GaN increases the stimulated emission (SE) threshold [3]. In this work, we demonstrate that the introduction of strong optical and carrier confinement into AlGaN/GaN heterostructures can significantly reduce the SE threshold. We also demonstrate SE in AlGaN epilayers with emission wavelengths as low as 327 nm at room temperature, illustrating that AlGaN is a suitable material for the development of deep ultraviolet laser diodes.

EXPERIMENT

The GaN, InGaN, and AlGaN epilayers studied were grown on (0001) sapphire substrates by MOCVD. The thickness of the $Al_xGa_{1-x}N$ layers were ~0.8 µm, and had alloy concentrations of x = 0.17 and 0.26. For the purpose of comparison, we used GaN and InGaN epilayers with thicknesses ranging from 100 nm to 7.2 µm. The sample growth parameters have been reported elsewhere [4, 5]. We also studied an AlGaN/GaN separate confinement heterostructure (SCH) grown by MBE on 6H-SiC. The active region of the SCH was a 70 Å thick GaN layer, which was sandwiched between a 600 Å $Al_{0.06}Ga_{0.94}N$ cladding layer and a 2300 Å $Al_{0.11}Ga_{0.89}N$

Mat. Res. Soc. Symp. Proc. Vol. 572 © 1999 Materials Research Society

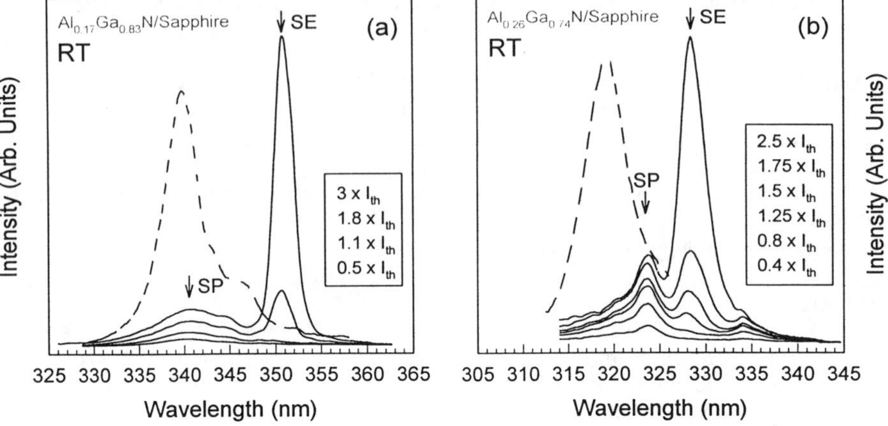

Fig. 1. RT SE spectra at several pump intensities above and below the SE threshold, I_{th}, for AlGaN layers with alloy concentrations of (a) 17% and (b) 26%. Low-power cw spontaneous emission is given by the dashed line.

waveguiding layer. The structure was deposited on top of a ~3 μm GaN buffer layer.

The experimental setup for the study of SE consisted of nanosecond tunable dye lasers pumped by either the doubled output of an injection seeded Nd:YAG laser (~7 ns pulsewidth and 10 Hz repetition rate at 532 nm) or the 308 nm XeCl line of an excimer laser (~8 ns pulsewidth and 10 Hz repetition rate). In the case of Nd:YAG pumping, the deep orange output of the dye laser was doubled to ~310 nm using a nonlinear crystal. The excimer-pumped dye laser emitted directly in the UV. All experiments were performed in the edge-emission geometry. The UV laser excitation source was focused to a line on the sample surface using a cylindrical lens [6]. A continuously variable ND filter was used to attenuate the laser power. The resultant emission was collected from the edge of the sample and focused onto the slits of a 1 meter spectrometer with a UV-enhanced gated CCD. The samples were mounted to the cold finger of a closed cycle helium refrigerator that allowed continuous temperature tuning from 10 K to 300 K. Photoluminescence (PL) was performed using the 244 nm line of an intracavity frequency doubled cw Ar$^+$ laser. PL was performed in a backscattering geometry to avoid distortion of the spectra due to reabsorption effects. For PL excitation (PLE) spectroscopy, the quasi-monochromatic light from a Xe lamp was dispersed by a 1/2 m spectrometer and used as the excitation source. The signal was collected from the sample and focused on the slits of a 1 meter double grating spectrometer coupled to a photomultiplier tube for detection.

RESULTS

The SE emission from AlGaN epilayers qualitatively resembles that from GaN epilayers. However, the emission wavelength lies in the deep UV. The RT emission spectra for AlGaN layers with alloy concentrations of 17% and 26% are shown in Figures 1 (a) and (b), respectively, for excitation densities above and below the SE threshold intensity I_{th}. The dashed

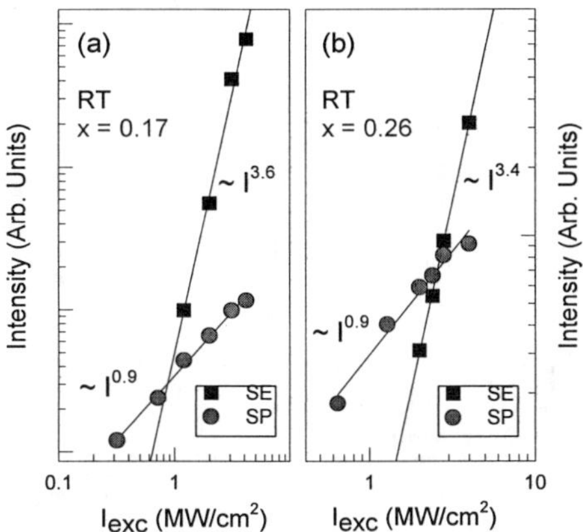

Fig. 2. Power dependence of the peaks marked in Fig. 1 at RT. The spontaneous emission peaks, denoted by circles, show an approximately linear increase with excitation intensity. SE exhibits a superlinear increase with power, and is represented by squares.

line indicates low power cw PL results. As seen from the figure, SE for both samples emerges out of the low energy wing of the spontaneous emission peak as the excitation density is increased. We believe that the SE observed from these epilayers (330 nm at RT) is the shortest ever reported in the literature for a semiconductor material. The spontaneous and stimulated emission peaks are separated by 10.5 and 8.5 nm, respectively, for the 17% and 26% epilayers, which is comparable to the spacing observed in GaN epilayers [7] and is considerably smaller than that seen in InGaN epilayers [8]. The redshift of the spontaneous emission seen under nanosecond excitation in Figs. 1 (a) and (b) compared to the cw spontaneous emission peaks is due to a reabsorption of the emitted radiation as it travels along the excitation path in the edge-emission experiments.

Figure 2 illustrates the emission intensity of the spontaneous and stimulated emission peaks as a function of excitation intensity I_{exc} for the AlGaN epilayers presented in Fig. 1. The spontaneous emission is seen to increase roughly linearly across the entire range of I_{exc}. Above I_{th}, we see the emergence of a new peak which exhibits a superlinear increase in emission intensity with I_{exc}, clearly indicating the onset of SE [9]. The values of I_{th} obtained from the AlGaN epilayers are slightly larger than those obtained from high quality GaN epilayers, and more than an order of magnitude greater than that of InGaN epilayers [3, 7, 9]. The high SE threshold for AlGaN epilayers makes the development of laser diodes using this material rather challenging.

Through this study, however, we show that AlGaN/GaN-based heterostructures can be used to produce lasing with both short emission wavelengths and low lasing thresholds. The SCH used in this study possesses a high degree of optical and carrier confinement. By pumping the sample at 335 nm, we obtained lasing at 358 nm (362 nm) at 10 K (RT) with a lasing threshold as low as 125 kW/cm^2. Figure 3 (a) shows the RT emission spectra from the SCH at several excitation intensities above and below I_{th}. The shorter wavelength peak at 341 nm is due to spontaneous emission from the x = 0.06 cladding region. As I_{exc} is increased, we see a narrow

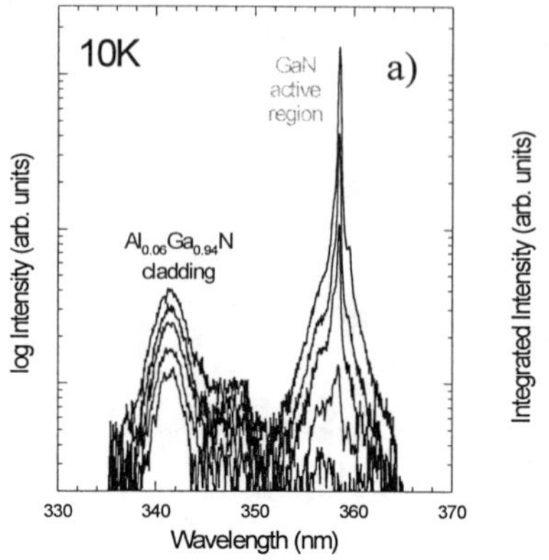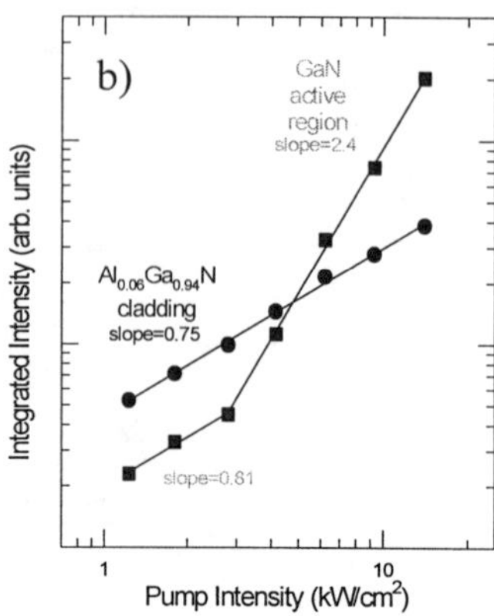

Fig. 3. a) 10 K power dependence of the SCH at several pump intensities above and below the lasing threshold. The peak at 341 nm is due to the cladding layer. b) Intensity-dependent behavior of the peaks shown in a). An approximately linear increase is seen for the cladding layer for all pump intensities and for the active region until the lasing threshold is reached. At higher intensities, a superlinear increase is observed at 358 nm along with spectral narrowing. The FWHM of the lasing peak is 3 Å.

lasing peak (3 Å FWHM) emerge at a wavelength of 358 nm. We did not observe any broadening of this peak as the temperature was raised up to RT. This indicates that the FWHM of the lasing peak is determined by the finesse of the cavity, which remains independent of temperature. In our previous study, we showed that the cavity is formed by microcracks caused by strain relaxation of AlGaN grown on SiC [10]. Fig. 3 (b) shows the detailed behavior of the emitted intensity as a function of I_{exc}. As can be seen from the graph, the cladding layer emission increases almost linearly with increasing I_{exc}. The emission from the active region behaves linearly until I_{th} is reached, after which a superlinear increase is observed. The RT SE threshold value for this sample was determined to be 125 kW/cm^2, representing a drastic reduction in comparison to GaN and AlGaN epilayers. This low threshold value is made possible by the carrier confinement and optical waveguiding properties of the structure. In fact the lasing threshold is comparable with the best InGaN epilayers which makes it suitable for the development of short wavelength LDs ($\lambda_{emission}$ < 370 nm at RT). It may be possible to further reduce the threshold by optimizing the sample parameters such as active layer thickness as well as the thickness and alloy concentrations of the cladding and waveguiding region.

CONCLUSION

We have systematically studied the stimulated emission properties of (In, Al) GaN thin films and heterostructures. The stimulated emission threshold of AlGaN epilayers was found to increase with increasing Al content compared to GaN, in contrast to InGaN epilayers, where an order of magnitude decrease is observed. Room temperature stimulated emission was observed at remarkably short wavelengths, demonstrating that AlGaN-based structures are a suitable material for deep ultraviolet laser diodes. Furthermore, we achieved a substantial reduction in the lasing threshold for AlGaN/GaN-based heterostructures. We showed that optical and carrier confinement will play the key roles in the reduction of the lasing threshold in these structures.

ACKNOWLEDGEMENTS

The work at Oklahoma State University was funded by BMDO, DARPA, and ONR.

REFERENCES

1. S. Nakamura and G. Fasol, *The Blue Laser Diode*, (Springer, Berlin, 1997).
2. I. Akasaki, S. Sota, H. Sakai, T. Tanaka, M. Koike, and H. Amano, Electron. Lett. **32**, 1105 (1996).
3. T. J. Schmidt, Yong-Hoon Cho, J. J. Song, and Wei Yang, Appl. Phys. Lett. **74**, 245 (1999).
4. S. Keller, A. C. Abare, M. S. Minsky, X. H. Wu, M. P. Mack, J. S. Speck, E. Hu, L. A. Coldren, U. K. Mishra, and S. P. DenBaars, Mater. Sci. Forum **264-268**, 1157 (1998).
5. S. Bidnyk, T. J. Schmidt, G. H. Park, and J. J. Song, Appl. Phys. Lett. **71**, 729 (1997).
6. X. H. Yang, T. J. Schmidt, W. Shan, J. J. Song, and B. Goldenberg, Appl. Phys. Lett. **66**, 1 (1995).
7. S. Bidnyk, T. J. Schmidt, B. D. Little, and J. J. Song, Appl. Phys. Lett. **74**, 1 (1999).
8. Yong-Hoon Cho, T. J Schmidt, S. Bidnyk, J. J. Song, S. Keller, U. K. Mishra, and S. P. DenBaars, Proc. MRS Fall **G6.54**, 161, Boston (1998).
9. S. Bidnyk, T. J. Schmidt, Y. H. Cho, G. H. Gainer, J. J. Song, S. Keller, U. K. Mishra, and S. P. DenBaars, Appl. Phys. Lett. **72**, 1623 (1998).
10. J. J. Song, A. J. Fischer, T. J. Schmidt, S. Bidnyk, and W. Shan, Nonlinear Optics **18** (2-1), 269 (1997).

ATOMIC SCALE ANALYSIS OF InGaN MULTI-QUANTUM WELLS

M. Benamara, Z.Liliental-Weber, W. Swider and J. Washburn,
E.O. Lawrence Berkeley National Laboratory, Berkeley CA 94720.
R. D. Dupuis, P. A. Grudowski and C. J. Eiting,
Microelectronics Research Center, University of Texas, Austin TX 78712.
J. W. Yang and M. A. Khan,
ECE Dept., University of South Carolina, Columbia, SC 29208.

Abstract

InGaN multiquantum wells grown by MOCVD on GaN have been investigated by transmission electron microscopy techniques and numerical analysis of high resolution (HREM) images. One objective of this research was to correlate the atomic structure and emission mechanisms of InGaN quantum well. The studied layers contained 13% or 20% In. It was shown that GaN/InGaN interfaces are rather rough and exhibit an oscillating contrast. Structural defects were found on these interfaces. The relative c-lattice parameter variation in the well was determined using numerical processing of HREM images. The lattice spacings appear to be larger than that expected from Vegard's law suggesting the presence of a biaxial strain. Further observations also revealed a redistribution of In within the well. Instead of a continous In-rich layer, quantum dots were often observed along the well with a regular spacing. The formation of these In-rich dots was not intented and their presence suggests either a periodic modulation of strain along the well or In-rich cluster formation.

Introduction

Recent years have witnessed a strong interest in III-nitride semiconductors and especially InGaN alloys because of the wide spectral range covered by these materials. These counpounds in epitaxial layers are known to be highly strained. This strain affects the structural and physical properties of the material through several mechanisms. It was recently shown by P. Perlin et al.[1] that strain induced piezoelectric fields are responsible for the emissions from InGaN quantum-wells (QW). Another explanation was given by Chichibu et al.[2] where emissions were suggested to arise from the recombination of excitons located at potential minima along the QW. This was supported by Krüger et al.[3] and by Grudowski et al.[4] whose photoluminescence measurements on InGaN QW showed a broadening of peaks. Thus, there is still some controversy about the emissions mechanisms of InGaN multi-quantum wells (MQW). Furthermore, the low miscibility of InN in GaN[5] and the large lattice mismatch between these two counpounds (11 %) affect indirectly the emissions through phase separation[6], segregation or defect generation[7]. In this paper, we report on an atomic scale study of the microstructure of InGaN MQW using High-Resolution Electron Microscopy (HREM).

Experiment

Several InGaN/GaN MQW structures were grown by metal organic chemical vapor deposition (MOCVD) on 1 μm thick GaN layer using conventional (0001) oriented sapphire substrates. Details of the growth conditions have been published elsewhere[4]. The structure of sample 1 consists of five periods of 2.5 nm $In_{0.13}Ga_{0.87}N$ separated by 50 nm $In_{0.03}Ga_{0.97}N$ barriers. The structure of sample 2 consists of twenty periods of 2.5 nm $In_{0.20}Ga_{0.80}N$

Mat. Res. Soc. Symp. Proc. Vol. 572 © 1999 Materials Research Society

separated by 50 nm GaN barriers. The TEM samples were prepared by standard procedure using ion-milling at liquid nitrogen temperature and examined using both Topcon 002B and Jeol ARM microscopes operating at 200 kV and 800 kV respectively. Before the observations, the samples were mounted on a gold covered grid and then dipped in a KOH (50%) solution for 5 min. This etch step was aimed at removing damage created during the ion milling.

For the numerical processing, HREM images were first digitized using a Kodak camera. An intensity profile for all bright spots ('blob') in the image is fitted using an apprpopriate two-dimensional function so that the exact coordinates of its center is accurately determined with sub-pixel resolution. Since each bright spot in the image corresponds to either an atomic column or an electron pathway (depending on the imaging conditions), the displacement of the atomic positions away from their bulk values can be determined for each of the unit cells along any crystallographic direction. This allows to maping the strain distribution. Thickness and composition have the same effect on the projected potential. Since the projected potential can be extracted and mapped from the intensity distribution within each unit cell, chemical fluctuations or thickness variations can be revealed. A detailed explanation of the procedure can be found in the original papers[8-9].

Result

Fig. 1 is a conventional image of sample 1 (five QW) structure taken along the [11$\overline{2}$0] direction. Although the In content in the wells is only 10% different from that in the barriers, the stack of layers is clear. Layers appear homogeneous and are easy to distinguish. The dark stripes correspond to the wells ($In_{0.13}Ga_{0.87}N$). A magnification of one the wells is shown in fig. 2. The layers are pseudomorphic to the underlying GaN layer and no misfit dislocations are present at the interfaces. Interfaces with barrier layers on either side are not sharp but rather are rough. An average thickness of layers was deduced from several parts of the sample and was equal to about 2.5 nm (5 basal planes). The striking feature of this image is that the layer does not appear homogeneous; it exhibits an oscillating contrast, which was also observed earlier[10].

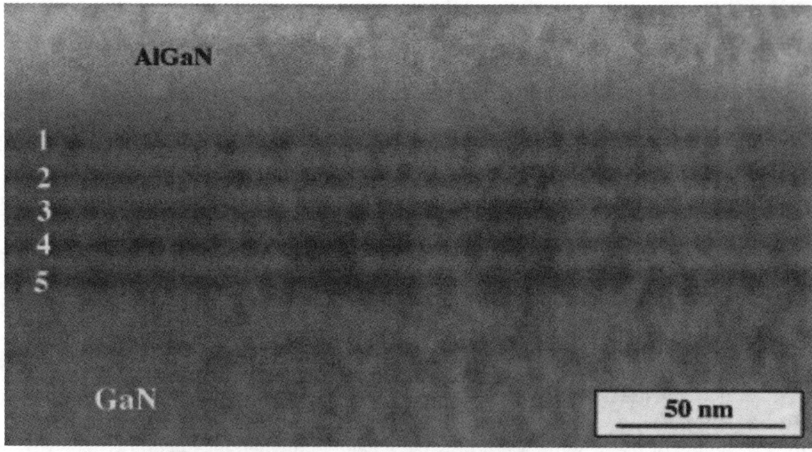

fig 1: TEM image showing the general structure of the 5 QW sample.

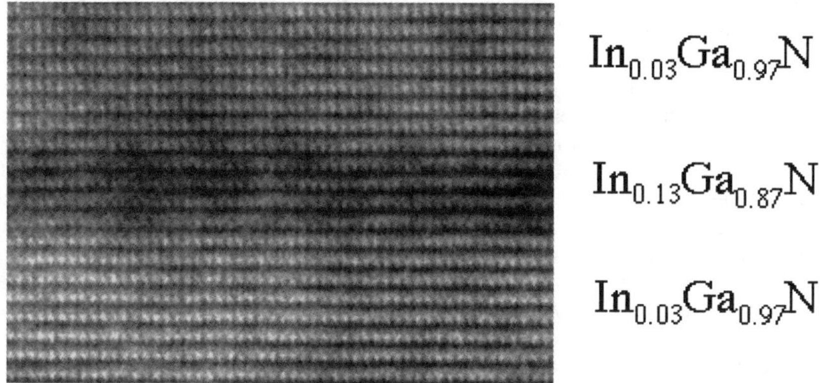

Fig 2: HREM image of one quantum well. The layer exibits a non-uniform contrast and interfaces between adjacent layers are rough.

Quantitative HREM was used to map strain profile across the well at different positions. The accuracy of the value was first determined by performing the same measurements in bulk GaN where lattice parameters are not supposed to vary. This accuracy depends on the pixel resolution at which the HREM image is digitized. A higher number of pixels results in a better accuracy. Since the unit cell in the image is not square but rectangular (c>a), the c lattice parameter corresponds to more pixels than the a parameter, and hence the determination of the c parameter is more accurate. In our case, the accuracy on a and c parameters in the bulk GaN is 0.8% and 0.5%, respectively; it allows us to detect any variation of the c-lattice parameter greater than 0.002 nm for average interatomic distance of 5.2 Å. Values of a and c lattice parameters perpendicular to the well at three different positions are represented in fig 3.

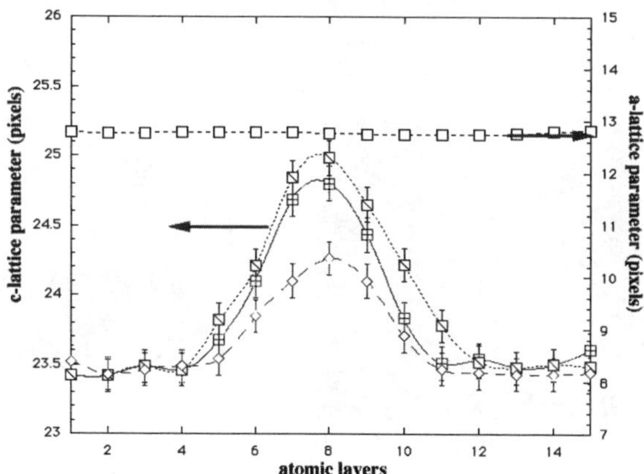

fig 3: Average strain profile across the In$_{0.03}$Ga$_{0.97}$N/In$_{0.13}$Ga$_{0.87}$N/In$_{0.03}$Ga$_{0.97}$N at three different positions. The lateral resolution is 5.2 Å.

These profiles were obtained by averaging the measured lattice parameter values in each column along each atomic rows parallel to the quantum well. It was noticed that the a parameter keeps the same value (3.19Å), which is not surprising since the well is only 25Å thick, but c exhibits a spike at the center of the well, corresponding to an expansion of the lattice parameter. The scale graduated in pixels gives the variation of the c-lattice parameter. The height and the width of the spike varies from place to place along the well as can be seen in fig.3. Its average amplitude is 2 %. The same measurement was conducted at the $In_{0.03}Ga_{0.97}N/GaN$ interfaces and the average spike height was found to be equal to 0.4%. Simple considerations implie that the c-lattice parameter of $In_{0.13}Ga_{0.87}N$ is equal to $c=c_{GaN} \times (1+2\%+0.4\%) = 5.31$ Å ± 0.02Å. An indium mole fraction x in the well can be extracted if one assumes the layer is relaxed. In such a case, the application of Vegard's law, i.e. $c=c_{GaN}(1-x)+c_{InN}(x)$ with $c_{GaN}=5.19$ Å and $c_{InN}=5.71$ Å, leads to an In content of about 23 % in the well for $c=5.31$Å, that is far above the intented 13 % In. But the a lattice parameter of a relaxed $In_{0.23}Ga_{0.77}N$ is $a_{0.23}=a_{GaN}(1-x)+a_{InN}(x)=3.27$Å, that is far above the measured value $a=3.19$Å. In case the $In_{0.13}Ga_{0.87}N$ layer would be relaxed, the lattice parameters in the well would be $c_0=5.26$Å and not 5.31 Å±0.02Å, and $a_0=a_{GaN}(1-x)+a_{InN}(x)=3.23$Å and not 3.19Å. The fact that $(c>c_0)$ and $(a<a_0)$ simultaneously lets us think that the layer is under biaxial compression. This observation is supported by the TEM experiments which showed that the layer is pseudomorphic. If we suppose that the In content in the well is the one intented (13%), the biaxial strain ε_c can be calculated and is equal to $\varepsilon_c=(c-c_0)/c_0=0.009+/-0.004$. The measured a and c lattice constants are linked to the Poisson ratio v by $\varepsilon_c/\varepsilon_a=((c-c_0)/c_0)/((a-a_0)/a_0)=-2v/(1-v)=0.7\pm0.3$. All the above numerical values lead to $v=0.24\pm0.08$. This value is in a good agreement with the one previously published [11].

Romano et al.[12] determined the value of the Poisson's ratio for GaN using X-ray diffraction. It is equal to $v=0.18\pm0.01$. Using this value, the combinaison of the above mentioned formulas gives the corresponding In content of 15.5 ± 0.4 % in the well. This value is close to the intented In concentration (13%).

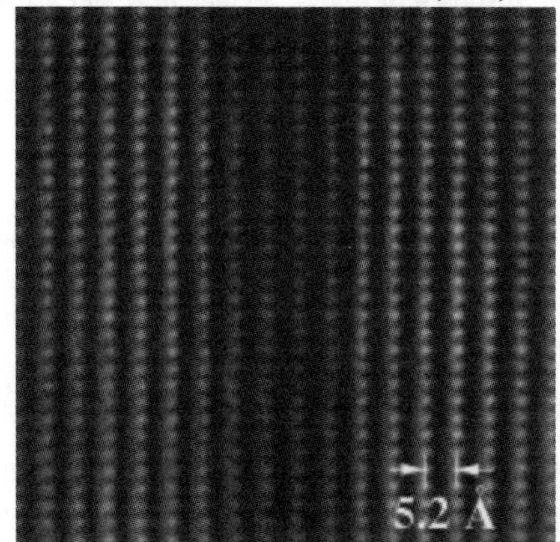
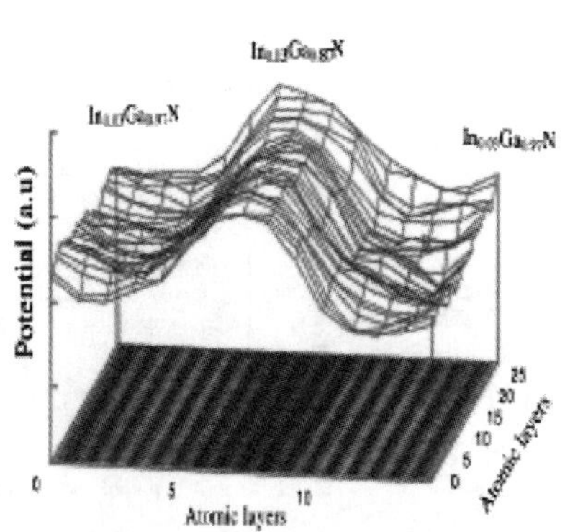

Fig. 4: (a) HREM image of $In_{0.03}Ga_{0.97}N/In_{0.13}Ga_{0.87}N/In_{0.03}Ga_{0.97}N$ quantum well and **(b)** the corresponding electron scattering potential map. The growth direction is from left to right.

Fig 4.b is a map of the electron projected potential that was extracted from the pattern change in the image fig. 4.a. Since the beam amplitudes of InN and GaN are similar when the crystal thickness is less than 20 nm, images were taken in thicker regions of the samples in order to extract chemical information from any pattern change. The scattering potential increases with sample thickness and with the average atomic weight of the atoms in each column. We can assume that the sample thickness does not change across the well because it is only 25 Å thick. Therefore, the scattering potential is directly related to any compositional change. Thus, the map in fig. 4.b can be seen as the representation of the Indium concentration across the well. It confirms the non-uniformity of Indium distribution within the well because variations of the potential reflects Indium content fluctuations[13]. The map shows also that interfaces are not sharp and the roughness is broader on the upper interface. As seen in fig 2, the succession of bright and dark areas along the well in Fig 2 suggests that a redistribution of In has taken place.

This feature is even more apparent when the In content increases from 13% to 20%. Fig. 5 is an HREM of sample 2. Quantum dot like structure with regular spacing is evident. The dots have an average diameter of 30 Å and their separation distance is about 40 Å. They have either a spheroidal or an elongated shape and their presence in certain regions implies that nucleation and growth are not the same all over the specimen. This regular arrangement of these clusters makes it unlikely that they are artifacts due to sample preparation. Only images of regions above 30 nm thick of samples were considered to avoid strain relaxation at edges of samples.

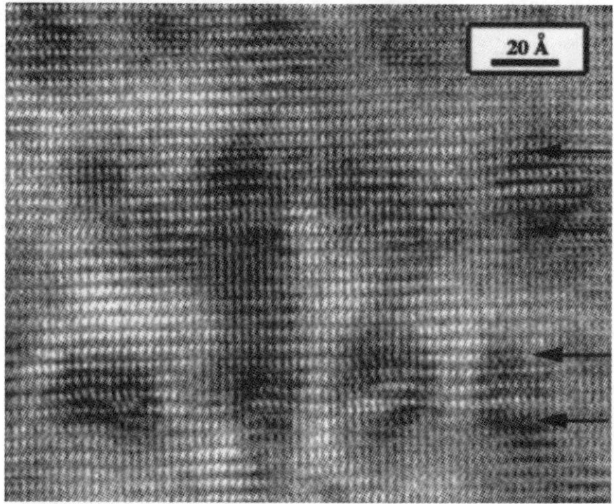

Fig 5: HREM taken along $[11\bar{2}0]$ of 20 QW sample. Note the regular arrangement of dots and the presence of strained regions and defects at $In_xGa_{1-x}N/GaN$. Arrows indicate interfaces.

The large contrast variation and abrupt modifications of pattern of adjacent unit cells in parts of the image prove that the structure is highly strained. The large difference in lattice parameter between wells and barriers could be responsible for the strained regions. Moreover, the periodic modulation of the c-lattice parameter along the well is a maximum inside the clusters. Formation of these cluters is likely to be related to an accumulation of In atoms.

In addition, a high density of dislocations and small dislocation loops in the (0001) plane were observed at interfaces quantum well/barrier. These loops appear at the $In_xGa_{1-x}N/GaN$ interfaces and not at the $GaN/In_xGa_{1-x}N$ ones. These loops may be the result of agglomeration of Indium atoms. Because of defect formation and modification of the pattern within the unit cell, no chemical mapping can be obtained from such sample areas.

Summary

InGaN multiquantum wells grown by MOCVD on GaN layers have been investigated by transmission electron microscopy techniques and numerical analysis of high resolution (HREM) images. It was shown that GaN/InGaN interfaces are rather rough and exhibit an oscillating contrast. Structural defects were found on the interfaces. Their density increases with the In mole fraction in the wells. The relative c-lattice parameter variation in the wells was determined using numerical processing of HREM images. This allowed us to determine strain in the wells and to estimate the Poisson's ratio. We showed that quantitative TEM is a suitable tool for determining independently the In composition in the well. Further observations also revealed a redistribution of In within the well. Instead of a continous In-rich layer, quantum dots were often observed along the well with a regular spacing. The formation of these In-rich dots was not intented and their presence suggests either a periodic modulation of strain along the well or In-rich cluster formation.

Acknowledgment

TEM work was supported by the Director, Office of Basic Science, Materials Science Division, U.S. Department of Energy, under the Contract No. DE-AC03-76SF00098. The use of the facility at the National Center for Electron Microscopy at Lawrence Berkeley National Laboratory is greatly appreciated. The work at The University of Texas was partially supported by DARPA under Grant MDA972-96-3-0014 monitored by R. Leheny and by ONR under grant N00014-95-1-1302 monitored by J. Zolper.

References

[1] P. Perlin, C. Kisielowski, V. Iota, B. A. Weinstein, L. Mattos, N. Shapiro, J. Kruger, E. R. Weber and J. Yang, Appl. Phys. Lett. 73 (19), (1998).

[2] S. Chichibu, T. Azuhata,T. Sota and S. Nakamura, Appl. Phys. Lett. 69, 4188 (1996).

[3] J. Krueger, C. Kisielowski, R. Klockenbrink, Sudhir G. S., Y. Kim, M. Rubin and E. R.Weber, MRS Symp. Proc. Vol. 468, 299 (1997).

[4] P. A. Grudowski, C. J. Eiting, J. Park, B. S. Shelton, D. J. H. Lambert and R. D. Dupuis, Appl. Phys. Lett. 71 (11), (1997).

[5] I. Ho and G. B. Stringfellow, Appl. Phys. Lett. 69 2701, (1996).

[6] N. A. El-Masry, E. L. Piner, S. X. Liu and S. M. Bedair, Appl. Phys. Lett. 72 (1), (1998).

[7] F. Ponce, D. Cherns, W. Goetz and R. S. Kern, MRS Symp. Proc. Vol. 482, 453 (1998).

[8] P. Schwander, C. Kisielowski, F. H. Baumann, Y. Kim and A. Ourmazd, Phys. Rev. Lett. 71, (25), 4150 (1993).

[9] C. Kisielowski, O. Schmidt and J. Wang, MRS Symp. Proc. Vol. 482, 369 (1998).

[10] Y. Narukawa, Y. Kawakami, M. Funato, S. Fujita, S. Fujita and S. Nakamura, Appl. Phys. Lett. 70 (8), 981 (1998).

[11] C. Kisielowski, J. Krüger, S. Ruvimov, T. Suski, J. W. Ager III, E. Kones, Z. Liliental-Weber, M. Rubin, E. R. Weber, M. D. Bremser and R. F. Davis, Phys. Rev. B, 54 (24), 17745 (1996).

[12] L. T. Romano, B. S. Krusor, M. D. Mc Cluskey, D. P. Bour and K. Nauka, Appl. Phys. Lett. 73 (13), 1757 (1998).

[13] C. Kisielowski, Z. Liliental-Weber and S. Nakamura, Jpn. J. Appl. Phys. Vol. 36, 6932 (1997).

TEM STUDY OF Mg-DOPED BULK GaN CRYSTALS

Z. LILIENTAL-WEBER, M. BENAMARA, S. RUVIMOV, J.H. MAZUR, J. WASHBURN,
I. GRZEGORY* and S. POROWSKI*

Materials Science Division, Lawrence Berkeley National Laboratory, Berkeley CA 94720,
62/203;
 * High Pressure Institut, Unipress, Warsaw, Poland

ABSTRACT

Transmission electron microscopy was applied to cross-sectioned samples to study surface
morphology, sample polarity and defect distribution in bulk GaN samples doped with Mg.
These crystals were grown from a Ga melt under high hydrostatic pressure of Nitrogen. It was
shown that the types of defects and their distribution along the c-axis depends strongly on
sample polarity. Based on this finding the growth rate along the c-axis for the two polar
directions was compared and shown to be approximately ten times larger for Ga polarity than
for N-polarity. In the part of the crystals with Ga polarity pyramidal defects with a base
consisting of high energy stacking faults were found. The parts of the crystals grown with N-
polarity were either defect free or contained regularly spaced stacking faults. Growth of these
regularly spaced cubic monolayers is polarity dependent; this structure was formed only for the
growth with N polarity and only for the crystals doped with Mg. Formation of this
superstructure is similar to the polytypoid structure formed in AlN crystals rich in oxygen. It is
also likely that oxygen can decorate the cubic monolayers and compensate Mg. This newly
observed structure may shed light on the difficulties of p-doping in GaN:Mg.

INTRODUCTION

Progress in GaN technology has been delayed for a long time because of difficulties in
obtaining p-doping. Only recently success of Amano et al [1] followed by Nakamura [2]
resulted in p-doping of GaN. GaN p-n junction blue light emitting diodes (LEDs) and lasers
have been obtained [3]. Despite this success Mg p-doping is still not fully understood.
Originally low-energy-electron-beam irradiation (LEEBI) [1] and recently thermal annealing
have been found to activate Mg and result in p-doping [2,3]. There are reports that Mg is not
distributed uniformly in the layer and has a high tendency to diffuse to the layer surface [4].

EXPERIMENTAL

In this paper structural studies of Mg doped bulk GaN crystals will be described. The GaN
crystals have been grown by the High Nitrogen Pressure Solution Method [5] from a solution of
liquid gallium containing 0.1-0.5 at.% of Mg [6]. Before the growth conditions are established,

Mat. Res. Soc. Symp. Proc. Vol. 572 © 1999 Materials Research Society

the solution was homogenized at a temperature of 1100°C for several hours. Then the temperature was increased up to the growth conditions (T = 1500 - 1600°C) and N2 pressure about 15 kbars was applied. Typical time of growth was in the range of 100 - 150 hours.

Crystals in the form of hexagonal platelets (similar to those for undoped crystals) reaching dimensions up to 8 mm were found in the cooler part of the crucible. This shows that the Mg impurity does not influence the relative growth rates to a significant degree and therefore the form of the hexagonal platelets is similar to that for undoped crystals. However, it was observed that the platelets with Mg are generally somewhat thicker suggesting that Mg impurity does accelerate the growth in the c-direction. Observation with the Nomarsky microscope shows that as in undoped samples one surface normal to the c-axis is more rough, with many surface steps while the opposite surface is always mirror like. Convergent beam electron diffraction (CBED) was applied to study polarity of these crystals [7]. Earlier studies of undoped GaN crystals showed that the polarity of the crystals can also be recognized by using mechano-chemical polishing with aqueous solution of KOH [8, 9]. Cross-section samples were prepared from crystals directly removed from the crucible and also from those etched chemically. Electron microscopy was applied in order to detect the presence of any structural defects, observation of surface morphology and determination of crystal polarity.

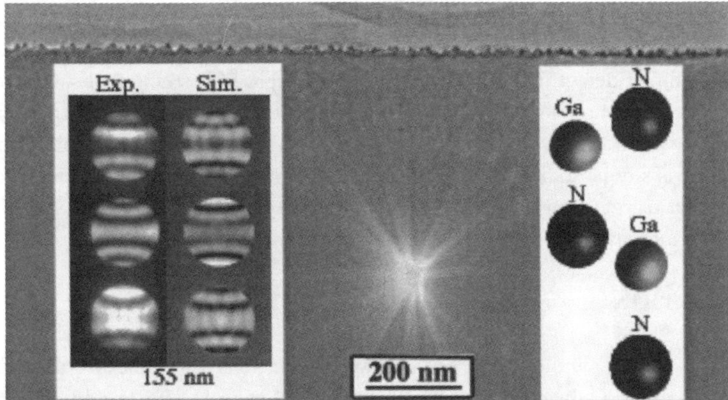

Fig. 1: Cross-section TEM micrograph of the rough side of an as-grown GaN:Mg platelet sample. The inset shows the experimental and calculated CBED patterns together with a hard-ball model showing atom arrangement along the c-axis.

Fig. 1 shows a cross-section electron micrograph from a crystal directly removed from the crucible. It can be seen that one side of the as grown crystal has a rough surface (saw-like) with an amplitude of 10 nm. The distance between the spikes was about 7-10 nm. The opposite side of this crystal did not show such undulation, the surface was practically atomically flat. Convergent beam electron diffraction (CBED) was applied to study polarity of these crystals and it turned out that the rough side of the crystal grew with N polarity (Fig. 1-inset). This is opposite to the undoped bulk crystals, where the rough side of the crystal grew with Ga polarity.

Mechano-chemical polishing of undoped crystals with an aqueous solution of KOH showed that the chemically active side grew with N polarity and the Ga-side was chemically inert [9,10]. Similar polishing applied to the Mg doped crystals showed that the previously rough crystal side became smooth while the mechanical damage remained on the opposite side. CBED studies confirmed that the Mg doped crystals behave similarly to undoped crystals in respect to the etching behavior, the crystal side with N polarity is chemically active.

Transmission electron microscopy (TEM) applied to undoped cross-section samples showed that not only surface morphology for N- and Ga-polarity is different but the part of the crystal on the N polar side remains practically defect free, while the part near the Ga polar side has high density of stacking faults distributed randomly (Fig. 2a). All three possible types of stacking faults have been found in these undoped crystals. Dislocation loops decorated by Ga precipitates were often attached to high energy stacking faults converting them locally to low energy stacking faults.

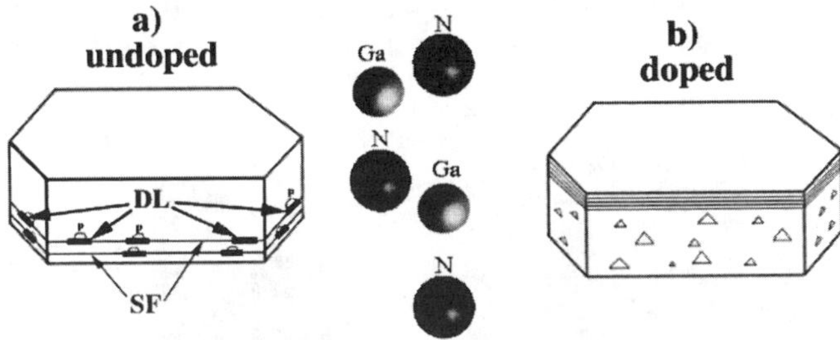

Fig. 2: Schematic drawing of defect distribution in the undoped bulk GaN crystals (a) and in the Mg doped crystals (b). The inset indicates crystal polarity. Stacking faults (SF) with attached dislocation loops (DL) decorated by Ga precipitates (P) are marked in the undoped crystals and stacking faults and pyramidal defects in the Mg doped crystals.

Dramatically different defect distribution was observed in the GaN samples doped by Mg. A large difference in crystal structural quality has also been observed for different growth runs. Based on their structural quality one can divide the crystals into three groups. The first group of crystals was practically free of any structural defects. In those crystals small precipitates could occasionally be observed on the Ga side.

The second group of crystals had a similar defect distribution as for undoped crystals, the side of the crystals with N-polarity was free of defects, while the opposite side of the platelet had a high density of defects (Fig. 3). It was noticed that the defect free area of the crystal was much thinner than the defective part of the crystal, opposite to the undoped crystals. The thickness of the defect free region along the c-axis direction was about 10% of the platelet thickness. It can be concluded that the growth of the crystal along the c-axis in Ga polarity is much faster than the growth with N-polarity.

The defects observed in Mg doped crystals were not simple stacking faults on c-planes, each defect had a pyramidal shape with the base of the pyramid facing the Ga side of the platelet.

High resolution transmission electron microscopy revealed that high energy stacking faults are formed on c-planes at the base of these pyramids and their sides are also stacking faults on planes inclined about 45° to their base (Fig. 3). Since these bulk crystals contain oxygen impurities (in the range of 10^{18} cm^{-3}), it is possible that these defects are "dome" shaped inversion domains as observed earlier in AlN crystals rich in oxygen, where the base of such pyramid forms a flat inversion domain and the sides are corrugated domains [10]. However, details of these defects are still under investigation. The fact that the sides of the pyramid can not be viewed edge-on in high resolution micrographs makes their interpretation more difficult.

In the third type of crystal the pyramidal type of defect was also present but the part of the crystal grown with N-polarity had a superstructure (Fig. 2b). Every 10 nm a monolayer with an enhanced contrast was observed (Fig. 4) creating an equidistant layer structure. High resolution imaging reveals the presence of stacking faults e.g. a unit of cubic material inserted equidistantly in the hexagonal material. This layer arrangement results in additional satellite spots dividing the (0001) reciprocal space into 20 equal spaces and giving 0.52x20=10.4 nm distances between the monolayers. The appearance of these monolayers is assumed to be related to Mg distribution

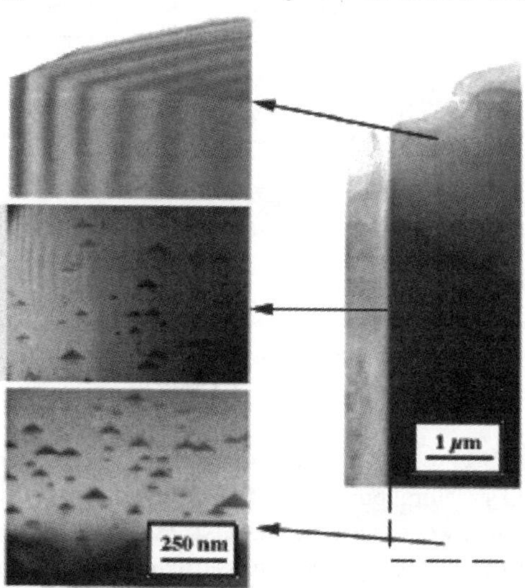

Fig. 3: TEM micrograph shows defect distribution in GaN:Mg crystals. Note that the upper part grown with N polarity is defect free.

in the sample. SIMS measurements for these crystals show Mg concentration 2×10^{20} cm^{-3} on the sample surface and flat Mg distribution in the range 6×10^{19} cm^{-3} in the remaining part of the crystal. No difference in Mg content was observed on the two sides of the crystal. Si and oxygen impurities were also detected in the crystals.

Based on theoretical predictions [11] Mg should substitute of Ga sites, the occupancy of Mg on Ga sites and N sites are expected to be different. Mg segregation is expected to be more

Based on theoretical predictions [11] Mg should substitute of Ga sites, the occupancy of Mg on Ga sites and N sites are expected to be different. Mg segregation is expected to be more likely on N-sites of the crystals. Predicted positions [12] for Mg occupancy on the N-side would likely lead to the formation of stacking faults. These structures are reminiscent of the polytype defects formed in AlN rich in oxygen [12,13]. In the AlN crystals with oxygen the formation of an octahedral metal layer surrounded by a mixture of N and oxygen atoms were expected. In our GaN crystals oxygen content is much lower (1018cm-3) but it cannot be excluded that oxygen decorates the stacking faults. The presence of oxygen would explain the fact that p-doping was not obtained in these crystals, oxygen most likely would be responsible for the observed compensation.

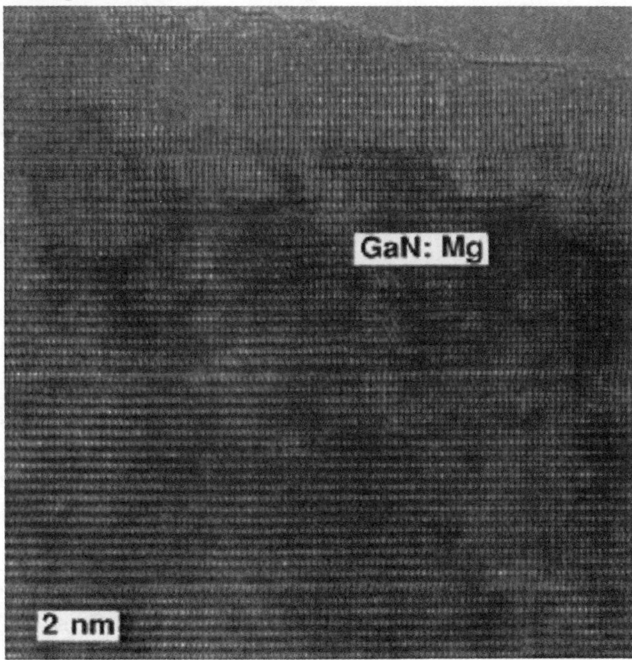

Fig. 4: High resolution image of low-energy stacking faults arranged equidistantly on the side of the GaN:Mg crystal which grew with N-polarity.

CONCLUSIONS

These studies showed that GaN bulk crystals grown from a Ga melt doped by Mg under hydrostatic pressure of nitrogen differ from undoped crystals. First of all the surface morphology of the as-grown crystals is reversed with respect to the crystal polarity in comparison with the undoped crystals. The N-side of the platelet is rough and the Ga site is smooth. The second important observation was that a much higher growth rate (about 10 times) was observed for the Ga polarity surface than for the N polarity surface. The third observation was related to defect distribution in these crystals. Some of them were almost structural defect pyramidal defects with the base on c-planes and the sides of these defects inclined approximately 45° are observed. A high energy stacking fault was observed as the base of these pyramidal defects. For some crystals regions grown with N polarity did not show any structural defects. However, in some crystals stacking faults (a cubic monolayer) equidistantly distributed were observed. This arrangement of stacking faults most likely is related to Mg segregation on N-sites and is very likely a new type of polytype like those observed earlier in AlN crystals rich in oxygen. Our crystals have much smaller oxygen concentration, but it is very likely that stacking faults are heavily decorated by oxygen. This would explain why p-doping was not obtained in these crystals since compensation by oxygen would take place [14].

ACKNOWLEDGMENT

This work was supported by the Director, Office of Basic Science, Materials Science Division, U.S. Department of Energy, under the Contract No. DE-AC03-76SF00098. The use of the facility at the National Center for Electron Microscopy at E.O. Lawrence Berkeley National Laboratory is greatly appreciated.

REFERENCES:

1.	H. Amano, M. Kito, K. Hiramatsu, and I. Akasaki, Inst. Phys. Conf. Ser. 106, 72 (1989).
2.	S. Nakamura, M. Senoh, and T. Mukai, Jpn. J. Appl. Phys. 31, L1708 (1991)
3.	S. Nakamura, Paper Plenary 1, presented at the 24th International Symposium on Compoud Semiconductors, San Diego CA, 8-11 September 1997.
4.	M.E. Lin, G. Xue, L. Zhou, J.E. Greeen and H. Morkoc, Appl. Phys. Lett., 63, 932 (1993).
5.	I. Grzegory, J. Jun, M. Bockowski, St. Krukowski, M. Wroblewski, B. Lucznik and S.Porowski, J. Phys. Chem. Solids 56, 639 (1995).
6.	S. Porowski, M. Bockowski, B .Lucznik, I. Grzegory, M. Wroblewski, H. Teisseyre, M. Leszczynski, E. Litwin-Staszewska, T .Suski, P. Trautman, K. Pakula and J.M. Baranowski, Acta Physica Polonica A vol., 92, 958 (1997).
7.	Z. Liliental-Weber, C. Kisielowski, S. Ruvimov, Y. Chen, J. Washburn, I. Grzegory, M. Bockowski, J. Jun, and S. Porowski, J. Electr. Mat. vol. 25, 1545 (1996).
8.	J.L. Weyher, S. Muller, I. Grzegory, and S. Porowski, J. Cryst. growth 182, 17 (1997).
9.	Z. Liliental-Weber, M. Benamara, O. Richter, W. Swider, J. Washburn, I. Grzegory, S. Porowski, J.W. Yang, and S. Nakamura, MRS Proc., vol. 512, 363 (1998).
10.	A.D. Westwood, R. A. Youngman, M.R. McCartney, A.N. Cormack, and M. R. Notis, J. Mater. Res. 10, 1287 (1995)
11.	C. Bungaro, K. Rapcewicz, and J. Bernholc, Rapid Communications 1999 in print.
12.	R. A. Youngman, A.D. Westwood, and M.R. McCartney, Mat. Res. Symp. Proc. 319 45 (1994)
13.	Y.Yan, M. Terauchi and M. Tanaka, Phil. Mag. A 77, 1027 (1998).
14.	C. G. Van de Walle, J. Neugebauer and C. Stampfl, in GaN and Related Semiconductors, edts. J.H. Edgar, S. Strite, I. Akasaki, H. Amano, and C. Wetzel, Emis Datareviews Series No. 23, An Inspec Publ (1999) p. 275.

DEFORMATION-INDUCED DISLOCATIONS IN 4H-SiC AND GaN

M. H. HONG[*], A. V. SAMANT[*], V. ORLOV[*], B. FARBER[*], C. KISIELOWSKI[**],
P. PIROUZ[*]

[*]Department of Materials Science and Engineering, Case Western Reserve University,
Cleveland, OH, 44106-7204.
[**]Lawrence Berkeley Laboratory, Berkeley, CA.

ABSTRACT

Bulk single crystals of 4H-SiC have been deformed in compression in the temperature range
550-1300°C, whereas a GaN thin film grown on a (0001) sapphire substrate was deformed by
Vickers indentation in the temperature range 25-800°C. The TEM observations of the deformed
crystals indicate that deformation-induced dislocations in 4H-SiC all lie on the (0001) basal plane
but depending on the deformation temperature, are one of two types. The dislocations induced by

deformation at temperatures above ~1100°C are complete, with a Burgers vector, **b**, of $\frac{1}{3}\langle 11\bar{2}0 \rangle$

but are all dissociated into two $\frac{1}{3}\langle 10\bar{1}0 \rangle$ partials bounding a ribbon of stacking fault. On the other

hand, the dislocations induced by deformation in the temperature range 550<T<~1100°C were
predominantly single leading partials each dragging a stacking fault behind them. From the width
of dissociated dislocations in the high-temperature deformed crystals, the stacking fault energy of
4H-SiC has been estimated to be 14.7±2.5 mJ/m^2. Vickers indentations of the [0001]-oriented
GaN film produced a dense array of dislocations along the three <11$\bar{2}$0> directions at all
temperatures. The dislocations were slightly curved with their curvature increasing as the
deformation temperature increased. Most of these dislocations were found to have a screw nature
with their **b** parallel to <11$\bar{2}$0>. Also, within the resolution of the weak-beam method, they were
not found to be dissociated. Tilting experiment show that these dislocations lie on the {1$\bar{1}$00}
prism plane rather than the easier (0001) glide plane.

INTRODUCTION

SiC, GaN, and alloys of the latter, are wide bandgap semiconductors showing considerable
promise for high-temperature and optoelectronic applications [1,2]. Recent advances in crystal
growth have resulted in the production of single-crystal, single polytype, SiC boules [3] and very-
nearly single crystalline GaN thin films [2]. In spite of their widespread interest, the deformation
behavior and microstructure of these materials have not been studied in sufficient detail.

The most common polytypes of SiC are 6H and 4H; these are now produced commercially in
boule form and are available as wafers. The stable ambient form of GaN has a 2H structure, and
thin films of the metastable 3C (cubic) polytype can also be grown by CVD on certain substrates.
All the polytypes of SiC and GaN are tetrahedrally coordinated and consist of two variants of a
basic tetrahedron: normal, T, and twinned, T'; each of these variants can occupy three spatial
positions, T_1, T_2, and T_3 (or T'_1, T'_2, and T'_3) (for details, see, e.g. [4, 5]). They predominantly
exhibit hexagonal symmetry, denoted by H in the Ramsdell notation [6], e.g. 2H (wurtzite), 4H,
6H or, in general, $2n$H (where n is an integer), or rhombohedral symmetry, denoted by R in the
Ramsdell notation, e.g. 15R or, in general, $(2n+1)$R. The 3C (zincblende) polytype is a subset of
the rhombohedral polytypes and is unique in that it consists of only one tetrahedral variant (either
all normal, T, or all twinned, T', tetrahedra) and additionally exhibits cubic (C) symmetry. Fig. 1
shows the <11$\bar{2}$0> projected structure of 4H-SiC and 2H-GaN with tetrahedral sequences
$...T_1T_2T'_1T'_3...$ (periodicity of four) and $...T_1T'_3...$ (periodicity of two), respectively. The
structure of each polytype can also be considered in terms of stacking of widely-spaced double
planes αA, βB and γC where α, β, γ represent basal planes of carbon or nitrogen (silicon or
gallium) atoms and A, B, and C represent interleaved parallel planes consisting of silicon or
gallium (carbon or nitrogen) atoms. Thus, the structure of 4H-SiC and 2H-GaN may also be

369

Mat. Res. Soc. Symp. Proc. Vol. 572 © **1999 Materials Research Society**

described in terms of the stacking sequences $...\alpha A\,\beta B\,\alpha A\,\gamma C...$ and $...\alpha A\,\gamma C...$, respectively. Note that all SiC and GaN polytypes are polar along the c-axis: the [0001] direction is distinct from the opposite $[000\bar{1}]$ direction.

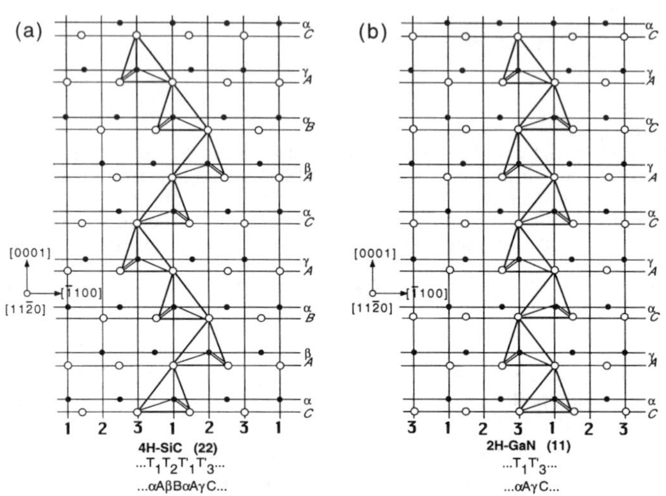

Fig. 1. The structure of (a) 4H-SiC and (b) 2H-GaN

The aim of this work is to investigate the deformation-induced microstructure of bulk 4H-SiC single crystals and a 2H-GaN thin film as a function of temperature using transmission electron microscopy (TEM). The dislocation structure is discussed based on the differences in the core structure of partials as well as the slip planes on which the dislocations lie.

EXPERIMENTAL PROCEDURES

The bulk 4H-SiC single crystals used in this study were grown by the modified sublimation technique [3]. Parallelepiped specimens with nominal dimensions 2×2×4 mm^3 were oriented and cut for single glide such that one pair of their lateral faces was parallel to { 1$\bar{1}$00 }, and the (0001) basal plane made an angle of 45° with respect to the compression axis. The samples were compressed at 550-1300 $°C$ in ultra-high purity argon to a strain of ~4-6%. On the other hand, the GaN film was grown by molecular beam epitaxy (MBE) on a (0001) sapphire substrate and was deformed by Vickers indentation in the temperature range 25-800°C. The indentation diagonals were aligned along the <11$\bar{2}$0> and $\langle 1\bar{1}00 \rangle$ directions on the (0001) surface of the film. From the deformed samples, 0.3 mm thick slices parallel to (0001) plane were sectioned with a diamond wheel cutter. Subsequently, the slices were ground with emery paper to a thickness of ~100 μm, then dimpled to a thickness of ~20 μm, and ion-milled to electron transparency at a voltage of 5 kV at an angle of ~15°. GaN thin films were prepared by back-side thinning with the foil surface normal to the [0001] direction from the sapphire substrate side. The thin TEM foils were examined in a Philips CM20 electron microscope operating at an accelerating voltage of 200 kV.

RESULTS AND DISCUSSION

The easy slip plane of hexagonal and rhombohedral polytypes is (0001) and the dislocations have a perfect $\mathbf{b}$=1/3<11$\bar{2}$0> Burgers vector [7]. As in other tetrahedrally-coordinated structures, these dislocations are dissociated into two partials with Burgers vectors $\mathbf{b}_l$=1/3<10$\bar{1}$0> and $\mathbf{b}_t$=1/3<01$\bar{1}$0>, where the subscripts l and t denote the leading and trailing partials, respectively. The dislocations dissociate as follows:

$$1/3<11\bar{2}0> = 1/3<10\bar{1}0> + 1/3<01\bar{1}0>$$

In tetrahedrally coordinated compounds, because of the polarity along the [0001] axis, the core of the perfect or partial basal dislocations consists of only one species, i.e. silicon (gallium) or carbon (nitrogen), and, because the partial dislocations belong to the glide plane, they are denoted as Si(g) [Ga(g)] or C(g) [N(g)] [8].

4H-SiC

Fig. 2 shows typical bright-field (BF) micrographs of 4H-SiC deformed in compression. Both micrographs in this figure were obtained using the $\mathbf{g}$= $\bar{1}$01$\bar{1}$ reflection near the [$\bar{1}$012] zone axis. Samant [9] has recently shown that plastic deformation of 4H-SiC above ~1100°C takes place by the activation of the (0001)<11$\bar{2}$0> slip system with the uncorrelated motion of partial dislocations. This is clearly shown in Fig. 2(a) where dissociated dislocations, consisting of a pair of leading/trailing partials bounding a ribbon of intrinsic stacking fault, are observed. Standard strain contrast experiments indicate that, as expected, the Burgers vectors of the partials are, respectively, parallel to <1$\bar{1}$00> and <10$\bar{1}$0> directions.

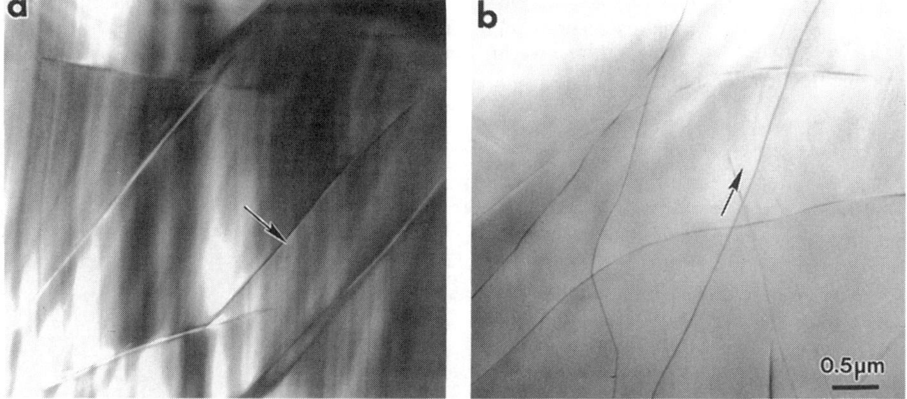

Fig. 2. TEM micrograph of 4H-SiC deformed in compression (a) at 1300 °C and (b) at 700 °C

From the width of partial separation, the stacking fault energy of 4H-SiC has been estimated to be 14.7 ± 2.5 mJ/m^2; this value is nearly five times larger than that ($2.9\pm0.5 mJ/m^2$) of 6H-SiC obtained by the same techniques [10].

Fig, 2(b) shows a BF micorgraph of dislocations induced by deformation at 700 °C. In this case, the microstructure is dominated by single leading partials without the associated trailing partials; each partial drags a stacking fault. The Burgers vectors of the single leading partials are all parallel to the <1$\bar{1}$00> directions, and LACBED experiments on three different segments have shown that they have a silicon core (i.e. they are Si(g) partials) [11]. Thus, it appears that in the 4H-SiC single crystals, deformation proceeds by nucleation and glide of single leading Si(g) partials at low-temperatures (<~1100 °C), whereas it proceeds by the generation and glide of total, although dissociated, dislocations at high temperatures (>~1100 °C). It has already been argued

[12] that only the leading partials, with a silicon core, nucleate at low temperatures ($<\sim 1100\,°C$), and a higher temperature ($>\sim 1100\,°C$) is required to nucleate the associated carbon-core trailing partials whereby the glide of leading/trailing pairs (i.e., dissociated dislocations) will carry out the plastic deformation: at a certain applied shear stress, the activation barrier for nucleation of trailing partials is higher than that of the leading partials. Since the formation of the same partial dislocation from the same source cannot occur more than once on the same glide plane, plastic deformation of the crystal can take place to a very limited extent at low temperatures. On the other hand, once thermal activation is sufficient to from the trailing partial, repeated dislocation multiplication can take place from the same source on the same planes, and plastic deformation will proceed by the glide of total dislocations. At these temperatures, the crystal will be ductile and large strains are obtained [12].

2H-GaN

Fig. 3 shows a (a) plan-view and (b) cross-sectional micrograph of the as-grown GaN specimen (before deformation). As shown by various authors (see, e.g., [13, 14]) there is a high density of threading dislocations in such films: some are those connected to misfit dislocation segments, and some generated to accommodate the tilt and twist misorientation of neighboring domains in the granular structure of the film. The plan-view micrograph in Fig. 3(a), with the inserted [0001] SADP, shows the typical cell structure of the film with the grain boundary dislocations defining the cell boundaries. They are mostly aligned along the $<11\bar{2}0>$ directions that are the Peierls valleys in non-cubic tetrahedrally-coordinated crystals. The cross-sectional micrograph in Fig.3(b) shows that the thickness of the GaN film deposited on the (0001) sapphire substrate is approximately 2 μm. The dislocations in this micrograph are mostly parallel to the c-axis, i.e., to the growth direction. Standard $\mathbf{g} \cdot \mathbf{b}=0$ invisibility criterion shows that most of the dislocations are **a**-type edge dislocations with a Burgers vector parallel to the $[11\bar{2}0]$ direction. The SADP obtained from an aperture covering both the 2H-GaN and the sapphire substrate shows the orientation relationship to be $[11\bar{2}0]_{GaN}//[1\bar{1}00]_{sap}$, $(1\bar{1}00)_{GaN}//(11\bar{2}0)_{sap}$.

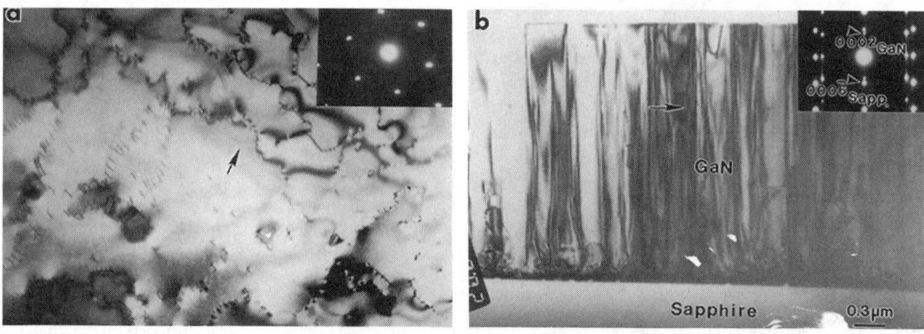

Fig. 3. (a) plan-view and (b) cross-sectional micrograph of the as-grown 2H-GaN

Fig. 4 shows micrographs taken under different reflections from the same region of a specimen indented at 450°C. From these, and micrographs taken under other reflections, the Burgers vectors of dislocations were determined. Practically all the dislocations appear under reflections of the type $\mathbf{g}=11\bar{2}0$ shown in Fig. 4(a). In this micrograph, the short segments are projections of threading dislocations (lying along the c-axis) in the film, while the dislocations denoted by symbols A and B, aligned along the $<1120>$ directions, are newly-generated (presumably by indentation) dislocations that are parallel to the (0001) plane of the film. Dislocations denoted by A are in contrast when imaged with reflections of the type $\mathbf{g}=1\bar{1}00$ (Fig. 4(b)), while they are out of contrast when imaged with reflections of the type $\mathbf{g}=\bar{1}010$ (Fig. 4(c)). On the other hand, dislocations denoted by B are out of contrast when imaged with reflections of the type $\mathbf{g} = 1\bar{1}00$ (Fig. 4(b)) and are in contrast with reflections of the type $\mathbf{g} = \bar{1}010$ (Fig. 4(c)). These and other micrographs indicate that dislocations A and B have Burgers vectors parallel to $[11\bar{2}0]$ and $[1\bar{2}10]$ directions, respectively. Since, the line directions of dislocations A and B are parallel to their Burgers vectors, it is concluded that all the indentation-induced dislocations have a perfect screw character.

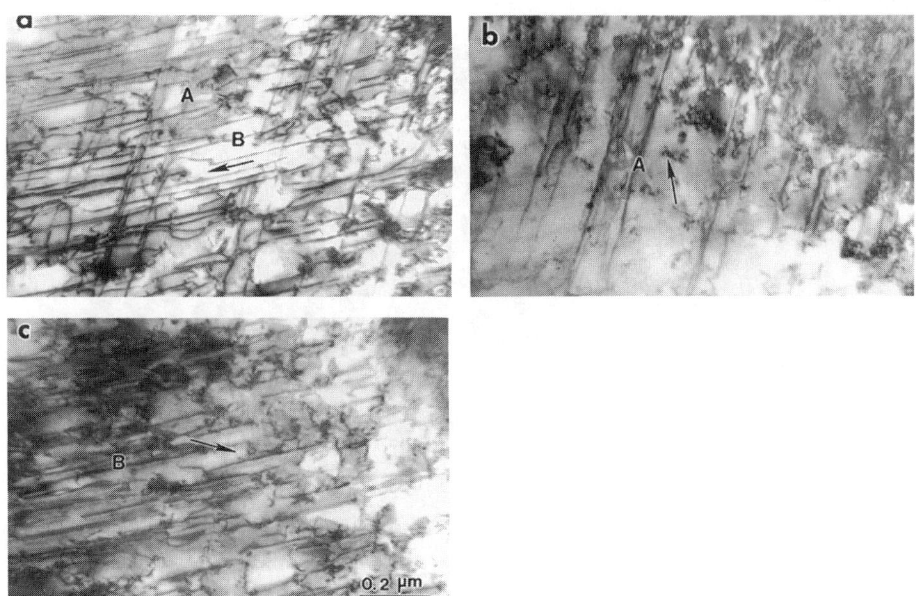

Fig. 4. TEM micrograph of a 2H-GaN indented at 450 °C; imaged by reflection (a) $\mathbf{g} = 11\bar{2}0$, (b) $\mathbf{g} = 1\bar{1}00$ and (c) $\mathbf{g} = \bar{1}010$

Since two different polytypes (2H and 3C) have often been observed in GaN, it would be expected that this material has a relatively low stacking fault energy and, consequently, that, as in 4H-SiC, all the basal dislocations in the GaN film would be dissociated into two partials. Indeed, recent weak-beam TEM of GaN powder deformed by pulverization shows that basal dislocations in this material are dissociated into two partials with a width of 3-8 nm, corresponding to a stacking fault energy of ~20 mJ/m^2 [15]. However, this is not the case in the present experiments; despite many attempts using the weak beam technique of TEM, no evidence of dissociation was found for any of the dislocations in the deformed film that were parallel to the basal plane. An example is shown in Fig. 5 which is a $\mathbf{g}/3\mathbf{g}$ weak-beam micrograph (with $\mathbf{g} = 11\bar{2}0$) from a GaN film indented at 300 °C; all the dislocations appear as single lines and not as pairs.

One possibility for the non-dissociation of basal $1/3 < 11\bar{2}0 >$ screw dislocations in Fig. 4 could be that they actually lie on the $\{ 1\bar{1}00 \}$ prism planes. Unlike the basal plane, the stacking fault energy on the prism planes is quite likely very high and the perfect dislocations, if dissociated at all on these planes, would have a small dissociation width (less than the resolution of the weak beam technique, ~2.5 nm). In order to check for this, extensive tilting experiments were carried out in the microscope and the changes in the shapes of individual kinked dislocations were noted with tilting. Figure 6 shows a typical example observed with the incident beam (a) nearly normal (tilted by ~5°) to the basal plane and (b) after tilting the foil approximately 35° away from position (a). Care was taken that the same region of the foil was imaged during the tilt. In (a), a super-kink pair is arrowed on an otherwise straight dislocation. In Fig. (b), the same super-kink pair has widened considerably after the foil has been tilted by ~35° about the $< 11\bar{2}0 >$ axis. This clearly indicates that the dislocation is not lying on the basal plane (in which case, the super-kink pair would have shrunk with the tilt) and the results are consistent with the fact that it lies on the prism plane. This sort of experiment was performed on five? kinked dislocation segments lying parallel to the basal plane and, in all cases, the dislocations were found to lie on one of the three { $1\bar{1}00$ } prism planes.

Fig. 5. Weak-beam TEM micrograph of the 2H-GaN deformed at 300°C using reflection $g=11\bar{2}0$ close to the [0001] zone axis

Two possibilities may account for these unexpected results. Under the complex stress field of the indentation, the $1/3<11\bar{2}0>$ screw dislocations may have nucleated on (0001) basal planes, where they would be dissociated; subsequently, they could have cross-slipped onto the prism planes by the Friedel-Escaig mechanism. More likely, however, the deformation–induced dislocations were probably nucleated on the prism planes from the sample surface because of large shear stresses on these planes induced by the indentation.

CONCLUSIONS

The deformation-induced dislocations in 4H-SiC are all basal dislocations and are divided into two types. For deformations performed above ~1100°C, the dislocations are predominantly dissociated perfect dislocations, while in crystals deformed below ~1100°C, the dislocations are predominantly single leading partials without their corresponding trailing partials. From the width of the dissociated dislocations, the stacking fault energy of 4H-SiC has been estimated to be 14.7 ± 2.5 mJ/m^2.

In the GaN film, Vickers indentation produced a dense array of dislocations on the { $1\bar{1}00$ } prism planes; these dislocations lie along the three $<11\bar{2}0>$ Peierls valleys on these planes. Most of these dislocations were found to have a screw character and, because of the high stacking fault energy on the prism planes, they were not dissociated.

ACKNOWLEDGEMENTS

This work was supported by grant number FG02-93ER45496 from the Department of Energy, and subcontract number 95-SPI-420757-CWRU from the Silicon Carbide consortium.

Thanks are due to Dr. Don Hobgood (previously of Northrop-Grumman) for providing a single crystal of 4H-SiC.

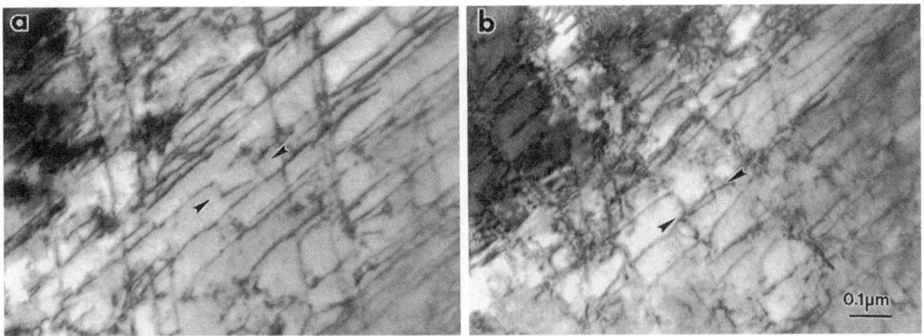

Fig. 6. TEM micrograph of deformed 2H-GaN (a) before and (b) after tilting around the [$11\bar{2}0$] axis

REFERENCES

1. W. J. Choyke, H. Matsunami and G. Pensl (edited), *Silicon Carbide, A Review of Fundamental Questions and Applications to Current Device Technology*, (Akademie Verlag GmbH **Volume I** and **II**, Berlin, 1997).
2. S. Nakamura and G. Fasol, *The Blue Laser Diode*, (Springer-Verlag, Berlin-Heidelberg, 1997).
3. Y. M. Tairov and V. F. Tsvetkov, J. Crystal Growth **43**, 209-212 (1978).
4. P. Pirouz and J. W. Yang, Ultramicroscopy **51**, 189-214 (1993).
5. P. Pirouz, Solid State Phenomena **56**, 107-132 (1997).
6. L. S. Ramsdell, Am. Mineralogist **32**, 64-82 (1947).
7. P. Pirouz, "Extended Defects in SiC and GaN Semiconductors", in *Proceedings of the International Conference on Silicon Carbide, III-Nitrides and Related Materials*, edited by G. Pensl, H. Morkoç, B. Monemar and E. Janzén (Trans Tech Publications Ltd. **264-268**, Zurich, Switzerland, 1998), *pp.* 399-408.
8. H. Alexander, P. Haasen, R. Labusch and W. Schröter, Foreword to J. Phys. (Paris) **40**, Colloque C6 (1979).
9. A. V. Samant, *Effect of Test Temperature and Strain-Rate on the Critical Resolved Shear Stress of Monocrystalline Alpha-SiC*, Ph.D. Thesis, Case Western Reserve University, 1999.
10. M. H. Hong, A. V. Samant and P. Pirouz, Phil. Mag. A (1999). In press.
11. X. J. Ning and P. Pirouz, J. Mater. Res. **11**, 884-894 (1996).
12. P. Pirouz, A. V. Samant, M. H. Hong, A. Moulin and L. P. Kubin, J. Mater. Res. (1999). In press.
13. B. Heying, X. H. Wu, S. Keller, Y. Li, D. Kapolnek, B. P. Keller, S. P. Denbaars and J. S. Speck, Appl. Phys. Lett. **68**, 643-645 (1996).
14. X. J. Ning, F. R. Chien, P. Pirouz, J. W. Yang and M. Asif Khan, J. Mater. Res. **11**, 580-592 (1996).
15. S. Takeuchi and K. Suzuki, Phil. Mag. Lett. (1999). In press.

Ca DOPANT SITE WITHIN ION IMPLANTED GaN LATTICE

H. Kobayashi* and W. M. Gibson
Department of Physics, the University at Albany, SUNY, Albany, NY 12222, USA

ABSTRACT

We have investigated the Ca dopant site within the GaN lattice using ion channeling in combination with Rutherford backscattering spectrometry (RBS), particle induced x ray emission (PIXE) and nuclear reaction analysis (NRA). Metalorganic chemical vapor deposition (MOCVD) grown GaN on c-plane sapphire substrates implanted with ^{40}Ca at a dose of 1×10^{15} cm^{-2} with post-implant annealing were investigated. The channeling results indicate that more than 80 % of Ca are near Ga sites even in as-implanted samples, however, they are displaced by ~ 0.2 Å from the Ga sites and that the Ca goes to the exact Ga sites after annealing at 1100°C. We think that the displaced Ca in the as-implanted samples are electrically compensated due to formation of complex defects with donor like point defects, such as Ca$_{Ga}$-V$_N$ and/or Ca$_{Ga}$-Ga$_N$, and that Ca$_{Ga}$ becomes electrically active when these complex defects are broken and the point defects diffuse away with annealing at 1100°C.

INTRODUCTION

GaN has attracted great interest not only for the fabrication of blue light emitting lasers but also for high power and high temperature devices because of its outstanding thermal and chemical stability [1, 2]. It has been reported that Ca implanted GaN becomes p-type from n-type with post-implant annealing at ~1100°C. The ionization level of Ca was estimated to be ~170 mV, which is as shallow as that of Mg [3]. Since Ca can be a p-type carrier without any co-implantation while Mg implantation requires P co-implantation to achieve p-type conductivity [4], Ca is expected to be a desirable p-type dopant. Ion implant damage and removal by annealing have been studied as a function of implantation dose and annealing temperature [5, 6]. However, most of these works concerned only electrical and optical properties or crystalline quality of the GaN itself, no work on structural information on impurity lattice location has been reported other than Si implanted GaN [7].

We have studied directly the lattice location of Ca implanted GaN using ion channeling combined with Rutherford backscattering spectrometry (RBS), particle induced x ray emission (PIXE) and nuclear reaction analysis (NRA) for Ga, Ca and N detection, respectively. We estimated the substitutionality of Ca in the GaN lattice from minimum channeling yield and studied whether the Ca is in Ga or N sites by comparing channeling angular distributions of Ca, Ga and N. We will also discuss displacements of Ca from lattice sites. In this paper, we will demonstrate the lattice location of Ca. We will also discuss the mechanism of p-type conductivity of Ca implanted GaN with post-implant annealing.

EXPERIMENTAL

2 µm-thick GaN layers were grown on c-plane sapphire substrates by metalorganic chemical vapor deposition (MOCVD) at 990°C with a 25 nm-thick GaN buffer layer grown at 520°C. ^{40}Ca ions were implanted into the undoped GaN layer at a dose of 1×10^{15} cm^{-2} at an energy of 200 keV. The implantation was performed at liquid nitrogen temperature. The maximum ^{40}Ca concentration

---

* present address : Research Center Sony Corporation, 134 Goudo-cho, Hodogaya-ku, Yokohama, Japan 240
  e-mail : hkobayas@src.sony.co.jp

Mat. Res. Soc. Symp. Proc. Vol. 572 © 1999 Materials Research Society

and projected range were estimated to be 9×10^{19} cm^{-3} and 110 nm, respectively, by TRIM calculations [8]. After ion implantation, the samples were annealed at 1100°C for 10 minutes with ramp-up and ramp-down times of 10 minutes each. The annealing was performed in flowing dry N$_2$ gas ambient (500 cc/min) using a conventional tube furnace made of quartz capable of operation up to 1200°C. The tube furnace was maintained at constant temperature, and ramp-up and ramp-down rates were controlled by inserting or removing the samples. Before annealing, samples were held at 300°C for 1 hour to eliminate water on the samples to avoid oxidation of the surfaces.

Ion beam analysis was performed using a 4 MV dynamitron accelerator at the University at Albany, SUNY. A 3 MeV He beam was used to obtain channeling yields of Ga and Ca by RBS and PIXE, respectively, and a 3.2 MeV proton beam was used for N detection by NRA. Nuclear resonant elastic scattering, ^{14}N (p, p) ^{14}N was used for N detection. A Si surface barrier detector (50 mm^2, FWHM = 14 keV) was placed at 171° from the beam direction for the RBS measurements. Figure 1 shows typical RBS energy spectra. The Ga signal window was set as shown in Fig.1, to correspond to the Ca implanted depth region. A minimum channeling yield (χ_{min}) of 1.4 % for the unimplanted GaN indicates the excellent crystalline quality of the GaN film. The χ_{min} of as-implanted samples and the annealed sample at 1100°C were 4.4 % and 1.9 %, respectively. This shows that the implant damage is considerably reduced by post implant annealing. However, the recovery is not perfect, suggesting that the optimum annealing temperature may be higher than 1100°C. Alternatively there may be some residual strain or dislocation defects in the implanted layer. Electrical measurements have shown that Ca dopant activation can be achieved at 1100°C. Samples annealed at 1150°C showed poor crystalline quality due to surface decomposition, so channeling angular scans were not done. To prevent such surface decomposition, either high N$_2$ pressure or an AlN encapsulation layer is necessary [9].

To investigate the implanted Ca lattice location, PIXE signals were measured by using a Si(Li) detector with a 4 μm-thick prolene window (20 mm^2, FWHM = 160 eV) that was placed at 135°. Figure 2 shows typical PIXE spectra. We placed 8 μm-thick aluminized mylar in front of the PIXE detector in order to reduce the intensity of the Ga L signal. The Ga L signal was reduced by a factor of 35 while more than 95% of the Ca K signal remained. This aluminized mylar was also useful to shield light from GaN luminescence produced by the ion bombardment.

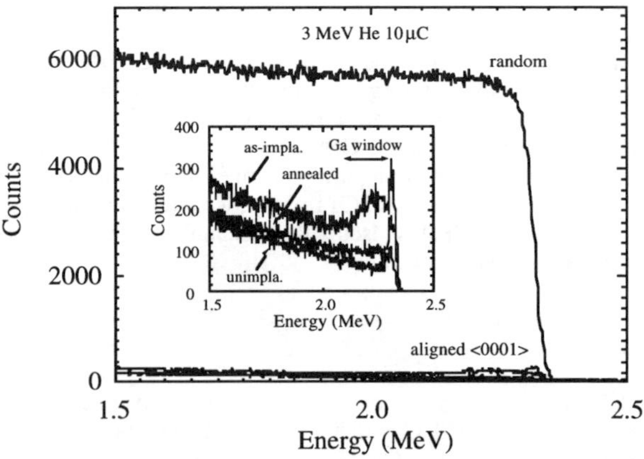

Fig. 1   RBS spectra for random and <0001> aligned GaN of as-implanted (1x10^{15} cm^{-2}), annealed (1100°C) and unimplanted samples.

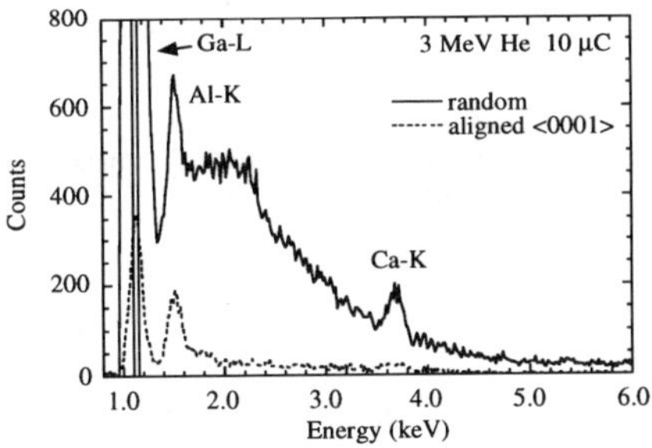

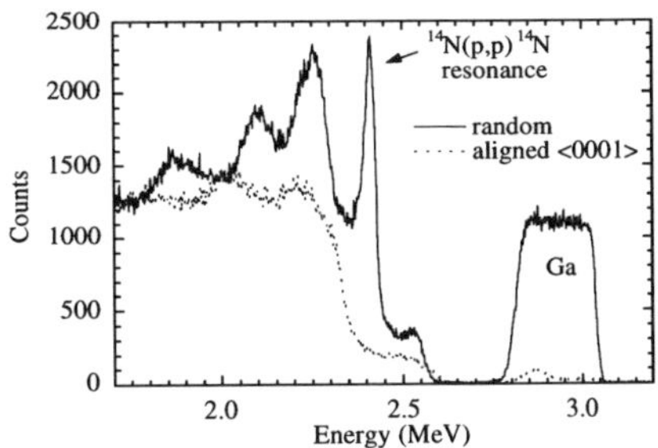

Fig. 2   PIXE spectra of the annealed sample.   Fig. 3   NRA spectra of the annealed sample.

Thus we minimized the dead time of the PIXE detector. In off-line data analysis, we obtained the Ca yield by curve fitting which took into account the Gaussian shaped Ca K signal and quadratic bremstrulung background. The same detector as used for the RBS measurement was used for N detection by NRA. The ^{14}N (p, p) ^{14}N reaction has a narrow resonance at 3.2 MeV with a resonance width of 20 keV. The reaction cross section is 35 times larger than that of RBS at this energy [10]. Figure 3 shows a typical NRA spectrum. A sharp peak due to the nuclear reaction can be seen. Since the resonance width corresponds to a depth width of 300 nm in GaN, we obtained the N signal only from the region where Ca was implanted, and could thus eliminate dechanneling contributions from deeper layers. To obtain the N yield precisely, we measured spectra at just below the resonance energy (off-resonance) to obtain the amount of continuum background under the resonance signal and subtracted the background from on-resonance spectra.

The samples were mounted on a two-axis goniometer and channeling angular distributions for Ca and Ga were obtained simultaneously. The channeling angular distributions of N were obtained in the same way. A total beam dose of 5 ~ 10 µC for each RBS, PIXE spectrum and 1 µC for the NRA spectrum was accumulated in a single angular scan. The beam spot size was 1.5 mm × 1.5 mm with divergence less than 1 mrad (0.06°).

RESULTS AND DISCUSSION

Channeling angular scans were performed in <0001> and $< 10\bar{1}1 >$ axial directions. Tilts through <0001> and $< 10\bar{1}1 >$ axes were made at angles of 10° and 26°, respectively, with respect to the {11$\bar{2}$0} plane to eliminate contribution of planer channeling at large tilt angles. Thus, we obtained good reproducibility of channeling distributions. Figure 4 shows the channeling angular distributions of an as-implanted sample. Channeling yields were normalized to the random yield which was obtained for each angular scan. The angular distributions of Ca have large dips even in the as-implanted sample in both the <0001> and the $< 10\bar{1}1 >$ directions as shown in Fig. 4. Table I summarizes χ_{min} of the as-implanted and the annealed samples. The fraction of impurity in substitutional site (f_{sub}) can be obtained by the following equation [11].

$$f_{sub} = \frac{1 - \chi_{min}(\text{impurity})}{1 - \chi_{min}(\text{host})}$$

We obtained f_{sub} of 0.86, 0.85 for the <0001>, $<10\bar{1}1>$, respectively. The fact that a large amount of Ca is already substitutional even in the as-implanted sample, suggests that dynamic annealing takes place effectively in GaN even for low temperature ion implantations [5, 12].

Figure 5 shows the channeling angular distribution of the annealed sample at 1100°C. The calculated f_{sub}'s are 0.82, 0.83 for the <0001>, $<10\bar{1}1>$, respectively, indicating that substitutionality of Ca in the annealed sample is almost same as that in the as-implanted sample.

The half angle of channeling dip can be expressed by the following equation [11],

$$\psi_{1/2} \propto \sqrt{\frac{Z_1 Z_2}{E}}$$

where Z_1 is the atomic number of incident atom, Z_2 is the average number of the constituent atoms in the row and E is energy of incident atom. Since the $<10\bar{1}1>$ atomic row of GaN consists of pure Ga and pure N rows, the half angle of Ga and N in the $<10\bar{1}1>$ are different. Therefore, it is possible to determine whether the Ca is in Ga or N sites by comparing half angles of Ca, Ga and N in the $<10\bar{1}1>$ direction. Table II shows a summary of $\psi_{1/2}$ values. The $\psi_{1/2}$'s were obtained from curve fitting procedures using a three-dimensional spline function. The accuracy of the half width measurement was estimated to be 0.06° based on reproducibility. We corrected the $\psi_{1/2}$'s of N obtained with 3.2 MeV protons to correspond to a 3 MeV incident He beam using the above equation so as that we can compare $\psi_{1/2}$'s. For the annealed sample, the $\psi_{1/2}$ of Ca in the $<10\bar{1}1>$ (0.49°) is almost the same as that of Ga (0.51°) and different from that of N (0.19°). Therefore, we conclude that most of the substitutional Ca is in the Ga site in the annealed sample.

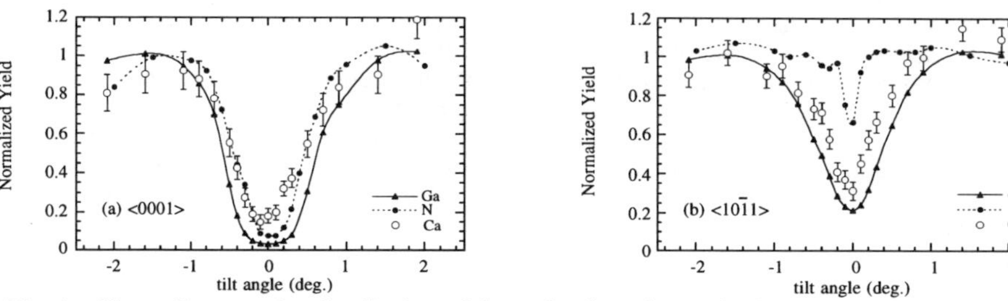

Fig. 4   Channeling angular distribution of the as-implanted sample along (a) the <0001> and (b) the $<10\bar{1}1>$ axial direction.

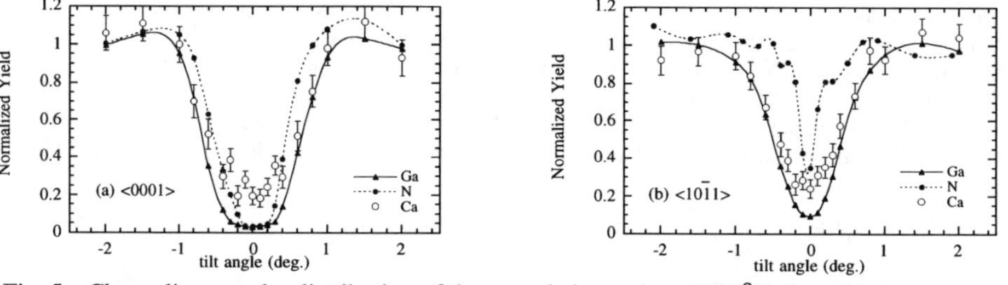

Fig. 5   Channeling angular distribution of the annealed sample at 1100°C along (a) the <0001> and (b) the $<10\bar{1}1>$ axial direction.

Table I. Summary of minimum channeling yield, χ_{min}.

Element	as-implanted		annealed	
	<0001>	<10$\bar{1}$1>	<0001>	<10$\bar{1}$1>
Ca	0.18	0.34	0.20	0.25
Ga	0.04	0.22	0.02	0.10
N	0.08	0.66	0.02	0.34

Table II. Summary of half angle of channeling dip, $\psi_{1/2}$ (degree). Corresponding values to 3 MeV He incident beam are shown in parentheses for N.

Element	as-implanted		annealed	
	<0001>	<10$\bar{1}$1>	<0001>	<10$\bar{1}$1>
Ca	0.47	0.31	0.68	0.49
Ga	0.61	0.50	0.67	0.51
N	0.49 (0.72)	0.10 (0.15)	0.50 (0.73)	0.13 (0.19)

On the other hand, for the as-implanted sample, the $\psi_{1/2}$'s of Ca are narrower than those of Ga in both the <0001> and the <10$\bar{1}$1> directions. This suggests that Ca is slightly displaced from the Ga sites in the as-implanted sample. The $\psi_{1/2}$ in the <10$\bar{1}$1> direction would also be narrowed if some fraction of the Ca is in the N site. However, the $\psi_{1/2}$ in the <0001> direction would not be narrowed. Since our data show that the $\psi_{1/2}$ in the <0001> direction is also narrowed similarly to the <10$\bar{1}$1> direction, we conclude that Ca atoms sit near the Ga sites but are slightly displaced. The projected displacement of impurity atoms from the host atoms (r_x) is related to the $\psi_{1/2}$'s by the following equation [11].

$$\frac{\psi_{1/2}(\text{impurity})}{\psi_{1/2}(\text{host})} = \frac{\{\ln[(Ca/r_x)^2 + 1]\}^{1/2}}{\{\ln[(Ca/\rho)^2 + 1]\}^{1/2}}$$

C is a constant of magnitude ~1.73, a is a screening distance and ρ is thermal vibration amplitude. For GaN, a is ~ 015 Å, ρ is ~ 0.10 Å [13]. We obtained r_x ~ 0.17 ± 0.05 Å in the <0001> and ~ 0.24 ± 0.05 Å in the <10$\bar{1}$1> directions, respectively, using the measured $\psi_{1/2}$'s of Ca and Ga for the as-implanted sample. This equation assumes an ordered lattice. We measured $\psi_{1/2}$'s of an unimplanted GaN sample and found that the $\psi_{1/2}$'s of the unimplanted sample and that of the as-implanted sample are consistent within error (0.06°). Therefore, contribution of the disorder in the as-implanted sample to the calculation is not serious.

From our results and electrical properties reported previously [3], in which Ca implanted GaN becomes p-type from n-type at an annealing temperature of 1100°C, we suggest the mechanism of electrical conductivity of Ca implanted GaN as follows:

(i) Ca is compensated electrically in the as-implanted GaN due to formation of complex defects with implantation induced donor like point defects, such as nitrogen vacancy (V_N) and/or antisite Ga (Ga_N), resulting in a complex in which Ca atoms are slightly displaced from exact Ga sites. GaN remains n-type due to residual donor like point defects.

(ii) After annealing at 1100°C, these complex defects are broken and donor like point defects are annihilated or diffuse away. After annealing, substitutional Ca_{Ga} works effectively as an acceptor, and GaN becomes p-type.

CONCLUSION

We have demonstrated the lattice location of Ca in ion implanted GaN with post-implant annealing using ion channeling. More than 80 % of Ca are near the Ga sites even in the as-implanted sample, however, they are slightly displaced from the Ga sites. Ca goes to the exact Ga sites after annealing at 1100°C. We think that Ca is electrically compensated due to formation of complex defects with donor like point defects, such as Ca_{Ga}-V_N and/or Ca_{Ga}-Ga_N in the as-implanted sample. Ca_{Ga} becomes electrically active when these complex defects are broken and the point defects diffuse away with annealing at 1100°C.

ACKNOWLEDGEMENTS

We would like to thank T. Asatsuma for the GaN deposition. This work was supported by Research Center, Sony Corporation.

REFERENCES

[1] J. C. Zolper and R. J. Shul, MRS Bull. **22**, 36 (1997)
[2] M. S. Shur and M. A. Khan MRS Bull. **22**, 44 (1997)
[3] J. C. Zolper, R. G. Wilson, S. J. Pearton, and R. A. Stall, Appl. Phys. Lett. **68**, 1945 (1996)
[4] S. J. Pearton, C. B. Vartuli, J. C. Zolper, C. Yuan, and R. A. Stall, Appl. Phys. Lett. **67**, 1435 (1995)
[5] H. H. Tan, J. S. Williams, J. Zou, D. J. H. Cockayne, S. J. Pearton, and R. A. Stall, Appl. Phys. Lett. **69**, 2364 (1996)
[6] H. H. Tan, J. S. Williams, J. Zou, D. J. H. Cockayne, S. J. Pearton, J. C. Zolper, and R. A. Stall, Appl. Phys. Lett. **72**, 1190 (1998)
[7] H. Kobayashi and W. M. Gibson, Appl. Phys. Lett. **73**, 1406 (1998)
[8] J. F. Ziegler, J. P. Biersack and U. Littmark, Stopping and Ranges of Ions in Matter (Pergamon, New York, 1988), Vol. I
[9] X. A. Cao, C. R. Abernathy, R. K. Singh, S. J. Pearton, M. Fu, V. Sarvepalli, J. A. Sekhar, J. C. Zolper, D. J. Rieger, J. Han, T. J. Drummond, R. J. Shul, and R. G. Wilson, Appl. Phys. Lett. **73**, 229 (1998)
[10] S. Bashkin, R. R. Carlson, and R. A. Douglas, Phys. Rev. **114**, 1552 (1959)
[11] L. C. Feldman, J. W. Mayer, and S. T. Picraux, *Materials Analysis by Ion Channeling* (Academic Press, New York, 1982)
[12] C. Liu, B. Mensching, K. Volz, and B. Rauschenbach, Appl. Phys. Lett. **71**, 2313 (1997)
[13] J. R. Tesmer and M. Nastasi, *Handbook of Modern Ion Beam Materials Analysis* (Materials Research Society, Pittsburgh, 1995)

GROWTH AND CHARACTERIZATION OF InGaN/GaN HETEROSTRUCTURES USING PLASMA-ASSISTED MOLECULAR BEAM EPITAXY

K. H. Shim, S.E. Hong, K.H. Kim, M. C. Paek, and K.I. Cho

Wide Band Gap Semiconductor Team, Microelectronics Technology Research Laboratory, Electronics and Telecommunications Research Institute
161 Kajung-Dong, Yusong-Gu, Taejon, Korea 305-350

Structural and optical properties of $In_{0.2}Ga_{0.8}N/GaN$ heterostructures grown by plasma-assisted molecular beam epitaxy have been investigated as a function of rf plasma power. Indium incorporation resulted in the higher rf power level suppressing 3D island growth with reduced introduction of defects in $In_{0.2}Ga_{0.8}N$ in comparison with GaN. Sharp morphology at interfaces and strong transitions in photoluminescence reveal the optimum rf power around 400 W in our experimental set up for the growth of $In_{0.2}Ga_{0.8}N/GaN$ heterostructures. Our experimental observations suggest that the presence of indium on surface modulates the rate of plasma stimulated desorption and diffusion, and reduces the formation of damaged subsurface.

INTRODUCTION

Recently, lots of achievements have been accomplished in the field of GaN-based nitride semiconductors by remarkable breakthroughs in epitaxial growth technology [1]. Plasma assisted molecular beam epitaxy (PAMBE) has demonstrated useful results along with its advantageous features such as low temperature processes and atomic layer growth. However, the development of GaN-based epilayers using PAMBE has been hindered by uncertainty about the effect of the plasma parameters on film quality and inadequate control of the energetic species in the nitrogen plasma. Meanwhile, the application of very low energy ion beams of a few tens of eV have focused on the modulation of surface reaction kinetics, strain relaxation, and island formation [2,3]. Despite the relevant role of plasma parameters, relatively few articles have focused on the study of plasma parameters in GaN epilayers using PAMBE [4,5]. A fundamental understanding of the interactions of energetic particles with the surface of GaN-based epi layers would enable the tradeoff between damage production and surface enhancement.

The objective of this study is to understand the effect of the rf power on the

Mat. Res. Soc. Symp. Proc. Vol. 572 © 1999 Materials Research Society

growth of GaN-based epilayers by exploring correlations between the structural/optical properties and the rf plasma power. Our experimental results on $In_{0.2}Ga_{0.8}N/GaN$ superlattice (SL) structures provide a framework for understanding the complicated role of plasma parameters on the growth of GaN-based heterostructures.

EXPERIMENT

GaN-based heterostructures were grown using PAMBE equipped with an inductively coupled rf plasma source (SVT Associates) [6]. After the conventional solvent cleaning and etching processes, 20-nm-thick AlN buffer layers were grown on sapphire (0001) at 500°C. Then GaN epilayers were grown at 720°C and followed by the growth of ten periods of $In_{0.2}Ga_{0.8}N/GaN(25/50Å)$ at 670°C. The thickness of InGaN was designed to be smaller than the critical thickness, 7.6 nm, estimated from a Matthew and Blakeslee model. The growth chamber maintained 7×10^{-5} torr for whole growth process. Langmuir probe technique was employed to analyze the distribution of energy and flux of nitrogen ions. The surface structure of epilayer was monitored by *in-situ* reflection of high-energy diffraction (RHEED) and the quality of obtained epilayers was characterized by atomic force microscopy (AFM), cross-sectional transmission electron microscopy (XTEM), and low-temperature photoluminescence (PL).

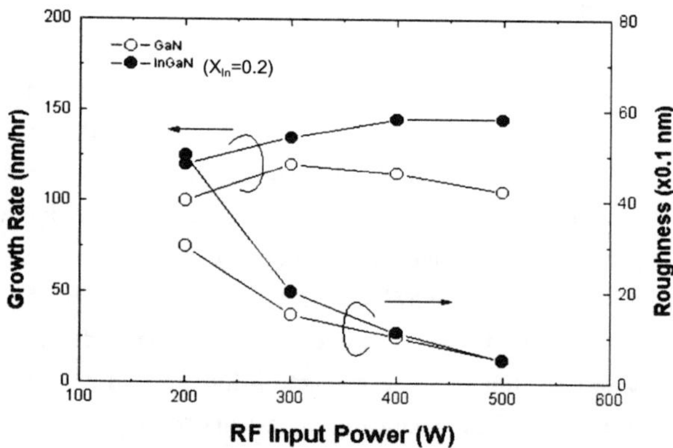

Fig. 1. Growth rate and roughness data plotted as a function of rf input powers for GaN, open circles, and $In_{0.2}Ga_{0.8}N$, closed circles.

RESULTS AND DISCUSSION

Fig. 1 represents the changes in both growth rate and root mean square (rms) roughness as a function of the input rf power of PAMBE. A typical, panoramic feature of PAMBE is well identified from the data in the figure 1. The decrease in growth rate at high rf power indicates that the flux of reactive nitrogen is sufficient at around 300 W. The surface roughness shows a significant decrease at low rf power regime and decreases monotonously at high rf power regime above 300 W. The changes in rms roughness revealed the evolution of surface morphology, corresponding to *in-situ* RHEED observation. However, Two distinctive features are noticeable in the InGaN growth: its growth rate saturates to 145 nm/hr at high rf power and its roughness is roughly two folds of GaN at low rf power around 200W. The origins causing such differences are discussed below in detail.

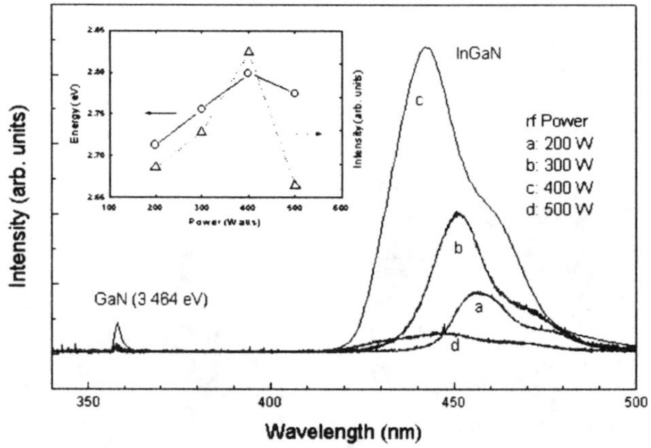

Fig. 2.   LT-PL spectra measured at 13 K from InGaN/GaN superlattice samples grown with the rf input powers of 200, 300, 400, and 500 W.

In an effort to investigate the effect of rf power on optical properties, $In_{0.2}Ga_{0.8}N$ /GaN SLs were prepared using different rf powers of 200, 300, 400, and 500 W.   Shown Fig. 2 represents PL spectra for the $In_{0.2}Ga_{0.8}N$/GaN samples, where the inset is a plot of energy and intensity of InGaN peaks. The spectrum c reveals optical properties much better than the others in terms of high emission efficiency and the peak position, 2.8 eV, of InGaN quantum wells corresponding to the energy levels of electrons and holes. The strong PL emission spectra were observed at around 300W for GaN [6]. Sharp and strong XRD peaks were observed from highly crystalline GaN epilayers grown with 300-400 W.   The increased energy and flux of nitrogen ions led to a preferred orientation structure and that the 2D planar growth enhanced at high rf power. The optimum rf power looks increased by ~100 W in InGaN.

Shown Figs 3 (a) and (b) are XTEM images taken from the sample grown with 400 W. Many angled-dislocations, noted as D, originate from threading dislocations at interfaces and facets of pyramid shape.   We observed that threading dislocations with a Burger's vector of g=<11-20>/3 exist as much as 10^{11} cm^{-2} at near to the interface, and that the dislocation density was tremendously decreased to ~7x10^{8} cm^{-2} at 0.4-1 μm.

The lattice constants of GaN were measured to be very close to that of bulk GaN, c=5.183 Å. The lattice constant of $In_{0.2}Ga_{0.8}N$ is under the compressive strain of 1.5%, although the original strain 2% between $In_{0.2}Ga_{0.8}N$ and GaN. According to TEM observations, dislocations were not newly generated at the interface of InGaN and GaN. This confirms that the relaxation in InGaN occurred partially in the limit of elastic deformation.

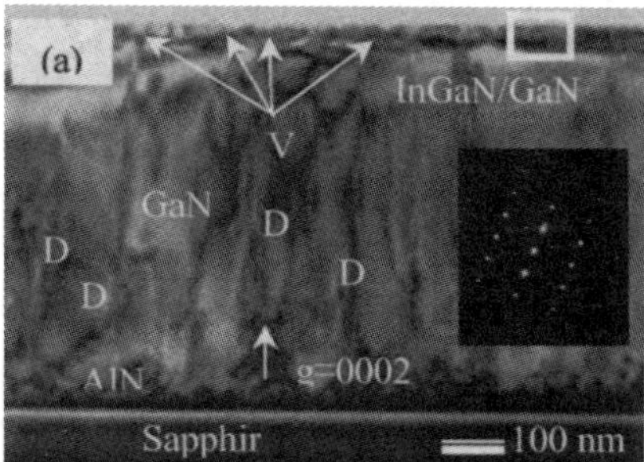

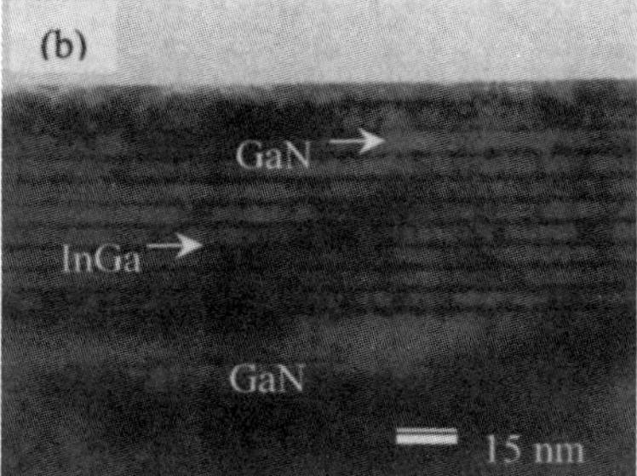

Fig. 3. (a) XTEM [11-20] images of InGaN/GaN SLs grown on sapphire substrates and (b) HR-XTEM image of the rectangle in (a). The micrographs were taken with g=0002 along [11-20] zone axis of hexagonal GaN. HR-XTEM image in (b) shows the sharp interface of InGaN/GaN with on dislocations newly generated due to the relaxation of lattice mismatch stress. The inset in (a) is the TED pattern showing spots from the sapphire substrate and the GaN epilayer.

Among various process regimes of ion bombardments, distinguished as a function of ion energy and ion/atom flux ratio, the operational conditions of PAMBE is located at a boundary dividing surface desorption and on-effect [7,8]. The previous data involving energetic particles often reveal discrepancies, which are particularly subject to the substrate materials, reactive particles, and plasma parameters affecting the energy and flux of ions. The enhanced diffusion of indium atoms on InGaN surface looks predominant at low rf power operation and resulted in very rough surface as shown in Fig.1. Therefore, the change caused by indium incorporation is attributed to the high surface diffusion and the large mass of indium.

The role of energetic particles in PAMBE can be understood by analyzing the changes in growth mode, which appears at the initial stage of thin film growth. The 2D-to-3D transition is unavoidable for the growth of heterostructures with large lattice mismatch between adjacent layers. The effects of ion-induced displacement on the nucleation rate can be estimated by adding a term for the breakup of the critical size nuclei. The forward reaction of nucleation which occurs by the direct impingement of Ga and the diffusion of single adatoms to critical-sized clusters is rate limited by thermal and plasma-stimulated desorption. Thus, the 3D nucleation depends mostly on the energy and flux of energetic particles at a certain temperature.

The optical properties of epilayers would be more likely depending on the ion energy. The displacement energy of bulk atoms is usually approximated as three times of the cohesive energy. By employing critical energies for the displacement of atoms at subsurface, the threshold energy of particles for defect formation at the subsurface are approximated as 16.5, 15, and 17 eV for AlN, GaN and InN, respectively [8]. Considering the Boltzman distribution with a dispersion of ~5 eV, the average ion energy less than ~14 eV looks safe for the growth of InN films without collision-induced defects. The rf powers ejecting ions with an average energy of 15 eV are 300 and 400 W for 1 and 2 sccm nitrogen gas flow, respectively. Consequently, the strong PL emission in Fig. 2 could be achieved from the InGaN/GaN grown with 400W. By raising the rf power from 400 W to 500 W, the plasma source becomes to supply ions with the energy ~21 eV and the flux increased by 30 %. The catastrophic degradation of the spectrum d in Fig. 2 presents that the plasma rf power must be controlled below 500 W.

The defects produced by ion bombardment may be annihilated at a higher temperature. However, the substrate temperature must be controlled below a certain limit to prevent surface decomposition. Nevertheless, our observations suggest that the energetic particles ejecting from a plasma source need to be removed completely or controlled properly for a specific application like the manipulation of surface roughening accompanied by stress relaxation.

CONCLUSIONS

From correlations between rf plasma power and the evolution of structural and optical properties of $In_{0.2}Ga_{0.8}N/GaN$ SLs, the incorporation of indium was observed to reduce the formation of damages, and correspondingly lead to enhance band-to-band optical transitions. It was important to control the rf power level below 400 W in order to suppress surface roughening without introducing defects at the subsurface of $In_{0.2}Ga_{0.8}N$ epilayers. The experimental results associated with indium incorporation could be explained in terms of the inelastic collision of energetic particles, stimulating kinetic reaction processes at surface as well as the creation of defects at subsurface.

ACKNOWLEDGMENTS

The support of the Ministry of Information and Communications of Korea is gratefully acknowledged. This work was partially initiated during Ph.D. research at University of Illinois at Urbana-Champaign. Authors thank J.B. Park at KMAC for his TEM analyses and Dr. C.S. Lee at KRISS for HR-XRD measurements.

REFERENCES

1. S. Nakamura, M. Senoh, S. Nagahama, N. Iwasa, T. Yamada, T. Matsushita, H. Kiyoku, Y. Sugimoto, T. Kozaki, H. Umemoto, M. Sano, and K. Chocho, Appl. Phys. Lett. **72**, 211(1998).
2. C.J. Tsai, H.A. Atwater, and T. Vreeland, Appl. Phys. Lett. **57**, 2305 (1990).
3. S.A. Barnett, C.H. Choi, and R. Kaspi, Mat. Res. Soc. Symp. Proc. 201 (1991).
4. M.S.H. Leung, R. Klockenbrink, C. Kisielowski, H. Fujii, J. Kruger, Sudkier G.S., A. Anders, Z. Liliental-Weber, M. Rudin, and E.R. Weber, Mater. Res. Soc. Symp. Proc. **449**, 221 (1997).
5. H. Fujii, C. Kisielowski, J. Krueger, M.S. Leung, R. Klockenbrink, M. Rubin, and E.R. Weber, Mat. Res.Soc. Symp. Proc. **449**, 227 (1997).
6. K.H. Shim, PhD thesis, University of Illinois at Urbana-Champaign, 1997.
7. M.V.R. Murty, H.A. Atwater, A.J. Kellock, and J.E.E. Baglin, Appl. Phys. Lett. **62**, 2566 (1993).
8. J.M.E. Harper, J.J.Cuomo, and H.T.G.Hentzell, Appl. Phys. Lett. **43**, 547 (1983).

PIEZOELECTRIC COEFFICIENTS OF ALUMINUM NITRIDE AND GALLIUM NITRIDE

C.M. Lueng*, H.L.W. Chan*, W.K. Fong**, C. Surya**, C.L. Choy*
*Department of Applied Physics and Materials Research Center,
 The Hong Kong Polytechnic University, Hung Hom, Kowloon, Hong Kong.
**Department of Electronic Engineering,
 The Hong Kong Polytechnic University, Hung Hom, Kowloon, Hong Kong.

ABSTRACT

Aluminum nitride (AlN) and gallium nitride (GaN) thin films have potential uses in high temperature, high frequency (e.g. microwave) acoustic devices. In this work, the piezoelectric coefficients of wurtzite AlN and GaN/AlN composite film grown on silicon substrates by molecular beam epitaxy were measured by a Mach-Zehnder type heterodyne interferometer. The effects of the substrate on the measured coefficients are discussed.

INTRODUCTION

Aluminum nitride (AlN) and gallium nitride (GaN) are III-V nitrides and the reported lattice parameters of AlN and GaN with wurtzite structure are: $a = 3.11$ Å, $c = 4.98$ Å for AlN [1] and $a = 3.189$ Å, $c = 5.185$ Å for GaN [1]. AlN and GaN have wide direct bandgaps of 6.2 [2] and 3.39 eV [3], repectively. Both of them have potential applications in devices working in high temperature and hostile environments [4]. Many different growth techniques have been used to prepare AlN and GaN films and molecular beam epitaxy (MBE) is one of the techniques that can produce high-quality epitaxial AlN and GaN films. Due to its superior properties, research in the physical properties and applications of AlN and GaN has attracted considerable interest [2]. However, to date there appears to be limited data on the piezoelectric coefficients of AlN as well as GaN. These parameters are important since both of them have potential use in microactuators, microwave acoustic and other microelectromechanical (MEM) devices [5].

EXPERIMENT

AlN film grown by MBE

AlN thin films were grown by using a SVTA BLT-N35 MBE system. The nitrogen molecules in a nitrogen gas stream were broken into atomic nitrogen by a SVTA RF plasma source operating at 13.56 MHz. Aluminum was evaporated by the K-cells. The atomic nitrogen and aluminum react to form AlN on a (111) silicon substrate.

The silicon substrate was cleaned by a degreasing process and etched in buffered HF. After the substrate had been transferred into the deposition chamber, the substrate surface was thermally cleaned at 940 °C for 1 hour. The substrate temperature was then decreased to about 600 °C and AlN thin film started to grow. The final thickness of the AlN film was about 450nm.

Sample geometry

In this experiment, two samples were studied. One is a 450nm thick AlN thin film while the other is a composite film, comprising a 140nm thick GaN film grown on a 30nm thick AlN

Mat. Res. Soc. Symp. Proc. Vol. 572 © 1999 Materials Research Society

buffer layer. Both of the films were grown on silicon substrates with (111) orientation. Figures 1 and 2 show the X-ray diffraction (XRD) patterns of the AlN and the composite film, respectively. The peak at 35.9° corresponds to the (0002) reflection of wurtzite AlN [6]. Another peak at 34.6° corresponds to the (0002) reflection of wurtzite GaN [7]. The XRD patterns confirmed that the films have the wurtzite structure, and the c axis is oriented along the normal of the substrate.

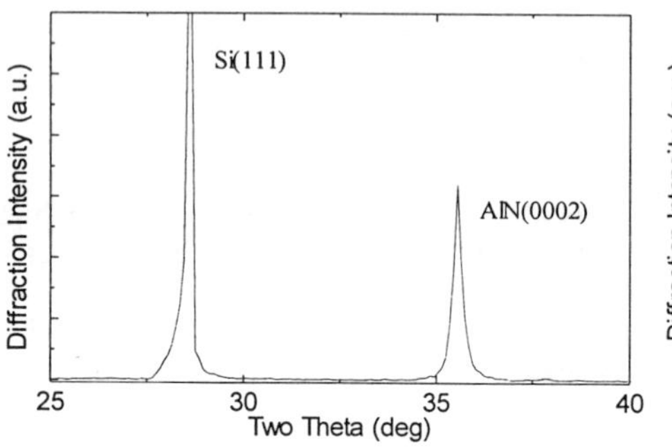

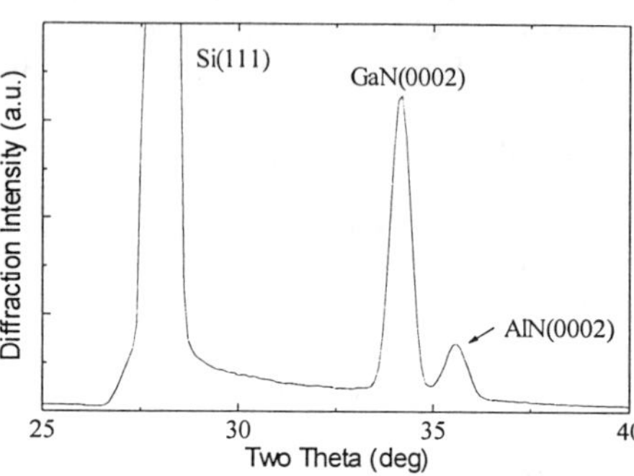

Figure 1: The XRD pattern of the AlN film.

Figure 2: The XRD pattern of the composite film.

Circular dots of aluminum of diameter 1mm were thermally evaporated at different positions on the top surface of the films. These dots act as top electrodes as well as mirrors to reflect the laser beam from the interferometer. The silicon substrate was attached to an aluminum block (connected to ground) fixed on a translation stage. An a.c. electric field was applied across the top electrode and the aluminum block. The thickness of the film changes due to the converse piezoelectric effect and the small displacement was measured using a Mach-Zehnder type heterodyne interferometer.

Measurement of d_{33} using laser interferometry

Figure 3 shows the schematic diagram of the Mach-Zehnder type heterodyne interferometer (SH-120 from B. M. Industries, France) used to measure the displacement. A linearly polarized laser beam, L (frequency f_L; wave number $k=2\pi/\lambda$, $\lambda = 632.8$ nm for a He-Ne laser) is split into a reference beam, R, and a probe beam, P, by a beam splitter (BS). R propagates through a Dove prism and a polarizing beam splitter (PBS) into a photodiode. The frequency of P is shifted by a frequency f_B (70 MHz) in a Bragg cell, and then this beam (now labeled S), is phase modulated by the surface displacement of the film sample, $x = u\cos(2\pi f_u t)$, where f_u is the vibration frequency and u is the displacement amplitude. For small u, only the sideband at $f_B + f_u$ is detected and its amplitude is

$$J_1(4\pi u/\lambda)/J_o(4\pi u/\lambda) \sim 2\pi u/\lambda = u/1007 \qquad (1)$$

where J_o and J_1 are the Bessel function of the zeroth and the first order, respectively. The ratio of amplitudes of the zeroth order (centerband) to the first order (sideband) of the Bessel function gives the absolute displacement of the sample surface. The ratio, $R' = J_1(4\pi u/\lambda)/J_o(4\pi u/\lambda)$ in dBm can be measured using a spectrum analyzer (HP3589). Let $R=10^{R'/20}$, the vibration displacement is [8]

$$u = 1007 * R' \qquad\qquad (2)$$

The d_{33}' coefficient (strain/applied field) of the film samples can be calculated as :

$$d_{33}' = (u/t)/(V/t) = u/V \qquad\qquad (3)$$

where t is the thickness of the sample and V is the voltage applied across the film sample. V was measured using an oscilloscope (Fig. 3) with a 50 Ω termination connected across the sample so as to ensure that the change in sample impedance with frequency does not cause a change in the voltage.

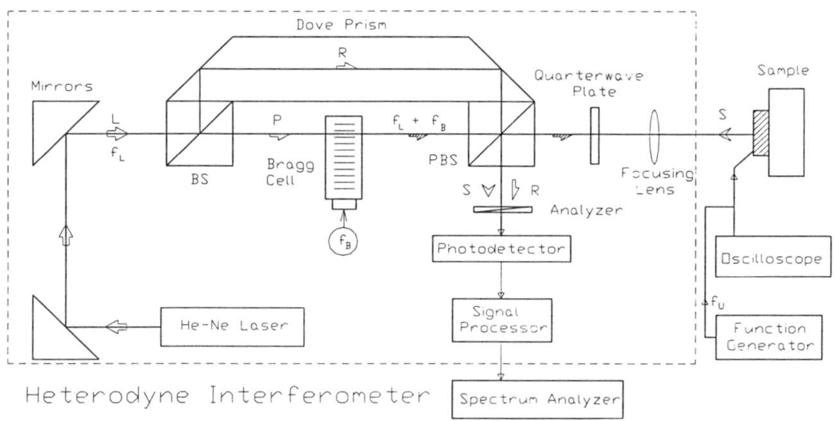

Figure 3: A Mach-Zehnder type heterodyne interferometer.

Sample mounting

When measuring vibration displacement using an interferometer, it is important to mount the sample properly. If the sample is not rigidly mounted on the aluminum block, other modes of vibration such as the bending mode may occur. The bending mode may have a vibration amplitude an order of magnitude larger than that of the thickness mode, hence a large error will result if the bending mode is also excited. One way to eliminate the bending effect is to reduce the size of the electrode and to glue the substrate to a rigid holder [9]. To ensure that no bending mode was present, the probe beam was scanned across the sample surface. As mentioned before, aluminum dots of diameter 1mm were deposited at different positions on the top surface of the sample. Measurements of the displacement amplitudes at these different positions gave essentially the same results, indicating that no bending mode was excited. As the electrode diameter was 1mm, the probe beam (about 100μm diameter) was also scanned across an Al dot to test whether bending vibration was excited within the 1mm dot. Again, constant vibration amplitudes are observed indicating that the dominant vibration is the thickness mode [10].

Substrate Clamping Effect

The constitutive equation for a piezoelectric material is [11]

$$S_i = s_{ij}T_j + d_{ki}E_k \qquad i,j = 1,\ldots,6 \qquad k = 1,\ldots,3 \qquad (4)$$

where S_i is the strain, T_j is the stress, s_{ij} is the compliance, d_{ki} is the piezoelectric coefficient and E_k is the applied electric field. When a thin piezoelectric film is grown on a thick substrate, the film is rigidly attached to the substrate and thus cannot vibrate freely. Hence, the measured coefficient cannot represent the true value for the film. However, it is possible to correct for the clamping effect, and the corrective factor depends on the symmetry and the orientation of the piezoelectric layer [12].

Since the film is clamped by a substrate, it cannot freely expand or contract in the interface. The assumption that strain S_1, S_2 and S_6 are equal to zero holds if the measurements are carried out at frequencies below the resonance frequencies of the film. If the top surface of the film is assumed to be free, T_3, T_4 and T_5 are zero on the free surface. For crystals with 6mm symmetry wurtzite structure [11], it is found that

$$T_1 = T_2 = \frac{-d_{31}E_3}{s_{11}^E + s_{12}^E} \qquad (5)$$

From the constitutive equation, we obtain

$$S_3 = 2s_{31}^E T_1 + d_{33}E_3 \qquad (6)$$

The measured (apparent) piezoelectric coefficient is $d_{33}'=S_3/E_3$, which can be related to the true coefficient d_{33} through Eq. (5) and (6):

$$d_{33}' = \frac{S_3}{E_3} = d_{33} - 2\left(\frac{d_{31}s_{13}^E}{s_{11}^E + s_{12}^E}\right) \qquad (7)$$

Therefore, the true d_{33} coefficient differs from the apparent coefficient by a corrective factor (the second term).

RESULTS

Measurements were made at the center of a 1mm diameter aluminum electrode located close to the center of the film. Figures 4 & 5 show the variation of the displacement with driving voltage at 10 kHz. From the slope of the line, the d_{33}' coefficients of the AlN and GaN/AlN composite film are found to be 3.9 pm/V and 2.65 pm/V, respectively. The correlation coefficients of the best fit straight lines for the AlN and the composite film are 0.99475 and 0.99651, respectively. The correction factor for AlN was calculated by using equation (7) and the compliance and d_{31} value given in ref. [13] and [14], respectively. The correction factor for AlN is 1.2 pm/V, hence the true piezoelectric coefficient d_{33} of AlN film is 5.1 pm/V. Figures 6 & 7 show the variation of the measured d_{33}' with frequency. It is seen that the value is approximately constant over the full range except at frequencies near 70 kHz for the AlN sample and 50 kHz for the GaN/AlN composite sample, where bending mode vibrations may have been excited. As it is difficult to estimate the substrate clamping effect in the GaN/AlN composite film, we are in a process of growing GaN films directly onto Si substrates. The results on these GaN films will be reported in the near future.

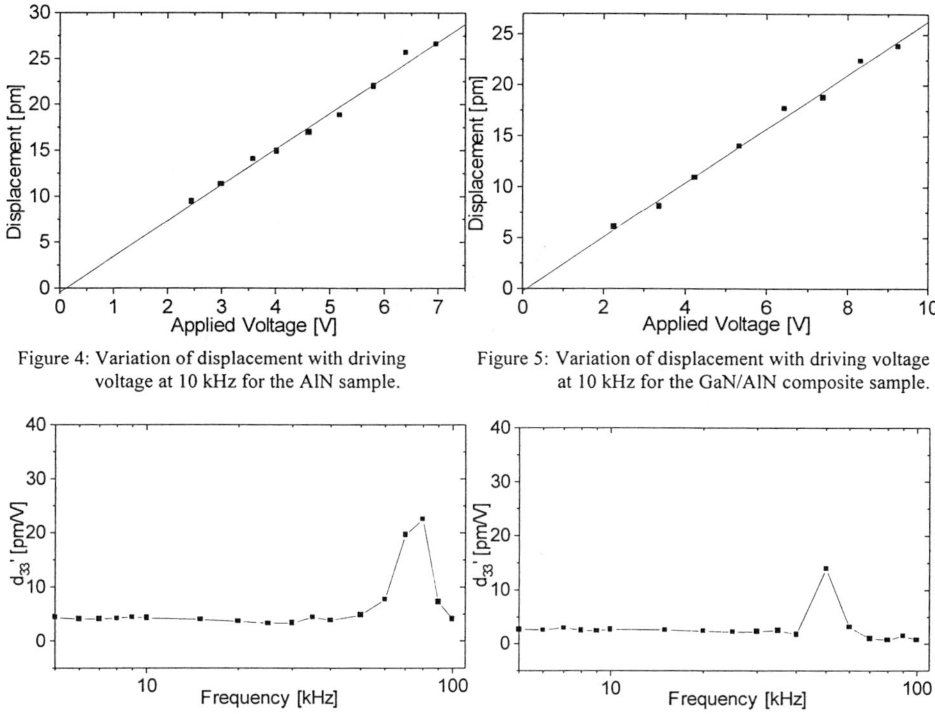

Figure 4: Variation of displacement with driving voltage at 10 kHz for the AlN sample.

Figure 5: Variation of displacement with driving voltage at 10 kHz for the GaN/AlN composite sample.

Figure 6: Variation of d_{33}' with frequency for the AlN sample.

Figure 7: Variation of d_{33}' with frequency for the GaN/AlN composite film sample.

CONCLUSIONS

In summary, the apparent piezoelectric coefficients d_{33}' of AlN and GaN/AlN composite films grown on Si substrates were found by a laser interferometric method to be 3.9 pm/V and 2.65 pm/V, respectively. The d_{33}' value of AlN film is approximately the same as that (4.0 pm/V) reported for AlN films grown by chemical vapour deposition [15]. By using a correction factor, the true d_{33} coefficient of AlN was found to be 5.1 pm/V and this value is similar to that (5.0 pm/V) reported in ref. [14].

REFERENCES

[1] Michael S. Shur and M. Asif Khan, Mat. Res. Bull., **22** (2), 44 (1997).
[2] S. Strite and H. Morkoç, J. Vac. Sci. Technol., **B10**, 1237 (1992).
[3] H. P. Maruska and J. J. Tietjen, Appl. Phys. Lett., **59**, 327 (1969).

[4] A. D. Bykhovski, V. V. Kaminski, M. S. Shur, Q. C. Chen and M. A. Khan, Appl. Phys. Lett., **68**, 818 (1996).

[5] S. N. Mohammad, Arnel A. Salvador and Hadis Morkoç, Proc. of the IEEE, **83**, 1306 (1995).

[6] Michihiro Miyauchi, Yukari Ishikawa and Noriyoshi Shibata, Jpn. J. Appl. Phys., **31**, L1714 (1992).

[7] K. Kubota, Y. Kobayashi, and K. Fujimoto, J. Appl. Phys., **66**, 2984 (1989).

[8] Z. Zhao, H. L. W. Chan and C. L. Choy, Ferroelectrics, **195**, 35 (1997).

[9] S. Muensit and I. L. Guy, Appl. Phys. Lett. **72**, 1896 (1998).

[10] A. L. Kholkin, Ch. Wütchrich, D. V. Taylor and N. Setter, Rev. Sci. Instrum., **67**, 1935 (1996).

[11] B. A. Auld, *Acoustic Fields and Waves in Solid Volume I*, 2nd ed. (Krieger Publishing Company, Malabar, Florida, 1990). p. 271, 370, 381.

[12] D. Royer and V. Kmetik, Electronics Letters, **28**, 1828 (1992).

[13] Lauroe E. McNeil, J. Am. Ceram. Soc., **76**, 1132 (1993).

[14] Tsubouchi,K., Sugai,K., and Mikoshiba,N., Proc. IEEE Ultrason. Symp., 375 (1981).

[15] Supasarote Muensit, PhD thesis, Macquarie University, 1998.

FAST AND SLOW UV-PHOTORESPONSE IN n-TYPE GaN

R. ROCHA[1], S. KOYNOV[1], P. BROGUEIRA[1], R. SCHWARZ[1], V. CHU[2], M. TOPF[3], D. MEISTER[3], and B.K. MEYER[3]

(1) Physics Department, Instituto Superior Técnico, P-1096 Lisbon, Portugal
(2) Instituto de Engenharia de Sistemas e Computadores, P-1000 Lisbon, Portugal
(3) I. Physics Department, University of Giessen, D-35392 Giessen, Germany

ABSTRACT

The photocurrent decay in n-type GaN films prepared by low-pressure chemical vapor deposition (LPCVD) was measured in the ms-to-s time range using steady-state UV light and in the μs time regime using short high-power pulses from higher harmonics of a Nd:YAG laser. A power law time dependence is observed with exponents ranging from -0.1 to -0.3, which is an indication of a broad distribution of trapping states inside the band gap. Combining Hall effect results and the magnitude of the initial slope of the photocurrent decay we estimate a mobility-lifetime product of 2.1×10^{-4} cm^2/V for photogenerated electrons at times below a few μs. Slow transients might be a handicap for applications of GaN in UV detectors.

1. INTRODUCTION

Applications of GaN-related materials in devices like UV-detectors or high-temperature transistors might be handicapped by slow current transients, as seen, for example, in the persistent photoconductivity effect (PPC) [1]. The PPC effect, which manifest itself as a prolonged non-exponential conductivity decay after exposure to sub-bandgap light, was attributed to internal strain or to charging and discharging of deep traps in AlGaN alloy films. Metastable changes in conductivity were observed up to several thousand seconds [2].

The samples we used for this study were prepared by low-pressure chemical vapor deposition without intentional doping. Hall measurements show that the carrier concentration is about 10^{18} cm^{-3} and the samples are n-type. Therefore the measured photocurrents were well below the dark current levels. Nevertheless, effects similar to those encountered in highly resistive AlGaN alloy films occur, like frequency-dependent photoconductivity and non-exponential current decay after pulsed excitation. We discuss whether both the slow (s) and the fast (μs) photocurrent decay can be described by changes of occupation of deep traps.

2. SLOW PHOTOCURRENT DECAY WITH CHOPPED LIGHT

The GaN samples were prepared by low-pressure chemical vapor deposition (LPCVD) from GaCl$_3$ and NH$_3$ sources on Al$_2$O$_3$ substrates [3]. The deposition temperature and process pressure were 965 $^{\circ}$C and 0.5 mbar, respectively. GaCl$_3$ was transported in a 50 sccm N$_2$ carrier gas heated to 70 $^{\circ}$C. The NH$_3$ flow was set at 100 sccm. This resulted in a growth rate of about 3 Å/s.

Mat. Res. Soc. Symp. Proc. Vol. 572 ©1999 Materials Research Society

Photocurrent decays were measured with coplanar Cr contacts. The 1 mm gap was illuminated either with chopped light from a 50 W Xe-lamp or with 5 ns pulses of the 532 nm line of a frequency-doubled Nd:YAG laser, or the fourth harmonic line at 266 nm. The applied voltage was kept between 1 and 3 V where ohmic behavior of the contacts was assured. The relatively large dark currents were compensated for electronically since the photocurrents were below the percent level.

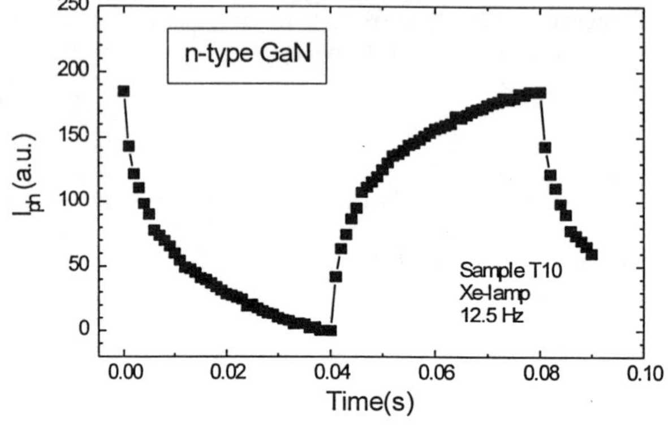

Fig. 1:
Modulated photocurrent with chopped UV light from a 50 W Xe lamp. The initial decay time constant is 2.1 ms.

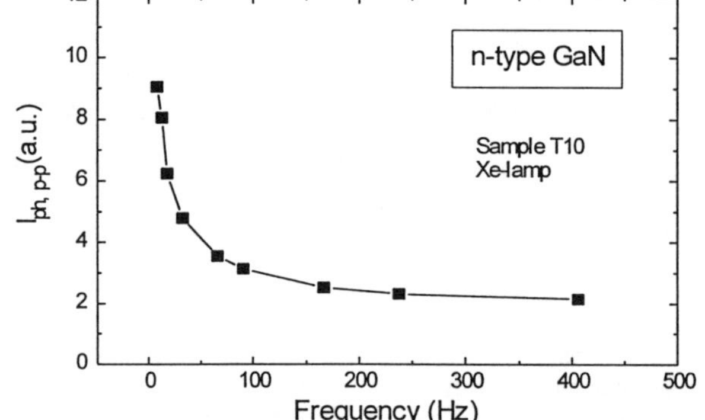

Fig. 2:
Frequency dependence of the peak-to-peak value of the photocurrent with chopped UV light.

As shown in the 12.5 Hz case in Fig. 1 the photocurrent never reaches a constant level due to the 1:1 duty cycle. Therefore, this experiment is not expected to reveal the very large time constants encountered in the PPC effect [2,4]. The initial decay time (response time) is about 2.1 ms in Fig. 1. Even though no saturation is reached one can estimate the exponent in a power law ansatz to the onset of the current decay to be about -0.1. We obtain the curve of Fig. 2 when the peak-to-peak photocurrent value is plotted as a function of the chopping frequency.

The complementary approach to the frequency domain is described in the next paragraph. However, the time domain studied below does not cover the times measured in this section. Also, we have to keep in mind that in the pulsed mode the average light bias, and the concomitant background carrier density, will be much lower. This could increase the photocurrent decay rate, i.e. reduce the observed response time, if it were governed by capture of free carriers in available shallow and deep traps.

3. FAST DECAY AFTER PULSED PHOTOEXCITATION

The initial photocurrents were of the order of the dark currents under pulsed excitation with 5 ns pulses from the 532 nm and 266 nm line of the Nd:YAG laser system. The power densities where about 14 and 5 MW/cm^2, respectively. The UV line lies slightly above the bandgap of 3.4 eV of GaN. However, the results shown in Fig. 3 for the 266 nm line do not differ significantly from the obviously strongly non-exponential time decay when the 532 nm line of the Nd:YAG laser or the 337 nm line of a pulsed nitrogen laser were used [5].

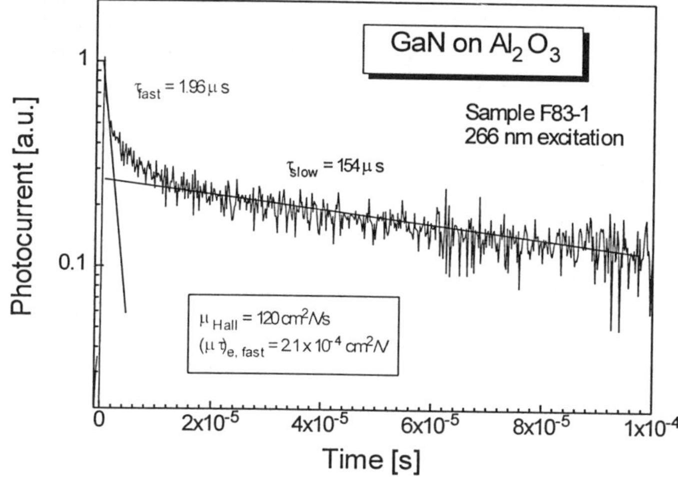

Fig. 3:
Photocurrent decay after pulsed laser excitation using 5 ns pulses at 266 nm wavelength.

Again, the photocurrent decay is strongly non-exponential with an initial decay time of 1.9 μs and a long tail with a time constant of 154 μs. The fast response time might be limited by the RC constant of the experimental set-up. The second time constant will depend on the time window chosen, and is, therefore, somewhat arbitrary. This is a typical feature of dispersive transport where the carrier mobility decreases with time due to trapped carriers which cannot contribute to band transport [6].

We can try to estimate a majority carrier $\mu\tau$-product (in our case $(\mu\tau)_e$ for electrons) if we combine the initial response time of Fig. 3 with the mobility value of 120 cm^2/Vs determined by Hall measurements. The result is $(\mu\tau)_e$ = 2.1x10^{-4} cm^2/V. This is the "effective" value at about 1 μs, it is expected to increase with time.

4. DISCUSSION

The three different time constants extracted from the photocurrent decays in Fig. 1 and 3 are quite consistent if we assume a time-dependent carrier drift mobility $\mu(t)$, as found in a number of disordered semiconductor materials [6]. This will lead to a power law decay in the photocurrent, $I_{ph}(t) \sim t^{-m}$, with an exponent m between 0 and 1. From Fig. 1 we find m = 0.1 and from Fig. 3 the range is from 0.27 to 0.33 for different samples and for different excitation wavelengths [5]. This would be compatible with a broad tail state distribution. Both, the presence of such defect states and the high dark conductivity of the films might be related to the heteroepitaxial growth of GaN on a foreign substrate (sapphire). Defect states could also be located at the surfaces and interfaces of crystallites that are seen in the AFM micrographs of our samples.

Without any further knowledge about the nature of trap states we can give an energy scale for those states if we transform the above time constants into trap energies using $E_{trap} = kT \, ln(v\tau)$, which is applicable for thermal emission. Under the assumption that electron traps are responsible for the above photocurrent transients we can sketch the location of the corresponding trap states as shown in Fig. 4, where we have chosen the phonon frequency v to be 10^{12} s^{-1}. The energy levels that correspond to 1.9 µs, 154 µs, and 2.1 ms are 0.36, 0.47, and 0.54 eV, respectively. A high defect density near the conduction band (CB) would be responsible for the position of the dark Fermi level E_F.

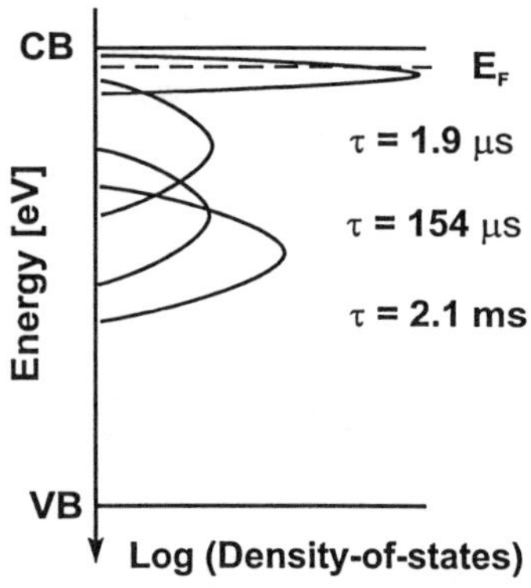

Fig. 4:
Sketch of the energy levels that correspond to the observed photocurrent decay times, assuming thermal reemission from electron traps. The magnitudes cannot be determined from the photocurrent decay alone and are therefore arbitrary.

From the n-type nature of the films we know that the photcurrent is carried by electrons. However, similarly to the case of electrons trapped near the conduction band as sketched in Fig. 4, we could also imagine that the time constants stem from reemission of holes trapped near the valence band, which would then recombine with free electrons or, with less probability, shallow-trapped electrons. The distinction should come from complementary experiments like photocapacitance measurements [7].

It is tempting to apply the Einstein relation to calculate the electron diffusion length L_e that would correspond to the above mentioned value of $(\mu\tau)_e = 2.1 \times 10^{-4}$ cm^2/V. We obtain L_e = 23 µm with an effective lifetime of 1.9 µs. This can be compared to results from the literature where an electron diffusion length of 0.2 µm and a lifetime of 0.1 ns was found in p-type GaN doped by Mg. The result was obtained from the analysis of electron beam induced current measurements [8]. There, however, electrons are the minority carriers, and the transport properties are low due to doping induced defects. We will be able to obtain an independent measure of the electron transport parameters in our case, once we have measured the steady-state photocurrent and the generation rate in absolute numbers.

5. CONCLUSION

In summary, our experiments hint to the existence of a broad state distribution inside the band gap of unintentionally doped n-type GaN samples which can explain the non-exponential photocurrent decay after pulsed laser excitation. The exponent of a power law decay is -0.1 in the case of chopped UV-light, where the average light bias is strong, and it is about -0.3 in the microsecond time regime measured with 5 ns pulses using both subbandgap and above bandgap laser light. We have also estimated the majority carrier diffusion length of electrons to be 23 μm at a response time of 1.9 μs.

Acknowledgment: S.K. would like to acknowledge a NATO fellowship. The work at IST is supported by the Fundação da Ciência e Tecnologia FCT (project no. PRAXIS/PCEX/P/FIS/7/98). All groups are members of the EC-funded network under the COPERNICUS programme (project no. ERB IC15 CT98 0819). R.S. and B.K.M. acknowledge financial support through the exchange program Acções Integradas of CRUP in Portugal and DAAD in Germany.

References

[1] M.D. McCluskey, N.M. Johnson, C.G. Van de Walle, D.P. Bour, M. Kneissl, and W. Walukiewicz, Phys. Rev. Lett. **80**, 4008 (1998).
[2] M.T. Hirsch, J.A. Wolk, W. Walukiewicz, and E.E. Haller, Appl. Phys. Lett. **71**, 1098 (1997).
[3] M. Topf, S. Koynov, S. Fischer, I. Dirnstorfer, W. Kriegseis, W. Burkhardt, and B.K. Meyer, Mat. Res. Soc. Symp. Proc. **449** (1997). S. Koynov, M. Topf, S. Fischer, B.K. Meyer, P. Radojkovic, E. Hartmann, Z. Liliental-Weber, J. Appl. Phys. **82**, 1 (1997).
[4] C.H. Qiu and J.I. Pankove, Appl. Phys. Lett. **70**, 1983 (1997).
[5] R. Schwarz, R. Rocha, P. Brogueira, S. Koynov, V. Chu, M. Topf, D. Meister, and B.K. Meyer, to be published.
[6] H. Antoniadis and E.A. Schiff, Phys. Rev. **B46**, 9842 (1992).
[7] Gyu-Chul Yi and Bruce W. Wessels, Appl. Phys. Lett. **68**, 3769 (1996).
[8] Z.Z. Bandic, P.M. Bridger, E.C. Piquette, and T.C. McGill, Appl. Phys. Lett. **73**, 3276 (1998).

EPITAXIAL GROWTH OF GaN THIN FILMS USING A HYBRID PULSED LASER DEPOSITION SYSTEM

Philippe Mérel *, Mohamed Chaker *, Henri Pépin* and Malek Tabbal **.
*INRS-Énergie et Matériaux, 1650 boul. Lionel Boulet, Varennes, Québec, Canada J3X 1S2.
merel@inrs-ener.uquebec.ca
**Department of Physics, American University of Beirut, P.O. Box: 11-0236, Bliss Street, Beirut, Lebanon.

ABSTRACT

A hybrid Pulsed Laser Deposition system was developed to perform epitaxial growth of GaN on sapphire(0001). This system combines the laser ablation of a cooled Ga target with a well-characterized atomic nitrogen source. Taking advantage of the flexibility of this unique deposition system, high quality GaN thin films were deposited by optimizing both the laser intensity and the nitrogen flux. To date, our best GaN films show a FWHM of the GaN(0002) rocking curve peak equal to 480 arcsec. This result has been obtained at a laser intensity of $I = 7 \times 10^7$ W/cm^2, a substrate temperature of 800°C and under Ga-rich growth conditions.

INTRODUCTION

Gallium nitride (GaN) and similar materials (AlN, InN) are the subject of intensive research because of their numerous applications in the fields of UV-visible light emitting/detecting devices and high power electronics [1]. The growth of GaN layers with a low density of defects has proved to be a challenge due to the lack of a high quality lattice matched substrate. Many conventional deposition techniques, like MOCVD [2] and MBE [3], have been used to grow these layers with a reasonable amount of success when combined with specific substrate preparation like the ELOG (Epitaxial Lateral Over Growth) process [4]. In this relatively new field, it is essential to search for alternative deposition techniques which could produce high quality thin films while offering a relatively simple process. Also, the comparison of the layers obtained using different growth methods can help to understand better the fundamental aspects of GaN heteroepitaxy.

One of the interesting features of Pulsed Laser Deposition (PLD) is that the ablated species (ions and neutrals) are available at high kinetic energy (10-100 eV) in the ablation plume [5]. Other advantages include its simplicity and its versatility. Also, PLD can be used with a reactive background gas leading to the growth of new compounds.

This work presents preliminary results on the epitaxial growth of GaN thin films using a hybrid Pulsed Laser Deposition (PLD) system combining the laser ablation of a cooled Ga target with an atomic nitrogen source developed at INRS-Énergie et Matériaux [6].

EXPERIMENT

The deposition system is presented in Fig. 1. The atomic nitrogen source is provided by a high-frequency (HF) plasma generated by an electromagnetic surface wave excited by a Ro-box [7] (applied field frequency $f = 13.56$ or 40.68 MHz) or a Surfatron [8] ($f = 440$ or 2450 MHz) wave launcher. The total HF absorbed power, P_A, is varied from 10 watts to 200 watts. The discharge is generated in a 4.5 mm inner diameter fused silica tube, 25 cm long, through which

Mat. Res. Soc. Symp. Proc. Vol. 572 © 1999 Materials Research Society

pure N_2 flows. The total gas flow, Q, and the pressure in the source, p_s, are in the range of 50 to 750 sccm and 1 to 30 Torr, respectively, as measured by flow meters and a Baratron pressure gauge.

The spatial afterglow region is next to the discharge, and it expands to fill a larger quartz vessel (diameter 3 cm, length 18 cm), which is connected at the other extremity to the deposition chamber. In our discharge conditions, the afterglow is separated into two different regions as one leaves the discharge. First comes 1) the early or "pink" afterglow, which is characterized by vibrationally excited N_2 molecules leading to self-ionization reactions (Penning effect), this gives way to 2) a late or "Lewis-Rayleigh" late afterglow where the dominant reaction

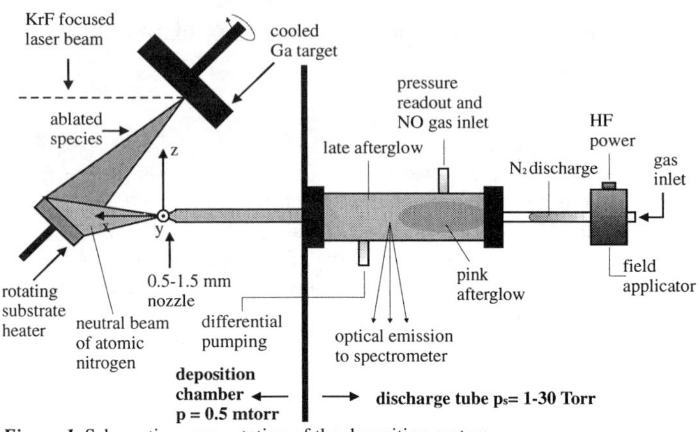

Figure 1. Schematic representation of the deposition system.

is volume recombination of nitrogen atoms. Measurement of the N-atom density through NO titration of the N-atoms is performed in the afterglow region where an Ar- 1.3% NO gas mixture can be introduced through a lateral gas inlet. The discharge and afterglow optical emissions are collected perpendicularly to the discharge tube. The final section of the nitrogen atomic source consists of a 30 cm long quartz tube (inner diameter of 6 mm) which extends into the deposition chamber trough a Swagelok seal. The tube is terminated by a nozzle. The pressure inside the deposition chamber, fixed at 0.5 mtorr, is controlled by both p_s and the nozzle diameter ϕ.

Inside the stainless steel deposition chamber, evacuated by a diffusion pump, the beam of a KrF excimer laser (248 nm, 100 mJ, 30Hz) is focused by a lens on a rotating Ga target (cooled to -30°C) located 4 cm away from the substrate. The lens is mounted on a translating stage so that the laser beam continuously sweeps the target over a one inch diameter. The rotation of the target combined with the laser beam motion ensures a uniform erosion pattern. A rotating heating stage holds the 1 inch sapphire(0001) substrate. A tungsten filament is used as the radiation source. In order to increase heat absorption and temperature uniformity, the backside of the substrate is coated with a 0.4 μm thick Mo layer deposited by magnetron sputtering. Prior to growth, the nitrogen source is used for the nitridation of the sapphire substrate at high temperature (T_s=950°C), thus favoring GaN nucleation.

After deposition, the crystalline quality of the GaN thin films is determined by triple-axis X-ray diffraction. The composition of the GaN layers is determined by EDX and XPS.

RESULTS

Atomic Nitrogen Source

The flexibility of our atomic nitrogen source enables us to measure the atomic nitrogen production as a function of various discharge conditions. For example, Fig. 2 presents the N-atom density measured in the late afterglow region as a function of the absorbed power, P_A, at different wave frequencies f (p_s=4 Torr). Whatever the frequency, [N] first increases rapidly with P_A and then saturates at a level that increases with f. This saturation value increases by approximately a factor of two when varying f from 13.56 MHz to 2450 MHz. One should further note from Fig. 2 that the absorbed power required to attain saturation decreases with increasing frequency, ranging from approximately 130 W at 13.56 MHz to 40 W at 2450 MHz. The initial rapid growth of [N] with P_A (observed before saturation) can be linked to an increase of the electron density [6].

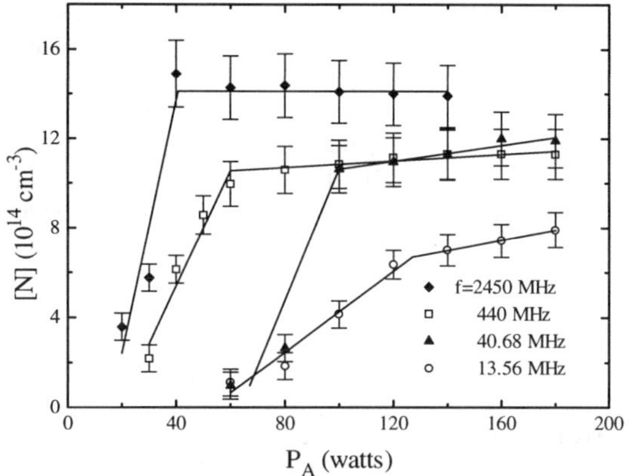

Figure 2. Nitrogen atom production, measured in the late afterglow, as a function of absorbed power at different applied field frequencies (p_s= 4 torr).

To gain insight into the saturation of the N-atom density with the increase of P_A, we turn to the gas temperature. We assume that the rotational temperature T_r of the nitrogen molecules in the discharge is in equilibrium with the gas temperature T_g. The value of T_r is obtained from the optical emission of the first positive system N_2 ($B^3\Pi_g$)-($A^3\Sigma^+_u$). We find that the observed saturation of the N-atom concentration occurs at a value of T_r near 500 K whatever the value of f. This suggests that, in the pressure range investigated, there is a threshold gas temperature above which the production of nitrogen atoms saturates. This can be understood by considering that the production of atomic nitrogen occurs through vibrational excitation of the N_2 molecules. Collisions between vibrationally excited molecules and high speed N_2 molecules destroy these excited states. The saturation level of the N-atom density is finally determined from the balance between vibrational excitation (which increases with the electron density) and vibrational energy loss (which increases with T_g).

GaN Layers

For GaN depositon, the plasma source of atomic nitrogen is operated at f = 440 MHz and P_A = 100 watts. During our first attempts of GaN thin film deposition, the optimal laser intensity was determined to be equal to 7×10^7 W/cm^2. Above this value, gallium droplets are observed at the surface of the thin film while working at a lower laser intensity greatly reduces the deposition rate. Figure 3 presents a ω-2θ diffraction pattern of a 0.5 μm thick sample deposited at T_s=800 °C. The atomic nitrogen source is operated with discharge conditions (p_s = 30 torr, Q = 750 sccm and ϕ = 0.5 mm) leading to an atomic nitrogen production percentage of

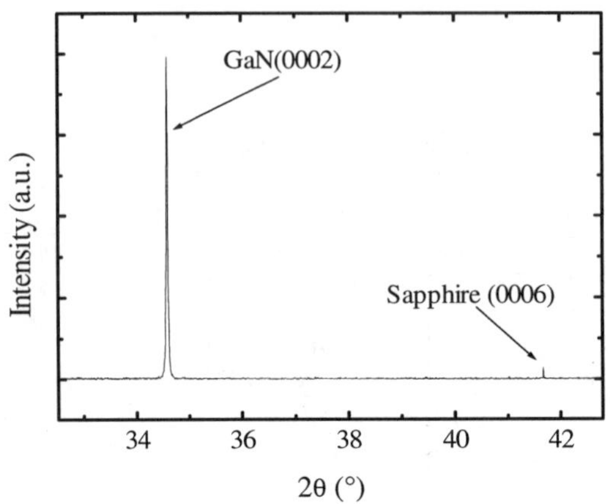

Figure 3. ω-2θ scan of a 0.5μm thick GaN layer showing the GaN(0002) and sapphire(0006) diffraction peaks (I=7x10^7 W/cm^2, T$_s$=800 °C). The sample is positioned in order to maximize the GaN(0002) peak intensity.

Table I. Properties of GaN layers deposited at two atomic nitrogen source pressures. (I=7x10^7 W/cm^2, T$_s$=800 °C),.

p_s (torr)	N-atom production [N]/[N$_2$] (%)	Thickness (μm)	Growth rate (μm/h)	N content (%)	FWHM GaN(0002) (arcsec)
30	0.4	0.5	0.08	48	480
4	1.7	0.7	0.25	55	936

this path, we estimate that the N-atom flux on the substrate increases by about one order of

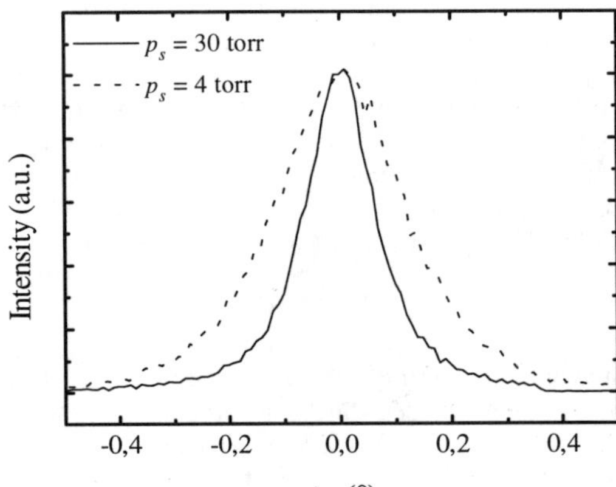

Figure 4. GaN(0002) rocking curve of GaN layers grown at two atomic nitrogen source pressures: p_s = 30 and 4 torr (I=7x10^7 W/cm^2, T$_s$=800 °C). The FWHM are 480 and 936 arcsec respectively.

[N]/[N$_2$]=0.4%. These parameters lead to a deposition rate of 0.08 μm/h. A well defined GaN(0002) peak is observed. From this, the growth orientation is determined to be GaN(0001) ‖ Sapphire(0001). The lattice parameter *c* of GaN is found to be 5.183Å compared to the tabulated value of 5.185Å.

GaN layers were also grown under nitrogen discharge conditions (p_s = 4 torr, Q = 370 sccm and φ = 1.5 mm) corresponding to a production percentage of [N]/[N$_2$]=1.7%. By taking into account the nitrogen production rate which has been determined 35 cm away from the substrate and the nitrogen atom recombination along

magnitude when the source pressure is varied from 30 torr to 4 torr. Table I presents the growth rates in these two cases. At p_s = 4 torr ([N]/[N$_2$]=1.7%), the growth rate is about 3 times higher than at p_s = 30 torr ([N]/[N$_2$]=0.4%). XPS and EDX analysis of the deposited samples also show a 55% nitrogen content for p_s = 4 torr compared to 48% at p_s = 30 torr. This corresponds to N-rich and Ga-rich growth conditions respectively.

Figure 4 shows rocking curves of the GaN(0002) peak of the two samples. The FWHM of this peak is reported in table I. We see that the sharpest peak, FWHM=480 arcsec, is obtained in

Ga-rich growth conditions (p_s = 30 torr) compared to FWHM = 936 arcsec when using N-rich growth conditions (p_s = 4 torr). The fact that Ga-rich conditions leads to higher quality GaN layers has been observed in Molecular Beam Epitaxy experiments [3]. A possible explanation of these results is presented in [9]. In this theoretical work, it is shown that the mobility of the Ga atoms on a GaN surface is greatly reduced under N rich conditions, a low mobility of Ga atoms resulting in a high density of defects in the growing layer.

CONCLUSIONS

An atomic nitrogen source based on a HF plasma has been developed for use in a hybrid PLD system for the epitaxial growth of GaN. The flexibility and ease of operation of this source enabled us to monitor systematically the N-atom concentration [N] as a function of the discharge wave frequency f under otherwise constant discharge conditions. Saturation of [N] as a function of f and P_A was observed when the temperature of the gas exceeds 500 K.

Coupling this nitrogen source with a PLD system, epitaxial growth of GaN films was achieved. We demonstrated that layers grown under Ga-rich conditions exhibit better crystalline quality than those obtained under N-rich conditions.

Work is currently in progress to further improve our understanding of the influence of deposition parameters such as the Ga and N fluxes, the substrate temperature and the presence of a GaN buffer layer on both the crystalline quality and the electrical characteristics of the PLD-GaN thin films.

ACKNOWLEDGMENTS

The authors gratefully acknowledge the financial support of MICRONET and Nortel. The help provided by Dr. James Webb of the National Research Council is also acknowledged.

REFERENCES

[1] H. Morkoç, S. Strite, G. B. Gao, M. E. Lin, B. Sverdlov and M. Burns, J. Appl. Phys. **76**, 1363 (1994).
[2] S. Nakamura, Y. Harada and M. seno, Appl. Phys. Lett. **58**, 2021 (1991).
[3]D. Schikora, M. Hanklen, D.J. As, K. Lischka, T. Litz, A. Woay, T. Buhrow and F. Henneberger, Phys. Rev **B54**, 8381 (1996).
[4] D. Kapolnek, S. Keller, R. Vetury, R. Underwood, P. Kozodoy, S. DenBaars, U. Mishra, Appl. Phys. Lett. **71**, 1204 (1997).
[5] D. H. Lowndes, D. B. Geohegan, A. A. Puretzky, D. P. Norton and C. M. Rouleau, Science **273**, 898 (1996).
[6] P. Mérel, M. Tabbal, M. Chaker, M. Moisan, A. Ricard, Plasma Sources Sci. Technol. **7**, 550 (1998).
[7] M. Moisan and Z. Zakrzewski, Rev. Sci. Instrum. **58**, 1895 (1987).
[8] M. Moisan, C. Beaudry and P. Leprince, Phys. Lett. **50A**, 125 (1974).
[9] T. Zywietz, J. Neugebauer and M. Scheffler, Appl. Phys. Lett. **73**, 488 (1998).

EPITAXIAL GROWTH OF ALN ON SI SUBSTRATES WITH INTERMEDIATE 3C-SIC AS BUFFER LAYERS

S. Q. Hong, H. M. Liaw, K. Linthicum*, R. F. Davis*, P. Fejes, S. Zollner, M. Kottke, and S. R. Wilson
Motorola Inc., Semiconductor Products Sector, 2100 E. Elliot Road, Tempe, ZA 85282
*Department of Materials Science and Engineering, North Carolina State University, Raleigh, NC 27695

ABSTRACT

Single crystalline AlN was successfully grown on a 3C-SiC coated Si (111) substrate by organometallic vapor phase epitaxy. The 3C-SiC film was obtained via the conversion of the Si near-surface region to SiC using gas-source molecular beam epitaxy. The quality of the AlN was mainly controlled by that of the SiC. The effects of Si pits and SiC hillocks formed during the conversion on subsequent AlN growth are discussed. Process optimization is suggested to improve the SiC buffer layer for subsequent AlN deposition.

INTRODUCTION

Silicon carbide (SiC) and aluminum nitride (AlN) are candidate materials for a wide variety of high-power, high-frequency, and high-temperature applications because of their unique materials properties as well as their excellent physical and chemical stability [1]. However, advances in both SiC and AlN technologies are hindered by the absence of large-area and high-quality single crystals for device fabrication. Successful epitaxial growth of 3C-SiC single crystals on Si substrates has been demonstrated since 1983 [2], despite the large mismatches in lattice constants and thermal expansion coefficients between the film and the substrate. Comparatively, heteroepitaxy of hexagonal SiC has received much less attention, though 4H- and 6H-SiC are actually of greater interest for power devices. This is due to their higher band gaps relative to the 3C material (3.265 and 3.023 eV, respectively, vs. 2.390 eV) and their commercial availability for on-going evaluation of device. It has been demonstrated [3] that single crystalline 6H-SiC can be grown at 1375°C on 2H-AlN coated sapphire (0001). The lattice mismatch between the carbide and the AlN layer is less than 1%. It therefore seems feasible to deposit hexagonal SiC on Si substrates if single-crystal 2H-AlN can be grown as an interlayer. The focus of this paper is the epitaxial growth and characterization of AlN single crystal films on Si(111) substrates. The AlN films were grown on thin 3C-SiC buffer layers previously formed by carbonization of the Si(111) substrates.

EXPERIMENTAL

Si (111) wafers oriented 3° off axis toward <110> direction were used as substrates. The surface of each Si wafer was converted to a thin 3C-SiC(111) layer via gas-source molecular beam epitaxy (MBE) prior to chemical vapor deposition (CVD) of an AlN film. Before conversion, the wafers were dipped in HF, rinsed with deionized H_2O, and placed on a SiC-coated graphite susceptor. The MBE reactor, with a base pressure of 10^{-9} Torr, was equipped with an in-situ reflection high-energy electron diffractometer (RHEED) for surface crystallographic analysis. The substrate was ramped from room temperature to 970°C at 7-8°C /min and kept at 970°C for 60 min. Ethylene (C_2H_4) was supplied from room temperature at a flow rate of 1.8 sccm. The SiC coated Si substrates were loaded in a CVD reactor. Trimethylalumimum (TMA) and ammonia (NH_3) were used for AlN deposition. The flow rates for TMA, NH_3, and additional H_2 were 26 µmol/min, 1500 sccm, and 3000 sccm, respectively. The AlN was grown at 1100°C for 90 min at a growth rate of 11 Å/min. For comparison, an AlN film was also directly deposited on Si(111) substrate under the same process conditions.

Mat. Res. Soc. Symp. Proc. Vol. 572 © 1999 Materials Research Society

Film thickness was measured by spectroscopic ellipsometry (SE). Crystallinity was examined with RHEED, X-ray diffraction (XRD), and electron diffraction. Scanning electron microscopy (SEM) and transmission electron microscopy (TEM) were employed to study surface morphology and defect formation. Auger electron spectrometry (AES) was conducted using a 5.0 keV primary beam at 30° angle of incidence to determine chemical compositions and the extent of C and O throughout the films. Sputtering during Auger depth profile analysis was accomplished by using a beam of 1.0 keV Xe^+ incident at 45°.

RESULTS AND DISCUSSION

Figure 1 is a RHEED pattern from the [110] azimuth of the converted SiC on the Si(111) substrate. It shows a single-domain (3x2) reconstruction of the cubic SiC. The formation of single crystalline 3C-SiC was confirmed by XRD as shown in Fig. 2. The two peaks at 2-theta =35.5° and 75.5° are SiC (111) and (222), respectively; the other two peaks represent Si (111) and (222). The Si (222) peak is actually Si (444) observed in second order. The d-spacing of the SiC (111) peak is 2.52Å corresponding to a cubic SiC lattice constant of 4.36Å, therefore the SiC lattice is fully relaxed.

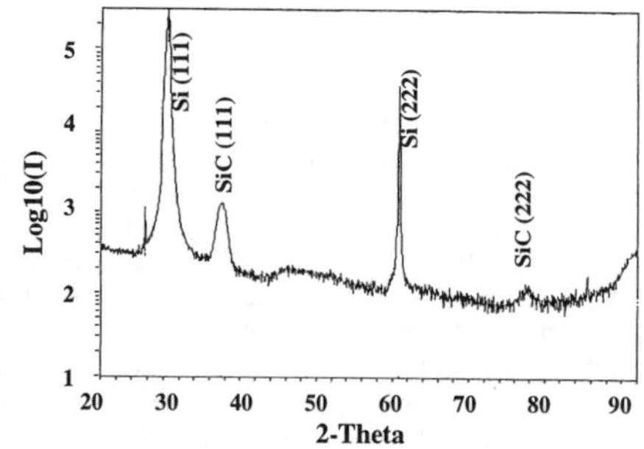

Figure 1　　RHEED pattern from the [110] azimuth of the converted SiC on the Si(111) substrate showing a single-domain (3x2) reconstruction of the cubic SiC.

Figure 2　　XRD theta-2theta scan of the single crystalline 3C-SiC film grown on the Si(111) substrate.

The surface of the converted SiC film contains many hillocks having an approximate diameter of 0.2-0.5 μm as revealed by SEM and shown in Fig. 3. Under the hillocks, {111} faceted pits appear in the Si substrate. It is known [4] that the pits are generated by Si outdiffusion from the substrate which reacts with C atoms forming SiC near pit openings and on pit walls. The formation of the pyramid-shaped pits is attributed to the low surface energy of the {111} Si facets. The polycrystalline SiC hillocks near the pit openings are an indication of strain relaxation [4]. The thickness of the converted layer determined by spectroscopic ellipsometry is approximately 3-5 nm. The higher thickness near hillocks/pits shown in Fig. 3 suggests an enhanced growth rate in the defective area. Further discussion regarding pit formation and its effect on subsequent 2H-AlN growth is presented below.

Figure 4 is an electron diffraction pattern of a cross-sectional AlN/SiC/Si sample, which was taken using the Si [0-11] zone axis, showing single crystals of AlN and the Si substrate. The diffraction from very thin SiC layer is too weak to be revealed in the same picture. Most of the diffraction spots are from hexagonal AlN and Si, with a small percentage resulting from double-diffraction. It is noted that the AlN spots are distorted indicating the existence of slightly misaligned grains despite its overall single-crystal nature. However, as revealed by RHEED and

XRD, AlN deposited directly on a Si substrate is polycrystalline (not shown). This result suggests the necessity of using SiC as a buffer layer for AlN growth under the process conditions of this research.

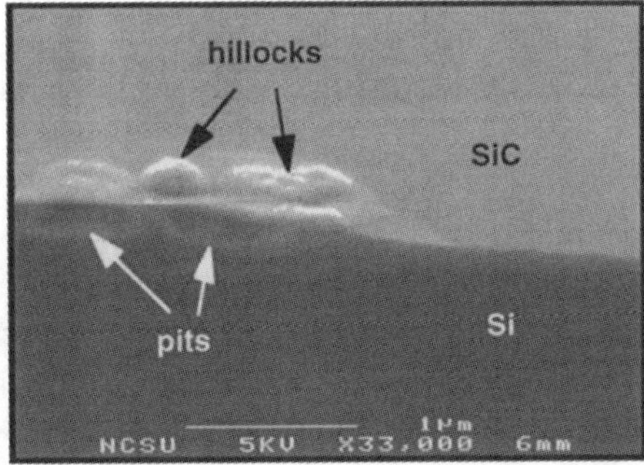

Figure 3   Cross-sectional SEM micrograph of a SiC/Si(111) sample showing the existence of hillocks and pits.

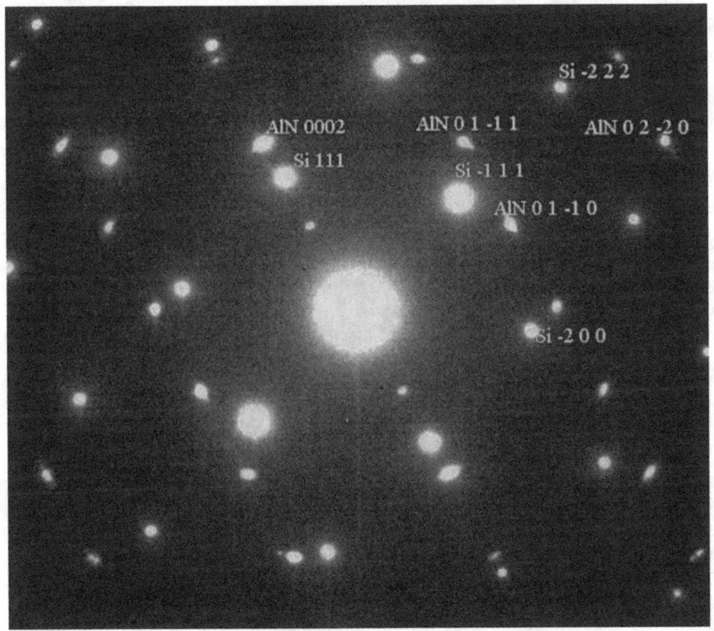

Figure 4   Electron diffraction pattern of a cross-sectional AlN/SiC/Si sample, showing the single crystalline AlN and Si substrate.

Figure 5 is the cross-sectional TEM micrograph of the same sample showing a hill-and-valley morphology on the surface of the AlN film. This morphology is believed to be associated with island growth and subsequent coalescence of the islands [5] and/or the formation of stacking mismatch boundaries at the steps on the SiC surface, as observed by Tanaka et al. [6]. Slight misalignment between adjacent islands as well as the formation of the mismatch boundaries can account for the distorted diffraction spots described above. A high density of dislocations are expected in the AlN but are not revealed due to the high contrast from the highly stressed film.

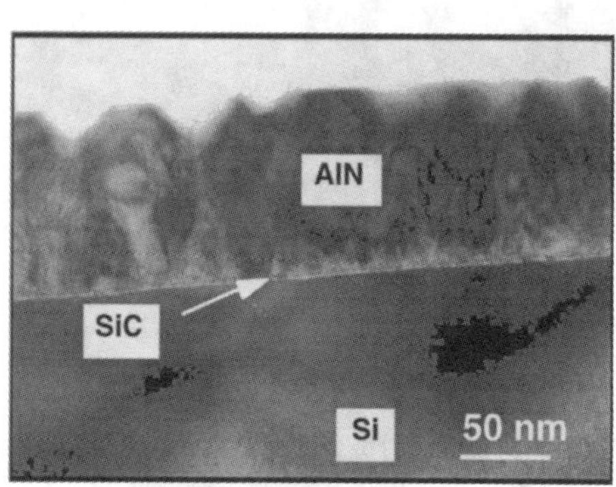

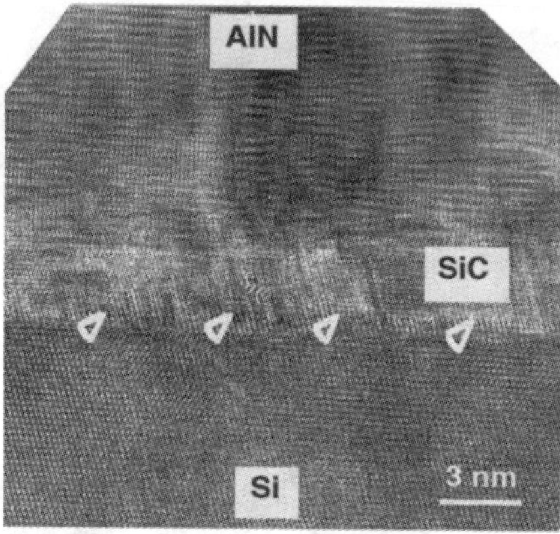

Figure 5   Cross-sectional TEM micrograph of a AlN/SiC/Si sample showing a hill-and-valley morphology on the surface of the AlN film.

Figure 6   High resolution TEM micrograph providing a lattice image of the AlN and the converted SiC layer.

A high resolution TEM micrograph in Fig. 6 provides a lattice image of the AlN and the converted SiC layer. The AlN-SiC interface is undulated because the surface of the converted SiC layer possesses a similar wave-like morphology. Therefore, the formation of the hill-and-valley morphology on the AlN surface hear is due to the island growth of both AlN and SiC. At the SiC-Si interface a high density of misfit dislocations are observed and indicated by arrows. The SiC film is completely relaxed as discussed in details in a separate paper [7].

An Auger depth profile from AlN into the Si substrate is presented in Fig. 7. The signal from carbon on the surface drops to the detection limit at about 7 min sputter time. There is no detectable C throughout the remaining AlN layer. It is believed that C is only on the AlN surface without extending into the film. The time required to sputter away the adventitious C is modified by surface roughness, especially when the incident sputtering beam is 45° with respect to wafer normal. Oxygen was detected throughout the AlN with heavier concentrations on the sample surface and at the AlN-SiC interface as would be expected. Incorporation of oxygen is not unusual during Al deposition as Al is a good oxygen getter. The thin SiC layer is not well resolved, but the Auger data suggests that there is C within the Si substrate. This could be associated with a SiC layer covering the walls of pyramidal pits under SiC-Si interface. These pits are a result of Si outdiffusion from the substrate during the SiC conversion at high temperatures as discussed earlier.

The Si pits and SiC hillocks formed during carbonization affect the quality of AlN layers. The polycrystalline hillocks near pit openings are a result of stress relaxation and are always unfavorable for subsequent AlN epitaxy. The effects of the Si pits fall into two categories. In the

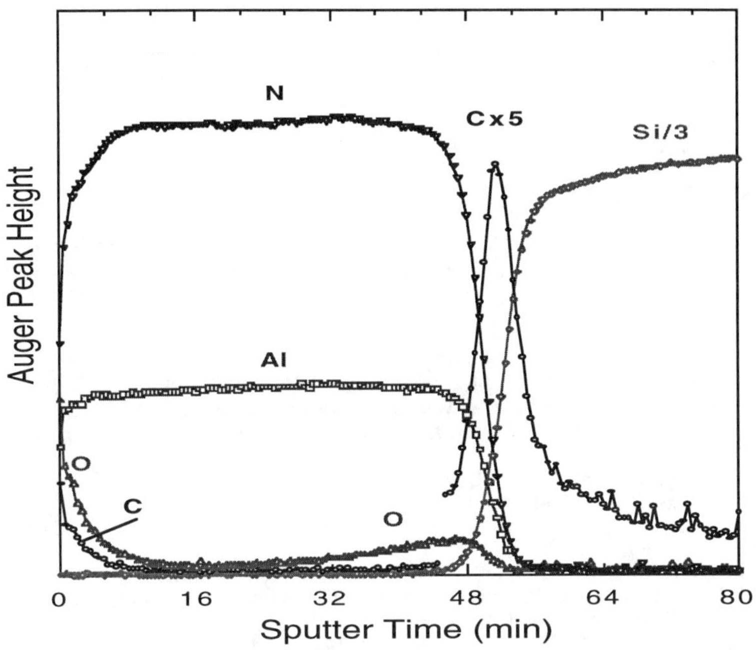

Figure 7   Auger depth profile of an AlN/SiC/Si(111) sample showing (a) adventitious C on the sample surface and (b) oxygen in the AlN with heavier concentrations on the sample surface and at the AlN-SiC interface.

early stage of carbonization, C atoms only partially cover the Si substrate. Si diffuses out from any of uncovered substrate regions resulting in the formation of {111} faceted voids. If the SiC nucleation density is low, the small pitted areas cannot be completely covered or sealed by the coalescence of C or SiC nuclei before a large pit opening is created. The continued dissociation of C_2H_4 results in the formation of SiC on pit walls. The subsequent AlN film then conforms to the morphology of the SiC layer. At high SiC nucleation densities, the originally pitted area is rapidly sealed resulting in the formation of SiC bridges over the voids generated by continued Si outdiffusion along {111} planes [8]. The subsequent AlN film appears to have a lower defect density in the latter case. In this research almost all the pits were found to be sealed [7].

The nucleation density is controllable by varying growth conditions. To increase the SiC nucleation density a low temperature is necessary in the early stage of conversion. Since the driving force for Si outdiffusion is the reduction in surface energy, a quick termination of the high temperature process after the SiC conversion is complete will further prevent the pit formation. Based on XPS results of the converted SiC surface more than 50% of the deposited carbon remained unreacted before AlN deposition. Moreover, wafer cleaning is also important as localized contamination, defects and remnant native oxide tend to enhance pit formation.

CONCLUSIONS

Monocrystalline films of AlN have been grown on Si (111) substrate using converted 3C-SiC as a buffer layer. Direct deposition of AlN on the Si substrate resulted in polycrystalline growth under the process conditions of this research. The quality of the AlN on the SiC is mainly limited by that of the SiC. Process optimization is necessary to improve the SiC quality, and especially to prevent Si pitting during SiC conversion.

ACKNOWLEDGMENTS

The authors would like to thank Semiconductor Research Corporation and Motorola for providing research funding. R. F. Davis was supported in part by Cobe Steel Ltd. University Professorship.

REFERENCES

[1] R. J. Trew, J. Yan, and P. M. Mock, Proc. IEEE **79-5**, 598 (1991)
[2] S. Nishino, J.A. Powell and H.A. Will, Appl. Phys. Letters **42**, 460 (1983).
[3] B.S. Sywe, Z.J. Yu, S. Burckhard, J. H. Edgar and J. Chaudhuri, J. Electrochem. Soc. **141**, 510 (1994).
[4] J. Schmitt, T. Troffer, K. Christiansen, S. Christiansen, R. Helbig, G. Pensl, H.P. Strunk, Materials Science Forum **264-268**, pt.1, 247 (1998).
[5] R. F. Davis, S. Tanaka, L. B. Rowland, R. S. Kern, Z. Sitar, S.K. Ailey, C. Wang, J. of Crystal Growth **164**, 132 (1996).
[6] S. Tanaka, R.S. Kern, J. Bentley and R.F. Davis, to be submitted.
[7] H. M. Liaw, S.Q. Hong, P. Fejes, D. Werho, H. Tompkins, S. Zollner, S. R. Wilson, K. Linthicum and R. F. Davis, this volume.
[8] J.P. Li and A.J. Steckl, J. Electrochem. Soc. **142**, p634 (1995).

SIMS AND CL CHARACTERIZATION OF MANGANESE-DOPED ALUMINUM NITRIDE FILMS

R.C. Tucceri*, C.D. Bland*, M.L. Caldwell*, M.H. Ervin**, N.P. Magtoto***, C.M. Spalding*, M.A. Wood**, H.H. Richardson*

*Department of Chemistry and Biochemistry, Ohio University, Athens, OH 45701, richards@helios.phy.ohiou.edu
**Army Research Laboratory/Shady Grove Industrial Park, 8705 Grovemont Circle Gathersburg, MD 20877-4117, mark_wood@mail.arl.mil
***Department of Chemistry, University of North Texas, Denton, TX, 76201, noelm@unt.edu

ABSTRACT

We have recently carried out MOCVD experiments that showed for the first time the doping of AlN thin films with manganese. Films of AlN that were doped with less than 0.1% of manganese showed emission bands at 427 nm, 488 nm and 600 nm in accordance with previous published excitation and emission spectra of manganese incorporated in bulk AlN. A film with a higher percentage of manganese (1.7%) grown on top of a pure AlN layer (underlayer) showed weak emission at 601 nm. A portion of the underlayer not covered during growth of the overlayer has a very strong emission band at 408 nm and a weaker band at 605 nm. SIMS and EDX analyses of this film revealed both carbon and oxygen contamination as well as diffusion of manganese into the AlN underlayer. The band at 408 nm is associated with direct gap emission from oxygen contaminated AlN and the band at 605 nm is from manganese incorporated by diffusion into the AlN underlayer.

INTRODUCTION

Group III nitrides have shown promise as being suitable materials for microelectronic and optoelectronic applications [1]. AlN attracts considerable interest because it possesses a very large band gap compared to the rest of the group III nitrides (6.2 eV) [2]. In addition, it has a high heat conductivity, excellent chemical and thermal stability; it will decompose rather than melt at 2400 °C [3]. Recently, rare earth doped III-V semiconductors have been the subject of great interest because of their strong visible luminescence [4,5]. However, very little attention has been paid to incorporation of transition metals into semiconductor hosts. Manganese has been incorporated in powder samples of AlN [6-8], with the tetravalent manganese ion, Mn^{+4}, occupying tetrahedral sites. The manganese activated AlN exhibits a maximum in the emission spectrum in the red region (~600 nm). Doping of AlN thin films with manganese has not been previously demonstrated. We present here for the first time, evidence that will show that manganese can be incorporated into a film of AlN grown on Si(100) obtained from a variety of surface analytical tools such as IR microscopy, SEM imaging, SIMS, EDX, XRF, CL, and XRD.

EXPERIMENTAL

Aluminum nitride (AlN) films were grown by metal organic chemical vapor deposition (MOCVD) in a high vacuum stainless steel reaction chamber. This system consists of a growth chamber, pumping unit, and a gas inlet. An Alcatel corrosive resistant turbo pump evacuates the chamber into the 10^{-5} Torr range. The source gases for the AlN were ammonia (NH_3) and trimethyl aluminum (TMA). Both gases were constantly maintained in the chamber at a ratio of 2.5:1, respectively. Silicon (Si) (100) substrates, ¼ in. by ½ in., were mounted onto a boron

Mat. Res. Soc. Symp. Proc. Vol. 572 © 1999 Materials Research Society

nitride ceramic heater and held in place by a molybdenum metal holder. A nichrome thermocouple was attached to the holder to determine the temperature during the deposition period. Pressures in the chamber were monitored with an ion gauge. The Si(100) substrates were flash annealed above 1200 °C to remove the surface oxide coating. The substrates were then kept at a constant temperature around 860 °C [9]. A stoichiometric ratio of 2.5:1 NH_3 to TMA was used for growing AlN films . Manganese decacarbonyl, ($[Mn(CO)_5]_2$), the dopant, was introduced into the flow reactor via a pulse valve. These films were characterized *ex situ* with IR reflectance microscopy, SEM imaging, XRF, XRD, CL, and SIMS.

RESULTS

With the use of IR microscopy, film thickness was determined by measuring the fringe spacing of the Fabry-Perot oscillations, while the chemical composition was analyzed by monitoring the peak frequency and bandwidth of the LO mode (880 – 935 cm^{-1}). The sharpness of LO mode observed in most of the films is related to the degree of crystalinity in the film [10]. The relatively fast growth rates (3-5 μm per hour) produced multiple nucleation centers that lead to a very rough surface. The film thickness ranged from 2 μm up to 27 μm.

Many samples show incorporation of manganese during the growth of AlN films. These films show light emission during electron bombardment (cathodoluminescence). The film referred to as sample 0209 is 12.18-μm thick, and its manganese content relative to aluminum is less than 0.1%. The CL spectrum of this film collected at room temperature exhibits few emission bands as shown in Figure 1. First amongst these bands is a slight shoulder that corresponds to an emission at 23419 cm^{-1} (427 nm). It is followed by a green band at 20491 cm^{-1} (488 nm), and then a red band at 16667 cm^{-1} (600 nm). The red band possesses multiple phonon states that have been described earlier by Karel and coworkers [7]. Figure 1 also includes the results of the curve fitting procedure in order to help locate the position of the multiple emission bands. Although these transitions have been shown to originate from Mn^{4+} ions in a tetrahedral field [7], only the emission from the red band has been observed previously. The blue and green transitions were observed only in the excitation spectrum.

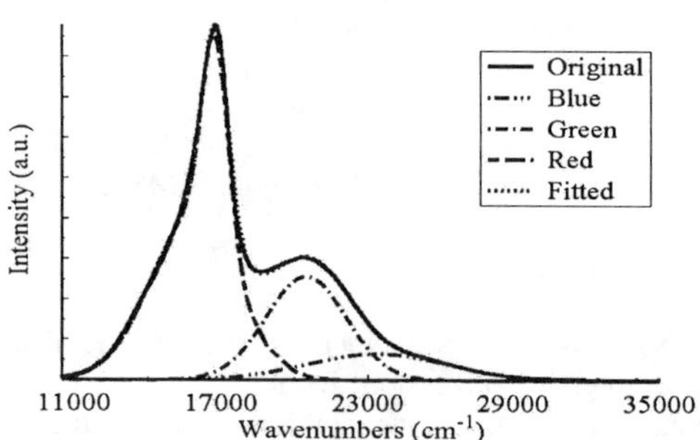

Figure 1. Deconvolved spectra and original spectrum for sample 0209.

A full characterization of sample 1020 was also carried out. The film had an overall thickness of 6.0 μm as calculated from IR reflectance microscopy data. In this sample, a manganese activated AlN layer (overlayer) was grown on top of a pure AlN layer.   SEM was used to clarify the morphology of the films. At 8000 times the magnification, the two film layers showed different structure. Figure 2 (panel a) displays an SEM image of the overlayer that exhibits a grain like appearance. Panel b is an SEM image of a portion of the sample not covered by the overlayer where a dendritic structure is observed. This portion of the film is referred to as the underlayer. The SEM images indicate that the films are rough. This is not surprising because the growth rates (on the order of 3-5 μm per hour) are fast enough to produce multiple nucleation sites.

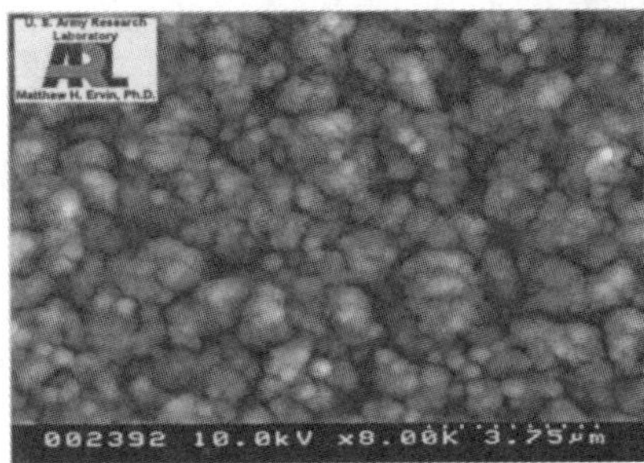

Figure 2. a) The overlayer of AlN (Mn incorporation) has a thickness of 6 μm. b) The underlayer containing only AlN has a thickness of 2.9 μm.

The elemental composition of the film was examined with x-ray fluorescence. Manganese was found in both layers but at varying relative percentages. The overlayer of AlN showed 1.7% manganese relative to aluminum, while the underlayer showed 0.6% manganese relative to aluminum. Relative percentages of the manganese were calculated from a standard material of aluminum and manganese. A sensitivity factor was then used to relate the intensities of the two elements. After finding that manganese is present in the film, image cathode luminescence (CL) was performed to determine the different regions of the film where visible light is emitted. Figure 3 shows the reversed CL image in which the dark band denotes the area where light is emitted. Three regions in the CL image are observed. The dark band in the center is the underlayer (0.6% Mn). Light areas to the left and right are the overlayer (1.7% Mn) and Si(100), respectively.

Figure 4 shows the EDX traces of both layers in sample 1020. Both the underlayer and overlayer show the presence of oxygen and carbon contaminations. The concentration of manganese, however, is below the limit of detection. The room temperature CL spectra from the overlayer and underlayer region are shown in figure 5. The overlayer region shows an emission band at 601 nm consistent with literature values for manganese activated AlN [8]. The CL spectrum from the underlayer region has an intense emission band at 408 nm and a weaker band at 605 nm. The band at 605 nm is most likely from the manganese that was incorporated in the AlN by diffusion from the overlayer region. Because the blue emission band is considerably stronger than the band at 605 nm, it cannot be attributed to Mn emission. Rather it could possibly be associated with contaminants like carbon or oxygen. Since emission bands due to carbon-doped AlN have been observed at wavelengths much lower than 400 nm [9], the band at 408 nm can be identified with oxygen contamination. Oxygen is introduced either as residual atmospheric gases (CO_2 and H_2O) or as CO from manganese decacarbonyl used as the doping agent.

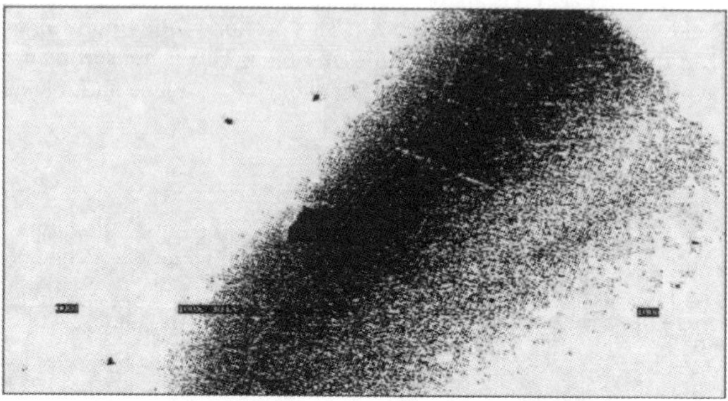

Figure 3. A cathode luminescence image of sample 1020 showing the overlayer (left white region), the underlayer (dark region), and the Si (100) substrate (right white region).

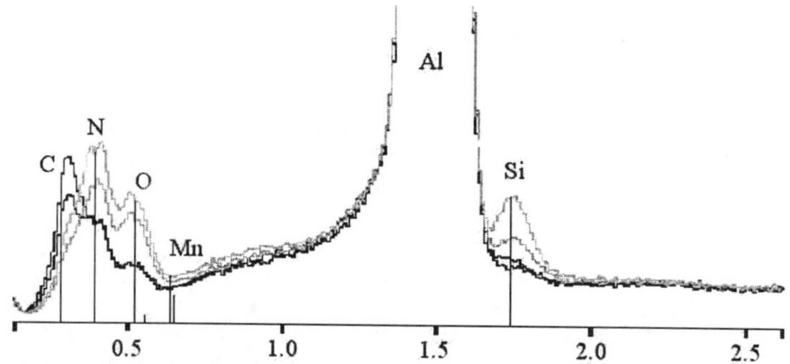

Figure 4. EDX spectra of sample 1020. The dark traces are spectra from the overlayer and the light traces are from the underlayer. The films show the presence of oxygen and carbon contaminants.

SIMS depth profiling was carried out in order to determine the elemental positions in sample 1020. The film was sputtered in the overlayer region of the sample. An oxygen ion beam at 7 kV was used to examine the surface. This made detection of oxygen contamination extremely difficult. The aluminum dimer peak was chosen because the monomer peak was off scale. In order to properly relate the amount of manganese in the profile to that of aluminum, the manganese trace is scaled up by a factor of 34. The results of this analysis are shown in figure 6. The depth profile reveals three layers. The top layer (labeled overlayer) in the depth profile is rich in manganese. Carbon is also found in this layer. It appears to be highly localized at the surface of the sample, and follows the same behavior as that of aluminum. This indicates that the observed carbon contamination is associated with the incomplete combustion of TMA. The depth profile of the next layer, labeled as underlayer, shows the presence of both Al and Mn. It is evident from Fig. 6 that the degree of manganese incorporation in the underlayer is significantly smaller than that of the overlayer. Since the underlayer is grown as pure AlN, it should not contain any manganese. Hence, the presence of manganese in this layer has to be the result of the diffusion of Mn from the overlayer. This lends support to the idea that the 605 nm emission band observed in the CL spectrum taken for the underlayer is from Mn that diffused from the overlayer. Finally, the third layer is identified as the Si(100) substrate. It is also interesting to point out that the silicon layer contains small amount of Mn, which indicates that Mn is able to diffuse down to a depth of about 7 μm. The absence of sharp, clearly defined interfaces as shown by the SIMS depth profile is indicative of a very rough sample.

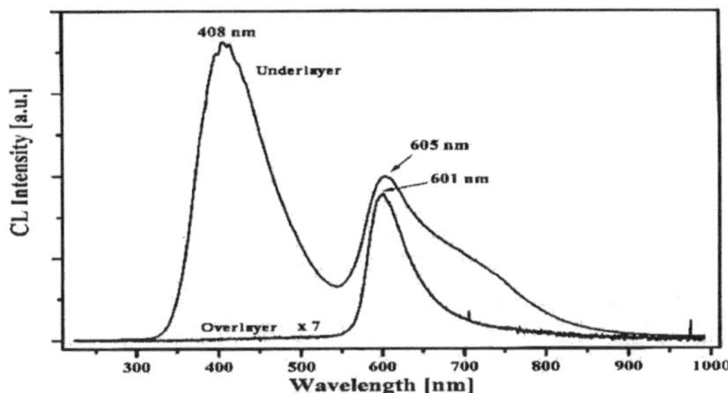

Figure 5. CL spectra of sample 1020 from the overlayer and the underlayer (top spectrum).

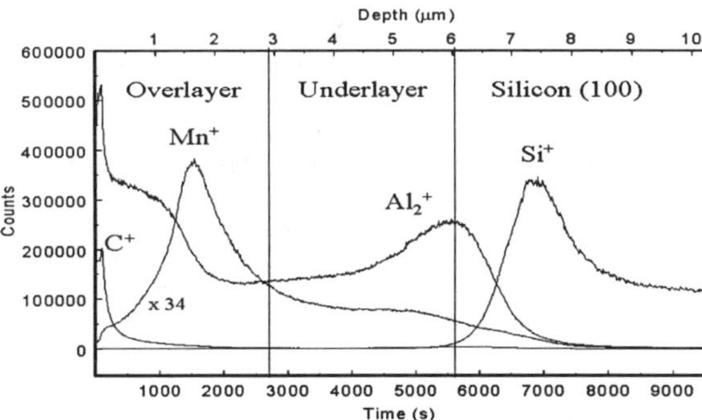

Figure 6. SIMS profile of sample 1020. The graph shows incorporation of manganese into the AlN film, and suggests a process where manganese diffuses from the overlayer to the underlayer.

CONCLUSIONS

The incorporation of a small amount of manganese (< 0.1%) in AlN gives rise to three emission bands at 427 nm, 488 nm and 600 nm in the CL spectrum. These emission wavelengths correspond to previously published excitation and emission data for manganese activated AlN. A film with a larger percentage of incorporated manganese (0.6%) exhibits strong emission at 408 nm and less intense emission at 605 nm. The blue emission band is the direct gap emission from oxygen-contaminated AlN, while the red emission band is from Mn incorporated in the AlN film. Manganese incorporation at 1.7% has emission only at 601 nm. EDX analysis demonstrates the presence of carbon and oxygen contamination. However, only oxygen contamination gives rise to emission at 408 nm. SIMS measurements reveal that manganese diffuses from the Mn-rich overlayer to the previously pure AlN underlayer.

ACKNOWLEDGEMENTS

This work is supported by a BMDO URISP grant N00014-96-1-0782 entitled "Growth, Doping and Contacts for Wide Band Gap Semiconductors". The authors would like to thank V. Jadwisienczak and H.J. Lozykowski at Ohio University for the CL spectra.

REFERENCES

1. J. L. Rouviere, M. Arlery, R. Niebuhr, K. H. Bachem, O. Briot, MRS Internet J. Nitride Semicond. Res. 1, 33 (1996).
2. F. Hasegawa, T. Takahashi, K. Kubo, Y. Nannichi, Jpn. J. Appl. Phys. 26, 1555 (1987).
3. D. V. Tsvetkov, A. S. Zubrilov, V. I. Nikolaev, V. A. Soloviev, V. A. Dmitriev, MRS Internet J. Nitride Semicond. Res. 1, 35 (1996).
4. A. J. Steckl, R. Birkhahn, Appl. Phys. Lett. 73, 1700 (1998).
5. H. J. Lozykowski, W. M. Jadwisienczak, Appl.Phys. Lett. 74, 1129 (1999).
6. F. Karel, J. Pastrňák, Czech. J. Phys B. 19, 79(1969).
7. F. Karel, J. Mareš, Czech. J. Phys. B. 22, 847 (1972).
8. F. Karel, J. Pastrňák, J. Hejduk, V. Losík, Phys. Stat. Sol. 15, 693 (1966).
9. A. D. Serra, N. P. Magtoto, D. C. Ingram, H. H. Richardson, Mat. Res. Soc. Symp. Proc. 482, 179 (1997).
10. U. Mazur, A. C. Cleary, J. Phys. Chem. 94, 189 (1990).
11. X. Tang, F. Hossain, K. Wongchotigul, M. G. Spencer, Appl. Phys. Lett. 72, 1501 (1998).

PHOTOLUMINESCENCE BETWEEN 3.36 eV AND 3.41 eV FROM GaN EPITAXIAL LAYERS

R. Seitz[*], C. Gaspar[*], T. Monteiro[*], E. Pereira[*], M.A. Poisson[**], B. Beaumont[***]

[*]Universidade de Aveiro, Departamento de Física, Aveiro, Portugal
[**]Thomson-CSF, Orsay, France
[***]CHREA-CNRS, Valbonne, France

1. INTRODUCTION

GaN, its alloys, QWs and MQWs have gained an important place among short-wavelength optical emitters and high temperature electronic devices [1,2]. The performance of such devices is limited by the presence of native and impurity defects. The understanding of the optical properties of the basic material allows us to improve its quality and thus increase the performance of these materials.

In non intentionally doped (nid) hexagonal good quality GaN layers grown on sapphire, 6H-SiC or Si, free exciton (FXC, FXB, FXA), donor bound exciton (DX), acceptor bound exciton (AX) and donor-acceptor pair (DAP) transitions have been reported by several authors [3, and references therein]. Besides these typical emissions, emission lines in the range 3.3 – 3.44 eV have been observed in nid and intentionally doped hexagonal GaN layers. However the nature of these recombinations is not completely clarified. Some authors assigned them to a superposition of LO phonon assisted transitions of DX and FX [3-8], excitons bound to neutral donors with deeper donor levels [5], band to impurity transitions and/or free to bound emission involving oxygen [9-11], DAP transitions [12-14], shallow bound excitons of cubic phases [15], excitons bound to structural defects [16-20] and Zn related recombinations [21].

In this work we analyse the luminescence between 3.36 eV and 3.41 eV of nid hexagonal GaN samples grown on sapphire. We found sample dependent emission lines with no DAP behaviour. From the data we are able to identify different kinds of recombination processes in the same spectral region.

2. EXPERIMENT

Hexagonal GaN layers of ca. 2 μm where grown by MOCVD on (0001) sapphire substrates. Steady state (SS) photoluminescence (PL) is excited by the 325 nm line of a He-Cd laser. The luminescence signal was dispersed by a SPEX 1704 1m monochromator and detected by a Hamamatsu photomultiplier. Excitation intensity was varied by neutral density (ND) filters. Time resolved (TR) measurements were carried out with a pulsed Xe lamp as an excitation source and a boxcar system for detection. The samples were mounted on a cold finger of a closed-cycle He cryostat and the temperature of the samples was varied between 12-300K.

Mat. Res. Soc. Symp. Proc. Vol. 572 © 1999 Materials Research Society

Samples are oxygen free and present a background concentration of approximately $5 \times 10^{17} \text{cm}^{-3}$.

3. RESULTS

Fig. 1 shows SSPL and TRPL of an oxygen free sample (sample A) where besides the excitonic emissions at 3.47 eV, DAP recombination at 3.27 eV and a yellow band (YB) recombination a set of lines can be observed in the range between 3.3 and 3.41 eV, namely at 3.400 eV, 3.342 and 3.328 eV (Fig. 2).

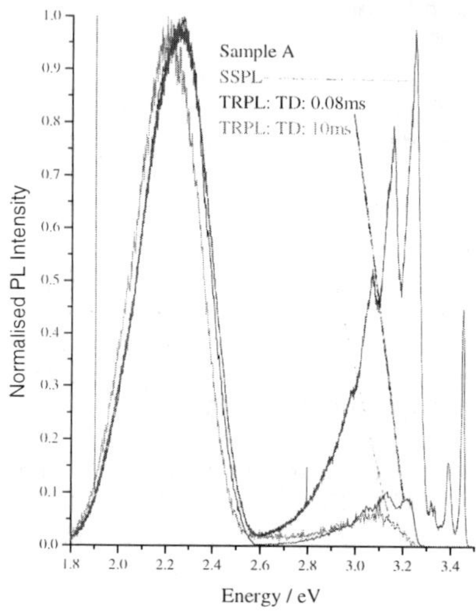

Fig. 1: SS and TRPL of sample A at 10K

Fig. 2: Temperature dependent PL spectra of sample A

From time dependent analysis of luminescence we are able to assume that all high energy side emissions are fast, shorter than microseconds as they do not appear in time resolved spectra. The TRPL spectra show that only the 3.27 eV DAP and the YB transitions have contributions with lifetimes above 10 μs. With increasing temperature the excitonic emissions shift to lower energies while the 3.400 eV emission shifts to higher energies (Fig. 2) indicating that they have different origin although they show a similar quenching. The 60 meV energy separation indicates that the 3.4 eV emission cannot be a LO assisted phonon replica of the excitonic recombinations. Also the relative intensity of the 3.4 eV band and excitonic emissions is position dependent.

There is no significant shift of the peak position of the 3.4 eV band with varying excitation density while under low levels of photoexcitation density a high energy shift of the excitonic emission is observed (Fig. 3).

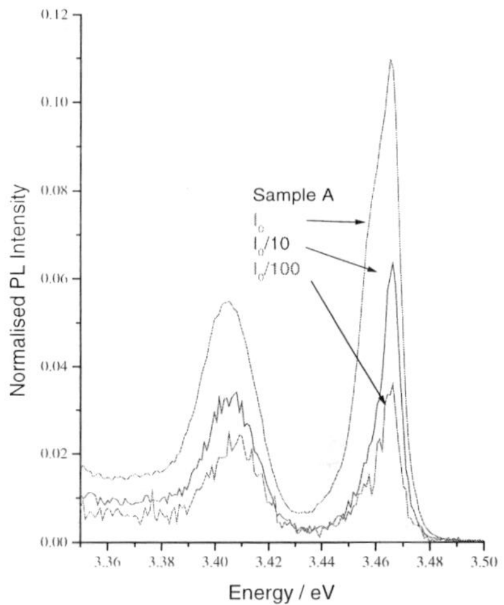

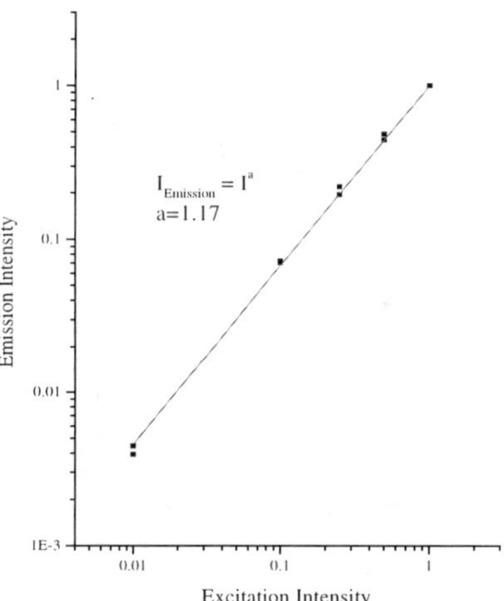

Fig. 3: Excitation intensity
dependence of PL of sample A

Fig. 4: Emission intensity for various
excitation intensities (sample A)

On the other hand a super-linear behaviour of the 3.400 eV band with increasing excitation intensity can be observed (Fig. 4)

Some of the nid samples present only excitonic lines and their vibronic replicas. However in some samples an overlap of several high energy emission lines (peaked between 3.3 to 3.43 eV) with LO phonon assisted replicas of excitonic transitions are observed as shown in Fig. 5 and 6 (sample B). Some of these lines show a smaller quenching than the excitonic emissions. It is interesting to note that a line that appears at 3.402 eV (very close in energy to the 3.400 eV emission of sample A) shifts to lower energies with increasing temperature (broken line in Fig. 5)

The line at 3.383 eV (sample B) shows a quite different evolution with temperature. In Fig. 7 and 8 a plot of luminescence intensity versus temperature is shown for the 3.400 eV line and DX emission of sample A and the 3.373 eV, 3.383 eV, 3.402 eV and DX line of sample B. The overall shape of the quenching curve of emissions of sample A and B are generally described by equations (1) and (2):

$$I(T) = I(0) [1 + C_i \exp (-E_{ai}/k_BT) + C_j \exp (-E_{aj}/k_BT)]^{-1} \qquad (1)$$

$$I(T) = I(0) [1 + C_k \exp (-E_{ak}/k_BT)]^{-1} \qquad (2)$$

where the weighting factors $C_\#$ express the weights of nonradiative dissociation channels and $E_\#$ stands for the thermal activation energy of the quenching processes. For the different lines these values are given in table 1.

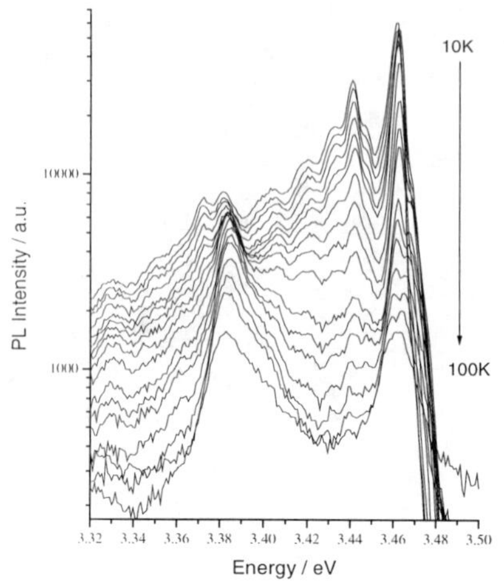

Fig. 5: Temperature dependent PL spectra of sample B

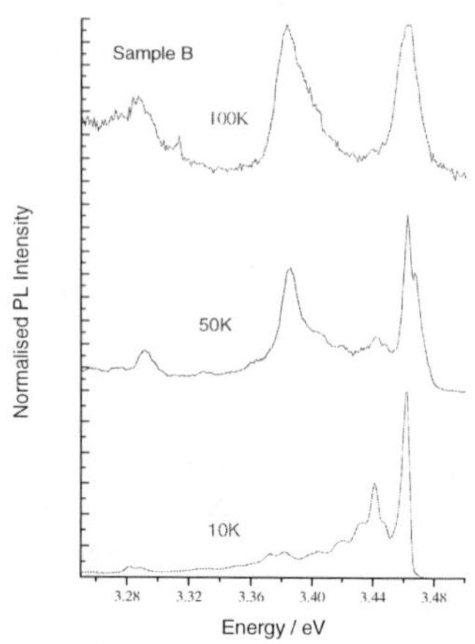

Fig. 6: PL spectra of sample B for 3 different temperatures

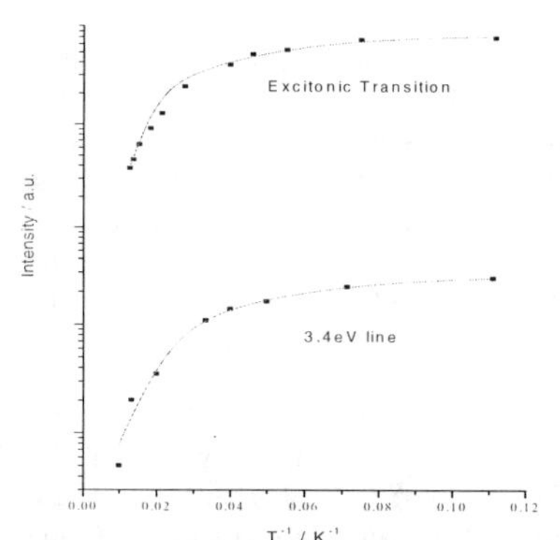

Fig. 7: Temperature quenching of sample A

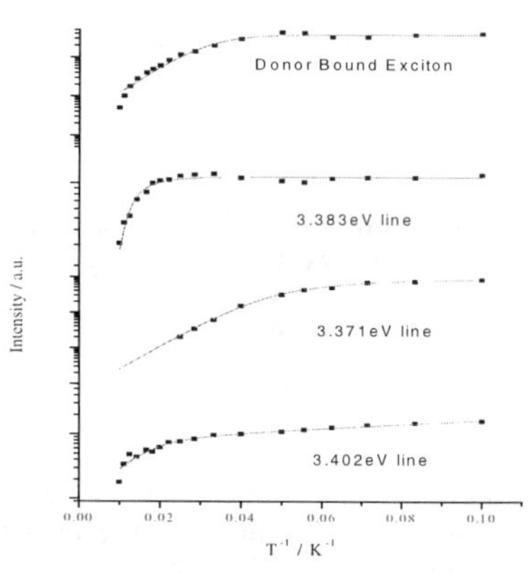

Fig. 8: Temperature quenching of sample B

SampleA					Sample B				
Line (eV)	C_1	C_2	E_{a1} (meV)	E_{a2} (meV)	Line (eV)	C_3	E_{a3} (meV)	C_4	E_{a4} (meV)
3.46 (exc.)	5.8	30	5.2	18.8	3.461 (DX)	140	13.6		
					3.402	5.0	5.6	340	44
3.400	5.0	233	3.5	17.4	3.383	68	29		
					3.373	592	10		

Table 1: Quenching parameters of the emission lines

4. Discussion

In the spectral region between 3.33 and 3.43 eV several sample dependent emission lines can be observed. The origin of the emission lines is discussed in detail for two samples.

4.1 Sample A

In this sample besides the characteristic 3.27 eV DAP emission no other lines reveal donor acceptor pair behaviour. In fact the 3.400 eV line shows a shift to higher energies with increasing temperature as recently reported in literature [11,12]. However no shift of the peak position to higher energies with increasing excitation density and no long lived luminescence (provided from distant pairs) was observed indicating that in our samples the 3.400 eV line is not a DAP recombination. It has been reported that this luminescence was attributed to oxygen related transitions [9] but as our nid samples are oxygen free as assessed by SIMS, this luminescence cannot be related to oxygen.

In our spectra no shoulder is observed at the 3.400 eV luminescence and no enhancement of the luminescence was observed when the sample was excited from the backside as has been mentioned in a recent report [19] indicating that no structural defects at the interface are involved in this recombination line.

While the exciton energy accompanies the band gap dependence on temperature the 3.400 eV line shifts to higher energies. This clearly shows that although both DX and the 3.400 eV line show similar temperature quenching they have different origins. A shift to higher energies usually indicates a free to bound or a bound to free transition. A free to bound transition occurs usually when at low temperature DAP emission is quenched due to ionisation of the donor. In our case no evidence of DAP is found. So we may assume that the 3.400 eV emission may be due to a bound to free transition. This would fit the following peak energy law [11]: $h\nu = E_g - E_d + nkT$ with n=1. This value accounts for the degeneracy of the top of the valence band. The donor level would be located about 100 meV below the conduction band. The similar quenching may suggest that this donor is populated from the DX transition.

4.2 Sample B

As in previous reports [3-8] some samples show only excitonic lines and their LO phonon replicas 92 meV apart. However, in some samples an overlap of several emission lines can be observed as in our sample B. In this sample the increase of the FX emission with temperature is clearly observed. Furthermore the quenching of the DX emission gives an activation energy equal to the spectroscopic separation between the FX and DX emissions (13.6 meV).

In this sample the 3.383 eV line is close to 92 meV below the free exciton. Previous results [7,8] clearly indicate a very small phonon coupling of the free exciton but in sample B the emission at 3.383 eV reaches the intensity of the free exciton emission at 100K. Therefore we can exclude a LO phonon replica. DAP transitions have been ascribed to the recombination mechanisms of a 3.383 eV line in Be doped samples [15,16]. Also a line at 3.382 eV has been ascribed to a strong phonon replica of a I_X

bound exciton peaked at 3.4735 eV[22]. However in our undoped sample B the 3.383 eV line does not change in peak position with temperature, photoexcitation density and delay times indicating that this emission is not a DAP recombination. On the other hand no I_X line was observed as in reference [22].

10 meV lower in energy (3.373 eV) another line is observed. This line shows a different quenching process than the other lines and cannot be ascribed to a LO phonon replica of the DX emission.

Although the 3.402 eV line appears very close in energy to the 3.400 eV emission of sample A and presents a similar quenching process the line in sample B shows a low energy shift with temperature while the line of sample A shifts to higher energies. So the two lines cannot be attributed to the same defect.

5. CONCLUSIONS

Emissions in GaN between the excitonic region and the 3.27 eV DAP transitions are very complex and sample dependent. We show that lines that appear in the same spectral region, or very close in energies in different samples present some common behaviour but also differ in fundamental characteristics so that they can be ascribed to different transitions. Even lines that might be LO phonon replicas due to their peak position are shown to be of different origin due to their different temperature behaviour.

These results show that different recombination channels are present due to donor and/or acceptor levels which are introduced in these samples.

REFERENCES

[1] S Nakamura, M Senoh, N Iwasa, S nagahama, Jpn. J. Appl. Phys. 34 (1995) L797

[2] S Nakamura, M Senoh, N Iwasa, S Nagahama, T Yamada, T Mukai, Jpn. J. Appl. Phys., 34 (1995) L1332

[3] Annamraju kasi Viswanath, Joo In Lee, Sungkyu Yu, Dongho Kim, Yoowho Choi, Chang-hee Hong, J. Appl. Phys., 84 (1998) 3848

[4] R Dingle, D d Sell, S E Stokowski, M Ilegems, Phys. Rev. B, 4 (1971) 1211

[5] W Götz, L T Romano, B S Krusor, N M Johnson, R J Molnar, Appl. Phys. Lett., 69 (1996) 242

[6] M Leroux, B Beaumont, N Grandjean, P Lorenzini, S Haffonz, P Vennéguès, J Massies, P Gibart, ICAM/E-MRS'97, Spring Meeting, Symposium L, L-III.1

[7] W Liu, M F Li, S J Xu, Kazuo Uchida, Koh Matsumoto, Semicond. Sci. Technol. 13 (1998) 769

[8] D Kovalev, B Averboukh, D Volm, B K Meyer, H Amano, I Akasaki, Phys. Rev. B, 54 (1996) 2518

[9] B C Chung, M Gershenzon, J. Appl. Phys. 72 (1992) 651

[10] M Smith, G D Chen, J Y Lin, H X Jiang, A Salvador, B N Sverdlov, A Botchkarev, H Morkoç, Appl. Phys. Lett., 66 (1995) 3474

[11] A V Andrianov, D E Lacklison, J W Orton, D J Dewship, S E Hooper, C T Foxon, Semicond. Sci. Techonol. 11 (1996) 366

[12] F Calle, F J Sanchez, J M G Tijero, M A Sanchez-Garcia, E Calleja, R Beresford, Semicond. Sci. Technol., 12 (1997) 1396

[13] F J sanchez, F Calle, M A sanchez-Garcia, E Calleja, E Munoz, C H Molloy, D J Somerford, J J Serrano, J M Blanco, Semicond. Sci. Technol., 13 (1998) 1130

[14] D J Dewsnip, A V Andrianov, I Harrison, J W Orton, D E Lacklison, G B Ren, S E Hooper, T S Cheng, C T Foxon, Semicond. Sci. Technol. 13 (1998) 500

[15] Stefan Strauf, Peter Michler, Jürgen Gutowski, Hartmut Selke, Udo Birkle, Sven Einfeldt, Detlef Hommel, Jour. Cryst. Growth 189/190 (1998) 682

[16] S Fischer, C. Wetzel, W. Walukiewicz, E E haller, Mat. Res. Soc. Symp. Proc. 395 (1996) 571

[17] L Eckey, J Ch. Holst, P Maxim, R Heitz, A Hoffmann, I Broser, B K Meyer, C Wetzel, E N Mokhov, P G Baranov, Appl. Phys. Lett. 68 (1996) 415

[18] C Wetzel, S Fischer, J Krüger, E E Haller, R J Molnar, T D Moustakas, E N Mokhov, P G Baranov, Appl. Phys. Lett., 68 (1996) 2556

[19] S Fischer, G Steude, D M Hofmann, F Kurth, F Anders, M Topf, B K Meyer, F Bertram, M Schmidt, J Christen, L Eckey, J Holst, A Hoffmann, B Mensching, B Rauschenbach, Jour. Cryst. Growth, 189/190 (1998) 556

[20] S T Kim, Y J Lee, S H chung, D C Moon, Semicon. Sci. Technol., 14 (1999) 156

[21] H Y An, O H Cha, J H Kim, G M Yang, K Y Lim, E –K Suh, H J Lee, J. Appl. Phys., 85 (1999) 2888

[22] D G Chtchekine, G D Gilliland, Z C Feng, S J Chua, D J Wolford, S E Ralph, M J Schurman, I Ferguson, MRS J. Nitride Semicond. Res. 4S1 G6.47 (1999)

ACKNOWLEDGMENTS

This work was partially supported by European community under contract BRPR-CT96-034. One of the authors (R. Seitz) gratefully acknowledges to JNICT by a maintenance grant BD/16284/98 (Praxis XXI Programm) .

DISORDER INDUCED IR ANOMALY IN HEXAGONAL AlGaN SHORT-PERIOD SUPERLATTICES AND ALLOYS

A. M. MINTAIROV *, A. S. VLASOV *, J. L. MERZ *, D. KORAKAKIS **, T. D. MOUSTAKAS **, A. O. OSINSKY ***, R. GASKA ***, M. B. SMIRNOV ****.

*EE Department, University of Notre Dame , Notre Dame, IN 46556
**ECE Department, University of Boston , Boston,MA, 02215
***APA Optics. 2950 N.E. 84th Lane, Blaine, MN, 55434
****Institute for Silicate Chemistry, Odoevskogo 24/2, 199155 St.Petersburg, Russia

ABSTRACT

We report an experimental (infrared reflectance spectroscopy) and theoretical study of the polar optical phonons in hexagonal ternary nitride compounds: AlN_m/GaN_n (n=2-8, m=4, 8) superlattices (SL) and spontaneously ordered $Al_xGa_{1-x}N$ (x=0.08-0.55) alloys. In infrared (IR) reflectivity spectra we revealed two modes having strong LO-TO splitting (20-150 cm^{-1}), and several modes, having a small (1-3 cm^{-1}) LO-TO splitting. All modes have a very high damping parameter ≥ 20 cm^{-1}. The unusual observation is the negative value of the oscillator strength for the weak IR mode at ~690 cm^{-1}, suggesting possible lattice instability, consistent with high damping observed. We found from lattice dynamical calculations that weak IR active modes correspond to modes localized at GaN-AlN interfaces. Our analysis has shown that an anomalous mode is induced by the disorder effects and arises due to strong overlapping of the LO-TO phonon branches of the bulk GaN and AlN. In SL samples the anomalous mode corresponds to phonons localized on interface inhomogenities.

INTRODUCTION

The study of the III-V nitrides GaN, AlN and their alloys has attracted considerable attention during the last several years due to their application in wide bandgap optoelectronics and microelectronics. The phonon properties of wurtzite GaN and AlN are intensively studied and well understood. Less is known about the phonon properties of ternary AlGaN. In the present paper we report the observation of an anomalous polar mode in IR reflectivity spectra of AlN/GaN short-period SLs and spontaneously ordered AlGaN alloys. This mode, observed at ~690 cm^{-1}, has negative oscillator strength. To our knowledge this is the first observation of such modes in crystals. Our analysis and lattice dynamical calculations show that this anomalous mode induced by GaN-AlN interface inhomogenities and disorder effects.

EXPERIMENTAL METHOD

Short period superlattice films were fabricated using a switched atomic layer metalorganic vapor deposition process as described in [1]. The films were deposited onto 1.6 µm GaN buffer layer, which was deposited on the basal plane of sapphire substrate. The superlattice (SL) consisted of a few hundred periods of (AlN_mGaN_n) unit cell made up of m monolayers of AlN and n monolayers of GaN. In the present work we studied superlattices having m=4, 8 and n=2, 4, 6, 8. The SL period have been determined by X-ray diffraction and transmission electron spectroscopy measurements [1].

427

Al$_x$Ga$_{1-x}$N alloys films having x=0.08, 0.22, 0.45, 0.55 and thickness 0.7, 1.4, 0.55, 0.45 μm, respectively, were grown by molecular beam epitaxy on the basal plane of sapphire substrates [2]. The X-ray data indicate SL ordering of the films [2].

The crystal symmetry of AlGaN SLs with even m+n and the alloy layers, grown on (0001) sapphire substrates, can be characterized by space group C_{3v}^1 [3]. This group has only ir active A$_1$ and E vibrational representations. We will distinguish the GaN- and AlN-type character of the phonon modes by figures 1 or 2. It should be noted that for alloys, the C_{3v}^1 space group follows from the fact that substitution of the Ga(Al) atoms in Al(Ga) sublattice (chemical disorder) eliminates improper translation (00$^1/_2$) of the C_{6v}^4 space group (bulk GaN and AlN), which replaces a sixfold screw axis by a threefold rotation one.

Polarized room temperature IR reflectance spectra were taken at oblique incidence (55°) with a Bruker IFS-66V spectrometer. In the simulations the reflectance coefficient was expressed in terms of dielectric functions ε_{xx} and ε_{zz} of constituent layers. The functions $\varepsilon_{\alpha\alpha}$ (α=x, z) were calculated from the standard expression:

$$\varepsilon_{\alpha\alpha}(\omega) = \varepsilon_\infty + \sum_j \frac{4\pi F_{\alpha j}}{\omega_{TO\alpha j}^2 - \omega^2 - i\omega\gamma_j}, \qquad (1)$$

Where $4\pi F_{\alpha j}$, $\omega_{TO\alpha j}$, γ_j and ε_∞ are mode oscillator strength, transversal frequency, damping, and high frequency dielectric function of the crystal. The subscript α shows the polarization of the mode. The oscillator strength was expressed via transversal and longitudinal phonon frequencies according to [4]. Phonon parameters for the sapphire substrate and buffer GaN layer entering (1), and the values of $4\pi F_{\alpha j}$ were taken from [5] and [6].

For AlGaN layers E(TO) and E(LO) phonon frequencies enter ε_{xx}, while A$_1$(LO) and A$_1$(TO) frequencies enter ε_{zz}. Only ε_{xx} contributes to s-polarized reflection, while both ε_{xx} and ε_{zz} contribute to p-polarization. Our calculations have shown that the p-polarized reflection coefficient very weakly depends on ε_{zz}. The only ε_{zz} parameter which affected IR reflectivity is the highest frequency A$_1$(LO2) phonon.

IR SPECTRA OF AlN/GaN SL AND ALLOYS

In Fig. 1 we present measured and calculated s-polarized (Fig.1,a-e) and p-polarized (f-j) IR spectra of three SL and two alloy samples. In most of the samples we can distinguish two main AlGaN restrahl bands - E(1) and E(2), corresponding to the main GaN- and AlN-type modes with frequencies of TO/LO components ~570/620 and ~640/840 cm^{-1}, respectively. In the p-polarization of the SL samples the E(2) band contains a sharp minimum at 740 cm^{-1} due to the A$_1$(LO) phonon of the buffer GaN layer. The minimum corresponding to the high frequency edge of E(2) band (~850 cm^{-1}) in s- and p-polarizations related to the E(x,y) and A$_1$(z) components of LO2 phonon, respectively, enabling us to measure its angular anisotropy.

In SL samples (Fig. 1, a,f and c,h) the fine structure of E(1) mode is revealed. It appears due to an additional weak (LO-TO splitting ~2 cm^{-1}) band e(1) near 600 cm^{-1}. We can also see that the E(1) band in alloy samples has asymmetry of its shape which can be an evidence of the e(1) mode contribution.

The most interesting feature in the reflectance spectra in Fig. 1 is the fine structure of the E(2) band observed in all samples. It consists of a small but distinct peak e(1a) at ~690 cm^{-1} followed by a very weak minimum e(2) at ~710 cm^{-1} and arises from two weak modes with small LO-TO splitting (1-2 cm^{-1}).

The unusual feature is the negative value of the oscillator strength for the e(1a) mode, i.e. peak in the spectra instead of the dip. This is demonstrated in Fig. 2. We see that only a

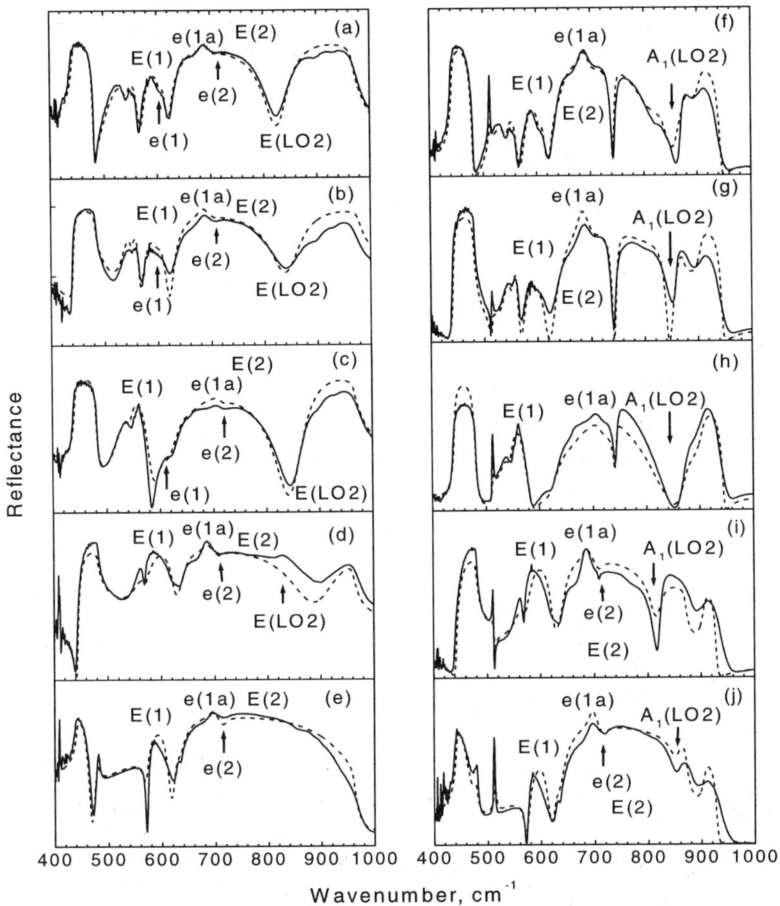

Fig. 1 IR spectra (a-e - s-polarization, f-j - p-polarization) of AlN$_m$GaN$_n$ SLs (nxm: 8x4 - a,f; 4x4 - b, g; 2x8 - c, h) and Al$_x$Ga$_{1-x}$N alloys (x: 0.22 - d, i; 0.45 - e, j)

negative value of the oscillator strength for the e(1a) mode leads to the characteristic peak at 690 cm^{-1} observed in the experiment. The important consequences of this anomaly is the high phonon damping ($\gamma_j \geq 20$ cm^{-1}) in AlGaN. It is clear from Fig. 2 that for smaller damping ($\gamma_j = 10$ cm^{-1}) the e(1a) reflection peak has at maximum unphysical value greater than unity. This corresponds to negative value of imaginary part of the dielectric function of the crystal Im(ε_{xx}), i.e. to the amplification of the electromagnetic waves and lattice instability. For $\gamma_j = 20$ cm^{-1} the broadening of the peak gives the positive value of Im(ε_{xx}) at 690 cm^{-1} and the value of the reflection coefficient less than unity.

In Fig. 3 we summarize the mode frequencies versus AlN content of all measured samples. The frequencies of the main E(TO1), E(LO1), E(TO2), as well as weak e(1a) and e(2) IR modes are close in SLs and alloys with the same AlN content.

As can be seen from Fig. 3 there is a the strong difference in the frequency of the $A_1(LO2)$ phonon in SLs and alloys indicating its different angular dispersion. In SLs the

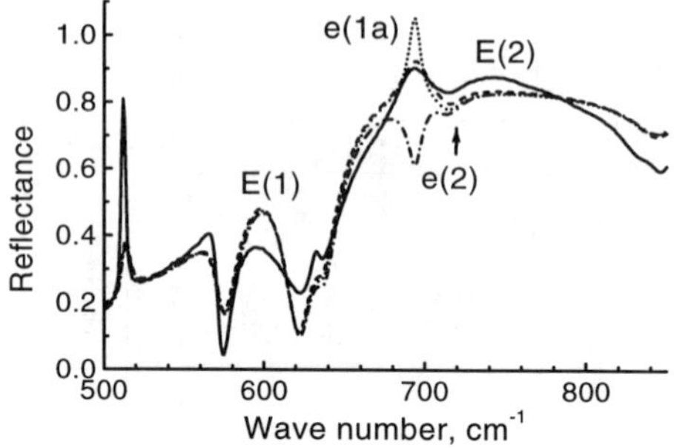

Fig. 2. Effect of anomalous e(1a) mode in IR spectra of AlGaN. Measured p-spectra of $Al_{0.55}Ga_{0.45}N$ alloy - solid curve; calculated curves (γ_j in cm^{-1}): dash-dotted - $F_{e(1a)}>0$, γ_j =10; dotted - $F_{e(1a)}<0$, γ_j =10; dashed $F_{e(1a)}<0$, γ_j =20 .

difference between frequencies of $A_1(LO2)$ and $E(LO2)$ is positive and has value ~50 cm^{-1}. In alloys it is negative and equals ~10 cm^{-1}. In SLs the angular dispersion of the LO2 mode frequency reflects the differences in the average macroscopic polarization field for the modes

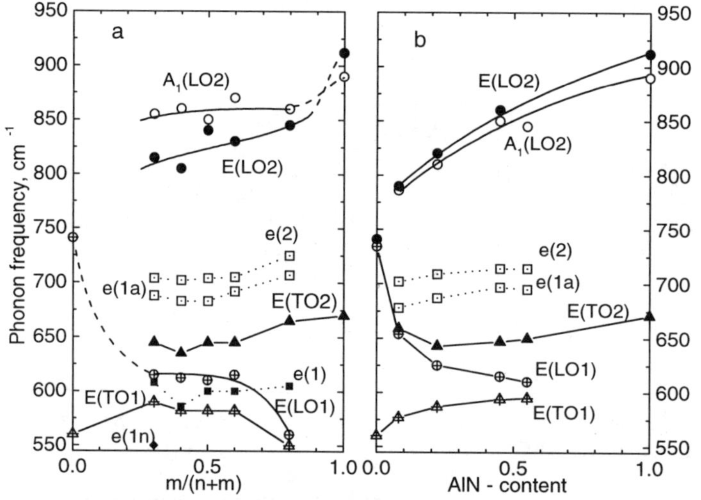

Fig. 3. Frequencies of the LO/TO components of the polar phonons of AlN_m/GaN_n SLs (a) and AlGaN alloys (b) versus AlN content. Only TO components frequencies are shown for e(1), e(2), and e(1a) modes because of their small LO-TO splitting value (~1-5 cm^{-1}).

propagating along and perpendicular to the layers [7]. This implies the existence of the GaN-AlN interfaces and has a purely long-range origin. In alloys the anisotropy of the LO2 phonon frequency reflects the anisotropy of the short-range atomic forces. As can be seen from Fig.3,b this anisotropy in alloys is the same as in bulk materials. These angular anisotropy effects for LO2 phonon are well reproduced in our microscopic lattice dynamical calculations (see below).

It can be seen from Fig. 3 that frequencies of the transversal modes E(TO1) and E(TO2) differ slightly   from those of the corresponding bulk materials compared with longitudinal ones - E(LO1) and E(LO2). For the GaN-type LO1-phonon a strong decreasing of the frequency in ternary AlGaN (down to 610 cm^{-1}) is unusual, because of its small dispersion in the bulk (670-740 cm^{-1}) [8]. However, the decreasing of the LO1 phonon frequency below the TO2 frequency is required the necessity of having a positive value of the oscillator strength for GaN-type phonons (lattice stability). If such a decrease involves short-range interaction, it occurs via a strong deformation of the Ga-N bond. If this happens, the local strains will act to prevent the bond deformation what can activate the anomalous e(1a) mode. This was confirmed by our lattice dynamical calculations.

LATTICE DYNAMICS OF ALN/GaN SL

We used lattice dynamical calculations of AlN_m/GaN_n SL optical phonon frequencies and IR activity. The model is based on the short-range interatomic potentials and rigid-ion Coulomb interactions developed for bulk GaN and AlN [8]. The positions of the atoms in the unit cell are calculated using the condition of the conservation of the bulk bond lengths and the value of the lattice parameter c determined from X-ray data.

The eigenvectors of GaN-type phonons calculated for a 4x4 SL for two different charge values of interface N atoms are presented in Fig. 4. In Fig. 4 we used the subscript l (for $l>1$) in our notations of modes, which is equal to a quantum number of the wavevector along the SL axis (see below). In model A the charges of the N interface atoms were equal to bulk GaN (-1.14) and AlN (-1.25) values for $AlGa_3N$ and Al_3GaN interfaces, respectively. In model B they were taken equal to the averaged value over the charge of the neighboring metallic atoms, i.e. -1.17 and -1.21, which qualitatively accounts for chemical bonding effects.

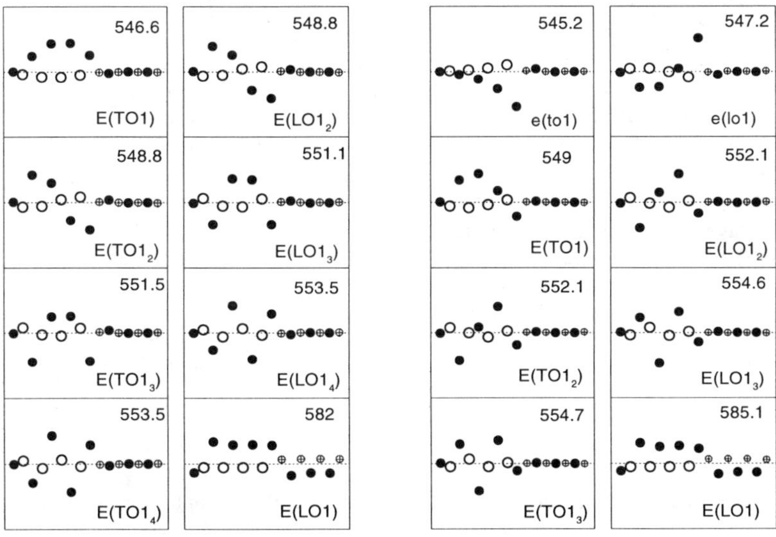

Fig. 4. Displacement patterns of the GaN-type E-modes in AlN_4GaN_4 SL calculated in model A (left) and model B (right).

We can see from Fig. 4 that in model A the confined $E(TO_l)$ modes are well described by the standing waves envelopes with wave vector $q = \frac{\pi}{ka}l$, where k is the number of monolayers, a is the thickness of the monolayer and $1 \leq l \leq k$. In this case the layer has one main IR active mode having $l=1$. In model B there is an additional IR active mode localized near the interface. In this case E(TO) mode is confined in k-1monolayers and has k-1 quantized components.

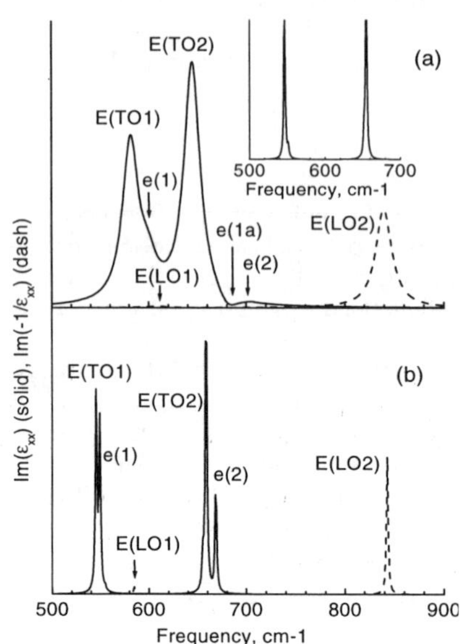

Fig. 5. Im (ε_{xx}) and Im($-1/\varepsilon_{xx}$) functions of AlN$_4$/GaN$_4$ SL: a - experimental; b - model B. Insert shows results for model A

Our observation of the weak e(1) and e(2) modes is qualitatively consistent with model B. This is demonstrated in Fig. 5 where we presented the transverse and longitudinal response functions of a 4x4 SL extracted from experimental IR spectra and calculated (using γ_j =3 cm^{-1}) in models A and B. The modes e(1) and e(2) localized at AlGa$_3$N and Al$_3$GaN interfaces are clearly revealed in model B. The differences in the calculated and observed frequencies indicates a deformation of bonds, which can be expected due to the presence of the anomalous mode.

From Fig.4 and 5 we see that in an ideal SL no anomalous mode exists because there is a strong decreasing of the E(LO1) phonon frequency down to 580 cm^{-1} due to its penetration in the AlN layer, which has purely electrostatic origin. In the calculation where we exclude this effect, putting the mass of AlN equal 1000, we obtain the E(LO1) phonon frequency of 670 cm^{-1} which is close to the frequency of the e(1a) mode. This shows that modes strongly localized in GaN bonds must have anomalous IR behavior in AlGaN material. While in alloys such localization is natural to assume due to disorder, in SLs the localization of modes can occur at interface inhomogeneities: fluctuation of layer thickness or layer intermixing. Our observation of the interface e(2) mode in alloys is consistent with their SL ordering. As at the GaN-AlN interface the reduction of the LO1 phonon frequency occurs via long-range interaction which does not involve bond deformation, we can suppose that SL ordering effects is the intrinsic feature of AlGaN.

An interesting consequence of the observed IR anomaly in AlGaN is the possibility of amplification of IR radiation if one can create conditions for decreasing phonon damping (see Fig. 2).

REFERENCES

1. M. Asif Khan, J. N. Kuzina, D. T. Olson, T. George, and W. T. Pike, Appl. Phys. Lett., **63**, 3471, (1993)

2. D. Korakakis, K. F. Ludwig, Jr., and T. D. Moustakas, Appl. Phys. Lett, **71**, 72 (1997).

3. Yu. E. Kitaev, M. F. Limonov, P. Tronc, and G. N. Yushin, Phys. Rev. B, **57**, (1998).

4. C. T. Kirk, Phys. Rev. B, **38**, 1255, 1988

5. A. S. Barker, Jr.and M. Ilegems, Phys.Rev. B **7**, 743, 1973

6. A. S. Barker, Jr., Phys. Rev. **132**, 1474 (1963)

7. M. Cardona, Superlattices and Microstructures, **7**, 180, 1990

8. V. Yu. Daviidov, Yu. V. Kitaev, I. N. Goncharuk, A. N. Smirnov, J. Graul, O. Seminchinova, D. Uffmann, M. B. Smirnov, A. P. Mirgorodsky and R. A. Evarestov, Phys. Rev. B **58**, 12899 1998.

NONDEGENERATE OPTICAL PUMP-PROBE SPECTROSCOPY OF HIGHLY EXCITED GROUP III NITRIDES

T.J. SCHMIDT,[a] J.J. SONG,[a] S. KELLER,[b] U.K. MISHRA,[b] S.P. DenBAARS,[b] and WEI YANG[c]

[a] Center for Laser and Photonics Research and Department of Physics
Oklahoma State University, Stillwater, Oklahoma 74078

[b] Electrical and Computer Engineering and Materials Departments
University of California, Santa Barbara, California 93106

[c] Honeywell Technology Center, 12001 State Highway 55, Plymouth, Minnesota 55441

ABSTRACT

We report the results of nondegenerate optical pump-probe absorption experiments performed on GaN and InGaN thin films and quantum wells under the conditions of strong optical band to band excitation. The evolution of the band edge in these materials was monitored as the number of photo-excited free carriers was increased beyond that required to achieve population inversion and observe stimulated emission. The band edge of InGaN is shown to exhibit markedly different high excitation behavior than that of GaN, explaining in part the reduction in stimulated emission threshold that typically accompanies the incorporation of indium into GaN to form InGaN. A comparison of the band edge absorption changes observed in pump-probe experiments to the gain spectra measured in variable-stripe gain experiments is also given.

INTRODUCTION

GaN and its respective alloys (InGaN and AlGaN) have been attracting an ever increasing amount of attention due to their physical hardness, inert nature, and large direct band gaps, making them promising materials for UV-Blue-Green light emitting devices and detectors.[1] Current technological advances have made high brightness light-emitting diodes (LED's) and cw laser diodes based on these materials a reality.[2] Their nonlinear properties are now becoming a focus for many research groups. Femtosecond four-wave-mixing (FWM) experiments have been used to study the dephasing times of the A and B free excitons in GaN,[3,4] and femtosecond pump-probe transient transmission experiments have been used to study the ultrafast carrier dynamics in InGaN expitaxial films.[5] Picosecond FWM experiments have shown strong optical nonlinearities below[6] and at the band edge of GaN,[7] as have nanosecond FWM experiments.[8] However, much information is still unknown about the optical phenomena exhibited by these materials at the high carrier densities at which practical devices operate. Recently, nanosecond pump-probe transmission experiments have shown exciton saturation due to resonant and below resonance optical excitation of the excitonic transitions of GaN.[7] Nanosecond pump-probe experiments with optical excitation above the band gap of GaN have also been reported[9,10] and have shown large values of induced transparency and induced absorption in the band gap region with increasing optical excitation. The time evolution of these band edge changes have been studied on femtosecond,[11] picosecond,[12] and nanosecond[13] time scales. Similar experiments have been performed on InGaN thin films and multiple quantum wells (MQWs).[14,15] A better understanding of the optical phenomena associated with high carrier concentrations in this material system is important, not only for general physical insight, but also as an aid in designing practical devices. We present here a direct comparison of the absorption properties of GaN and InGaN-based structures in their highly excited state. The magnitude of the nonlinearities studied in this work suggests the possibility of new photonic devices based on the group III nitrides as optical switches.

EXPERIMENT

The GaN samples used in this work were nominally undoped single-crystal films grown by metalorganic chemical vapor deposition (MOCVD) on (0001) oriented sapphire substrates. Thin AlN buffer layers, approximately 50 nm thick, were deposited on the substrate at 775 °C before the growth of the GaN epilayers. The GaN layers were then deposited at 1040 °C directly on the AlN buffer layers. A

Mat. Res. Soc. Symp. Proc. Vol. 572 © 1999 Materials Research Society

GaN layer thickness of 0.38 μm was used for the data presented here. The $In_{0.18}Ga_{0.82}N$ layer used in this study was grown by MOCVD at 800 °C on a 1.8 μm thick GaN layer deposited at 1060 °C on (0001) oriented sapphire . The $In_{0.18}Ga_{0.82}N$ layer was 0.1 μm thick and was capped by a 0.05 μm GaN layer. The InGaN/GaN MQW was grown by MOCVD on a 1.8 μm thick GaN buffer layer grown on a (0001) oriented sapphire substrate. The active region was made up of 12 quantum wells consisting of 3 nm thick InGaN wells and 4.5 nm thick Si doped (n ~ $2x10^{18}$ cm^{-3}) GaN barriers. The average In composition in the wells was ~ 18 %. The structure was capped by a 0.1 μm thick $Al_{0.07}Ga_{0.93}N$ layer. A detailed description of the growth conditions is given elsewhere.[16] The average In composition was measured using high-resolution x-ray diffraction and assuming Vegard's law. We note that the actual InN fraction could be smaller due to systematic overestimation when assuming Vegard's law in this strained material system.[17]

The nondegenerate optical pump-probe experiments were performed using frequency doubled radiation from a nanosecond dye laser as a UV pumping source and broadband fluorescence from a dye solution as the probe source. The experimental system for the GaN studies consisted of an amplified dye laser pumped by the second harmonic of an injection seeded Nd:YAG laser operating at 10 Hz. The deep red radiation from the dye laser was frequency doubled in a nonlinear crystal to produce the near UV wavelengths used to synchronously pump the GaN layers above their band gaps and the dye solution. The UV fluorescence from the dye solution was collected and focused onto the GaN layer coincidental with the pump beam. The broadband transmitted probe was then collected and focused on the slits of a 1 meter spectrometer and spectrally analyzed using a UV enhanced, gated CCD. A pump wavelength of 337 nm (3.678 eV) was used for the GaN data presented in this report. For the InGaN epilayer and MQW experiments, the third harmonic of the Nd:YAG laser (3.49 eV) was used in place of the dye laser to synchronously pump the individual samples and the dye solution. The sample temperature was varied between 10 K and room temperature through the use of a closed cycle refrigerator.

RESULTS

Figure 1(a) shows the 10 K absorption spectra near the band gap for a 0.38 μm GaN sample subjected to several different pump power densities (I_{exc}). We note here that the unpumped absorption spectra agree very well with published cw absorption values for the same sample.[18] Fig. 1(c) shows the measured absorption changes with respect to the unpumped spectra for the pump densities given in Fig. 1(a), where $\Delta\alpha = \alpha(I_{exc}) - \alpha(0)$. Induced transparency ($\Delta\alpha$ negative) in the excitonic region and induced absorption ($\Delta\alpha$ positive) in the below-gap region are clearly seen with increasing pump density. Fig. 1(b) shows room temperature (RT) absorption spectra for several different pump densities. We note that at RT, the A and B free exciton features seen in Fig. 1(a) have broadened and merged into one and the resulting induced transparency with increasing I_{exc} is about one third that at 10 K. The below-gap induced absorption is seen in Fig. 1(d) to be approximately half that at 10 K. The decrease in the free exciton absorption with increasing I_{exc} is attributed to many body effects, such as exciton screening by free carriers, causing a diminution of their oscillator strength.[19] Lattice heating has been proposed as the origin of the below-gap induced absorption,[8,11] but is not consistent with observations in other nitride materials (see the following section). It's origin, therefore, is still not well understood. We note that the observed induced absorption is spectrally located in the region in which stimulated emission (SE) is observed and net optical gain is expected, indicating a complex relation exists between induced absorption and gain in MOCVD-grown GaN thin films. We also note that these samples are optically thick (αL ~ 4 for the pump wavelength and sample thickness, $L = 0.38$ μm, used in this study), so the pump intensity is appreciably diminished as it traverses the sample thickness. Therefore, the resulting values of $\Delta\alpha$ presented here are lower limit values of the actual change with pump density.

Figures 2(a) and 2(b) show the absorption spectra of the InGaN reference layer near the fundamental absorption edge at 10 K and RT, respectively. The oscillatory structure is a result of thin film interference. With increasing excitation density of the pump pulse, the absorption coefficient in the band tail region is seen to decrease significantly. This bleaching was observed to saturate for I_{exc} exceeding ~ 2 MW/cm^2 at 10 K and RT. Similar behavior is observed for the MQW sample (not shown),

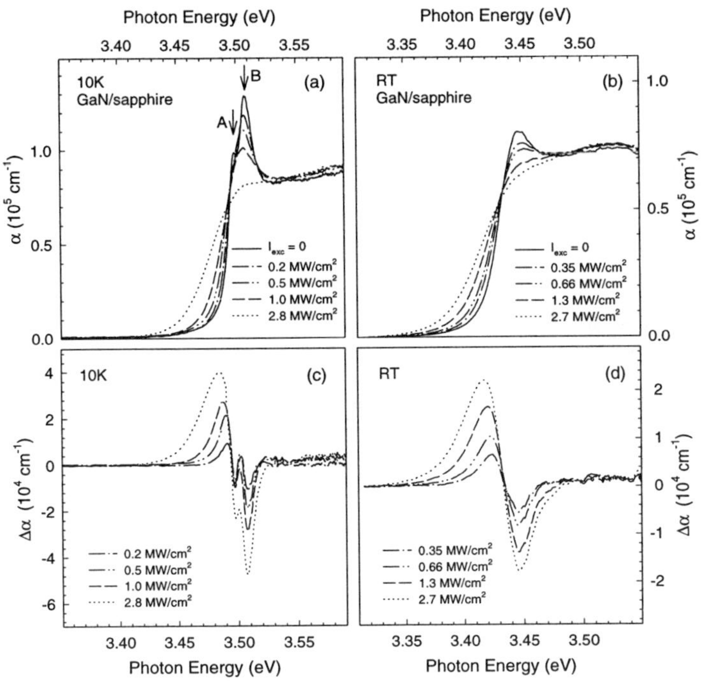

Figure 1: (a) 10 K and (b) room temperature absorption spectra near the fundamental absorption edge as a function of optical excitation density (I_{exc}) for a 0.38 μm thick GaN layer grown by MOCVD on (0001) oriented sapphire. (c), (d) Differential absorption spectra as a function of I_{exc}, $\Delta\alpha = \alpha(I_{exc}) - \alpha(0)$, for the spectra in (a) and (b), respectively. The A and B free exciton transitions are clearly seen in the unpumped 10 K spectrum. Complete exciton saturation is seen for I_{exc} approaching 3 MW/cm^2 at both 10 K and room temperature. Induced transparency in the excitonic region and induced absorption in the below-gap region are clearly seen with increasing I_{exc} at both temperatures. The excitation wavelength was 337 nm (3.678 eV).

the only difference being a larger spectral region exhibiting absorption bleaching due to the larger band tailing exhibited by the MQW sample. The differential absorption spectra are also shown in Figs. 2(c) and 2(d) for clarity. We note that the induced transparency associated with the absorption bleaching is quite large, exceeding 3×10^4 cm^{-1} at both 10 K and RT. The spectral region in which SE is observed is also indicated in Figs. 2(c) and 2(d). Clear features in the induced absorption bleaching spectra are seen to coincide with these spectral regions and are attributed to net optical amplification (gain) of the probe pulse. Again, similar behavior was observed for the InGaN/GaN MQW sample. Induced absorption was not observed for either InGaN based structure. This lack of induced absorption in the gain region of InGaN-based structures explains in part the typical reduction in SE threshold that accompanies the incorporation of indium into GaN to form InGaN.[20]

The modal gain spectra measured by the variable-stripe method of Shaklee and Leheny[21] are shown in Figs. 3(a) and 3(b) for the InGaN/GaN MQW and InGaN thin film, respectively. The spectra were taken for excitation lengths less than 200 μm to minimize re-absorption induced distortions in the spectra.[22,23] The excitation densities in Figs. 3(a) and 3(b) are given with respect to the SE threshold (I_{th}) measured for long (> 2 mm) excitation lengths. A clear blueshift in the gain peak with increasing optical excitation is seen for the MQW. This blueshift was observed to stop for $I_{exc} > 12 \times I_{th}$. Further increases in I_{exc} resulted only in an increase in the modal gain maximum. The large shift in the gain maximum of

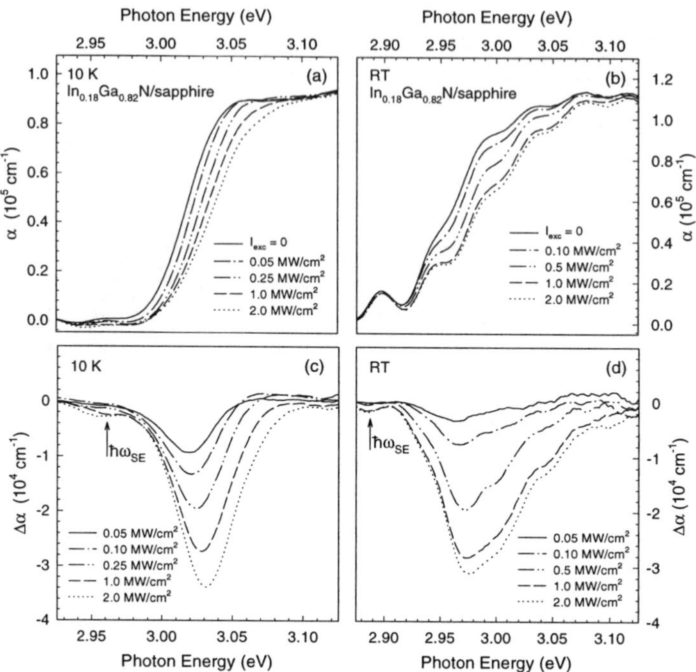

Figure 2: (a) 10 K (b) room temperature optical absorption spectra of an InGaN thin film near the fundamental absorption edge as a function of above-gap optical excitation. (c), (d) Differential absorption spectra, $\Delta\alpha(I_{exc}) = \alpha(I_{exc}) - \alpha(0)$, as a function of above-gap optical excitation for the spectra shown in (a) and (b), respectively. The oscillatory structure in (a) and (b) is a result of thin film interference. The excitation wavelength was 355 nm (3.492 eV). The SE energy is indicated for completeness.

the MQW to higher energy with increasing I_{exc} is consistent with band filling of localized states in the InGaN active layers. Similar behavior was observed at room temperature. The blueshift in the gain spectra of the InGaN thin film is seen to be considerably smaller than that of the MQW. It was also observed to stop at considerably lower excitation densities. The modal gain spectra of both samples are seen to correspond spectrally with the low energy tail of the band tail state absorption bleaching spectra, as shown in Figs. 3(c) and 3(d).

Figures 3(c) and 3(d) show a comparison of the absorption bleaching spectra of the InGaN/GaN MQW and the InGaN thin film for several excitation densities below and above the SE threshold. An interesting difference between the two structures is the behavior of the absorption bleaching as the SE threshold is exceeded. As the pump density is increased, the bleaching is observed to increase for both samples. As the SE threshold is exceeded, though, the bleaching of the MQW tail states is seen to *decrease* significantly with increasing excitation, while the bleaching of the epilayer's tail states continues to increase with increasing excitation density. This is clearly seen in Figs. 3(c) and 3(d), where the dotted and dashed lines show the bleaching spectra for excitation densities below the SE threshold and the solid lines show the bleaching spectra for excitation densities above the SE threshold. The SE thresholds for the MQW and epilayer were measured to be ~ 350 and 300 kW/cm^2, respectively, for the experimental conditions. The bleaching behavior of the MQW is attributed to the opening up of the bottleneck in the carrier relaxation process due to the fast depopulation of carriers participating in the SE process.[24] For the InGaN thin film, the recombination lifetime was found to always be significantly

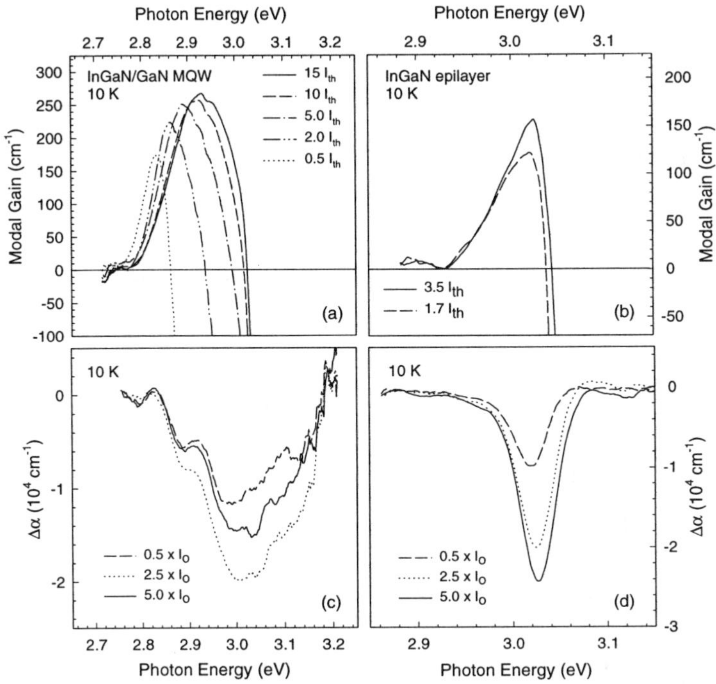

Figure 3: 10 K modal gain spectra of (a) an InGaN/GaN MQW and (b) an InGaN epilayer as a function of above-gap optical excitation. The excitation densities, I_{exc}, are given with respect to the SE thresholds, I_{th}, measured for long (> 2 mm) excitation lengths. A clear blueshift in the gain maximum and gain/absorption crossover point is seen with increasing optical excitation for the MQW sample. This trend is much less obvious for the InGaN epilayer. (c), (d) 10 K nanosecond nondegenerate pump-probe differential absorption spectra for the structures in (a) and (b), respectively, showing absorption bleaching ($\Delta\alpha$ negative) of band tail states with increasing I_{exc}. $\Delta\alpha(I_{exc}) = \alpha(I_{exc}) - \alpha(0)$ and $I_o = 100$ kW/cm². The SE threshold for the MQW and epilayer was found to be ~ 350 and 300 kW/cm², respectively, for the experimental conditions of (c) and (d). An excitation wavelength of 355 nm (3.492 eV) was used for the data shown in (a) through (d).

shorter than the pump pulse, leading to no observable change in the relaxation dynamics of the higher energy band tail states as the SE threshold is exceeded.

CONCLUSIONS

In summary, nanosecond nondegenerate optical pump-probe absorption experiments have been performed on GaN thin films, InGaN thin films, and InGaN/GaN MQWs grown by MOCVD on (0001) oriented sapphire in order to further understand the optical phenomena associated with high carrier densities in this material system. The evolution of the band edge was monitored as the number of free carriers was increased by photo-excitation. Exciton saturation was observed in GaN with increasing carrier concentration, with a resulting decrease in the absorption coefficient approaching $\Delta\alpha = -4 \times 10^4$ cm⁻¹ at 10 K and -2×10^4 cm⁻¹ at RT. In addition, large below-gap induced absorption exceeding $\Delta\alpha = 4 \times 10^4$ cm⁻¹ at 10 K and 2×10^4 cm⁻¹ at RT was observed in GaN as the pump density was increased to over 3 MW/cm² . The exciton saturation is explained by many body effects, such as screening by excess free carriers. The origin of the below-gap induced absorption is still not well understood, but is most likely a result of the high density of defects in MOCVD-grown GaN films. In contrast, InGaN based structures were observed to only exhibit strong bleaching of band tail states with increasing above-gap optical excitation. The induced transparency associated with the absorption bleaching was found to exceed 3×10^4 cm^{-1} at both 10 K and RT. The magnitude of the bleaching in the InGaN/GaN MQW was found to be significantly affected by the onset of SE, indicating the carriers responsible for bleaching and SE share the same recombination channels in these structures. Optical gain was observed on the low energy tail of the absorption bleaching spectra for both InGaN structures. The large values of induced transparency/absorption studied in this work suggest the potential of new optoelectronic applications, such as optical switching, for the group III nitrides.

ACKNOWLEDGMENTS

The work at Oklahoma State University was funded by BMDO, DARPA, and ONR.

REFERENCES

1. S.N. Mohammad and H. Morkoc, Prog. Quant. Electr. **20**, 361 (1996), and references therein.
2. S. Nakamura, M. Senoh, N. Iwasa, S. Nagahama, T. Yamada, and T. Mukai, Jpn. J. Appl. Phys., Part 1 **34**, L1332 (1995). S. Nakamura, M. Senoh, S. Nagahama, N. Iwasa,, T. Yamada, T. Matsushita, Y. Sugimoto, and H. Kiyoku, Appl. Phys. Lett. **70**, 868 (1997). S. Nakamura, MRS Internet J. Nitride Semicond. Res. **4S1**, G1.1 (1999).
3. A.J. Fischer, W. Shan, G.H. Park, J.J. Song, D.S. Kim, D.S. Yee, R. Horning, and B. Goldenberg, Phys. Rev. B **56**, 1077 (1997).
4. R. Zimmermann, M. Hofmann, D. Weber, J. Mobius, A. Euteneuer, W.W. Ruhle, E.O. Gobel, B.K. Meyer, H. Amano, and Akasaki, MRS Internet. J. Nitride Semicond. Res. **2**, Article 24 (1997).
5. C.K. Sun, F. Vallee, S. Keller, J.E. Bowers, and S.P. DenBaars, Appl. Phys. Lett. **70**, 2004 (1997).
6. B. Taheri, J. Hays, J.J. Song, and B. Goldenberg, Appl. Phys. Lett. **68**, 587 (1996).
7. H. Haag, P. Gilliot, D. Ohlmann, R. Levy, O. Briot, R.L. Aulombard, MRS Internet J. Nitride Semicond. Res. **2**, Article 21 (1997).
8. H. Haag, P. Gilliot, R. Levy, B. Honerlage, O. Briot, S. Ruffenach-Clur, and R.L. Aulombard, Appl. Phys. Lett. **74**, 1436 (1999).
9. T.J. Schmidt, J.J. Song, Y.C. Chang, R. Horning, and B. Goldenberg, Appl. Phys. Lett. **72**, 1504 (1998).
10. T.J. Schmidt, Y.C. Chang, and J.J. Song, SPIE Conf. Proceedings Series **3419**, 61 (1998).
11. A.J. Fischer, B.D. Little, T.J. Schmidt, J.J. Song, R. Horning, and B. Goldenberg, SPIE Conf. Proceedings Series **3624-25** (1999).
12. H. Haag, P. Gilliot, R. Levy, B. Honerlage, O. Briot, S. Ruffenach-Clur, and R.L. Aulombard, Phys. Rev. B **59**, 2254 (1999).
13. Theodore J. Schmidt, Ph.D. thesis, Oklahoma State University, 1998.
14. T.J. Schmidt, Y.H. Cho, G.H. Gainer, J.J. Song, U.K. Mishra, and S.P. DenBaars, Appl. Phys. Lett. **73**, 1892 (1998).
15. T.J. Schmidt, Y.H. Cho, S. Bidnyk, J.J. Song, S. Keller, U.K. Mishra, and S.P. DenBaars, SPIE Conf. Proceedings Series **3625-7** (1999).
16. B.P. Keller, S. Keller, D. Kapolnek, W.N. Jiang, X.F. Wu, H. Masui, X.H. Wu, B. Heying, J.S. Speck, U.K. Mishra, and S.P. DenBaars, J. Electron. Mater. **24**, 1707 (1995). S. Keller, A.C. Abare, M.S. Minski, X.H. Wu, M.P. Mack, J.S. Speck, B. Hu, L.A. Coldren, U.K. Mishra, and S.P. DenBaars, Materials Science Forum, **264-268**, 1157 (1998).
17. T. Takeuchi, H. Takeuchi, S. Sota, H. Sakai, H. Amano, and I. Akasaki, Jpn. J. Appl. Phys. **36**, L177 (1997). M.D. McCluskey, C.G. Van deWalle, C.P. Master, L.T. Romano, and N.M. Johnson, Appl. Phys. Lett. **72**, 2725 (1998).
18. A.J. Fischer, W. Shan, J.J. Song, Y.C. Chang, R. Horning, and B. Goldenberg, Appl. Phys. Lett. **71**, 1 (1997).
19. C. Klingshirn and H. Haug, Phys. Rep. **70**, 315 (1981), and references therein.
20. T.J. Schmidt, Y.H. Cho, J.J. Song, and Wei Yang, Appl. Phys. Lett. **74**, 245 (1999).
21. K.L. Shaklee and R.F. Leheny, Appl. Phys. Lett. **18**, 475 (1971).
22. T.J. Schmidt, S. Bidnyk, Y.H. Cho, A.J. Fischer, J.J. Song, S. Keller, U.K. Mishra, and S.P. DenBaars, Appl. Phys. Lett. **73**, 3689 (1998).
23. T.J. Schmidt, S. Bidnyk, Y.H. Cho, A.J. Fischer, J.J. Song, S. Keller, U.K. Mishra, and S.P. DenBaars, MRS Internet J. Nitride Semicond. Res. **4S1**, G6.54 (1999).
24. T. Breitkopf, H. Kalt, C. Klingshirn, and A. Reznitsky, J. Opt. Soc. Am. B. **13**, 1251 (1996).

STUDY OF NEAR-THRESHOLD GAIN MECHANISMS IN MOCVD-GROWN GaN EPILAYERS AND InGaN/GaN HETEROSTRUCTURES

S. BIDNYK *, T. J. SCHMIDT *, B. D. LITTLE *, J. J. SONG *
*Oklahoma State Univ., Center for Laser and Photonics Research and Dept. of Physics, Stillwater, OK.

ABSTRACT

We report the results of an experimental study on near-threshold gain mechanisms in optically pumped GaN epilayers and InGaN/GaN heterostructures at temperatures as high as 700 K. We show that the dominant near-threshold gain mechanism in GaN epilayers is inelastic exciton-exciton scattering for temperatures below ~ 150 K, characterized by band-filling phenomena and a relatively low stimulated emission (SE) threshold. An analysis of both the temperature dependence of the SE threshold and the relative shift between stimulated and band-edge related emission indicates electron-hole plasma is the dominant gain mechanism for temperatures exceeding 150 K. The dominant mechanism for SE in InGaN epilayers and InGaN/GaN multiple quantum wells was found to be the recombination of carriers localized at potential fluctuations resulting from nonuniform indium incorporation. The SE spectra from InGaN epilayers and multiple quantum wells were comprised of extremely narrow emission lines and no spectral broadening of the lines was observed as the temperature was raised from 10 K to over 550 K. Based on the presented results, we suggest a method for significantly reducing the carrier densities needed to achieve population inversion in GaN, allowing for the development of GaN-active-medium laser diodes.

INTRODUCTION

The first results on stimulated emission (SE) in GaN at low temperatures were reported in the literature more than a quarter of a century ago.[1] GaN-based structures have been shown to be chemically stable and able to withstand high temperatures.[2] Recently commercialized bright blue light emitting and laser diodes are all based on InGaN/GaN heterostructures.[3] Understanding the gain mechanisms in this material is extremely important from both a fundamental physics and a device optimization standpoint. There have been several studies performed explaining the origin of SE in GaN epilayers and InGaN multiquantum wells (MQWs) at various temperatures. However, the results reported in the literature are often contradictory.[4-7] In this work we systematically studied the behavior of SE in GaN epilayers and InGaN/GaN MQWs over a wide temperature range. We demonstrate that at temperatures below 150 K excitonic-related gain mechanisms dominate the near-threshold emission behavior in GaN epilayers. For temperatures above 150 K we show that EHP recombination is the mechanism responsible for SE in GaN. We explain the SE behavior for the InGaN/GaN MQWs in terms of localized states recombination, thus showing that SE in InGaN/GaN MQWs and GaN epilayers are distinctly different.

EXPERIMENT

The GaN epilayers and InGaN/GaN MQW samples used in this study were grown by metalorganic chemical vapor deposition (MOCVD). Sample structures and growth parameters are described in Refs. 8 and 9. The samples were mounted on a copper heat sink attached to a custom-built wide temperature range cryostat/heater system. The SE part of this study was

Mat. Res. Soc. Symp. Proc. Vol. 572 © 1999 Materials Research Society

performed in an edge emission geometry. In order to avoid distortion of the spontaneous emission spectra due to re-absorption processes, the laser beam was focused on the sample surface and spontaneous emission was collected from a direction near normal to the sample surface.[8]

RESULTS AND DISCUSSIONS

GaN epilayers

To determine if the SE threshold density occurs above or below the Mott density (the critical density beyond which no excitons can exist), we studied the temperature behavior of the SE threshold in GaN epilayers grown on sapphire and SiC substrates over a wide temperature range (20 to 700 K), as shown in Fig. 1. For temperatures above 200 K, the SE thresholds roughly followed an exponential dependence: $I_{th} = I_0 \exp(T/T_0)$, with $T_0 \cong 170$ K (Ref. 2). This exponential behavior of the SE threshold is qualitatively similar to that observed in other material structures. However, as the temperature is reduced to below 200 K, a significant (faster than exponential) reduction in the SE threshold was observed in GaN epilayers. This is associated with a change in the gain mechanism indicating a drastic increase in the SE efficiency at low temperatures. Recently, Fischer *et al.* convincingly demonstrated the presence of excitonic resonances in GaN epilayers well above RT through optical absorption measurements.[10] Excitons in GaN epilayers cannot be easily ionized due to the relatively large exciton binding energy. However, at near-SE-threshold (near-I_{th}) pump densities the picture is not straightforward due to screening of the Coulomb-interaction by the photo generated free carriers. In general, the existence of excitons depends on the strength of the Coulomb-interaction which in turn depends on the density and distribution of carriers among bound and unbound states. It has been predicted theoretically that in a material system with a relatively large exciton binding energy, one could expect inelastic ex-ex scattering to have the lowest SE threshold at low temperatures.[11] Schmidt *et al.* performed pump-probe experiments and confirmed the presence of excitons at pump

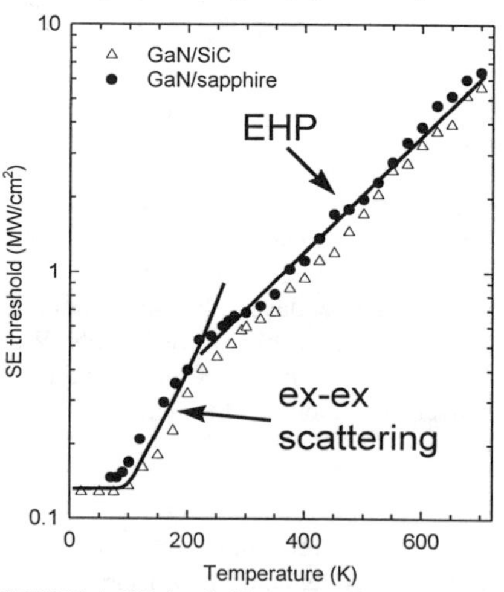

FIGURE 1. SE threshold as a function of temperature for GaN thin films grown on SiC (open triangles) and sapphire (filled circles).

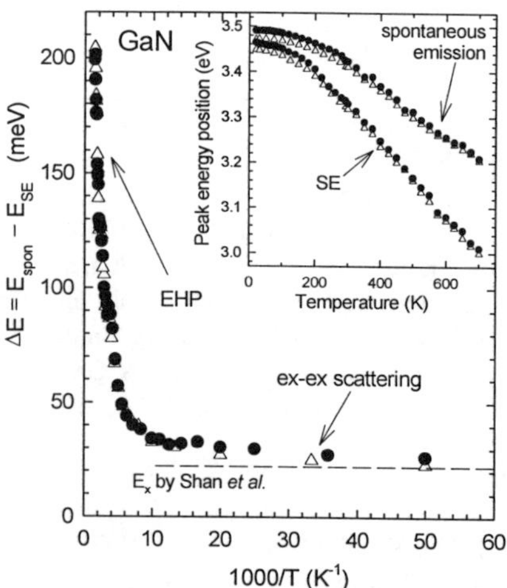

FIGURE 2. Energy difference (ΔE) between spontaneous and SE peaks as a function of temperature for GaN thin films grown on SiC (open triangles) and sapphire (filled circles).

densities above the SE threshold at 10 K in reflection spectra.[12] These observations support the idea that excitons persist above the densities required to observe SE in GaN at cryogenic temperatures.

The effects of excitons on the SE process can be better understood by studying the temperature and power dependence of the SE peak position. We measured the energy position of the SE peak at near-I_{th} pump densities and the spontaneous emission peak position using low power cw PL techniques in the temperature range of 20 to 700 K for samples grown on sapphire and SiC substrates, as shown in the inset of Fig. 2. The position of the spontaneous and SE peaks in the GaN epilayers is influenced by residual strain resulting from thermal-expansion mismatch between the epilayers and the substrates. This difference in energy position for the two samples is largest at low temperature and gradually decreases as the temperature is increased.

To avoid strain-related complications, we restricted ourselves to an analysis of the relative energy separation between the spontaneous and SE peaks, $\Delta E = E_{spon} - E_{SE}$, as depicted in Fig. 2. As we approach low temperatures (T<150 K), ΔE asymptotically approaches the exciton binding energy ($E_x = 21$ meV) measured by photoreflectance.[13] However, at temperatures above 150 K, ΔE monotonically increases and reaches values as high as 200 meV at 700 K. The behavior of the energy difference between the spontaneous and SE peaks at low temperatures (<150 K) is well explained by inelastic ex-ex scattering. In the case of ex-ex scattering the energy difference between the two peaks can be estimated from:[14]

$$\Delta E = E_{spon} - E_{SE}^{ex-ex} = \left(E_g - E_x \right) - \left(E_g - 2E_x - E_k^{e-h} \right) = E_x + E_k^{e-h}, \qquad (1)$$

where E_g is the bandgap energy and E_k^{e-h} is the kinetic energy of the unbound electron-hole pair created during the excitonic collision. At low excitation densities and low temperatures, one can consider the bands to be empty. The unbound electron-hole pairs created during the process have small kinetic energy ($E_k^{e-h} \approx 0$). Thus, ΔE approaches E_x as T $\rightarrow$ 0 K, as shown in Fig. 2.

For high temperatures (T>150 K), the energy difference between the spontaneous and SE peaks gradually increases from ~35 meV to a few hundred meV. Both the large energy difference and the relatively high SE thresholds in this temperature range (Fig. 1.) point to EHP recombination as the dominant gain mechanism. In EHP recombination, a large number of excited carriers cause band-gap renormalization effects resulting in large values of ΔE. Under such high excitation conditions, excitons are dissociated by many body interactions.[4] As further evidence to support EHP in this temperature range, we point out that excitons have not been observed in GaN at highly elevated temperatures, even under extremely low excitation conditions.[10] We therefore conclude that EHP recombination is responsible for gain in GaN thin films at these elevated temperatures. Since no significant change in the behavior of the SE threshold and SE peak position was observed for temperatures between 150 K and 700 K (Fig. 1 and Fig. 2), we conclude EHP recombination to be the dominant gain mechanism for temperatures exceeding 150 K.

At temperatures below 150 K, the effect of the kinetic energy E_k^{e-h} on ex-ex scattering recombination process could be observed in the excitation density dependence of ΔE. As the excitation intensity or temperature is increased, the bottom of the bands become filled. Thus, unbound electron-hole pairs created in the process of ex-ex collision must have higher energies, and the kinetic energy E_k^{e-h} can no longer be neglected. The inset in Fig. 3 shows the power dependence of ΔE at three different temperatures near the point when the gain mechanism experiences a transition from inelastic ex-ex scattering to EHP recombination. For temperatures

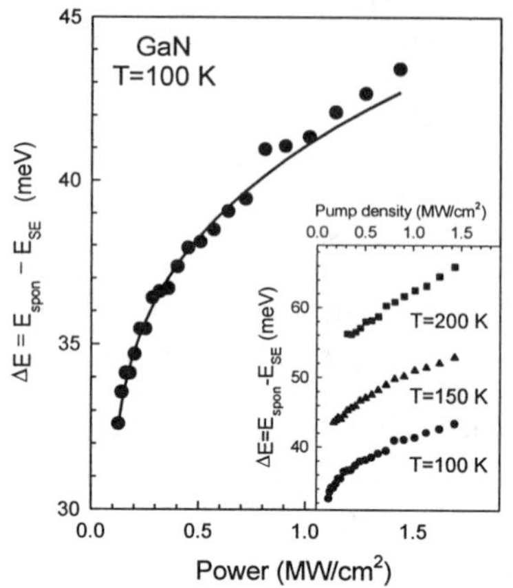

FIGURE 3. Energy separation (ΔE) between the spontaneous and SE peaks as a function of excitation density at 100 K. The solid line represents a theoretical fit of the experimental data (open circles) in Eq. (2). The inset shows the change in the behavior of ΔE at different temperatures.

below 150 K, we observed a rapid increase of ΔE at near-I_{th} pump densities, as shown in the inset of Fig. 3. This shift is most likely associated with band-filling which causes increased values of E_k^{e-h}. For temperatures above 150 K, this strong near-I_{th} shift in ΔE is not observed. At high temperatures (T >150 K) SE originates from EHP recombination and the gradual increase in ΔE with increasing excitation density is caused by band-gap renormalization effects. For one photon pumping and elliptical bands, the band-filling effect associated with ex-ex scattering gives a calculated line shift E_k^{e-h} proportional to $\left(I - I_{th}\right)^{1/3}$ (see Refs. 14 and 15). The substitution of this expression into Eq. (1) yields:

$$\Delta E = E_x + a\left(I - I_{th}\right)^{1/3}, \qquad (2)$$

A fit of the experimental data (open circles) taken at 100 K into Eq. (2) is shown in Fig. 3 by a solid line. From the fit, we obtained the values of exciton binding energy E_x= 28 meV and the SE threshold I_{th} = 100 kW/cm^2, which is in a reasonable agreement with experimental results and supports the idea of ex-ex scattering being the dominant SE mechanism for temperatures below 150 K.

Since ex-ex scattering has a lower SE threshold than recombination from an EHP (Fig. 1), it would be advantageous from a device standpoint if SE was dominated by excitonic effects at RT and beyond. This could be achieved, for example, by introducing 2-D spatial confinement of carriers. By tailoring the width of a GaN active layer sandwiched between AlGaN confinement layers, one would expect a significant increase in the exciton binding energy. For increased values of exciton binding energy, a strong reduction of the homogeneous broadening due to reduced Fröhlich interactions is expected.[16] This could potentially extend the ex-ex scattering gain mechanism to RT. The SE threshold for such structures would be significantly reduced due to carrier confinement and a shift in the dominant near-I_{th} gain mechanism to that of ex-ex scattering.

Stimulated emission in InGaN/GaN multiple quantum wells

In this section, the results of SE studies from InGaN/GaN MQWs are presented. The InGaN/GaN MQW samples were pumped under the same experimental conditions described above for the GaN epilayers. Figure 4 shows an emission spectrum from the MQW sample pumped above the SE threshold at RT. The spontaneous emission peak appears to be broad with a full width at half maximum (FWHM) of approximately 20 nm. A drastic spectral narrowing occurs when we excite the sample above the SE threshold, which generates narrow SE peak(s) with FWHM of less than 0.1 nm. Note that the FWHM of the SE peak in high-quality GaN epilayers is about 2 nm at RT. Such narrow SE lines in InGaN/GaN MQWs with inhomogeneous In incorporation can only be explained in terms of deeply localized states. Large localization

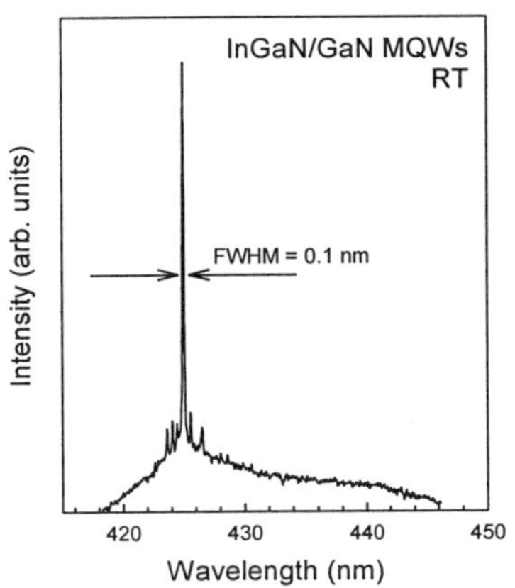

FIGURE 4. Emission spectra from InGaN/GaN MQW sample at RT. The sample is pumped 1.1 times the SE threshold of 55 kW/cm^2.

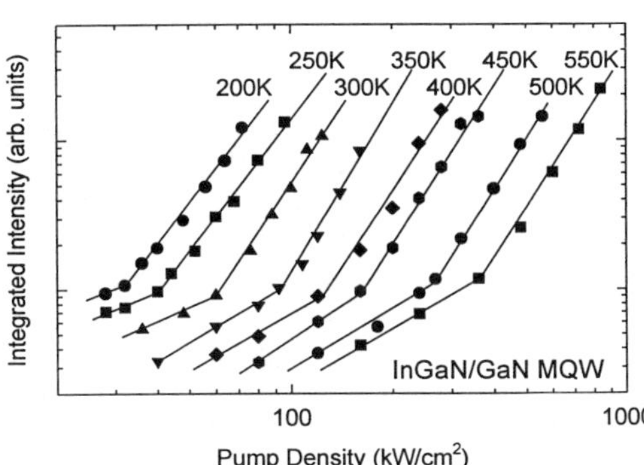

FIGURE 5. Integrated intensity of InGaN/GaN MQW emission as a function of pump density for different temperatures. The slope change from 0.8-1.3 to 2.2-3.0 indicates the transition from spontaneous emission to SE.

introduces discrete atomic-like levels which are observed as narrow peaks in SE spectra. To further support this idea, we performed a study of SE in the wide temperature range. No noticeable broadening of the SE peak(s) was observed when the temperature was varied over a range of several hundred degrees. The low temperature sensitivity of the FWHM is consistent with that expected from strongly localized carriers. The details of this study are reported elsewhere.[9] The SE threshold for the InGaN MQWs is observed to be an order of magnitude lower than for a GaN thin film. It is likely that such a low SE threshold in the MQW is due to the strong localization of carriers at potential fluctuations in the InGaN active layers.

We also found that an increase in temperature leads to a decrease in PL intensity. This indicates the onset of efficient losses and a decrease in quantum efficiency of the MQW. At high temperatures, only a small fraction of carriers reach the conduction band minima, and most of them recombine non-radiatively. The modal gain depends only on radiatively recombining carriers. Therefore, the temperature increase efficiently decreases modal gain and leads to an increase in the SE threshold. To evaluate the number of carriers that recombine radiatively, we studied the integrated photoluminescence intensity as a function of excitation power for different temperatures, as shown in Fig. 5. For the temperature range studied, we found that under low excitation densities, the integrated intensity I_{integ} from the sample increases almost linearly with pump density I_p (i.e. $I_{integ} \propto I_p^\gamma$, where $\gamma=0.8$-1.3), whereas at high excitation densities, this dependence becomes superlinear (i.e. $I_{integ} \propto I_p^\beta$, where $\beta=2.2$-3.0). The excitation pump power at which the slope of I_{integ} changes corresponds to the SE threshold at each temperature. Interestingly, the slopes of I_{integ} below and above the SE threshold do not significantly change over the temperature range involved in this study. This indicates that recombination of deeply localized carriers is the dominant SE mechanism in InGaN/GaN MQWs in the entire temperature range studied.

5. CONCLUSIONS

In conclusion, we have studied the gain mechanisms in GaN epilayers and InGaN/GaN MQWs in the temperature range of 20 to 700 K. We observed that for temperatures below 150 K the dominant near-threshold gain mechanism in GaN epilayers is inelastic exciton-exciton scattering, characterized by a low stimulated emission threshold. For temperatures exceeding 150 K, the dominant gain mechanism was shown to be electron-hole plasma recombination, characterized by a relatively high SE threshold and a large separation between spontaneous and stimulated emission peaks. The emission spectra for the InGaN-based structures were found to be drastically different than that of GaN over the entire temperature range studied. SE spectra in InGaN/GaN MQWs were comprised of extremely narrow lines and no broadening of the lines was observed as the temperature was raised by several hundred degrees. The InGaN/GaN MQWs exhibited an order of magnitude lower SE threshold than that of GaN epilayers. The SE behavior for the InGaN/GaN MQWs is explained in terms of carrier localization at potential fluctuations in the InGaN layers. The temperature sensitivity of the SE threshold of InGaN/GaN samples was measured and compared with GaN epilayers. This study demonstrates that the SE mechanisms in InGaN/GaN MQWs and GaN are distinctly different.

ACKNOLEDGEMENTS

We gratefully acknowledge the support of ONR, DARPA, AFSOR, and NSF.

REFERENCES

1. R. Dingle, K. L. Shaklee, R. F. Leheny, and R. B. Zetterstrom, Appl. Phys. Lett. **19**, 5 (1971)
2. S. Bidnyk, B. D. Little, T. J. Schmidt, Y. H. Cho, J. Krasinski, J. J. Song, B. Goldenberg, W. Yang, W. G. Perry, M. D. Bremser, and R. F. Davis, J. Appl. Phys. **85**, 1792 (1999)
3. S. Nakamura and G. Fasol, *The Blue Laser Diode*, (Springer, Berlin, 1997)
4. H. Amano and I. Akasaki, *Proc. Topical Workshop on III-V Nitrides*, 193, Nagoya, Japan (1995)
5. I. M. Catalano, A. Cingolani, M. Ferrara, M. Lugarà and A. Minafra, Solid State Comm. **25**, 349 (1978)
6. J. Holst, L. Eckey, A. Hoffman, I. Broser, B. Schöttker, D. J. As, D. Schikora, and K. Lischka, Appl. Phys. Lett. **72**, 1439 (1998)
7. T. J. Schmidt, S. Bidnyk, Yong-Hoon Cho, A. J. Fischer, J. J. Song, S. Keller, U. K. Mishra, and S. P. DenBaars, Appl. Phys. Lett. **73**, 3689 (1998), *and references therein.*
8. S. Bidnyk, T. J. Schmidt, B. D. Little, and J. J. Song, Appl. Phys. Lett. **74**, 1 (1999)
9. S. Bidnyk, T. J. Schmidt, Y. H. Cho, G. H. Gainer, J. J. Song, S. Keller, U. K. Mishra, and S. P. DenBaars, Appl. Phys. Lett. **72**, 1623 (1998)
10. A. J. Fischer, W. Shan, J. J. Song, Y. C. Chang, R. Horning, and B. Goldenberg, Appl. Phys. Lett. **71**, 1981 (1997)
11. I. Galbraith and S. W. Koch, J. Crystal Growth **159**, 667 (1996)
12. T. J. Schmidt, J. J. Song, Y. C. Chang, R. Horning, and B. Goldenberg, Appl. Phys. Lett. **72** (1998)
13. W. Shan, B. D. Little, A. J. Fischer, J. J. Song, B. Goldenberg, W. G. Perry, M. D. Bremser, and R. F. Davis, 16369 (1996)
14. R. Levy and J. B. Grun, Phys. Stat. Sol. (a) **22**, 11 (1974)
15. X. H. Yang, J. M. Hays, W. Shan, J. J. Song, and E. Cantwell, Appl. Phys. Lett. **62**, 1071 (1992)
16. H. Jeon, J. Ding, A. V. Nurmikko, H. Luo, N. Samarth, and J. K. Furdyna, Appl. Phys. Let. **57**, 2413 (1990)

ELECTRON TRANSPORT IN THE III-V NITRIDE ALLOYS

B. E. FOUTZ*, S. K. O'LEARY**, M. S. SHUR***, and L. F. EASTMAN*
* School of Electrical Engineering, Cornell University, Ithaca, New York 14853
** Faculty of Engineering, University of Regina, Regina, Saskatchewan, Canada S4S 0A2
*** Department of Electrical, Computer, and Systems Engineering, Rensselaer Polytechnic
Institute, Troy, New York 12180-3590

ABSTRACT

We study electron transport in the alloys of aluminum nitride and gallium nitride and
alloys of indium nitride and gallium nitride. In particular, employing Monte Carlo simulations we determine the velocity-field characteristics associated with these alloys for various
alloy compositions. We also determine the dependence of the low-field mobility on the alloy
composition. We find that while the low-field mobility is a strong function of the alloy composition, the peak and saturation drift velocities exhibit a more mild dependence. Transient
electron transport is also considered. We find that the velocity overshoot characteristic is
a strong function of the alloy composition. The device implications of these results are discussed.

INTRODUCTION

The III-V nitride semiconductors, gallium nitride (GaN), aluminum nitride (AlN), and
indium nitride (InN), offer considerable potential for electronic and optoelectronic device
applications [1]. Many nitride based devices employ alloys of these materials. For example,
alloys of AlN and GaN are used in field-effect transistors and photodetectors [2] and alloys
of AlN and GaN and alloys of InN and GaN are used in lasers [3]. These device applications
have fueled considerable interest in the fundamental properties these alloys, and thus these
alloys have been the focus of considerable attention in recent years.

In order to analyze and improve the design of electronic devices fabricated with alloys of
the III-V nitrides, a thorough understanding of the electron transport within these materials
is necessary. While electron transport in bulk GaN [4, 5, 6], AlN [7, 8], and InN [9, 10], has
been extensively examined, the transport of electrons in the III-V nitride alloys has yet to be
the focus of much attention. In a recent paper, however, Albrecht *et al.* [11] employed Monte
Carlo simulations to investigate electron transport in alloys of AlN and GaN. It was found
that while the low-field mobility decreases dramatically with the Al content, the saturation
drift velocity is found to be relatively insensitive.

In this paper, we further study electron transport in the alloys of the III-V nitrides. In
particular, using Monte Carlo simulations we study electron transport in both alloys of AlN
and GaN ($Al_xGa_{1-x}N$) and alloys of InN and GaN ($In_xGa_{1-x}N$). Steady-state and transient electron transport are considered in this analysis, and the device implications of these
results are considered.

SIMULATIONS

The approach adopted here is similar to that employed by Bhapkar and Shur for the
treatment of wurtzite GaN [6]. In particular, a three-valley model for the conduction band

445

Mat. Res. Soc. Symp. Proc. Vol. 572 © 1999 Materials Research Society

is employed. Non-parabolicity is considered in all valleys, the non-parabolicity being treated through the application of the Kane model. We assume that all donors are ionized, and that the free electron concentration is equal to the dopant concentration. For each simulation, the motion of one thousand electrons is examined. The scattering mechanisms considered are (1) ionized impurity, (2) polar optical phonon, (3) piezoelectric, and (4) acoustic deformation potential. Intervalley and alloy scattering are also considered, alloy scattering being incorporated using the expression of Ridley [12]. Electron degeneracy and electron screening effects are also accounted for.

The material parameters corresponding to GaN, AlN, and InN, are tabulated in Table I of Foutz et $al.$ [13]. The alloy material parameters are determined through linear interpolation. The alloy scattering potential, Δ, is nominally set to 1 eV [14]. For the purposes of our analysis, the GaN, AlN, and InN band structures of Lambrecht and Segall [15, 16, 17] are adopted, the corresponding band structural parameters being tabulated in Tables II, III, and IV of Foutz et $al.$ [13]. Owing to the uncertainty in the band structures, we ascribe an effective mass equal to the free electron mass to all of the upper conduction band valleys.

RESULTS

We first consider steady-state electron transport. In Figure 1, we plot the steady-state velocity-field characteristic associated with $Al_xGa_{1-x}N$ for various alloy compositions. For all cases, the temperature is set to 300 K and the doping concentration is set to 10^{17} cm^{-3}. We see that the velocity-field characteristic is changed rather dramatically as the alloy composition is varied. In particular, it is seen that the peak drift velocity diminishes, the field at which the peak in the velocity-field characteristic occurs increases, the low-field mobility decreases, and the sharpness of the peak softens as the Al content is increased. While Albrecht et $al.$ [11] found that the saturation drift velocity is insensitive to the alloy content, we find that the alloyed materials considered in Figure 1, $Al_{0.2}Ga_{0.8}N$ and $Al_{0.4}Ga_{0.6}N$, exhibit lower saturation drift velocities than that exhibited by either pure AlN or pure GaN. This is attributable to our considerably greater upper valley effective mass and alloy scattering potential selections [14]. The steady-state velocity-field characteristic associated with $In_xGa_{1-x}N$, for various alloy compositions, is depicted in Figure 2.

The dependence of the low-field mobility on the alloy composition, corresponding to both $Al_xGa_{1-x}N$ and $In_xGa_{1-x}N$, is depicted in Figure 3, these results being extracted from our low-field Monte Carlo simulations of electron transport. Once again, in all cases we set the temperature to 300 K and the doping concentration to 10^{17} cm^{-3}. As was observed by Albrecht et $al.$ [11], the low-field mobility is a strong function of the alloy composition. In particular, small amounts of Al in $Al_xGa_{1-x}N$ dramatically reduce the low-field mobility. Similarly, it has been found that small amounts of Ga added to InN dramatically reduce the corresponding low-field mobility; this cannot be seen from Figure 3, but can be inferred from the fact that pure InN is found to have a low-field mobility of 2900 cm^2/V·s. These results suggest that while alloy scattering plays a critical role in determining the low-field electron transport in these materials, it plays a less important role in determining high-field electron transport. Thus, III-V nitride alloys might be more suitable for short channel devices where the electric fields are higher and alloy scattering will not significantly degrade the device performance.

Steady-state electron transport is the dominant transport mechanism for devices with

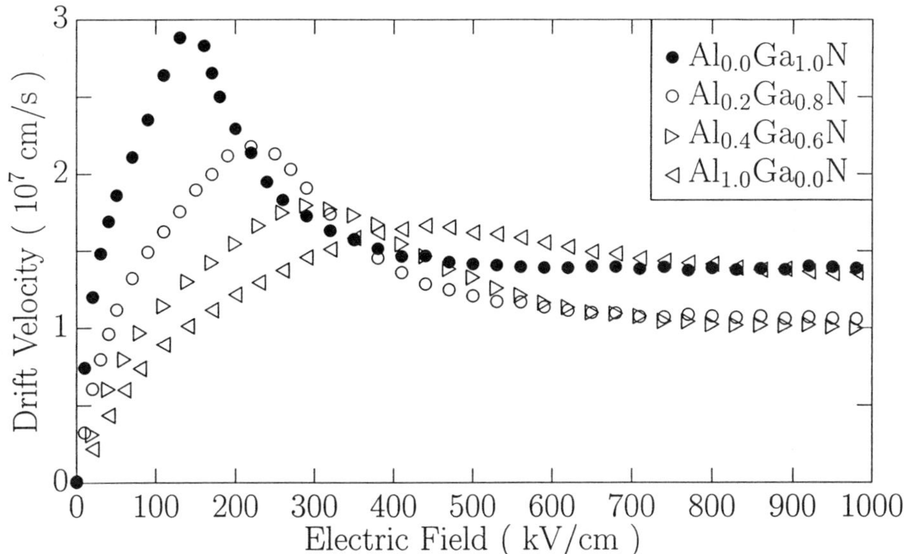

Figure 1: The velocity-field characteristic associated with $Al_xGa_{1-x}N$ for various alloy compositions. For all cases, the temperature is set to 300 K, the doping concentration is set to 10^{17} cm^{-3}, and the alloy scattering potential is set to 1 eV.

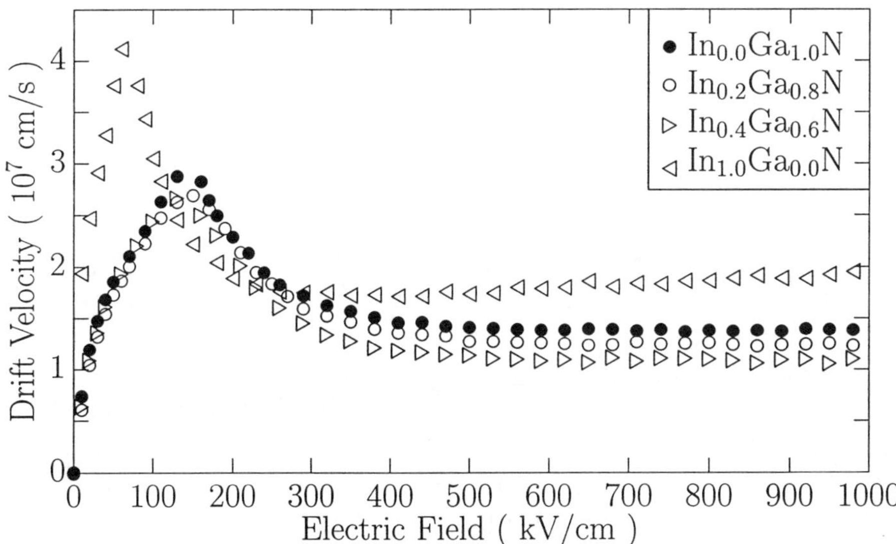

Figure 2: The velocity-field characteristic associated with $In_xGa_{1-x}N$ for various alloy compositions. For all cases, the temperature is set to 300 K, the doping concentration is set to 10^{17} cm^{-3}, and the alloy scattering potential is set to 1 eV.

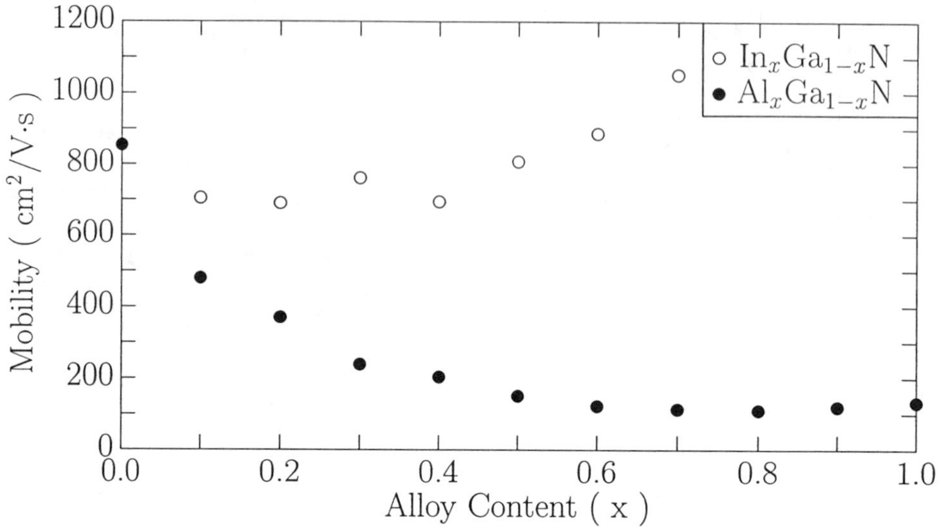

Figure 3: The low-field mobility for various alloy compositions for $Al_xGa_{1-x}N$ and $In_xGa_{1-x}N$, as determined from our low-field Monte Carlo simulations of electron transport. For all cases, the temperature is set to 300 K, the doping concentration is set to 10^{17} cm^{-3}, and the alloy scattering potential is set to 1 eV. For $In_xGa_{1-x}N$, the low-field mobility at x=0.8, 0.9, and 1.0, is found to be 1500, 2000, and 2900 cm^2/V·s. Note that the data points coincide for the case of x=0.0.

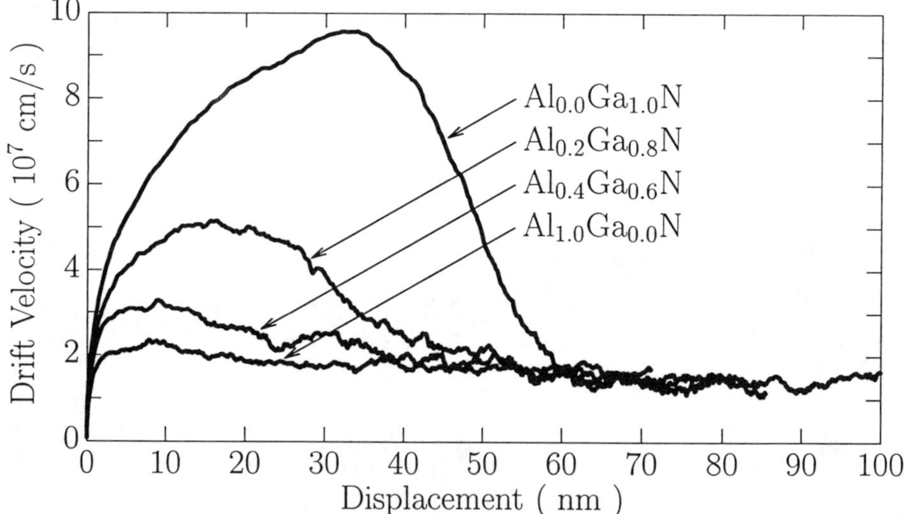

Figure 4: The average electron velocity as a function of displacement for various alloy compositions for $Al_xGa_{1-x}N$. For all cases, the temperature is set to 300 K, the doping concentration is set to 10^{17} cm^{-3}, and the alloy scattering potential is set to 1 eV. The applied electric field is 500 kV/cm in all cases.

larger dimensions. For devices with smaller dimensions, however, transient electron transport must also be considered when evaluating device performance. In studying transient electron transport in the III-V nitride alloys, we follow the approach of Foutz *et al.* [18] and consider the response of electrons to the sudden application of a constant electric field. We assume that the electrons are in equilibrium prior to the application of the electric field. In Figure 4, we plot the average electron velocity of electrons in $Al_xGa_{1-x}N$ as a function of displacement for various alloy compositions; we assume that the electrons are initially distributed about zero displacement. We have applied a constant electric field of 500 kV/cm, and set the temperature to 300 K and the doping concentration to 10^{17} cm^{-3} in all cases [19]. We see that the transient response varies dramatically with the Al content. In particular, while the peak average electron velocity of pure GaN overshoots, by a considerable margin, its corresponding steady-state velocity, this overshoot is dramatically reduced when the Al content is increased. This would suggest that electronic devices fabricated with the III-V nitrides should not use alloying if one wishes transient electron transport to enhance the resultant device performance.

CONCLUSIONS

In conclusion, we have studied steady-state and transient electron transport in the III-V nitride alloys. We have found that while the low-field mobility is a strong function of the alloy composition, the peak and saturation drift velocities exhibit a more mild dependence. Velocity overshoot effects are found to substantially diminish with alloy composition. These results suggest that III-V nitride alloyed devices are more suitable for short channel devices where the electric fields are high and the degradation in device performance due to alloy scattering is not significant. We have also found that III-V nitride based alloys should not be used in devices which hope to exploit transient electron transport effects.

ACKNOWLEDGEMENTS

The authors wish to thank the Office of Naval Research for financial support under their MURI program: Grant # N00014-96-1-1223; Project Monitor: J. C. Zolper. One of the authors (S. K. O.) gratefully acknowledges support from the Natural Sciences and Engineering Research Council of Canada.

REFERENCES

[1] S. N. Mohammad and H. Morkoç, Prog. Quant. Electron. **20**, 361 (1996).

[2] M. S. Shur and M. A. Khan, Mater. Res. Bull. **22** (2), 44 (1997).

[3] S. Nakamura, Mater. Res. Bull. **22** (2), 29 (1997).

[4] M. A. Littlejohn, J. R. Hauser, and T. H. Glisson, Appl. Phys. Lett. **26**, 625 (1975).

[5] M. Shur, B. Gelmont, and M. A. Khan, J. Electron. Mater. **25**, 777 (1996).

[6] U. V. Bhapkar and M. S. Shur, J. Appl. Phys. **82**, 1649 (1997).

[7] S. K. O'Leary, B. E. Foutz, M. S. Shur, U. V. Bhapkar, and L. F. Eastman, Solid State Commun. **105**, 621 (1998).

[8] J. D. Albrecht, R. P. Wang, P. P. Ruden, M. Farahmand, and K. F. Brennan, J. Appl. Phys. **83**, 1446 (1998).

[9] S. K. O'Leary, B. E. Foutz, M. S. Shur, U. V. Bhapkar, and L. F. Eastman, J. Appl. Phys. **83**, 826 (1998).

[10] E. Bellotti, B. K. Doshi, K. F. Brennan, J. D. Albrecht, and P. P. Ruden, J. Appl. Phys. **85**, 916 (1999).

[11] J. D. Albrecht, R. Wang, P. P. Ruden, M. Farahmand, E. Bellotti, and K. F. Brennan, Mater. Res. Symp. Proc. **482**, 815 (1998).

[12] B. K. Ridley, *Quantum Processes in Semiconductors* (Oxford, New York, 1982).

[13] B. E. Foutz, S. K. O'Leary, M. S. Shur, and L. F. Eastman, J. Appl. Phys. (in press).

[14] As was pointed out by Albrecht *et al.* [11], the selection of a specific numerical value for Δ may be the the subject of considerable debate. Albrecht *et al.* [11] chose $\Delta = 0.01$ eV for their nominal selection.

[15] W. R. L. Lambrecht and B. Segall, in *Properties of Group III Nitrides*, No. 11 *EMIS Datareviews Series*, edited by J. H. Edgar (Inspec, London, 1994), pg. 141.

[16] W. R. L. Lambrecht and B. Segall, in *Properties of Group III Nitrides*, No. 11 *EMIS Datareviews Series*, edited by J. H. Edgar (Inspec, London, 1994), pg. 135.

[17] W. R. L. Lambrecht and B. Segall, in *Properties of Group III Nitrides*, No. 11 *EMIS Datareviews Series*, edited by J. H. Edgar (Inspec, London, 1994), pg. 151.

[18] B. E. Foutz, L. F. Eastman, U. V. Bhapkar, and M. S. Shur, Appl. Phys. Lett. **70**, 2849 (1997).

[19] The analysis of Foutz *et al.* [13] demonstrates that velocity overshoot effects can occur over substantially enhanced distances when the electric field is lower. In particular, for an electric field selection of 210 kV/cm, Figure 2a of Foutz *et al.* [13] demonstrates that velocity overshoot effects in pure GaN are exhibited over distances in excess of 0.3 μm. We chose 500 kV/cm as AlN does not exhibit velocity overshoot until the electric field in excess of 450 kV/cm, as was demonstrated by Foutz *et al.* [13].

HIGH-QUALITY GaN GROWN BY
MOLECULAR BEAM EPITAXY ON Ge(001)

H. SIEGLE*,**, Y. KIM*,**, SUDHIR G. S. *,**, J. KRÜGER *,**, P. PERLIN**, J.
W. AGER III**, C. KISIELOWSKI***, E. R. WEBER*
*Department of Materials Science and Mineral Engineering, UC Berkeley, Berkeley,
CA 94720
** Lawrence Berkeley National Laboratory, Materials Science Division, Berkeley,
CA 94720
*** Lawrence Berkeley National Laboratory, National Center for Electron Microscopy,
Berkeley, California 94720

ABSTRACT

We report on growth of GaN on Germanium as an alternative substrate material. The
GaN films were deposited on Ge(001) substrates by plasma-assisted molecular beam
epitaxy. Atomic force microscopy, x-ray diffraction, photoluminescence, and Raman
spectroscopy were used to characterize the structural and optical properties of the films.
We observed that the Ga/N ratio plays a crucial role in determining the phase purity and
crystal quality. Under N-rich conditions the films were phase-mixed, containing cubic and
hexagonal GaN, while in the Ga-rich regime they were purily hexagonal. The latter
samples show bandedge luminescence with linewidths as small as 31 meV at low
temperatures.

INTRODUCTION

The availability of suitable substrate materials is a key requirement for the epitaxial
growth of GaN thin films. The lack of GaN substrates of appropriate size and structural
perfection has motivated the search for alternative substrates for heteroepitaxial growth.
The most common substrates, sapphire, SiC and GaAs, exhibit large differences in lattice
constants and thermal expansion coefficients to GaN, resulting in large amounts of stress
and high defect concentrations in the epilayers. [1,2] Ge in contrast possess the same
thermal expansion coefficient as GaN. [3] Since most of the stress in heteroepitaxial GaN
layers arises from the ill-matched expansion coefficients, growth on Ge would lower the
amount of stress in the GaN epilayers and thus improve their structural, electrical, and
optical quality. Besides, Ge can easily be doped and therefore could serve as a
backside/bottom contact.

Mat. Res. Soc. Symp. Proc. Vol. 572 © 1999 Materials Research Society

We demonstrate that Ge provide an alternative substrate material for GaN growth. The GaN layers were deposited on Ge(001) substrates by plasma-assisted molecular beam epitaxy (MBE). Their structural and optical properties were characterized by atomic force microscopy, x-ray diffraction, Raman and Photoluminescence spectroscopy.

EXPERIMENT

A Riber 1000 MBE system was used for the epitaxial growth. An effusion cell provided elemental gallium by evaporation and activated nitrogen was produced by a constricted glow discharge (CGD) plasma source with pure nitrogen gas (99.9995%). Details of the plasma source are given elsewhere. [4]

Intrinsic Ge (001) wafers with miscuts of 6 degree toward <110> served as substrates. Prior growth they were degreased by boiling in acetone and isopropyl alcohol and were finally rinsed in deionized water to remove Ge oxides. The substrates were then heated up to 680°C in ultra high vacuum for thermal desorption of surface contaminants. In contrast to growth on sapphire, the Ge substrates were not nitridated. Subsequently, a GaN buffer layer was deposited on the substrate at 600°C for 5 minutes. Finally, the GaN main layer were grown for 2 hours at 680°C. The Ga/N flux ratio at the deposition of the buffer as well as the main layer was varied by changing the nitrogen flow rate while the Ga cell temperature was kept constant at 880°C. We found best results with 5 sccm N flow at a background pressure of 3.5 mTorr for the buffer layer growth and 35 sccm for the main layer growth. The resulting thickness of the GaN epilayer was about 1 μm.

The surface morphology was tested using contact-mode atomic force microscopy (AFM). The full width at half maximum (FWHM) of the rocking curve and the a- and c-lattice parameters were measured with a Siemens D-5000 diffractometer containing a four-bounce Ge monochromator. Room-temperature micro-Raman spectra were recorded with the 488 nm line of an Ar^{2+} laser in backscattering geometry. A notch-filter suppressed elastically scattered light while the Raman signal was analyzed by a single 0.5 m spectrometer equipped with a charged coupled device detector. To get information on the layers' optical properties we performed low-temperature photoluminescence (PL) spectroscopy. PL at 4 K was excited by a 50 mW HeCd laser, diffracted by a 0.85 m double-grating monochromator and detected by a UV-sensitive photomultiplier. Secondary ion mass spectroscopy was carried out on selected layers to measure their compositions.

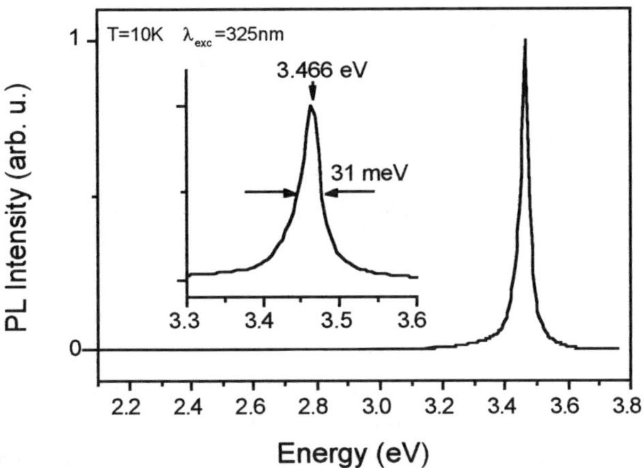

Fig. 1: *Typical low-temperature PL spectrum of GaN grown on Ge after excitation at 325 nm (3.81 eV). The near-bandgap luminescence exhibits a relatively narrow linewidth of 31 meV and peaks at the stress-free position of 3.466 eV.*

RESULTS AND DISCUSSION

Figure 1 displays a typical PL spectrum of the GaN layers grown on Ge substrates taken at low temperatures after excitation at 325 nm (3.81 eV). The spectrum is dominated by the hexagonal near-bandgap, excitonic luminescence at 3.466 eV. No cubic PL can be observed indicating high hexagonal phase purity. Defect-related broad bands, as e.g. the yellow luminescence, do not appear in the spectrum either. Along with the narrow linewidth of 31 meV, this demonstrates the good optical quality of the layers which is superior to growth on silicon with best linewidths of around 40 meV [5] and 100 meV [6], respectively. The PL line position at 3.466 eV corresponds exactly to the stress-free value [1].

As can be seen from Fig. 2 which shows an atomic-force

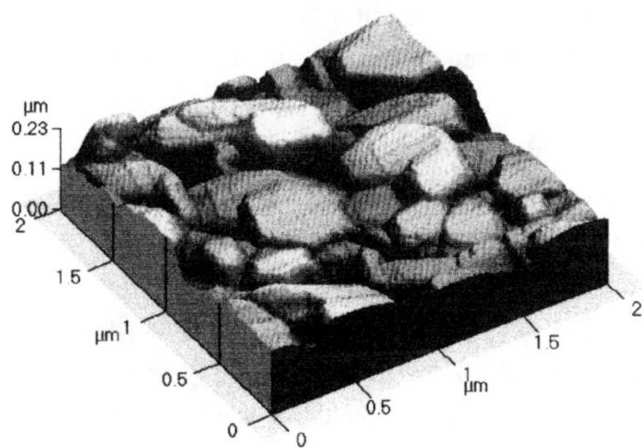

Fig. 2: *AFM picture of a 2×2 μm² area of GaN/Ge. The root-mean-square roughness is 20.6 nm.*

microscopy (AFM) image taken of a 2×2 μm^2 area, the growth mode is three-dimensional, typical for MBE-GaN because of the low growth temperature. We determined a root-mean-square (rms) roughness of 20.6 nm. This value is in the range we normally observe from GaN deposited on sapphire, [7].

To get information about the structural properties of our layers we performed micro-Raman spectroscopy analysis. This technique allows us to determine the stress and the free-carrier concentrations in the layers and gives us at least a qualitative impression of the crystalline quality by measuring the linewidths of the Raman modes and by looking for disorder-activated scattering. For details about the use of Raman spectroscopy as a characterization tool in general, the reader is referred to [8].

In Fig. 3a we plot a typical Raman spectrum taken in $z(..)\underline{z}$ geometry of an 1 μm thick layer grown under optimized Ga-rich conditions. The spectrum is dominated by the hexagonal E_2(high) mode, which is the only mode allowed in this scattering geometry assuming the c-axis to be perpendicular to the surface. We found linewidths in the range of 5 cm^{-1} comparable to the values known from hexagonal GaN deposited on sapphire, [9]. The absence of the A_1(LO) mode indicates a very high free-carrier concentration in the layer exceeding $1\times10^{19} cm^{-3}$ [10]. Beside point defects, which might be caused by non-perfect growth conditions, we expected the incorporation of Ge atoms in-diffusing from the substrate during the growth process to be the main cause for the high free carrier concentration. Similar to Si, Ge is a donor in GaN. To verify our assumption we performed SIMS measurements on selected GaN layers. The SIMS profiles confirmed the indiffusion of Ge into the GaN epilayers.

The E_2(high) Raman mode reacts sensitively on pressure [11]. Since it is a nonpolar mode its frequency neither depends on the propagation direction relative to the c-axis, as observed from the A_1 and E_1 modes forming the so-called quasi modes [12] nor does the mode shifts due to interaction with free carriers. Therefore, its frequency is a direct measure of stress in the layer. We found in all our layers the E_2 to be located at the stress-free frequency of 567 cm^{-1} [13] in agreement with our PL results presented in Fig. 1. Our x-ray measurements yield similar results. The lattice parameter has been determined to 5.1748 Å.

Similar to the growth of GaN on GaAs [14,15] we observed that the Ga/N flux ratio plays a crucial role in determining the phase purity of the layers. When growing in the Ga-rich regime we found purily hexagonal GaN layers, as displayed in the PL and Raman spectra of Figs. 1 and 3. With N-rich conditions the layers become phase mixed, containing both cubic 3C-GaN as well as hexagonal 2H-GaN. On the right side of Fig. 3, a Raman spectrum of such a phase-mixed layer is shown. One can clearly see the cubic TO mode beside the hexagonal E_2 mode. Similar results were found in the PL spectra, in

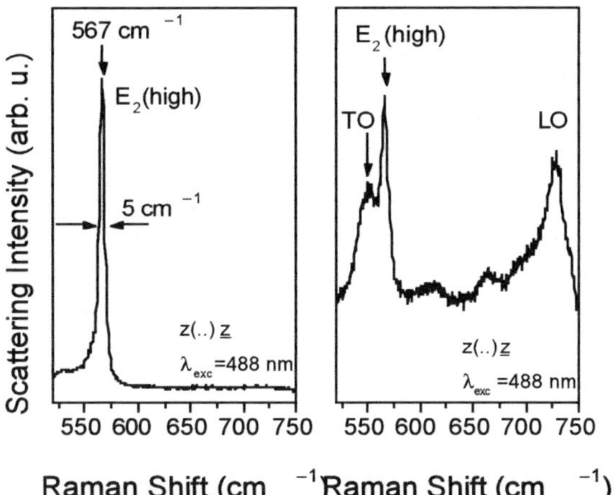

Fig. 3a,b: *Typical room-temperature Raman spectrum taken in z(..)$\underline{z}$ geometry of an 1 µm thick GaN/Ge layer grown under optimized Ga-rich conditions (left, Fig. 3a) and of a phase-mixed layer (right, Fig. 3b).*

which 200 meV below the hexagonal near-bandgap luminescence the corresponding cubic PL occurred. Together with the coexistence of both phases in the layers the crystal quality deteriorates, indicated by broader Raman lines and the appearance of disorder-activated Raman scattering [16]. The absence of long-range order in disordered materials yields to a breakdown of the $q = \Delta k \approx 0$ selection, where q is the wavevector of the phonon and Δk the difference between incident and scattered photon wavevector [17]. Thus, the Raman spectrum reflects the vibrational density of states.

SUMMARY

Summarizing our results, we achieved growth of high-quality GaN on Ge(001) substrates. Optimization of the growth conditions resulted in stress-free GaN layers, as indicated by PL and Raman spectroscopy. We found a strong dependency of the material's quality on the Ga/N flux rate. When grown under Ga-rich conditions, the layer exhibit a comparatively high optical quality, as indicated by the near-bandedge luminescence linewidth of 31 meV and the absence of defect-related bands. Under N-rich growth conditions, the GaN layers are comprised of both the hexagonal and the cubic phase. In all cases, the layers exhibit a rather large free electron density which has to be attributed to the in-diffusion of Ge from the substrate. Also, the current crystalline quality calls for further optimization of the growth conditions.

ACKNOWLEDGMENTS

The authors thank M. Straßburg for carrying out the SIMS measurements. One of the authors (H.S.) acknowledges the support of a DAAD fellowship. This work was supported by the Office of Energy Research, Office of Basic Energy Sciences, Division of Materials Sciences of the U.S. Department of Energy under Contract No. DE-AC03-76SF00098.

REFERENCES

1. C. Kisielowski, J. Krüger, S. Ruvimov, T. Suski, J. W. Ager III, E. Jones, Z. Lilienthal-Weber, M. Rubin, E. R. Weber, M. D. Bremser, R. F. Davis, Phys. Rev. B **54**, 17745 (1996)
2. J. Krüger, N.Shapiro, S.Subramanya, Y.Kim, H.Siegle, P.Perlin, E,R.Weber, W.S. Wong, T.Sands, N.W.Cheung, R. J. Molnar, this MRS meeting
3. Landolt-Börnstein Tables, edited by O. Madelung, M. Schulz, and H. Weiss (Springer, Berlin 1982)
4. A. Anders, N. Newman, M. Rubin, M. Dickinson, E. Jones, P. Phatak, and A. Gassmann, Rev. Sci. Instrum. **67**, 905 (1996)
5. Y. Nakada, I. Aksenov, H. Okumura, Appl. Phys. Lett. **73**, 827 (1998)
6. B. Yang, A. Trampert, O. Brandt, B. Jenichen, and K. H. Ploog, J. Appl. Phys. **83**, 3800 (1998)
7. H. Fujii, C. Kisielowski, J. Krüger, R. Klockenbrink, M. S. H. Leung, Sudhir G. S., H. Sohn, M. Rubin, and E. R. Weber; Mat.Res.Soc.Symp.Proc. **449**, 227 (1997)
8. Light Scattering in Solids I-VI, edited by M. Cardona and G. Güntherodt, Topics Appl. Phys. (Springer, Berlin, Heidelberg)
9. H. Siegle, L. Eckey, A. Hoffmann, C. Thomsen, B. K. Meyer, D. Schikora, M. Hankeln, K. Lischka, Solid State Communications **96**, 943 (1995)
10. H. Harima, T. Inoue, S. Nakashima, H. Okumura, Y. Ishida, S. Yoshida, H. Hamaguchi, Paper presented on the ICNS'97, October 27-31, 1997, Tokushima, Japan
11. H. Siegle, A. R. Goñi, C. Thomsen, C. Ulrich, K. Syassen, B. Schöttker, D. J. As, D. Schikora, Materials Research Society, Symposium Proceedings Vol. 468, ed. C. R. Abernathy, H. Amano, J. C. Zolper, pp. 225 (1997)
12. L. Filippidis, H. Siegle, A. Hoffmann, C. Thomsen, K. Karch, and F. Bechstedt, phys. stat. sol. (b) **198**, 621 (1996)
13. H. Siegle, A. Hoffmann, L. Eckey, C. Thomsen, J. Christen, F. Bertram, B. Schmidt, K. Hiramatsu, Appl. Phys. Lett. **71**, 2490 (1997)
14. O. Brandt, H. Yang, B. Jenichen, Y. Suzuki, L. Däweritz, and K. H. Ploog, Phys. Rev. B **52**, R2253 (1995)
15. D. Schikora, M. Hankeln, D. J. As, K. Lischka, T. Litz, A. Waag, T. Buhrow, F. Henneberger, Phys. Rev. B **54**, R8381 (1996)
16. H. Siegle, G. Kaczmarczyk, L. Filippidis, P. Thurian, A. Hoffmann, C. Thomsen, Zeitschrift für Physikalische Chemie **200**, 187 (1997)
17. M. H. Brodsky, in Light Scattering in Solids I, edited by M. Cardona (Springer, Berlin, 1975), pp. 205

CARRIER RECOMBINATION DYNAMICS OF $Al_xGa_{1-x}N$ EPILAYERS GROWN BY MOCVD

Yong-Hoon Cho*, G. H. Gainer*, J. B. Lam*, J. J. Song*, W. Yang**, and S. A. McPherson**

* Center for Laser and Photonics Research and Department of Physics
  Oklahoma State University, Stillwater, OK 74078
**Honeywell Technology Center, 12001 State Highway 55, Plymouth, MN 55441

ABSTRACT

We present a comprehensive study of the optical characteristics of $Al_xGa_{1-x}N$ epilayers by means of photoluminescence (PL), PL excitation, and time-resolved PL spectroscopy. All $Al_xGa_{1-x}N$ epilayers were grown by metalorganic chemical vapor deposition and the Al mole fraction (x) was varied from 0 to 0.6. We observed that (i) the full width at half maximum of the PL emission, (ii) the energy difference between the PL emission peak energy and the PLE absorption edge, and (iii) the effective lifetime increase with increasing x. These facts indicate that degree of band-gap fluctuation due to a spatially inhomogeneous Al alloy content distribution increases with increasing x. We observed anomalous temperature-induced emission shift behavior for $Al_xGa_{1-x}N$ epilayers, specifically, an S-shaped (decrease-increase-decrease) temperature dependence of the peak energy with increasing temperature. This anomalous temperature-dependent emission behavior was enhanced as the Al mole fraction was increased. Since the band-gap fluctuation in $Al_xGa_{1-x}N$ epilayers due to inhomogeneous spatial variations of the Al content increases with increasing Al content, we believe that band-gap fluctuation causes the PL peak energy to deviate from the typical temperature dependence of the energy gap shrinkage. Therefore, the anomalous temperature-induced emission shift can be attributed to energy tail states due to alloy potential inhomogeneities in the $Al_xGa_{1-x}N$ epilayers with large Al content.

INTRODUCTION

Much interest has been focused on III-V nitride compound semiconductors and their heterostructures due to their potential applications such as short-wavelength light emitting devices [1,2] solar-blind ultraviolet detectors [3], and high power and high temperature devices [4,5]. In particular, the ternary compound $Al_xGa_{1-x}N$ has the potential for use in light emitting and detecting devices covering nearly the entire deep-ultraviolet (UV) region of the spectrum (190 – 350 nm). The recombination mechanism for InGaN-based structures has been widely investigated by several authors and detailed emission properties from localized states was discussed for both spontaneous and stimulated emission of InGaN/GaN quantum structures [6,7]. However, the detailed spontaneous emission properties of the AlGaN-based structures have not yet been investigated.

In this work, we report the results on optical properties of $Al_xGa_{1-x}N$ epilayers as a function of Al content x ($0 \leq x \leq 0.6$) by means of photoluminescence (PL), PL excitation (PLE), and time-resolved PL (TRPL) spectroscopy. By studying alloy epilayers, we can avoid (or minimize) other ambiguous effects such as strain-induced piezoelectric polarization, quantum confinement, layer thickness variations, and interface-related defects usually involved in the case of quantum structures. We observed that the full width at half maximum (FWHM) of the PL emission, the energy difference between the PL emission peak energy and the PLE absorption

Mat. Res. Soc. Symp. Proc. Vol. 572 © 1999 Materials Research Society

edge, and the lifetime increase with increasing x from 0.17 to 0.60, indicating that the degree of Al alloy potential fluctuations increases with increasing x. From the temperature-dependent PL measurements, we observed that the PL emission from $Al_xGa_{1-x}N$ with high x did not follow the typical temperature dependence of the energy gap shrinkage.

EXPERIMENT

The $Al_xGa_{1-x}N$ epilayers used in this work were grown by low-pressure metalorganic chemical vapor deposition (MOCVD) on (0001) oriented sapphire at a growth temperature of 1050 °C. Prior to $Al_xGa_{1-x}N$ growth, a thin ~ 5-nm-thick AlN buffer layer was deposited on the sapphire at a temperature of 625 °C. Triethylgallium, triethylaluminum, and ammonia were used as precursors in the $Al_xGa_{1-x}N$ growth. The $Al_xGa_{1-x}N$ layer thickness was about 1 μm. In order to evaluate the Al alloy composition of the $Al_xGa_{1-x}N$ thin films, the samples were analyzed with a high-resolution x-ray diffractometer using Cu $K\alpha_1$ radiation. The angular distances between the $Al_xGa_{1-x}N$ and AlN (or GaN) peaks were obtained by ω-2θ scans. PL experiments were performed using the 244 nm line of an intracavity doubled cw Ar^+ laser as an excitation source. PL and PLE experiments were also carried out using the quasimonochromatic light emission from a Xe lamp dispersed by a ½ m monochromator as the excitation source. Both the PL and PLE experiments used a photomultiplier tube in conjunction with a 1 m double spectrometer as a

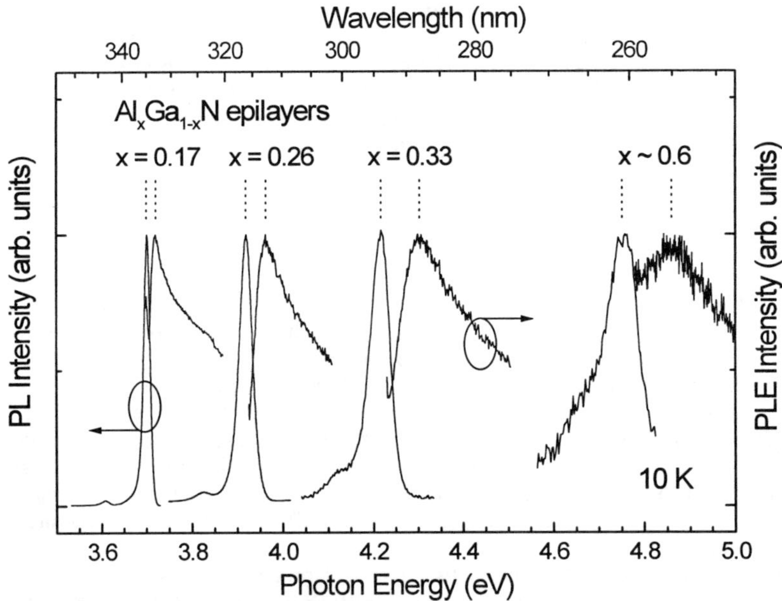

Figure 1. 10K PL and PLE spectra for $Al_xGa_{1-x}N$ epilayers with Al content x = 0.17, 0.26, 0.33, and 0.60. The PL spectra were measured using second-order diffraction through a monochromator and all the spectra were normalized. Note that the Stokes shift (the energy difference between the PL emission peak energy and the PLE absorption edge) increases with increasing x, indicating that the degree of potential fluctuations increases with increasing x.

detector. TRPL measurements were carried out using a picosecond pulsed laser system consisting of a cavity-dumped dye laser synchronously pumped by a frequency-doubled modelocked Nd:YAG laser for sample excitation and a streak camera for detection. We observed room temperature UV stimulated emission from optically pumped $Al_xGa_{1-x}N$ epilayers with Al content as high as 26 %. Detailed stimulated emission properties were reported elsewhere [8].

RESULTS AND DISCUSSIONS

Figure 1 shows 10 K PL and PLE spectra for $Al_xGa_{1-x}N$ epilayers with Al content $x = 0.17$, 0.26, 0.33, and 0.60 with PL peak energies of ~ 3.70, 3.92, 4.22, and 4.76 eV, respectively. The PL spectra were measured using second-order diffraction through a monochromator and all the spectra were normalized. The decrease in PLE signal above the PLE peak position with increasing excitation energy is due to the decrease in the excitation intensity of a Xe lamp source. Note that the FWHM of the PL emission and the Stokes shift (the energy difference between the PL emission peak energy and the PLE absorption edge) monotonically increase with varying x from 0 to 0.60, as shown in Fig. 1.

TRPL and time-integrated PL (TIPL) results measured at 10 K are shown in Fig. 2 for GaN, $Al_{0.17}Ga_{0.83}N$, and $Al_{0.33}Ga_{0.67}N$ epilayers. In the case of the GaN epilayer, the lifetime of the free exciton (FX) is about 30 ps while that of the bound exciton (BX) is about 40 ps. In the case of both the $Al_{0.17}Ga_{0.83}N$, and $Al_{0.33}Ga_{0.67}N$ epilayers, in contrast, the measured lifetime increases with decreasing emission energy, and hence, the peak energy of the emission shifts to the low energy side as time proceeds. This behavior is most likely due to alloy potential fluctuations. We observed that the FWHM is about 4, 17, and 48 meV and the overall lifetime is about 30, 250, and 450 ps for the GaN, $Al_{0.17}Ga_{0.83}N$, and $Al_{0.33}Ga_{0.67}N$ epilayers, respectively. These facts indicate that the alloy potential fluctuation of the $Al_{0.33}Ga_{0.67}N$ epilayer is larger than that of the $Al_{0.17}Ga_{0.83}N$ epilayer. These facts obtained from Figs. 1 and 2 indicate that the degree

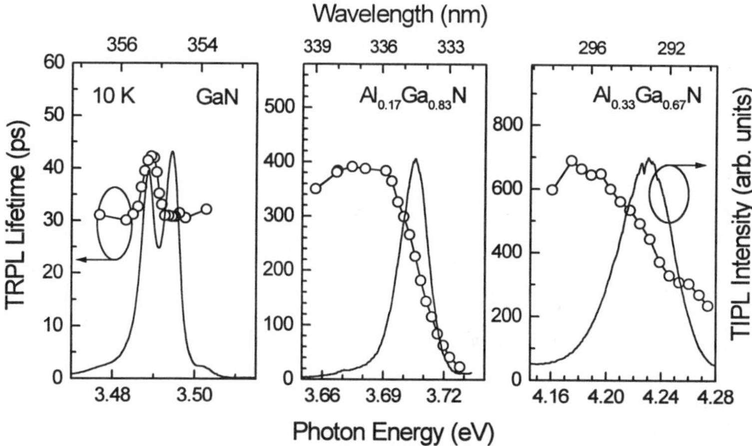

Figure. 2. 10 K time-resolved and time-integrated PL results for (a) GaN, (b) $Al_{0.17}Ga_{0.83}N$, and (c) $Al_{0.33}Ga_{0.67}N$ epilayers. A PL FWHM of about 4, 17, and 48 meV and an overall lifetime of about 35, 230, and 450 ps were obtained for GaN, $Al_{0.17}Ga_{0.83}N$, and $Al_{0.33}Ga_{0.67}N$ epilayers, respectively.

of Al alloy potential fluctuations increases with increasing x.

Generally, the PL peak energy (E_{PL}) follows the well-known temperature dependence of the energy gap shrinkage: $E_g(T) = E_g(0) - \alpha T^2/(\beta + T)$, where $E_g(T)$ is the band-gap transition energy at temperature T, and α and β are known as the Varshni thermal coefficients [9]. The parameters $\alpha = 8.32 \times 10^{-4}$ eV/K and $\beta = 835.6$ K for the GaN Γ_9^V - Γ_7^C transition were previously extracted by photoreflectance studies [10]. The temperature-dependent PL peak shift for the GaN and $Al_xGa_{1-x}N$ layers with small x value ($x < 0.1$) was consistent with the estimated energy decrease, while the PL emission from $Al_xGa_{1-x}N$ with higher x value did not follow the typical temperature dependence of the energy gap shrinkage. The anomalous behavior of photon energy as a function of temperature in the $Al_{0.17}Ga_{0.83}N$ and $Al_{0.33}Ga_{0.67}N$ epilayers is shown in Fig. 3. For the $Al_{0.17}Ga_{0.83}N$ ($Al_{0.33}Ga_{0.67}N$) epilayer, with increasing temperature up to 20 (90) K (region I), an initial small decrease in E_{PL} was observed, followed by an increase in E_{PL} in the temperature range of 20 – 70 K (90 – 150 K) (region II) and finally a decrease again in the temperature above 70 (150) K (region III). In the case of the $Al_{0.33}Ga_{0.67}N$ epilayer, E_{PL} more clearly shows the S-shaped (redshift-blueshift-redshift) behavior with increasing temperature. As

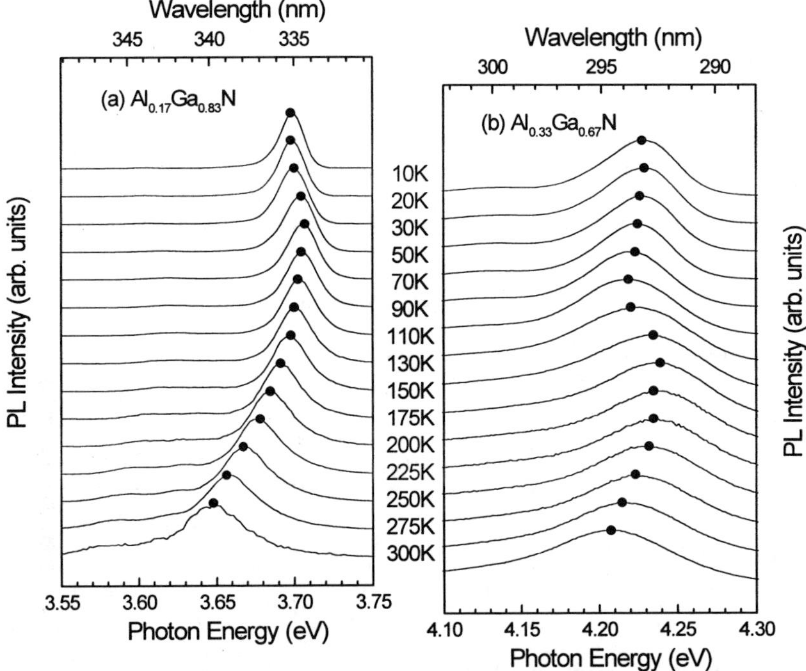

Figure 3. Evolution of the PL spectra for the (a) $Al_{0.17}Ga_{0.83}N$ and (b) $Al_{0.33}Ga_{0.67}N$ epilayers over a temperature range from 10 to 300 K. The main emission peak of both samples (closed circles) shows an S-shaped shift with increasing temperature. All spectra are normalized and shifted in the vertical direction for clarity. For the $Al_{0.17}Ga_{0.83}N$ ($Al_{0.33}Ga_{0.67}N$) epilayer, with increasing temperature up to 20 (90) K, an initial small decrease in peak energy was observed, followed by an increase in energy in the temperature range of 20 – 70 K (90 – 150 K) and finally a decrease again in the temperature above 70 (150) K.

the temperature increases from 10 to 90 K for the $Al_{0.33}Ga_{0.67}N$ epilayer, E_{PL} redshifts 8.6 meV. This value is similar to the expected band-gap shrinkage of ~ 7.2 meV for the GaN epilayer over this temperature range [10]. For a further increase in temperature from 90 to 150 K, the PL peak blueshifts 20.2 meV for the $Al_{0.33}Ga_{0.67}N$ epilayer. If we consider the temperature-induced band-gap shrinkage of ~ 11.7 meV expected for the case of GaN, the actual blueshift of the PL peak with respect to the band-edge is about 31.9 meV over this temperature range. When the temperature is further increased above 150 K, the peak position redshifts again. From the observed redshift of 31.7 meV combined with the expected band-gap shrinkage of ~ 47 meV between 150 and 300 K, we estimate an actual blueshift of the PL peak relative to the band-edge to be about 15.3 meV in this temperature range.

Consequently, the PL emission in region I exhibits a characteristic luminescence behavior from lower energy tail states caused by an inhomogeneous spatial band-gap distribution in the alloy materials. In region II, the carriers populate higher-energy tail states. In region III, the emission is due to thermally populated carriers at higher-energy states and follows the typical temperature dependence of the energy gap again. We note that the PL peak energy deviation from the $E_g(0) - \alpha T^2/(\beta + T)$ function and the temperature ranges of regions I and II increase with increasing Al content of the $Al_xGa_{1-x}N$ epilayers. Since the band-gap fluctuation in $Al_xGa_{1-x}N$ epilayers due to alloy potential variations of the Al content increases with increasing Al content, we believe that band-gap fluctuation causes the PL peak energy to deviate from the typical temperature dependence of the energy gap shrinkage. Therefore, the temperature-induced emission shift observed in this work for $Al_xGa_{1-x}N$ epilayers with large Al content is attributed to energy tail states due to inhomogeneous alloy potential fluctuations. We note, however, that we did not observe an actual "effective" redshift with respect to the fundamental energy gap in temperature region I, in contrast to the case of InGaN/GaN multiple quantum wells (MQWs) [11]. In the InGaN/GaN MQWs, the redshift energy was about five times larger than the expected band-gap shrinkage over the temperature range I and was attributed to the carrier relaxation process into the localized states in the InGaN/GaN MQWs [11]. Further detailed studies are in progress to clarify the relationship between the observed anomalous temperature-induced emission shift and temperature-dependent carrier dynamics in the $Al_xGa_{1-x}N$ alloy.

CONCLUSIONS

We have investigated the optical characteristics of MOCVD-grown $Al_xGa_{1-x}N$ ($0 \le x \le 0.6$) epilayers by means of PL, PLE, and TRPL spectroscopy. We observed that (i) the FWHM of the PL emission, (ii) the Stokes shift between the $Al_xGa_{1-x}N$ PL peak energy and the band-edge obtained from PLE spectra, and (iii) the effective lifetime increase with increasing x. These facts indicate that the degree of band-gap fluctuation due to a spatially inhomogeneous Al alloy content distribution increases with increasing x. We observed anomalous temperature-induced emission shift behavior for $Al_xGa_{1-x}N$ epilayers, specifically, an S-shaped (decrease-increase-decrease) temperature dependence of the peak energy with increasing temperature. Both the PL peak energy deviation from the typical energy gap temperature dependence and the temperature range in which the anomalous emission shift occurs increase with increasing Al content of the $Al_xGa_{1-x}N$ epilayers. This is attributed to spatially inhomogeneous Al content variations causing more band-gap fluctuation with increasing Al content. Therefore, the anomalous temperature-induced emission shift is related to thermal population of energy tail states due to alloy potential inhomogeneities in the $Al_xGa_{1-x}N$ epilayers.

ACKNOWLEDGMENTS

This work was supported by AFOSR, BMDO, ONR, and DARPA.

REFERENCES

1. S. Nakamura, M. Senoh, N. Iwasa, S. Nagahama, T. Yamada, and T. Mukai, Jpn. J. Appl. Phys. **34**, L1332 (1995).

2. S. Nakamura, M. Senoh, S. Nagahama, N. Iwasa, T. Yamada, T. Matsushita, Y. Sugimoto, and H. Kiyoku, Appl. Phys. Lett. **69**, 4056 (1996).

3. B. W. Lim, Q. C. Chen, J. Y. Yang, and M. A. Khan, Appl. Phys. Lett. **68**, 3761 (1996).

4. Y. F. Wu, B. P. Keller, S. Keller, D. Kapolnek, P. Kozodoy, S. P. DenBaars, and U. K. Mishra, Appl. Phys. Lett. **69**, 1438 (1996).

5. X. H. Yang, T. J. Schmidt, W. Shan, J. J. Song, and B. Goldenberg, Appl. Phys. Lett. **66**, 1 (1995).

6. Y. H. Cho, G. H. Gainer, A. J. Fischer, J. J. Song, S. Keller, U. K. Mishra, and S. P. Denbaars, Appl. Phys. Lett. **73**, 1370 (1998).

7. T. J. Schmidt, Y. H. Cho, G. H. Gainer, J. J. Song, S. Keller, U. K. Mishra, and S. P. Denbaars, Appl. Phys. Lett. **73**, 560 (1998).

8. T. J. Schmidt, Y. H. Cho, J. J. Song, and W. Yang, Appl. Phys. Lett. **74**, 245 (1999).

9. Y. P. Varshni, Physica **34**, 149 (1967).

10. W. Shan, T. J. Schmidt, X. H. Yang, S. J. Hwang, J. J. Song, and B. Goldenberg, Appl. Phys. Lett. **66**, 985 (1995).

11. Y. H. Cho, G. H. Gainer, A. J. Fischer, J. J. Song, S. Keller, U. K. Mishra, and S. P. DenBaars, Appl. Phys. Lett. **73**, 1370 (1998).

COMPARATIVE STUDY OF GAN GROWTH PROCESS BY MOVPE

Jingxi Sun, J.M. Redwing*, T. F. Kuech
Department of Chemical Engineering, University of Wisconsin-Madison, Madison, WI53706
*Epitronics, Phoenix, AZ85027

ABSTRACT

A comparative study of two different MOVPE reactors used for GaN growth is presented. Computational fluid dynamics (CFD) was used to determine common gas phase and fluid flow behaviors within these reactors. This paper focuses on the common thermal fluid features of these two MOVPE reactors with different geometries and operating pressures that can grow device-quality GaN-based materials. Our study clearly shows that several growth conditions must be achieved in order to grow high quality GaN materials. The high-temperature gas flow zone must be limited to a very thin flow sheet above the susceptor, while the bulk gas phase temperature must be very low to prevent extensive pre-deposition reactions. These conditions lead to higher growth rates and improved material quality. A certain range of gas flow velocity inside the high-temperature gas flow zone is also required in order to minimize the residence time and improve the growth uniformity. These conditions can be achieved by the use of either a novel reactor structure such as a two-flow approach or by specific flow conditions. The quantitative ranges of flow velocities, gas phase temperature, and residence time required in these reactors to achieve high quality material and uniform growth are given.

INTRODUCTION

Gallium nitride (GaN)-based materials have been used successfully for fabricating short-wavelength optoelectronic as well as high-temperature and high-power devices. Metalorganic vapor phase epitaxy (MOVPE) is the primary technique to grow these thin films and related devices. However, the growth of GaN-based materials has been problematic. High growth temperatures, gas phase pre-deposition reactions, and subsequent gas flow complexity complicate the MOVPE growth of GaN. These complicating factors have impeded the design and scale-up of MOVPE reactors for GaN-based material growth. The current understanding of GaN MOVPE growth is limited to the influence of specific reactor geometries and growth conditions on the growth rate and uniformity. The understanding of the dominant reaction pathways and their interaction with transport phenomena has been insufficient for design and optimization of MOVPE reactors for GaN growth.

In order to investigate the growth mechanism, the MOVPE of GaN has been extensively studied using various numerical modeling techniques [1]. Based on theoretical calculation and analysis of experimental data, kinetic mechanisms of differing complexity, which include a number of gas phase and surface reactions, have been proposed in these modeling studies. Typically, the proposed mechanism is substantiated by showing the consistency with a very limited set of experimental observations, usually growth rate data. Due to the complexity of the proposed growth mechanism and very limited growth-based data available for comparison, it is very difficult for these modeling investigations to provide general information which can be directly applied to practical technology, such as the design, scale-up and optimization of reactors capable of producing deposition rate and composition uniformity necessary for fabrication of various GaN-based devices.

Mat. Res. Soc. Symp. Proc. Vol. 572 © 1999 Materials Research Society

In this study, we studied the MOVPE of GaN-based materials by employing an alternative modeling strategy. Since trimethylgallium (TMG) is very dilute in the $NH_3 : H_2$ mixture, the gas phase reactions have almost no affect on the thermal fluid flow behaviors. However, the growth chemistry pathways are significantly determined by the thermal flow conditions, and can lead to deleterious pre-deposition reactions for GaN growth. Therefore, we performed, without any gas and surface reactions involved, a comparative study of the thermal flow behavior within two different reactors that can grow device quality GaN-based materials. We identified some common features of thermal flow inside these two different reactors. Based on these thermal flow features, we are further studying the growth chemistry in a step-wise fashion, with an increasing chemical complexity in the gas phase at each step, in an effort to discover how these observed common thermal flow features determine the growth pathways. Compared to a typical growth chemistry study, this step-wise growth chemistry investigation provides significant insight into the primary reactions in the GaN MOVPE processes. Based on combined modeling and experimental effort, we aim at providing a general model of GaN growth. In this paper, we will address our study on thermal flow behaviors inside GaN reactors. Unlike the typical III-V reactor, the GaN reactor involves much higher growth temperature and more extensive pre-deposition gas phase reactions, therefore the primary focus of this investigation are the flow velocity distribution, temperature profile and residence time.

REACTORS & EXPERIMENT

Schematic diagrams of the two reactors used in this study are shown in Fig.1 and Fig.2. The first reactor we studied is a working vertical reactor, the second one is the two-flow reactor invented by Nakamura [2]. We have detailed operating conditions for the vertical reactor, while our simulation study of the two-flow reactor is based on the incomplete experimental data distributed in several publications and patents by Nakamura [2][3][4]. Therefore, we present our detailed study on the vertical reactor. The simulation of the two-flow reactor is used as a comparison to verify the features present in the vertical reactor. For the vertical reactor, trimethylgallium (TMG), in a hydrogen carrier gas, is supplied through the inner tube while ammonia and hydrogen are supplied in the outer tube. The outer wall of the reactor is water-cooled. The graphite susceptor is inductively heated. The inner diameter of the reactor is 85 mm and the outer diameters of the two inlet tubes are 6.4 mm and 25.4 mm respectively. The opening of the inner tube is widened at the end above the susceptor, which is 70 mm in diameter. This reactor is operated at a pressure of 100 Torr and a susceptor temperature of 1100°C. The coolant and the inlet gas temperature are assumed to be at room temperature. The films were

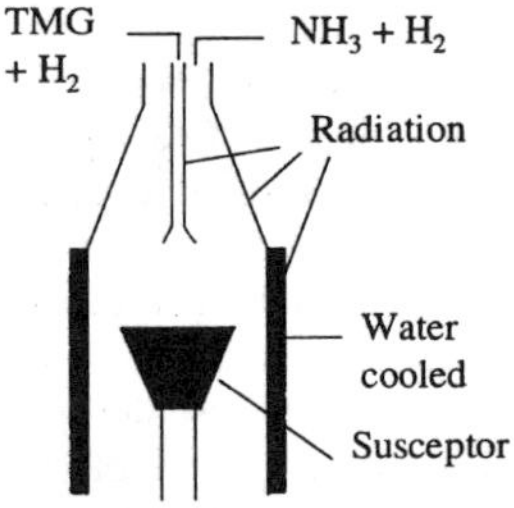

Figure 1. Diagram of the vertical reactor

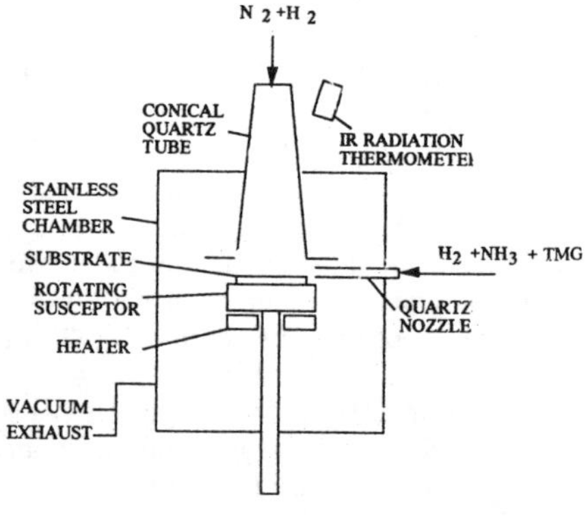

Figure 2. Diagram of the two-flow reactor ([2])

grown on 2-inch c-plane sapphire. The two-flow reactor is operated at 760 torr and a susceptor temperature of 1000°C. The growth reactants (TMG + NH$_3$ + H$_2$) are supplied from a horizontal inlet, and a vertical pushing gas (H$_2$ + N$_2$) is added in order to confine the reactants to the growing surface [2]

MODELING

The modeling study was performed using a multi-purpose computational fluid dynamics code, which includes conjugate heat transfer and radiation [5,6]. The fundamental equations of continuity, momentum, and energy balances and species conservation are used to describe the system [7]. The reactor model is based on numerical solution of the nonlinear, coupled partial differential equations representing the conservation of momentum, energy, total mass and individual species using the finite element method (FEM). The FEM transport model solves for flow and heat transfer (including conduction in the walls, convection, and radiation) for the reactor configuration. The thermal diffusion component that drives high molecular species away from hot regions, towards cold regions, has also been included because of the high growth temperature.

RESULTS

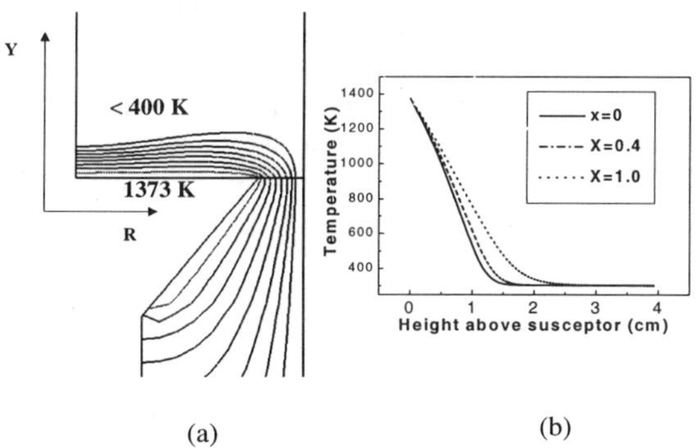

Figure 3. Temperature profiles for the vertical reactor. (a) Temperature contour, (b) Temperature distribution above susceptor at three radial position x = R/R$_{wafer}$.

The detailed results for the vertical reactor are initially presented, followed by brief results on the two-flow reactor for comparison. Figure 3 and 4 show the temperature and flow profile inside the vertical reactor under the standard operating conditions. From Fig. 3(a), we can see that the high-temperature flow zone is confined to a thin gas flow sheet above the susceptor. The temperature decrease from growth temperature to ~ 400K within a very short distance away from the growth surface, resulting in a very high temperature gradient above the growth surface. The bulk of the gas phase temperature remains around the inlet temperature, which is about room temperature. Figure 4 (a) show the corresponding streamlines. Typically, there are recirculations above the susceptor due to the high input velocity of the inner TMG feed tube [8]. These recirculations could lead to trapping of particulate matter, which may subsequently deposit onto the film. The removal of recirculations for this case is accomplished by changing the geometry. TMG flow velocity is lowered with a much wider opening at the end. This modification improves the uniformity and quality of the deposited film on the substrate. The radial velocity becomes increasingly uniform when the flow approaches the growth surface, which is also a necessary requirement for the deposited film uniformity. In order to make comparison between the two investigated reactors, we quantify the above thermal fluid behavior.

465

From Fig. 3 (b) and Fig. 4(b), we can see that the thickness of the high-temperature zone is about 1-2 cm, while the gas flow velocity inside the high temperature flow zone is about 0.01-0.6 m/sec.

The simulation geometry for the two-flow reactor and a typical temperature profile is shown in Figure 5. By comparing to the corresponding temperature profiles in

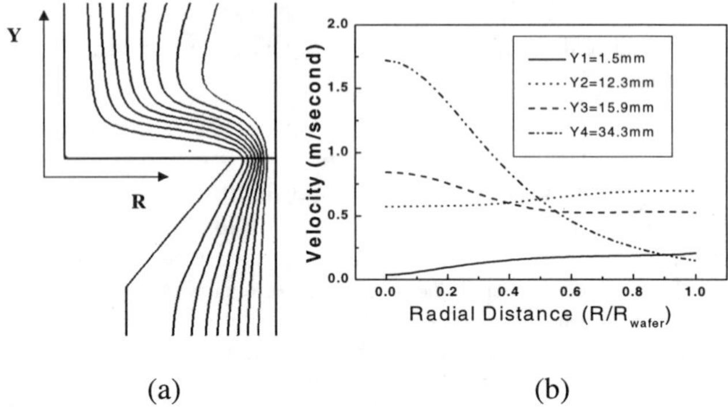

(a)   (b)

Figure 4. Flow profile for the vertical reactor. (a): Streamline, (b): Velocity radial distribution above susceptor.

the vertical reactor, similar behaviors can be identified. A very thin high-temperature flow sheet is also formed above the susceptor. The thickness of the high-temperature flow sheet is about 1.0-1.5 cm. The gas temperature outside this flow sheet is below 400K. Similar to the vertical reactor, a uniform flow velocity distribution is observed above the growth surface. The gas flow velocity above the susceptor from our simulation is in the range of the reported value [4], 0.02-0.5 m/sec.

We further investigate these common thermal fluid behaviors by studying thermal diffusion and residence time within the vertical reactor. The high temperature gradient above the growth surface leads to high thermal diffusion coefficients. High thermal diffusion effects can reduce the concentration of higher molecular weight species near the growth front. Figure 6 indicates the impact of a change in thermal diffusion in the TMG concentration profile above the substrate. The TMG concentration on the substrate surface is significantly reduced by the high thermal diffusion effect in this reactor. Since the reactant residence time affects the extent of gas phase reactions, and hence the quality of the deposited film, the residence time is a very useful parameter to study the thermal fluid behavior for GaN MOVPE reactors. We studied the residence time by integrating the trace of released particles from the wide opened end of the inner TMG tube until the particle leave the susceptor edge. Figure 7 shows the particle trace and radial distribution of the residence times. The molar ratio of TMG to ammonia at every particle releasing point is also

Figure 5. Temperature contour for the two-flow reactor.

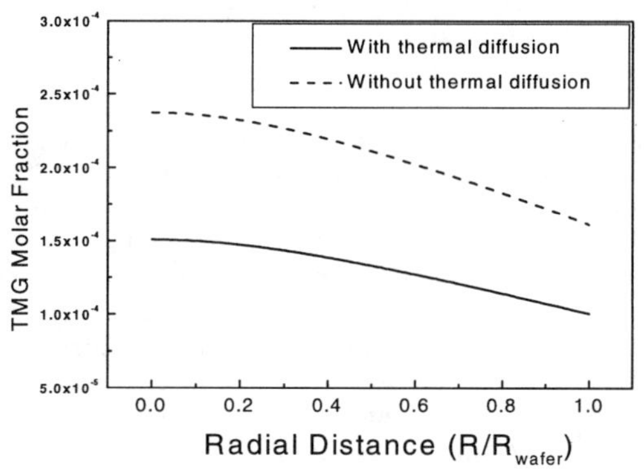

Figure 6. The effect of thermal diffusion on the radial molar distribution of TMG on the substrate

included. From these results, we can see that the total residence time for TMG within this reactor is less than 0.3 second. The residence time decreases at higher ratio of TMG to NH3.

DISCUSSION

The most commonly used MOVPE precursors for GaN, trimethylgallium (TMG) and ammonia, react very rapidly to form an adduct, even at room temperature. These adducts and subsequent adduct-related reactions lead to changes in the nature and concentration of reactants. The gas phase depletion of the actual growth precursors can result in poor growth uniformity, and most probably material quality. In addition, these adduct-related reactions take place away from the heated susceptor. Additional parameters, such as the design of reactor inlet mixing critically influences the

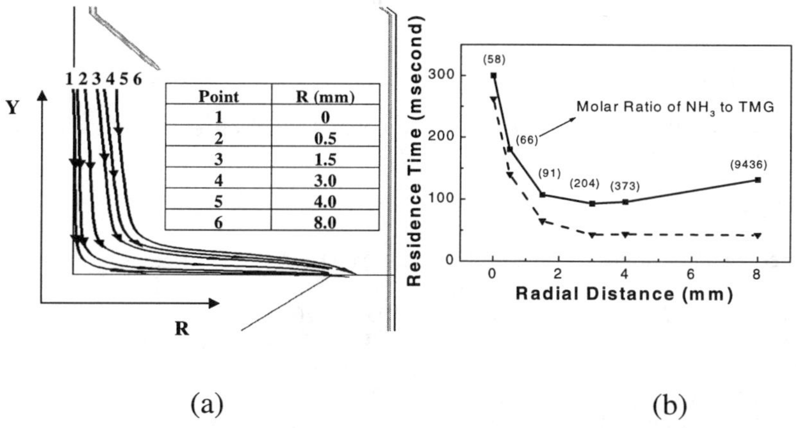

Point	R (mm)
1	0
2	0.5
3	1.5
4	3.0
5	4.0
6	8.0

(a)   (b)

Figure 7. (a) Particle releasing trace, (b) Residence time distribution. < are the total residence time, while 6 are the residence time inside high-temperature zone (T > 400 K).

epitaxial growth.

From our simulation results, we can identify some common thermal flow features that are observed both in a working vertical reactor and the two-flow reactor. A thin high-temperature flow sheet of ~ 1 - 2 cm thick is formed above the susceptor, while the bulk of the gas phase temperature is below 400K. Within the high-temperature flow sheet, the flow velocity is very uniform and in a certain range of 0.01-0.6 m/sec. Since these two reactors have totally different geometries and are operated under different pressures (100 Torr vs. 760 Torr), we believe that these common features might be correlated to growth of device quality GaN-based materials. Prevention of the adduct-related gas phase reactions plays a critical role in the GaN growth. Reduction in the gas phase temperature and minimization of residence time are very effective ways to prevent these pre-deposition reactions. The bulk gas phase temperature inside these two reactors is kept at a low temperature of 300 ~ 400K. This temperature profile will minimize the pre-deposition reactions, or delay these reactions until the high-temperature flow zone.   Since the high-temperature flow zone is restricted to a thin flow sheet right above the susceptor, the reactants react vigorously within this high-temperature zone with a very short residence time, leading to film deposition on the substrate. There is a small residence time of the reactants where the NH_3 and TMG are well mixed within the high temperature flow zone. This small contact time reduces the extent of the pre-deposition reactions. This specific temperature profile is achieved by a high TMG flow from the inner input tube in the vertical reactor, and by the pushing gas flow in the two-flow reactor.

CONCLUSIONS

A comparative study of GaN growth by MOVPE was performed to identify the common thermal fluid behaviors for MOVPE reactors that are being used to grow device quality GaN-based materials. The comparison of the temperature and flow profiles for two high-performance reactors has revealed several common features for the thermal flow behaviors inside these reactors, which can be served as general engineering guidelines for design and optimization of MOVPE reactors for GaN growth.

ACKNOWLEDGEMENTS

This work is supported by the ONR-MURI on compliant substrates and the EPRI-DARPA initiative on high power devices.

REFERENCES

1. H. Jurgensen, D. Schmitz, G. Strauch, E. Woelk, M. Dauelsberg, L. Kadinski, Y. N. Markarov, MIJ-NSR, (USA), Vol. 1 Art. 26 (1996).
2. S. Nakamura, Jap. J. Appl. Phys. Vol. 30 (8), (1991) 1620
3. Nakamura, T. Mukai, and M. Senoh, Appl. Phys. Lett. Vol.64, (1994) p1687.
4. S. Nakamura, United States Patent, No. 5334227, (1994).
5. H. M. Manasevit, J. Crystal Growth, 13/14, (1972) p306.
6 . CFDACE theory manual, CFDRC incorporation, Version 5 (1998).
7. R. B. Bird, W. E. Stewart and E. N. Lightfoot, Transport Phenomenon, Wiley, New York (1960).
8. S. A. Safvi, J. M. Redwing, M. A. Tisher, T. F. Kuech, J. of Electrochem. Soc. 144(5) (1997) 1789.

Part V

GaN Devices and Processing

AlGaN Microwave Power HFETs on Insulating SiC Substrates

Gerry Sullivan, Ed Gertner, Richard Pittman, Mary Chen, Richard Pierson, Aiden Higgins; Rockwell Science Center, Thousand Oaks, CA: Qisheng Chen; APA Optics, Blaine, MN: Jin-Wu Yang; University of South Carolina, Columbia, S.C.: R. Peter Smith, Raul Perez, Abdur Khan; JPL/Caltech, Pasadena, CA: Joan Redwing, Brian McDermott; Epitronics/ATMI, Danbury, CT: Karim Boutros, Spectrolab, Sylmar, CA.

ABSTRACT

AlGaN HFETs are attractive devices for high power microwave applications. Sapphire, which is the typical substrate for AlGaN epitaxy, has a very poor thermal conductivity, and the power performance of AlGaN HFETs fabricated on sapphire substrates is severely limited due to this large thermal impedance. We report on HFETs fabricated on high thermal conductivity electrically insulating SiC substrates. Excellent RF power performance for large area devices is demonstrated. On-wafer CW measurements at 10 GHz of a 320 micron wide FET resulted in an RF power density of 2.8 Watts/mm, and measurements of a 1280 micron wide FET resulted in a total power of 2.3 Watts. On-wafer pulsed measurements, at 8 GHz, of a 1280 micron wide FET resulted in a total power of 3.9 Watts. Design of a hybrid microwave amplifier will be discussed.

INTRODUCTION

Wide band gap semiconductors, such as AlGaN and SiC, are being developed for a variety of applications. These materials have properties that make them very attractive for microwave power generation. In Table 1, some of the materials properties of SiC and GaN are compared to Si and GaAs. As can be seen, the band gaps and breakdown fields of GaN and SiC are similar, and are much larger than those of GaAs and Si. The maximum electron velocity in GaN is predicted to be a bit larger than that in SiC.   The low field electron mobility in bulk samples of SiC and GaN is also similar. At an AlGaN / GaN heterojunction, however, the electron mobility can be much higher. Room temperature Hall mobilities in excess of 2000 cm^2/Vs have been measured in AlGaN / GaN heterojunction structures, with corresponding sheet charge densities of 5 $x10^{12}$ cm^{-2}. This much larger mobility is a distinct advantage for AlGaN / GaN HFETs, relative to SiC devices.

In power transistors, efficient thermal management is critically important. Table 1 shows that the thermal conductivity of SiC is much higher that that of GaN. This listed thermal conductivity for SiC (4.9 W/cm K) is the value for very high purity, lightly N-type material. Currently, insulating SiC has a somewhat lower conductivity (3.3 W/cm K [1]), but this value should improve as the SiC substrate quality improves. Sapphire, which is the typical substrate for AlGaN epitaxy, has a thermal conductivity of only 0.2 W/cm K. This very poor thermal conductivity severely limits the performance of RF power HFETs on sapphire substrates. Growth of AlGaN HFET structures onto electrically insulating SiC substrates should combine the superior electrical properties of the AlGaN / GaN heterostructure with the superior thermal properties of bulk SiC, resulting in high performance microwave power transistors.

Mat. Res. Soc. Symp. Proc. Vol. 572 © 1999 Materials Research Society

Table 1. Selected materials properties of SiC and GaN, compared to Si and GaAs. (a-electron mobility in bulk, lightly N-type GaN; b- electron mobility at an AlGaN / GaN heterojunction)

Property/Material	Si	GaAs	4H-SiC	GaN
Bandgap (eV)	1.1	1.4	3.3	3.4
Breakdown field [$\times 10^5$ V/cm]	2	4	8	8
Electron mobility [cm^2/Vs]	1400	8500	800	900 [a] 2000 [b]
Maximum velocity [$\times 10^7$ cm/s] (E=8 E5 V/cm)	0.5	0.7	1.0	1.4
Thermal conductivity [W/cm K]	1.5	0.5	4.9	1.3

EXPERIMENT

The nitrides display a large piezoelectric effect, relative to other III-V materials [2]. For the (0001) orientation typically used for epitaxial growth, the piezoelectric effect is a major source of charge at AlGaN / GaN heterojunctions. Figure 1 shows that the electron charge density densities in undoped AlGaN / GaN heterostructures increase linearly with increasing AlGaN alloy composition, and densities well in excess of 1 $\times 10^{13}$ cm^{-2} can be achieved. This sheet charge density is several times larger than is typically seen in AlGaAs / GaAs heterostructures. Because the mobility of electrons in AlGaN / GaN heterostructures are several times lower than the mobility of electrons in AlGaAs / GaAs, the total current for a given gate width is very similar for the two materials systems. The drain breakdown voltage for the nitride structure is many times larger than for the arsenide structure, for similarly sized transistors.

The structures reported on in this paper were all grown by low pressure MOCVD using the standard metalorganic precursors for the group III material, ammonia for the column V material, and silane for the dopant. Adding intentional donors to the AlGaN layer increases the electron sheet charge density, as expected. Figure 2 shows the measured Hall mobility versus sheet charge density for a variety of structures grown on SiC substrates, some of which contain intentional doping in the AlGaN. The structure with a 300K sheet charge density of 4.9 $\times 10^{12}$ cm^{-2} and a mobility of 2150 cm^2/Vs, has a 77K sheet charge density of 4.9 $\times 10^{12}$ cm^{-2} and a mobility in excess of 12,000 cm^2/Vs. This large low temperature mobility attests to the high quality of the heterointerface.

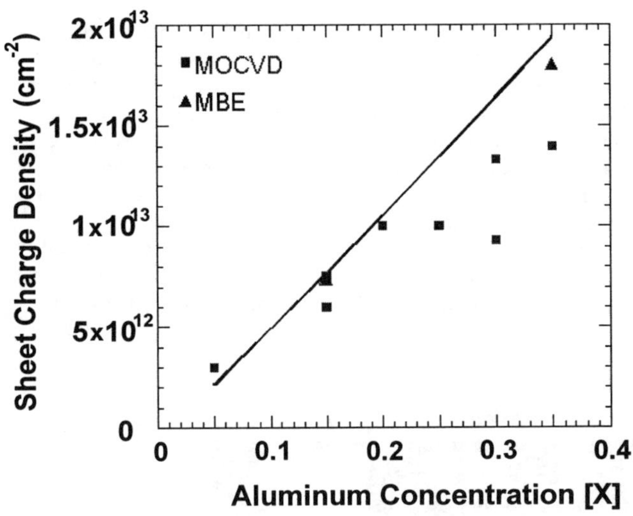

Figure 1. Two-dimensional electron sheet charge density at AlGaN / GaN heterostructures, as a function of the alloy composition of the AlGaN [2].

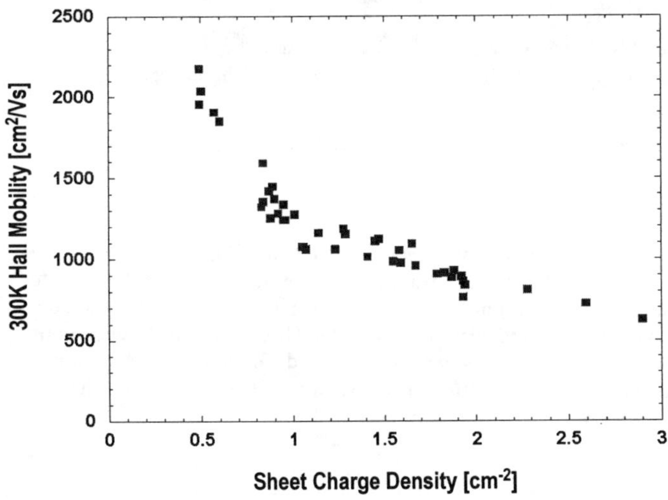

Figure 2. Room temperature Hall mobility versus sheet charge density for a variety of AlGaN/GaN structures grown by MOCVD on SiC substrates at Rockwell.

The epitaxial layer structure of the devices reported on in this paper consisted of a 50 Angstrom undoped AlGaN cap, a 100 Angstrom AlGaN region doped with 5 x10^{18} donors/cm^2, a 50 Angstrom undoped spacer layer, a one micron thick GaN layer, and a 1000 Angstrom thick AlN nucleation layer. The aluminum alloy concentration in the AlGaN is 25%. This structure is grown on a 350 micron thick electrically insulating 4H SiC substrate. The structure has a room temperature sheet charge density of about 1 x10^{13} cm^{-2}, and a Hall mobility of 1350 cm^2/Vs.

HFETs were fabricated in this layer structure. Implantation was used for isolation. Ohmic contacts consist of Ti-Al annealed at 950 $^{\circ}$C. The optically defined Pt-Au gate is 0.6 microns long. A microphotograph of the layout of the 1280 micron wide device reported on in this paper is shown in Figure 3. It consists of 16 gates, each of which is 80 microns wide. Successive sources are interconnected using air bridges.

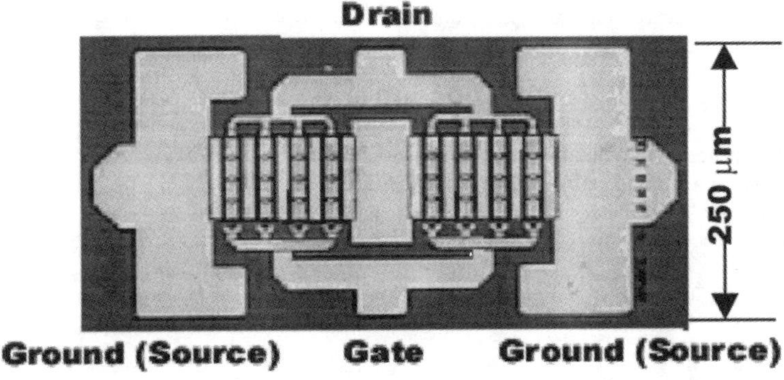

Figure 3. Microphotograph of a 1280 micron wide AlGaN HFET.

RESULTS

Shown in Figure 4 are the transistor characteristics for HFETs on sapphire and SiC. The thermally induced collapse of the drain current at high power that occurs for the device on sapphire is largely absent for the device on SiC. This demonstrated the efficacy of the SiC at conducting the heat away from the HFET. The difference in the absolute magnitude of the drain currents is not due to the different substrates, but is related to differences in the layer structure during the two different growths. The HFET on SiC has a current density of about 1 amp/mm.

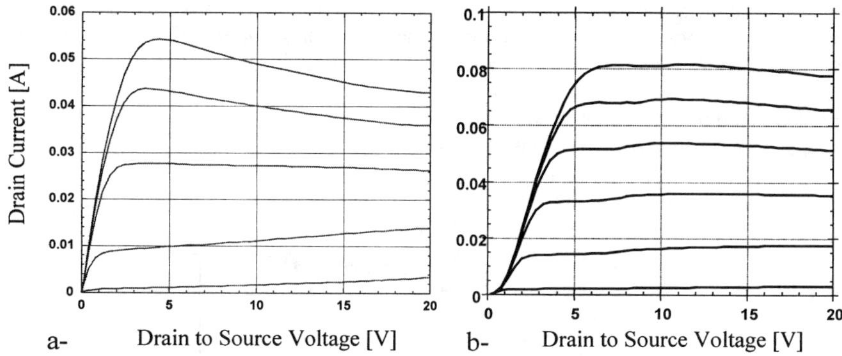

Figure 4. Comparison of the transistor characteristics of an 80 micron wide AlGaN HFET on a sapphire substrate (a) and a SiC substrate (b).

The transconductance versus gate voltage for the HFET on SiC shown in Figure 4b is shown in Figure 5. The peak transconductance is 240 mS/mm, and the pinch-off voltage is about -4.2 Volts.

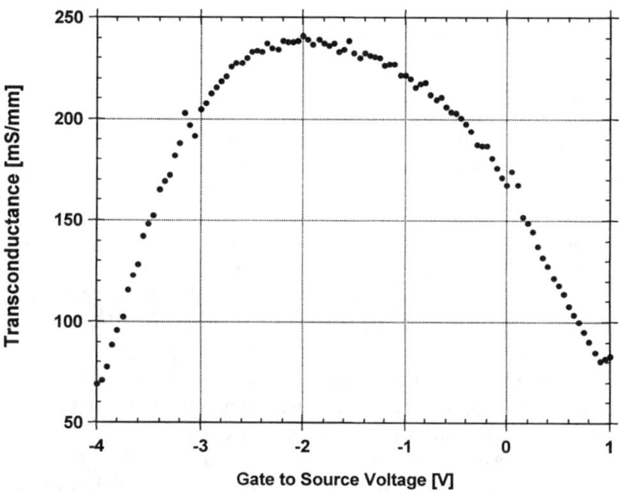

Figure 5. Transconductance versus gate voltage for the 80 micron wide HFET on a SiC substrate shown in Figure 4b. The drain bias was 7 Volts, and the source to drain spacing is 2 microns.

The 10 GHz gain compressions characteristics for the HFETs on SiC are shown in Figure 6. These measurements were done on the wafer using controlled impedance probes. The small signal gain is in excess of 10 dB. With increasing drain bias, the output power saturates as expected. Increasing the drain bias voltage does little to improve the maximum output power. This is believed to be due to the high thermal impedance for a wafer resting on a metal chuck.

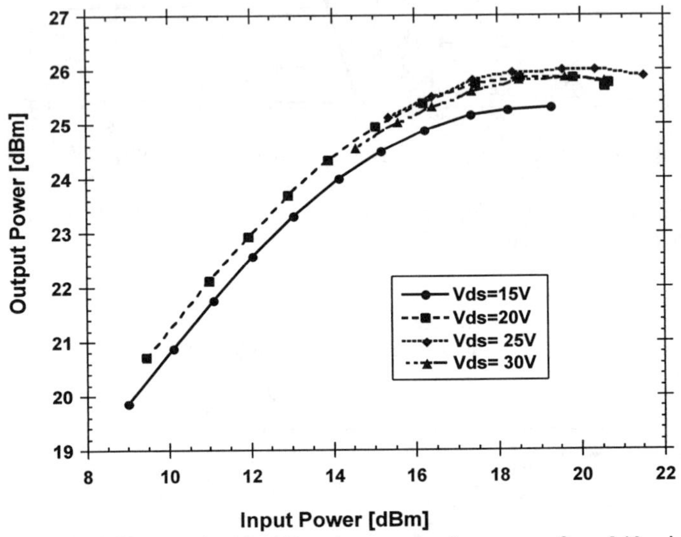

Figure 6. CW on-wafer 10 GHz gain compression curves for a 240 micron wide HFET on a SiC substrate. The source to drain spacing is 3 microns.

On-wafer CW 10 GHz power testing was done on these HFETs on SiC substrates. To reduce the thermal impedance, the SiC substrate was attached to the metal chuck using thermal grease. A power density of 2.8 Watts/mm was measured in a 320 micron wide FET. A total power of 2.3 Watts was measured in a 1280 micron wide FET.

Pulsed testing of devices is a way to measure the devices while they are still in wafer form, and to estimate the performance of the devices once they have been properly packaged with efficient thermal management. The pulsed measurements were done by applying a 10 microsecond DC pulse to the drain with a 0.1% duty cycle, and a 20 microsecond 8 GHz pulse to the gate which symmetrically overlapped the drain bias pulse. Output tuners were used to match the fundamental frequency. As shown in Figure 7, a 1280 micron wide device with 3 micron source-to-drain spacing produced a maximum power of 3.9Watts (3W/mm) with 7.7 dB gain and 26% P.A.E. The drain bias was 32Volts, and the gate bias was -1.5V. A somewhat lower drain bias (25V) resulted in a somewhat higher 30% P.A.E., producing 3.3 Watts (2.6W/mm) with 8.5 dB gain.

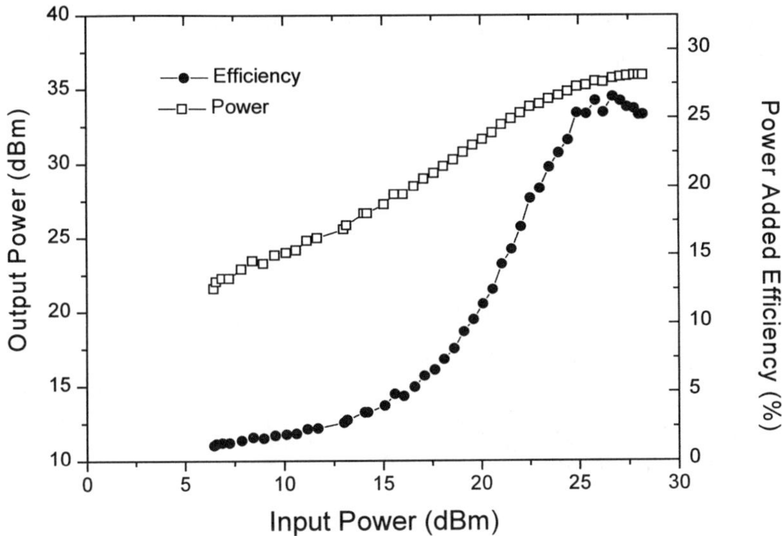

Figure 7. Pulsed on-wafer 8 GHz gain compression curves for a 1280 micron wide HFET on a SiC substrate. The source-to-drain spacing is 3 microns, V_{ds} is 32V and V_{gs} is -1.5V.

Large signal 10 GHz measurements (Load pull) were done on these transistors to determine optimal RF impedances. As shown in Figure 8, the input impedance is quite low, as expected for a large power transistor with a comparatively long optically-defined gate. Reducing the gate length using electron beam lithography will move the input impedance closer to 50 Ohms, making the input matching easier and more efficient. This low input impedance will probably limit the size of practical power HFETs to a few millimeters of gate width.

The input impedance has also been rotated clock-wise around the Smith chart, relative to a similarly sized GaAs HFET, so that its impedance is almost entirely real. This rotation for the AlGaN HFET is believed to be a reflection of the lack of through-substrate via holes. The ground connection for the AlGaN HFET's source, instead of being a low inductance via, is a long front-side metal line. This extra impedance between the source and ground is very undesirable, and degrades the RF performance of the existing HFETs.

The optimal output impedance for the 1280 micron wide HFET is high, relative to a similarly sized GaAs HFET, reflecting the large drain bias possible with the AlGaN HFET. The optimal impedance for maximum power moved even closer to 50 Ohms with increasing drain bias, as expected. The output impedance mismatch which degrades the total power by one dB is a fairly large circle, making the output matching for these transistors fairly simple.

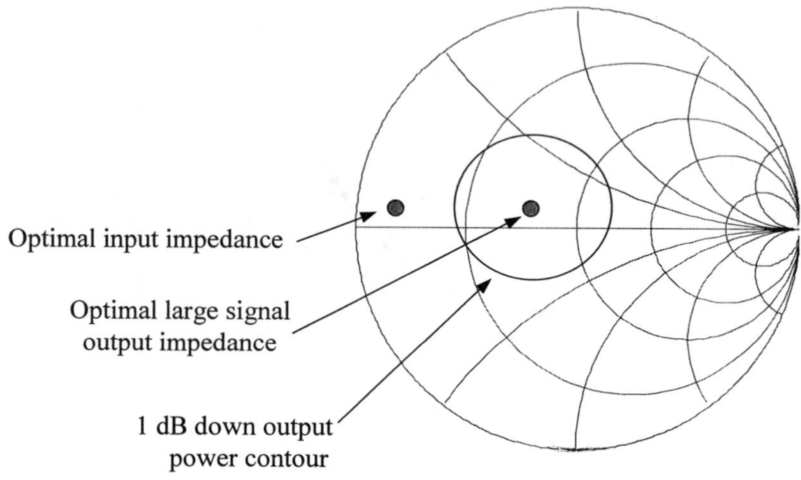

Figure 8. Large signal on-wafer impedance measurements of 1280 micron wide HFETs, optimizing for maximum power. The drain bias is 25 Volts, and the source-to-drain spacing is 3 microns.

Shown in Figure 9 is a 9 GHz hybrid amplifier designed using two of these 1280 micron wide HFETs. The passive matching components are fabricated on 250 micron thick GaAs substrates using microstrip waveguide format. The input and output are matched to 50 Ohms. As can be seen, the input matching requires considerably more circuitry than does the output matching, reflecting the greater impedance mismatch on the gates of the HFETs relative to the drains. Also, the total area required for the passive matching components is much larger that that required for the active area of the transistors. This relatively large area for the passive matching components at 10 GHz is a strong argument against fabricating MMICs, rather than hybrids, given the current high cost and small wafer size of electrically insulating SiC. At higher frequencies, the performance degradation associated with a hybrid design will force MMIC or flip-chip designs to be used, in spite of the costs.

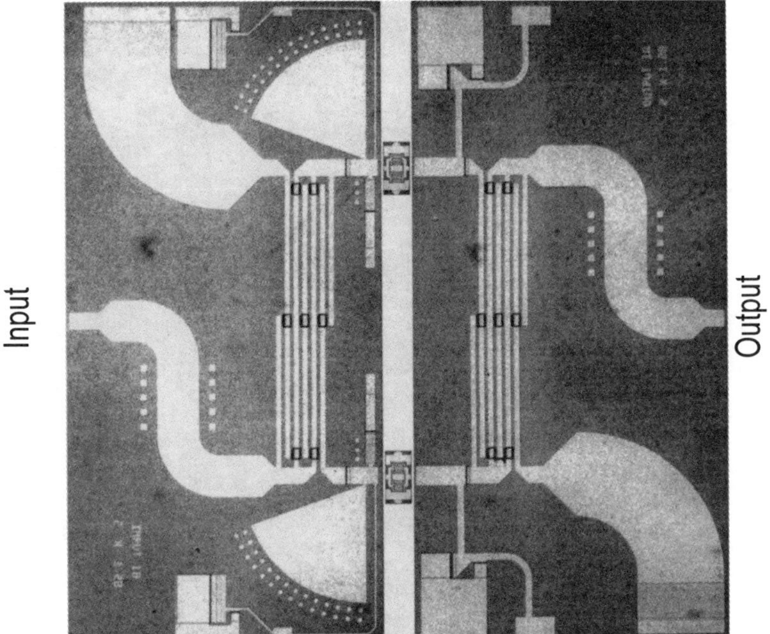

Figure 9. 9 GHz Hybrid amplifier using two transistors, each of which is 1280 microns wide.

CONCLUSIONS

AlGaN / GaN HFETs are the device of choice for microwave power applications. For these power applications, the improved thermal conductivity of electrically insulating SiC substrates provides a huge advantage in performance, relative to HFETs on sapphire substrates. At 10 GHz, hybrid amplifiers are the cost-effective architecture, with the large area passive matching components being fabricated on inexpensive substrates.

ACKNOWLEDGEMENTS

We gratefully acknowledge BMDO (Dr. Kepi Wu) for supporting this work.

REFERENCES

1- S.T. Allen, W.L. Pribble, R.A. Sadler, T.S. Alcorn, Z. Ring and J.W. Palmour, Recent Progress in SiC Microwave MESFETs, Spring MRS Conference, Talk Y1.2, San Francisco, 1999.

2- E. T. Yu, X. Z. Dang, P. M. Asbeck, S. S. Lau and G. J. Sullivan, "Spontaneous and piezoelectric polarization effects in III-V nitride heterostructures", to appear in J. Vac. Sci. and Tech., June/July 1999.

Recessed gate GaN MESFETs fabricated by the photoelectrochemical etching process

Won Sang Lee,* Yoon Ho Choi,* Ki Woong Chung,*
Moo Whan Shin,** and Dong Chan Moon***

* Device & Materials Lab, LG Corporate Institute of Technology, 16 Woomyeon-Dong,
Seocho-gu, Seoul, Korea 137-724
**Department of Inorganic Materials Eng., Myongji University,
38-2 Nam-Dong, Yongin-Si, Kyunggi-Do, Korea
***Department of Electronic Materials Eng., Kwang Woon University,
447-1 Wolgye-Dong, Seoul, Korea

Abstract

A new photo-electrochemical etching method was developed and used to fabricate GaN MESFETs. The etching process uses photoresist for masking illumination and the etchant is KOH based. The etching rate with 1.0 mol% of KOH for n-GaN is as high as 1600 Å/min under the Hg illumination of 35 mW/cm^2. The MESFET saturates at $V_{DS} = 4$ V and pinches off at $V_{GS} = -3$ V. The maximum drain current of the device is 230 mA/mm at 300 K and the value is remained almost same for 500 K operation. The characteristic frequencies, f_T and f_{max}, are 6.35 GHz and 10.25 GHz, respectively. Insensitivity of the device performance to temperature was attributed to the defect-related high activation energy of dopants for ionization and band-bending at the subgrain boundaries in GaN thin films.

1. INTRODUCTION

Wide bandgap semiconductors based upon the III-Nitride system, which are mainly developed for optical application, have many properties which are ideal for electronic devices for high temperature, high frequency, high power, and radiation hard application. A variety of high frequency electronic devices can be fabricated from GaN-based semiconductors; these devices are predicted to offer superior DC and RF performance compared to more conventional Si and GaAs devices from the view point of high power. Theoretical calculations for GaN MESFET predict output power density near 5 W/mm, power-added efficiency higher than 50 %, and linear power gain about 20 dB for an optimized device structure [1]. In addition, GaN can allow for the AlGaN/GaN heterostructures which was demonstrated to produce two-dimensional electron gas (2 DEG) and thus makes possible several novel devices that can operate at frequencies beyond the capability of SiC devices. Also, the 2 DEG permits low resistances and low noise performance not possible with SiC [2].

Despite the excellent electronic properties of GaN, the fabrication of GaN MESFETs with novel designs have been hindered by the chemical inertness of this material. Thus, the   most   successful etching process so far has been dry etching method including reactive ion etching (RIE) [3],   electron cyclotron resonance (ECR) RIE [4], and inductively coupled plasma (ICP) RIE [5]. However, these techniques are known to result in ion-induced damage on the etch surface which is highly undesirable for the high frequency and high power operation of devices. Wet chemical etching is a desirable substitute for dry etching method by providing low damage on the surface of the active region. There has been a report on the photoelectrochemical wet etching process for GaN films using KOH solution under the illumination of Hg arc lamp [6]. However, the experiment was performed using Ti metal mask which is not normally used in conventional wet etching process and is not suitable for the fabrication of microelectronic devices. In this paper, we report on the photoelectrochemical etching

Mat. Res. Soc. Symp. Proc. Vol. 572 © 1999 Materials Research Society

process for GaN thin films using photoresist mask. For the first time, the etching method is demonstrated to be well applicable to the fabrication of recessed gate GaN MESFET. The I_{DS}, g_m, and V_{th} of the device at different operating temperatures and f_T and f_{max} characteristics are discussed as well.

2. EXPERIMENTS

There are many ways of etching the wide bandgap semiconductors to make transistors. The dry processing was used mainly because no convenientchemical solution for good ething characteristics is available. Once we can provide the chemical solution, the photoelectrochemical etching process is preferred, in particular, by the need for a recessed gate GaN MESFET. The etching mask in this experiments is a i-line photoresist (THMR-iP1800, TOK), eliminating the hard baking process which degrades the adhesion of photoresist to the surface and hence leads to easy lift-off process. The sample for this experiment consists of 1.3 μm thick undoped ($3 \sim 4 \times 10^{16}/cm^3$) GaN buffer layer, 500 Å thick Si doped ($2 \times 10^{17}/cm^3$) n-GaN active layer and 300 Å thick Si doped ($2 \times 10^{18}/cm^3$) n^+-GaN cap layer grown on c-plane sapphire substrate by metal-organic chemical vapor deposition (MOCVD) method. GaN epi layer quality was measured by PL, XRD, and Hall measurement of 300 K. Dry etching for device isolation, i.e. mesa etching, used an ECR dry etcher. Etching rate was varied with RF bias, microwave power, and processing chamber pressure. Prior to the gate recess etching process, the source and the drain ohmic contact with a contact resistance of 4×10^{-6} Ωcm^2 was obtained by the deposition of Ti/Al (=300 Å/2000 Å) bilayer followed by a rapid thermal annealing (RTA) at 700 ℃ for 10 seconds. The recess wet etching rate was examined as functions of the mole percentage of the KOH solution and the etching time. After an extraction of optimized conditions, the photoelectrochemical etching method was directly employed for the fabrication of a recessed gate GaN MESFET. An n^+ cap layer with a thickness of 300 Å was etched out for the recessed gate structure. The schottky contact was formed by a Pt/Au (=400 Å/4000 Å) alloy and it showed a good blocking capability. The gate length and the gate width of the device were 0.7 μm and 100 μm, respectively. The spacing between the gate and source or drain was 5 μm. The morphology of the etched surface was characterized by Scanning Electron Microscopy (SEM) and Atomic Force Microscopy (AFM). The dc performance of the device was characterized at different device operating temp. From the S-parameter measurement in a frequency range of $1 \sim 18$ GHz with a microwave network analyzer, HP 8510C, we measured the f_T and f_{max}.

3. RESULTS AND DISCUSSION

Fig. 1 shows the photoluminescence and X-ray diffraction of n-GaN epi layer at room temperature. FWHM of band edge emission, (0002) θ scan, (1012) Φ scan are 43 meV and 7.2 arcmin and 22.5 arcmin, respectively.

Carrier concentration and mobility has changed by Si mole fraction. When the Si mole fraction varied range of 0.5 $\sim$ 7 nmol/min, carrier concentration, mobility are increased of $2 \times 10^{17}/cm^3 \sim 5.6 \times 10^{18}/cm^3$ and decreased of 430 $cm^2/V.sec \sim 135$ $cm^2/V.sec$, respectively. Optimum condition of GaN dry etching is Ar:5, BCl_3:4, Cl_2:3, Pressure:3, RF:450 V, Mw:300 W, He:3 and its etching rate has 950 Å/min.

Ohmic contact metal used Ti/Al system, contact resistance was 5×10^{-6} Ωcm^2 at 700 °C 10 sec RTP annealing. Fig. 2 shows the photoelectrochemical etching depth as functions of etching time and the concentration of KOH in the etchant solution. The etching rate is linearly increased with etching time. The etching rate is shown to increase as the KOH concentration is increased. The etching rates for the 0.5 mol % and 1.0 mol % of KOH in the solution are about 460 Å/min and 1600 Å/min, respectively. Note that the etching rate observed in this experiments is comparable with the etching rate achieved in a typical RIE method. The illumination intensity of Hg arc lamp during the etching process was 35 mW/cm^2. Fig. 3 shows the morphology of the etched surface displaying a well defined etched edge. The sample in this figure was etched (DC bias 2.5 V, UV Power of 35 mW) to a depth of about 2000 Å. This figure demonstrates that the photoresist can be used as a suitable pattern mask in this etching process.

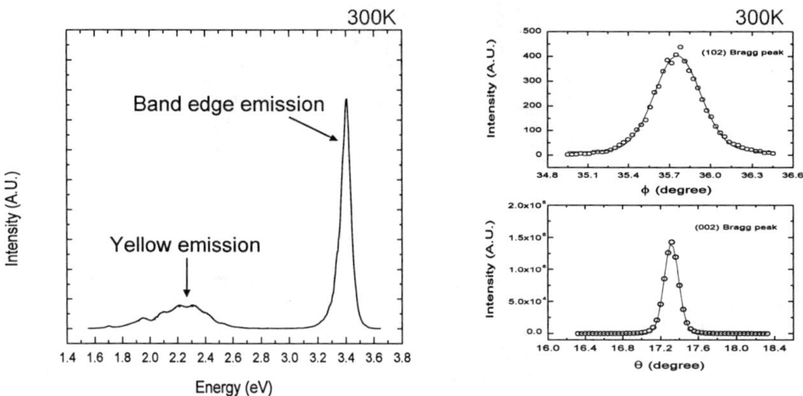

Fig. 1 Photoluminescence and Xray diffraction of n-GaN epi layer at 300K

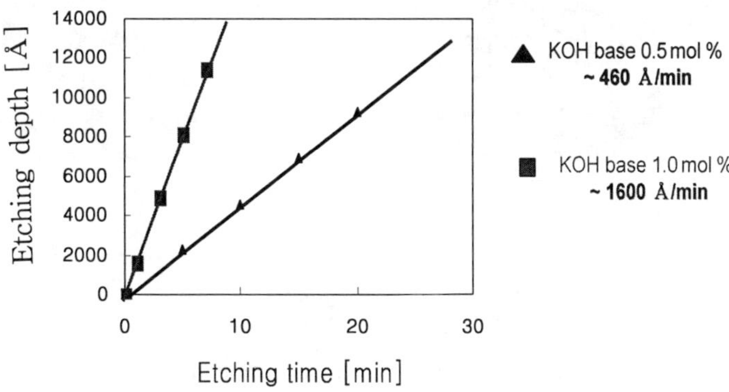

Fig. 2 Photoelectrochemical etching depth

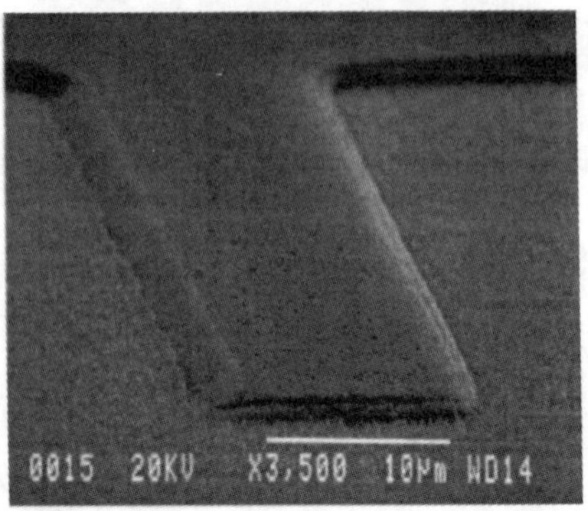

Fig. 3 Etched surface by SEM

The surface roughness was examined by AFM and the result is shown in Fig. 4. The rms roughness from the etched GaN surface is about 37 Å. The rms roughness for the pre-etched GaN surface is measured to be 3.2 Å. The surface morphology has been changed by the etching, but the value of etching-induced roughness is low enough to fabricate a MESFET structure with a gate length of 0.7 μm and a channel thickness of 0.05 μm.

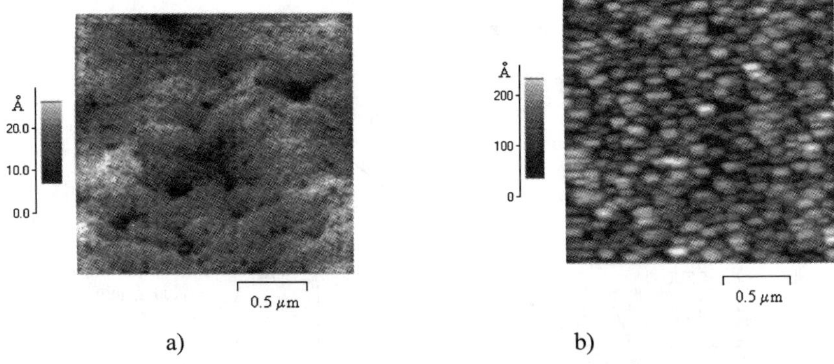

Fig. 4 Surface roughness by AFM
a) as-grown b) etched surface

The optimized etching conditions were directly applied to the fabrication of recessed gate GaN MESFET and Fig. 5 shows the room temperature and high temperature current-voltage characteristics of the device. The gate voltage step is - 1 V and the saturation of the source-drain current occurs at Vds = 4 V and the pinch-off voltage at Vgs = - 3 V. The maximum drain-source current of the device operating at 300 K is about 240 mA/mm and no significant change is observed at 500 K. The current level is somewhat lower than the value expected from the device design, reasons for which will be discussed later.

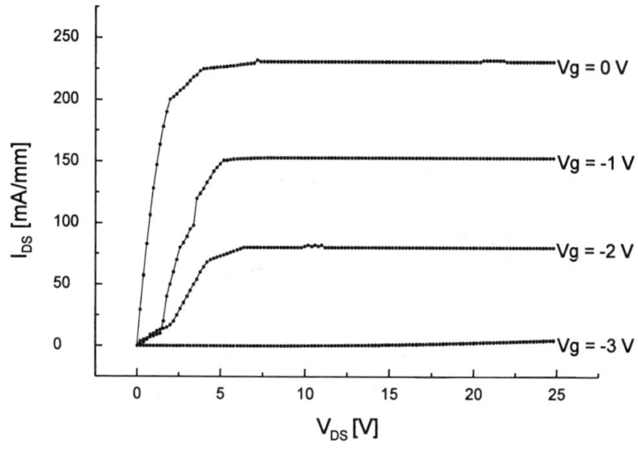

a)

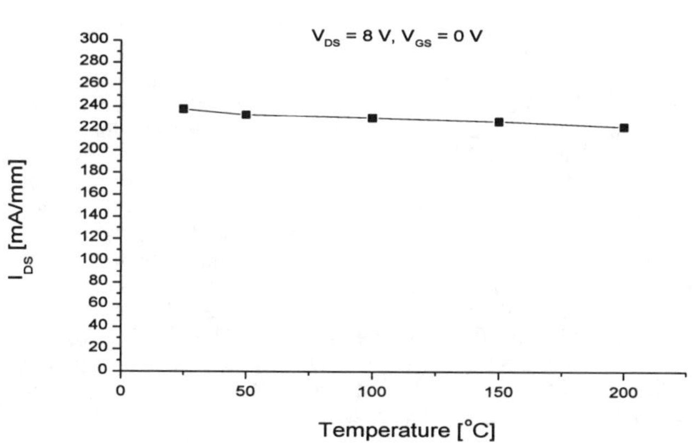

b)

Fig. 5 Current-voltage characteristics of GaN MESFET
a) I_{ds} vs. V_{ds} with V_{gs} (V_{gs} varies from 0 V to -3 V by -1 V from the top curve,
b) Drain current change to the operating temperature
(V_{ds} = 8 V, V_{gs} = 0 V, temperature from room temperature to 200 °C)

The measured RF performance is in Fig. 6. The cut-off frequency and maximum oscillation frequency are 6.25 GHz and 10.25 GHz at Vgs = 0 V, Vds = 8 V, respectively.

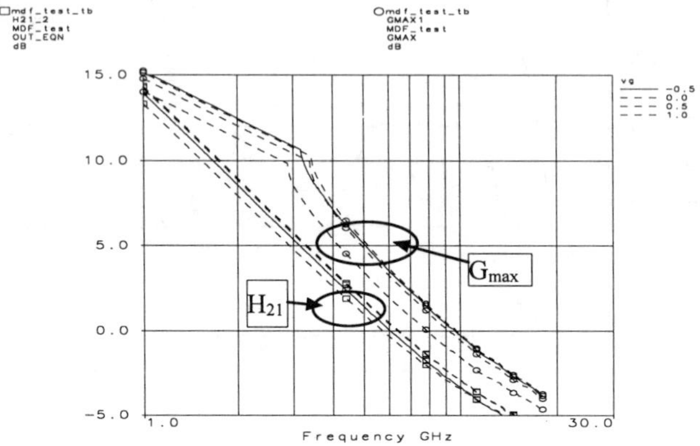

Fig. 6 f_T and f_{max} characteristics of GaN MESFET

Theoretically, the current is supposed to decrease with the operating temperatures due to the reduction of the electron velocity and mobility to the temperatures. One explanation for no change in the drain current level is a high activation energy of dopants for ionization. It has been reported that the activation energy of dopants in thin films with two dimensional defects is known to be higher than that without the defects [6]. When the activation is high and the dopants are in a "freeze-out region" [7], the number of the effective free carriers increases with temperature. Therefore, it is possible that the insensitivity of the drain current with temperature be attributed to a presumable freeze-out of dopants in the active region of device, otherwise, the drain current should be decreased due to the lower mobility and electron velocity at high temperatures. Another explanation for the insensitivity is trapping of carriers at the subgrain boundaries. GaN thin films are known to possess high density of subgrain boundaries that are responsible for the band-bending 8]. The driving force for electrons to cross the transverse boundaries is larger when the device operates at higher temperatures. Thus, the mobility of electrons at higher temperature could be higher than that at lower temperature as is the observed abnormal behavior in the mobility-doping relationship [9].

4. CONCLUSION

For the first time, the photoelectrochemical etching process using a photoresist mask was employed for the fabrication of recessed gate GaN MESFETs. The maximum etching rate with 1.0 mol % of KOH for n-GaN was 1600 Å/min. The fabricated GaN MESFET shows a current saturation with the maximum current of 240 mA/mm at room temperature. Insensitivity of the change in the drain current to the temperature was attributed to the defect-related high activation energy of dopants and band-bending. The developed photoelectric wet etching process is likely to be used widely in the fabrication of devices using GaN, the chemically inert semiconductor.

REFERENCES

1. R. J. Trew and M. W. Shin, Int. J. High Speed Electronics and Systems,6, 211 (1995).
2. M. S. Shur, A. Khan, B. Gelmont, R. J. Trew, and M. W. Shin, Ins. Phys. Conf. Series No. 141: Chapter 4, 419 (1994).
3. M. E. Lin, Z. F. Fan, Z. Ma, L. H. Allen, and H. Morcoc, Appl. Phys. Lett. 64, 887 (1994).
4. R. J. Shul, G. B. Aclellan, S. A. Casalnuovo, D. J. Reiger, S. J. Pearton, C. Constantine, C. Barrat, R. F. Karlicek. C. Tran, and M. Schurman, Appl. Phys. Lett., 69, 1119 (1996).
5. J. R. Milehan, S. J. Pearton, C. R. Abernathy, J. D. Mackenzie, R. J. Shul, and S. P. Kilcoyene, J. Vac. Sci. Technol., A.14, 836(1996).
6. M. W. Shin, R. J. Trew, and G. L. Bilbro, IEEE Electron Device Lett., 15, 292 (1994).
7. M. W. Shin, R. J. Trew, G. L. Bilbro, D. L. Dreifus, and A. J. Tessmer, J. Mater. Sci., 6, 111 (1995).
8. F. A. Fonce, MRS Bulletin, 22, 51 (1997).
9. Y. Park, B. Kim, I. Kim, and E. Oh, Proc. '98 Korea-Japan Joint Workshop on SWSODM, 54 (1998).

CURRENT-VOLTAGE CHARACTERISTICS OF UNGATED AlGaN/GaN HETEROSTRUCTURES

J.D. Albrecht[†], P.P. Ruden[†], S.C. Binari[‡], K. Ikossi-Anastasiou[‡],
M.G. Ancona[‡], R.L. Henry[‡], D.D. Koleske[‡], and A.E. Wickenden[‡]

[†]ECE Department, University of Minnesota, Minneapolis, MN 55455
[‡]Electronics Science and Technology Div., Naval Research Laboratory, Washington, DC 20375

ABSTRACT

Results of a systematic study of the current vs. voltage characteristics of ungated AlGaN/ GaN heterostructures grown on sapphire substrates are presented. It is experimentally observed that the saturation current nearly doubles as the source-to-drain channel lengths decrease from 11.8 to 1.7μm. The average electric field at which current saturation occurs is 10 to 30kV/cm, i.e. much less than the electron velocity saturation field. The experimental data is interpreted in the framework of a new model that takes into account the non-uniformity of the electron density in the channel, electron velocity saturation, and thermal effects. The temperature dependent electron transport characteristics of the model are based on Monte Carlo simulations of electron transport in GaN. It is shown that appreciable contact resistance, which leads to partial channel depletion near the source, and significant self-heating of the devices under high drain-to-source bias are the main reasons for the observed current saturation. The effective ambient temperature in the channel of the devices is calculated from a two-dimensional thermal model of heat dissipation through the sapphire substrate. Equilibrium channel carrier concentrations and low-field mobilities are determined from Hall effect data. The ungated structures are demonstrated to provide much useful materials and process characterization data for the development of AlGaN/GaN heterostructure field effect transistors.

INTRODUCTION

AlGaN/GaN heterostructures have excellent potential for the realization of high-power, high-frequency heterostructure field effect transistors (HFETs) [1,2,3]. The GaN channel material is characterized by favorable electron transport parameters, such as high mobility and high peak velocity. In addition, polarization charges at the hetero-interface can neutralize a large electron charge density, thus allowing for high channel carrier concentrations. However, most epitaxial III-nitride structures are currently grown on sapphire substrates. Sapphire has a relatively low thermal conductivity and, consequently, the performance of AlGaN/GaN HFETs that are heatsunk through the sapphire substrate has been shown to be limited by deleterious thermal effects at high voltages and currents [4]. Furthermore, the fabrication technology for these devices is still relatively immature. Hence, contact resistances are often large and can seriously impair device performance [4].

Mat. Res. Soc. Symp. Proc. Vol. 572 © 1999 Materials Research Society

Ungated HFET structures are of interest as simple characterization structures used in, e.g., transmission line method (TLM) determinations of the equilibrium sheet resistivity and contact resistance, and as saturated resistors in some circuit application. Here, we present a systematic theoretical and experimental examination of ungated AlGaN/GaN heterostructures grown on sapphire substrates.

DEVICE STRUCTURE

The III-nitrides used in this study were grown by MOCVD on sapphire substrates. The material structure consisted of an AlN nucleation layer, followed by 3μm of undoped GaN and 300Å of Si doped $Al_{0.3}Ga_{0.7}N$. Ohmic contacts were formed using alloyed Ti/Al/Ni/Au. Device isolation was accomplished by implantation with nitrogen. The contact resistances varied considerably with annealing conditions. The impact of the different contact resistance values on the current vs. voltage characteristics of the ungated structures will

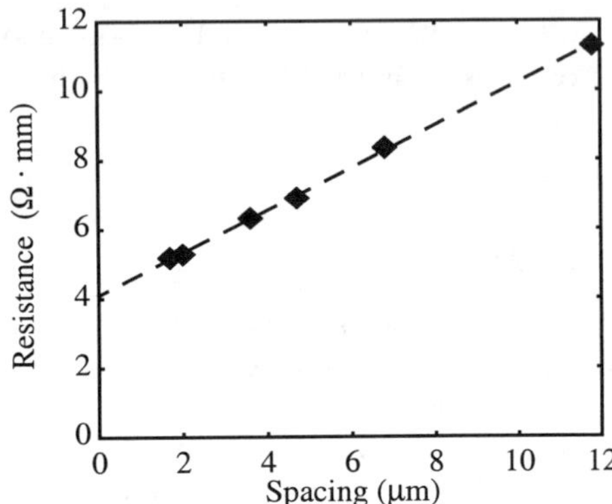

Figure 1: Measured low-field resistances vs. device length for ungated AlGaN/GaN heterostructures.

be examined in more detail below. We will first focus on a set of devices, all from the same wafer, with relatively low contact resistance. The devices tested consist of 75μm wide ungated channels with 1.7, 2.0, 3.6, 4.7, 6.8, and 11.8μm source and drain contact spacings. The measured low-voltage resistances are plotted as a function of the contact spacing in Figure 1. Simple TLM analysis indicates that the sheet resistivity for this heterostructure is 610Ω/□ and the specific contact resistance is 2.0Ω·mm.

DEVICE MODEL AND KEY PARAMETERS

The device model developed for this investigation is a charge-control/gradual-channel approximation model that incorporates salient results of previous Monte Carlo electron transport simulations for GaN as a function of the ambient temperature [5]. The measured low-field TLM data shown in Figure 1, together with independent Hall measurements of the electron mobility, provide information about the equilibrium channel electron concentration, $n_{so}=1/q\mu_{LF}\rho_{\square}$, where μ_{LF} is the low-field electron mobility and q is the electron charge. Under applied drain-to-source bias, V_{DS}, the local electron concentration in the channel can be related to the difference between the surface potential, $\phi_s(x)$, and the channel potential, $V_{ch}(x)$. This relationship can be expressed as,

$$qn_s(x) = \frac{\varepsilon}{d_{eff}}[\phi_s(x) - V_{ch}(x)] + qn_{so} \tag{1}$$

where ε is the material permittivity. The surface-to-channel distance, d_{eff}, is the thickness of the AlGaN layer plus the effective thickness of the two-dimensional electron gas (~20Å), which is determined from a self-consistent calculation of the quasi-two-dimensional subband structure as discussed in ref. [4] and references therein. The channel coordinate, x, ranges from zero at the source-end to L_{DS} at the drain-end. The surface potential, referenced to the source potential, can be assumed to vary linearly between the source and drain contacts: $\phi_s(x) = (V_{DS}/L_{DS}) \cdot x$.

The low-field mobility, the saturation (peak) velocity, and the critical (peak) field for velocity saturation, results of Monte Carlo simulations for electron transport in GaN have been parameterized as functions of temperature, doping concentration, and compensation [5]. The velocity vs. electric field curve for field strengths below the critical field, F_c, is well approximated with the simple analytic expression used in ref. [4]. Expressing this velocity vs. field relationship in terms of a field dependent mobility, we obtain,

$$\mu(T) = \begin{cases} \mu_{LF}(T)\left[1 + \left(\dfrac{\mu_{LF}(T)}{v_c(T)} - \dfrac{1}{F_c(T)}\right)\dfrac{dV_{ch}}{dx}\right]^{-1} & \text{for} \quad \dfrac{dV_{ch}}{dx} \leq F_c \\[2em] v_c(T)\Big/\left(\dfrac{dV_{ch}}{dx}\right) & \text{for} \quad \dfrac{dV_{ch}}{dx} > F_c, \end{cases}$$

(2)

where v_c is the electron drift velocity at F_c. The drain-to-source drift current can now be calculated from,

$$I = qn_s(x)\mu\frac{dV_{ch}}{dx},$$

(3)

with the boundary conditions for the channel potential given by

$$V_{ch}(0) = IR_c$$

(4)

and $V_{ch}(L_{DS}) = V_{DS} - IR_c,$

(5)

where R_c is the contact resistance. Integration of (3) over the length of the channel yields after some manipulation the following implicit relationship between I and V_{DS}:

$$\frac{\varepsilon}{d_{eff}}V_{DS} = q[n_s(L_{DS}) - n_s(0)] + \frac{IL_{DS}}{\mu_{LF}V_{DS}}\ln\left|\frac{qn_s(L_{DS}) + I\left(\dfrac{1}{\mu_{LF}F_c} - \dfrac{L_{DS}}{\mu_{LF}V_{DS}} - \dfrac{1}{v_c}\right)}{qn_s(0) + I\left(\dfrac{1}{\mu_{LF}F_c} - \dfrac{L_{DS}}{\mu_{LF}V_{DS}} - \dfrac{1}{v_c}\right)}\right|.$$

(6)

Because of the voltage drops across the source and drain contacts under bias, the electron concentration is reduced below the equilibrium value at the source-end of the channel and enhanced above the equilibrium value at the drain-end. The non-uniform electron concentration implies a non-uniform longitudinal electric field. This model is similar in character to a model developed in refs. [6 and 7] for GaAs FETs.

Lastly, as indicated in equation (2) above, the transport parameters depend on the lattice temperature. The power dissipated in the channel raises the ambient temperature due to the non-zero thermal impedance of the structure. The devices are heatsunk through the sapphire substrate. The thermal impedance characterizing the heat transfer from the channel to the heatsink was determined by executing a separate, two-dimensional heat-flow model. Obviously, a two-dimensional model of the thermal conductance is only a rough approximation to the real, three-dimensional problem. In addition, it was found that the thermal impedance depends on the layout of the contact pattern and on the effectiveness with which these contacts are heatsunk by the probes. However, for the parameter range of interest, a satisfactory approximation to the effective thermal impedance obtained from these calculations (taking into account the temperature dependences of the thermal conductivities of sapphire and of the III-nitrides) is given by $R_{th} = (23.2 - 0.26 \cdot (L_{DS} - 2)) \cdot (1 + (T - T_{heatsink})/T_o)$ K·mm/W, with $T_o = 725 - 5.8 \cdot (L_{DS} - 2)$. Here L_{DS} is given in μm. The channel temperature, T, is related to the power dissipated:

$$T = T_{heatsink} + R_{th} I V_{DS} \tag{7}$$

with $T_{heatsink} = 300K$.

The thermal effects in the current calculation as a function of applied voltage are incorporated by solving the system of coupled equations given by (6) and (7) to obtain a self-consistent solution. The contact resistances are treated as temperature independent in all of the calculations presented.

RESULTS AND DISCUSSION

Figure 2 shows the measured and calculated current vs. voltage characteristics of the un-gated AlGaN/GaN heterostructures for various channel lengths. The curves are shown as broken lines where the calculated channel temperature exceeds 650K. The agreement between the measured and calculated currents is very good, in particular for the longer devices. To explore the or-

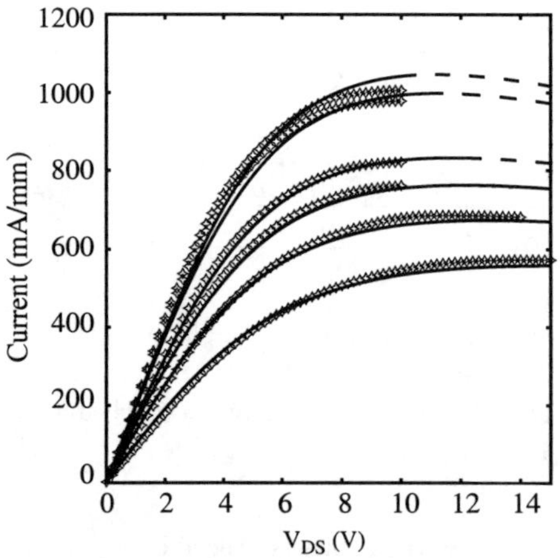

Figure 2: Measured (points) and calculated (solid lines) current-voltage characteristics. The channel lengths are L_{DS} = 1.7 (top curve), 2.0, 3.6, 4.7, 6.8, and 11.8μm (lowest curve).

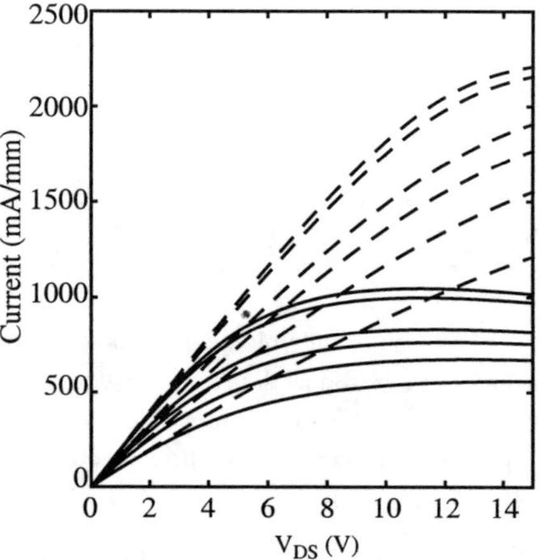

Figure 3: Comparison of the calculated currents for the same devices as in Fig. 2 with (solid lines) and without (broken lines) thermal effects taken into account.

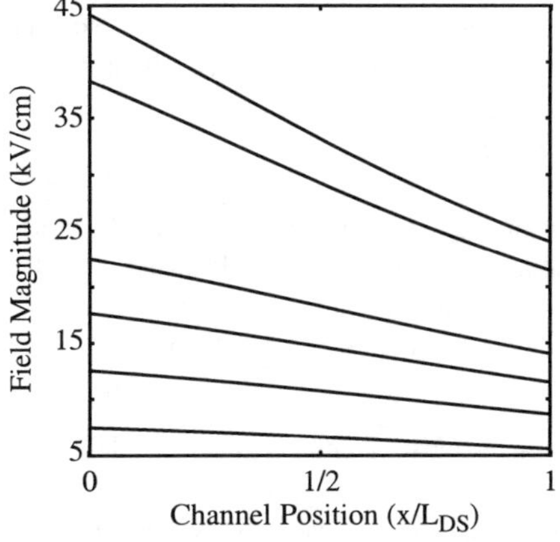

Figure 4: Local longitudinal electric field corresponding to the current-voltage characteristics shown in Fig. 2 for V_{DS} = 10.0 V. The top curve corresponds to the shortest device, the bottom curve to the longest.

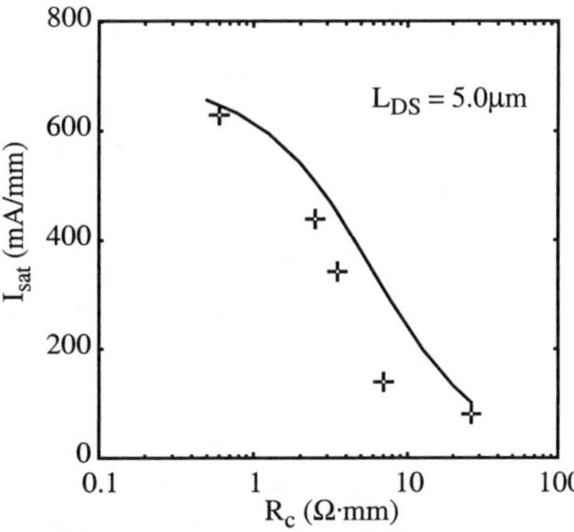

Figure 5: Calculated saturation currents as a function of the contact resistance for a series of ungated AlGaN/GaN heterostructures with L_{DS} = 5μm. Also shown are experimental results from devices with different contact resistances due to different annealing conditions.

igin of the observed current saturation, Figure 3 displays the currents calculated assuming perfect heatsinking (R_{th} = 0), together with the calculated results of Figure 2. Clearly, the thermal effects are critical in leading to the measured saturation currents that increase with decreasing channel length. The electron velocity depends most strongly on the temperature at low fields. Electron velocity saturation alone leads to saturation currents that are independent of the channel length (approximately equal to 2300mA/mm in our model), although the voltage at which saturation is reached increases with channel length. The conclusion that the carrier velocity is actually relatively low in the devices studied here is supported by a calculation of the longitudinal electric field in the channel. This field can be obtained by integrating the current equation (3) numerically. Figure 4 displays the longitudinal electric fields for the whole set of devices versus the normalized channel coordinate, x/L_{DS}, for an applied bias of 10V. It is apparent that even in the short devices the electric field at the source-end barely reaches 45kV/cm, well below the field where the velocity peaks [5]. Evident from Figure 4 is the partial carrier depletion near the source-end of the channel. This effect is larger in devices with short channels because the contact resistances play a greater role. In a separate calculation, the effect of diffusion implied by the non-uniform carrier density was examined and it was found to be very small for these ungated devices [8].

From the results obtained, it is to be expected that the saturation current will depend on the value of the contact resistances. A larger source contact resistance implies a stronger tendency for the channel to be pinched off near the source-end and, consequently, a larger longitudinal electric field and carrier velocity in that region when current saturation is reached. To examine the effect of the contact resistance, the saturation currents were calculated as a function of the specific contact resistance for devices with L_{DS} = 5μm. The result is shown in Figure 5 Also displayed are experimental data obtained from devices on a single wafer (although a different wafer from the one considered above) that underwent different annealing procedures. The contact resistances varied over a relatively wide range when annealing conditions were changed from 830C for 2sec (highest contact resistance) to 860C for 10sec (lowest contact resistance) [4]. Good qualitative agreement between the calculation and the experimental data is observed. However, for the case of high contact resistance there are quantitative discrepancies which may be attributable to variations in the equilibrium channel carrier concentrations with process conditions. The calculations assumed no variation in carrier concentration, transport parameters, or thermal impedance. A field analysis similar to the one shown in Figure 4 demonstrates that electron velocity saturation ensures current saturation in devices with very high contact resistance and that thermal effects are relatively less important.

In summary, calculated current-voltage characteristics of ungated AlGaN/GaN heterostructures have been presented and found to be in good agreement with experimental data. The calculations reveal that the electric field and the carrier concentration remain fairly uniform and that thermal degradation of the electron velocity is the main cause of the observed current saturation in devices with relatively low contact resistance. Devices with high contact resistance on the other hand show more pronounced field and carrier density non-uniformity. This work also indicates that further study of the gate-to-source and gate-to-drain regions in AlGaN/GaN HFETs is needed to fully understand the observed device performance. A numerical model similar to the one presented here, which includes diffusion, will be the subject of future work.

ACKNOWLEDGMENT

The work at the University of Minnesota was supported in part by Hughes Research Laboratories and by the National Science Foundation. The work at NRL was partially supported by the Office of Naval Research.

REFERENCES

1. N.X. Nguyen, C. Nguyen, and D.E. Grider, Electronics Lett. 34, 811 (1998).
2. G.J. Sullivan, J.A. Higgins, M.Y. Chen, J.W. Yang, Q. Chen, R.L. Pierson, and B.T. McDermott, Electrics Lett. 34, 922 (1998).
3. Y.-F. Wu, B. P. Keller, P. Fini, S. Keller, T. J. Jenkins, L. T. Kehias, S. P. Denbaars, and U. K. Mishra, IEEE Electron Device Lett. 19, no.2, 50, (1998).
4. P.P. Ruden, J.D. Albrecht, A. Sutandi, S.C. Binari, K. Ikossi-Anastasiou, M.G. Ancona, R.L. Henry, D.D. Koleske, and A.E. Wickenden, MRS Internet J. Nitride Semicond. Res. 4S1, G6.35 (1999).
5. J.D. Albrecht, R.P. Wang, P.P. Ruden, M. Farahmand, and K.F. Brennan, J. Appl. Phys., 83, 4777 (1998).
6. T. Harin, K. Takahashi, and Y. Shibata, IEEE Trans. Electron Devices ED-30, 1743 (1983).
7. J. Baek. M.S. Shur, K.W. Lee, and T. Vu, IEEE Trans. Electron Devices ED-32, 2426 (1985).
8. J.D. Albrecht, P.P. Ruden, to be published.

Hydrostatic and uniaxial stress dependence and photo induced effects on the channel conductance of n-AlGaN/GaN heterostructures grown on sapphire substrates

A. K. Fung[a], C. Cai[a], P. P. Ruden[a], M. I. Nathan[a], M. Y. Chen[b], B. T. McDermott[b], G. J. Sullivan[b], J. M. Van Hove[c], K. Boutros[d], J. Redwing[d], J. W. Yang[e], Q. Chen[e], M. A. Khan[e], W. Schaff[f], M. Murphy[f]
[a]Department of Electrical and Computer Engineering, University of Minnesota, Minneapolis, MN 55455, afung@ece.umn.edu, nathan@ece.umn.edu
[b]Rockwell International Science Center, Thousand Oaks, CA 91358
[c]SVT Associates, Eden Prairie, MN 55344
[d]Epitronics/ATMI, Phoenix, Arizona 85027
[e]APA Optics, Blaine, Minnesota 55449
[f]School of Electrical Engineering, Cornell University, Ithaca, NY 14853

Abstract

We measure the hydrostatic stress, uniaxial stress, and photo induced dependence of the channel conductance of two-dimensional electron gas AlGaN/GaN heterostructures grown on c-axis sapphire. The structures examined are grown by nitrogen-plasma molecular beam epitaxy and metal organic chemical vapor deposition. Electrical conductance measurements are made with four point probes on Hall bar samples. Both, hydrostatic stress and uniaxial stress result in changes in the conductance. Moreover, these changes in conductance have long settling times after the stress is applied and may be due to deep level defects, the energy levels of which change with stress. Stress coefficients extracted from the samples are partially attributed to deep level defects and to the piezoelectric effect resulting from different piezoelectric coefficients of GaN and AlN. Photo induced changes of the two-dimensional electron gas are also observed. We find that pulsed illumination produces long transient times in the conductance. These transients are reduced by thermal heating in some samples. However, they can still be present at 153°C.

Introduction

Stress experiments are useful in the characterization of semiconductors. They can be used to determine piezoelectric constants[1] and they can reveal the presence of deep defects having energy levels that change with pressure.[2] Recent efforts have utilized applied static stress/strain to piezoelectrically modify the two-dimensional (2D) electron concentrations at AlGaAs/GaAs[3] and AlGaN/GaN[4] heterostructure interfaces. The changes in channel conductance can be related to differences in the piezoelectric constants and to differences in the stress/strain profiles of the materials that constitute the heterostructures. Pressure has also been used to quantify the amount of energy shift of deep levels through the changes in conductance[2] and photoluminesce.[5] By the application of piezoelectric effects, and through the understanding of the behavior of defects on the properties of semiconductors, novel and improved device structures can be implemented.[6,7]

In this study we report on hydrostatic and uniaxial stress effects in the 2D electron gas channel conductance of AlGaN/GaN heterostructures. We find that the channel conductance exhibits a slow (on the order of 10^2 to 10^3 seconds) change with the application of pressure. This slow transient follows a very fast change that occurs on the time scale of the application or removal of pressure (a few seconds). We also examine the photo response of the samples in attempts to gain more information about the cause of these long pressure induced transients.

Mat. Res. Soc. Symp. Proc. Vol. 572 © 1999 Materials Research Society

We find that the photo responses also have long transients and that increasing the temperature of the sample can reduce them. However, the long transients can still persist even at 153°C.

Experiment

The four different epitaxial layers used in this study are grown by four different labs with either nitrogen-plasma source molecular beam epitaxy (MBE) or metal organic chemical vapor deposition (MOCVD). Here we will designate the MBE wafers as 1(a) and 1(b) and the MOCVD wafers as 2(a) and 2(b). Schematic layer structures of the different wafers are shown in Fig. 1. Nominal non-illuminated room temperature Hall mobility (cm^2/Vs)/electron sheet carrier concentration (cm^{-2}) are $580/5.40x10^{12}$, $723/1.38x10^{13}$, $933/3.50x10^{12}$ and 1520 $/1.08x10^{13}$ for wafers 1(a), 1(b), 2(a) and 2(b), respectively. Standard microfabrication techniques are used to produce Hall bars for our experiment.

Hydrostatic pressure up to 0.9GPa is applied to the samples in this study in a Unipress cell through a 1:1 mixture of hexane:pentane. Temperature in the cell is monitored with a type-T copper-constantan thermocouple. Temperature changes in the cell follow a damped response to the ambient lab temperature and can vary by as much as ±0.5°C.

Compressive uniaxial pressure up to 0.6GPa is applied to a rectangularly prepared sample by placing one end of the sample flush against a fixed aluminum platform and by applying pressure to the other end of the sample with a tungsten anvil. The rectangular sample dimensions are typically 0.85mm x 0.43mm x 2.5mm with the long edge parallel to the direction of applied stress. Details of the apparatus and method can be found elsewhere.[3,8]

Photoconductivity measurements are performed with various excitation sources, light emitting diodes with wavelengths of 470 and 928nm, a halogen lamp, and a mercury-gas-fluorescent white lamp. Electrical measurements are performed with a HP4156A dc semiconductor analyzer. Conductance measurements are obtained using four point contacts on Hall bars.

5nm $Al_{0.15}Ga_{0.85}N$	5nm GaN
25nm $Al_{0.15}Ga_{0.85}N$ $n = 2x10^{18}\ cm^{-3}$	20nm $Al_{0.2}Ga_{0.8}N$
5nm $Al_{0.15}Ga_{0.85}N$	1µm GaN
1µm GaN	5nm AlN
Sapphire Substrate Wafer 1(a)	Sapphire Substrate Wafer 1(b)

30nm $Al_{0.15}Ga_{0.85}N$ $n = 2x10^{18}\ cm^{-3}$	30nm $Al_{0.15}Ga_{0.85}N$
10nm $Al_{0.15}Ga_{0.85}N$	100nm GaN $n = 2x10^{17}\ cm^{-3}$
5µm GaN	1 µm GaN
40nm AlN	Sapphire Substrate Wafer 2(b)
Sapphire Substrate Wafer 2(a)	

Figure 1: Schematic epitaxial layer structures of wafers used in this study. Wafer 1(a) and 1(b) are grown by MBE, wafer 2(a) and 2(b) are grown by MOCVD.

Results

The application of hydrostatic stress result in changes in conductance of 1.0%/GPa and 3.5%/GPa for wafers 1(a) and 2(a), respectively. Compressive uniaxial stress produce changes in the conductance on wafer 1(a) of −0.25%/GPa and −0.63%/GPa for stress in the [11$\bar{2}$0] and [$\bar{1}$010] GaN directions, respectively. These values are determined from the steady state values after the pressure induced transient on the conductance and are the result of at least two effects: the piezoelectric effect and a change in the ionization of deep level defects. For the piezoelectric effect, the presence of inversion domains[9] can modify the measured change in conductance since the sign of the piezoelectric polarization depends on the orientation of the (hexagonal) growth axis. Figure 2 shows time dependence of the conductance of wafer 1(a) for an applied compressive uniaxial stress. The time domain behavior of the conductance shows the presence of at least two processes, which result in a fast initial change in conductance and a long decay in conductance with applied stress. The piezoelectric polarization at each interface of dissimilar materials and their neutralization by available free carriers from the contacts or the semiconductor bulk occur essentially instantaneously as do stress induced energy level shifts of deep level defects. Both processes occur in a much shorter time than the time step of each data point and will follow the applied pressure without time delay. In Fig. 2, at the leading edge where stress is applied, the loading process is done over a period of a few seconds so as not to overload the sample. After the stress is applied it can be seen that there is a slow process that affects the conductance of the sample. The measured conductance is determined by the number of carriers and their mobility. Although the mobility of the electrons can change with the change in electron density due to screening and other effects, slow trapping and detrapping of carriers with pressure most likely cause the long time constants. These slow trapping times may be due to energy barriers associated with certain types of defects. Also, band discontinuities of the different materials in the heterostructure and band bending from space charge may additionally impede and slow movement of free carriers in response to applied

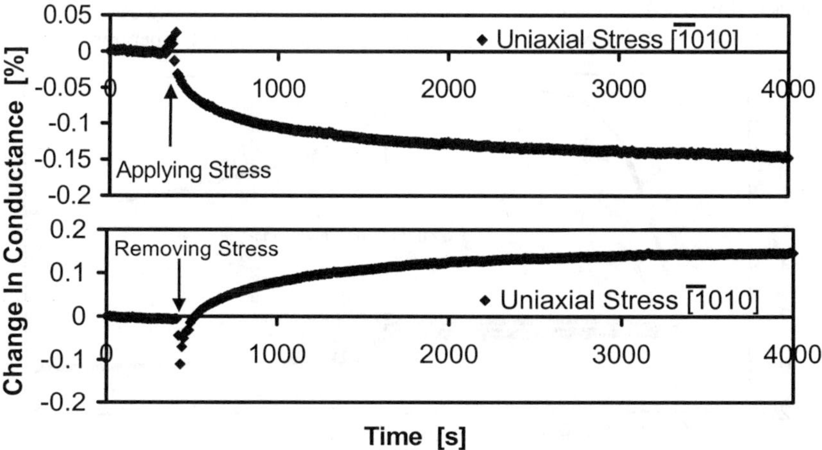

Figure 2: Time dependence of conductance for compressive uniaxial stress on wafer 1(a) along the [$\bar{1}$010] GaN direction. Application of stress goes from 0.38GPa to 0.64GPa. The removal of stress goes from 0.64GPa to 0.26 GPa.

pressure. In an attempt to further understand the slow transient behavior of the conductance with pressure we also examine the photoconductive response of the samples under various illumination conditions.

Figure 3 shows the pulsed illumination response for all four wafers examined in this study. All the samples show slow conductance time transients with illumination. This is not surprising; persistent photoconductivity (PPC) has been widely reported for bulk GaN[10,11] and AlGaN/GaN heterostructures.[12,13] In our samples, we find that the conductance also changes with infrared illumination (928nm). Wafers 1(a), 1(b) and 2(b) show an increase in conductance with illumination whereas wafer 2(a) shows a slight decrease in conductance. The increase in conductance may be caused by photo ionization of donor atoms or deep-donor-like defect levels. The decrease in conductance can be due to the same processes that have been reported to cause optical quenching.[14,15] For wafer 2(a) prior to illumination, the sample may be in a state such that thermally or optically generated electrons from the valence band are present in the conduction band with their corresponding valence band holes at least partly trapped in gap states that have long lifetimes.[11] Then, upon illumination, the trapped holes in the gap states within the photon energy range of the infrared LED (1.34eV) are freed into the valence band where they can recombine with conduction band electrons directly or via recombination centers in the band gap, resulting in a decrease in conductance.[14,15] To elucidate the slow conductance transients resulting from illumination we examine the effect of increasing the temperature on optical quenching and on the PPC of our samples.

Since heating reduces both optical quenching and PPC and thermal noise makes it difficult to observe these effects otherwise. It is necessary to increase the difference between the initial (without optical quenching) and final (with optical quenching) conductive states for optical quenching and also for the initial (illuminated) and final (non-illuminated) conductive states for PPC. To generate electron-hole pairs we use a mercury-vapor-fluorescent lamp for all the samples examined. To optically quench (reduce) the conductance for wafer 2(a) while the fluorescent lamp is on, we use a halogen lamp. With increasing temperatures we find that for wafer 2(a) optical quenching is reduced for fixed photon fluxes. Furthermore, at increasing temperatures the effect of the halogen illumination increases the conductance rather than decreases it (Fig. 4(a)). In addition to reducing optical quenching, heating can decrease the PPC transient times for wafer 2(a). The transient times at higher temperatures are shorter

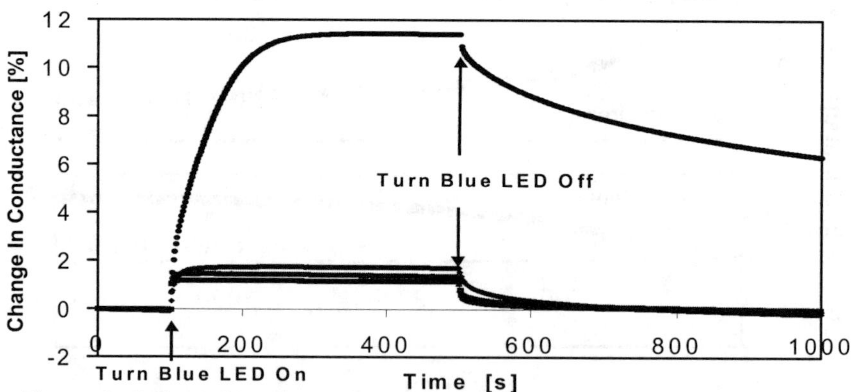

Figure 3: Pulsed illumination response for all four wafers studied in this experiment. Long conductance transients are observed. Listing wafers from greatest to smallest change in conductance at 300 seconds are wafer 2(a), 2(b), 1(b) and 1(a).

than at 27°C, however they can still persist up to 153°C (see Fig. 4(b)). Heating also reduces the amount of photo-induced changes in the conductance for wafer 1(a). The change in conductance for this sample from the illuminated state to the non-illuminated state at different temperatures and a fixed fluorescent illumination intensity are −0.9%, −0.35% and −0.25% for 27°C, 90°C and 153°C, respectively. This effect can be the result of thermal ionization of states in the band gap, leaving gap states emptier, such that it becomes less likely for photons with energy less than the band gap to generate free carriers. Transient times are not reported for wafer 1(a) since they could not be accurately quantified due to thermal noise. We also note that the thermal coefficients for the electrical conductance of wafers 1(a) and 2(a) are linearly interpolated to be −0.24%/°C and −0.46%/°C, respectively, between 27°C and 153°C with fluorescent illumination. This can be due to a decrease in mobility and possibly to pyroelectric effects that produce fixed polarization charges at the 2D electron gas interface that repel electrons from the conduction path.

Conclusion

We have measured the hydrostatic and uniaxial stress dependence and the photo response of AlGaN/GaN heterostructures. Hydrostatic stress results in changes in conductance of 1.0%/GPa and 3.5%/GPa for wafers grown by MBE and MOCVD, respectively. For the MBE wafer, uniaxial stress coefficients for the change in conductance are −0.25%/GPa and

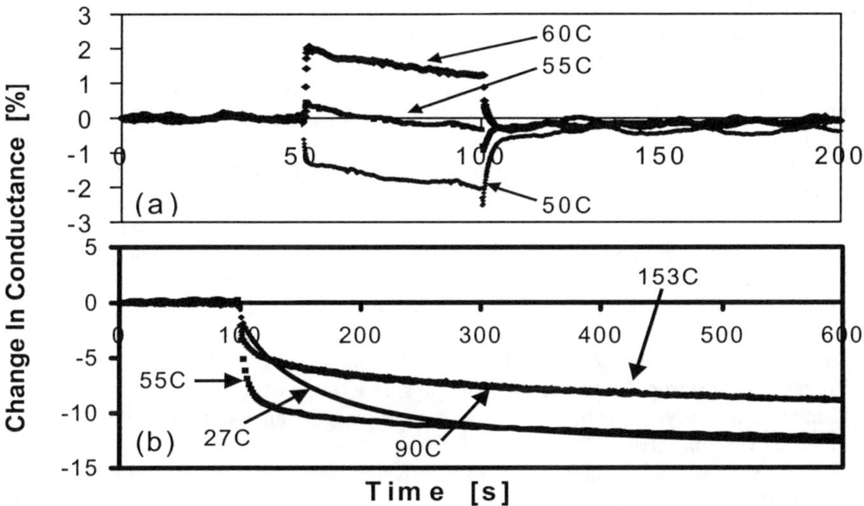

Figure 4: (a) Optical quenching as a funcion of temperature for wafer 2(a). The sample is illuminated with a fluorescent lamp at all times. A halogen lamp is used to quench the conductance when it is turned on between 50 and 100 seconds. (b) Persistent photoconductivity as a function of temperature for wafer 2(a). The sample is illuminated with a fluorescent lamp prior to 0 seconds. The lamp is turned off at 100 seconds. PPC response is nearly identical at 90C and 153C.

-0.63%/GPa for stress in the $[11\bar{2}0]$ and $[\bar{1}010]$ GaN directions, respectively. These hydrostatic and uniaxial stress coefficients are attributed to the piezoelectric effect and stress induced changes in ionization of defect levels. We can not quantitatively separate these two effects. The application of stress results in conductance transients on the order of 10^2 to 10^3 seconds. Because of this, traps in the band gap due to defects are suspected. Photoconductivity measurements show that the conductance of the MBE sample increases with light having energy less than the band gap and may be due to the photo ionization of donors or deep level electron traps. For the MOCVD sample, photoconductivity measurements show optical quenching. This may be the result of hole trap levels in the band gap. Both the MBE and MOCVD samples show persistent photoconductivity with long transients. The transients are reduced with an increase in temperature, however even at 153°C long transients can still be observed.

Acknowledgments

We thank Brian Ishaug and Xiaobo Zhang for providing the optical spectra of the light sources used in this experiment and John Albrecht for helpful discussions. This work was supported in part by NSF-ECS9612539, NSF-DMR and ONR N/N00014-9525758.

References

1. G. Arlt and P. Quadflieg, Phys. Status Solidi 25, 323 (1968)
2. J. Mercy, C. Bousquet, J. Robert, A. Raymond, G. Gregoris, J. Beerens, J. Portal, and P. Frijlink, Surf. Sci. 142, 298 (1984)
3. A. Fung, J. Albrecht, M. Nathan, and P. Ruden, J. Appl. Phys. 84, 3741 (1998)
4. R. Gaska, J. Yang, A. Bykhovski, M. Shur, V. Kaminski, and S. Soloviov, Appl. Phys. Lett. 72, 64 (1998)
5. S. Kim, I. Herman, J. Tuchman, K. Doverspike, L. Rowland, and D. Gaskill, Appl. Phys. Lett. 67, 380 (1995)
6. O. Ambacher, J. Smart, J. Shealy, N. Weimann, K. Chu, M. Murphy, W. Schaff, L. Eastman, R. Dimitrov, L. Wittmer, M. Stutzmann, W. Rieger, and J. Hilsenbeck, J. Appl. Phys. 85, 3222 (1999)
7. W. Yang, T. Nohava, S. Krishnankutty, R. Torreano, S. McPherson, and H. Marsh, Appl. Phys. Lett. 73, 978 (1998)
8. S. Lu, C. Meng, F. Wiiliamson, and M. Nathan, J. Appl. Phys. 69, 8241 (1991)
9. L. Romano, J. Northrup, M. O'Keefe, Appl. Phys. Lett. 69, 2394 (1996)
10. E. Munoz, E. Monroy, J. Garrido, I. Izpura, F. Sanchez, M. Sanchez-Garcia, B. Beaumount, and P. Gibart, Appl. Phys. Lett. 71, 870 (1997)
11. C. Qiu, W. Melton, M. Leksono, J. Pankove, B. Keller, and S. DenBaars, Appl. Phys. Lett. 69, 1282 (1996)
12. J. Li, J. Lin, H. Jiang, M. Khan, and Q. Chen, J. Vac. Sci. Tech. B 15, 1117 (1997)
13. X. Dang, C. Wang, E. Yu, K. Boutros, and J. Redwing, Appl. Phys. Lett. 72, 2745 (1998)
14. Z. Huang, D. Mott, P. Shu, R. Zhang, J. Chen, and D. Wickenden, J. Appl. Phys. 82, 2707 (1997)
15. M. Hirsch, O. Seifert, O. Kirfel, J. Parisi, J. Wolk, W. Walukiewicz, E. Haller, O. Ambacher, M. Stutzmann, Mat. Res. Soc. Symp. Pro. 482, 531 (1998)

THE INFLUENCE OF SPONTANEOUS AND PIEZOELECTRIC POLARIZATION ON NOVEL ALGAN/GAN/INGAN DEVICE STRUCTURES

B. E. FOUTZ, M. J. MURPHY, O. AMBACHER, V. TILAK, J. A. SMART, J. R. SHEALY, W. J. SCHAFF, and L. F. EASTMAN
School of Electrical Engineering, Cornell University, Ithaca, New York 14853

ABSTRACT

The strong spontaneous polarization and piezoelectric effects in the wurtzite III-nitride semiconductors lead to new possibilities for device design. In typical heterojunction field effect transistors these effects are used to create large electron concentrations at the AlGaN/GaN interface. However, we examine several other possible device structures which include heterojunctions of AlGaN, GaN, and InGaN. For example, we find the strong electric fields present in these structures allow us to create quantum wells greater than 1 eV deep. Both Ga-faced and N-faced materials are explored. The two-dimensional electron gas concentrations in these structures are found using a self-consistent 1-D Schrödinger-Poisson solver modified to incorporate the effects of spontaneous and piezoelectric polarization. The boundary conditions at the heterojunction interfaces and at the surface and substrate are discussed in detail. Electron concentrations are compared with those obtained experimentally through capacitance-voltage and Hall effect measurements.

INTRODUCTION

The group III-nitrides possess a large spontaneous and piezoelectric polarization. The presence of this strong polarization is supported by both theoretical calculations of its existence [1] and the large electron concentrations which result at the AlGaN/GaN heterojunctions in transistor structures. Simple models [2, 3, 4] have been used to calculate the electron concentration at a single heterointerface and support the hypothesis that the 2D electron gas found at the interface is induced by polarization effects. However, to determine the conduction band profile and electron concentrations in more complex device structures more sophisticated and generalized methods are needed. One such method is the use of a self-consistent one-dimensional Schrödinger-Poisson solver. The solver used in this study, CBAND (for conduction band solver), has been modified to incorporate the effects of spontaneous and piezoelectric polarization. Specifically, the polarization is included in the solver through the use of thin layers of space charge at each heterointerface of the structure. In this paper we describe the theory used to calculate the conduction band profile and electron sheet charge in nitride based heterostructures and then compare these results with experimental data. Finally, we examine CBAND results for two novel device structures.

THEORY

The polarization present in the group III-nitrides is due to the lack of inversion symmetry along the c-axis of the wurtzite crystal structure. In relaxed material there exists a built-in or spontaneous polarization [1]. This polarization points toward the substrate for Ga-face material and points toward the surface in N-face material. (For Ga-face material the

	AlN	GaN	InN
P_{SP} (C/m^2)	-0.081	-0.029	-0.032
e_{33} (C/m^2)	1.46	0.73	0.97
e_{31} (C/m^2)	-0.60	-0.49	-0.57
C_{13} (GPa)	108	103	92
C_{33} (GPa)	373	405	224
a_0 (Å)	3.112	3.189	3.54

Table I: The constants used to calculate the polarization in III-nitride layers. P_{SP} is the spontaneous polarization. e_{33} and e_{31} are piezoelectric constants. C_{13} and C_{33} are elastic deformation constants and a_0 is the lattice constant.

positive direction is toward the surface by convention.) The polarization in the material can be changed by placing it under strain. This change in polarization is commonly called the piezoelectric polarization and is given by

$$P_{PE} = 2\frac{a - a_0}{a_0}\left(e_{31} - e_{33}\frac{C_{13}}{C_{33}}\right),$$

(1)

where a is the lattice constant under strain, and a_0 is the lattice constant of the relaxed material. The constants e_{31} and e_{33} are piezoelectric constants and C_{13} and C_{33} are elastic deformation constants. The total polarization in a given layer is simply the sum of the spontaneous and piezoelectric polarization, i.e., $P = P_{SP} + P_{PE}$. The constants used in our calculation are from Bernardini et al. [1] and Wright [5] and are shown in Table I. For alloys, the constants are linearly interpolated. At a heterojunction there is usually a change in the polarization on each side. This abrupt change in polarization causes a bound sheet charge. In general, the bound sheet charge is the polarization of the bottom layer minus the polarization of the top layer, $\sigma = P\,(\text{bottom}) - P\,(\text{top})$.

We now examine the particular situation present in AlGaN/GaN field effect transistors. We assume the thin AlGaN layer is pseudomorphically lattice matched to the thick GaN layer below it. Figure 1 shows the situation for both Ga-face and N-face material. In a Ga-faced structure, a positive bound charge is created at the deeper interface which causes the formation of a 2-D electron gas at the lower interface. In N-face material, the positive bound charge is present at the upper interface and the 2-D electron gas will form there. With the proper structure geometry, it may be possible to form a 2-D hole gas at the other interface [6], however, this has yet to be reported experimentally.

With a theoretical understanding of polarization we now wish to predict the effect this polarization will have on conduction band diagrams and electron concentrations for a variety of device structures. For this purpose we use a 1-D Schrödinger-Poisson solver. First the Schrödinger's equation is solved to find the electron concentration and then Poisson's equation is solved to determine the conduction band profile. The new conduction band profile will modify the solution of the electron concentration, so the process is repeated until there is little change in the solution, i.e., the solution is self-consistent. Our computer program which solves this self-consistent problem, called CBAND, is very similar to other self-consistent solvers [7]. The program however, must be modified to incorporate the effects of the spontaneous and piezoelectric polarization. This is accomplished by adding thin layers of charge at the heterojunction interfaces equivalent to the bound sheet charge caused by the

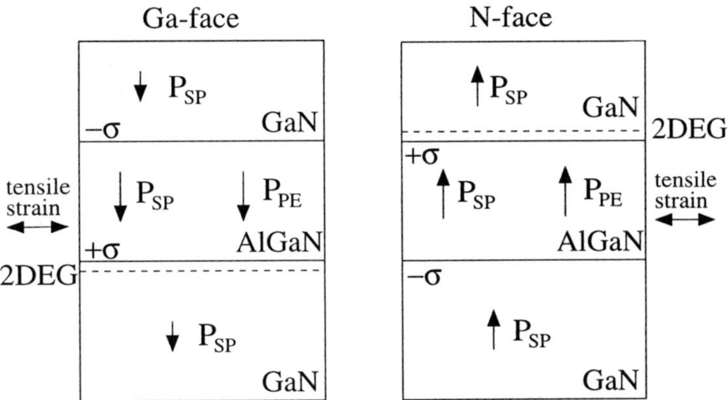

Figure 1: The direction of polarization and the location of the 2DEG in Ga-face and N-face AlGaN HFETs. In both cases, the AlGaN layer is under tensile strain leading to both a spontaneous and piezoelectric component to the polarization. For Ga-face materal the direction of polarization causes the formation of a 2DEG at the lower interface. In N-face material the direction of polarization is reversed causing the 2DEG to form at the upper interface.

polarization. The charge in these layers is added to the space charge used to solve Poisson's equation.

It is also necessary to specify boundary conditions for the conduction band at the surface, heterojunction, and substrate. In our structures we assumed a Schottky barrier contact at the surface, pinning the conduction band to a fixed value. For Al_xGaN, we assume the barrier is $\Phi_B = 0.84 + 1.3x$ eV[2]. At the heterointerface, we use the conduction band discontinuity, ΔE_c, given by Wei and Zunger [8]. It turns out, for reasonable values of Φ_B and ΔE_c, there is little impact on the resulting electron concentrations. This is the case, since the electron concentration is dominated by the polarization induced bound charge. This can be seen, for example, from the analytical expression given by Asbeck, Eq.(2) of [3] (or Eq.(15) of [2]). At the substrate the conduction band is simply set to one half of the band gap of the material next to the substrate. It turns out that the choice of conduction band level at the substrate also has very little impact on the 2-D electron gas concentrations since the substrate is typically far away from the heterojunction.

RESULTS

In Figure 2a the calculated solution for the conduction band diagram and electron concentration for a typical AlGaN/GaN FET is shown. We assume the material is Ga-face, and do not include the top GaN layer that is shown in Figure 1. The AlGaN barrier layer is 300 Å thick and contains 30% aluminum. In this case, the bound sheet charge, σ, is 1.68×10^{13} cm^{-2}. Therefore a 6 Å layer containing 2.8×10^{20} cm^{-3} positive space charge is included at the interface. The conduction band discontinuity at the interface is 0.378 eV and the Schottky barrier at the surface is pinned at 1.23 eV. The calculation uses a 1 micron GaN buffer and the conduction band is pinned to 1.7 eV at the substrate boundary. From the curvature of the conduction band we find the maximum electric field, located at the

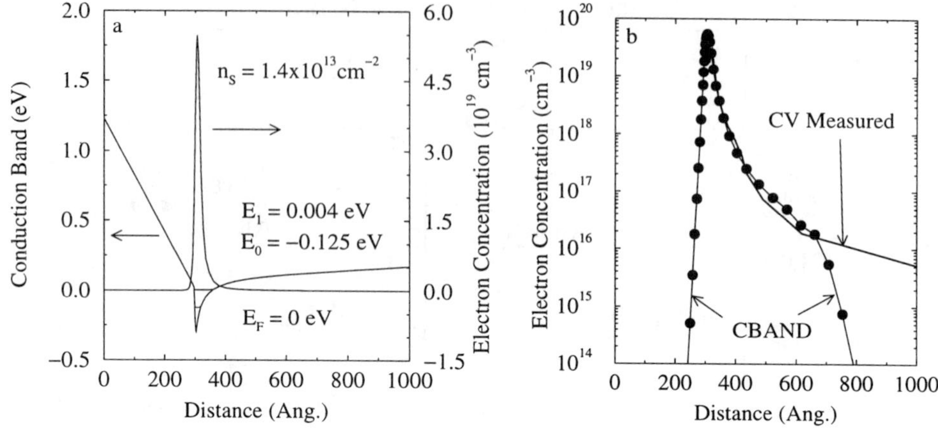

Figure 2: The calculated conduction band diagram and electron concentration for a typical Al-GaN/GaN HFET. The 2DEG sheet density and the location of the subbands with respect to the Fermi level are shown in (a). Figure (b) shows a comparison of the electron concentration calculated by CBAND along with an experimentally measured profile determined by the capacitance-voltage technique.

heterojunction, is about 2 MV/cm. The total electron sheet concentration is 1.4×10^{13} cm^{-2} and the maximum electron density is 5.4×10^{19} cm^{-3} with 22% of the electron sheet density in the AlGaN layer. The electron concentration profile calculated by CBAND can be compared with an experimentally determined profile through the capacitance-voltage technique. As shown in Figure 2b, there is good agreement between the profiles over three orders of magnitude in carrier concentration.

Next we compare the electron concentration of grown structures with our calculations. Figure 3 depicts the measured electron concentration as a function of the aluminum content in the barrier. The boxed items show the data for undoped structures grown by MBE [9] and the circles show data for undoped structures grown by MOCVD. To demonstrate that both spontaneous and piezoelectric polarization is important to explain all the sheet charge found in AlGaN/GaN devices, calculated results are shown for three situations. The three curves show the calculated sheet density for a 200 Å AlGaN barrier with only the effects of piezoelectric polarization included, only the effects of spontaneous polarization, and with both. Most of the experimental data lies between the calculation for the spontaneous polarization only and the calculation that includes both spontaneous and piezoelectric polarization. Since the CBAND sheet densities can be considered an upper bound it is clear that both spontaneous and piezoelectric polarizations are important in determining the total sheet charge.

It is also fairly straight forward to use CBAND to investigate the conduction band profiles and electron concentrations in novel device structures. In Figure 4 we show the results of calculations for two possible device layers. Figure 4a shows how an AlGaN buffer can be used to increase the confinement of the 2D gas by providing a greater than 1 eV conduction band wall on either side of the 2D gas. In this structure a five percent AlGaN buffer is used. When the material is abruptly changed to GaN the change in polarization induces a strong

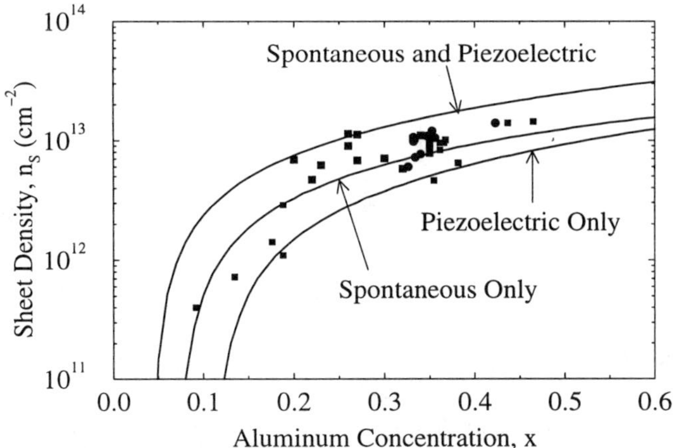

Figure 3: A comparison of experimentally determined sheet charge as a function of aluminum concentration of the AlGaN barrier with that calculated by CBAND. The calculated curves show results only including the effects of piezoelectric polarization, spontaneous polarization only, and both.

electric field which bends the conduction band down sharply. In this case a 300 Å GaN layer is used. Then in the top barrier layer, a 30% aluminum layer, the direction of the electric field is reversed and the conduction band is lifted sharply upward. The electron concentration in this structure is predicted to be 1.2×10^{13} cm^{-2} and may increase performance due to the increased confinement of the electrons. This may be especially important during transistor operation, when the electron energy can rise dramatically due to the large electric field along the channel.

We can further increase the sheet density of devices by using an InGaN buffer. The conduction band diagram is shown in Figure 4b. Using 20% indium in the buffer increases the strain and piezoelectric polarization of the the AlGaN barrier enough to increase the sheet charge to 3.5×10^{13} cm^{-2}. Increasing the indium content of the buffer continues to increase the 2-D gas sheet concentration, however, at some point strain relaxation will occur.

CONCLUSIONS

In this paper the spontaneous and piezoelectric properties of the group III-nitrides have been outlined. These effects have been incoporated into a self-consistent 1-D Schrödinger-Poisson solver which calculates conduction band diagrams and electron concentration profiles in group III-nitride device structures. It has been shown that both the spontaneous and piezoelectric polarization are important for determining sheet carrier concentration observed in grown devices.

ACKNOWLEDGEMENTS

The authors wish to thank the Office of Naval Research for financial support under their MURI program: Grant # N00014-96-1-1223; Project Monitor Dr. J. Zolper. Dr. Ambacher

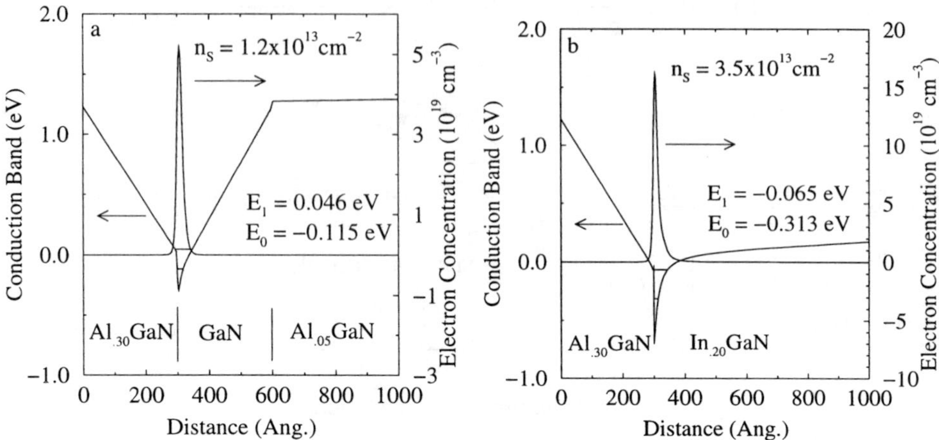

Figure 4: Two novel device structures calculated by CBAND. Figure (a) shows the results for a structure which uses an AlGaN buffer to increase the 2DEG confinement. A structure which includes indium in the buffer to increase the strain and piezoelectric polarization in the AlGaN is shown in (b).

would like to thanks the Alexander von Humboldt Stiftung for a Feodor Lynen fellowship.

REFERENCES

[1] F. Bernardini, V. Fiorentini, and D. Vanderbilt, Phys. Rev. B **56**, R10024 (1997).

[2] O. Ambacher, J. Smart, J. R. Shealy, N. G. Weimann, K. Chu, M. Murphy, W. J. Schaff, and L. F. Eastman, R. Dimitrov, L. Wittmer, M. Stutzmann, W. Reiger, and J. Hilsenbeck, J. Appl. Phys. **85**, 3222 (1999).

[3] P. M. Asbeck, E. T. Yu, S. S. Lau, G. J. Sullivan, J. Van Hove, and J. Redwing, Elec. Lett. **33**, 1230 (1997).

[4] A. D. Bykhovski, R. Gaska, M. S. Shur, Appl. Phys. Lett. **73**, 3577 (1998).

[5] A. F. Wright, J. Appl. Phys. **82** 2833 (1997).

[6] F. D. Sala, A. D. Carlo, P. Lugli, F. Bernardini, V. Fiorentini, R. Scholz, J.-M. Jancu, Appl. Phys. Lett. **74**, 2002 (1999).

[7] I-H. Tan, G. L. Snider, L. D. Chang, and E. L. Hu, J. Appl. Phys. **68** 4071 (1990).

[8] S. Wei, A. Zunger, Appl. Phys. Lett. **69**, 2719 (1996).

[9] M. J. Murphy, B. E. Foutz, K. Chu, H. Hu, W. Yeo, W. J. Schaff, O. Ambacher, L. F. Eastman, T. J. Eustis, R. Dimitrov, M. Stutzmann, W. Rieger, MRS Internet J. Nitride Semicond. Res. **4S1**, G8.4 (1999).

PIEZOELECTRIC SCATTERING IN LARGE-BANDGAP SEMICONDUCTORS AND LOW-DIMENSIONAL HETEROSTRUCTURES

B.K. RIDLEY*, N.A. ZAKHLENIUK, C.R. BENNETT, M. BABIKER, and D.R. ANDERSON
Department of Physics, University of Essex, Colchester, CO4 3SQ, UK
* also Department of Electrical Engineering, Cornell University, USA

ABSTRACT

We develop a rigorous theory of piezoacoustic phonon limited electron transport in bulk GaN and GaN-based heterostructures. Within the Boltzmann equation approach we derive a new expression for the momentum relaxation rate and show that the Pauli principle restrictions are comparable in importance to a screening effect at temperatures up to 150 K provided that the electron density is large. This is of particular importance for electrons in GaN/AlN-based quantum wells where very high electron densities initiated by the piezoelectric effect have recently been reported. Variations of the piezoacoustic phonon limited electron mobility with the lattice temperature and with the electron density for a zinc-blende and wurtzite GaN are presented.

INTRODUCTION

Large-bandgap semiconductors, such as GaN and AlN, have important piezoelectric properties due to the large values of their piezoelectric tensor components [1]. As a consequence of this, the electron interaction with piezoacoustic phonons provides a significant contribution to the transport coefficients of the electrons within a very wide temperature range. The importance of piezoacoustic scattering is well known and has been studied both experimentally and theoretically. Usually piezoacoustic scattering dominates at low temperatures, but this is not the case in the context of large-bandgap semiconductors where even near the room temperature this scattering can be comparable with polar optical and deformation acoustic phonon scattering [2, 3]. This situation appears to be ripe for investigations on piezoelectric scattering in low-dimensional heterostructures which rigorously and consistently take into account specific features of the nitride large-bandgap semiconductors.

In this paper we study the piezoacoustic mobility of two-dimensional (2D) electrons confined in an infinite potential quantum-well (QW) formed from a GaN/AlN double heterostructure. Both the zinc-blende (ZB) and wurtzite (WZ) crystal structures of GaN are considered. The effect of the screening on the electron-piezoacoustic-phonon interaction is incorporated via the Lindhard dielectric function. For comparison we also calculate the electron mobility of the three-dimensional (3D) electron gas in bulk GaN.

Mat. Res. Soc. Symp. Proc. Vol. 572 © 1999 Materials Research Society

THEORY

In order to calculate the electron kinetic coefficients it is necessary to solve the Boltzman kinetic equation for the distribution function $F(\vec{k})$,

$$-\frac{e\vec{E}}{\hbar}\frac{dF(\vec{k})}{d\vec{k}} = \sum_{\vec{k'}}\left[W_{\vec{k'}\vec{k}}F(\vec{k'})[1-F(\vec{k})] - W_{\vec{k}\vec{k'}}F(\vec{k})[1-F(\vec{k'})]\right], \tag{1}$$

where $W_{\vec{k}\vec{k'}}$ is the probability of scattering between electron states with wavevectors $\vec{k}$ and $\vec{k'}$ due to the phonon emission and absorption processes, $\vec{E}$ is an external electric field, and e is the electronic charge. Because we will consider an electron gas with an arbitrary degree of degeneracy, the Pauli principle restrictions (PPR) are explicitly taken into account in Eq. (1). Expressing $F(\vec{k})$ as a sum of a symmetric part $F_o(\varepsilon)$ and an antisymmetric part $F_1(\vec{k})$, we obtain a solution for the antisymmetric part in the form:

$$F_1(\vec{k}) = \frac{\hbar}{m^*}e\tau(\varepsilon)\frac{dF_o(\varepsilon)}{d\varepsilon}(\vec{k}\cdot\vec{E}), \tag{2}$$

where m^* is an electron effective mass, $F_o(\varepsilon) = \left[\exp\{(\varepsilon - \varepsilon_F)/k_oT_o\} + 1\right]^{-1}$ is an equilibrium Fermi distribution function, ε_F is the Fermi energy, T_o is the lattice temperature, k_o is the Boltzmann constant, and $\tau(\varepsilon)$ is the momentum relaxation time which is given by the expressions

$$\frac{1}{\tau(\varepsilon)} = \frac{2\pi}{\hbar}\sum_{\vec{q}}\frac{1}{\varepsilon_{el}^2(\vec{q'})}|C(\vec{q})|^2|G_{ep}(\vec{q})|^2\frac{\vec{k}\cdot\vec{q'}}{k^2}\Phi(\varepsilon,\hbar\omega_q)\delta(\varepsilon_{\vec{k}-\vec{q}}-\varepsilon), \tag{3}$$

$$\Phi(\varepsilon,\hbar\omega_q) = \left[\frac{(N_q+1)[1-F_o(\varepsilon-\hbar\omega_q)]}{1-F_o(\varepsilon)} + \frac{N_q[1-F_o(\varepsilon+\hbar\omega_q)]}{1-F_o(\varepsilon)}\right]. \tag{4}$$

Here $C(\vec{q})$ and $G_{ep}(\vec{q})$ are, respectively, the coupling constant and the form-factor for the electron-phonon interaction, $\varepsilon_{el}(\vec{q'})$ is the static Lindhard dielectric function [4] which accounts for the electronic screening of the electron-phonon interaction, $\omega_q = c_\alpha q$ is the phonon frequency at wavevector $\vec{q}$, c_α is the longitudinal ($\alpha = L$) or transverse ($\alpha = T$) sound velocity, $N_{\vec{q}} = \left[\exp(\hbar\omega_q/k_oT_o)-1\right]^{-1}$ is the phonon distribution function. The expressions obtained above are general for 3D and 2D electron gases interacting with bulk deformation acoustic or piezoacoustic phonons. In the 3D case $\vec{k}$ is a three-dimensional electron wavevector and $\vec{q'} \equiv \vec{q}$, while in the 2D case $\vec{k}$ is a two-dimensional electron wavevector and $\vec{q'} = \vec{q}_\parallel$ with $\vec{q}_\parallel$ the in-plane component of the phonon wavevector $\vec{q}$. The form-factor $G_{ep}(\vec{q}) = 1$ for a 3D gas and $|G_{ep}(q_\perp)|^2 = \pi^4(q_\perp d/2)^{-2}\sin^2(q_\perp d/2)/[\pi^2 - (q_\perp d/2)^2]^2$ for 2D electrons occupying only the first subband of the infinite potential QW, where $q_\perp$ is the component of $\vec{q}$ normal to the

QW and d is the QW width. A careful analysis of the expression for $\Phi(\varepsilon, \hbar\omega_q)$ in Eq.(4), which after using expressions for $F_o(\varepsilon)$ and N_q can be presented in an equivalent form as

$$\Phi(\varepsilon, \hbar\omega_q) = (2N_q + 1)\cosh^2[(\varepsilon - \varepsilon_F)/2k_oT_o]/[\cosh^2[(\varepsilon - \varepsilon_F)/2k_oT_o] + \sinh^2(\hbar\omega_q/2k_oT_o)],$$

shows that despite the fact that the phonon energy $\hbar\omega_q$ is always smaller than the electron energy ε, which is of the order of ε_F, the phonon energy must be retained in the argument of the functions $F_o(\varepsilon \pm \hbar\omega_q)$ provided that the lattice temperature is low, $k_oT_o < \sqrt{8m^*c_\alpha^2\varepsilon_F}$. A numerical estimation of the right-hand side of this inequality gives respectively 30 K for bulk GaN with electron density $n_0=10^{18}$ cm^{-3} and 140 K for 2D electrons of density $n_0=3\times10^{13}$ cm^2 (such high electron densities have recently been reported [5] in a GaN/AlN heterostructure and were attributed to the doping effect of the intrinsic piezoelectric field). The term in question in Eq. (3) takes into account the effect of PPR on the momentum relaxation rate and its presence means that the relaxation rate of an individual electron depends on the electron density n_0 (even if the screening is ignored) through the Fermi energy. At high lattice temperatures, when $k_oT_o > \sqrt{8m^*c_\alpha^2\varepsilon_F}$, we obtain $\Phi(\varepsilon, \hbar\omega_q) \approx (2N_q + 1) \approx 2k_oT_o/\hbar\omega_q$ and $\tau(\varepsilon)$ depends on n_0 only in the presence of the screening. This brings qualitatively new features into the temperature and density dependences of the electron mobilities. It is important to note that the expression obtained in Eq. (3) for the momentum relaxation rate is only an approximate solution within the Boltzmann equation approach. It becomes exact only in the test particle approach [6, 7] in the Bloch-Gruneisen regime. The details of its derivation will be reported elsewhere.

The coupling coefficient $|C(\vec{q})|^2 = \left[\hbar(e\hat{h})^2/2\rho Vc_\alpha q\right]H_\alpha(\vec{e})$ for the piezoacoustic phonons depends in general on the phonon wave polarization α and on the orientation of the vector $\vec{e} \equiv \vec{q}/q$ with respect to the crystal axes [8] (here ρ is a material density, V is a sample volume, and $\hat{h}$ is a component of the piezoelectric tensor). The explicit form of the function $H_\alpha(\vec{e})$ depends on the crystal symmetry. For the ZB symmetry $\hat{h} = h_{14}$, $H_L(\vec{e}) = (6q_xq_yq_z/q^3)^2$ and $H_T(\vec{e}) = 4[q_x^2q_y^2/q^4 + q_x^2q_z^2/q^4 + q_y^2q_z^2/q^4] - H_L(\vec{e})$ are the longitudinal and transverse components, respectively. For the crystal with WZ symmetry $\hat{h} = h_{33}$, $H_L(\vec{e}) = (q_z/q)^2[1 - (h_x/h_{33})(q_x^2 + q_y^2)/q^2]^2$ with the notation $h_x = h_{33} - h_{31} - 2h_{15}$, and $H_T(\vec{e}) = [(q_x^2 + q_y^2)/q^2][h_{15}/h_{33} + (h_x/h_{33})q_z^2/q^2]^2$. The usual approximation is to use the angular average of $H_\alpha(\vec{e})$. This approach is justified in the case of bulk semiconductors, but it requires a special investigation in the case of low-dimensional heterostructures. The physical reason for this is that for 2D electrons there is no momentum conservation along the quantization direction. As a consequence of this, the phonon wavevector components $\vec{q}_\perp$ and $\vec{q}_\parallel$ play principally different roles in the electron-phonon interaction. Using the bulk-like expression for the coupling coefficient may lead to the wrong estimate of the corresponding interaction strength. It is necessary first to express $H_\alpha(\vec{e})$ as a functions of $\vec{q}_\perp$ and $\vec{q}_\parallel$ and then average it over the orientations of the in-plane wavevector $\vec{q}_{\parallel}$ [9]. As a result, the coupling coefficient does not depend on the orientation of $\vec{q}_{\parallel}$, but it depends on the ratio of $q_{\perp}/q_{\parallel}$ and also on the crystal orientation of the QW planes. Here we will consider three different orientations of the QW for the ZB structure, when the 2D planes are parallel to the (001), (111), and (110) planes, respectively, and one specific orientation of the QW for the WZ structure: namely when the 2D planes are normal to the [0001] axis which is the usual growth direction.

The low field electron mobilities are calculated using Eqs. (2) - (4) and we obtain

$$\mu_p = \frac{e}{m^* n_o} \left(\frac{m^* k_o T_o}{\pi \hbar^2} \right)^{\frac{p}{2}} \left(\frac{8}{9\pi} \right)^{\frac{p}{2}-1} \int\limits_o^\infty x^{\frac{p}{2}} \tau_p(x) F_o(x)[1 - F_o(x)]dx, \tag{5}$$

where p=3 for 3D electrons and p=2 for 2D electrons, $x = \varepsilon / k_o T_o$. The Fermi energy at a given lattice temperature T_o and a given electron density n_o is defined by the equation

$$2\pi^2 \hbar^3 n_o / (2m^* k_o T_o)^{3/2} = \int\limits_o^\infty x^{1/2}[\exp(x - z_o) + 1]^{-1} dx \quad \text{for the 3D electron gas, where}$$

$z_o = \varepsilon_F / k_o T_o$, and for the 2D electrons $z_o = \ln[\exp(\pi \hbar^2 n_o / m^* k_o T_o) - 1]$.

RESULTS AND DISCUSSION

First we evaluate the mobility of 2D electrons as a function of lattice temperature for a given sheet density. This is shown in the Fig. 1 for ZB and WZ. For ZB the mobility is almost independent of the orientation of the plane of the QW. The following material parameters were used in our calculations: $\rho = 6.1 \, g/cm^3$, $c_L = 6.56 \times 10^5 \, cm/s$, $c_T = 2.68 \times 10^5 \, cm/s$, $\varepsilon_s = 9.5$ for both ZB and WZ structures, $m^* = 0.21 m_o$, $h_{14} = 6.66 \times 10^7 \, V/cm$ for ZB, and $m^* = 0.23 m_o$, $h_{33} = 10.86 \times 10^7 \, V/cm$, $h_{31} = -3.91 \times 10^7 \, V/cm$, $h_{15} = -3.57 \times 10^7 \, V/cm$ for the WZ structure. The QW width is d=50 Å and the electron sheet density is $n_o = 10^{13} \, cm^{-2}$.

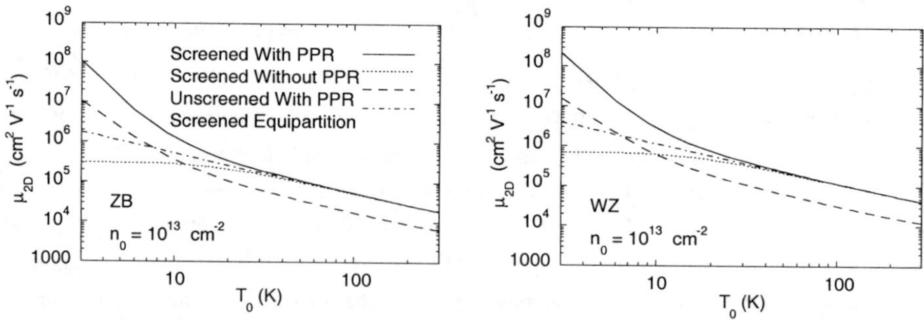

FIG. 1. Temperature dependence of the 2D mobility in WZ and ZB GaN/AlN heterostructures using different models for the momentum relaxation rate.

As is seen in the Fig. 1 the PPR plays an important role at temperatures below 50K. Screening is also important and it leads to an increase in the mobility in comparison with the unscreened case. But at T_0 less than 20K the effect of the PPR is much stronger than the screening effect. For comparison we also show the calculated mobility using the equipartition approximation for the phonon distribution function. This approximation can be used only at higher temperatures to give the correct value of the mobility.

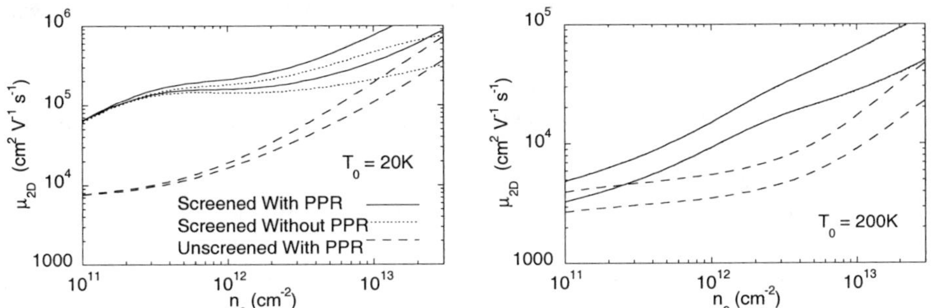

FIG. 2. Mobility of 2D electrons in WZ and ZB GaN/AlN heterostructures as a function of the electron density at two different temperatures: T=200K and T=20K using different models for the momentum relaxation rate. The upper curve in each figure for each pair corresponds to WZ GaN.

In Fig. 2 we display the calculated mobility of 2D electrons for WZ and ZB GaN as a function of the electron density at T_0=20K and T_0=200K. At low temperatures the PPR is important if the density is bigger than 10^{12} cm^{-2}. At high temperatures the PPR as expected is not important, but the effect of screening is still considerable.

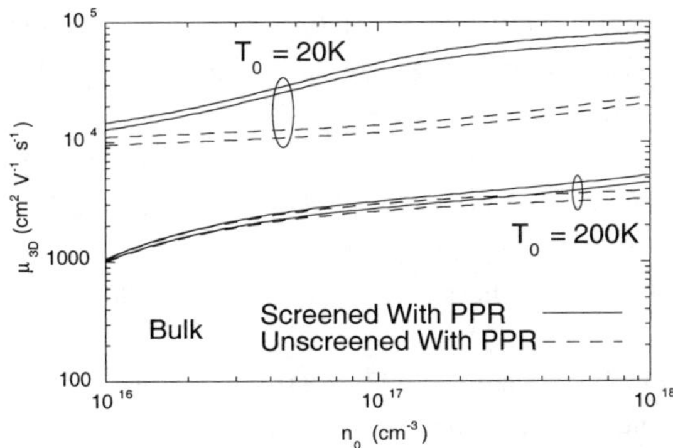

FIG. 3. Mobility of 3D electrons in bulk GaN as a function of the electron density. The upper curve for each pair corresponds to WZ GaN.

In Fig. 3 the same dependences are presented for 3D electrons. Because the Fermi energy of a 3D electron gas of density $n_0=10^{18}$ cm^{-3} is smaller than that of 2D electrons of density $n_0=10^{13}$ cm^{-2}, the PPR for 3D electrons is important at sufficiently lower temperatures. For the range of densities n_0 shown in Fig. 3 these temperatures are below 20K and the PPR is not important here. There is a remarkable difference between effect of the screening in 2D and 3D cases. In the 2D case the screening becomes weaker when n_0 increases while in the 3D case this is vice versa. The reason is that the screening length in the 2D case does not depend on n_0 while in the 3D case it does depend [4]. At the same time the increase in n_0 leads to an increase of the Fermi energy and consequently to increase of the wavevectors of the phonons interacting with electrons. It follows from the Lindhard function [4] that in the 2D case this results in a weakening of the screening effect.

CONCLUSION

The main conclusion which can be drawn from the analysis above is that in GaN at temperatures below 150K both the effects of screening and the PPR should be taken into account in order to obtain the correct result for the piezoacoustic phonon limited mobility in 2D case if the electron density is 10^{13} cm^{-2} or higher. This also true for the deformation acoustic phonon scattering. This conclusion becomes important in light of recent data about high electron sheet densities in GaN/AlN heterostructures because of internal piezoelectric field doping. Also, we note that WZ material has a slightly higher mobility than ZB material and that the mobility is not sensitive to the orientation of the plane for 2D electrons.

REFERENCES

[1] A. Bykhovski, B. Gelmont, M. Shur, and K. Khan, J. Appl. Phys. **77**, p. 1616 (1995).

[2] R. Oberhuber, G. Zander, and P. Vogl, Appl. Phys. Letters **73**, p. 818 (1998).

[3] N.A. Zakhleniuk, C.R. Bennett, B.K. Ridley, and M.Babiker, Appl. Phys. Letters **73**, 2485 (1988).

[4] For a 3D electron gas the Lindhard dielectric function can be found in G.D. Mahan, *Many-Particle Physics*, Plenum Press, New York, 1990, p. 438, and for a 2D gas in T. Ando, A.B. Fowler, and F. Stern, Rev. Mod. Phys. **54**, p.450 (1982).

[5] R. Gaska, M.S. Shur, A.D. Bykhovski, A.O. Orlov, and G.L. Snider, Appl. Phys. Lett. **74**, p. 287 (1999).

[6] P.J. Price, Solid State Commun. **51**, p. 607 (1984).

[7] V.E. Gantmakher and Y.B. Levinson, *Carrier Scattering in Metals and Semiconductors*, North-Holland, Amsterdam, 1987, Chapters 2 and 4.

[8] J.D. Zook, Phys. Rev. **136**, p.869 (1964).

[9] V. Karpus, Semicond. **21**, p.1180 (1987).

ACTIVATION CHARACTERISTICS OF DONOR AND ACCEPTOR IMPLANTS IN GaN

X. A. Cao,* S. J. Pearton,* R. K. Singh,* R. G. Wilson,** J. A. Sekhar,*** J. C. Zolper,**** J. Han,***** D. J. Rieger,***** R. J. Shul,***** H. J. Guo ****** S. J. Pennycook,****** and J. M. Zavada*******

*Department of Materials Science and Engineering, University of Florida, Gainesville, FL 32611, USA
**Consultant, Stevenson Ranch, CA 91381, USA
***Micropyretics Heaters International, Inc. Cincinnati, OH 45215, USA
****Office of Naval Research, Arlington, VA 22217, USA
*****Sandia National Laboratories, Albuquerque, NM 87185, USA
******Oak Ridge National Laboratory, Solid State Division, Oak Ridge, TN 37831,USA
*******European Research Office, USARDSG, London, England

ABSTRACT

The ionization levels of different donor and acceptor species implanted into GaN were measured by temperature-dependent Hall data after high temperature (1400 °C) annealing. The values obtained were 28 meV (Si), 48 meV (S), 50 meV (Te) for the donors, and 170 meV for Mg acceptor. P-type conductivity was not achieved with either Be or C implantation. Basically all of the implanted species show no distribution during activation annealing. For high implant doses (5×10^{15} cm^{-2}) a high concentration of extended defects remains after 1100 °C anneals, but higher temperatures (1400 °C) produces a significant improvement in crystalline quality in the implanted region.

INTRODUCTION

Ion implantation is an effective technology for selected-area doping or isolation of GaN-based devices.[1-3] As reviewed previously by Zolper,[1] implantation of donors at high dose ($>5\times10^{14}$ cm^{-2}) can be used to decrease source and drain access resistance in field effect transistors (FETs), at lower doses to create channel regions for FETs, while sequential implantation of both acceptors and donors may be used to fabricate p-n junctions. Two different device structures have been demonstrated using the latter method, namely a junction field-effect transistor[2] and a planar, homojunction light-emitting diode (LED).[4]

At high implant doses ($\geq 5\times10^{14}$ cm^{-2}) it is clear that conventional rapid thermal annealing (RTA) at 1100-1200 °C can activate the dopants but not remove the ion-induced structural damage.[5] At higher annealing temperatures ($\geq$1400 °C), the equilibrium N_2 pressure over GaN is >1000 bar,[6] and only two methods have proven effective in preventing surface decomposition. The first is use of high N_2 pressures (15 kbar),[7] and the second is deposition of AlN encapsulation layers[8] (at 1400 °C the equilibrium N_2 pressure above AlN is only 10^{-8} bar). The latter is clearly more convenient.

What is needed is a better understanding of the optimum implanted species for creation of n- and p-type regions in GaN, based on the highest achievable electron or hole concentrations, the residual damage and the redistribution during high temperature annealing. In this paper we report on the high temperature activation characteristics of donor implant species (Si, S and Te) and

513

Mat. Res. Soc. Symp. Proc. Vol. 572 © 1999 Materials Research Society

acceptor species (Mg and C). The redistribution of all these dopants plus Se and Be was investigated by using secondary ion mass spectrometry (SIMS). Finally, the efficiency of high temperature RTA for removing lattice damage in implanted GaN was examined by transmission electron microscopy (TEM).

EXPERIMENTAL

Layers of GaN 2-3 μm thick were grown at ~1040 °C on c-plane Al_2O_3 by atmospheric pressure Metal Organic Chemical Vapor Deposition (MOCVD), using triethylgallium and ammonia. From x-ray diffraction and photoluminescence measurements we know this material is typical of the current state-of-the-art heteroepitaxial GaN.

The samples were implanted at 25 °C with 150 keV $^{24}Mg^+$, 80 keV $^9Be^+$, 80 keV $^{12}C^+$, 200 keV $^{32}S^+$, 300 keV $^{80}Se^+$, or 600 keV $^{128}Te^+$ ions at doses of $3-5\times10^{14}$ cm^{-2}, and with 150 keV $^{28}Si^+$ at a dose of 5×10^{15} cm^{-2}. This puts the projected range, R_p, of the implanted species at least 1500 Å into the GaN in all cases, avoiding effects due to near-surface point defect injection. The samples were capped with ~1000 Å of reactively sputtered AlN, and annealed at temperatures of 900-1500 °C under a N_2 ambient in the Zapper^{TM} furnace described previously.[9] The dwell time at the peak temperature was ~10 secs. After annealing, the AlN was selectively etched in aqueous KOH at 80 °C.[10] For measurement of the electrical properties, HgIn ohmic contacts were alloyed to the corners of 3×3 mm^2 sections, and Hall effect data was recorded at 25 °C in all cases. The atomic distributions before and after annealing were measured by SIMS, and the data quantified using the as-implanted sample as a standard. In addition, the Si-implanted samples were also examined by plan-view TEM, since Si remains the standard implant species for GaN n-type doping, and it has a representative mass number that allows comparison to damage expected with S, Ca and Mg.

RESULTS AND DISCUSSION

Figure 1 shows an Arrhenius plot of sheet carrier concentration in Si^+ implanted material. In the AlN encapsulated samples activation occurs with an activation energy of ~5.2 eV before saturating at ~1400 °C. We interpret this activation energy as the average required to move the interstitial Si atom to a vacant substitutional site by short-range diffusion and to simultaneously remove compensating point defects so that the Si is electrically active. Note that at 1500 °C the sheet electron density decreases, and this was accompanied by a decrease in carrier mobility. This increase in compensation is consistent with Si beginning to occupy both Ga sites (where it is a donor), and N sites (where it is an acceptor). This is commonly observed with Si implantation in other III-V materials.[11] The peak n-type doping level we obtained is ~5×10^{20} cm^{-3}, and the corresponding activation efficiency is 90%. This very high doping level produces extremely good specific contact resistances for W and WSi_x metallization, with values $\leq10^{-6}$ $\Omega\cdot$cm^2 after annealing in the range 600-900 °C. This demonstrates the efficiency of the implantation approach for reducing contact resistances in GaN electronic devices.

There is also interest in the group VI donors, S, Se and Te, which do not have the potential drawback of being amphoteric like Si in GaN. Figure 2 shows the Arrhenius plots of S^+ and Te^+ activation in GaN. The sheet carrier concentrations show activation energy of 3.2 eV and 1.5 eV for the temperature range between 1000-1200 °C for S^+ and Te^+ respectively. These are significantly lower than that for Si, but their physical origins should be basically the same. The sheet electron density essentially saturates above 1200 °C for S implantation, with maximum volume density of ~5×10^{18} cm^{-3}. However the electron density does not saturate even at 1400 °C for the Te case. It is likely that because of the much greater atomic weight of [128]Te, even higher annealing temperatures would be required to remove all its associated lattice damage, and that the activation characteristics are still being dominated by this defect removal process. Even though implanted Si^+ at the same dose showed evidence of site-switching and self-compensation, it still produces a higher peak doping level than the non-amphoteric donors S and Te. This suggests the group VI donors do not have any advantage over Si for creation of n-type layers in GaN. From temperature-dependent Hall measurements, we find the ionization levels of Si, S and Te are 28 meV, 48 meV and 50 meV respectively, so that the donors are fully ionized at room temperature. This was verified by elevated temperature Hall measurements (up to 150 °C), where no additional increase in free carrier concentration was observed. In this data, we assumed a compensation level of ~25%, as is typical for GaN implanted material.[6]

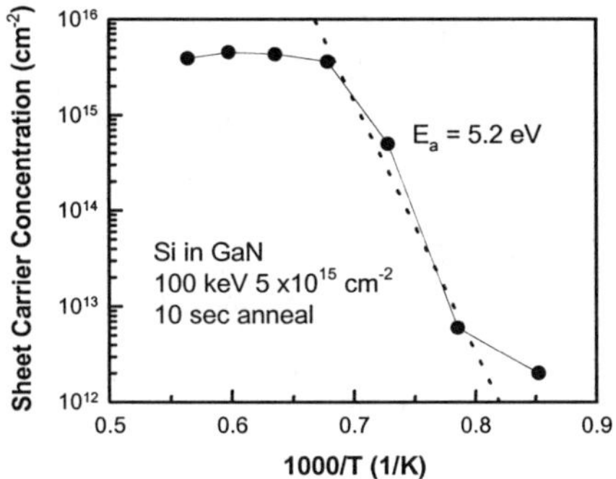

Fig. 1. Arrhenius plot of sheet electron concentration versus inverse anneal temperature for Si^+ implanted GaN.

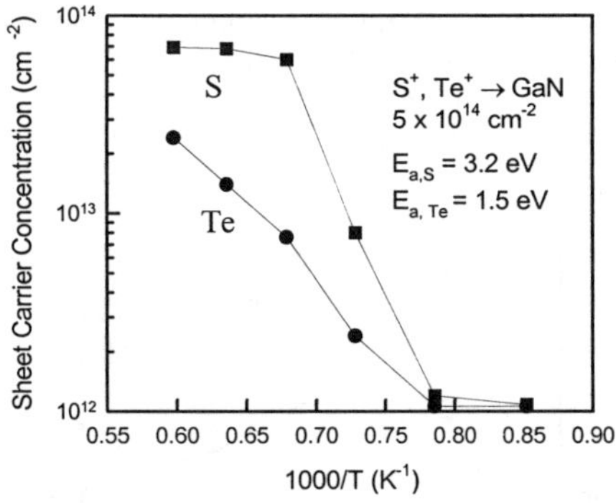

Fig. 2. Sheet carrier densities in S^+ or Te^+ implanted GaN as a function of annealing temperature.

The effects of post-implant annealing temperature on the sheet carrier concentrations in Mg^+ and C^+ implanted GaN are shown in Figure 3. There are two important features of the data: first, we did not achieve p-type conductivity with carbon, and second only ~1% of the Mg produces a hole at 25 °C. Carbon has been predicted previously to have a strong self-compensation effect,[12] and it has been found to produce p-type conductivity only in metal organic molecular beam epitaxy where its incorporation on a N-site is favorable.[13] Based on an ionization level of ~170 meV, the hole density in uncompensated Mg-doped GaN would be calculated to be ~10% of the Mg acceptor concentration when measured at 25 °C. In our case we see an order of magnitude less holes than predicted. This should be related to the existing n-type carrier background in the material and perhaps to residual lattice damage which is also n-type in GaN, At the highest annealing temperature (1400 °C), the hole density falls, which could be due to Mg coming out of the solution or to the creation of further compensating defects in the GaN.

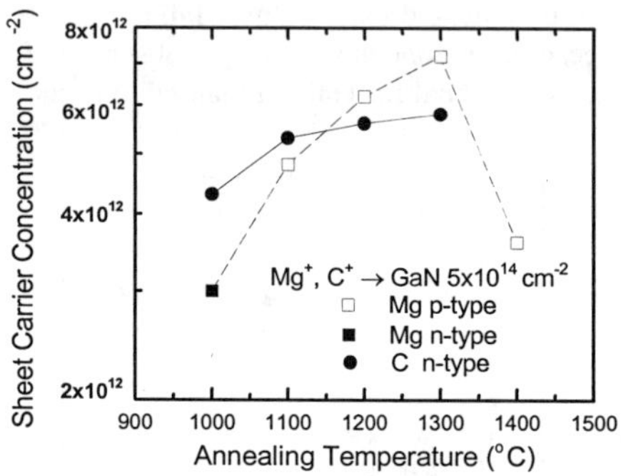

Fig. 3. Sheet carrier densities in Mg$^+$ or C$^+$ implanted GaN as a function of annealing temperature.

Figure 4 shows the SIMS profiles of Si as-implanted and 1400 °C annealed samples. There is little redistribution of the Si at 1400 °C, with $D_{Si} \leq 10^{-13}$ cm^2·s^{-1} at this temperature calculated from the change in width at half-maximum. Similarly, we found that S, Se, Te, Mg and C are all extremely slow diffusers when implanted into GaN, with $D_{eff} \leq 2 \times 10^{-13}$ cm^2·sec^{-1} at 1450 °C. The extremely stable nature of dopants in GaN means that junction placement should be quite precise and there will be fewer problems with lateral diffusion of the source/drain regions towards the gate. This is promising for the fabrication of GaN-based power devices, which require creation of doped well or source/drain regions by implantation.

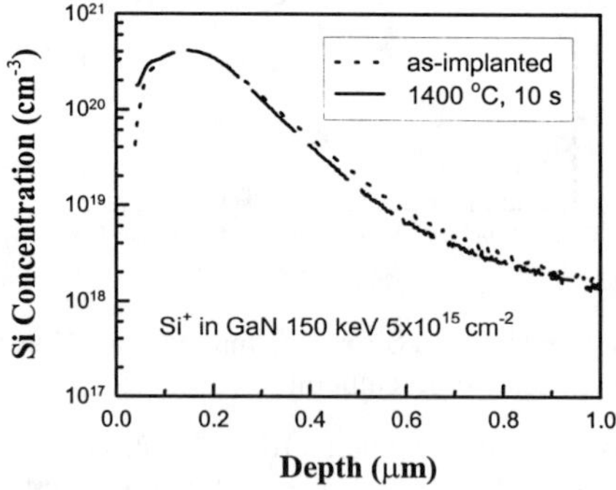

Fig. 4. SIMS profiles of implanted Si in GaN (150 keV, 5×10^{15} cm^{-2}) before and after annealing at 1400 °C for 10 s.

In the particular case of implanted Be, there was an initial broadening of the profile at 900 °C (figure 5), corresponding to an effective diffusivity of ~5×10^{-13} cm^2·sec^{-1} at this temperature. However there was no subsequent redistribution at temperatures up to 1200 °C. It appears that in GaN, the interstitial Be undergoes a type of transient-enhanced diffusion until these excess point defects are removed by annealing, at which stage the Be is basically immobile.

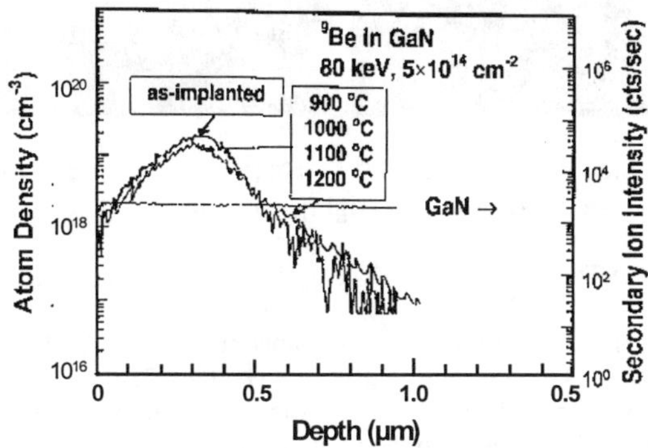

Fig. 5. SIMS profiles of implanted Be in GaN (80 keV, 5×10^{14} cm^{-2}) before and after annealing at different temperatures for 10 s.

Figure 6 shows a plan view TEM from a Si-implanted sample after annealing at 1100 °C (left) and 1400 °C (right) for 10 secs. This high dose implant (150 keV, 5×10^{15} cm^{-2}) represents a worst-case scenario in terms of damage removal. The sample still contains a high density of extended defects (~10^{10} cm^{-2}) after 1100 °C annealing. We ascribe these defects to the formation of dislocation loops in the incompletely repaired lattice. By sharp contrast, annealing at 1400 °C for 10 secs brings a substantial reduction in the implant-induced defects. The lower density of defects (~10^9 cm^{-2}) could be ascribed to the threading dislocations arising from lattice-mismatch in the heteroepitaxy. This appears to correlate well with the fact that the highest electron mobility and carrier density in these samples was observed for 1400 °C annealing. Clearly the ultra-high temperature annealing is required to completely remove lattice damage in GaN implanted with high doses. However it may not be needed in lower dose material (≤5×10^{13} cm^{-2}) where the amount of damage created is correspondingly less.

SUMMARY AND CONCLUSIONS

The redistribution and activation of a wide variety of possible donor and acceptor species in GaN was examined. S and Te are found to produce lower room-temperature n-type doping levels than Si at the same dose, while only Mg was found to produce p-type doping (C and Be implanted samples remained n-type). None of the implanted species showed measurable diffusion at 1450 °C, but Be did display an apparent defect-assisted redistribution at lower temperature (900 °C). For high implant dose (e.g. 5×10^{15} cm^{-2} for Si$^+$), annealing at 1100 °C is insufficient to remove the lattice damage, while 1400 °C produces much lower defect densities.

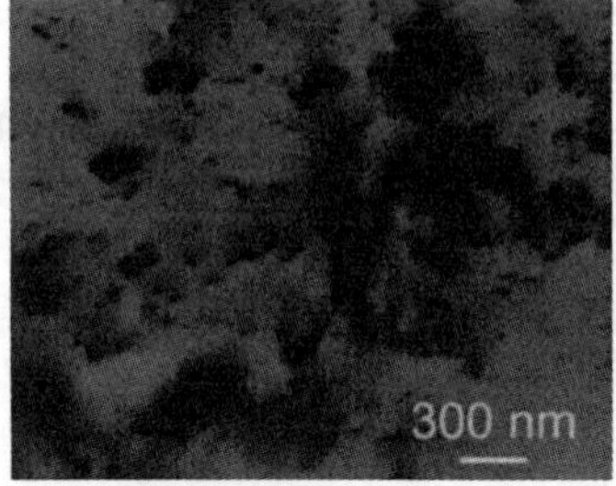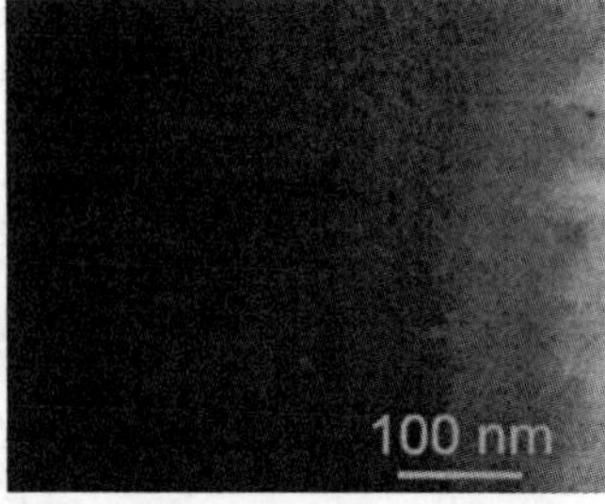

Fig. 6. TEM plan view from Si^+ implanted GaN (5×10^{15} cm^{-2}, 150 keV) after 1100 °C (left) and 1400 °C (right), 10 s annealing.

ACKNOWLEDGEMENTS

The work at UF is partially supported by grants from DARPA/EPRI (D. Radack and J. Melcher), MDA 972-98-1-0006, and NSF (L. Hess), DMR-9732865. The work of R. G. Wilson is partially supported by a grant from ARO. Sandia is a multiprogram laboratory operated by Sandia Corporation, a Lockheed-Martin company, for the US Department of Energy under contract No. DEAC04-94 AL 85000.

REFERENCES

1. J. C. Zolper, *GaN and Related Materials,* edited by S. J. Pearton, p. 371 (Gordon and Breach, NY 1997).
2. J. C. Zolper and R. J. Shul, MRS Bulletin **22**, 36 (1997).
3. S. C. Binari, L. B. Rowland, W. Kruppa, G. Kelner, K. Doverspike and D. K. Gaskill, Electron. Lett. **30**, 1248 (1994).
4. H. P. Maruska (unpublished, 1997).
5. H. H. Tan, J. S. Williams, J. Zou, D. J. H. Cockayne, S. J. Pearton and R. A. Stall, Appl. Phys. Lett. **69**, 2364 (1996).
6. S. Porowski and I. Grzegory, *GaN and Related Materials*, edited by S. J. Pearton, p. 295 (Gordon and Breach, NY 1997).
7. J. C. Zolper, J. Han, S. B. Van Deusen, M. H. Crawford, R. M. Biefeld, J. Jun, T. Suski, J. M. Baranowski and S. J. Pearton, Mat. Res. Soc. Symp. Proc. **482,** 609 (1998).
8. J. C. Zolper, D. J. Reiger, A. G. Baca, S. J. Pearton, J. W. Lee and R. A. Stall, Appl. Phys. Lett. **69**, 538 (1996).
9. M. Fu, V. Sarvepalli, R. K. Singh, C. R. Abernathy, X. A. Cao, S. J. Pearton and J. A. Sekhar, Mat. Res. Soc. Symp. Proc. **483,** 345 (1998).
10. J. R. Mileham, S. J. Pearton, C. R. Abernathy, J. D. MacKenzie, R. J. Shul and S. P. Kilcoyne, Appl. Phys. Lett. **67,** 1119 (1995).
11. S. J. Pearton, J. S. Williams, K. T. Short, S. T. Johnson, D. C. Jacobsen, J. M. Poate, J. M. Gibson and D. O. Boerma, J. Appl. Phys. **65,** 1089 (1989).
12. P. Bogulawski, E. L. Briggs and J. Bernholc, Phys. Rev. B **51,** 17255 (1995).
13. C. R. Abernathy, J. D. MacKenzie, S. J. Pearton and W. S. Hobson, Appl. Phys. Lett. **66,** 1969 (1995).

TRANSMUTATION DOPING OF III-NITRIDES

GALINA POPOVICI
Rockbit International Inc, 7601 Will Rogers Blvd, Fort Worth, TX 76140
gpopovici@rbi-gearhart.com, rrmihai@showme.missouri.edu

Abstract

Transmutation doping of (In, Ga, Al)N compounds by neutron irradiation is a promising and totally unexplored field to date. It is much more effective than that of Si due to large neutron capture cross section and abundance of In, Al and Ga isotopes participating in reaction. This should make the irradiation possible in low-flux reactors and result in smaller radiation damage. Annealing of the radiation damage seems feasible.

Doping is a controlled introduction of impurity atoms in appropriate sites and states in the crystal lattice. It is designed to change the electrical, optical and/or other properties of semiconductors in a controllable manner. There are four methods of doping: during growth, by ion implantation, by diffusion and by transmutation.

In the case of nitrides and other wide band gap III-V semiconductors, doping is generally performed simultaneously with the growth. Noteworthy progress in the growth and doping of high quality epitaxial III-nitride films by a variety of methods has recently been achieved.[1,2] It is well known that it is easy to obtain GaN and InN of n-type through unintentional doping. The source of the unintentional doping is still controversial.[1,2,3] Unintentionally doped InN has high electron concentrations up to 10^{18}- 10^{19} cm^{-3}. Controlled n-type conductivity in GaN and diluted InGaN alloys is generally achieved by Si doping during growth [3] or by ion implantation.[4] Ways to successfully and reproducibly dope AlN have not been found.[1] Under many conditions of growth and doping AlN remains an insulator with resistivity of 10^9 to 10^{12} Ωcm.[1]

Ion implantation can be a useful method of GaN doping. Since p- and n-type, and also highly resistive layers, have been realized, the method can be used to fabricate of all-implanted devices.[5, 6] However, much more research should be done to achieve good reproducibility.

Transmutation doping is used industrially for obtaining n-type silicon. Samples of pure Si are irradiated with neutrons in nuclear reactors. One of the isotopes of Si can absorb thermal neutrons and undergo subsequent β-decay. As a result, Si transmutes into phosphorus. The material is then annealed for healing the radiation damage related to the unavoidable presence of fast neutrons. It is possible to control precisely the concentration of P atoms by controlling the neutron fluence. Transmutation of B into Li was studied for diamond doping [7, 8] and showed promising results.

Data on relevant transmutation reactions for AlN, GaN and InN are given in **Table I.** Nitrogen has a very small cross section of neutron absorption. Even if it absorbs a neutron, it changes into another stable isotope of N. Transmutations of Al, Ga and In occurs with the absorption of a neutron and subsequent β-decay, except for a weak channel of transition of In into Cd, when the neutron absorption is followed by the capture of a K-

Mat. Res. Soc. Symp. Proc. Vol. 572 © 1999 Materials Research Society

shell electron or by the emission of a positron. [9,10] Absorption of the neutron puts the mother nucleus in an excited state, from which it decays with a certain half life, given as $t_{1/2}$ in Table I. In all cases given in Table I, the daughter isotopes are stable. The last column in Table I gives the concentration of the transmuted atoms N_t for one hour of neutron irradiation at a thermal neutron flux of 10^{13} cm^{-2} s^{-1}. [a)] It was calculated through the relation:

$$N_t = N\,a\,S\,F\,t$$

where   N is the concentration of mother atoms, a is the isotopic abundance of those atoms, S is the cross section for neutron capture (averaged over thermal neutron spectrum), F is the thermal neutron flux, t is the irradiation time.

Table I. Data on transmutation by neutron capture of Si, Al, Ga and In [11]

Mother Nucleus (Z, m)	Abundance (%)	Cross Section (barn)	Daughter Nucleus (Z, m)	$t_{1/2}$	Energy (MeV)	N_t in 1 h (cm^{-3})
Si (14, 30)	3.1	0.11	P (15, 31)	2.6 h	1.49	4.5×10^{12}
Al (13, 27)	100	0.23	Si (14, 28)	2.3 min	2.82	4.0×10^{14}
Ga (31, 69)	60.1	1.8	Ge (32, 70)	21 min	1.65	1.6×10^{15}
Ga (31, 71)	39.9	0.15	Ge (32, 72)	14.1 h	1.5-3.15	1.0×10^{14}
In (49, 115)	95.7	72	Sn (50, 116)	54 min	0.6-11.6	1×10^{16}
In (49, 115)	95.7	42	Sn (50, 116)	14 s	3.3	0.94×10^{16}
In (49, 113)	4,3	3	Sn (50, 114)	1.2 min	~ kV	3.0×10^{13}
In (49, 113)	4.3	0.16	Cd (48, 114)	49.5 days	~ kV	0.16×10^{13}

Z is the atomic number, m is the atomic mass, cross sections are in barns
(1 barn = 10^{-24} cm^{-2}).
Concentrations of atoms used in the calculation:
N(Si) = 0.37×10^{23} cm^{-3}
N(Al) = 0.48×10^{23} cm^{-3}
N(Ga) = 0.43×10^{23} cm^{-3}
N(In) = 0.65×10^{22} cm^{-3}

As one can see from Table I, neutron capture for In, Ga, and Al is more effective than that for Si by two orders of magnitude. Therefore, neutron fluences for obtaining the same dopant concentration will be about two orders of magnitude smaller. This will result in smaller radiation damage and will make the irradiation possible in low-flux reactors.

All three transmutated atoms Si, Ge and Sn are amphoteric impurities and can enter both sublattices, III and V. But it is unlikely that these atoms will enter the nitrogen

---

sublattice on   account of the small radius of the nitrogen atom (see **Table II**). Most probably Si, Ge and Sn will enter the III-sublattice. All three dopants will thus be donors.

Table II. Atomic radii of the elements.

Atom	Al	Ga	N	Si	Ge	Sn
Atomic radius (μm)	0.143	0.122	0.070	0.117	0.122	0.140

The recoil energy of transmutated atoms will be in the range used for ion implantation (50-250 keV). Though the energy of the nuclear reaction is of the order of MeV, from the conservation of impulse it follows that most of energy is taken by the electron.

Successful implantation doping of GaN layers has shown that annealing of radiation damage can be effective. [4, 5, 6, 12] AlN has a high radiation resistance and can be annealed to high temperatures due to the fact that AlN is a much more stable compound than GaN and InN.[1] To our knowledge there are no papers on implantation or irradiation of InN layers. However, transmutation doping of InN may be efficient. In has a high efficiency of transmutation reaction, three order of magnitude higher than that of Si. It should result in a short time of irradiation and consequently in a weak radiation damage.

Transmutation doping of III-N semiconductors is promising and to date totally unexplored field. Doping of GaN by transmutation of Ga into Ge should be feasible. Since doping of AlN by more conventional methods was unsuccessful, an attempt to obtain n-type AlN by transmutation doping would be interesting from both experimental and theoretical points of view. InN has the highest efficiency of transmutation. This should result in a weaker radiation damage. InN transmutation doping thus may also be achieved.

REFERENCES

[1] G. Popovici, H. Morkoc, and S. N. Mohammad, *Deposition and Properties of III-Nitrides by Molecular Beam Epitaxy* in " Group III nitride semiconductor compounds, Physics and Applications" Edited by Bernard Gil, (Claredon, Oxford UK), p.19-69, 1998.

[2] G. Popovici and H. Morkoc, *Growth and Doping of, and Defects in III-Nitrides* in " "III-V Nitride Semiconductors and Optoelectronic Devices" v.2 " edit by S. Pearton, to be published in 1999.

[3] W. Kim, A. E. Botchkarev, A. Salvador, G. Popovici, H. Tang and H. Morkoc,. J. Appl. Phys. 82, 219 (1997).

[4] S. J. Pearton, C. B. Vartuli, J. C. Zolper, C. Yuan and R. A. Stall, Appl. Phys. Lett, 67,1435 (1995).

[5] M. Rubin, N. Newman, J. C. Chen, T. C. Fu, and J. T. Ross, Appl. Phys. Letters, 64, 64 (1994).

[6] C. J. C. Zolper, R. G. Wilson, S. J. Pearton, and R. A. Stall, Appl. Phys. Lett, 68,1945 (1996).

[7] G. Popovici , A. A. Melnikov , V.S. Varichenko, S. Khasawinah, T. Sung, M. A. Prelas, A. B. Denisenko, N. M. Penina, V.A. Martinovich, E. N. Drozdova , A .M. Zaitsev, and W. R. Fahrner, Diamond Relat.Mater.5, 761 (1996).

[8] S.A. Khasawinah, G. Popovici, J. Farmer, T. Sung, M. A. Prelas, J. Chamberlain, and H. White, J. Mater. Research, 10, 2523 (1995).

[9] R.L. Heat, Table of the Isotopes, in CRC Handbook of Chemistry and Physics, Ed. R.C. Weast, CRC Press, West Palm Beach, 1978, pp.B-270 to B-354.

[10] F.W.Walker, J.R.Parrington, F. Feiner, Nuclides and Isotopes, 14-th Edition, GN Nuclear Energy, General Electric Company, 1989.

[11] The data compiled by M. Popovici, Missouri University Research Reactor.

[12] H. H. Tan, J. S. Williams, J. S. Zou, d. J. H. Cockayne, S. J. Pearton, R. A. Stall, Appl. Phys. Lett. 69, 2364 (1996).

High Barrier Height n-GaN Schottky diodes with a barrier height of 1.3 eV by using sputtered copper metal

W.C. Lai[*], M. Yokoyama[*], C.Y. Chang[**], J.D. Guo[***], J.S. Tsang[***], S.H. Chan[***] and S.M. Sze[*]

[*]National Cheng Kong University, Department of Electrical Engineering, Tainan, Taiwan, R.O.C.
[**]National Chiao Tung University, Institute of Electronic, Hsinchu 30050, Taiwan, R.O.C.
[***]National Nano Device Laboratories, Hsinchu 30050, Taiwan, R. O. C.

ABSTRACT

Copper Schottky diodes on n-type GaN grown by metal-organic chemical vapor deposition were achieved and investigated. Ti/Al was used as the ohmic contact. The copper metal is deposited by the Sputter system. The barrier height was determined to be as high as Φ_B =1.13eV by current-voltage (I-V) method and corrected to be Φ_B =1.35eV as considered the ideality factor, n, with the value of 1.2. By the capacitance-voltage (C-V) method, the barrier height is determined to be Φ_B =1.41eV. Both results indicate that the sputtered copper metal is a high barrier height Schottky metal for n-type GaN.

INTRODUCTION

GaN-based materials have been intensively studied recently for the blue emission diodes. Besides that, they have the potential in the applications of high temperature, high frequency and high power electronic devices due to its wide band gap and chemical stability characteristics. For some electronic devices like MESFET, it is necessary to fabricate a Schottky contact with a high breakdown voltage and a low reverse leakage current, up to a high reverse voltage. The metals with high work functions have been reported by many researchers to be used as a Schottky contact for GaN[1-9]. In this study, the Cu metal deposited by the sputter system was used as a Schottky contact to fabricate a Cu/n-GaN Schottky diode. The Schottky barrier height of the copper metal for n-GaN grown on the Al_2O_3 substrate was investigated by using the I-V and C-V measurements.

Mat. Res. Soc. Symp. Proc. Vol. 572 © 1999 Materials Research Society

EXPERIMENT

The samples used for this study were grown by the low pressure metal-organic vapor phase epitaxy (MOVPE) system. Trimethylgallium (TMGa) and NH_3 were used as source materials, and SiH_4 diluted in H_2 was used as the n-type dopant source. The sapphire substrate was first heated at 1100°C for 20min in the stream of H_2. A 25nm thick GaN layer was deposited as the buffer layer at 525°C. Then, the substrate was heated up to 1050°C and a 2µm thick Si doped GaN layer was grown on the GaN buffer layer. The carrier concentration of the Si doped GaN film was $5 \times 10^{17} cm^{-3}$ measured by the Hall measurement. The GaN samples were first cleaned with organic solvents, followed by etching in a $HCl:H_2O$ solution, and then were loaded in an E-gun evaporator equipped with the cryo pump. Ti (50nm) and Al (100nm) were evaporated using the conventional lift-off technique. After lift-off process, the sample was annealed at 550°C for 5min in nitrogen ambient in order to obtain ohmic characteristics. Then, Cu (100nm) was deposited on the GaN film by sputtering and the pattern of the Cu contact was formed by using the wet etching with a solution of 10% HNO_3. Prior to the sputtering process, the sample was cleaned in boiling aqua regia for 5 min and dipped in $HCl:H_2O$ for 1min. The area of the Schottky contact was 9.5×10^{-5} cm^2. The current-voltage (I-V) and capacitance-voltage (C-V) characteristics were measured with HP4145 semiconductor parameter analyzer and Keithly 590 C-V analyzer.

RESULT AND DISCUSSION

The typical I-V characteristics of the Cu/GaN Schottky diode measured at room temperature are shown in Fig.1. An increase of the forward current was observed above the applied voltage of 0.4V. The I-V characteristic function of the Schottky diode is described by the following equations[11]:

$$J=J_0(\exp(qV/nkT)-1) \dots\dots\dots\dots\dots\dots\dots(1)$$
$$J_0=A^{**}T^2\exp(-q\phi_{bn0}/kT)\dots\dots\dots\dots\dots\dots(2)$$

where J_0 is the saturation current density, n the ideality factor, A^{**} the effective Richardson constant, k the Boltzmann's constant and ϕ_{bn0} the measured barrier height. The value of Φ_{bn0} can be deduced from the I-V measurement as the effective Richardson constant is decided. The theoretical value of A^{**} is 26 A/cm^{-2}K^{-2} [1]. From the forward bias log I-V plot as shown in Fig.1, the values of the Schottky barrier height and the ideality factor are 1.13eV and 1.2, respectively.

To eliminate the effect of the series resistance of the diodes, the Norde's method was used to calculate the Schottky barrier height[12]. From this method, the Schottky barrier height of the Cu/GaN can be calculated by using a plot of the following function:

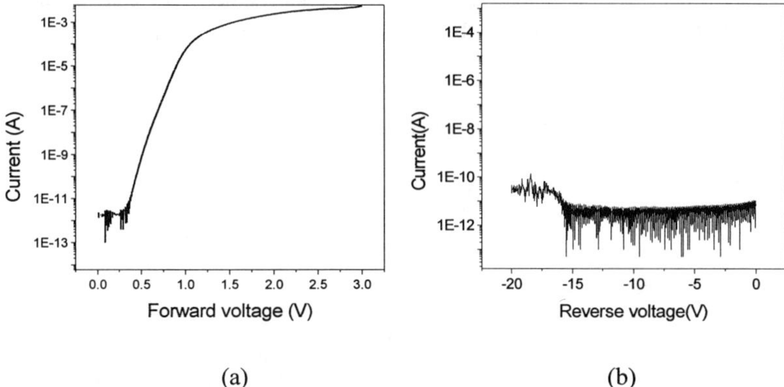

(a)   (b)

Fig. 1 The I-V characteristic plot of the Cu/GaN Schottky diode at room temperature (a) forward biases (b) reverse biases.

$$F(V) = \frac{V}{2} - \frac{kT}{q} \ln(\frac{I}{AA**T^2}) \quad\dots\dots\dots\dots\dots\dots\dots\dots\dots\dots\dots\dots\dots(3)$$

From the plot of the F(V) vs. V suggested in Fig.2, the minimum point $F(V_m)$of F(V) can be find by setting dF/dV=0. And then the corresponding voltage V_m can be obtained. Therefore, the Schottky barrier height can be calculated by the following equation[13]:

$$\phi_B = F(V_m) - (\frac{1}{2} - \frac{1}{n})V_m - (\frac{2-n}{n})\frac{kT}{q} \quad\dots\dots\dots\dots\dots\dots\dots\dots\dots\dots\dots(4)$$

where ϕ_B is Schottky barrier height, V_m is correspond voltage of the minimum point of F(V) and n is the ideal factor. From Eq.(4), the calculated Schottky barrier high of the Cu/GaN is 1.15eV.

Besides the I-V methods, the Schottky barrier height of Cu/GaN can also be measured by using the C-V characteristic method. Fig.(3) shows the C-V characteristic curve and the $1/C^2-$ V plot as a function of DC bias changing from $-5V$ to $0V$ in steps of 0.02V, with a small signal of 100kHz. The capacitance increases gently with the increase of the applied voltage, which is shown in Fig.3(a). That is because of the shrinkage of the depletion region of the Cu/GaN Schottky contact. The built in potential of this Cu/GaN Schottky barrier is obtained from the $1/C^2$-voltage plot via $V_{bi}=V_0+kT/q$, where V_0 is the voltage intercept. V_0 is determined to be 1.37eV which is fitted and marked in Fig.3(b). Neglecting image force lowering, the barrier height, Φ_{bn}(C-V) is given by

$$\Phi_{bn} (C\text{-}V) = qV_{bi} + (E_c\text{-}E_f) \quad\dots\dots\dots\dots\dots\dots\dots\dots\dots\dots\dots\dots\dots(5)$$

Where $(E_c\text{-}E_f)$ is the energy difference between the conduction band and the Fermi level and given by $(E_c-E_f) = (kT/q)\ln(N_c/N_d)$. N_c is the effective state density in the conduction band and N_d is the donor concentration of the film. The theoretical value of N_c is defined as

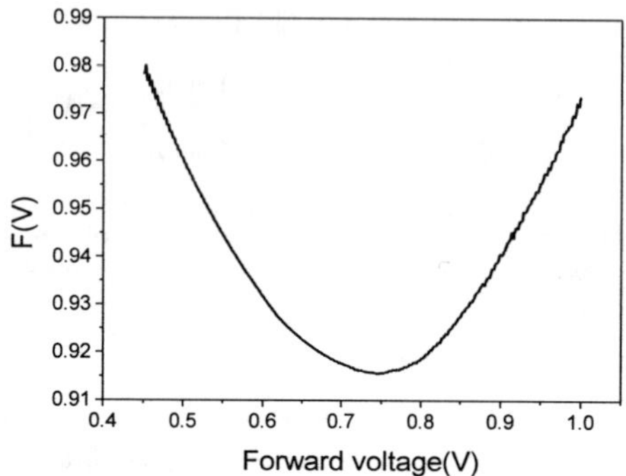

Fig.2 The norde plot of the Cu/GaN Schottky Diode

$2(2\pi m_n{}^*kT/h^2)^{3/2}$. The typical value of the effective mass, $m_n{}^*$, is equal to $0.22m_0$. As mentioned above, the doping concentration of GaN measured by the Hall measurement is $5 \times 10^{17} cm^{-3}$. Therefore, the values of (E_c-E_f) and the barrier height calculated from Eq.5 are 0.04 eV and 1.41eV, respectively.

The value of the barrier height measured from the C-V method is larger than that measured by the I-V method. This is because the transport mechanism of these diodes is not purely due to the thermionic emission. For these diodes, the barrier height of Φ_{bno} is voltage or electric field sensitive and Φ_{bn} is not.[10] The two barrier heights are related but are not identical quantities. Because the ideality factor of the I-V characteristic is not 1 for the Cu/n-GaN Schottky diode, the Schottky barrier height of the copper metal on n-GaN must be correlated with a more fundamental barrier height, Φ_{bf}, at zero electric field by the following equation[14]:

$$\Phi_{bf}=n\Phi_{bn0}-(n-1) kT/q \ln(N_c/N_d) \quad\ldots\ldots\ldots\ldots\ldots(6)$$

where N_c is the effective conduction-band density of the states. When the junction is at zero electric field, i.e., the flat band condition, there should be no tunnelling or image force lowering. The corrected barrier height calculated from Eq.(6) is 1.35eV. The quantitative Φ_{bf} is, therefore, more appropriate value to compare with Φ_{bn}. The corrected barrier height, Φ_{bf}, now can be considered as the barrier height of a Schottky diode with a transport mechanism of thermionic emission and without any image force lowering effect.[10]

The results of the barrier height calculated from above methods are summarized in Table I. The Schottky barrier height of the Cu/GaN diodes measured from I-V and C-V methods are different from that predicted by the Schottky-Mott model, which indicates that the barrier height is equal to the difference between the metal work function of Cu and the electron affinity of GaN (χ=4.2eV). Such a high Schottky barrier height of the Cu/GaN contact obviously can not be explained by the Schottky–Mott model and it may result from the

Table I. Schottky barrier height of Cu/GaN diode

	Ideality n	Φ_{bn0}(I-V) [Eq.(1)]	Φ_b(Norde Plot) [Eq. (4)]	Φ_{bn}(C-V) [Eq. (5)]	Φ_{bf} [Eq. (6)]
Cu/GaN Schottky diode	1.2	1.13 eV	1.15 eV	1.41 eV	1.35 eV

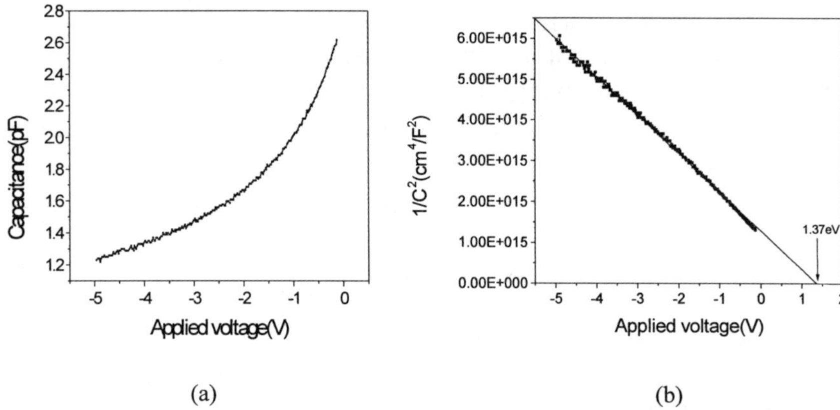

(a)   (b)

Fig. 3 (a) C-V characteristics and (b) $1/C^2$ plot of Cu/GaN Schottky diode.

interface reaction or the Fermi level pinning effect which is also found in the GaAs-based or InP-based materials. The details of the Schottky contact Cu on GaN, such as the reaction between Cu and GaN, are currenty extensively studied and will be published later.

In conclusion, the Cu/GaN Schottky diode with a high barrier height has been achieved and investigated. The Cu metal was sputter-deposited on n-GaN. The barrier height of the Cu/GaN Schottky diodes measured by the I-V and C-V method are 1.15 and 1.41eV, respectively. The contact mechanism of Cu/GaN can not be completely explained by the Schottky –Mott model and the mechanism is suggested to be the surface fermi-level pinning effect.

The authors wish to acknowledge the support a portion for this work by the National Sciences Council (NSC-88-2215-E009-015) and Epistar Corporation .

Reference

1) P. Hacke, T. Detchprohm, K. Hiramatsu and N. Sawaki: Appl. Phys. Lett. 63 (1993) 2676.

2) S. C. Binari, H. B. Dietrich, G. Kelner, L. B. Rowland, K. Doverspike and D. K. Gaskill:

Electron. Lett. 30 (1994) 909.

3) M. R. H. Khan, T. Detchprohm, P. Hacke, K. Hiramatasu and N. Sawaki: J. Phys. D 28 (1995) 1169.

4) J. G. Guo, M. S. Feng, R. J. Guo, F. M. Pan and C. Y. Chang: Appl. Phys. Lett. 67 (1995) 2657.

5) L. Wang, M. Y. Nathan, T. –H. Lim, M. A. Khan and Q. Chen: Appl. Phys. Lett. 68 (1996) 1267.

6) A. T. Ping, A. C. Scmitz, M. A. Khan and I. Adesida: Electron. Lett. 32 (1996) 68.

7) S. N. Mohammad, Z. F. Fan, A. E. Botchkarev, W. Kim, O. Aktas, A. SalVador and H. Morkoc: Electron. Lett. 32 (1996) 598.

8) K. Suzue, S. N. Mohammad, Z. F. Fan, W. Kim, O. Akats, A. E. Botchkarev and H. Morkoc: J. Appl. Phys. 80 (1996) 4467.

9) E. V. Kalinina, N. I. Kuznetsou, V. A. Dmitriev, K. G. Irvine, and C. H. Carter Jr.: J. Electron. Mater. 25 (1996) 831.

10) Q. Z. Liu, L. S. Yu, S. S. Lau, J. M. Redwing, N. R. Perkins and T. F. Kuoch: Appl. Phys. Lett. 70 (1997) 1275.

11) S. M. Sze: Physics of Semiconductor Devices (John Wiley & Sons, New York, 1981) 2nd ed., Chap. 5.

12) H. Norde: J. Appl. Phys. 50 (1979) 5052.

13) D. K. Schroder: Semiconductor Material And Device Characterization (John Wiley & Sons, New York, 1990), 153-154.

14) L. F. Wagner, R. W. Young, and A. Sugerman: IEEE Electron. Device Lett. ED-4, (1983) 320.

IIIB- Nitride Semiconductors for High Temperature Electronic Applications

X. BAI, D. M. HILL and M. E. KORDESCH,
Department of Physics and Astronomy, kordesch@helios.phy.ohiou.edu,
Ohio University, Athens OH 45701

ABSTRACT

Thin films of ScN and YN were grown on silicon, quartz and sapphire using metal evaporation and an RF atomic nitrogen source. YN decomposes on contact with water vapor, and only AlN capped films could be stabilized. ScN is stable in air and water, and thin films of this material deposited at temperatures between 300 and 900 $^{\circ}$C show a substrate-dependent film texture. Typical growth rates were ~ 0.1 nm/second with a 300W N discharge at about 0.1 mTorr Nitrogen pressure. Structural characterization by x-ray diffraction, infrared transmission spectroscopy and Hall effect measurements on n-type ScN and the fabrication of p-n junctions of n– type ScN with silicon are presented.

INTRODUCTION

The IIIB metals Scandium, Yttrium and Lanthanum are less commonly used for wide bandgap nitride semiconductors compared to the IIIA metals Aluminum, Gallium and Indium. The IIIB metals have one d electron rather than one p electron in the outer shell. The IIIB-nitrides have bandgaps in the 2-2.4 eV range, crystallize in the rock salt structure, and melt above 2600 $^{\circ}$C. The films are yellow to deep red in color [1]. Several groups have investigated ScN, both for its mechanical properties in analogy to the well-known coating material TiN [2,3], and both experimental and theoretical investigations of its electronic properties [4-6]. The cubic ScN lattice is a good match for the IIIA nitrides; because of the high melting temperatures of both Sc and ScN this material might ultimately be used to replace InN in IIIA-V semiconductor alloys for high temperature applications.

We report the results of vacuum evaporation of scandium metal in an atomic N ambient for producing thin films of ScN, and diode structures made from n-type ScN and p-type Si. YN nitride films decomposed in humid air in a matter of hours to days.

EXPERIMENT

The vacuum system is based on a stainless steel 6-way cross with 150 mm outside diameter copper sealed flanges. The vertical ports are used for sample insertion (top), the evaporator feedthroughs (bottom). The four horizontal ports are used for the SVTA Inc. radio frequency atomic nitrogen source, viewports, shutter, main chamber pump and quartz crystal thin film thickness monitor. The system has a 100 l/sec turbo pump roughing system, and a 1500 l/sec cryopump for use during deposition, because the atomic nitrogen caused corrosion and oil decomposition problems with our conventional pumping system. Base pressure in the unbaked system was in the 10^{-7} Torr range in several hours.

The substrates were clamped to a pyrolytic boron nitride-coated graphite heater, the substrate temperature was monitored with an optical pyrometer. A four-conductor (6mm diameter copper conductors) feedthrough was used to support and make contact to two tungsten

Mat. Res. Soc. Symp. Proc. Vol. 572 © 1999 Materials Research Society

boats (6mm wide tungsten foil 0.25mm thick with a central dimple) for evaporation of two
metals at one time. Typical deposition currents were in the 100-125 Ampere range. The
nitrogen source was operated on pure nitrogen at about 2×10^{-4} Torr. Sc deposition rates were
controlled manually in the 0.05-0.3 nm/sec range.

Substrates were washed in alcohol and preheated in the chamber to degas the substrates
for about 1 hr. Small (2-3 mm diameter) chips of Sc or Y metal (99.99%) were placed in the
boats for evaporation. Hall effect measurements were made with a commercial Hall effect set-up
(Keithly Instruments).

A schematic of the experimental set-up is shown in the figure 1.

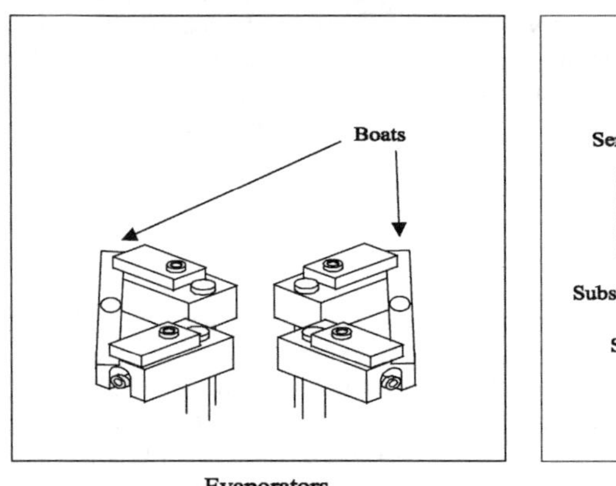

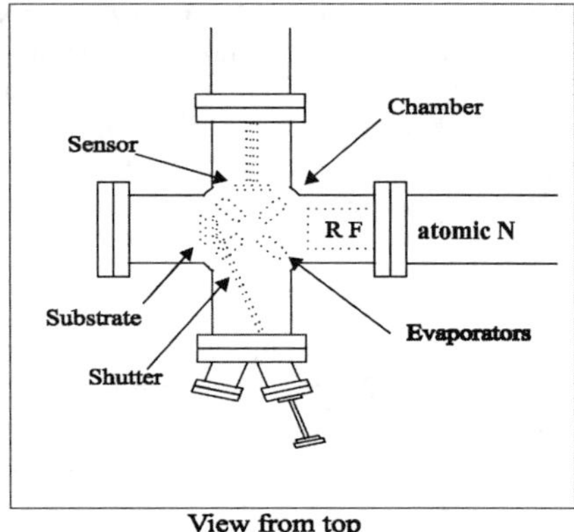

Figure 1. Schematic of experimental set-up.

RESULTS

Several dozens films were grown on various substrates: quartz microscope slides (Quartz
Scientific, Inc. Fairport Harbor OH) and silicon (100). Typically, a low metal evaporation rate
was used, because higher rates were very difficult to control, and resulted in Sc- rich ScN or
metal capped films. Films where the deposition resulted in a Sc metal overlayers were usually
discarded immediately, and no further ScN growth was attempted. ScN and YN "skin" or
"crust" formed on the metal in the evaporation boat. Well-formed bulk cubic crystals were
observed in SEM observation of the evaporation remnants. While still in the chamber, YN films
showed a deep red color, ScN films were yellow to greenish-yellow. Upon Removal from the
vacuum system, the YN films became transparent in a matter of hours. YN capped with AlN
was more stable, but decomposed from the edges of the film inward with the same eventual
outcome.

Sc metal films could be evaporated in this system, which showed only Sc metal lines in
x-ray diffraction, with no detectable Sc-oxide.

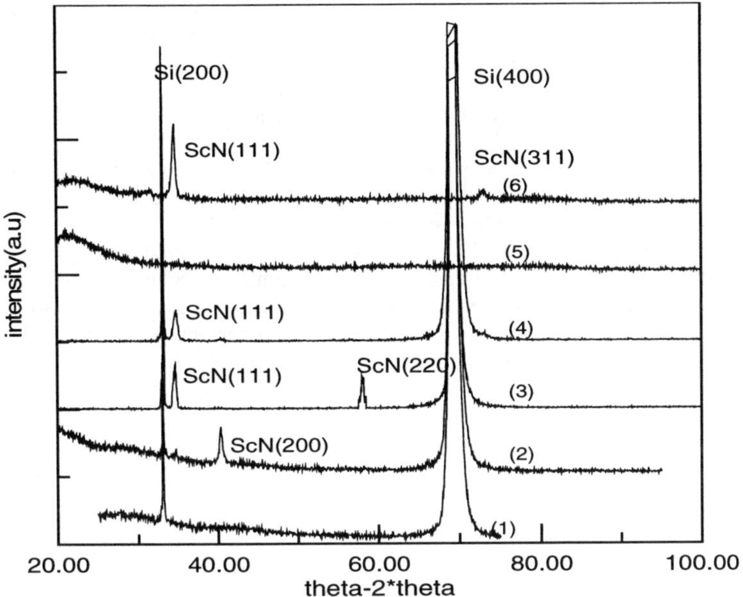

Figure 2: Typical θ-2θ x-ray diffraction curves for ScN on various substrates. Nominal deposition parameters: Curves (1) and (5), Silicon and quartz substrates, respectively. Curve (2) 240 nm ScN, deposited at 650 °C, 0.05 nm/sec, (3) 400 nm ScN, 700 °C 0.2 nm/sec, (4) 170 nm ScN, 700 °C, 0.3 -0.4nm/sec. (6) 220 nm ScN on quartz, 650 °C, 0.03 nm/sec.

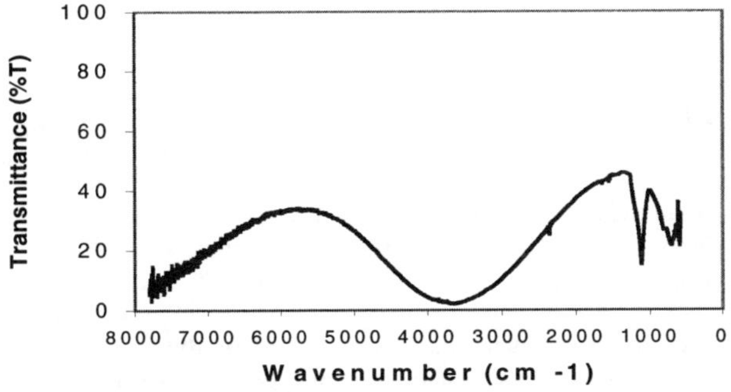

Figure 3. Infrared Transmission of a ScN/quartz film. The LO phonon mode is observed at 1104 cm^{-1} . The sinusoidal oscillations 2000-8000 cm^{-1} are interference fringes.

Infrared transmission measurements were made on these films, originally to determine the thickness, however the refractive index of scandium nitride is not known to sufficient accuracy for thickness measurements using interference fringes. The thermal conductivity of ScN is reportedly in the range for AlN, and the IIIA nitrides, which are quite high. The phonon modes for these materials have recently been reported [7], and the ScN LO mode value observed is consistent with the magnitude of the IIIA-nitrides.

Several types of metal contacts were sputter-deposited onto ScN thin films for Hall measurements and to form simple schottky-diode contacts. The Hall data are reported in Table 1 below. No successful schottky diodes were made.

Diode structures were formed by deposition of ScN on conductive p-type Si(100) (Boron doped, 0.2-1.0 -cm, from Virginia Semiconductor). Indium metal contacts were pressed onto the surface of the ScN film manually, the silicon contact was made by pressing In onto a freshly scrapped area of the film where silicon was exposed. Gold contacts were also used. The I-V curves for indium and Au contacted diodes are shown in figure 4.

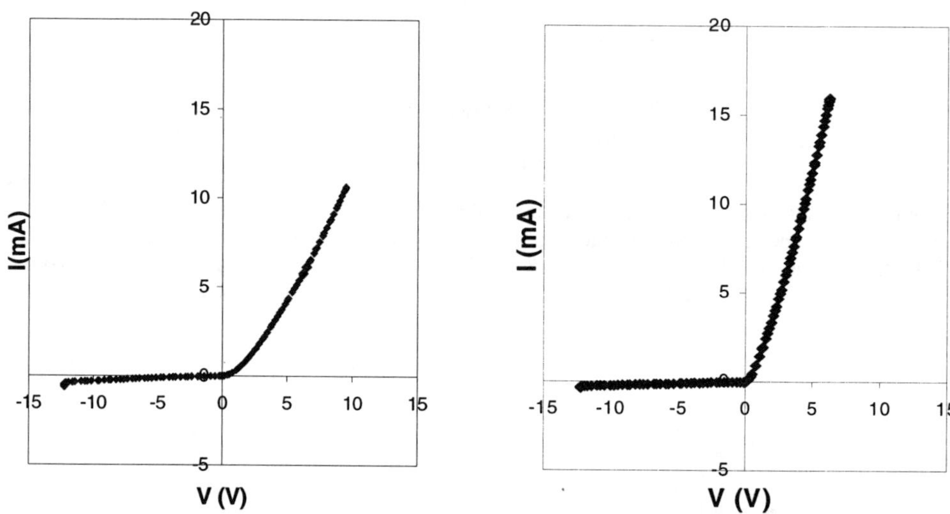

Figure 4: LEFT: 100 nm ScN on p-Si(100), gold contacts..
         RIGHT: 130 nm ScN on p-type Si(100), indium contacts.

Hall effect measurements were made on several of the ScN films. Films that were sufficiently conductive for this measurement were n type. It was suggested in ref. [4] that Mg might be suitable for p-doping of ScN. Magnesium metal was co-evaporated in the second boat at a rate corresponding to about 1% Mg. The Mg content was checked with x-ray fluorescence and is in the 1% range. The Mg-doped ScN sample does show a positive Hall coeffcient. ScN p-n junctions are planned.

The data are summarized in the table below.

Table 1. Hall Effect Data

Sample (substrate)	Thickness (nm)	Contact	Resistivity ($\Omega\cdot$cm)	Hall Coefficient (cm^3/gs)	Carrier Density (cm^{-3})
980727A (quartz *)	130	Ti	6.39E-2	-4.66E0	1.34E18
980728B (quartz *)	180	Ti	6.03E-1	-3.57E0	1.75E18
990129 (quartz **) before anneal	500	Ni/Au	2.97E-2	2.47E1	2.50E17
990129 (quartz **) annealed	500	Ni/Au	1.18E-1	2.53E1	2.47E17

Anneal condition: N_2 , temperature: 447°C, duration: 5min.

DISCUSSION

ScN thin films grown on silicon and quartz show similar orientation as observed for reactively sputtered ScN on MgO in ref.[2,3]. On silicon we have grown both (100) and (111) textured films, on quartz, the (100) orientation predominates. The as-grown films with good conductivity are n-type, suggesting oxygen as the n-dopant, in analogy to GaN. In this case, we expect that the source may be water that desorbs from the system walls during the deposition. In several cases, insulating films were obtained, although similar in color, the bandgap measurements (in the 2.5-3 eV range) on these films (on quartz) indicates some Sc-oxide in the film.

Initial attempts show that Mg may indeed be a p-type dopant for ScN.

CONCLUSIONS

ScN thin films were grown by thermal Sc evaporation in an atomic nitrogen environment. Films were grown on silicon and quartz. Conductive n-ScN/p-Si(100) diodes were tested. YN films were unstable in humid air, and were not investigated further. The results for ScN thin films suggest that diodes and possibly other electronic devices made from ScN junctions may survive operation at elevated temperatures. Although not investigated further in this study, the transition from ScN to Sc-metal is very simple in this growth method, so that Sc contacts for high temperature use should also be considered.

ACKNOWLEDGMENTS

We would like to thank John Dismukes for helpful discussions. This work supported by BMDO through ONR Grants N00014-95-1-0298, and N00014-96-1-0782, -1183.

REFERENCES

1. Gmelin's Handbuch der anorganischen Chemie, 8 Aufl., Deutsche Chem. Gesell., R.J. Meyer, ed., Leipzig-Berlin, Verlag Chemie, 1924-. Volume SE C2, Nitride, Vergleichende Angaben.,pg 146ff.

2. D. Gall, I. Petrov, N. Hellgren, L. Hultman J.E. Sundgren and J.E. Greene, J. Appl. Phys. **84**, 6034 (1998).
3. D. Gall, I. Petrov, L.D. Madsen, J.E. Sundgren and J.E. Greene, J. Vac. Sci. Technol. **A16**, 2411 (1998).
4. J.P. Dismukes and T.D. Moustakas, Proceedings of the III-V Nitrieds Materials and Processing Symposium, The Electrochemical Society, Pennington NJ, 1996.
5. A.G. Petukhov, W.R.L. Lambrecht and B. Segall, Phys. Rev. B. **53**, 4324 (1996).
6. K. Kunze and J.F. Harrison, J. Amer. Chem. Soc. **112,** 3812 (1990).
7. H.Harima, T.Inoue, S. Nakashima, H. Okumura, Y. Ishida, S. Yoshida, T. Koizumi, H.Grille, and F. Bechstedt, Appl. Phys. Lett. **74**, 191 (1999).

PHOTO-ASSISTED RIE of GaN in $BCl_3/Cl_2/N_2$

N. Medelci, A. Tempez, I. Berishev, D. Starikov, and A. Bensaoula
Nitride Materials and Devices Laboratory, SVEC-University of Houston, Houston, TX

ABSTRACT

Gallium nitride (GaN) has been under intense investigation due to its unique qualities (wide band gap, chemical and temperature stability) for optoelectronic and high temperature/high power applications. To this end, reactive ion etching (RIE) experiments were performed on GaN thin films using $BCl_3/Cl_2/Ar$. These resulted in etch rates of 1400 Å/min at -400 V dc bias[1]. However, rough etched surfaces, nitrogen surface depletion and high chlorine content were observed. In order to remedy these shortcomings, a photo-assisted RIE process using a filtered Xe lamp beam was developed, resulting in higher etch rates but again in nitrogen depleted surfaces[2]. Preliminary results on using nitrogen instead of argon in the process chemistry show a big improvement in photo-asssisted etch rates (50%) and Ga/N ratio (0.78 versus 1.25). In this paper, the effects of epilayer doping, dc bias, nitrogen flow rate and photo-irradiation flux on GaN etch rates, surface morphology and composition are presented. Finally, preliminary results on the use of a KrF excimer laser beam in the GaN photo-assisted RIE process are presented.

INTRODUCTION

Wide band gap III-V nitrides are emerging as the materials of choice for high temperature and high power electronics and blue-UV emitters and detectors. Synthesis of gallium nitride (GaN) is being developed for the fabrication of light emitting diodes, laser diodes and flat panel displays. These characteristics, resulting from their strong chemical strength, become a drawback for their processing and thus for their industrial development.

Conventional reactive ion etching (RIE) using halogen-based chemistries achieved relatively low etch rates[3-6]. High density plasma techniques such as electron cyclotron resonance (ECR), inductively coupled plasma (ICP) and magnetron RIE lead to higher etch rates[7-13]. However, these ion-assisted methods cannot avoid ion bombardment damage and surface roughening at high RF powers. In addition, nitrogen depletion is usually associated with high ion energies. Vertical sidewalls have been realized with chemically assisted ion beam etching (CAIBE) using HCl[14]. Alternative dry etching methods for low lattice damage are low energy electron enhanced etching (LE4)[15] and photo-assisted etching. The last technique has already been demonstrated for GaAs and Si[16-17]. Photo-assisted etching of GaN has also been tested using an ArF excimer laser in a HCl ambient[18]. However, the preliminary etch rate was low (optimized etch rates have not yet been reported).

Encouraging results for RIE of GaN have been reported using $BCl_3/Cl_2/Ar$, with etch rates reaching up to 1,200 Å/min at 200 W RF power[1]. Nevertheless, nitrogen depletion, which increases with increasing dc self-bias, was observed after etching. Moreover, higher etch rates are desirable. Therefore, a photo-assisted RIE process using the same chemistry was developed, resulting in higher etch rates but still nitrogen depleted surfaces[2]. Substitution of Ar with N_2 in the gas chemistry was tested with the aim of reducing the surface nitrogen depletion. Encouraging results for both etch rates and nitrogen surface content were obtained. In this paper, photo-assisted RIE of GaN in $BCl_3/Cl_2/N_2$ using both a filtered Xe lamp and a

Mat. Res. Soc. Symp. Proc. Vol. 572 © 1999 Materials Research Society

KrF excimer laser are evaluated and compared to standard RIE. Doping, nitrogen flow rate and lamp optical power density effects on etch rates, surface morphology and composition are investigated.

EXPERIMENTAL DETAILS

The gallium nitride films were grown on sapphire wafers at 750 °C by molecular beam epitaxy[19]. The GaN samples were patterned using photoresist (Shipley 1813). For a given set of etching conditions, two unassisted samples (one blank and one patterned) and two light-exposed ones were processed in the same run. Etching was performed in an Oxford Plasma Technology 80 μp RIE reactor. The base pressure of the reactor was 2×10^{-6} Torr. A xenon lamp with an emission spectrum ranging from 200 to 900 nm was used. The light beam traveled through a glass window (UV filter), an IR filter, reflected on a mirror and impinged at normal incidence onto the samples with a spot size of 2 cm in diameter through a quartz window. The maximum light power density was 38 mW/cm^2. A KrF excimer laser (248 nm) was also used as a light source in the photo-assisted RIE process.

Gallium nitride thin films were etched in 10 sccm BCl_3/10 sccm Cl_2/N_2 at 30 mTorr. The RF power was varied from 100 to 400 W, which corresponds to a dc self-bias range of -200 to -480 V. Etch rates were measured on the patterned samples using a Tencor 250-alpha step profilometer. The GaN surface morphology and etch profiles were checked with a JEOL JSM-5410 scanning electron microscope (SEM). XPS was employed to investigate any resulting surface stoichiometry modification from etching. For this purpose, a Perkin-Elmer PHI ESCA system was used. Mg Kα radiation was utilized as the source of excitation. The energy scale was calibrated from the C 1s line at 285 eV from adventitious carbon on the surface. The base pressure during the analysis was in the low 10^{-10} Torr range. The surface composition was determined from integration of B 1s, C 1s, N 1s, O 2s, Ga 2p$^{3/2}$ and Cl 2p peak area, using the sensitivity factors 0.171, 0.314, 0.499, 0.733, 2.751 and 0.954 for each element, respectively. In the case of GaN grown on an insulating substrate, surface charging build-up was compensated by a flow of low energy electrons. The detection angle was 45° with respect to the sample surface.

RESULTS AND DISCUSSION

Etch Rate Results

GaN thin films were etched in $BCl_3/Cl_2/N_2$ at 100 to 400 W RF powers (-220 to -480 V dc self-biases). The standard RIE rate increased from 400 Å/min at -220 V up to 1330 Å/min at -470 V (Figure 1). The etch rate at a given dc self-bias under filtered Xe lamp light irradiation was higher when compared to the standard RIE rate. The photo-assisted etch rate also increased as a function of dc self-bias. The highest etch rate, 3240 Å/min was reached at -470 V dc self-bias. The highest standard RIE rate obtained at -470 V could be obtained at -250 V with the use of photo-radiation. The photo-assisted to standard rate ratios are in the 1.9-4.4 range with the highest photo-enhancement observed at 200 W RF power (-300 V dc self-bias).

The effects of nitrogen flow rate, Xe lamp beam power density and GaN doping level on etch rates have also been investigated for both the standard and photo-assisted RIE processes. Both n and p-type GaN etch rates decrease with increasing carrier concentration,

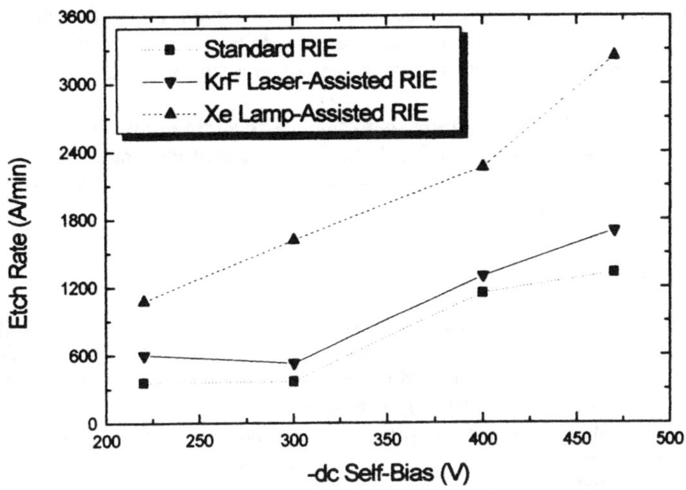

Figure 1. Photo-assisted (both filtered Xe lamp and KrF excimer laser) and standard etch rates and ratios of GaN as a function of dc self-bias voltage. The GaN films were etched in 10 sccm Cl_2/10 sccm BCl_3/10 sccm N_2 at 30 mTorr and 38 mW/cm^2 optical power density.

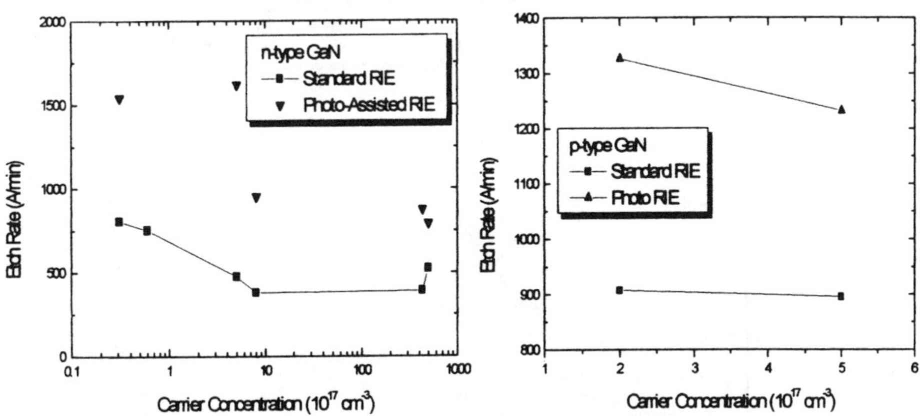

Figure 2. Photo-assisted and standard etch rates of n and p-type GaN as a function of carrier concentration. The GaN films were etched in 10 sccm Cl_2/10 sccm BCl_3/10 sccm N_2 at 30 mTorr and 38 mW/cm^2 optical power density.

particularly under filtered Xe lamp-assisted RIE conditions, even though they are higher than the standard ones (Figure 2). Optimum N_2 flow rate was found to be in the 10-15 sccm range. As the Xe lamp beam power density increases, we notice a sharp increase in the etch rate at 30 mW/cm^2.

Etching under the same conditions but using a KrF excimer laser beam (248 nm) has also been performed. As shown in Figure 1, the laser-assisted etch rates (maximum: 1700 Å/min at -470 V) and photo-assisted to standard rate ratios (maximum: 1.7 at -220 V) were lower than those obtained with the filtered Xe lamp.

Surface Morphology and Composition

The GaN surface morphology as examined by SEM exhibits 'etch pits' at high dc bias (-470 V). We notice however that, at all dc biases, the etched surfaces under illumination are smoother than the standard etched ones. Moreover, we observe that the photo-assisted etched surface at –300 V is smoother than the as-grown surface.

Using XPS analysis, the surface composition change due to etching was investigated. In the standard RIE process, the surface Ga/N versus dc self-bias does not show a defined trend (Figure 3). At –300 V, the surface is more depleted in N than the starting GaN surface. For -200V and –480 V, the surface Ga/N ratio is similar to that of the as-deposited surface (1.55). However, after etching at -380V, the surface is even closer to stoichiometry than the starting GaN. When the GaN sample is illuminated during the etch process with the filtered Xe lamp beam, the surface Ga/N ratio decreases as the dc self-bias increases, reaching 1.37. at

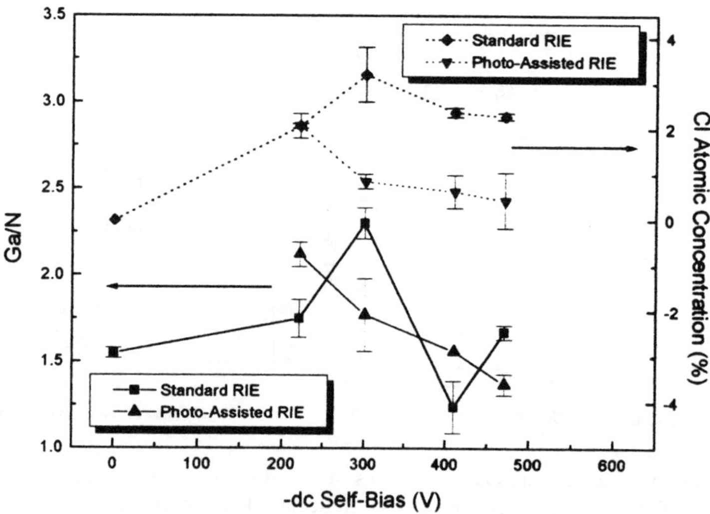

Figure 3. Surface XPS Ga to N ratios and Cl atomic concentrations as a function of dc bias.

-480V. When the sample is illuminated with the KrF excimer laser beam, the variation of the Ga/N ratio with the dc self-bias follows the fluctuations of the ratio obtained with standard RIE.

The standard RIE process always results in a surface containing 2% or more of chlorine (Figure 3). The lowest chlorine residual content is obtained at -220V dc self-bias and the highest content at -300V. The photo-assisted etching process (for both KrF excimer laser and filtered Xenon lamp) always leads to a surface as clean or cleaner than the standard RIE process. In the case of the filtered Xe lamp, the surface residual chlorine content decreases as the dc self-bias increases. At -300V dc self-bias, the chlorine content is below 0.8%. When using the KrF excimer laser, the photo-assisted process is only better than standard RIE for the highest dc self-biases (-420 and -480V in our study).

It is difficult to explain how irradiation influences and enhances the etching process without knowledge of the surface composition during etch and real time monitoring of the etch products. It is believed that irradiation may excite the adsorbed species (reactants and products) and thus enhance product desorption. This would also lead to higher coverage of the material surface with reactive species. The availability of N radicals in the plasma and the photo-excitation of the surface resident species might enhance surface diffusion and accelerate the formation of more volatile end products. In addition, photo-excited surface Ga atoms with unsatisfied bonds react more readily with N atoms. Such process should result in higher surface turn-over and etch rates. Finally, photo-generated carriers take part in the photochemistry, as shown by the decrease in etch rate for the p-type material as doping level increases.

CONCLUSIONS

Photo-assisted reactive ion etching of GaN using both a filtered Xe lamp and a KrF excimer laser in $BCl_3/Cl_2/N_2$ was investigated. The standard and photo-assisted RIE etch rates always increase with increasing dc self-bias. Higher etch rates are observed when the materials are exposed to both a filtered Xe lamp and a KrF excimer laser beam, reaching 3240 and 1700 Å/min at –470 V dc bias, respectively. Higher photo-assisted to standard etch ratios were observed for the filtered Xe lamp case, with a maximum (4.4) observed at –300V dc bias. Use of illumination lowers surface chlorine atomic concentration. Etch rates decrease as GaN doping level increases for both p and n-type materials. Hence, the photo-assisted RIE process of GaN in $BCl_3/Cl_2/N_2$ results in higher etch rates, and smoother and cleaner etched surfaces when compared to the standard process performed using the same conditions.

ACKNOWLEDGMENTS

This work was supported by funds from a NASA cooperative agreement #NCC8-127 to SVEC, a Texas Advanced Research Program Grant # 1-1-27764, and a Texas Advanced Technology Program Grant # 1-1-32061. This material is also based upon work supported by the U.S. Civilian Research and Development foundation under Award No. REI-247. This work made use of TCSUH/MRSEC Shared facilities supported by the State of Texas through the Texas Center for Superconductivity at the University of Houston and by the National Science Foundation under Award Number DMR-9632667. The authors would also like to thank Darren Tucker for helping with the RIE system and Susan Street for performing the SEM analysis.

REFERENCES

[1] N. Medelci , A. Tempez, E. Kim, N. Badi, D. Starikov, I. Berichev, and A. Bensaoula, Mat. Res. Soc. Symp. Proc. **512**, 285 (1998).

[2] A. Tempez, N. Medelci, N. Badi, D. Starikov, I. Berishev, and A. Bensaoula, *"Photoenhanced reactive ion etching of III-V nitrides in $BCl_3/Cl_2/Ar/N_2$ plasmas"*, to be published in J. Vac. Sci. and Technol. A (1999).

[3] I. Adesida, A. Mahajan, E. Andideh, M. Asif Khan, D. T. Olsen, and J. N. Kuznia, Appl. Phys. Lett. **63**(20), 2777 (1993).

[4] Heon Lee, David B. Oberman and James S. Harris Jr., Appl. Phys. Lett. **67**(12), 1754 (1995).

[5] M. E. Lin, Z. F. Fan, Z. Ma, L. H. Allen, and H. Morkoç, Appl. Phys. Lett. **64**(7), 887 (1994).

[6] A. T. Ping, I. Adesida, M. Asif Khan and J. N. Kuznia, Electronics Letters, Vol. **30**, No. 22, 1895 (1994).

[7] C. B. Vartuli, S. J. Pearton, J. W. Lee, J. D. McKenzie, C.R. Abernathy, and R. J. Shul, J. Vac. Sci. and Technol. A **15**(3), 638 (1997).

[8] G. F. McLane, T. Monahan, D. W. Eckart, S. J. Pearton, and C. R. Abernathy, J. Vac. Sci. and Technol. A **14**(3), 1046 (1996).

[9] C. B. Vartuli, S. J. Pearton, C.R. Abernathy, and R. J. Shul, A. J. Howard, S. P. Kilcoyne, J. E. Parmeter and M. Hagerott-Crawford, J. Vac. Sci. and Technol. A **14**(3), 1011 (1996).

[10] G. F. McLane, L. Casas, S. J. Pearton, and C. R. Abernathy, Appl. Phys. Lett. **66**(24), 3328 (1995).

[11] G. F. McLane, L. Casas, R. T. Lareau, D. W. Eckart, C. B. Vartuli, S. J. Pearton, and C. R. Abernathy, J. Vac. and Sci. Technol. A **13**(3), 724 (1995).

[12] Hyun Cho, C. B. Vartuli, S. M. Donovan, C.R. Abernathy, S. J. Pearton, R. J. Shul, C. Constantine, J. Vac. Sci. and Technol. A **16**(3), 1631 (1998).

[13] R. J. Shul, C. G. Willinson, M. M. Bridges, J. Han, J. W. Lee, S. J. Pearton, C.R. Abernathy, J. D. McKenzie, S. M. Donovan, L. Zhang, and L. F. Lester, J. Vac. Sci. and Technol. A **16**(3), 1621 (1998).

[14] A. T. Ping, I. Adesida, M. Asif Khan, Appl. Phys. Lett. **67**(9), 1250 (1995).

[15] H. P. Gillis, D. A. Choutov, K. P. Martin, M. D. Bremser, and R. F. Davis, Journal of Electronics Materials Vol. **26** No. 3 (1997).

[16] F.A. Houle, Phys. Rev. B **19**, 10120 (1989).

[17] S. Takatani, S. Yamamoto, H. Takawaza, and K. Mochiji, J. Vac. Sci. and Technol. B **13**(6) 2340 (1995).

[18] R.T. Leonard, S. M. Bedair, Appl. Phys. Lett. **68**(6), 794 (1996).

[19] I. Berishev, E. Kim, and A. Bensaoula, J. Vac. Sci. and Technol. A **16**(5), 2791 (1998).

CORRELATION OF DRAIN CURRENT PULSED RESPONSE WITH MICROWAVE POWER OUTPUT IN AlGaN/GaN HEMTs

S. C. Binari, K. Ikossi-Anastasiou, W. Kruppa, H. B. Dietrich, G. Kelner, R. L. Henry, D. D. Koleske, and A. E. Wickenden
Naval Research Laboratory, Washington, D. C., 20375

ABSTRACT

The drain-current response to short (<1μs) gate pulses has been measured for a series of GaN HEMT wafers that have similar dc and small-signal characteristics. This response has been found to correlate well with the measured microwave power output. For example, for devices where the pulsed drain current is greater than 70% of the dc value, output power densities of up to 2.3 W/mm are attained. This is in contrast with 0.5 W/mm measured for devices with low pulse response (less than 20% of the dc value). These results, which can be explained by the presence of traps in the device structure, provide a convenient test which is predictive of power performance.

INTRODUCTION

GaN-based microwave power transistors have set the state-of-the art for output power density [1] and have the potential to replace GaAs-based transistors for a number of high-power applications. The rapid advances made in GaN-based devices stem from the electronic properties of the GaN-based material system. However, due to the relative immaturity of the materials growth and device processing, the measured microwave power output is frequently limited by trapping effects [2]. Trapping effects have been shown to play a role in the operation of many GaN-based FETs. In addition to the limitations they impose on microwave power output, traps have been shown to result in drain-current transients subsequent to a gate or drain voltage pulse [2], drain current collapse after the application of a high drain bias [3], and transconductance and output resistance frequency dispersion [4]. One of the objectives of this work is to develop a test or measurement technique that can serve as a useful predictor of device microwave power performance. The earlier this test can be performed in the device processing sequence, the more useful it will be.

During the development of GaAs field-effect transistors, significant attention was directed toward the understanding and minimization of trapping effects. In the GaAs technology, the drain current response to gate- and drain-voltage pulses was extensively utilized [5, 6] as a means of investigating trapping effects. It was also shown that pulsed current-voltage characteristics can serve as a indicator of microwave power performance [7,8]. In the work presented here, the measured microwave power output in GaN-based HEMTs has been found to correlate well with the drain-current pulsed response.

MATERIALS GROWTH AND DEVICE FABRICATION

The device cross section is shown in Fig. 1. All HEMTs were fabricated with 4 to 6 μm source-drain spacings and a nominal gate length of 1 μm. The gate consists of two gate fingers with a total gate width of 150 μm. The ohmic contacts were Ti/Al/Ni/Au and had a contact resistance in the range of 1-2 Ω-mm. Pt/Au was used for the gate metallization and the devices were isolated with N implantation.

Mat. Res. Soc. Symp. Proc. Vol. 572 © 1999 Materials Research Society

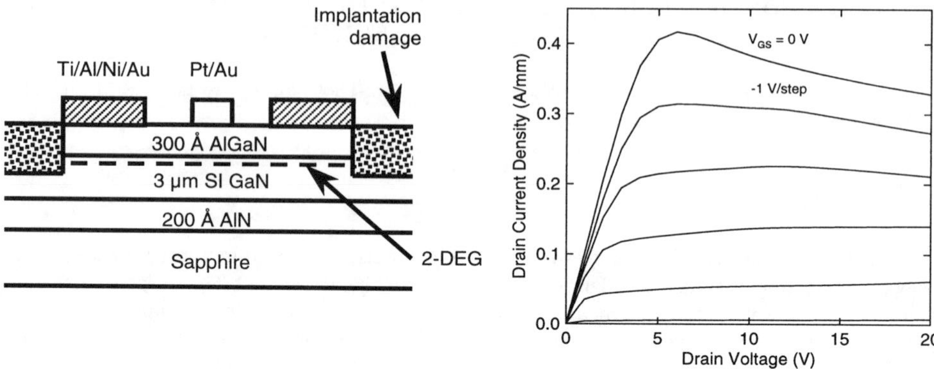

Fig. 1. Device cross section.   **Fig. 2.** Representative drain characteristics.

The epitaxial layers used in this work were grown in an inductively-heated, water-cooled, vertical MOCVD reactor. Triethylaluminum, trimethylgallium, and NH_3 were used as the reactant sources, and Si_2H_6 was the Si dopant source. The AlGaN/GaN HEMT structures consisted of a 3 µm thick undoped, high-resistivity GaN buffer layer grown on top of a thin (200 Å) low-temperature nucleation layer. The buffer layer was employed to spatially remove the active part of the device from the higher-defect-density material near the substrate interface. An AlGaN layer with a total thickness of 300 Å was grown on top of the GaN buffer. Five epitaxial layer designs were used in this work. The salient features of these wafers, including the average values measured across a wafer for sheet resistance, mobility, sheet carrier concentration, I_{max} (the maximum drain current at a forward gate current of 0.1 mA/mm), and the threshold voltage, V_{th}, are summarized in Table I. Wafers 1 and 5 were undoped, wafers 2-4 had Si-doped AlGaN layers with ~30 Å undoped AlGaN spacer layers, and wafers 3 and 4 had a ~50 Å undoped AlGaN cap.

HEMT Wafer #	Al fraction	Si_2H_6 flow (sccm)	R_{sh} ($\Omega/\square$)	Mobility (cm^2/V-s)	n_{sh} (10^{13} cm^{-2})	I_{max} (mA/mm)	V_{th} (V)
1	0.3	0	760	800	1.0	480	-4.5
2	0.3	0.26	690	830	1.1	550	-4.0
3	0.3	0.26	590	845	1.3	520	-4.5
4	0.3	0.52	610	870	1.2	560	-5.5
5	0.4	0	700	725	1.2	500	-4.5

Table I. Materials and device electrical characteristics.

RESULTS

The current-voltage characteristics, drain current pulsed response, and the microwave power output were measured for the HEMTs described above. Drain characteristics representative of the devices studied here are shown in Fig. 2. The negative slope at the larger drain voltages are due to thermal effects. The transfer characteristics for a group of devices (that

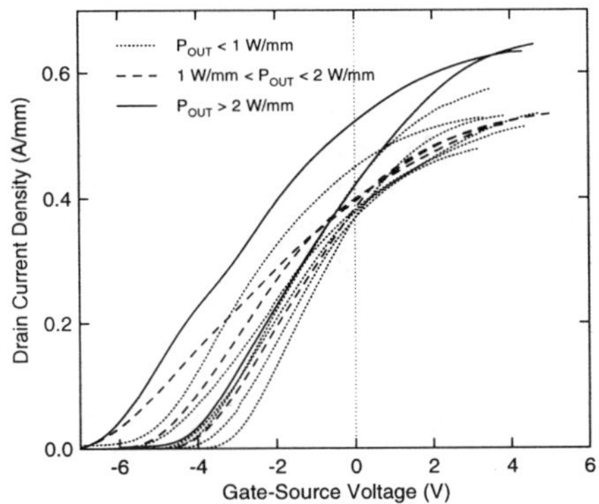

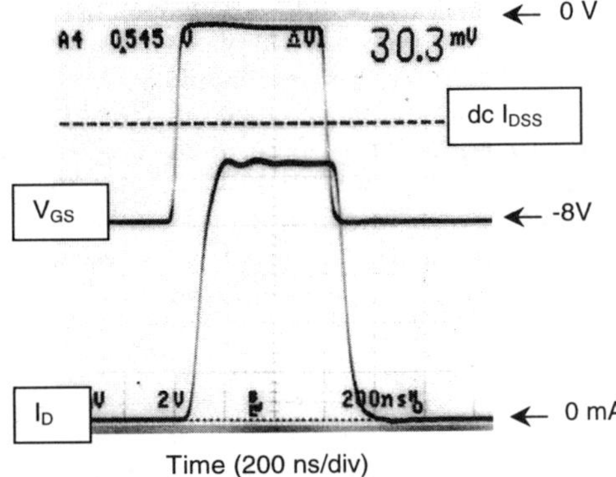

Fig. 3. Transfer characteristics for $V_{DS} = 7$ V. The maximum forward gate current was 0.1 mA/mm. The line style used represents the different output power levels measured.

Fig. 4. Drain current response to a gate voltage pulse. The gate voltage is pulsed from a level less than V_{th} to 0 V.

were subsequently measured for pulsed-response and power output) are shown in Fig. 3. The maximum drain current level, I_{max}, is within the range of 0.5 to 0.65 A/mm.

An estimate of the maximum power that can be obtained from the transistor is $\Delta I \Delta V / 8$, where ΔI is the available current swing ($= I_{max}$), and ΔV is the knee voltage subtracted from the gate-drain breakdown voltage at pinch-off. Based on the dc values of current and voltage, the estimated power output from this group of wafers should be comparable. However, as described below, the power output level varies by more than a factor of 5.

The drain current response to a gate voltage pulse was also measured for this group of devices. An example of this is shown in Fig. 4. For this case, the gate voltage is pulsed from a level of -8V to 0V and V_{DS} was 1V. The pulse width was 0.6 μs and the duty cycle was 1%. The drain current was measured with a current probe. The dc value of I_{DSS} ($V_{GS} = 0$) is also shown in the figure. This device exhibits a high pulse-to-dc current ratio, 90%. Other devices that were measured exhibited a much lower ratio; as low as 5%.

The choice of the starting value of V_{GS} has a significant effect on the observed drain current pulse. In general, to see a significant difference between the pulsed and dc current level, the starting V_{GS} must be at the threshold voltage or lower. This variation is shown in Fig. 5. In addition, values of V_{DS} from 0.1 to 10V were investigated and it was determined that V_{DS} typically had a minimal effect on the current ratio. This behavior is shown in Fig. 5, where the starting V_{GS} is varied from -8 to 0V for two different values of V_{DS}.

The microwave power output for these devices was measured using on-wafer microwave probes. The input and output tuning as well as the gate and drain bias were adjusted to maximize the output power. The measurement frequency was 2 GHz, at which the small-signal gain was between 10 and 13 dB. Output power levels as high as 2.3 W/mm and as low as 0.4 W/mm were measured. The power output is plotted as a function of drain current pulse response in Fig. 6, and a strong correlation is observed.

The results described above were obtained under normal room lighting conditions. The effect of ultraviolet illumination on both the microwave and pulsed current measurement was investigated. It was generally observed that ultraviolet illumination significantly increased the power output and pulse response ratio for HEMTs where these values were low, but had a minimal effect for devices where these values were high. For example, for devices that performed poorly in normal lighting, the current ratio was found to increase from 0.12 to 0.65 and the microwave power output was found to increase by 4 dB with ultraviolet illumination. For devices that performed well, the pulse ratio increased from 0.84 to 0.90 and the power output increased by <0.5 dB with the application of ultraviolet illumination.

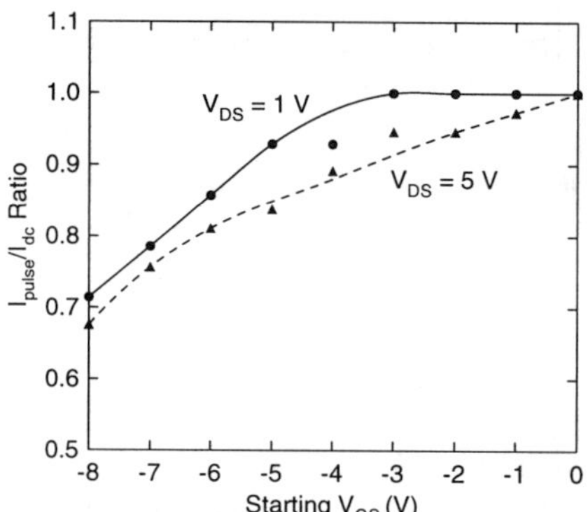

Fig. 5. Variation of pulse response with starting gate-source voltage.

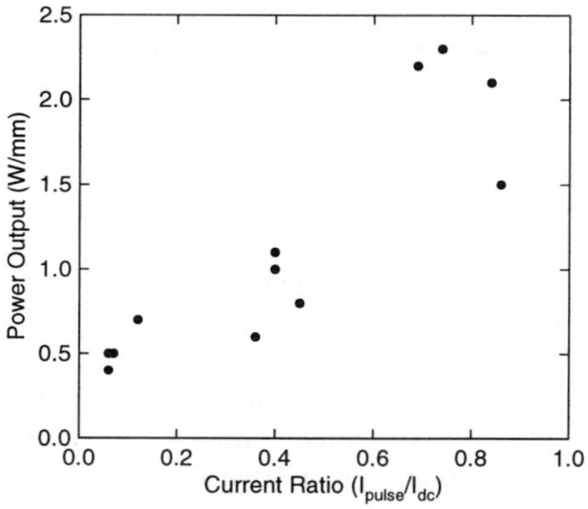

Fig. 6. Correlation of microwave power output at 2 GHz with drain current response.

DISCUSSION & CONCLUSIONS

The measured results can be explained by the presence of traps in the device structure. The pulse measurement reported here is related to gate lag measurements done originally on GaAs MESFETs to assess their performance potential in digital circuits. In those cases, the gate lag phenomenon was usually attributed to surface states in the access regions between the metal contacts which acted as electron traps. Although in the present case, additional study is necessary to establish with certainty that surface states are involved in the observed phenomena, the lack of a gate recess or a systematic passivation procedure, makes this a likely possibility. A plausible mechanism consists of electrons being trapped at the access region surface with a negative gate voltage and causing a lifting of the quantum well below, thereby reducing its free electron density. When the gate voltage switches to 0 V, thereby opening the channel under the gate, the trapped electrons cannot respond quickly, and the 2 DEG density below increases only slowly to its full value.

In the power output tests, devices with more surface states will have lower values of output power, since the maximum drain current is limited by the steady-state trap occupancy of the surface states which causes a reduction in the electron density in the quantum well. Although further tests are necessary to fully characterize and then hopefully eliminate the mechanism causing the limitation in output power, the observed correlation between gate-pulse response and power performance yields a simple screening procedure which can be utilized early in the device fabrication process.

ACKNOWLEDGMENTS

This work was supported by the Office of Naval Research.

REFERENCES

1. S. T. Sheppard, K. Doverspike, W. L. Pribble, S. T. Allen, J. W. Palmour, L. T. Kehias, and T. J. Jenkins, to be published, IEEE EDL, Apr. 1999.

2. S. C. Binari, H. B. Dietrich, W. Kruppa, G. Kelner, N. S. Saks, A. Edwards, J. M. Redwing, A. E. Wickenden, and D. D. Koleske, Proc. Inter. Conf. Nitride Semicond., pp.476-478, 1997.

3. S. C. Binari, W. Kruppa, H. B. Dietrich, G. Kelner, A. E. Wickenden, and J. A. Freitas, Jr., Solid-State Electron. **41**, pp. 1549-1554, (1997).

4. W. Kruppa, S. C. Binari, and K. Doverspike, Electronics Lett. **31**, pp. 1951-2, (1995).

5. M. Rocci, Physica **129B**, pp. 119-138, (1985).

6. R. Yeats, D. C. D'Avanzo, K. Chan, N. Fernandez, T. W. Taylor, and C. Vogel, IEEE IEDM Digest, pp. 842-845, (1988).

7. A. Platzker, A. Palevsky, S. Nash, W. Struble, and Y. Tajima, IEEE MTT-S Digest, pp. 1137-1140, (1990).

8. J. C. Huang, G. Jackson, S. Shanfield, W. Hoke, P. Lyman, D. Atwood, P. Saledas, M. Schindler, Y. Tajima, A. Platzker, D. Masse, and H. Statz, IEEE MTT-S Digest, pp. 713-716, (1991).

PHOTOIONIZATION SPECTRA OF TRAPS RESPONSIBLE FOR CURRENT COLLAPSE IN GaN MESFETS

P.B. KLEIN, J.A. FREITAS, Jr., AND S.C. BINARI
Naval Research Laboratory, Washington DC 20375-5347, klein@bloch.nrl.navy.mil

ABSTRACT

Current collapse in GaN MESFETS is believed to result from the trapping of carriers in the high resistivity GaN layer, and can be reversed by the application of light.   Light photoionizes (or photoneutralizes) the carriers, releasing them from the traps and restoring all or part of the original I-V characteristics of the device.   In these investigations we have taken advantage of this effect to characterize the traps responsible for current collapse in an n-channel GaN MESFET.   At fixed source-drain voltage, the incremental light-induced drain current, above that measured in the dark, and normalized per incident photon, is measured as a function of wavelength.   The resulting photoionization spectrum reflects two absorption thresholds corresponding to two distinct electron traps.   Because of the nature of the measurement, these traps can be identified as those responsible for current collapse in the device.

INTRODUCTION

Significant progress has been made over the last several years in understanding the physical characteristics of the group-III nitride material system, as well as in developing optoelectronic and electronic devices based on these materials. The material parameters of this system promise the possibility of convenient sources and detectors in the UV and blue portions of the spectrum, as well as electronic devices capable of operating at high power, high temperature and in adverse environments. Indeed, blue-green LED's , as well as blue lasers operating for several thousand hours are now commercially available, in spite of the fact that the materials used to fabricate these devices contain relatively high concentrations of defects. Similarly for electronic devices, FETs have been successfully produced in GaN as well as in AlGaN/GaN HEMT structures. While the properties of these FETs continue to improve, trap-related phenomena still hinder the reproducible fabrication of high quality electronic devices.

Two of the most commonplace trapping phenomena are persistent photoconductivity (PPC) and "current collapse". PPC is the optical excitation of photoconductivity in a material that exists for times long after the optical excitation source is removed. This process is often associated with the presence of a metastable deep defect, such as the DX center in AlGaAs, or can be induced by the transport of photoexcited carriers across macroscopic potential barriers [1]. It has been noted with respect to the AlGaAs/GaAs system that while PPC poses no direct problem for FETs, its presence does indicate the possibility of other transient device instabilities associated with charge trapping, such as shifts in the threshold voltage [2]. "Current collapse" refers to the trapping of charge at deep level centers in the structure, resulting in a dramatic reduction in the current flowing through the device, and hence in its output power.

Current collapse is initiated by subjecting the device to a high electric field, such as that experienced by ramping the drain voltage up to a large value. This is thought to inject hot carriers from the channel into regions of the structure where they can be trapped by deep defects. The resulting buildup of charge forms a depletion region in the channel which tends to pinch off

Mat. Res. Soc. Symp. Proc. Vol. 572 © 1999 Materials Research Society

the channel and reduce the source-drain current. Khan et al. [3] studied collapse in an AlGaN/GaN heterostructure insulated gate FET (HIGFET) and, following analogous work in similar AlGaAs/GaAs structures [4], suggested that the trapping occurred in the AlGaN gate insulator. Similarly, Binari et al. [5] observed this effect in GaN MESFETS. In that case, the carrier trapping was assumed to occur in the semi-insulating GaN layer below the active n-type channel layer. In both of these studies it was observed that the current collapse could be reversed by the application of light, and that the effectiveness of the light illumination diminished monotonically with increasing wavelength.

It is the purpose of this investigation to take advantage of the light sensitivity of the current collapse phenomenon in order to begin to identify the traps responsible for this effect in GaN MESFETs. While current interest is more focused on the more efficient HFET structures, it is important to understand these processes in the simpler MESFET, where only the GaN material contributes. With channel carriers trapped on deep level centers in the structure, the optical reversal of current collapse should be viewed as a photoionization (or photoneutralization, depending on the charge state of the trap) of the trapped carrier from the trap. Consequently, the dependence of the light-induced increase in the drain current on the illumination wavelength should reflect the photoionization spectrum of the trap. This spectrum, and its associated photoionization threshold, can serve as an identifier for the trap involved in the current collapse process.

EXPERIMENT

Details of the MESFET design and characterization are described in Ref. 5. The device layout for the MESFET is shown in Fig. 1 and a cross-section is shown in Fig. 2. The FET was fabricated with a source-drain spacing of 5μm, a gate width of 150μm and a gate length of 1.5μm. The active channel was grown on top of a thick, undoped semi-insulating buffer layer in order to spatially remove the active part of the device from the higher-defect-density material near the sapphire substrate. Hall measurements at 300K indicated a channel carrier concentration of $2 \times 10^{17} cm^{-3}$ and a mobility of 410 cm^2/V-sec. The current collapse measurements were carried out by obtaining drain current characteristics of the device with light illumination and in the dark. These were determined using an HP4145B semiconductor parameter analyzer, which measures the drain characteristics with a single sweep of V_{DS} for each gate bias.

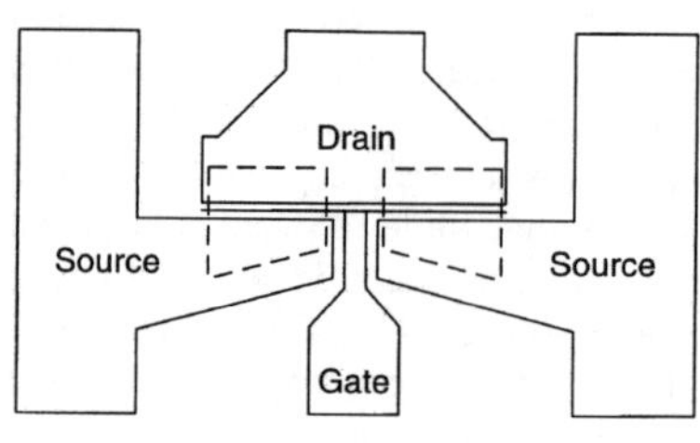

Fig. 1. MESFET layout. Solid lines indicate source, drain and gate metallizations. Dotted lines indicate implant isolation regions.

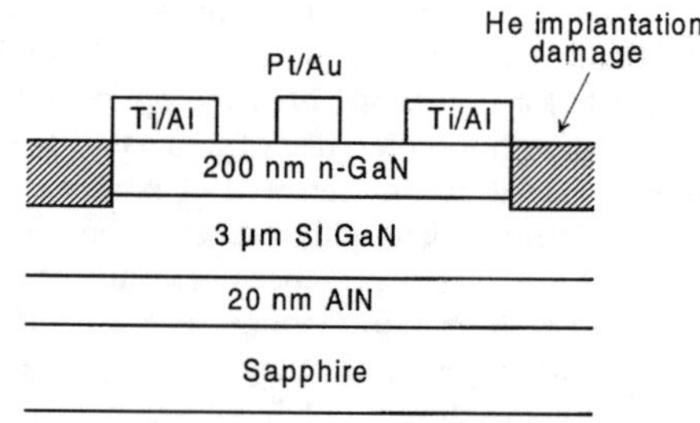

Fig. 2. MESFET cross-section.

For reproducible measurements of current collapse, it was found necessary to initialize the device before each measurement by proximity illumination with a blue GaN LED (1mW), which emptied all or most of the traps. The measurements were carried out near zero gate bias by setting the system to record two consecutive I-V curves with V_{GS} at 0V and -10mV, with a 5 sec delay between scans. After the first I-V curve was recorded in the dark, an electronically activated shutter was opened, allowing the device to be measured during the second scan under optical illumination. Monochromatic light was obtained from a 75W Xe arc lamp (PTI model A1010) and a Spex 1680B 0.22m. double monochromator with a 1200 gpm holographic grating, with the spectrometer bandpass set to approximately 3.5nm. The light was collected with a spherical mirror and focused onto the device with an output power of generally tens of μW in an image area of about 20mm^2. I-V curves were obtained for light-on and light-off (dark) conditions, and the *fractional* increase in drain current taken under illumination (at wavelength λ) over that taken in the dark, $(I_\lambda\text{-}I_{dark})/I_{dark}$, was measured at a predetermined fixed drain voltage (taken at V_{DS}=5V). This light-induced increase in I_{DS} was normalized at each wavelength by the incident photon flux Φ (photons/cm^2/sec). The resultant quantity, defined by $R(\lambda) \equiv [(I_\lambda\text{-}I_{dark})/I_{dark}]/\Phi(\lambda)$ is a measure of the number of traps emptied by photoionization per incident photon. The spectral dependence of $R(\lambda)$ reflects the photoionization spectrum [6] of the trap, thus allowing us to probe the traps responsible for the current collapse.

RESULTS

The experimentally determined spectrum R(hv), plotted as a function of incident photon energy (hv = hc/λ), is shown as the filled circles in Fig. 3. Two broad absorptions are observed. In addition, a rise in the drain current is observed near the GaN bandgap, and is assumed to result from photoexcited carriers injected into the channel. As observed by Binari et al. [5], a monotonic decrease in the effectiveness of light to reverse current collapse is observed with increasing wavelength. This optically-induced reversal appears to result from the photoionization of two distinct trapping centers, labeled Trap 1 and Trap 2. It should be emphasized again that the absorptions seen here represent optical transitions *from traps that are intimately connected with the current collapse phenomenon.*

The dotted line in the figure represents a best fit of the data to the functional form for the photoionization cross-section of a deep-level defect [7,8], where a vertical, "forbidden" optical transition is assumed: $\sigma(hv) \propto (hv\text{-}E_{th})^{3/2}/(hv)^3$, where E_{th} is the absorption threshold. A similar form for the analogous "allowed" transition [8], which employs an additional fitting parameter, does not significantly improve the fit. The failure of this fitting procedure appears to result from the fact that the observed absorptions are particularly broad. Large optical linewidths associated with deep centers in semiconductors often reveal the presence of strong coupling of the electronic states of the deep center to the vibrational states of the lattice [6]. Such coupling is usually associated with a significant lattice distortion at the defect site. As sketched in Fig. 4, at 300K, absorption between the trap and the conduction band can take place from vibrational excited states as well as from the lowest trap state, thus resulting in a Gaussian broadening of the absorption.

In such cases, photoionization data must be fitted by a convolution of the photoionization cross-section with a Gaussian broadening function. This procedure has been successfully

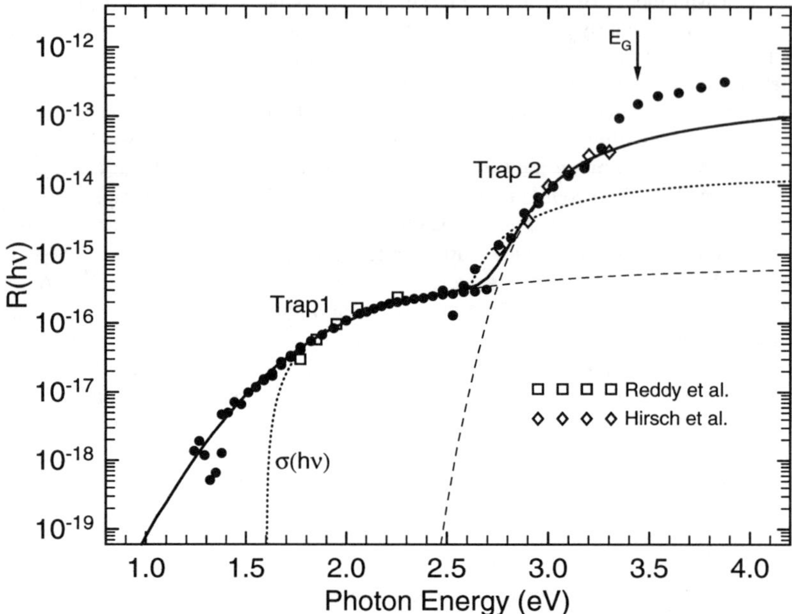

Fig. 3. The filled circles represent the spectral dependence of the optical reversal of current collapse, R(hν), that reflects the photoionization spectrum of the traps causing this effect. The dotted line represents the best fit of the data to a standard deep level photoionization spectrum, while the solid line is a fit employing the convolution of the photoionization spectrum with a Gaussian broadening function. The open squares and diamonds represent data from recent photoconductivity studies (individually scaled).

employed in studies of the DX center in AlGaAs [9] and the EL2 center in semi-insulating GaAs [10], and has recently been applied to account for capacitance transient spectra of the deep "E2" center in n-GaN [11]. Using the approach of Mooney et al. [9], the solid line in Fig. 3 represents such a fit of the current collapse data, and accounts quite well for the spectral dependence of R(hν). The dashed lines are the Trap 1 and Trap 2 components of this fit. For each absorption, an amplitude, an energy threshold and a Gaussian broadening parameter are deduced. For Trap 1, $A_1=1.5\times10^{-16}$, $E_1=1.8$ eV, and $\sigma_1=0.26$ eV.   For Trap 2, $A_2=1.4\times10^{-14}$, $E_2=2.85$ eV, and $\sigma_2=0.10$ eV. Note that these broadening parameters correspond to Gaussians with FWHM of approximately 0.6eV and 0.25eV, respectively, which verifies the large breadth of the absorptions.   Unlike the unbroadened photoionization cross-section, there is significant absorption at energies lower than the threshold energy. As can be seen from the diagram in Fig. 5, these two threshold energies correspond to two very deep traps. It is also important to note that these optical threshold energies exceed the trap depths that might be determined from a DLTS measurement of the same center, for example, by the lattice relaxation energy (or Franck-Condon energy) d_{FC} indicated in Fig. 5.

The data in Fig. 3 is also compared to two recent investigations of photoconductivity in GaN. Hirsch et al. [12] studied the spectral dependence of PPC in MOCVD grown, unintentionally doped n-GaN, and deduced the photoionization spectrum (diamonds) shown (scaled) in Fig. 3 from the time dependence of photoconductivity buildup in their samples. Similarly, Reddy et al. [13] also measured PPC in undoped, MBE-grown GaN, but in a slightly different spectral region: their photo-current results are shown (scaled) as the squares in Fig. 3. That these data agree with our current measurements for Trap 2 and Trap 1, respectively, suggests the possibility that we are observing the same traps, but it is curious that each of these studies reports a single trap. The signal-to-noise of the data of Hirsch et al. would suggest that they would have been unable to detect Trap 1, but it is not clear whether Reddy et al. investigated the energy range of Trap 2, or whether that trap simply does not exist in MBE-grown material. If the latter were found to be true it would be a significant point, as the trap is a major contributor to current collapse in GaN-based devices.

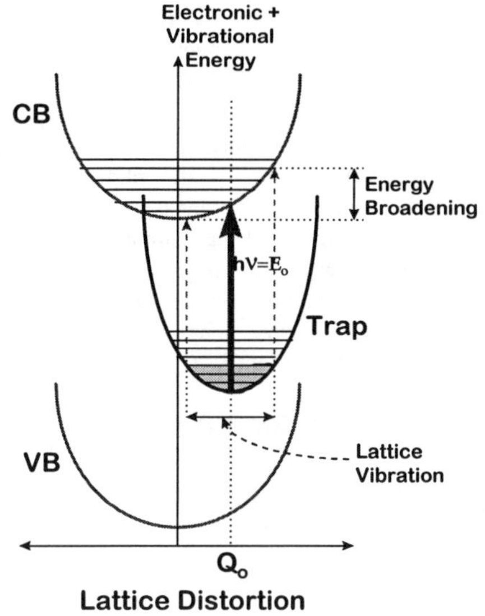

Fig. 4. Broadening of absorption lines from a strongly lattice-coupled deep level defect.

We have also studied the dependence of the light-induced reversal of current collapse on the excitation intensity of the light source. It was found that at 400nm (3.1eV), the optical response from Trap 2 is almost fully saturated (i.e. most of the traps are empty) at the highest power used, corresponding to about 1mW/cm^2. For Trap 1, studied at 600nm (2.07eV), the highest excitation power was found to be within an order of magnitude of saturation values. Saturation at such low power levels suggests either a low trap density, a large photoionization cross-section, or a small capture cross-section, although these parameters cannot yet be unraveled by the present measurements. However, Hirsch et al. have observed very slow photoconductivity buildups associated with the deep center that appears to be identical to Trap 2 in the current work. From these results they concluded that the capture cross-section for this trap must be extremely small, in agreement with our current observations.

CONCLUSIONS

In these studies we have taken advantage of the optical reversibility of the current collapse phenomenon in order to probe the deep traps responsible for this process in a GaN MESFET. In addition to a near-bandedge effect believed due to the injection of photoexcited carriers into the channel, two broad, below-gap absorptions were observed, which we have associated with the photoionization of trapped electrons at two distinct deep centers in the semi-insulating GaN layer. These absorptions were much too broad to fit with the standard spectral dependence for a deep-center photoionization   cross-section. Assuming strong coupling of the deep trap with the lattice, the data was well accounted for by a convolution of the photoionization cross section and a Gaussian broadening function.   The resulting threshold energies reflect two very deep traps associated with current collapse, with absorption thresholds at 1.8 eV and 2.85 eV. These traps may have been observed in recent PPC studies as well.

ACKNOWLEDGEMENTS

This work has been supported in part by the Office of Naval Research.

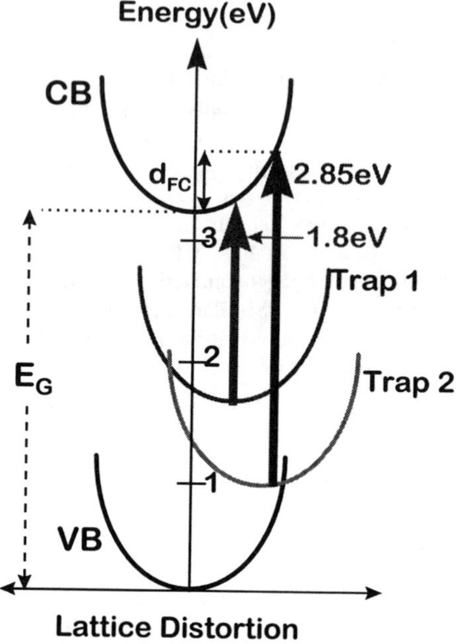

Fig. 5. Threshold absorption energies of Trap 1 and Trap 2 relative to the GaN band edges, as a function of local lattice distortion at the defect site.

REFERENCES

1. H.J. Quiesser and D.E. Theodorou, Phys. Rev. Lett. **43**, 401 (1979).

2. T.N. Theis and S.L. Wright, Appl. Phys. Lett. **48**, 1374 (1986).

3. M.A. Khan, M.S. Shur, Q.C. Chen and J.N. Kuznia, Electron. Lett. **30**, 2175 (1994).

4. R. Fisher, T.J. Drummond, W. Kopp, H. Morkoc, K. Lee and M. Shur, Electron. Lett. **19**,789 (1983).

5. S.C. Binari, W. Kruppa, H.B. Dietrich, G. Kelner, A.E. Wickenden and J.A. Freitas, Jr., Solid State Electron. **41**, 1549 (1997).

6. B.K. Ridley, *Quantum Processes in Semiconductors,* 3rd ed. (Clarendon Press, Oxford, 1993).

7. G. Lucovsky, Solid State Commun. **3**, 299 (1965).

8. J.C. Inkson, J. Phys. C **14**, 1093 (1981).

9. P.M. Mooney, G.A. Northrup, T.N. Morgan and H.G. Grimmeiss, Phys. Rev. B **37**, 8298 (1988).

10. A. Chantre, G. Vincent and D. Bois, Phys. Rev. B **23**, 5335 (1981).

11. P. Hacke, P. Ramvall, S. Tanaka, Y. Aoyagi, A. Kuramata, K. Horino and H. Munekata, Appl. Phys. Lett. **74**, 543 (1999).

12. Michele T. Hirsch, J.A. Wolk, W. Walukiewicz and E.E. Haller, Appl. Phys. Lett. **71**, 1098 (1997).

13. C.V. Reddy, K. Balakrishnan, H. Okumura and S. Yoshida, Appl. Phys. Lett. **73**, 244 (1998).

AUTHOR INDEX

Gertner, E., 471
Gibson, W.M., 377
Glans, P-A., 39
Goldstein, J., 253
Gong, M., 117
Grau, M., 259
Gregory, R.B., 33, 231
Grehk, T.M., 39
Griffin, J., 45
Grudowski, P.A., 357
Grzegory, I., 363
Guo, H.J., 513
Guo, J.D., 523
Gurray, A., 173
Gutmann, R.J., 63
Gwilliam, R.M., 129

Hageman, P.R., 345
Hall, W.B., 23
Han, J., 513
Hanson, T., 23
Harris, C.I., 197
Hecht, C., 149
Henkel, T., 117
Henry, R.L., 489, 541
Heuken, M., 321
Higgins, A., 471
Hill, D.M., 529
Hjelm, M., 69
Hofmann, D., 259, 275
Holland, O.W., 33
Homewood, K.P., 129
Hong, M.H., 369
Hong, S.E., 383
Hong, S.Q., 219, 407
Hopkins, R.H., 245
Huang, J., 207
Hubbard, C.W., 93

Iakimov, T., 265
Ibbetson, J.P., 315
Ikossi-Anastasiou, K., 489, 541
Ila, D., 123
Irvine, K.G., 167

Jacobs, K., 327
Jacobsson, H., 265
Janzén, E., 265
Je, J.H., 141
Johansson, L.I., 39
Johnson, C.M., 129, 135
Jones, K.A., 185, 339
Juergensen, H., 321

Käckell, P., 69
Kang, S.C., 141
Karlsson, S., 197
Keller, S., 351, 433
Kelner, G., 541
Khan, A., 471
Khan, M.A., 357
Khan, M.A., 495

Khatri, S., 23
Khlebnikov, I., 57
Kim, J.G., 333
Kim, K.H., 383
Kim, Y., 289, 295, 451
Kisielowski, C., 369, 451
Klein, P.B., 547
Knights, A.P., 129
Kobayashi, H., 377
Kobayashi, N., 117
Kölbl, M., 271
Koleske, D.D., 489, 541
Konkar, A., 231
Konstantinov, A.O., 197
Korakakis, D., 427
Kordesch, M.E., 529
Kordina, O., 167
Kornegay, K., 45, 173
Kottke, M., 407
Koynov, S., 395
Krüger, J., 289, 295, 451
Kruppa, W., 541
Kuech, T.F., 463
Kum, B.H., 141
Kwon, Y.H., 351

Lai, W.C., 523
Lam, J.B., 351, 457
Larkin, D.J., 123
Larsen, P.K., 345
Lee, W.S., 481
Li, P., 301
Li, Y., 3
Liaw, H.M., 219, 407
Liliental-Weber, Z., 295, 357, 363
Lin, C., 207
Lindner, J.K.N., 111
Linthicum, K.J., 219, 307, 407
Linville, R.J., 281
Lischka, K., 225
Little, B.D., 351, 439
Long, F.H., 201
Lourenço, M.A., 129
Lowney, D., 327
Lu, W.J., 87
Luenenbuerger, M., 321
Lueng, C.M., 389

Macfarlane, P.J., 51
MacMillan, M.F., 23
Madangarli, V., 57, 75
Magtoto, N.P., 413
Mani, S.S., 23
Marchand, H., 315
Masri, P., 213
Masuda, Y., 191
Matin, M., 3
Matsumoto, K., 179, 191
Mazur, J.H., 363
McDermott, B., 471
McDermott, B.T., 495
McGlothlin, H.M., 3

SUBJECT INDEX